Mechanics

Graduate Texts In Physics

Graduate Texts in Physics publishes core learning/teaching material for graduate- and advanced-level undergraduate courses on topics of current and emerging fields within physics, both pure and applied. These textbooks serve students at the MS- or PhD-level and their instructors as comprehensive sources of principles, definitions, derivations, experiments and applications (as relevant) for their mastery and teaching, respectively. International in scope and relevance, the textbooks correspond to course syllabi sufficiently to serve as required reading. Their didactic style, comprehensiveness and coverage of fundamental material also make them suitable as introductions or references for scientists entering, or requiring timely knowledge of, a research field.

Series Editors

Professor Richard Needs
Cavendish Laboratory
JJ Thomson Avenue
Cambridge CB3 0HE, UK
E-mail: rn11@cam.ac.uk

Professor William T. Rhodes
Florida Atlantic University
Imaging Technology Center
Department of Electrical Engineering
777 Glades Road SE, Room 456
Boca Raton, FL 33431, USA
E-mail: wrhodes@fau.edu

Professor H. Eugene Stanley
Boston University
Center for Polymer Studies
Department of Physics
590 Commonwealth Avenue, Room 204B
Boston, MA 02215, USA
E-mail: hes@bu.edu

Florian Scheck

Mechanics

From Newton's Laws to Deterministic Chaos

Fifth Edition

Springer

Prof. Dr. Florian Scheck
Universität Mainz
Institut für Physik, Theoretische Elementarteilchenphysik
Staudinger Weg 7
55099 Mainz
Germany
scheck@uni-mainz.de

Title of the original German edition: *Mechanik, 8. Auflage*
(Springer-Lehrbuch)
ISBN 978-3-540-71377-7

ISSN 1868-4513 e-ISSN 1868-4521
ISBN 978-3-642-26046-9 e-ISBN 978-3-642-05370-2
DOI 10.1007/978-3-642-05370-2
Springer Heidelberg Dordrecht London New York

Cover design: Integra Software Services Pvt. Ltd., Pondicherry

Printed on acid-free paper.

Springer is a part of Springer Science+Business Media (springer.com)

Preface

Purpose and Emphasis. Mechanics not only is the oldest branch of physics but was and still is the basis for all of theoretical physics. Quantum mechanics can hardly be understood, perhaps cannot even be formulated, without a good knowledge of general mechanics. Field theories such as electrodynamics borrow their formal framework and many of their building principles from mechanics. In short, throughout the many modern developments of physics where one frequently turns back to the principles of classical mechanics its model character is felt. For this reason it is not surprising that the *presentation* of mechanics reflects to some extent the development of modern physics and that today this classical branch of theoretical physics is taught rather differently than at the time of Arnold Sommerfeld, in the 1920s, or even in the 1950s, when more emphasis was put on the theory and the applications of partial-differential equations. Today, *symmetries* and *invariance principles*, the *structure of the space–time continuum*, and the *geometrical structure* of mechanics play an important role. The beginner should realize that mechanics is not primarily the art of describing block-and-tackles, collisions of billiard balls, constrained motions of the cylinder in a washing machine, or bicycle riding. However fascinating such systems may be, mechanics is primarily the field where one learns to develop general principles from which equations of motion may be derived, to understand the importance of symmetries for the dynamics, and, last but not least, to get some practice in using theoretical tools and concepts that are essential for all branches of physics.

Besides its role as a basis for much of theoretical physics and as a training ground for physical concepts, mechanics is a fascinating field in itself. It is not easy to master, for the beginner, because it has many different facets and its structure is less homogeneous than, say, that of electrodynamics. On a first assault one usually does not fully realize both its charm and its difficulty. Indeed, on returning to various aspects of mechanics, in the course of one's studies, one will be surprised to discover again and again that it has new facets and new secrets. And finally, one should be aware of the fact that mechanics is not a closed subject, lost forever in the archives of the nineteenth century. As the reader will realize in Chap. 6, if he or she has not realized it already, mechanics is an exciting field of research with many important questions of qualitative dynamics remaining unanswered.

Structure of the Book and a Reading Guide. Although many people prefer to skip prefaces, I suggest that the reader, if he or she is one of them, make an exception for once and read at least this section and the next. The short introductions at the beginning of each chapter are also recommended because they give a summary of the chapter's content.

Chapter 1 starts from Newton's equations and develops the elementary dynamics of one-, two-, and many-body systems for unconstrained systems. This is the basic material that could be the subject of an introductory course on theoretical physics or could serve as a text for an integrated (experimental and theoretical) course.

Chapter 2 is the "classical" part of general mechanics describing the principles of canonical mechanics following Euler, Lagrange, Hamilton, and Jacobi. Most of the material is a MUST. Nevertheless, the sections on the symplectic structure of mechanics (Sect. 2.28) and on perturbation theory (Sects. 2.38–2.40) may be skipped on a first reading.

Chapter 3 describes a particularly beautiful application of classical mechanics: the theory of spinning tops. The rigid body provides an important and highly nontrivial example of a motion manifold that is not a simple Euclidean space $\mathbb{R}^{2f}$, where f is the number of degrees of freedom. Its rotational part is the manifold of SO(3), the rotation group in three real dimensions. Thus, the rigid body illustrates a Lie group of great importance in physics within a framework that is simple and transparent.

Chapter 4 deals with relativistic kinematics and dynamics of pointlike objects and develops the elements of special relativity. This may be the most difficult part of the book, as far as the physics is concerned, and one may wish to return to it when studying electrodynamics.

Chapter 5 is the most challenging in terms of the mathematics. It develops the basic tools of differential geometry that are needed to formulate mechanics in this setting. Mechanics is then described in geometrical terms and its underlying structure is worked out. This chapter is conceived such that it may help to bridge the gap between the more "physical" texts on mechanics and the modern mathematical literature on this subject. Although it may be skipped on a first reading, the tools and the language developed here are essential if one wishes to follow the modern literature on qualitative dynamics.

Chapter 6 provides an introduction to one of the most fascinating recent developments of classical dynamics: stability and deterministic chaos. It defines and illustrates all important concepts that are needed to understand the onset of chaotic motion and the quantitative analysis of unordered motions. It culminates in a few examples of chaotic motion in celestial mechanics.

Chapter 7, finally, gives a short introduction to continuous systems, i.e. systems with an infinite number of degrees of freedom.

Exercises and Practical Examples. In addition to the exercises that follow Chaps. 1–6, the book contains a number of practical examples in the form of exercises followed by complete solutions. Most of these are meant to be worked out on a personal computer, thereby widening the range of problems that can be solved with elementary means, beyond the analytically integrable ones. I have tried to choose examples simple enough that they can be made to work even on a programmable pocket computer and in a spirit, I hope, that will keep the reader from getting lost in the labyrinth of computional games.

Length of this Book. Clearly there is much more material here than can be covered in one semester. The book is designed for a two-semester course (i.e., typically, an introductory course followed by a course on general mechanics). Even then, a certain choice of topics will have to be made. However, the text is sufficiently self-contained that it may be useful for complementary reading and individual study.

Mathematical Prerequisites. A physicist must acquire a certain flexibility in the use of mathematics. On the one hand, it is impossible to carry out all steps in a deduction or a proof, since otherwise one will not get very far with the physics one wishes to study. On the other hand, it is indispensable to know analysis and linear algebra in some depth, so as to be able to fill in the missing links in a logical deduction. Like many other branches of physics, mechanics makes use of many and various disciplines of mathematics, and one cannot expect to have all the tools ready before beginning its study. In this book I adopt the following, somewhat generous attitude towards mathematics. In many places, the details are worked out to a large extent; in others I refer to well-known material of linear algebra and analysis. In some cases the reader might have to return to a good text in mathematics or else, ideally, derive certain results for him- or herself. In this connection it might also be helpful to consult the appendix at the end of the book.

General Comments and Acknowledgements. This fifth English edition follows closely the eigth German edition (volume 1 of a series of five textbooks). As compared to the third English edition published in 1999, there are a number revisions and additions. Some of these are the following. In Chap. 1 more motivation for the introduction of phase space at this early stage is given. A paragraph on the notion of hodograph is added which emphasizes the special nature of Keplerian bound orbits. Chap. 2 is supplemented by some extensions and further explanations, specifically in relation with Legendre transformation. Also, a new section on a generalized version of Noether's theorem was added, together with some enlightening examples. In Chap. 3 more examples are given for inertia tensors and the use of Steiner's theorem. Here and in Chap. 4 the symbolic "bra" and "ket" notation is introduced in characterizing vectors and their duals. The present, fifth edition differs from the previous, fourth edition of 2005 by a few corrections and some additions in response to specific questions asked by students and other readers.

The book contains the solutions to all exercises, as well as some historical notes on scientists who made important contributions to mechanics and to the mathematics on which it rests. The index of names, in addition to the subject index, may also be helpful in locating quickly specific items in mechanics.

This book was inspired by a two-semester course on general mechanics that I have taught on and off over the last decades at the Johannes Gutenberg University at Mainz and by seminars on geometrical aspects of mechanics. I thank my collaborators, colleagues, and students for stimulating questions, helpful remarks, and profitable discussions. I was happy to realize that the German original, since its first appearance in October 1988, has become a standard text at German speaking universities and I can only hope that it will continue to be equally successful in its English version. I am grateful for the many encouraging reactions and suggestions I have received over the years. Among those to whom I owe special gratitude are P. Hagedorn, K. Hepp, D. Kastler, H. Leutwyler, L. Okun, N. Papadopoulos, J.M. Richard, G. Schuster, J. Smith, M. Stingl, N. Straumann, W. Thirring, E. Vogt, and V. Vento. Special thanks are due to my former student R. Schöpf who collaborated on the earlier version of the solutions to the exercises. I thank J. Wisdom for his kind permission to use four of his figures illustrating chaotic motions in the solar system, and P. Beckmann who provided the impressive illustrations for the logistic equation and who advised me on what to say about them.

The excellent cooperation with the team of Springer-Verlag is gratefully acknowledged. Last but not least, I owe special thanks to Dörte for her patience and encouragement.

As with the German edition, I dedicate this book to all those students who wish to study mechanics at some depth. If it helps to make them aware of the fascination of this beautiful field and of physics in general then one of my goals in writing this book is reached.

Mainz, March 2010 *Florian Scheck*

As in the past, I will keep track of possible errata on a page attached to my home page. The latter can be accessed via http://wwwthep.physik.uni-mainz.de/site/.

Contents

1. Elementary Newtonian Mechanics

This chapter deals with the kinematics and the dynamics of a finite number of mass points that are subject to internal, and possibly external, forces, but whose motions are not further constrained by additional conditions on the coordinates. (The mathematical pendulum will be an exception). Constraints such as requiring some mass points to follow given curves in space, to keep their relative distance fixed, or the like, are introduced in Chap. 2. Unconstrained mechanical systems can be studied directly by means of Newton's equations and do not require the introduction of new, generalized coordinates that incorporate the constraints and are dynamically independent. This is what is meant by "elementary" in the heading of this chapter – though some of its content is not elementary at all. In particular, at an early stage, we shall discover an intimate relationship between invariance properties under coordinate transformations and conservation laws of the theory, which will turn out to be a basic, constructive element for all of mechanics and which, for that matter, will be felt like a *cantus firmus*[1] throughout the whole of theoretical physics. The first, somewhat deeper analysis of these relations already leads one to consider the nature of the spatial and temporal manifolds that carry mechanical motions, thereby entering a discussion that is of central importance in present-day physics at both the smallest and the largest dimensions.

We also introduce the notion of *phase space*, i.e. the description of physical motions in an abstract space spanned by coordinates and corresponding momenta, and thus prepare the ground for canonical mechanics in the formulation of Hamilton and Jacobi.

We begin with Newton's fundamental laws, which we interpret and translate into precise analytical statements. They are then illustrated by a number of examples and some important applications.

1.1 Newton's Laws (1687) and Their Interpretation

We begin by stating Newton's fundamental laws in a formulation that is close to the original one. They are as follows:

[1] *cantus firmus*: a preexisting melody, such as a plainchant excerpt, which underlies a polyphonic musical composition.

F. Scheck, *Mechanics*, Graduate Texts in Physics, 5th ed.,
DOI 10.1007/978-3-642-05370-2_1, © Springer-Verlag Berlin Heidelberg 2010

I. Every body continues in its state of rest or of uniform rectilinear motion, except if it is compelled by forces acting on it to change that state.
II. The change of motion is proportional to the applied force and takes place in the direction of the straight line along which that force acts.
III. To every action there is always an equal and contrary reaction; or, the mutual actions of any two bodies are always equal and oppositely directed along the same straight line.

In order to understand these fundamental laws and to learn how to translate them into precise analytical expressions we first need to interpret them and to go through a number of definitions. On the one hand we must clarify what is meant by notions such as "body", "state of motion", "applied force", etc. On the other hand we wish to collect a few (provisional) statements and assumptions about the space-time continuum in which mechanical motions take place. This will enable us to translate Newton's laws into local equations, which can then be tested, in a quantitative manner, by comparison with experiment.

Initially, "bodies" will be taken to be *mass points*, i.e. pointlike *particles* of mass m. These are objects that have no spatial extension but do carry a finite mass. While this idealization is certainly plausible for an elementary particle like the electron, in studying collisions on a billiard table, or relative motions in the planetary system, it is not clear, a priori, whether the billiard balls, the sun, or the planets can be taken to be massive but pointlike, i.e. without spatial extension. For the moment and in order to give at least a preliminary answer, we anticipate two results that will be discussed and proved later.

(i) To any finite mass distribution (i.e. a mass distribution that can be completely enclosed by a sphere of finite radius), or to any finite system of mass points, one can assign a center of gravity to which the resultant of all external forces applies. This center behaves like a pointlike particle of mass M, under the action of that resultant, M being the total mass of the system (see Sects. 1.9 and 3.8).

(ii) A finite mass distribution of total mass M that looks the same in every direction (one says it is *spherically symmetric*) creates a force field in the outer, mass-free space that is identical to that of a pointlike particle of mass M located at its center of symmetry (Sect. 1.30). A spherical sun acts on a planet that does not penetrate it like a mass point situated at its center. In turn, the planet can be treated as a pointlike mass, too, as long as it is spherically symmetric.

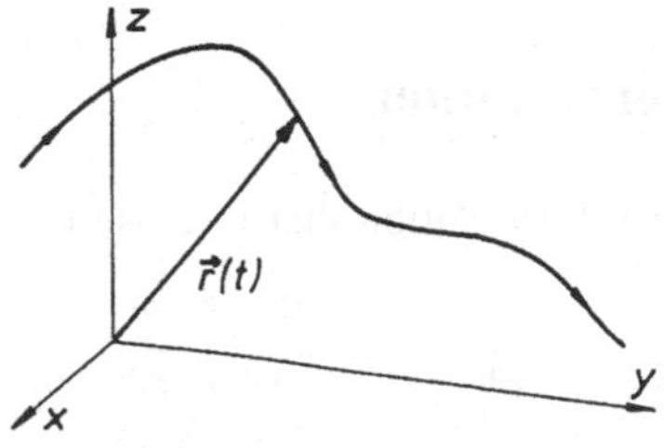

Fig. 1.1. Example of an orbit with accelerated motion. While the orbit curve is a coordinate independent, geometric object, its description by the position vector $\boldsymbol{r}(t)$ depends on the choice of origin and coordinates

In Law I, motion, or state of motion, refers to the trajectory $\boldsymbol{r}(t)$ of the mass point in coordinate space $\mathbb{R}^3$, where $\boldsymbol{r}$ describes its position at a given time t. Figure 1.1 shows an example for an arbitrary trajectory in three-dimensional space. State of rest means that $\dot{\boldsymbol{r}}(t) = 0$ for all times t, while uniform rectilinear motion is that motion along a straight line with constant velocity. It is important to realize that motion is always *relative* motion of (at least two) physical systems. For instance, a particle moves relative to an observer (i.e. a measuring apparatus). It is only meaningful to talk about the relative positions of particle A and particle B, or about the position of particle A with respect to an observer, at any fixed time.

Experimental experience allows us to assume that the space in which the physical motion of a mass point takes place is *homogeneous* and *isotropic* and that it has the structure of a three-dimensional Euclidean space $\mathbb{E}^3$. Homogeneous here means that no point of $\mathbb{E}^3$ is singled out in any respect. Isotropic in turn means that there is no preferred direction either (more on this will be said in Sect. 1.14). Thus the space of motions of the particle is an *affine space*, in agreement with physical intuition: giving the position $\boldsymbol{x}(t) \in \mathbb{E}^3$ of a particle at time t is not meaningful, while giving this $\boldsymbol{x}(t)$ relative to the position $\boldsymbol{y}(t)$ of an observer (at the same time) is. If we endow the affine space with an origin, e.g. by relating all positions to a given observer, the space is made into the real three-dimensional vector space $\mathbb{R}^3$. This is a metric space on which scalar and cross products of vectors are defined as usual and for which base systems can be chosen in a variety of ways.

In nonrelativistic physics *time* plays a special role. Daily experience tells us that time appears to be universal in the sense that it runs uniformly without being influenced by physical events. In order to sharpen this statement one may think of any moving particle as being accompanied on its journey by its own clock, which measures what is called the particle's *proper time* τ. On his clock an observer B then measures the time

$$t^{(B)} = \alpha^{(B)}\tau + \beta^{(B)} . \tag{1.1}$$

Here $\alpha^{(B)}$ is a positive constant indicating the (relative) unit that B chooses in measuring time, while $\beta^{(B)}$ indicates where B has chosen his origin of time, relative to that of the moving clock.

Equation (1.1) can also be written in the form of a differential equation,

$$\frac{\mathrm{d}^2 t^{(B)}}{\mathrm{d}\tau^2} = 0 , \tag{1.2}$$

which is independent of the constants $\alpha^{(B)}$ and $\beta^{(B)}$. While (1.1) relates the proper time to that of a specific observer, (1.2) contains the statement that is of interest here for *all* possible observers. We conclude that time is described by a one-dimensional affine space, or, after having chosen an origin, by the real line $\mathbb{R}$. For the sake of clarity we shall sometimes also write $\mathbb{R}_t$ ("t" for "time").[2]

[2] It would be premature to conclude that the space–time of nonrelativistic physics is simply $\mathbb{R}^3 \times \mathbb{R}_t$ as long as one does not know the symmetry structure that is imposed on it by the dynamics. We return to this question in Sect. 1.14. In Sect. 4.7 we analyze the analogous situation in relativistic mechanics.

The trajectory $\boldsymbol{r}(t)$ is often described in terms of a specific coordinate system. It may be expressed by means of Cartesian coordinates,

$$\boldsymbol{r}(t) = \big(x(t), y(t), z(t)\big) ,$$

or by means of spherical coordinates

$$\boldsymbol{r}(t) : \{r(t), \varphi(t), \theta(t)\} ,$$

or any other coordinates that are adapted to the system one is studying.

Examples of motions in space are:

(i) $\boldsymbol{r}(t) = (v_x t + x_0,\ 0,\ v_z t + z_0 - gt^2/2)$ in Cartesian coordinates. This describes, in the x-direction, uniform motion with constant velocity v_x, a state of rest in the y-direction, and, in the z-direction, the superposition of the uniform motion with velocity v_z and free fall in the gravitational field of the earth.

(ii) $\boldsymbol{r}(t) = \big(x(t) = R\cos(\omega t + \phi_0),\ y(t) = R\sin(\omega t + \phi_0),\ 0\big)$.

(iii) $\boldsymbol{r}(t) : \big(r(t) = R,\ \varphi(t) = \phi_0 + \omega t,\ 0\big)$.

Examples (ii) and (iii) represent the same motion in different coordinates: the trajectory is a circle of radius R in the (x, y)-plane that the particle follows with constant angular velocity ω.

From the knowledge of the function $\boldsymbol{r}(t)$ follow the *velocity*

$$\boldsymbol{v}(t) \stackrel{\text{def}}{=} \frac{\mathrm{d}}{\mathrm{d}t}\boldsymbol{r}(t) \equiv \dot{\boldsymbol{r}} \tag{1.3}$$

and the *acceleration*

$$\boldsymbol{a}(t) \stackrel{\text{def}}{=} \frac{\mathrm{d}}{\mathrm{d}t}\boldsymbol{v}(t) \equiv \dot{\boldsymbol{v}} = \ddot{\boldsymbol{r}} . \tag{1.4}$$

In Example (i) above, $\boldsymbol{v} = (v_x,\ 0,\ v_z - gt)$ and $\boldsymbol{a} = (0,\ 0,\ -g)$. In Examples (ii) and (iii) we have $\boldsymbol{v} = \omega R\big(-\sin(\omega t + \phi_0),\ \cos(\omega t + \phi_0),\ 0\big)$ and $\boldsymbol{a} = \omega^2 R\big(-\cos(\omega t + \phi_0),\ -\sin(\omega t + \phi_0),\ 0\big)$, i.e. $\boldsymbol{v}$ has magnitude ωR and direction tangent to the circle of motion. The acceleration has magnitude $\omega^2 R$ and is directed towards the center of that circle.

The velocity vector is a tangent vector to the trajectory and therefore lies in the tangent space of the manifold of position vectors, at the point $\boldsymbol{r}$. If $\boldsymbol{r} \in \mathbb{R}^3$, this tangent space is also an $\mathbb{R}^3$ and can be identified with the space of positions. There are cases, however, where we have to distinguish between the position space and its tangent spaces. A similar remark applies to the acceleration vector.

1.2 Uniform Rectilinear Motion and Inertial Systems

Definition. *Uniform rectilinear motion* is a state of motion with constant velocity and therefore vanishing acceleration, $\ddot{\boldsymbol{r}} = 0$.

The trajectory has the general form

$$r(t) = r^0 + v^0 t \, , \tag{1.5}$$

where r^0 denotes the initial position, v^0 the initial velocity, $r^0 = r(t = 0)$, and $v^0 = v(t = 0)$. The velocity is constant and the acceleration is zero at all times:

$$\begin{aligned} v(t) &\equiv \dot{r}(t) = v^0 \, , \\ a(t) &\equiv \ddot{r}(t) = 0 \, . \end{aligned} \tag{1.6}$$

We remark that (1.6) are differential equations characteristic for uniform motion. A specific solution is only defined if the *initial conditions* $r(0) = r^0$, $v(0) = v^0$ are given. Equation (1.6) is a linear, homogeneous system of differential equations of second order; v^0 and r^0 are integration constants that can be freely chosen.

Law I states that (1.5) with arbitrary constants r^0 and v^0 is the characteristic state of motion of a mechanical body to which no forces are applied. This statement supposes that we have already chosen a certain frame of reference, or a class of frames, in coordinate space. Indeed, if all force-free motions are described by the differential equation $\ddot{r} = 0$ in the reference frame $\mathbf{K}_0$, this is not true in a frame **K** that is *accelerated* with respect to $\mathbf{K}_0$, (see Sect. 1.25 for the case of rotating frames). In **K** there will appear fictitious forces such as the centrifugal and the Coriolis forces, and, as a consequence, force-free motion will look very complicated. There exist, in fact, specific frames of reference with respect to which force-free motion is always uniform and rectilinear. They are defined as follows.

Definition. Reference frames with respect to which Law I has the analytical form $\ddot{r}(t) = 0$ are called *inertial frames.*

In fact, the first of Newton's laws *defines* the class of inertial frames. This is the reason why it is important in its own right and is more than just a special case of Law II. With respect to inertial frames the second law then has the form

$$m\ddot{r}(t) = K \, ,$$

where K is the resultant of the forces applied to the body. Thus Newton's second law takes a particularly simple form in inertial systems. If one chooses to describe the motion by means of reference frames that are accelerated themselves, this fundamental law will take a more complicated form although it describes the same physical situation. Besides the resultant K there will appear additional, fictitious forces that depend on the momentary acceleration of the noninertial system.

The inertial systems are particularly important because they single out the group of those transformations of space and time for which the equations of motion (i.e. the equations that follow from Newton's laws) are *form invariant* (i.e. the structure of the equations remains the same). In Sect. 1.13 we shall construct the class of all inertial frames. The following proposition is particularly important in this connection.

1.3 Inertial Frames in Relative Motion

> Let **K** be an inertial frame. Any frame **K**′ that moves with constant velocity $\boldsymbol{w}$ relative to **K** is also inertial (see Fig. 1.2).

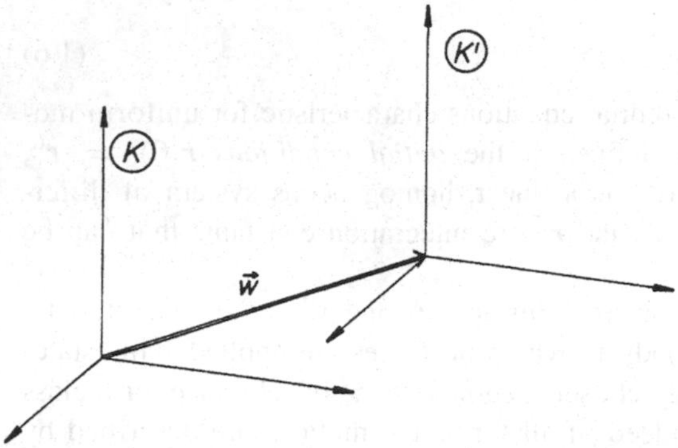

Fig. 1.2. If **K** is an inertial system, then so is every **K**′ whose axes are parallel to those of **K** and which moves with constant velocity $\boldsymbol{w}$ relative to **K**

Proof. The position vector $\boldsymbol{r}(t)$ with respect to **K** becomes $\boldsymbol{r}'(t) = \boldsymbol{r}(t) - \boldsymbol{w}t$ with respect to **K**′. Since $\boldsymbol{w}$ is constant, there follows $\ddot{\boldsymbol{r}}'(t) = \ddot{\boldsymbol{r}}(t) = 0$. All force-free motions satisfy the same differential equation (1.6) in either reference system, both of which are therefore inertial frames. □

The individual solution (1.5) looks different in **K** than in **K**′: if the systems coincide at $t = 0$, the initial condition $(\boldsymbol{r}^0, \boldsymbol{v}^0)$ with respect to the first is equivalent to the initial condition $(\boldsymbol{r}^0, \boldsymbol{v}^0 - \boldsymbol{w})$ with respect to the second.

1.4 Momentum and Force

In Law II we identify "motion" with the *momentum*:

$$\boldsymbol{p}(t) \stackrel{\text{def}}{=} m\dot{\boldsymbol{r}}(t) \equiv m\boldsymbol{v}\,, \tag{1.7}$$

i.e. the product of *inertial mass* and momentary velocity. The second law, when expressed as a formula, then reads[3]

$$\frac{\mathrm{d}}{\mathrm{d}t}\boldsymbol{p}(t) = \boldsymbol{K}(\boldsymbol{r}, \dot{\boldsymbol{r}}, t) \tag{1.8a}$$

[3] We have interpreted "change of motion" as the time derivative of the momentum. Law II does not say this so clearly.

or, if the inertial mass is independent of the state of motion,

$$m\ddot{\boldsymbol{r}}(t) = \boldsymbol{K}(\boldsymbol{r}, \dot{\boldsymbol{r}}, t) . \tag{1.8b}$$

If the second form (1.8b) applies, the proportionality factor m, the inertial mass of the body, can be determined, with respect to a sample body of reference mass m_1, by alternatively exposing the body and the sample to the same force field and by comparing the resultant accelerations: their ratio fulfills $m/m_1 = |\ddot{\boldsymbol{r}}^{(1)}|/|\ddot{\boldsymbol{r}}|$.

The mass of macroscopic bodies can be changed by adding or removing matter. In nonrelativistic physics at macroscopic dimensions mass is an additive quantity; that is, if one joins two bodies of masses m_1 and m_2, their union has mass $(m_1 + m_2)$. Another way of expressing this fact is to say that mass is an *extensive* quantity.

In the realm of physics at microscopic dimensions one finds that mass is an invariant, characteristic property. Every electron has the mass $m_e = 9.11 \times 10^{-31}$ kg, a hydrogen atom has a fixed mass, which is the same for any other hydrogen atom, all photons are strictly massless, etc.

The relationship (1.7) holds only as long as the velocity is small as compared to the speed of light $c \simeq 3 \times 10^8$ m/s. If this is not the case the momentum is given by a more complicated formula, viz.

$$\boldsymbol{p}(t) = \frac{m}{\sqrt{1 - \boldsymbol{v}^2(t)/c^2}} \boldsymbol{v}(t) , \tag{1.9}$$

where c is the velocity of light (see Chap. 4). For $|\boldsymbol{v}| \ll c$ the expressions (1.9) and (1.7) differ by terms of order $\mathcal{O}(|\boldsymbol{v}|^2/c^2)$. For these reasons – mass being an invariant property of elementary particles, and its role in the limit of small velocity $\boldsymbol{v} \to \boldsymbol{0}$ – one also calls the quantity m the *rest mass* of the particle. In the older literature, when considering the quotient $m/\sqrt{1 - \boldsymbol{v}^2(t)/c^2}$, one sometimes talked about this as the moving, velocity dependent mass. It is advisable, however, to avoid this distinction altogether because it blurs the invariant nature of rest mass and hides an essential difference between relativistic and nonrelavistic kinematics. In talking about mass we will always have in mind the invariant rest mass.

We assume the force $\boldsymbol{K}(\boldsymbol{r}, \dot{\boldsymbol{r}}, t)$ to be given a priori. More precisely we are talking about a *force field*, i.e. a vector-valued function over the space of coordinates and, if the forces are velocity dependent, the space of velocities. At every point of this six-dimensional space where $\boldsymbol{K}$ is defined this function gives the force that acts on the mass point at time t. Such force fields, in general, stem from other physical bodies, which act as their sources. Force fields are vector fields. This means that different forces that are applied at the same point in space, at a given time, must be added vectorially.

In Law III the notion "action" stands for the (internal) force that one body exerts on another. Consider a system of finitely many mass points with masses m_i and position vectors $\boldsymbol{r}_i(t)$, $i = 1, 2, \ldots, n$. Let $\boldsymbol{F}_{ik}$ be the force that particle i exerts on particle k. One then has $\boldsymbol{F}_{ik} = -\boldsymbol{F}_{ki}$. Forces of this kind are called *internal forces* of the n-particle system. This distinction is necessary if one wishes to describe the interaction with further, possibly very heavy, external objects by means of *external forces*. This is meaningful, for instance, whenever the reaction on the external objects is negligible. One should keep in mind that the distinction between internal and external forces is artificial and is made only for practical reasons. The source of an external force can always be defined to be part of the system, thus converting the external to an internal force. Conversely, the example discussed in Sect. 1.7 below shows that the two-body problem with internal forces can be reduced to an effective one-body problem through separation of the center-of-mass motion, where a fictitious particle of mass $\mu = m_1 m_2/(m_1 + m_2)$ moves in the field of an external force.

1.5 Typical Forces. A Remark About Units

The two most important fundamental forces of nature are the gravitational force and the Coulomb force. The other fundamental forces known to us, i.e. those describing the strong and the weak interactions of elementary particles, have very small ranges of about 10^{-15} m and 10^{-18} m, respectively. Therefore, they play no role in mechanics at laboratory scales or in the planetary system.

The gravitational force is always attractive and has the form

$$\boldsymbol{F}_{ki} = -G m_i m_k \frac{\boldsymbol{r}_i - \boldsymbol{r}_k}{|\boldsymbol{r}_i - \boldsymbol{r}_k|^3} \,. \tag{1.10}$$

This is the force that particle k with mass m_k applies to particle i whose mass is m_i. It points along the straight line that connects the two, is directed from i to k, and is inversely proportional to the square of the distance between i and k. G is Newton's gravitational constant. Apart from G (1.10) contains the *gravitational masses* (heavy masses or weights) m_i and m_k. These are to be understood as parameters characterizing the strength of the interaction. Experiment tells us that gravitational and inertial masses are proportional to one another ("all bodies fall at the same speed"), i.e. that they are essentially of the same nature. This highly remarkable property of gravitation is the starting point for Einstein's equivalence principle and for the theory of general relativity. If read as the *gravitational* mass, m_i determines the strength of the coupling of particle i to the force field created by particle k. If understood as being the *inertial* mass, it determines the local acceleration in a given force field. (The third of Newton's laws ensures that the situation is symmetric in i and k, so that the discussion of particle k in the field of particle i is exactly the same.)

In the case of the Coulomb force, matters are different: here the strength is determined by the *electric charges* e_i and e_k of the two particles,

$$\boldsymbol{F}_{ki} = \kappa_C e_i e_k \frac{\boldsymbol{r}_i - \boldsymbol{r}_k}{|\boldsymbol{r}_i - \boldsymbol{r}_k|^3} \,, \tag{1.11}$$

which are not correlated (for macroscopic bodies) to their masses. A ball made of iron with given mass may be uncharged or may carry positive or negative charges. The strength as well as the sign of the force are determined by the charges. For sign e_i = sign e_k it is repulsive; for sign e_i = −sign e_k it is attractive. If one changes the magnitude of e_k, for instance, the strength varies proportionally to e_k. The accelerations induced by this force, however, are determined by the inertial masses as before. The parameter κ_C is a constant that depends on the units used (see below).

Apart from these fundamental forces we consider many more forms of forces that may occur or may be created in the macroscopic world of the laboratory. Specific examples are the *harmonic force*, which is always attractive and whose magnitude is proportional to the distance (Hooke's law), or those force fields which arise from the variety of electric and magnetic fields that can be created by all kinds of arrangements of conducting elements and coils. Therefore it is meaningful to regard the force field on the right-hand side of (1.8) as an independent element of the theory that can be chosen at will. The equation of motion (1.8) describes, in differential form, how the particle of mass m will move under the influence of the force field. If the situation is such that the particle does not disturb the source of the force field in any noticeable way (in the case of gravitation this is true whenever $m \ll M_{\text{source}}$) the particle may be taken as a probe: by measuring its accelerations one can locally scan the force field. If this is not a good approximation, Law III becomes important and one should proceed as in Sect. 1.7 below.

We conclude this section with a remark about units. To begin with, it is clear that we must define units for three observable quantities: time, length in coordinate space, and mass. We denote their dimensions by T, L, and M, respectively:

$$[t] = T\,, \qquad [r] = L\,, \qquad [m] = M\,,$$

the symbol $[x]$ meaning the physical dimension of the quantity x. The dimensions and measuring units for all other quantities that occur in mechanics can be reduced to these basic units and are therefore fixed once a choice is made for them. For instance, we have

$$\begin{aligned}
&\text{momentum}: \; [p] = MLT^{-1}\,,\\
&\text{force}: \; [K] = MLT^{-2}\,,\\
&\text{energy} = \text{force} \times \text{displacement}: \; [E] = ML^2T^{-2}\,,\\
&\text{pressure} = \text{force/area}: \; [b] = ML^{-1}T^{-2}\,.
\end{aligned}$$

For example, one can choose to measure time in seconds, length in centimeters, and mass in grams. The unit of force is then $1\,\text{g}\,\text{cm}\,\text{s}^{-2} = 1\,\text{dyn}$, the energy unit is $1\,\text{g}\,\text{cm}^2\,\text{s}^{-2} = 1\,\text{erg}$, etc. However, one should follow the International System of Units (SI), which was agreed on and fixed by law for use in the engineering sciences and for the purposes of daily life. In this system time is measured in seconds, length in meters, and mass in kilograms, so that one obtains the following derived units:

$$\begin{aligned}
&\text{force: } 1\,\text{kg}\,\text{m}\,\text{s}^{-2} = 1\,\text{Newton}\ (= 10^5\,\text{dyn})\,,\\
&\text{energy: } 1\,\text{kg}\,\text{m}^2\,\text{s}^{-2} = 1\,\text{Joule}\ (= 10^7\,\text{erg})\,,\\
&\text{pressure: } 1\,\text{kg}\,\text{m}^{-1}\,\text{s}^{-2} = 1\,\text{Pascal} = 1\,\text{Newton/m}^2\,.
\end{aligned}$$

If one identifies gravitational and inertial mass, one finds the following value for Newton's gravitational constant from experiment:

$$G = (6.67428 \pm 0.00067) \times 10^{-11}\,\text{m}^3\,\text{kg}^{-1}\,\text{s}^{-2}\,.$$

For the Coulomb force the factor κ_C in (1.11) can be chosen to be 1. (This is the choice in the Gaussian system of electrodynamics.) With this choice electric charge is a derived quantity and has dimension

$$[e] = M^{1/2} L^{3/2} T^{-1} \quad (\kappa_C = 1)\,.$$

If instead one wishes to define a unit for charge on its own, or, equivalently, a unit for another electromagnetic quantity such as voltage or current, one must choose the constant κ_C accordingly. The SI unit of current is 1 ampere. This fixes the unit of charge, and the constant in (1.11) must then be chosen to be

$$\kappa_C = \frac{1}{4\pi\varepsilon_0} = c^2 \times 10^{-7}\,,$$

where $\varepsilon_0 = 10^7/4\pi c^2$ and c is the speed of light, see Eq. (4.1) below.

1.6 Space, Time, and Forces

At this point it may be useful to give a provisional summary of our discussion of Newton's laws I–III. The first law shows the uniform rectilinear motion (1.5) to be the natural form of motion of every body that is not subject to any forces. If we send such a body from A to B it chooses the shortest connection between these points, a straight line. As one may talk in a physically meaningful manner only about motion relative to an observer, Law I raises the question in which frames of reference does the law actually hold. In fact, Law I defines the important class of inertial systems. Only with respect to these does Law II assume the simple form (1.8b).

The space that supports the motions described by Newton's equations is a three-dimensional Euclidean space, i.e. a real space where we are allowed to use the well-known Euclidean geometry. A priori this is an affine space. By choosing an origin we make it a real vector space, here $\mathbb{R}^3$. Important properties of the space of physical motions are its homogeneity ("it looks the same everywhere") and its isotropy ("all directions are equally good"). Time is one-dimensional; it is represented by points on the real line. In particular, there is an ordering relation which classifies times into "earlier" and "later", past and future.

Combining the momentary position of a particle and the time at which it takes on that position, we obtain an *event* $(\boldsymbol{x}(t), t) \in \mathbb{R}^3 \times \mathbb{R}_t$, a point in the combined space–time continuum. This definition is particularly important for relativistic physics, which exhibits a deeper symmetry between space and time, as we shall see later.

In comparing (1.2) and (1.8b) notice the asymmetry between the space and the time variables of a particle. Let τ again be the proper time as in (1.1), and t the time measured by an observer. For the sake of simplicity we choose the same unit for both, i.e. we set $\alpha^{(B)} = 1$. Equation (1.2) tells us that time runs uniformly and does not depend on the actual position of the particle nor on the forces which are applied to it. In contrast, the equation of motion (1.8) describes as a function of time the set of all possible trajectories that the particle can move on when it is subject to the given force field. Another way of expressing this asymmetry is this: $\boldsymbol{r}(t)$ is the *dynamical variable.* Its temporal evolution is determined by the forces, i.e. by the dynamics. The time variable, on the other hand, plays the role of a *parameter* in nonrelativistic mechanics, somewhat like the length function in the description of a curve in space. This difference in the assignment of the variables' roles is characteristic of the nonrelativistic description of systems of mass points. It does not hold for continuum mechanics or for any other field theory. It is modified also in physics obeying special relativity, where space and time hold more symmetric roles.

Having clarified the notions in terms of which Newton's laws are formulated, we now turn to an important application: the two-body problem with internal forces.

1.7 The Two-Body System with Internal Forces

1.7.1 Center-of-Mass and Relative Motion

In terms of the coordinates $\boldsymbol{r}_1$, $\boldsymbol{r}_2$ of the two particles whose masses are m_1 and m_2, the equations of motion read

$$m_1 \ddot{\boldsymbol{r}}_1 = \boldsymbol{F}_{21} \,, \quad m_2 \ddot{\boldsymbol{r}}_2 = \boldsymbol{F}_{12} = -\boldsymbol{F}_{21} \,. \tag{1.12}$$

The force that particle "2" exerts on particle "1" is denoted by $\boldsymbol{F}_{21}$. We will adopt this notation throughout: $\boldsymbol{F}_{ki}$ is the force field that is created by particle number k and is felt by particle number i. Taking the sum of these we obtain the equation $m_1 \ddot{\boldsymbol{r}}_1 + m_2 \ddot{\boldsymbol{r}}_2 = 0$, which is valid at all times. We define the center-of-mass coordinates

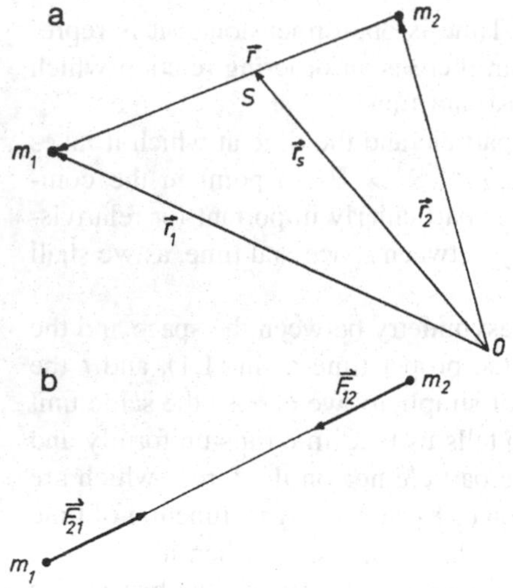

Fig. 1.3. (**a**) Definition of center-of-mass and relative coordinates for the two-body system. The relative coordinate $\boldsymbol{r}$ is independent of the choice of origin. (**b**) Force and reaction force in the two-body system. These gives rise to a central force in the equation of motion for the relative coordinate

$$\boldsymbol{r}_S \stackrel{\text{def}}{=} \frac{1}{m_1+m_2}\left(m_1\boldsymbol{r}_1 + m_2\boldsymbol{r}_2\right); \tag{1.13}$$

this means that $\ddot{\boldsymbol{r}}_S = 0$, i.e. the center of mass moves at a constant velocity. The dynamics proper is to be found in the *relative motion.* Define

$$\boldsymbol{r} \stackrel{\text{def}}{=} \boldsymbol{r}_1 - \boldsymbol{r}_2 \,. \tag{1.14}$$

By inverting (1.13) and (1.14) we have (see also Fig. 1.3)

$$\boldsymbol{r}_1 = \boldsymbol{r}_S + \frac{m_2}{m_1+m_2}\boldsymbol{r}\,, \quad \boldsymbol{r}_2 = \boldsymbol{r}_S - \frac{m_1}{m_1+m_2}\boldsymbol{r}\,. \tag{1.15}$$

Inserting these in (1.12) and using $\ddot{\boldsymbol{r}}_S = 0$, we find that the equation of motion in the relative coordinates becomes

$$\mu\ddot{\boldsymbol{r}} = \boldsymbol{F}_{21} \,. \tag{1.16}$$

The mass parameter

$$\mu \stackrel{\text{def}}{=} \frac{m_1 m_2}{m_1+m_2}$$

is called the *reduced mass.* By separating the center of mass we have reduced the two-body problem to the motion of *one* particle with mass μ.

1.7.2 Example: The Gravitational Force Between Two Celestial Bodies (Kepler's Problem)

In the case of the gravitational interaction (1.10), (1.12) becomes

$$\begin{aligned} m_1\ddot{\boldsymbol{r}}_1 &= -G\frac{m_1 m_2}{r^2}\,\frac{\boldsymbol{r}_1-\boldsymbol{r}_2}{r} = -G\frac{m_1 m_2}{r^2}\,\frac{\boldsymbol{r}}{r}\,, \\ m_2\ddot{\boldsymbol{r}}_2 &= -G\frac{m_1 m_2}{r^2}\,\frac{\boldsymbol{r}_2-\boldsymbol{r}_1}{r} = +G\frac{m_1 m_2}{r^2}\,\frac{\boldsymbol{r}}{r}\,, \end{aligned} \tag{1.17}$$

where $\boldsymbol{r} = \boldsymbol{r}_1 - \boldsymbol{r}_2$ and $r = |\boldsymbol{r}|$, from which follow the equations of motion in center-of-mass and relative coordinates

$$\ddot{\boldsymbol{r}}_S = 0 \quad\text{and}\quad \mu\ddot{\boldsymbol{r}} = -G\frac{m_1 m_2}{r^2}\,\frac{\boldsymbol{r}}{r}\,.$$

We can read off the behavior of the system from these equations: the center of mass moves uniformly along a straight line (or remains at rest). The relative motion is identical to the motion of a *single*, fictitious particle of mass μ under the action of the force

$$-G\frac{m_1 m_2}{r^2}\,\frac{\boldsymbol{r}}{r}\,.$$

Since this is a *central force*, i.e. one which always points towards the origin or away from it, it can be derived from a *potential* $U(r) = -A/r$ with $A = Gm_1m_2$. This can be seen as follows.

Central forces have the general form $\boldsymbol{F}(\boldsymbol{r}) = f(r)\hat{\boldsymbol{r}}$, where $\hat{\boldsymbol{r}} = \boldsymbol{r}/r$ and $f(r)$ is a scalar function that should be (at least) continuous in the variable $r = |\boldsymbol{r}|$. Define then

$$U(r) - U\big(r_0\big) = -\int_{r_0}^{r} f(r')\mathrm{d}r'\,,$$

where r_0 is an arbitrary reference value and where $U(r_0)$ is a constant. If we take the gradient of this expression this constant does not contribute and we obtain

$$\begin{aligned} \boldsymbol{\nabla} U(r) &= \frac{\mathrm{d}U(r)}{\mathrm{d}r}\boldsymbol{\nabla} r = -f(r)\boldsymbol{\nabla}\sqrt{x^2+y^2+z^2} \\ &= -f(r)\boldsymbol{r}/r\,. \end{aligned}$$

Thus, $\boldsymbol{F}(\boldsymbol{r}) = -\boldsymbol{\nabla} U(r)$. In the case of central forces the *orbital angular momentum*

$$\boldsymbol{l} \stackrel{\text{def}}{=} \mu\boldsymbol{r}\times\dot{\boldsymbol{r}}$$

is conserved; both its magnitude and its direction are constants in time. This follows from the observation that the acceleration is proportional to $\boldsymbol{r}$: $\mathrm{d}\boldsymbol{l}/\mathrm{d}t = \mu\boldsymbol{r}\times\ddot{\boldsymbol{r}} = 0$.

As a consequence, the motion takes place entirely in a plane perpendicular to $\boldsymbol{l}$, namely the one spanned by $\boldsymbol{r}^0$ and $\boldsymbol{v}^0$. Since the motion is planar, it is convenient to introduce polar coordinates in that plane, viz.

$$x(t) = r(t)\cos\phi(t)\,, \quad y(t) = r(t)\sin\phi(t)\,, \tag{1.18}$$

so that the components of the angular momentum are

$$l_x = l_y = 0\,, \quad l_z = \mu r^2\dot{\phi} \equiv l = \text{const}\,,$$

and, finally,

$$\dot{\phi} = l/\mu r^2\,. \tag{1.19a}$$

Furthermore the total energy, i.e. the sum of kinetic and potential energy, is conserved. In order to show this start from the equation of motion for a particle in the force field of a more general potential $U(\boldsymbol{r})$

$$\mu\ddot{\boldsymbol{r}} = -\boldsymbol{\nabla}U(\boldsymbol{r})\,.$$

This is an equation relating two vector fields, the acceleration multiplied by the reduced mass on the left-hand side, and the gradient field of the scalar function $U(\boldsymbol{r})$ on the right-hand side. Take the scalar product of these vector fields with the velocity $\dot{\boldsymbol{r}}$ to obtain the scalar equation

$$\mu\dot{\boldsymbol{r}}\cdot\ddot{\boldsymbol{r}} = -\dot{\boldsymbol{r}}\cdot\boldsymbol{\nabla}U(\boldsymbol{r})\,.$$

The left-hand side is the time derivative of $(\mu/2)\dot{\boldsymbol{r}}^2$. On the right-hand side, and with the decomposition $\boldsymbol{r} = \{x, y, z\}$, one has

$$\dot{\boldsymbol{r}}\cdot\boldsymbol{\nabla}U(\boldsymbol{r}) = \frac{\mathrm{d}x}{\mathrm{d}t}\frac{\partial U(\boldsymbol{r})}{\partial x} + \frac{\mathrm{d}y}{\mathrm{d}t}\frac{\partial U(\boldsymbol{r})}{\partial y} + \frac{\mathrm{d}z}{\mathrm{d}t}\frac{\partial U(\boldsymbol{r})}{\partial z}$$

which is nothing but the *total* time derivative of the function $U(\boldsymbol{r}(t))$ along smooth curves $\boldsymbol{r}(t)$ in $\mathbb{R}^3$. If these are solutions of the equation of motion, i.e. if they fulfill $\mu\dot{\boldsymbol{r}}\cdot\ddot{\boldsymbol{r}} = -\dot{\boldsymbol{r}}\cdot\boldsymbol{\nabla}U(\boldsymbol{r})$, one obtains

$$\mu\dot{\boldsymbol{r}}\cdot\ddot{\boldsymbol{r}} + \dot{\boldsymbol{r}}\cdot\boldsymbol{\nabla}U(\boldsymbol{r}) = \frac{\mathrm{d}}{\mathrm{d}t}\left(\frac{1}{2}\mu\dot{\boldsymbol{r}}^2 + U(\boldsymbol{r})\right) = 0\,, \quad \text{hence}$$

$$\frac{\mathrm{d}E}{\mathrm{d}t} = 0\,, \quad \text{where} \quad E = \frac{1}{2}\mu\dot{\boldsymbol{r}}^2 + U(\boldsymbol{r})\,.$$

Thus, even though in general $E(\boldsymbol{r}, \dot{\boldsymbol{r}})$ is a function of the position $\boldsymbol{r}$ and the velocity $\dot{\boldsymbol{r}}$, it is constant when evaluated along any solution of the equation of motion. Later on we shall call this kind of time derivative, taken along a solution, the *orbital derivative.*

For the problem that we are studying in this section this result implies that

$$E = \tfrac{1}{2}\mu\boldsymbol{v}^2 + U(r) = \tfrac{1}{2}\mu\left(\dot{r}^2 + r^2\dot{\phi}^2\right) + U(r) = \text{const}\,. \tag{1.20}$$

We can extract $\dot{r}$ as a function of r from (1.20) and (1.19):

$$\dot{r} = \sqrt{\frac{2\left(E - U(r)\right)}{\mu} - \frac{l^2}{\mu^2 r^2}}\,. \tag{1.19b}$$

Eqs. (1.19a) and (1.19b) form a system of two coupled ordinary differential equations of first order. They were obtained from two conservation laws, the conservation of the modulus of the angular momentum and the conservation of the total energy of the relative motion. Although this coupled system is soluble, see Sect. 1.29 and Practical Example 6 below, the procedure is somewhat cumbersome. It is simpler to work out a parametric form of the solutions by obtaining the radial variable as a function of the azimuth, $r = r(\phi)$ (thereby losing information on the evolution of $\boldsymbol{r}(t)$ as a function of time, though).

By "dividing" (1.19b) by (1.19a) and making use of $\mathrm{d}r/\mathrm{d}\phi = (\mathrm{d}r/\mathrm{d}t)/(\mathrm{d}\phi/\mathrm{d}t)$, we find that

$$\frac{1}{r^2}\frac{\mathrm{d}r}{\mathrm{d}\phi} = \sqrt{\frac{2\mu\big(E - U(r)\big)}{l^2} - \frac{1}{r^2}}\;.$$

This differential equation is of a type that can always be integrated. This means that its solution is reducible to ordinary integrations. It belongs to the class of ordinary *differential equations with separable variables*, cf. Sect. 1.22 below, for which general methods of solution exist. In the present example, where $U(r) = -A/r$, there is a trick that allows to obtain solutions directly, without doing any integrals. It goes as follows.

Setting $U(r) = -A/r$ and replacing $r(\phi)$ by the function $\sigma(\phi) = 1/r(\phi)$, we obtain the differential equation

$$-\frac{\mathrm{d}\sigma}{\mathrm{d}\phi} = \sqrt{\frac{2\mu(E + A\sigma)}{l^2} - \sigma^2}\;,$$

where we have made use of $\mathrm{d}\sigma/\mathrm{d}\phi = -r^{-2}\mathrm{d}r/\mathrm{d}\phi$.

It is convenient to define the following constants:

$$p \stackrel{\text{def}}{=} \frac{l^2}{A\mu}\;, \qquad \varepsilon \stackrel{\text{def}}{=} \sqrt{1 + \frac{2El^2}{\mu A^2}}\;.$$

The parameter p has the dimension of length, while ε is dimensionless. Indeed, A has the same physical dimension as an energy times a length. Likewise l^2 has the dimension (energy×mass×length^2). Hence, $l^2/(A\mu)$ is a length, ε is dimensionless. Inserting these definitions the differential equation becomes

$$\left(\frac{\mathrm{d}\sigma}{\mathrm{d}\phi}\right)^2 + \left(\sigma - \frac{1}{p}\right)^2 = \frac{\varepsilon^2}{p^2}\;.$$

This equation is solved by substituting $\sigma - 1/p = (\varepsilon/p)\cos\phi$. Rewritten in terms of the original variable $r(\phi)$ the general solution of the Kepler problem is

$$r(\phi) = \frac{p}{1 + \varepsilon\cos\phi}\;. \tag{1.21}$$

Before proceeding to analyze these solutions we remark that (1.19a) is a consequence of the conservation of angular momentum and is therefore valid for any central force. The quantity $r^2\dot{\phi}/2$ is the surface velocity at which the radius vector moves over the plane of motion. Indeed, if $\boldsymbol{r}$ changes by the amount $\mathrm{d}\boldsymbol{r}$, the radius sweeps out the area $\mathrm{d}F = |\boldsymbol{r} \times \mathrm{d}\boldsymbol{r}|/2$. Thus, per unit of time,

$$\frac{\mathrm{d}F}{\mathrm{d}t} = \frac{1}{2}|\boldsymbol{r} \times \dot{\boldsymbol{r}}| = \frac{l}{2\mu} = \frac{1}{2}r^2\dot{\phi} = \text{const} . \tag{1.22}$$

This is the content of *Kepler's second law* (1609):

> The radius vector from the sun to the planets sweeps out equal areas in equal times.

We note under which conditions this statement holds true: it applies to any central force but only in the two-body problem; for the motion of a planet it is valid to the extent the interaction with the other planets is negligible compared to the action of the sun.

In studying the explicit form of the solutions (1.21) it is useful to introduce Cartesian coordinates (x, y) in the plane of the orbit. Equation (1.21) is then turned into a quadratic form in x and y, and the nature of the Kepler orbits is made more evident: they are conics. One sets

$$x = r\cos\phi + c , \quad y = r\sin\phi ,$$

and chooses the constant c so that in the equation

$$r^2 = (x - c)^2 + y^2 = [p - \varepsilon r\cos\phi]^2 = [p - \varepsilon(x - c)]^2$$

the terms linear in x cancel. As long as $\varepsilon \neq 1$, this is achieved by the choice

$$c = \frac{\varepsilon p}{1 - \varepsilon^2} .$$

Finally, with the definition

$$a \stackrel{\text{def}}{=} \frac{p}{1 - \varepsilon^2} ,$$

the function (1.21) becomes

$$\frac{x^2}{a^2} + \frac{y^2}{a^2 - c^2} = 1 , \tag{1.21$'$}$$

i.e. an equation of second order containing only the squares of x and y. Here two distinct cases are possible.

(i) $\varepsilon > 1$, i.e. $c^2 > a^2$. In this case (1.21′) describes a hyperbola. The center of the force field lies at one of the foci. For the attractive case (A and p are positive) the branch of the hyperbola that opens toward the force center is the physical one. This applies to the case of gravitational interaction (cf. Fig. 1.4).

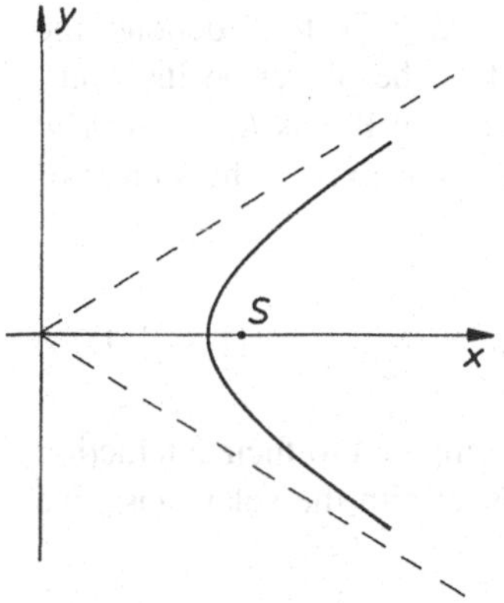

Fig. 1.4. If in the Kepler problem the energy E (1.20) is positive, the orbits of relative motion are branches of hyperbolas. The figure shows the relevant branch in the case of an attractive force

These concave branches describe the orbits of meteorites whose total energy is positive. Physically speaking, this means that they have enough kinetic energy to escape from the attractive gravitational field to infinity.

The branch turning away from the force center is the relevant one when the force is repulsive, i.e. if A and p are negative. This situation occurs in the scattering of two electric point charges with equal signs.

(ii) $\varepsilon < 1$, i.e. $c < a$. In this case the energy E is negative. This implies that the particle cannot escape from the force field: its orbits must be finite everywhere. Indeed, (1.21′) now describes an ellipse (cf. Fig. 1.5) with

$$\text{semimajor axis } a = \frac{p}{1-\varepsilon^2} = \frac{A}{2(-E)} ,$$

$$\text{semiminor axis } b = \sqrt{a^2 - c^2} = \sqrt{pa} = \frac{l}{\sqrt{2\mu(-E)}} .$$

The orbit is a *finite* orbit. It is closed and therefore periodic. This is *Kepler's first law*: the planets move on ellipses with the sun at one focus. This law holds true only for the gravitational interaction of two bodies. All finite orbits are closed and are ellipses (or circles). In Sect. 1.24 we return to this question and illustrate it with a few examples for interactions close to, but different from, the gravitational

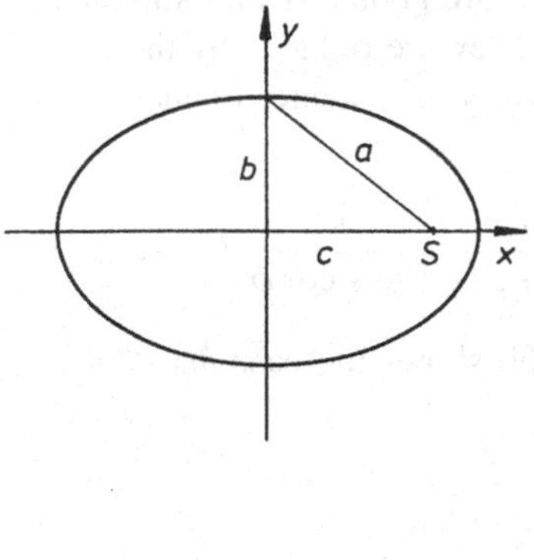

Fig. 1.5. If the energy (1.20) is negative, the orbit is an ellipse. The system is bound and cannot escape to infinity

case. The area of the orbital ellipse is $F = \pi ab = \pi a\sqrt{ap}$. If T denotes the period of revolution (the time of one complete circuit of the planet on its orbit), the area law (1.22) says that $F = Tl/2\mu$. A consequence of this is *Kepler's third law* (1615), which relates the third power of the semimajor axis to the square of the period, viz.

$$\frac{a^3}{T^2} = \frac{A}{(2\pi)^2\mu} = \text{const} = \frac{G(m_1+m_2)}{(2\pi)^2} . \tag{1.23}$$

If one neglects the mutual interactions of the planets compared to their interaction with the sun and if their masses are small in comparison with the solar mass, we obtain:

> For all planets of a given planetary system the ratio of the cubes of the semimajor axes to the squares of the periods is the same.

Of course, the special case of circular orbits is contained in (1.21′). It occurs when $\varepsilon = 0$, i.e. when $E = -\mu A^2/2l^2$, in which case the radius of the orbit has the constant value $a = l^2/\mu A$.

The case $\varepsilon = 1$ is a special case which we have so far excluded. Like the circular orbit it is a singular case. The energy is exactly zero, $E = 0$. This means that the particle escapes to infinity but reaches infinity with vanishing kinetic energy. The orbit is given by

$$y^2 + 2px - 2pc - p^2 = 0 ,$$

where c may be chosen at will, e.g. $c = 0$. The orbit is a parabola.

So far we have studied the relative motion of two celestial bodies. It remains to transcribe this motion back to the true coordinates by means of (1.15). As an example we show this for the finite orbits (ii). Choosing the center of mass as the origin, one has

$$\boldsymbol{S}_1 = \frac{m_2}{m_1+m_2}\boldsymbol{r} , \quad \boldsymbol{S}_2 = -\frac{m_1}{m_1+m_2}\boldsymbol{r} .$$

The celestial bodies 1 and 2 move along ellipses that are geometrically similar to the one along which the relative coordinate moves. They are reduced by the scale factors $m_2/(m_1+m_2)$ and $m_1/(m_1+m_2)$, respectively. The center of mass S is a common focus of these ellipses:

$$S_1(\phi) = \frac{m_2}{m_1+m_2}\,\frac{p}{1+\varepsilon\cos\phi} , \quad S_2(\phi) = \frac{m_1}{m_1+m_2}\,\frac{p}{1+\varepsilon\cos\phi} \quad \left(S_i \equiv |\boldsymbol{S}_i|\right) .$$

Figure 1.6a shows the case of equal masses; Fig. 1.6b shows the case $m_1 \ll m_2$.

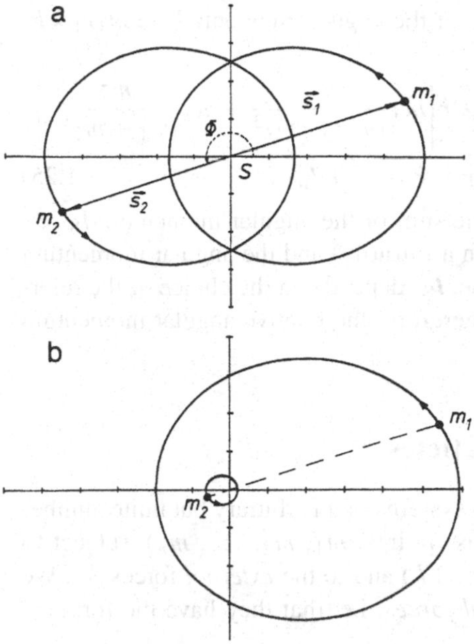

Fig. 1.6. Upon transformation of the relative motion of Fig. 1.5 to the real motion of the two celestial bodies, both move on ellipses about the center of mass S, which is at one of the foci. (**a**) shows the situation for equal masses $m_1 = m_2$; (**b**) shows the case $m_2 \gg m_1$. Cf. Practical Example 1.1

1.7.3 Center-of-Mass and Relative Momentum in the Two-Body System

As we have seen, the equations of motion can be separated in center-of-mass and relative coordinates. Similarly, the sum of the momenta and the sum of the angular momenta can be split into parts pertaining to the center-of-mass motion and parts pertaining to the relative motion. In particular, the total kinetic energy is equal to the sum of the kinetic energies contained in the center-of-mass and relative motions, respectively. These facts are important in formulating conservation laws.

Let $\boldsymbol{P}$ be the momentum of the center of mass and $\boldsymbol{p}$ the momentum of the relative motion. We then have, in more detail,

$$\boldsymbol{P} \stackrel{\text{def}}{=} (m_1 + m_2)\dot{\boldsymbol{r}}_S = m_1\dot{\boldsymbol{r}}_1 + m_2\dot{\boldsymbol{r}}_2 = \boldsymbol{p}_1 + \boldsymbol{p}_2$$

$$\boldsymbol{p} \stackrel{\text{def}}{=} \mu\dot{\boldsymbol{r}} = \frac{1}{m_1 + m_2}(m_2\boldsymbol{p}_1 - m_1\boldsymbol{p}_2)\,,$$

or, by inverting these equations,

$$\boldsymbol{p}_1 = \boldsymbol{p} + \frac{m_1}{m_1 + m_2}\boldsymbol{P}\,; \quad \boldsymbol{p}_2 = -\boldsymbol{p} + \frac{m_2}{m_1 + m_2}\boldsymbol{P}\,.$$

The total kinetic energy is

$$T_1 + T_2 = \frac{\boldsymbol{p}_1^2}{2m_1} + \frac{\boldsymbol{p}_2^2}{2m_2} = \frac{\boldsymbol{p}^2}{2\mu} + \frac{\boldsymbol{P}^2}{2(m_1 + m_2)}\,. \tag{1.24}$$

Thus, the kinetic energy can be written as the sum of the kinetic energy of relative motion, $\boldsymbol{p}^2/2\mu$, and the kinetic energy of the center of mass, $\boldsymbol{P}^2/2(m_1+m_2)$. We note that there are no mixed terms in $\boldsymbol{p}$ and $\boldsymbol{P}$.

In a similar way we analyze the sum $\boldsymbol{L}$ of the angular momenta $\boldsymbol{l}_1 = m_1\boldsymbol{r}_1 \times \dot{\boldsymbol{r}}_1$ and $\boldsymbol{l}_2 = m_2\boldsymbol{r}_2 \times \dot{\boldsymbol{r}}_2$. One finds that

$$\boldsymbol{L} = \boldsymbol{l}_1 + \boldsymbol{l}_2 = \boldsymbol{r}_S \times \dot{\boldsymbol{r}}_S(m_1+m_2) + \boldsymbol{r} \times \dot{\boldsymbol{r}}\left[m_1\frac{m_2^2}{(m_1+m_2)^2} + m_2\frac{m_1^2}{(m_1+m_2)^2}\right]$$
$$= (m_1+m_2)\boldsymbol{r}_S \times \dot{\boldsymbol{r}}_S + \mu\boldsymbol{r} \times \dot{\boldsymbol{r}} \equiv \boldsymbol{l}_S + \boldsymbol{l}_{\text{rel}} . \tag{1.25}$$

The total angular momentum splits into the sum of the angular momentum $\boldsymbol{l}_S$ relative to the origin O (which can be chosen arbitrarily) and the angular momentum of the relative motion $\boldsymbol{l}_{\text{rel}}$. The first of these, $\boldsymbol{l}_S$, depends on the choice of the reference system; the second, $\boldsymbol{l}_{\text{rel}}$, does not. Therefore, the relative angular momentum is the relevant dynamical quantity.

1.8 Systems of Finitely Many Particles

These notions and definitions generalize to systems of an arbitrary but finite number of particles as follows. We consider n mass points $(m_1, m_2, \ldots, m_n)$, subject to the *internal* forces $\boldsymbol{F}_{ik}$ (acting between i and k) and to the *external* forces $\boldsymbol{K}_i$. We assume that the internal forces are *central forces*, i.e. that they have the form

$$\boldsymbol{F}_{ik} = F_{ik}(r_{ik})\frac{\boldsymbol{r}_k - \boldsymbol{r}_i}{r_{ik}} \qquad \left(r_{ik} \stackrel{\text{def}}{=} |\boldsymbol{r}_i - \boldsymbol{r}_k|\right) , \tag{1.26}$$

where $F_{ik}(r) = F_{ki}(r)$ is a scalar and continuous function of the distance r. (In Sect. 1.15 we shall deal with a somewhat more general case.) Central forces can be derived from potentials

$$U_{ik}(r) = -\int_{r_0}^{r} F_{ik}(r')\mathrm{d}r' , \tag{1.27}$$

and we have $\boldsymbol{F}_{ik} = -\boldsymbol{\nabla}_k U_{ik}(r)$, where

$$r = \sqrt{\left(x^{(i)} - x^{(k)}\right)^2 + \left(y^{(i)} - y^{(k)}\right)^2 + \left(z^{(i)} - z^{(k)}\right)^2} ,$$

and the gradient is given by

$$\boldsymbol{\nabla}_k = \left(\frac{\partial}{\partial x^{(k)}}, \frac{\partial}{\partial y^{(k)}}, \frac{\partial}{\partial z^{(k)}}\right) .$$

(Remember that F_{ik} is the force that i exerts on k.) The equations of motion read

$$\begin{aligned}
m_1\ddot{\boldsymbol{r}}_1 &= \boldsymbol{F}_{21} + \boldsymbol{F}_{31} + \ldots + \boldsymbol{F}_{n1} + \boldsymbol{K}_1 , \\
m_2\ddot{\boldsymbol{r}}_2 &= \boldsymbol{F}_{12} + \boldsymbol{F}_{32} + \ldots + \boldsymbol{F}_{n2} + \boldsymbol{K}_2 , \\
&\vdots \\
m_n\ddot{\boldsymbol{r}}_n &= \boldsymbol{F}_{1n} + \boldsymbol{F}_{2n} + \ldots + \boldsymbol{F}_{n-1n} + \boldsymbol{K}_n , \quad \text{or}
\end{aligned} \tag{1.28}$$

$$m_i\ddot{\boldsymbol{r}}_i = \sum_{k\neq i}^{n} \boldsymbol{F}_{ki} + \boldsymbol{K}_i , \quad \text{with} \quad \boldsymbol{F}_{ki} = -\boldsymbol{F}_{ik} .$$

With these assumptions one proves the following assertions.

1.9 The Principle of Center-of-Mass Motion

The center of mass S of the n-particle system behaves like a single particle of mass $M = \sum_{i=1}^{n} m_i$ acted upon by the resultant of the external forces:

$$M\ddot{\boldsymbol{r}}_S = \sum_{i=1}^{n} \boldsymbol{K}_i \,, \quad \text{where} \quad \boldsymbol{r}_S \stackrel{\text{def}}{=} \frac{1}{M}\sum_{i=1}^{n} m_i \boldsymbol{r}_i \,. \tag{1.29}$$

This principle is proved by summing the equations (1.28) over all particles. The internal forces cancel in pairs because $\boldsymbol{F}_{ki} = -\boldsymbol{F}_{ik}$, from Newton's third law.

1.10 The Principle of Angular-Momentum Conservation

The time derivative of the total angular momentum equals the sum of all external torques:

$$\frac{\mathrm{d}}{\mathrm{d}t}\left(\sum_{i=1}^{n} \boldsymbol{l}_i\right) = \sum_{j=1}^{n} \boldsymbol{r}_j \times \boldsymbol{K}_j \,. \tag{1.30}$$

Proof. For a fixed particle with index i

$$m_i \boldsymbol{r}_i \times \ddot{\boldsymbol{r}}_i = \sum_{k \neq i} F_{ki}\left(r_{ki}\right) \frac{\boldsymbol{r}_i \times (\boldsymbol{r}_i - \boldsymbol{r}_k)}{r_{ik}} + \boldsymbol{r}_i \times \boldsymbol{K}_i \,.$$

The left-hand side is equal to

$$m_i \frac{\mathrm{d}}{\mathrm{d}t}\left(\boldsymbol{r}_i \times \dot{\boldsymbol{r}}_i\right) = \frac{\mathrm{d}}{\mathrm{d}t}\boldsymbol{l}_i \,.$$

Taking the sum over all i yields the result (1.30). The internal forces cancel pairwise because the cross product is antisymmetric while the scalar function $F_{ik}(r_{ik})$ is symmetric in i and k. □

1.11 The Principle of Energy Conservation

The time derivative of the total internal energy is equal to the total power (work per unit time) of the external forces, viz.

$$\frac{\mathrm{d}}{\mathrm{d}t}(T+U)=\sum_{i=1}^{n}\left(\boldsymbol{v}_i\cdot\boldsymbol{K}_i\right),\quad\text{where}$$

$$T=\frac{1}{2}\sum_{i=1}^{n}m_i\dot{\boldsymbol{r}}_i^2\equiv\sum T_i\quad\text{and}\tag{1.31}$$

$$U=\sum_{i=1}^{n}\sum_{k=i+1}^{n}U_{ik}\left(r_{ik}\right)\equiv U\left(\boldsymbol{r}_1,\ldots,\boldsymbol{r}_n\right).$$

Proof. For fixed i one has

$$m_i\ddot{\boldsymbol{r}}_i=-\boldsymbol{\nabla}_i\sum_{k\neq i}U_{ik}\left(r_{ik}\right)+\boldsymbol{K}_i\ .$$

Taking the scalar product of this equation with $\dot{\boldsymbol{r}}_i$ yields

$$m_i\ddot{\boldsymbol{r}}_i\cdot\dot{\boldsymbol{r}}_i=\frac{1}{2}\frac{\mathrm{d}}{\mathrm{d}t}\left(m_i\dot{\boldsymbol{r}}_i^2\right)=-\dot{\boldsymbol{r}}_i\cdot\boldsymbol{\nabla}_i\sum_{k\neq i}U_{ik}\left(r_{ik}\right)+\dot{\boldsymbol{r}}_i\cdot\boldsymbol{K}_i\ .$$

Now we take the sum over all particles

$$\frac{\mathrm{d}}{\mathrm{d}t}\left(\sum_i\frac{1}{2}m_i\dot{\boldsymbol{r}}_i^2\right)=-\sum_{i=1}^{n}\sum_{\substack{k=1\\k\neq i}}^{n}\dot{\boldsymbol{r}}_i\cdot\boldsymbol{\nabla}_iU_{ik}\left(r_{ik}\right)+\sum_{i=1}^{n}\dot{\boldsymbol{r}}_i\cdot\boldsymbol{K}_i$$

and isolate the terms $i=a,\ k=b$ and $i=b,\ k=a$, with $b>a$, of the double sum on the right-hand side. Their sum is

$$\dot{\boldsymbol{r}}_a\cdot\boldsymbol{\nabla}_aU_{ab}+\dot{\boldsymbol{r}}_b\cdot\boldsymbol{\nabla}_bU_{ba}=\left[\dot{\boldsymbol{r}}_a\cdot\boldsymbol{\nabla}_a+\dot{\boldsymbol{r}}_b\cdot\boldsymbol{\nabla}_b\right]U_{ab}=\frac{\mathrm{d}}{\mathrm{d}t}U_{ab}\,,$$

because $U_{ab}=U_{ba}$. From this it follows that

$$\frac{\mathrm{d}}{\mathrm{d}t}\left[\sum_{i=1}^{n}\frac{1}{2}m_i\dot{\boldsymbol{r}}_i^2+\sum_{i=1}^{n}\sum_{k=i+1}^{n}U_{ik}\left(r_{ik}\right)\right]=\sum_{j=1}^{n}\dot{\boldsymbol{r}}_j\cdot\boldsymbol{K}_j\ .\quad\square$$

We consider next an important special case: the *closed* n-particle system.

1.12 The Closed n-Particle System

A system is said to be *closed* if all external forces vanish. Proposition 1.9 reduces now to

$$M\ddot{\boldsymbol{r}}_S = 0\,, \quad \text{or} \quad M\dot{\boldsymbol{r}}_S = \sum_{i=1}^{n} m_i \dot{\boldsymbol{r}}_i \stackrel{\text{def}}{=} \boldsymbol{P} = \text{const}\,, \quad \text{and}$$

$$\boldsymbol{r}_S(t) = \frac{1}{M}\boldsymbol{P}t + \boldsymbol{r}_S(0) \quad \text{with} \quad \boldsymbol{P} = \sum_{i=1}^{n} \boldsymbol{p}_i = \text{const}\,.$$

This is the principle of *conservation of momentum*: the total momentum of a closed system is conserved.

Proposition 1.10 reads

$$\sum_{i=1}^{n} \boldsymbol{r}_i \times \boldsymbol{p}_i \equiv \sum_{i=1}^{n} \boldsymbol{l}_i \equiv \boldsymbol{L} = \text{const}\,.$$

The total angular momentum is also an integral of the motion.

Proposition 1.11 finally becomes

$$T + U = \sum_{i=1}^{n} \frac{\boldsymbol{p}_i^2}{2m_i} + \sum_{k>i} U_{ik}(r_{ik}) \equiv E = \text{const}\,.$$

In summary, the closed n-particle system is characterized by 10 integrals or constants of the motion, viz.

$\boldsymbol{P}$, the total momentum; $\boldsymbol{P} = \text{const}$	*Momentum Conservation*
$\boldsymbol{r}_S(t) - \frac{1}{M}\boldsymbol{P}t = \boldsymbol{r}_S(0)$	*Center-of-Mass Principle*
$E = T + U = \frac{\boldsymbol{P}^2}{2M} + T_{\text{rel}} + U$	*Conservation of Energy*
$\boldsymbol{L} = \sum_{i=1}^{n} \boldsymbol{l}_i = \boldsymbol{r}_S \times \boldsymbol{P} + \boldsymbol{l}_{\text{rel}}$	*Conservation of Angular Momentum*

The quantities $\{\boldsymbol{r}_S(0), \boldsymbol{P}, \boldsymbol{L}, E\}$ form the *ten classical constants of the motion* of the closed n-particle system.

This remarkable result calls for questions and some comments:

(i) Perhaps the most obvious question is whether the existence of ten integrals of the motion guarantees integrability of the equations of motion, and if so, for which number n of particles it does so. The answer may seem surprising at this point: a closed two particle system whith central forces is indeed integrable, the general closed three particle system is not. In other terms, while the constants of the motion guarantee integrability for $n = 2$, this is not true for $n \geq 3$. The reason for this observation is that, in addition to be conserved, the integrals of the motion must

fulfill certain conditions of compatibility. We shall come back to this question in Sect. 2.37.
(ii) Why are there just ten such integrals? The answer to this question touches upon a profound relationship between invariance of a physical system under space-time coordinate transformations and conservation laws. It turns out that the most general affine transformation that relates one inertial system to another depends on the same number ten of real parameters. This is what is worked in the next section. Here and in Chap. 2 it will become clear that there is a one-to-one correspondence between these parameters and the ten integrals of the motion.

1.13 Galilei Transformations

It is not difficult to verify that the most general affine transformation g that maps inertial frames onto inertial frames must have the following form:

$$\boxed{\begin{aligned} &\boldsymbol{r} \underset{g}{\mapsto} \boldsymbol{r}' = \mathbf{R}\boldsymbol{r} + \boldsymbol{w}t + \boldsymbol{a} \quad \text{with} \quad \mathbf{R} \in O(3),\ \det \mathbf{R} = +1 \ \text{or} \ -1\,, \\ &t \underset{g}{\mapsto} t' = \lambda t + s \quad \text{with} \quad \lambda = +1 \ \text{or} \ -1\,. \end{aligned}} \tag{1.32}$$

Here $\mathbf{R}$ is a rotation, $\boldsymbol{w}$ a constant velocity vector, $\boldsymbol{a}$ a constant vector of dimension length. We analyze this transformation by splitting it into several steps, as follows.

1. A shift of the origin by the constant vector $\boldsymbol{a}$:

$$\boldsymbol{r}' = \boldsymbol{r} + \boldsymbol{a}\,.$$

2. Uniform motion of $\mathbf{K}'$ relative to $\mathbf{K}$, with constant velocity, such that $\mathbf{K}$ and $\mathbf{K}'$ coincide at time $t = 0$:

$$\boldsymbol{r}' = \boldsymbol{r} + \boldsymbol{w}t\,.$$

3. A rotation whereby the system $\mathbf{K}'$ is rotated away from $\mathbf{K}$ in such a way that their origins are the same, as shown in Fig. 1.7, $\boldsymbol{r}' = \mathbf{R}\boldsymbol{r}$. Let

$$\boldsymbol{r} = (x \equiv r_1,\ y \equiv r_2,\ z \equiv r_3) \quad \boldsymbol{r}' = (x' \equiv r_1',\ y' \equiv r_2',\ z' \equiv r_3')\,.$$

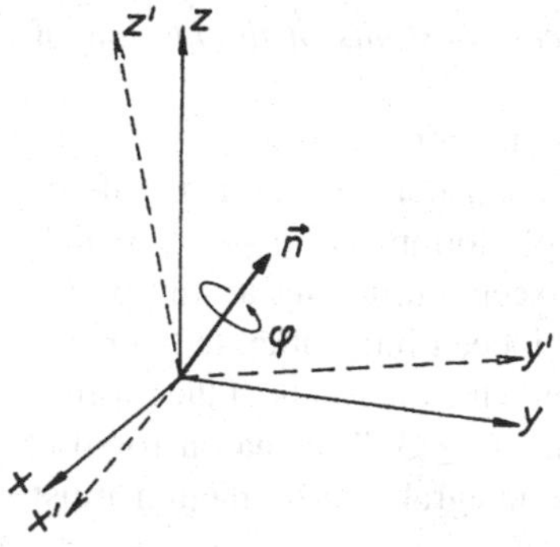

Fig. 1.7. Two Cartesian coordinate systems that are connected by a rotation about the direction $\hat{n}$ by an angle φ

When written in components, $\boldsymbol{r}' = \mathbf{R}\boldsymbol{r}$ is equivalent to

$$r_i' = \sum_{k=1}^{3} R_{ik} r_k \,, \quad i = 1, 2, 3 \,.$$

We must have $\boldsymbol{r}'^2 = \boldsymbol{r}^2$ (this is the defining condition for the rotation group), i.e.

$$\begin{aligned}
&\sum_{i=1}^{3} r_i' r_i' = \sum_{k=1}^{3}\sum_{l=1}^{3}\sum_{i=1}^{3} R_{ik} R_{il} r_k r_l \stackrel{!}{=} \sum_{k=1}^{3} r_k r_k \,, \quad \text{and thus} \\
&\sum_{i=1}^{3} R_{ik} R_{il} \stackrel{!}{=} \delta_{kl} \,, \quad \text{or} \quad \sum_{i=1}^{3} (\mathbf{R}^T)_{ki} R_{il} \stackrel{!}{=} \delta_{kl} \,.
\end{aligned} \tag{1.33}$$

$\mathbf{R}$ is a real orthogonal 3×3 matrix. Equation (1.33) implies $(\det \mathbf{R})^2 = 1$, i.e. $\det \mathbf{R} = +1$ or -1. Equation (1.33) yields 6 conditions for the 9 matrix elements of $\mathbf{R}$. Therefore $\mathbf{R}$ depends on 3 free parameters, for example a direction $\hat{\boldsymbol{n}}$ about which $\mathbf{K}'$ is rotated with respect to $\mathbf{K}$ and which is given by its polar angles (θ, ϕ) and the angle φ by which $\mathbf{K}$ must be rotated in order to rearch $\mathbf{K}'$ (see Fig. 1.7).

4. A shift of the time origin by the fixed amount s:

$$t' = t + s \,.$$

Collecting all steps we see that the general transformation

$$\begin{pmatrix} \boldsymbol{r} \\ t \end{pmatrix} \underset{g}{\longmapsto} \begin{pmatrix} \boldsymbol{r}' = \mathbf{R}\boldsymbol{r} + \boldsymbol{w}t + \boldsymbol{a} \\ t' = \lambda t + s \end{pmatrix} \tag{1.34}$$

with, initially, $\det \mathbf{R} = +1$ and $\lambda = +1$, depends on 10 real parameters, viz.

$$g = g(\underbrace{\varphi, \hat{\boldsymbol{n}}}_{\mathbf{R}}, \boldsymbol{w}, \boldsymbol{a}, s) \,.$$

There are as many parameters in the Galilei transformation as there are constants of the motion in the closed n-particle system. The transformations g form a group, the *proper, orthochronous Galilei group* $\mathrm{G}_+^{\uparrow}$[4]. In order to show this, we consider first the composition of two subsequent transformations of this kind. We have

$$\begin{aligned}
\boldsymbol{r}_1 &= \mathbf{R}^{(1)}\boldsymbol{r}_0 + \boldsymbol{w}^{(1)} t_0 + \boldsymbol{a}^{(1)} \,; \quad t_1 = t_0 + s^{(1)} \,, \\
\boldsymbol{r}_2 &= \mathbf{R}^{(2)}\boldsymbol{r}_1 + \boldsymbol{w}^{(2)} t_1 + \boldsymbol{a}^{(2)} \,; \quad t_2 = t_1 + s^{(2)} \,.
\end{aligned}$$

Writing the transformation from $\boldsymbol{r}_0$ to $\boldsymbol{r}_2$ in the same way,

$$\boldsymbol{r}_2 = \mathbf{R}^{(3)}\boldsymbol{r}_0 + \boldsymbol{w}^{(3)} t_0 + \boldsymbol{a}^{(3)} \,, \quad t_3 = t_0 + s^{(3)} \,,$$

we read off the following relations

[4] The arrow pointing "upwards" stands for the choice $\lambda = +1$; that is, the time direction remains unchanged. The plus sign stands for the choice $\det \mathrm{R} = +1$.

$$\begin{aligned}
\mathbf{R}^{(3)} &= \mathbf{R}^{(2)} \cdot \mathbf{R}^{(1)} \\
\boldsymbol{w}^{(3)} &= \mathbf{R}^{(2)} \boldsymbol{w}^{(1)} + \boldsymbol{w}^{(2)} , \\
\boldsymbol{a}^{(3)} &= \mathbf{R}^{(2)} \boldsymbol{a}^{(1)} + s^{(1)} \boldsymbol{w}^{(2)} + \boldsymbol{a}^{(2)} , \\
s^{(3)} &= s^{(2)} + s^{(1)} .
\end{aligned} \tag{1.35}$$

One now shows explicitly that these transformations do form a group by verifying that they satisfy the *group axioms*:

1. There is an operation defining the composition of two Galilei transforms:
$$g\big(\mathbf{R}^{(2)}, \boldsymbol{w}^{(2)}, \boldsymbol{a}^{(2)}, s^{(2)}\big) g\big(\mathbf{R}^{(1)}, \boldsymbol{w}^{(1)}, \boldsymbol{a}^{(1)}, s^{(1)}\big) = g\big(\mathbf{R}^{(3)}, \boldsymbol{w}^{(3)}, \boldsymbol{a}^{(3)}, s^{(3)}\big) .$$
This is precisely what we verified in (1.35).
2. This composition is an associative operation: $g_3(g_2 g_1) = (g_3 g_2) g_1$. This is so because both addition and matrix multiplication have this property.
3. There exists a unit element, $\boldsymbol{E} = g(\mathbb{1}, \boldsymbol{0}, \boldsymbol{0}, 0)$, with the property $g_i \boldsymbol{E} = \boldsymbol{E} g_i = g_i$ for all $g_i \in \mathrm{G}_+^\uparrow$.
4. For every $g \in \mathrm{G}_+^\uparrow$ there is an inverse transformation g^{-1} such that $g \cdot g^{-1} = \boldsymbol{E}$. This is seen as follows. Let $g = g(\mathbf{R}, \boldsymbol{w}, \boldsymbol{a}, s)$. From (1.35) one sees that $g^{-1} = g(\mathbf{R}^\mathrm{T}, -\mathbf{R}^\mathrm{T}\boldsymbol{w}, s\mathbf{R}^\mathrm{T}\boldsymbol{w} - \mathbf{R}^\mathrm{T}\boldsymbol{a}, -s)$ is its inverse. Indeed, one verifies
$$\begin{aligned}
&g(\mathbf{R}^\mathrm{T}, -\mathbf{R}^\mathrm{T}\boldsymbol{w}, s\mathbf{R}^\mathrm{T}\boldsymbol{w} - \mathbf{R}^\mathrm{T}\boldsymbol{a}, -s)\, g(\mathbf{R}, \boldsymbol{w}, \boldsymbol{a}, s) \\
&= g(\mathbf{R}^\mathrm{T}\mathbf{R}, \mathbf{R}^\mathrm{T}\boldsymbol{w} - \mathbf{R}^\mathrm{T}\boldsymbol{w}, \mathbf{R}^\mathrm{T}\boldsymbol{a} - s\mathbf{R}^\mathrm{T}\boldsymbol{w} + s\mathbf{R}^\mathrm{T}\boldsymbol{w} - \mathbf{R}^\mathrm{T}\boldsymbol{a}, -s + s) \\
&= g(\mathbb{1}, \boldsymbol{0}, \boldsymbol{0}, 0) .
\end{aligned}$$

It will become clear later on that there is a deeper connection between the ten parameters of the proper, orthochronous Galilei group and the constants of the motion of the closed n-particle system of Sect. 1.12 and that it is therefore no accident that there are exactly ten such integrals. We shall learn that the invariance of a mechanical system under

(i) time translations $t \mapsto t' = t + s$ implies the conservation of total energy E of the system;
(ii) space translations $\boldsymbol{r} \mapsto \boldsymbol{r}' = \boldsymbol{r} + \boldsymbol{a}$ implies conservation of total momentum $\boldsymbol{P}$ of the system. The components of $\boldsymbol{a}$ correspond to the components of $\boldsymbol{P}$ in the sense that if the system is invariant only under translations along a fixed direction, then only the projection of $\boldsymbol{P}$ onto that direction is conserved;
(iii) rotations $\boldsymbol{r} \mapsto \boldsymbol{r}' = \mathbf{R}(\varphi)\boldsymbol{r}$ about a fixed direction implies the conservation of the projection of the total angular momentum $\boldsymbol{L}$ onto that direction.

The assertions (i–iii) are the content of a theorem by Emmy Noether, which will be proved and discussed in Sect. 2.19 and, in a somewhat more general form in 2.41.

Finally, one easily convinces oneself that in the center-of-mass motion the quantity

$$\boldsymbol{r}_\mathrm{S}(0) = \boldsymbol{r}_\mathrm{S}(t) - \frac{\boldsymbol{P}}{M} t$$

stays invariant under the transformations $\boldsymbol{r} \mapsto \boldsymbol{r}' = \boldsymbol{r} + \boldsymbol{w}t$.

We conclude this section by considering the choices $\det \mathbf{R} = -1$ and/or $\lambda = -1$ that we have so far excluded. In the Galilei transformation (1.34) the choice $\lambda = -1$ corresponds to a reflection of the time direction, or *time reversal.* Whether or not physical phenomena are invariant under this transformation is a question whose importance goes far beyond mechanics. One easily confirms that all examples considered until now are indeed invariant. This is so because the equations of motion contain only the acceleration $\ddot{\boldsymbol{r}}$, which is invariant by itself, and functions of $\boldsymbol{r}$:

$$\ddot{\boldsymbol{r}} + \boldsymbol{f}(\boldsymbol{r}) = 0 \, .$$

By $t \mapsto -t$ the velocity changes sign, $\dot{\boldsymbol{r}} \mapsto -\dot{\boldsymbol{r}}$. Therefore, the momentum $\boldsymbol{p}$ and also the angular momentum $\boldsymbol{l}$ change sign. The effect of time reversal is equivalent to *reversal of motion.* All physical orbits can be run over in either direction, forward or backward.

There are examples of physical systems, however, that are not invariant under time reversal. These are systems which contain frictional forces proportional to the velocity and whose equations of motion have the form

$$\ddot{\boldsymbol{r}} + K\dot{\boldsymbol{r}} + \boldsymbol{f}(\boldsymbol{r}) = 0 \, .$$

With time reversal the damping caused by the second term in this equation would be changed to an amplification of the motion, i.e. to a different physical process.

The choice $\det \mathbf{R} = -1$ means that the rotation $\mathbf{R}$ contains a space reflection. Indeed, every $\mathbf{R}$ with $\det \mathbf{R} = -1$ can be written as the product of *space reflection* (or *parity*) $\mathbf{P}$:

$$\mathbf{P} \stackrel{\text{def}}{=} \begin{pmatrix} -1 & 0 & 0 \\ 0 & -1 & 0 \\ 0 & 0 & -1 \end{pmatrix} ,$$

and a rotation matrix $\bar{\mathbf{R}}$ with $\det \bar{\mathbf{R}} = +1$, $\mathbf{R} = \mathbf{P} \cdot \bar{\mathbf{R}}$. Note that $\mathbf{P}$ turns a coordinate system with right-handed orientation into one with left-handed orientation.

1.14 Space and Time with Galilei Invariance

(i) The invariance of mechanical laws under translations ($\boldsymbol{a}$) is a manifestation of the *homogeneity* of the physical, three-dimensional space; invariance under rotations ($\mathbf{R}$) is an expression of its *isotropy.* Here we wish to discuss these relations a little further. Imagine that we observe the motion of the sun and its planets from an inertial frame $\mathbf{K}_0$. In that frame we establish the equations of motion and, by solving them, obtain the orbits as a function of time. Another observer who uses a frame $\mathbf{K}$ that is shifted and rotated compared to $\mathbf{K}_0$ will describe the same planetary system by means of the same equations of motion. The explicit solutions will look different in his system, though, because he sees the same physics taking place at a different point in space and with a different spatial orientation. However, the equations of motion that the system obeys, i.e. the basic differential equations, are the same in either frame. Of course, the observer in $\mathbf{K}$ may also choose his time zero differently from the one in $\mathbf{K}_0$, without changing anything in the physics that takes place. It is in this sense that space and time are homogeneous and space, in addition, is isotropic. Finally, it is also admissible to let the two systems $\mathbf{K}$ and $\mathbf{K}_0$ move with *constant* velocity $\boldsymbol{w}$ relative to each other. The equations of motion depend only on differences of coordinate vectors $(\boldsymbol{x}^{(i)} - \boldsymbol{x}^{(k)})$ and therefore do not change. In other words, physical motion is always *relative* motion.

So far we have used the *passive* interpretation of Galilei transformations: the physical system (the sun and its planets) are given and we observe it from different inertial frames. Of course, one can also choose the *active* interpretation, that is, choose a fixed inertial system and ask the question whether the laws of planetary motion are the same, independent of where the motion takes place, of how the orbits are oriented in space, and of whether the center of mass is at rest with respect to the observer or moves at a constant velocity $\boldsymbol{w}$.

[Another way of expressing the passive interpretation is this: an observer located at a point A of the universe will abstract the same fundamental laws from the motion of celestial bodies as another observer who is located at a point B of the universe. For the active interpretation, on the other hand, one would ask a physicist at B to carry out the same experiments as a physicist whose laboratory is based at A. If they obtain the same results and reach the same conclusions, under the conditions on the relative position (or motion) of their reference frames defined above, physics is Galilei invariant.]

(ii) Suppose we consider two physically connected events (a) and (b), the first of which takes place at position $\boldsymbol{x}^{(a)}$ at time t^a, while the second takes place at position $\boldsymbol{x}^{(b)}$ at time t^b. For example, we throw a stone in the gravitational field of the earth such that at t^a it departs from $\boldsymbol{x}^{(a)}$ with a certain initial velocity and arrives at $\boldsymbol{x}^{(b)}$ at time t^b. We parametrize the orbit $\boldsymbol{x}$ that connects $\boldsymbol{x}^{(a)}$ and $\boldsymbol{x}^{(b)}$ and likewise the time variable by

$$\begin{aligned} \boldsymbol{x} &= \boldsymbol{x}(\tau) \quad \text{with} \quad \boldsymbol{x}^{(a)} = \boldsymbol{x}(\tau_a)\,, \quad \boldsymbol{x}^{(b)} = \boldsymbol{x}(\tau_b)\,, \\ t &= t(\tau) \quad \text{with} \quad t^a = t(\tau_a)\,, \quad t^b = t(\tau_b)\,, \end{aligned}$$

where τ is a scalar parameter (the proper time). The time that a comoving clock will show has no preferred zero. Furthermore, it can be measured in arbitrary units. The most general relation between t and τ is then $t(\tau) = \alpha\tau + \beta$ with α and β real constants. Expressed in the form of a differential equation this means that $\mathrm{d}^2 t/\mathrm{d}\tau^2 = 0$. Similarly, the orbit $\boldsymbol{x}(\tau)$ obeys the differential equation

$$\frac{\mathrm{d}^2\boldsymbol{x}}{\mathrm{d}\tau^2} + \boldsymbol{f}(\boldsymbol{r})\left(\frac{\mathrm{d}t}{\mathrm{d}\tau}\right)^2 = 0\,,$$

with $\mathrm{d}t/\mathrm{d}\tau = \alpha$ and where $\boldsymbol{f}$ is minus the force divided by the mass. The comparison of these differential equations shows the asymmetry between space and time that we noted earlier. Under Galilei transformations, $t(\tau) = \alpha\tau + \beta$ becomes $t'(\tau) = \alpha\tau + \beta + s$; that is, time differences such as $(t^a - t^b)$ remain unchanged. Time $t(\tau)$ runs linearly in τ, independently of the inertial frame one has chosen.

In this sense the time variable of nonrelativistic mechanics has an absolute character. No such statement applies to the spatial coordinates, as will be clear from the following reasoning.

We follow the same physical motion as above, from two different inertial frames **K** (coordinates $\boldsymbol{x}, t$) and $\mathbf{K}'$ (coordinates $\boldsymbol{x}', t'$). If they have the same orientation, they are related by a Galilei transformation $g \in \mathrm{G}_+^\uparrow$, so that

$$
\begin{aligned}
t'^a - t'^b &= t^a - t^b \,, \\
\left(\boldsymbol{x}'^{(a)} - \boldsymbol{x}'^{(b)}\right)^2 &= \left(\mathbf{R}\left(\boldsymbol{x}^{(a)} - \boldsymbol{x}^{(b)}\right) + \boldsymbol{w}\left(t^a - t^b\right)\right)^2 \\
&= \left(\left(\boldsymbol{x}^{(a)} - \boldsymbol{x}^{(b)}\right) + \mathbf{R}^{-1}\boldsymbol{w}\left(t^a - t^b\right)\right)^2 .
\end{aligned}
$$

(The last equation follows because the vectors z and $\mathbf{R}z$ have the same length.) In particular, the transformation law for the velocities is

$$
\boldsymbol{v}' = \mathbf{R}\left(\boldsymbol{v} + \mathbf{R}^{-1}\boldsymbol{w}\right) \quad \text{and} \quad \boldsymbol{v}^2 = (\boldsymbol{v}' - \boldsymbol{w})^2 \,.
$$

In observing the same physical process and measuring the distance between points (a) and (b), observers in **K** and $\mathbf{K}'$ reach different conclusions. Thus, unlike the time axis, orbital space does not have a universal character.

The reason for the difference in the results obtained in measuring a distance is easy to understand: the two systems move relative to one another with constant velocity $\boldsymbol{w}$. From the last equation we see that the velocities at corresponding space points differ. In particular, the initial velocities at point (a), i.e. the initial conditions, are not the same. Therefore, calculating the distance between (a) and (b) from the observed velocity and the time difference gives different answers in **K** and in $\mathbf{K}'$. (On the other hand, if we chose the initial velocities in (a) to be the same with respect to **K** and to $\mathbf{K}'$, we would indeed find the same distance. However, these would be two different processes.) The main conclusion is that, while it is meaningful to talk about the spatial distance of two events taking place at the *same* time, it is not meaningful to compare distances of events taking place at *different* times. Such distances depend on the inertial frame one is using. In Sect. 4.7 we shall establish the geometrical structure of space–time that follows from these considerations.

1.15 Conservative Force Fields

In our discussion of the n-particle system (Sects. 1.8–1.12) we had assumed the internal forces to be central forces and hence to be potential forces. Here we wish to discuss the somewhat more general case of *conservative* forces.

Conservative forces are defined as follows. Any force field that can be represented as the (negative) gradient field of a time-independent, potential energy $U(\boldsymbol{r})$,

$$
\boldsymbol{F} = -\boldsymbol{\nabla} U(\boldsymbol{r}) \,,
$$

is called *conservative.* This definition is equivalent to the statement that the work done by such forces along a path from $\boldsymbol{r}_0$ to $\boldsymbol{r}$ depends only on the starting point and on the end point but is independent of the shape of that path. More precisely, a force field is conservative precisely when the path integral

$$\int_{\boldsymbol{r}_0}^{\boldsymbol{r}} (\boldsymbol{F} \cdot \mathrm{d}\boldsymbol{s}) = -\big(U(\boldsymbol{r}) - U(\boldsymbol{r}_0)\big)$$

depends on $\boldsymbol{r}$ and $\boldsymbol{r}_0$ only. As already indicated, the integral can then be expressed as the difference of the potential energies in $\boldsymbol{r}$ and $\boldsymbol{r}_0$. In particular, the balance of the work done or gained along a closed path is zero if the force is conservative, viz.

$$\oint_\tau (\boldsymbol{F} \cdot \mathrm{d}\boldsymbol{s}) = 0\,,$$

for any closed path τ.

What are the conditions for a force field to be conservative, i.e. to be derivable from a potential? If there is a potential U (which must be at least C^2), the equality of the mixed second derivatives $\partial^2 U/\partial y \partial x = \partial^2 U/\partial x \partial y$ (cyclic in x, y, z) implies the relations

$$\frac{\partial F_y}{\partial x} - \frac{\partial F_x}{\partial y} = 0 \quad \text{(plus cyclic permutations)}\,.$$

Thus the *curl* of $\boldsymbol{F}(\boldsymbol{r})$, i.e.

$$\operatorname{curl} \boldsymbol{F} \stackrel{\text{def}}{=} \left(\frac{\partial F_z}{\partial y} - \frac{\partial F_y}{\partial z}, \frac{\partial F_x}{\partial z} - \frac{\partial F_z}{\partial x}, \frac{\partial F_y}{\partial x} - \frac{\partial F_x}{\partial y}\right),$$

must vanish. This is a *necessary* condition, which is *sufficient* only if the domain over which the function $U(\boldsymbol{r})$ is defined and where $\operatorname{curl} \boldsymbol{F}$ vanishes is singly connected. Singly connected means that every closed path that lies entirely in the domain can be contracted to a point without ever meeting points that do not belong to the domain. Let τ be a smooth, closed path, let S be the surface enclosed by it, and let $\hat{n}$ be the local normal to this surface. Stokes' theorem of vector analysis then states that the work done by the force $\boldsymbol{F}$ along the path τ equals the surface integral over S of the normal component of its curl:

$$\oint_\tau (\boldsymbol{F} \cdot \mathrm{d}\boldsymbol{s}) = \iint_S \mathrm{d}f (\operatorname{curl} \boldsymbol{F}) \cdot \hat{n}\,.$$

This formula shows the relationship between the condition $\operatorname{curl} \boldsymbol{F} = 0$ and the definition of a conservative force field: the integral on the left-hand side vanishes, for all closed paths, only if $\operatorname{curl} \boldsymbol{F}$ vanishes everywhere.

We consider two examples, for the sake of illustration.

Example (i) A central force has vanishing curl, since

$$\left(\text{curl } f(r)\boldsymbol{r}\right)_x = \frac{\mathrm{d}f}{\mathrm{d}r}\left(\frac{\partial r}{\partial y}z - \frac{\partial r}{\partial z}y\right)$$
$$= \frac{\mathrm{d}f}{\mathrm{d}r}\frac{1}{r}(yz - zy) = 0 \quad \text{(plus cyclic permutations)} .$$

Example (ii) The curl of the following force field does not vanish everywhere:

$$F_x = -B\frac{y}{\varrho^2}\,, \quad F_y = +B\frac{x}{\varrho^2}\,, \quad F_z = 0\,,$$

where

$$\varrho = x^2 + y^2 \quad \text{and} \quad B = \text{const} .$$

(This is the magnetic field around a straight, conducting wire.) It vanishes only outside the z-axis, i.e. in $\mathbb{R}^3$ from which the z-axis ($x = 0,\ y = 0$) has been cut out. Indeed, as long as $(x, y) \neq (0, 0)$ we have

$$(\text{curl } \boldsymbol{F})_x = (\text{curl } \boldsymbol{F})_y = 0\,,$$
$$(\text{curl } \boldsymbol{F})_z = B\left(\frac{1}{\varrho^2} - \frac{2x^2}{\varrho^4} + \frac{1}{\varrho^2} - \frac{2y^2}{\varrho^4}\right) = 0\,.$$

For $x = y = \varrho = 0$, however, the z-component does not vanish. An equivalent statement is that the closed integral $\oint(\boldsymbol{F}\cdot \mathrm{d}\boldsymbol{s})$ vanishes for all paths that do not enclose the z-axis. For a path that winds around the z-axis once one finds that

$$\oint(\boldsymbol{F}\cdot \mathrm{d}\boldsymbol{s}) = 2\pi B\,.$$

This is shown as follows. Choose a circle of radius R around the origin that lies in the (x, y)-plane. Any other path that winds around the z-axis once can be deformed continuously to this circle without changing the value of the integral. Choose then cylindrical coordinates ($x = \varrho\cos\phi,\ y = \varrho\sin\phi,\ z$). Then $\boldsymbol{F} = (B/\varrho)\hat{\boldsymbol{e}}_\phi$ and $\mathrm{d}\boldsymbol{s} = \varrho\,\mathrm{d}\phi\hat{\boldsymbol{e}}_\phi$, where $\hat{\boldsymbol{e}}_\phi = -\hat{\boldsymbol{e}}_x\sin\phi + \hat{\boldsymbol{e}}_y\cos\phi$, and $\oint(\boldsymbol{F}\cdot \mathrm{d}\boldsymbol{s}) = B\int_0^{2\pi}\mathrm{d}\phi = 2\pi B$. A path winding around the z-axis n times would give the result $2\pi nB$.

Yet, in this example one can define a potential, viz.

$$U(\boldsymbol{r}) = -B\arctan(y/x) = -B\phi\,.$$

This function is unique over any partial domain of $\mathbb{R}^3$ that avoids the z-axis. However, as soon as the domain contains the z-axis this function ceases to be unique, in spite of the fact that curl $\boldsymbol{F}$ vanishes everywhere outside the z-axis. Clearly, such a domain is no longer singly connected.

1.16 One-Dimensional Motion of a Point Particle

Let q be the coordinate, p the corresponding momentum, and $F(q)$ the force. We then have

$$\dot{q} = \frac{1}{m} p ; \quad \dot{p} = F(q) . \tag{1.36}$$

This is another way of writing the equation of motion. The first equation repeats the definition of the momentum, $F(q)$ on the right-hand side of the second equation is the force field (in one dimension).

The kinetic energy is $T = m\dot{q}^2/2 = p^2/2m$. The function $F(q)$ is assumed to be continuous. In one dimension there is always a potential energy $U(q) = -\int_{q_0}^{q} F(q')\mathrm{d}q'$ such that $F(q) = -\mathrm{d}U(q)/\mathrm{d}q$. The total energy $E = T + U$ is conserved:

$$\frac{\mathrm{d}E}{\mathrm{d}t} = \frac{\mathrm{d}(T+U)}{\mathrm{d}t} = 0 .$$

Indeed, calculating the derivatives of T and U one has

$$\frac{\mathrm{d}(T+U)}{\mathrm{d}t} = m\dot{q}\ddot{q} + \dot{q}\frac{\mathrm{d}U}{\mathrm{d}t} = \dot{q}\,\{m\ddot{q} - F(q)\} = 0 ,$$

where use was made of the equation of motion $m\ddot{q} = F(q)$. Note that the time derivative is not arbitrary but is taken along *solutions* $q(t)$ of the equation of motion. Such solutions are also called *orbits* of the system and, therefore, the time derivative which is relevant here, is called the *orbital derivative.*

Take as an example the harmonic force $F(q) = -\kappa p$, with κ a real positive constant, i.e. a force that is linear in the coordinate q and tends to drive the system back to the equilibrium position $q_0 = 0$ (Hooke's law),

$$\dot{q} = \frac{1}{m} p , \qquad \dot{p} = -\kappa q .$$

In this example kinetic and potential energy are

$$T = \frac{1}{2} m\dot{q}^2 = \frac{p^2}{2m} , \quad U(q) = \frac{1}{2}\kappa q^2 .$$

Consider a particular solution of the equation of motion $m\ddot{q} = -\kappa q$, for instance the one that starts at $(q = -a, p = 0)$ at time $t = 0$,

$$q(t) = -a\cos\left(\sqrt{\kappa/m}\, t\right) , \quad p(t) = a\sqrt{m\kappa}\,\sin\left(\sqrt{\kappa/m}\, t\right) .$$

The spatial motion which is actually seen by an observer is the oscillatory function $q(t) = -a\cos(\sqrt{\kappa/m}\, t)$ in coordinate space. Although this is a simple function of time, it would need many words to describe the temporal evolution of the particle's trajectory to a third party. Such a description could go as follows: "The particle starts at $q = -a$, where its kinetic energy is zero, its potential energy is maximal and equal to the total energy $U(q = a) = E = (1/2)\kappa a^2$. It accelerates, as it is driven to the origin, from initial momentum zero to $p = a\sqrt{m\kappa}$ at the time it passes the origin. At this moment its potential energy is zero, its kinetic energy $T_{\text{kin}} = E$ is maximal. Beyond that point the particle is slowed down until it reaches its maximal position $q = a$ where its momentum vanishes again. After that time the momentum changes sign, increases in magnitude until the particle passes the origin, then decreases until the particle reaches its initial position. From then on the motion repeats periodically, the period being $T = 2\pi\sqrt{m}/\sqrt{\kappa}$."

The physics of the particle's motion becomes much simpler to describe if one is ready to accept a small step of abstraction: Instead of studying the coordinate function $q(t)$ in its one-dimensional manifold $\mathbb{R}$ alone, imagine a *two*-dimensional space with abscissa q and ordinate p,

$$\left\{\mathbb{R}^2 \,, \text{ with coordinates } (q, p)\right\} ,$$

and draw the solutions $(q(t), p(t))$ as curves in that space, parametrized by time t. In the example we have chosen these are periodic motions, hence closed curves. Now, the actual physical motion becomes obvious and, in fact, quite simple: In the two-dimensional space spanned by q and p the particle moves on a closed curve, reflecting the alternating behaviour of position and momentum, as well as of kinetic and potential energies. Figures of this kind are shown below in Sect. 1.17.

These elementary considerations and the example we have given may be helpful in motivating the following definitions.

We introduce a compact notation for the equations (1.36) by means of the following *definitions.* With

$$\underset{\sim}{x} = \left\{x_1 \stackrel{\text{def}}{=} q,\ x_2 \stackrel{\text{def}}{=} p\right\} ; \quad \underset{\sim}{\mathcal{F}} = \left\{\mathcal{F}_1 \stackrel{\text{def}}{=} \frac{1}{m} p, \quad \mathcal{F}_2 \stackrel{\text{def}}{=} F(q)\right\} .$$

the equations (1.36) are packed into one single differential equation for a two-component variable

$$\dot{\underset{\sim}{x}} = \underset{\sim}{\mathcal{F}}(\underset{\sim}{x}, t) \,. \tag{1.37}$$

The solutions $x_1(t) = \varphi(t)$ and $x_2(t) = m\dot{\varphi}(t)$ of this differential equation are called *phase portraits.* The energy function $E(q, p) = E(\varphi(t), \dot{\varphi}(t))$, when taken along the phase curves, is constant.

The $\underset{\sim}{x}$ are points of a *phase space* $\mathbb{P}$ whose dimension is $\dim \mathbb{P} = 2$. One should note that the abscissa q and the ordinate p, a priori, are independent variables that span the phase space. The ordinate p becomes a function of q only along solution curves of (1.36) or (1.37). The physical motion "flows" across the phase space. To illustrate this new picture of mechanical processes we consider two more examples.

1.17 Examples of Motion in One Dimension

1.17.1 The Harmonic Oscillator

The harmonic oscillator is defined by its force law $F(q) = -m\omega^2 q$. The applied force is proportional to the elongation and is directed so that it always drives the particle back to the origin. The potential energy is then

$$U(q) = \tfrac{1}{2} m\omega^2 \left(q^2 - q_0^2\right) , \tag{1.38}$$

where q_0 can be chosen to be zero, without loss of generality. One has

$$\dot{\underset{\sim}{x}} = \underset{\sim}{\mathcal{F}}(\underset{\sim}{x}) \quad \text{with} \quad x_1 = q\,, \quad x_2 = p\,, \quad \text{and}$$

$$\mathcal{F}_1 = \frac{1}{m} p = \frac{1}{m} x_2\,, \quad \mathcal{F}_2 = F(q) = -m\omega^2 x_1\,,$$

so that the equations of motion (1.37) read explicitly

$$\dot{x}_1 = \frac{1}{m} x_2\,, \quad \dot{x}_2 = -m\omega^2 x_1\,.$$

The total energy is conserved and has the form

$$E = \frac{x_2^2}{2m} + \frac{1}{2} m\omega^2 x_1^2 = \text{const}\,.$$

One can hide the constants m and ω by redefining the space, the momentum, and time variables as follows:

$$z_1(\tau) \stackrel{\text{def}}{=} \omega\sqrt{m}\, x_1(t)\,, \quad z_2(\tau) \stackrel{\text{def}}{=} \frac{1}{\sqrt{m}} x_2(t)\,, \quad \tau \stackrel{\text{def}}{=} \omega t\,.$$

This transformation makes the energy a simple quadratic form,

$$E = \tfrac{1}{2}\left[z_1^2 + z_2^2\right] ,$$

while time is measured in units of the inverse circular frequency $\omega^{-1} = T/2\pi$. One obtains the system of equations

$$\frac{\mathrm{d}z_1(\tau)}{\mathrm{d}\tau} = z_2(\tau)\,, \quad \frac{\mathrm{d}z_2(\tau)}{\mathrm{d}\tau} = -z_1(\tau)\,. \tag{1.39}$$

It is not difficult to guess their solution for the initial conditions $z_1(\tau = 0) = z_1^0$, $z_2(\tau = 0) = z_2^0$. It is

$$z_1(\tau) = \sqrt{\left(z_1^0\right)^2 + \left(z_2^0\right)^2}\, \cos(\tau - \varphi)\,,$$

$$z_2(\tau) = -\sqrt{\left(z_1^0\right)^2 + \left(z_2^0\right)^2}\, \sin(\tau - \varphi)\,, \quad \text{where}$$

$$\sin\varphi = z_2^0 \Big/ \sqrt{\left(z_1^0\right)^2 + \left(z_2^0\right)^2}\,, \quad \cos\varphi = z_1^0 \Big/ \sqrt{\left(z_1^0\right)^2 + \left(z_2^0\right)^2}\,.$$

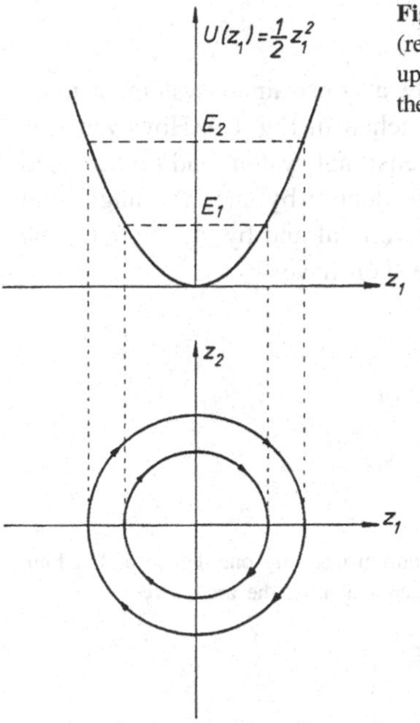

Fig. 1.8. The harmonic oscillator in one dimension: z_1 is the (reduced) position, z_2 the (reduced) momentum variable. The upper part shows the potential energy; the lower part shows the phase portraits for two values of the energy

The motion corresponding to a fixed value of the energy E becomes particularly clear if followed in *phase space* (z_1, z_2). The solution curves in phase space are called *phase portraits.* In our example they are circles of radius $\sqrt{2E}$, on which the system moves clockwise. The example is completely symmetric in coordinate and momentum variables. Figure 1.8 shows in its upper part the potential as a function of z_1 as well as two typical values of the energy. In the lower part it shows the phase portraits corresponding to these energies.

Note what we have gained in describing the motion in phase space rather than in coordinate space only. True, the coordinate space of the harmonic oscillator is directly "visible". However, if we try to describe the temporal evolution of a specific solution $q(t)$ in any detail (i.e. the swinging back and forth, with alternating accelerations and decelerations, etc.), we will need many words for a process that is basically so simple. Adopting the phase-space description of the oscillator, on the other hand, means a first step of abstraction because one interprets the momentum as a new, independent variable, a quantity that is measurable but not directly "visible". The details of the motion become more transparent and are very simple to describe: the oscillation is now a closed curve (lower part of Fig. 1.8) from which one directly reads off the time variation of the position and momentum and therefore also that of the potential and kinetic energy.

The transformation to the new variables $z_1 = \omega\sqrt{m}q$ and $z_2 = p/\sqrt{m}$ shows that in the present example the phase portraits are topologically equivalent to circles along which the oscillator moves with constant angular velocity ω.

1.17.2 The Planar Mathematical Pendulum

Strictly speaking, the planar pendulum is already a constrained system: a mass point moves on a circle of constant radius, as sketched in Fig. 1.9. However, it is so simple that we may treat it like a free one-dimensional system and do not need the full formalism of constrained motion yet. We denote by $\varphi(t)$ the angle that measures the deviation of the pendulum from the vertical and by $s(t) = l\varphi(t)$ the length of the corresponding arc on the circle. We then have

$$T = \tfrac{1}{2}m\dot{s}^2 = \tfrac{1}{2}ml^2\dot{\varphi}^2 ,$$

$$U = \int_0^s mg\sin\varphi' \, ds' = mgl\int_0^\varphi \sin\varphi' \, d\varphi' , \quad \text{or}$$

$$U = -mgl[\cos\varphi - 1] .$$

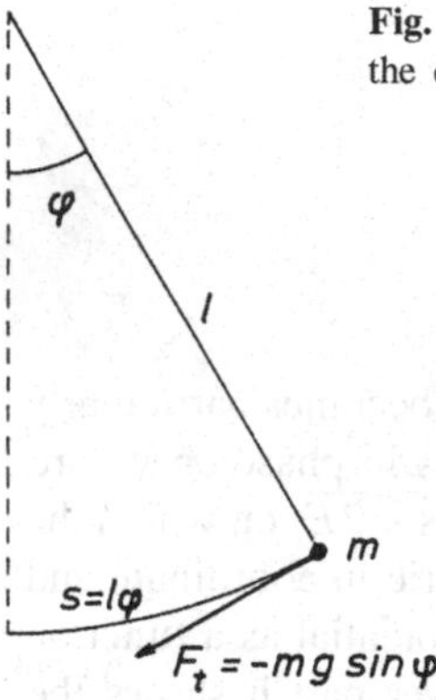

Fig. 1.9. The plane mathematical pendulum has only one degree of freedom: the deviation φ from the vertical, or, equivalently, the arc $s = l\varphi$

We introduce the constants

$$\varepsilon \stackrel{\text{def}}{=} \frac{E}{mgl} = \frac{1}{2\omega^2}\dot{\varphi}^2 + 1 - \cos\varphi \quad \text{with} \quad \omega^2 \stackrel{\text{def}}{=} \frac{g}{l} .$$

As in Sect. 1.17.1 we set $z_1 = \varphi$, $\tau = \omega t$, and $z_2 = \dot{\varphi}/\omega$. Then $\varepsilon = z_2^2/2 + 1 - \cos z_1$, while the equation of motion $ml\ddot{\varphi} = -mg\sin\varphi$ reads, in the new variables,

$$\frac{dz_1}{d\tau} = z_2(\tau) , \quad \frac{dz_2}{d\tau} = -\sin z_1(\tau) . \tag{1.40}$$

In the limit of small deviations from the vertical one has $\sin z_1 = z_1 + O(z_1^3)$ and (1.40) reduces to the system (1.39) of the oscillator. In Fig. 1.10 we sketch the potential $U(z_1)$ and some phase portraits. For values of ε below 2 the picture is qualitatively similar to that of the oscillator (see Fig. 1.8). The smaller ε, the

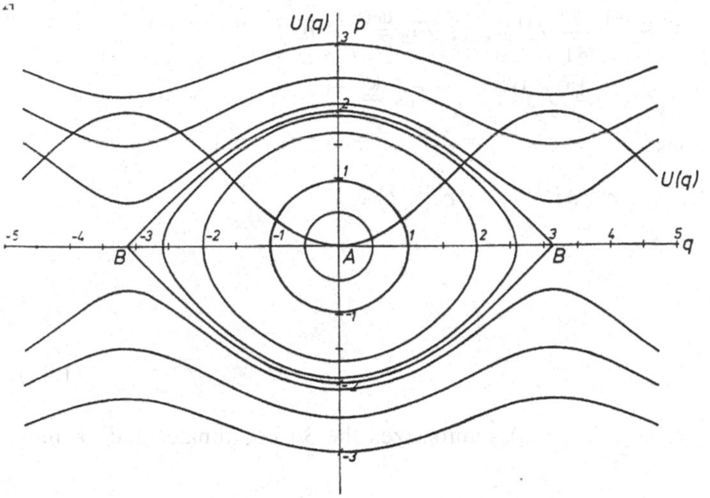

Fig. 1.10. The potential energy $U(q) = 1 - \cos q$ of the plane pendulum, as well as a few phase portraits $p = \sqrt{2(\varepsilon - 1 + \cos q)}$, as a function of q and for several values of the reduced energy $\varepsilon = E/mgl$. Note that in the text $q \equiv z_1$, $p \equiv z_2$. The values of ε can be read off the ordinate

closer this similarity. For $\varepsilon > 2$ the pendulum always swings in one direction, either clockwise or anticlockwise. The boundary $\varepsilon = 2$ between these qualitatively different domains is a singular value and corresponds to the motion where the pendulum reaches the uppermost position but cannot swing beyond it. In Sect. 1.23 we shall show that the pendulum reaches the upper extremum, which is also an unstable equilibrium position, only after infinite time. This singular orbit is called the *separatrix*; it separates the domain of oscillatory solutions from that of rotating solutions.

Note that in Fig. 1.10 only the interval $q \in [-\pi, +\pi]$ is physically relevant. Beyond these points the picture repeats itself such that one should cut the figure at the points marked B and glue the obtained strip on a cylinder.

1.18 Phase Space for the n-Particle System (in $\mathbb{R}^3$)

In Sect. 1.16 we developed the representation of one-dimensional mechanical systems in phase space. It is not difficult to generalize this to higher-dimensional systems such as the n-particle system over $\mathbb{R}^3$. For this purpose we set

$$x_1 \stackrel{\text{def}}{=} x^{(1)}\,, \quad x_2 \stackrel{\text{def}}{=} y^{(1)}\,, \quad x_3 \stackrel{\text{def}}{=} z^{(1)}\,, \quad x_4 \stackrel{\text{def}}{=} x^{(2)}\,, \ldots$$
$$x_{3n} \stackrel{\text{def}}{=} z^{(n)}\,, \quad x_{3n+1} \stackrel{\text{def}}{=} p_x^{(1)}\,, \quad x_{3n+2} \stackrel{\text{def}}{=} p_y^{(1)}\,, \ldots x_{6n} \stackrel{\text{def}}{=} p_z^{(n)}\,.$$

This allows us to write the equations of motion in the same compact form (1.37) provided one defines

$$\mathcal{F}_1 \stackrel{\text{def}}{=} \frac{1}{m_1} p_x^{(1)}\,, \quad \mathcal{F}_2 \stackrel{\text{def}}{=} \frac{1}{m_1} p_y^{(1)}\,, \ldots, \mathcal{F}_{3n} \stackrel{\text{def}}{=} \frac{1}{m_n} p_z^{(n)}\,,$$

$$\mathcal{F}_{3n+1} \stackrel{\text{def}}{=} \mathcal{F}_x^{(1)}\,, \quad \mathcal{F}_{3n+2} \stackrel{\text{def}}{=} \mathcal{F}_y^{(1)}\,, \ldots, \mathcal{F}_{6n} \stackrel{\text{def}}{=} \mathcal{F}_z^{(n)}\,.$$

The original equations

$$\dot{\boldsymbol{p}}^{(i)} = \boldsymbol{F}^{(i)}\left(\boldsymbol{r}^{(1)}, \ldots, \boldsymbol{r}^{(n)}, \dot{\boldsymbol{r}}^{(1)}, \ldots, \dot{\boldsymbol{r}}^{(n)}, t\right),$$

$$\dot{\boldsymbol{r}}^{(i)} = \frac{1}{m_i} \boldsymbol{p}^{(i)}$$

then read

$$\boxed{\dot{\underset{\sim}{x}} = \underset{\sim}{\mathcal{F}}(\underset{\sim}{x}, t)\,.} \tag{1.41}$$

The variable $\underset{\sim}{x} = (x_1, x_2, \ldots, x_{6n})$ summarizes the $3n$ coordinates and $3n$ momenta

$$\boldsymbol{r}^{(i)} = \left(x^{(i)}, y^{(i)}, z^{(i)}\right), \quad \boldsymbol{p}^{(i)} = \left(p_x^{(i)}, p_y^{(i)}, p_z^{(i)}\right), \quad i = 1, \ldots, n\,.$$

The n-particle system has $3n$ coordinates or *degrees of freedom*, $f = 3n$. (The number of degrees of freedom, i.e. the number of independent coordinate variables, will always be denoted by f.)

$\underset{\sim}{x}$ is a point in phase space whose dimension is $\dim \mathbb{P} = 2f$ ($= 6n$, here). This compact notation is more than a formal trick: one can prove a number of important properties for first-order differential equations such as (1.41) that do not depend on the dimension of the system, i.e. the number of components it has.

Remark: Very much like in coordinate space alone, in treating specific problems of mechanics one should choose sets of coordinates in phase space which are optimally adapted to the system one is studying. For instance, returning to the n-body problem considered above, a good choice is the set of *Jacobi coordinates* for which an interesting example may be found in Exercise 2.24 and its solution. The idea is to introduce relative and center-of-mass coordinates and momenta for subsystems of increasing particle number. Like in the two-body system this allows to identify the physically relevant degrees of freedom and to separate them from the center-of-mass motion. A system of relevance for celestial mechanics is the three-body problem where one of the bodies is much heavier than the other two.

1.19 Existence and Uniqueness of the Solutions of $\dot{\underset{\sim}{x}} = \underset{\sim}{\mathcal{F}}(\underset{\sim}{x}, t)$

A striking feature of the phase portraits in Figs. 1.8 and 1.10 is that no two phase curves ever intersect. (Point B of Fig. 1.10 seems an exception: the separatrix arriving from above, the one departing towards the bottom, and the unstable equilibrium meet at B. In reality they do not intersect because B is reached at different times in the three cases; see below.) This makes sense on physical grounds: if two phase portraits did intersect, on arriving at the point of intersection the system would have the choice between two possible ways of continuing its evolution. The description by means of (1.41) would be incomplete. As phase portraits, indeed, do not intersect, a single point $\underset{\sim}{y} \in \mathbb{P}$ together with (1.41) fixes the whole portrait. This point $\underset{\sim}{y}$, which defines the positions and momenta (or velocities), can be understood as the *initial condition* that is assumed at a given time $t = s$. This condition defines how the system will continue to evolve locally.

The theory of ordinary differential equations gives precise information about the existence and uniqueness of solutions for (1.41), provided the function $\underset{\sim}{\mathcal{F}}(\underset{\sim}{x}, t)$ fulfills certain regularity conditions. This information is of immediate relevance for physical orbits that are described by Newton's equations. We quote the following basic theorem but refer to the literature for its proof (see e.g. Arnol'd 1992).

Let $\underset{\sim}{\mathcal{F}}(\underset{\sim}{x}, t)$ with $\underset{\sim}{x} \in \mathbb{P}$ and $t \in \mathbb{R}$ be continuous and, with respect to $\underset{\sim}{x}$, continuously differentiable. Then, for any $\underset{\sim}{z} \in \mathbb{P}$ and any $s \in \mathbb{R}$ there is a neighborhood U of $\underset{\sim}{z}$ and an interval I around s such that for all $\underset{\sim}{y} \in U$ there is precisely one curve $\underset{\sim}{x}(t, s, \underset{\sim}{y})$ with t in I that fulfills the following conditions:

$$\begin{aligned} &\text{(i)} \quad \frac{\partial}{\partial t}\underset{\sim}{x}(t, s, \underset{\sim}{y}) = \underset{\sim}{\mathcal{F}}[\underset{\sim}{x}(t, s, \underset{\sim}{y}), t] \,, \\ &\text{(ii)} \quad \underset{\sim}{x}(t = s, s, \underset{\sim}{y}) = \underset{\sim}{y} \,, \\ &\text{(iii)} \quad \underset{\sim}{x}(t, s, \underset{\sim}{y}) \quad \text{has continuous derivatives in } t, s, \text{ and } \underset{\sim}{y}. \end{aligned} \tag{1.42}$$

$\underset{\sim}{y}$ is the initial point in phase space from which the system starts at time $t = s$. The solution $\underset{\sim}{x}(t, s, \underset{\sim}{y})$ is called the *integral curve of the vector field* $\underset{\sim}{\mathcal{F}}(\underset{\sim}{x}, t)$.

For later purposes (see Chap. 5) we note that $\underset{\sim}{\mathcal{F}}(\underset{\sim}{x}, t) \equiv \underset{\sim}{\mathcal{F}}_t(\underset{\sim}{x})$ can be understood to be a vector field that associates to any $\underset{\sim}{x}$ the velocity vector $\dot{\underset{\sim}{x}} = \underset{\sim}{\mathcal{F}}_t(\underset{\sim}{x})$. This picture is a useful tool for approximate constructions of solution curves in phase space in those cases where one does not have closed expressions for the solutions. This can be done by graphical means by sketching the velocity field and drawing curves to which this field is tangent. Alternatively, one may choose to perform a numerical integration of the equation of motion thereby obtaining solutions as chains of small arcs in phase space.

1.20 Physical Consequences of the Existence and Uniqueness Theorem

Systems described by the equations of motion (1.41) have the following important properties:

(i) They are *finite dimensional*, i.e. every state of the system is completely determined by a point $\underset{\sim}{z}$ in $\mathbb{P}$. The phase space has dimension $2f$, where f is the number of degrees of freedom.
(ii) They are *differential systems*, i.e. the equations of motion are differential equations of finite order.
(iii) They are *deterministic*, i.e. the initial positions and momenta determine the solution locally (depending on the maximal neighborhood U and maximal interval I) in a unique way. In particular, this means that two phase curves do not intersect (in U and I).

Suppose we know all solutions corresponding to all possible initial conditions,

$$\underset{\sim}{x}(t,s,\underset{\sim}{y}) \equiv \underset{\sim}{\Phi}_{t,s}(\underset{\sim}{y}) \,. \tag{1.43}$$

This two-parameter set of solutions defines a mapping of $\mathbb{P}$ onto $\mathbb{P}$, $\underset{\sim}{y} \mapsto \underset{\sim}{x} = \underset{\sim}{\Phi}_{t,s}(\underset{\sim}{y})$. This mapping is unique, and both it and its inverse are differentiable. The set $\underset{\sim}{\Phi}_{t,s}(\underset{\sim}{y})$ is called the *flow* in phase space $\mathbb{P}$.

Consider a system whose initial configuration at time s is $\underset{\sim}{y} \in \mathbb{P}$. The flow describes how the system will evolve from there under the action of its dynamics. At time t it takes on the configuration $\underset{\sim}{x}$, where t may be later or earlier than s. In the first case we find the future evolution of the system, in the second we reconstruct its past. As is customary in mathematics, let the symbol $\circ$ denote the composition of two maps. For example,

$$x \underset{f}{\mapsto} y = f(x) \underset{g}{\mapsto} z = g(y) \quad \text{or} \quad x \underset{g\circ f}{\longmapsto} z = g\big(f(x)\big) \,.$$

With the times r, s, t in the interval I we then have

$$\underset{\sim}{\Phi}_{t,s} \circ \underset{\sim}{\Phi}_{s,r} = \underset{\sim}{\Phi}_{t,r} \,, \qquad \underset{\sim}{\Phi}_{s,s} = 1 \,,$$

$$\frac{\partial}{\partial t}\underset{\sim}{\Phi}_{t,s} = \underset{\sim}{F}_t \circ \underset{\sim}{\Phi}_{t,s} \quad \text{with} \quad \underset{\sim}{F}_t \stackrel{\text{def}}{=} \left.\frac{\partial}{\partial t}\underset{\sim}{\Phi}_{t,s}\right|_{s=t} \,.$$

For *autonomous* systems, i.e. for systems where $\underset{\sim}{\mathcal{F}}$ does not depend explicitly on time, we have

$$\underset{\sim}{x}(t+r,\, s+r,\, \underset{\sim}{y}) = \underset{\sim}{x}(t,s,\underset{\sim}{y}) \,, \quad \text{or} \quad \underset{\sim}{\Phi}_{t+r,s+r} = \underset{\sim}{\Phi}_{t,s} \equiv \underset{\sim}{\Phi}_{t-s} \,. \tag{1.44}$$

In other words, such systems are invariant under time translations.

Proof. Let $t' = t + r$, $s' = s + r$. As $\partial/\partial t = \partial/\partial t'$, we have

$$\frac{\partial}{\partial t}\underset{\sim}{x}(t + r \equiv t' ; s + r \equiv s', \underset{\sim}{y}) = \underset{\sim}{\mathcal{F}}\big(\underset{\sim}{x}(t', s', \underset{\sim}{y})\big)$$

with the initial condition

$$\underset{\sim}{x}(s', s', \underset{\sim}{y}) = \underset{\sim}{x}(s + r, s + r, \underset{\sim}{y}) = \underset{\sim}{y} .$$

Compare this with the solution of

$$\frac{\partial}{\partial t}\underset{\sim}{x}(t, s, \underset{\sim}{y}) = \underset{\sim}{\mathcal{F}}\big(\underset{\sim}{x}(t, s, \underset{\sim}{y})\big) \quad \text{with} \quad \underset{\sim}{x}(s, s, \underset{\sim}{y}) = \underset{\sim}{y} .$$

From the existence and uniqueness theorem follows

$$\underset{\sim}{x}(t+r, s+r, \underset{\sim}{y}) = \underset{\sim}{x}(t, s, \underset{\sim}{y}) . \qquad \square$$

In principle, for a complete description of the solutions of (1.41) we should add the time variable as an additional, orthogonal coordinate to the phase space $\mathbb{P}$. If we do this we obtain what is called the *extended phase space* $\mathbb{P} \times \mathbb{R}_t$, whose dimension is $(2f+1)$ and thus is an odd integer. As time flows monotonously and is not influenced by the dynamics, the special solution $(\underset{\sim}{x}(t), t)$ in extended phase space $\mathbb{P} \times \mathbb{R}_t$ contains no new information compared to its projection $\underset{\sim}{x}(t)$ onto phase space $\mathbb{P}$ alone. Similarly, the projection of the original flow $\{\phi_{t,s}(\underset{\sim}{y}), t\}$ in extended phase space $\mathbb{P} \times \mathbb{R}_t$ onto $\mathbb{P}$ is sufficient to give an almost complete image of the mechanical system one is considering.

Figure 1.10, which shows typical phase portraits for the planar pendulum, yields a particularly instructive illustration of the existence and uniqueness theorem. Given an arbitrary point $\underset{\sim}{y} = (q, p)$, at arbitrary time s, the entire portrait passing through this point is fixed completely. Clearly, one should think of this figure as a three-dimensional one, by supplementing it by a time axis. For example, a phase curve whose portrait (i.e. its projection onto the (q, p)-plane) is approximately a circle in this three-dimensional space will wind around the time axis like a spiral (make your own drawing!). The point B, at first, seems an exception: the separatrix (A) corresponding to the pendulum being tossed from its stable equilibrium position so as to reach the highest position without "swinging through", the separatrix (B), which starts from the highest point essentially without initial velocity, and the unstable equilibrium (C) seem to coincide. This is no contradiction to Theorem 1.19, though, because (A) reaches the point B only at $t = +\infty$, (B) leaves at $t = -\infty$, while (C) is there at any finite t.

We summarize once more the most important consequences of Theorem 1.19. At any point in time the state of the mechanical system is determined completely by the $2f$ real numbers $(q_1, \ldots, q_f; p_1, \ldots, p_f)$. We say that it is finite dimensional. The differential equation (1.41) contains the whole dynamics of the system. The flow, i.e. the set of all solutions of (1.41), transports the system from all possible initial conditions to various new positions in phase space. This transport, when read as a map from $\mathbb{P}$ onto $\mathbb{P}$, is bijective (i.e. it is one to one) and is differentiable in either direction. The flow conserves the differential structure of the dynamics. Finally, systems described by (1.41) are deterministic: the complete knowledge of the momentary configuration (positions and momenta) fixes uniquely all future and past configurations, as long as the vector field is regular, as assumed for the theorem.[5]

1.21 Linear Systems

Linear systems are defined by $\underset{\sim}{\mathcal{F}} = \boldsymbol{A}\underset{\sim}{x} + \underset{\sim}{b}$. They form a particularly simple class of mechanical systems obeying (1.41). We distinguish them as follows.

1.21.1 Linear, Homogeneous Systems

Here the inhomogeneity $\underset{\sim}{b}$ is absent, so that

$$\dot{\underset{\sim}{x}} = \boldsymbol{A}\underset{\sim}{x}, \quad \text{where} \quad \boldsymbol{A} = \{a_{ik}\},$$

or, written in components,

$$\dot{x}_i = \sum_k a_{ik} x_k . \tag{1.45}$$

Example. The harmonic oscillator is described by a linear, homogeneous equation of the type (1.41), viz.

$$\left.\begin{aligned} \dot{x}_1 &= \frac{1}{m} x_2 \\ \dot{x}_2 &= -m\omega^2 x_1 \end{aligned}\right\} \quad \text{or} \quad \begin{pmatrix} \dot{x}_1 \\ \dot{x}_2 \end{pmatrix} = \begin{pmatrix} 0 & \frac{1}{m} \\ -m\omega^2 & 0 \end{pmatrix} \begin{pmatrix} x_1 \\ x_2 \end{pmatrix} . \tag{1.46}$$

The explicit solutions of Sect. 1.17.1 can also be written as follows:

$$x_1(t) = x_1^0 \cos\tau + x_2^0/m\omega \sin\tau , \quad x_2(t) = -x_1^0 m\omega \sin\tau + x_2^0 \cos\tau .$$

Set $\tau = \omega(t-s)$ and

$$\underset{\sim}{y} = \begin{pmatrix} y_1 \\ y_2 \end{pmatrix} \equiv \begin{pmatrix} x_1^0 \\ x_2^0 \end{pmatrix} .$$

Then

$$\underset{\sim}{x}(t) \equiv \underset{\sim}{x}(t,s,\underset{\sim}{y}) = \underset{\sim}{\Phi}_{t,s}(\underset{\sim}{y}) = \boldsymbol{M}(t,s) \cdot \underset{\sim}{y} , \quad \text{with}$$

$$\boldsymbol{M}(t,s) = \begin{pmatrix} \cos\omega(t-s) & \frac{1}{m\omega}\sin\omega(t-s) \\ -m\omega\sin\omega(t-s) & \cos\omega(t-s) \end{pmatrix} . \tag{1.47}$$

One confirms that $\underset{\sim}{\phi}_{t,s}$ and $\boldsymbol{M}(t,s)$ depend only on the difference $(t-s)$. This must be so because we are dealing with an autonomous system. It is interesting to note that the matrix $\boldsymbol{M}$ has determinant 1. We shall return to this observation later.

[5] Note that the existence and uniqueness is guaranteed only locally (in space and time). Only in exceptional cases does the theorem allow one to predict the long-term behavior of the system. Global behavior of dynamical systems is discussed in Sect. 6.3. Some results can also be obtained from energy estimates in connection with the virial, cf. Sect. 1.31 below.

1.21.2 Linear, Inhomogeneous Systems

These have the general form

$$\dot{\underset{\sim}{x}} = \boldsymbol{A}\underset{\sim}{x} + \underset{\sim}{b} \,. \tag{1.48}$$

Example. Lorentz force with homogeneous fields. A particle of charge e in external electric and magnetic fields is subject to the force

$$\boldsymbol{K} = \frac{e}{c}\dot{\boldsymbol{r}} \times \boldsymbol{B} + e\boldsymbol{E} \,. \tag{1.49}$$

In the compact notation we have

$$x_1 = x \,, \; x_2 = y \,, \; x_3 = z \,, \; x_4 = p_x \,, \; x_5 = p_y \,, \; x_6 = p_z \,.$$

Let the magnetic field point in the z-direction, $\boldsymbol{B} = B\hat{\boldsymbol{e}}_z$, i.e. $\dot{\boldsymbol{r}} \times \boldsymbol{B} = (\dot{y}B, \, -\dot{x}B, 0)$. Setting $K = eB/mc$ we then have $\dot{\underset{\sim}{x}} = \boldsymbol{A}\underset{\sim}{x} + \underset{\sim}{b}$, with

$$\boldsymbol{A} = \begin{pmatrix} 0 & 0 & 0 & 1/m & 0 & 0 \\ 0 & 0 & 0 & 0 & 1/m & 0 \\ 0 & 0 & 0 & 0 & 0 & 1/m \\ 0 & 0 & 0 & 0 & K & 0 \\ 0 & 0 & 0 & -K & 0 & 0 \\ 0 & 0 & 0 & 0 & 0 & 0 \end{pmatrix} , \quad \underset{\sim}{b} = e \begin{pmatrix} 0 \\ 0 \\ 0 \\ E_x \\ E_y \\ E_z \end{pmatrix} . \tag{1.50}$$

For a complete treatment of linear systems we refer to the mathematical literature (see e.g. Arnol'd 1992). Some aspects will be dealt with in Sects. 6.2.2 and 6.2.3 in the framework of linearization of vector fields. A further, important example is contained in Practical Example 2.1 (*small oscillations*).

1.22 Integrating One-Dimensional Equations of Motion

The equation of motion for a one-dimensional, autonomous system reads $m\ddot{q} = K(q)$. If $K(q)$ is a continuous function it possesses a potential energy

$$U(q) = -\int_{q_0}^{q} K(q')\,\mathrm{d}q' \,,$$

so that the law of energy conservation takes the form

$$\tfrac{1}{2}m\dot{q}^2 + U(q) = E = \text{const} \,.$$

From this follows a first-order differential equation for $q(t)$:

$$\frac{\mathrm{d}q}{\mathrm{d}t} = \sqrt{\frac{2}{m}\left(E - U(q)\right)} \,. \tag{1.51}$$

This is a particularly simple example for a differential equation with *separable variables* whose general form is

$$\frac{\mathrm{d}y}{\mathrm{d}x} = \frac{g(y)}{f(x)} \tag{1.52}$$

and for which the following proposition holds (see e.g. Arnol'd 1992).

Theorem. Assume the functions $f(x)$ and $g(y)$ to be continuously differentiable in a neighborhood of the points x_0 and y_0, respectively, where they do not vanish, $f(x_0) \neq 0$, $g(y_0) \neq 0$. The differential equation (1.52) then has a unique solution $y = F(x)$ in the neighborhood of x_0 that fulfills the initial condition $y_0 = F(x_0)$ as well as the relation

$$\int_{x_0}^{x} \frac{\mathrm{d}x'}{f(x')} = \int_{y_0}^{F(x)} \frac{\mathrm{d}y'}{g(y')} . \tag{1.53}$$

When applied to (1.51) this means that

$$t - t_0 = \sqrt{\frac{m}{2}} \int_{q_0}^{q(t)} \frac{\mathrm{d}q'}{\sqrt{E - U(q')}} ; \tag{1.54}$$

that is, we obtain an equation which yields the solution if the quadrature on the right-hand side can be carried out. The fact that there was an integral of the motion (here the law of energy conservation) allowed us to reduce the second-order equation of motion to a first-order differential equation that is solved by simple quadrature.

Equations (1.54) and (1.51) can also be used for a qualitative discussion of the motion: since $T + U = E$ and since T must be $T \geq 0$, we must always have $E \geq U(q)$. Consider, for instance, a potential that has a local minimum at $q = q_0$, as sketched in Fig. 1.11. At the points A, B, and C, $E = U(q)$. Therefore, solutions with that energy E must lie either between A and B, or beyond C, $q_A \leq q(t) \leq q_B$, or $q(t) \geq q_C$.

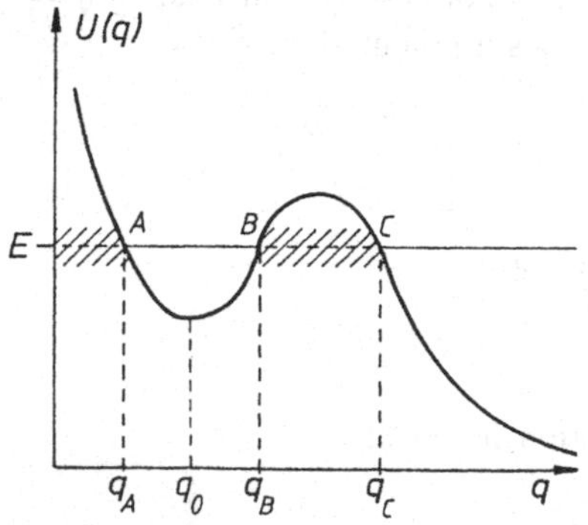

Fig. 1.11. Example of potential energy in one dimension. From energy conservation the kinetic energy must vanish in A, B, and C, for a given total energy E. The hatched areas are excluded for the position variable q

As an example, we consider the first case. Here we obtain *finite* orbits; the points A and B are *turning points* where the velocity $\dot{q}$ passes through zero, according to (1.51). The motion is periodic, its period of oscillation being given by $T(E) = 2\times$ (running time from A to B). Thus

$$T(E) = \sqrt{2m} \int_{q_A(E)}^{q_B(E)} \frac{dq'}{\sqrt{E - U(q')}} . \tag{1.55}$$

1.23 Example: The Planar Pendulum for Arbitrary Deviations from the Vertical

Figure 1.12 shows the maximal deviation $\varphi_0 \le \pi$. According to Sect. 1.17.2 the potential energy is $U(\varphi) = mgl(1-\cos\varphi)$. For $\varphi = \varphi_0$ the kinetic energy vanishes, so that the total energy is given by

$$E = mgl\bigl(1 - \cos\varphi_0\bigr) = mgl(1 - \cos\varphi) + \tfrac{1}{2}ml^2\dot{\varphi}^2 .$$

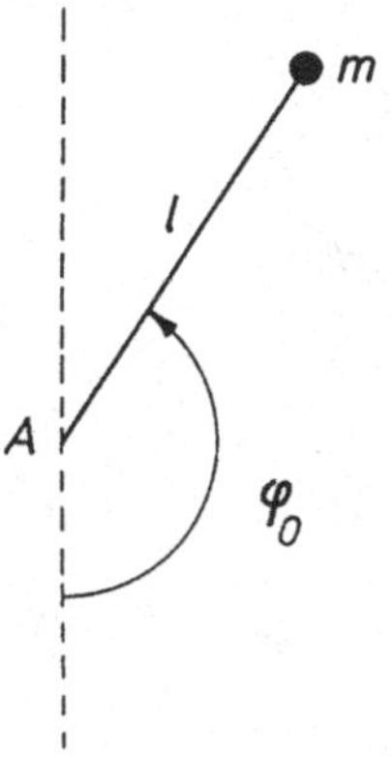

Fig. 1.12. Plane mathematical pendulum for an arbitrary deviation $\varphi_0 \in [0, \pi]$

The period is obtained from (1.55), replacing the arc $s = l\varphi$ by φ:

$$T = 2\sqrt{2m} \int_0^{\varphi_0} l\, d\varphi / \sqrt{mgl(\cos\varphi - \cos\varphi_0)} . \tag{1.56}$$

With $\cos\varphi = 1 - 2\sin^2(\varphi/2)$ this becomes

$$T = 2\sqrt{2}\sqrt{\frac{l}{g}} \int_0^{\varphi_0} d\varphi / \sqrt{\cos\varphi - \cos\varphi_0}$$

$$= 2\sqrt{\frac{l}{g}} \int_0^{\varphi_0} d\varphi / \sqrt{\sin^2(\varphi_0/2) - \sin^2(\varphi/2)} . \tag{1.56'}$$

Substituting the variable φ as follows:

$$\sin\alpha \stackrel{\text{def}}{=} \frac{\sin(\varphi/2)}{\sin(\varphi_0/2)} ,$$

one obtains

$$\begin{aligned} \mathrm{d}\varphi &= 2\,\mathrm{d}\alpha \sin(\varphi_0/2))\sqrt{1-\sin^2\alpha}\Big/\sqrt{1-\sin^2(\varphi_0/2)\sin^2\alpha} , \\ \varphi &= 0 \to \alpha = 0 , \\ \varphi &= \varphi_0 \to \alpha = \pi/2 , \end{aligned}$$

and therefore

$$T = 4\sqrt{\frac{l}{g}}K\big[\sin(\varphi_0/2)\big] , \tag{1.57}$$

where $K(z) = \int_0^{\pi/2} \mathrm{d}\alpha/\sqrt{1-z^2\sin^2\alpha}$ denotes the complete elliptic integral of the first kind (see e.g. Abramowitz, Stegun 1965).

For small and medium-sized deviations from the vertical, one can expand in terms of $z = \sin(\varphi_0/2)$ or directly in terms of $\varphi_0/2$:

$$\begin{aligned} \left(1-z^2\sin^2\alpha\right)^{-1/2} &\simeq 1 + \sin^2\alpha\frac{z^2}{2} + \sin^4\alpha\frac{3z^4}{8} \\ &\simeq 1 + \tfrac{1}{2}\sin^2\alpha\left(\tfrac{1}{4}\varphi_0^2 - \tfrac{1}{48}\varphi_0^4\right) + \tfrac{3}{8}\sin^4\alpha\tfrac{1}{16}\varphi_0^4 . \end{aligned}$$

The integrals that this expansion leads to are elementary, viz.

$$\int_0^{\pi/2} \sin^{2n} x\,\mathrm{d}x = \frac{\pi}{2n!}\left(n-\frac{1}{2}\right)\left(n-\frac{3}{2}\right)\cdots\frac{1}{2} , \quad (n = 1, 2, \ldots) .$$

Thus, one obtains

$$\begin{aligned} K(z) &\simeq \frac{\pi}{2}\left[1 + \frac{1}{4}z^2 + \frac{9}{64}z^4\right] \\ &\simeq \frac{\pi}{2}\left[1 + \frac{1}{16}\varphi_0^2 + \left(\frac{9}{64} - \frac{1}{12}\right)\frac{\varphi_0^4}{16}\right] , \end{aligned}$$

and, finally,

$$T \simeq 2\pi\sqrt{\frac{l}{g}}\left[1 + \frac{1}{16}\varphi_0^2 + \frac{11}{3072}\varphi_0^4\right] . \tag{1.58}$$

The quality of this expansion can be judged from a numerical comparison of successive terms as shown in Table 1.1.

The behavior of T (1.57) in the neighborhood of $\varphi_0 = \pi$ can be studied separately. For that purpose one calculates the time t_Δ that the pendulum takes to

Table 1.1. Deviation from the harmonic approximation

φ_0	$\frac{1}{16}\varphi_0^2$	$\frac{11}{3072}\varphi_0^4$
10°	0.002	3×10^{-6}
20°	0.0076	1×10^{-4}
45°	0.039	1.4×10^{-3}

swing from $\varphi = \pi - \Delta$ to $\varphi = \varphi_0 = \pi - \varepsilon$, where $\varepsilon \ll \Delta$, cf. Fig. 1.13. Introduce $x = \pi - \varphi$ as a new variable and let $T^{(0)} = 2\pi\sqrt{l/q}$. Then

$$\frac{t_\Delta}{T^{(0)}} = \frac{1}{\pi\sqrt{2}}\int_\varepsilon^\Delta \frac{dx}{\sqrt{\cos\varepsilon - \cos x}} \simeq \frac{1}{\pi}\int_\varepsilon^\Delta \frac{dx}{\sqrt{x^2 - \varepsilon^2}} = \frac{1}{\pi}\ln 2\frac{\Delta}{\varepsilon}\,, \tag{1.59}$$

where we have approximated $\cos x$ by $1 - x^2/2$. For $\varphi_0 \to \pi$, i.e. for $\varepsilon \to 0$, t_Δ tends to infinity logarithmically. The pendulum reaches the upper (unstable) equilibrium only after infinite time.

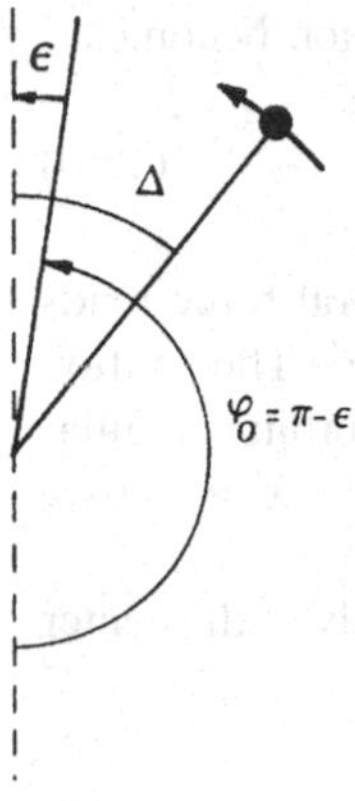

Fig. 1.13. The plane pendulum for large deviations, say $\varphi_0 = \pi - \varepsilon$, where ε is small compared to 1. In the text we calculate the time t_Δ the pendulum needs to swing from $\varphi = \pi - \Delta$ to the maximal value φ_0. One finds that t_Δ goes to infinity like $-\ln\varepsilon$, as one lets ε tend to zero

It is interesting to note that the limiting case $E = 2mgl$ (unstable equilibrium or separatrix) can again be integrated by elementary means. Returning to the notation of Sect. 1.17.2, the variable $z_1 = \varphi$ now obeys the differential equation

$$\frac{1}{2}\left(\frac{dz_1}{d\tau}\right)^2 + (1 - \cos z_1) = 2\,, \quad \text{or} \quad \frac{dz_1}{d\tau} = \sqrt{2(1 + \cos z_1)}\,.$$

Setting $u \stackrel{\text{def}}{=} \tan(z_1/2)$, we find the following differential equation for u:

$$du/\sqrt{u^2+1} = d\tau\,,$$

which can be integrated directly. For example, the solution that starts at $z_1 = 0$ at time $\tau = 0$ fulfills

$$\int_0^u \mathrm{d}u'/\sqrt{u'^2+1} = \int_0^\tau \mathrm{d}\tau' , \quad \text{and hence} \quad \ln\left(u+\sqrt{u^2+1}\right) = \tau .$$

With $u = (e^\tau - e^{-\tau})/2$, the solution for z_1 is obtained as follows:

$$z_1(\tau) = 2 \arctan(\sinh \tau) .$$

If we again choose $z_1 = \pi - \varepsilon$, i.e. $u = \cot \varepsilon/2 \simeq 2/\varepsilon$, we have $u + \sqrt{u^2+1} \simeq 4/\varepsilon$ and $\tau(\varepsilon) \simeq \ln(4/\varepsilon)$. The time to swing from $z_1 = 0$ to $z_1 = \pi$ diverges logarithmically.

1.24 Example: The Two-Body System with a Central Force

Another important example is the two-body system (over $\mathbb{R}^3$) with a central force, to which we now turn. It can be analyzed in close analogy to the one-dimensional problem of Sect. 1.22.

The general analysis of the two-body system was given in Sect. 1.7. Since the force is supposed to be a central force (assumed to be continuous), it can be derived from a spherically symmetric potential $U(r)$. The equation of motion becomes

$$\mu \ddot{\boldsymbol{r}} = -\nabla U(r) , \quad \text{with} \quad \mu = \frac{m_1 m_2}{m_1 + m_2} ; \tag{1.60}$$

$\boldsymbol{r} = \boldsymbol{r}_1 - \boldsymbol{r}_2$ is again the relative coordinate and $r = |\boldsymbol{r}|$. If the central force reads $\boldsymbol{F} = F(r)\hat{\boldsymbol{r}}$, the corresponding potential is $U(r) = -\int_{r_0}^{r} F(r')\mathrm{d}r'$. The motion takes place in the plane perpendicular to the conserved relative orbital angular momentum $\boldsymbol{l}_{\mathrm{rel}} = \boldsymbol{r} \times \boldsymbol{p}$. Introducing polar coordinates in that plane, $x = r\cos\varphi$ and $y = r \sin\varphi$, one has $\dot{\boldsymbol{r}}^2 = \dot{r}^2 + r^2\dot{\varphi}^2$.

The energy of relative motion is conserved because no forces apply to the center of mass and therefore total momentum is conserved:

$$T_{\mathrm{S}} + E = \frac{\boldsymbol{P}^2}{2M} + \frac{\mu}{2}(\dot{r}^2 + r^2\dot{\varphi}^2) + U(r) = \text{const} . \tag{1.61}$$

Thus, with $l \equiv |\boldsymbol{l}| = \mu r^2 \dot{\varphi}$,

$$E = \frac{1}{2}\mu\dot{r}^2 + \frac{l^2}{2\mu r^2} + U(r) = \text{const} . \tag{1.62}$$

$T_r \stackrel{\text{def}}{=} \mu\dot{r}^2/2$ is the kinetic energy of radial motion, whereas the term $l^2/2\mu r^2 = \mu r^2\dot{\varphi}^2/2$ can be read as the kinetic energy of the rotatory motion, or as the potential energy pertaining to the centrifugal force,

$$\boldsymbol{Z} = -\nabla\left(\frac{1}{2}\mu r^2\dot{\varphi}^2\right) = -\hat{\boldsymbol{r}}\frac{\partial}{\partial r}\left(\frac{1}{2}\mu r^2\dot{\varphi}^2\right) = -\mu r\dot{\varphi}^2\hat{\boldsymbol{r}} = -\frac{\mu}{r}v_r^2\hat{\boldsymbol{r}} .$$

From angular-momentum conservation

$$l = \mu r^2 \dot{\varphi} = \text{const} , \tag{1.63}$$

and from energy conservation (1.62) one obtains differential equations for $r(t)$ and $\varphi(t)$:

$$\frac{dr}{dt} = \sqrt{\frac{2}{\mu}(E - U(r)) - \frac{l^2}{\mu^2 r^2}} \equiv \sqrt{\frac{2}{\mu}(E - U_{\text{eff}}(r))} , \tag{1.64}$$

$$\frac{d\varphi}{dt} = \frac{1}{\mu r^2} , \quad \text{with} \tag{1.65}$$

$$U_{\text{eff}}(r) \stackrel{\text{def}}{=} U(r) + \frac{l^2}{2\mu r^2} , \tag{1.66}$$

where the latter, $U_{\text{eff}}(r)$, can be interpreted as an *effective potential*. When written in this form the analogy to the truly one-dimensional case of (1.51) is clearly visible. Like (1.51) the equation of motion (1.64) can be solved by separation of variables, yielding r as a function of time t. This must then be inserted into (1.65), whose integration yields the function $\varphi(t)$. Another way of solving the system of equations (1.64) and (1.65) is to eliminate the explicit time dependence by dividing the second by the first and by solving the resulting differential equation for r as a function of φ, viz.

$$\frac{d\varphi}{dr} = \frac{1}{r^2 \sqrt{2\mu(E - U_{\text{eff}})}} . \tag{1.67}$$

This equation is again separable, and one has

$$\varphi - \varphi_0 = l \int_{r_0}^{r(\varphi)} \frac{dr}{r^2 \sqrt{2\mu(E - U_{\text{eff}})}} . \tag{1.68}$$

Writing $E = T_r + U_{\text{eff}}(r)$, the positivity of T_r again implies that $E \geq U_{\text{eff}}(r)$. Thus, if $r(t)$ reaches a point r_1, where $E = U_{\text{eff}}(r_1)$, the radial velocity $\dot{r}(r_1)$ vanishes. Unlike the case of one-dimensional motion this does not mean (for $l \neq 0$) that the particle really comes to rest and then returns. It rather means that it has reached a point of greatest distance from, or of closest approach to, the force center. The first is called *perihelion* or, more generally, *pericenter*, the second is called *aphelion* or *apocenter*. It is true that the particle has no radial velocity at r_1 but, as long as $l \neq 0$, it still has a nonvanishing angular velocity.

There are various cases to be distinguished.

(i) $r(t) \geq r_{\min} \equiv r_P$ ("P" for "perihelion"). Here the motion is not finite; the particle comes from infinity, passes through perihelion, and disappears again towards infinity. For an *attractive* potential the orbit may look like the examples sketched in Fig. 1.14. For a *repulsive* potential it will have the shape shown in Fig. 1.15. In the former case the particle revolves about the force center once or several times; in the latter it is repelled by the force center and will therefore be scattered.

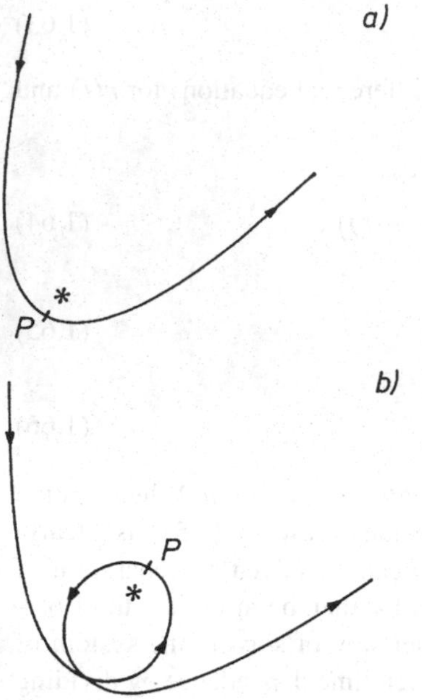

Fig. 1.14. Various infinite orbits for an attractive potential energy. P is the point of closest approach (pericenter or perihelion)

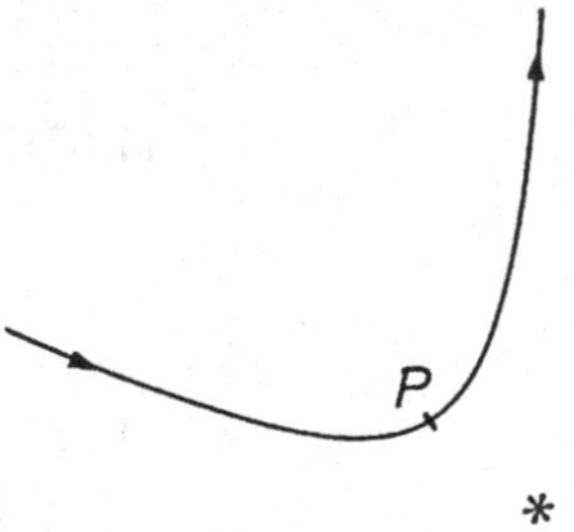

Fig. 1.15. Typical infinite orbit for a repulsive central potential

(ii) $r_{\text{min}} \equiv r_{\text{P}} \leq r(t) \leq r_{\text{max}} \equiv r_{\text{A}}$ ("A" for "aphelion"). In this case the entire orbit is confined to the circular annulus between the circles with radii r_{P} and r_{A}. In order to construct the whole orbit it is sufficient to know that portion of the orbit which is comprised between an aphelion and the perihelion immediately succeeding it (see the sketch in Fig. 1.16). Indeed, it is not difficult to realize that the orbit is symmetric with respect to both the line SA and the line SP of Fig. 1.16. To see this, consider two polar angles $\Delta\varphi$ and $-\Delta\varphi$, with $\Delta\varphi = \varphi - \varphi_A$, that define directions symmetric with respect to SA, see Fig. 1.17, with

$$\Delta\varphi = l \int_{r_{\text{A}}}^{r(\varphi)} \frac{\mathrm{d}r}{r^2\sqrt{2\mu(E - U_{\text{eff}})}} .$$

One has

$$U_{\text{eff}}(r) = U_{\text{eff}}(r_{\text{A}}) + \big(U_{\text{eff}}(r) - U_{\text{eff}}(r_{\text{A}})\big) = E + \big(U_{\text{eff}}(r) - U_{\text{eff}}(r_{\text{A}})\big)$$

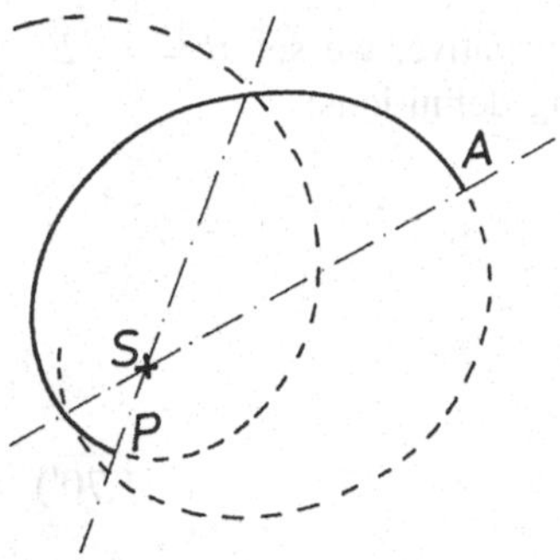

Fig. 1.16. Bound, or finite, orbit for an attractive central potential. The orbit has two symmetry axes: the line SA from the force center S to the apocenter, and the line SP from S to the pericenter. Thus, the entire rosette orbit can be constructed from the branch PA of the orbit. (The curve shown here is the example discussed below, with $\alpha = 1.3$, $b = 1.5$.)

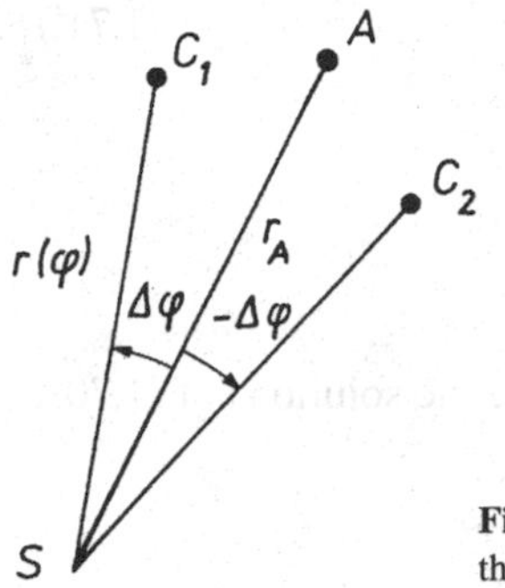

Fig. 1.17. Two symmetric positions before and after passage through the apocenter

and therefore

$$\Delta\varphi = l \int_{r_A}^{r(\varphi)} \frac{dr}{r^2\sqrt{2\mu\big(U_{eff}(r_A) - U_{eff}(r)\big)}} . \tag{1.69}$$

Instead of moving from A to C_1, by choosing the other sign of the square root in (1.69), the system may equally well move from A to C_2. From (1.67) this means that one changes the direction of motion, or, according to (1.64) and (1.65), that the direction of time is reversed. As $r(\varphi)$ is the same for $+\Delta\varphi$ and $-\Delta\varphi$, we conclude that if $C_1 = \{r(\varphi), \varphi = \varphi_A + \Delta\varphi\}$ is a point on the orbit, so is $C_2 = \{r(\varphi), \varphi = \varphi_A - \Delta\varphi\}$. A similar reasoning holds for P. This proves the symmetry stated above.

We illustrate these results by means of the following example.

Example. A central potential of the type $U(r) = -a/r^\alpha$. Let (r, φ) be the polar coordinates in the plane of the orbit. Then

$$\frac{dr}{dt} = \pm\sqrt{\frac{2E}{\mu} - \frac{2U(r)}{\mu} - \frac{l^2}{\mu^2 r^2}} , \tag{1.70}$$

$$\frac{d\varphi}{dt} = \frac{l}{\mu r^2} . \tag{1.71}$$

Since we consider only finite orbits for which E is negative, we set $B \stackrel{\text{def}}{=} -E$. We introduce dimensionless variables by the following definitions:

$$\varrho(\tau) \stackrel{\text{def}}{=} \frac{\sqrt{\mu B}}{l} r(t), \quad \tau \stackrel{\text{def}}{=} \frac{B}{l} t .$$

The equations of motion (1.70) and (1.71) then read

$$\frac{\mathrm{d}\varrho}{\mathrm{d}\tau} = \pm \sqrt{\frac{2b}{\varrho^\alpha} - \frac{1}{\varrho^2} - 2}, \tag{1.70$'$}$$

$$\frac{\mathrm{d}\varphi}{\mathrm{d}\tau} = \frac{1}{\varrho^2}, \tag{1.71$'$}$$

where we have set

$$b \stackrel{\text{def}}{=} \frac{a}{B} \left(\frac{\sqrt{\mu B}}{l} \right)^\alpha .$$

The value $\alpha = 1$ defines the Kepler problem, in which case the solutions of (1.70′) and (1.71′) read

$$\varrho(\varphi) = 1/b\big(1 + \varepsilon \cos(\varphi - \varphi_0)\big) \quad \text{with} \quad \varepsilon = \sqrt{1 - 2/b^2} .$$

The constant φ_0 can be chosen at will, e.g. $\varphi_0 = 0$. Figures 1.18–1.22 show the orbit $\varrho(\varphi)$ for various values of the parameters α and b. Figure 1.18 shows two Kepler ellipses with $b = 1.5$ and $b = 3$. Figures 1.19, 1.20 illustrate the situation

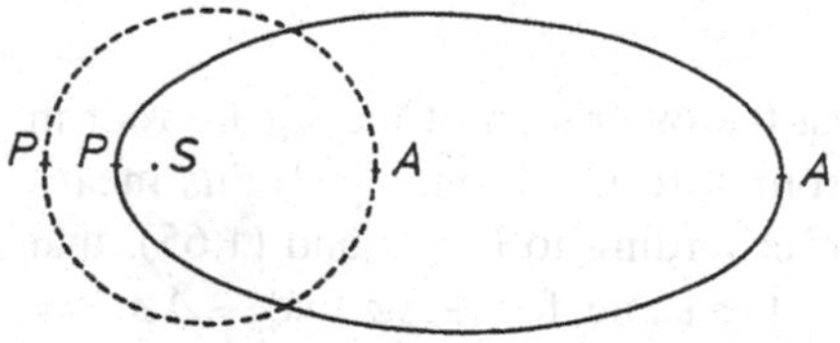

Fig. 1.18. Two Kepler ellipses ($\alpha = 1$) with different eccentricities. Cf. Practical Example 1.4

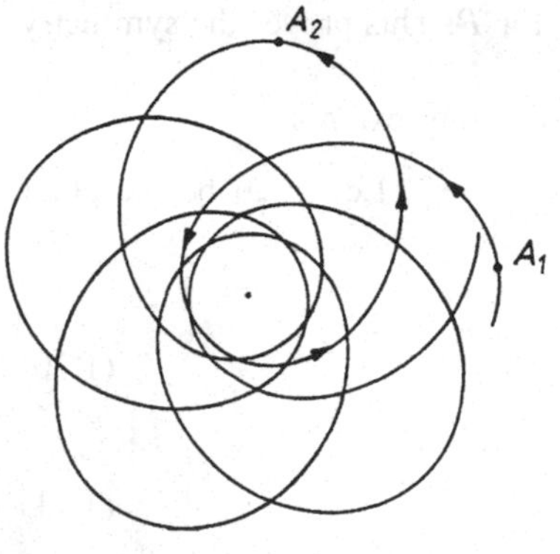

Fig. 1.19. Rosette orbit for the potential $U(r) = -a/r^\alpha$ with $\alpha = 1.3$ and $b = 1.5$

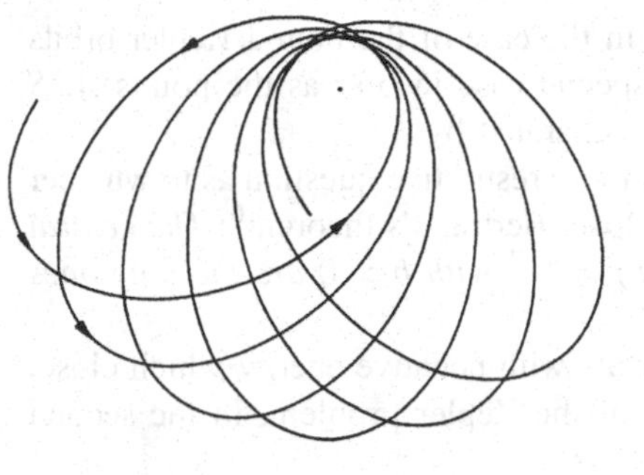

Fig. 1.20. A rosette orbit as in Fig. 1.19 but with $\alpha = 1.1,\ b = 2.0$

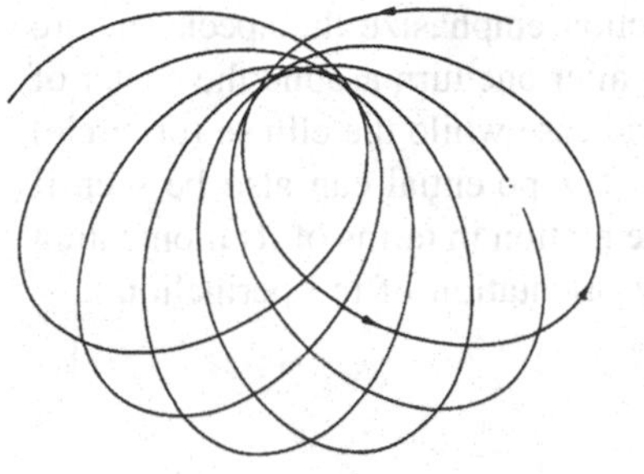

Fig. 1.21. Example of a rosette orbit that "stays behind", with $\alpha = 0.9,\ b = 2$

for $\alpha > 1$ where the orbit "advances" compared with the Kepler ellipse. Similarly, Figs. 1.21, 1.22, valid for $\alpha < 1$, show it "staying behind" with respect to the Kepler case. In either case, after one turn, the perihelion is shifted compared with the Kepler case ($\alpha = 1$) either forward ($\alpha > 1$) or backward ($\alpha < 1$). In the former case there is more attraction at perihelion compared to the Kepler ellipse, in the latter, less, thus causing the rosette-shaped orbit to advance or to stay behind, respectively.

Remark: From the above exercise it seems plausible that finite orbits which *close* after a finite number of revolutions about the origin are the exception rather than the rule. For this to happen the angle φ between the straight lines SA and SP of Fig. 1.16 must be a rational number times 2π, $\varphi = (n/p)(2\pi)$, $n, p \in \mathbb{N}$, where p is the number of branches PA of the orbit needed to close it, and n is

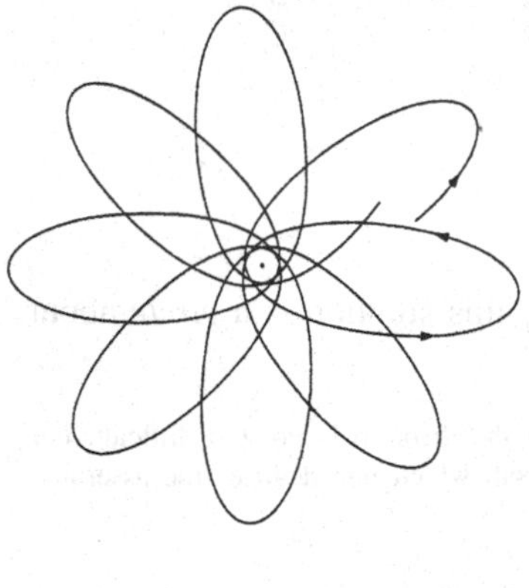

Fig. 1.22. A rosette orbit as in Fig. 1.21 but with $\alpha = 0.8$, $b = 3$

the number of turns about S. For example, in the case of the bound Kepler orbits we have $p = 2$ and $n = 1$. This is a very special case insofar as the points A, S and P lie on a straight line, with A and P separated by S.

The following theorem answers the even more restrictive question as to whether *all* finite orbits in a given central potential close: Bertrand's theorem[6]: *The central potentials $U(r) = \alpha/r$ with $\alpha < 0$ and $U(r) = br^2$ with $b > 0$ are the only ones in which* all *finite (i.e. bound) orbits close.*

As we know in the first case, it is the orbits with negative energy which close; they are the well-known ellipses or circles of the Kepler problem. In the second case all orbits are closed and elliptical.

Remarks: The examples studied in this section emphasize the special nature of the Kepler problem whose bound orbits close after one turn around the center of force. The rosette-like orbit represents the generic case while the ellipse (or circle) is the exception. This property of the attractive $1/r$-potential can also be seen if instead of the plane of motion in $\mathbb{R}^3$ we study the motion in terms of its momentum $\boldsymbol{p} = (p_x, p_y)^T$. The solution (1.21) for abitrary orientation of the perihelion

$$r(t) = \frac{p}{1 + \varepsilon \cos(\phi(t) - \phi_0)} ,$$

when decomposed in terms of Cartesian coordinates (x, y) in the plane of motion, reads

$$x(t) = \frac{p}{1 + \varepsilon \cos(\phi(t) - \phi_0)} \cos(\phi(t) - \phi_0) ,$$
$$y(t) = \frac{p}{1 + \varepsilon \cos(\phi(t) - \phi_0)} \sin(\phi(t) - \phi_0) .$$

The derivatives of $x(t)$ and of $y(t)$ with respect to time are

$$\dot{x}(t) = -p \frac{\sin(\phi - \phi_0)}{[1 + \varepsilon \cos(\phi(t) - \phi_0)]^2} \dot{\phi} = -\frac{1}{p}(r^2 \dot{\phi}) \sin(\phi - \phi_0) ,$$
$$\dot{y}(t) = p \frac{\cos(\phi - \phi_0) + \varepsilon}{[1 + \varepsilon \cos(\phi(t) - \phi_0)]^2} \dot{\phi} = \frac{1}{p}(r^2 \dot{\phi}) \big[\cos(\phi - \phi_0) + \varepsilon\big] .$$

Upon multiplication with the reduced mass μ, making use of the conservation law (1.19a) for ℓ, the modulus of the angular momentum, $\ell = \mu r^2 \dot{\phi}$, and inserting the definition $p = \ell^2/(A\mu)$, one obtains

$$p_x = \mu \dot{x} = -\frac{A\mu}{\ell} \sin(\phi - \phi_0) ,$$
$$p_y = \mu \dot{y} = \frac{A\mu}{\ell} \big\{\cos(\phi - \phi_0) + \varepsilon\big\} .$$

In a two-dimensional space spanned by p_x and p_y this solution is a *circle* about the point

[6] J. Bertrand (1873): R. Acad. Sci. **77**, p.849. The proof of the theorem is not too difficult. For example, Arnol'd proposes a sequence of five problems from which one deduces the assertion, (Arnol'd, 1989).

$$\left(0, \varepsilon(A\mu/\ell)\right) = \left(0, \sqrt{(A\mu/\ell)^2 + 2\mu E}\right),$$

where we have inserted the definition of the excentricity, $\varepsilon = \sqrt{1 + 2E\ell^2/\mu A^2}$. The radius of this circle is $R = A\mu/\ell$. The bound orbits in the space spanned by p_x and p_y are called *hodographs*. In the case of the Kepler problem they are always circles.

This remarkable result is related to another constant of the motion, the Hermann-Bernoulli-Laplace-Lenz vector, that applies to the $1/r$ potential. We will show this in the framework of canonical mechanics in Exercise 2.31.

1.25 Rotating Reference Systems: Coriolis and Centrifugal Forces

Let **K** be an inertial system and **K**′ another system that coincides with **K** at time $t = 0$ and rotates with angular velocity $\omega = |\boldsymbol{\omega}|$ about the direction $\hat{\boldsymbol{\omega}} = \boldsymbol{\omega}/\omega$, as shown in Fig. 1.23. Clearly, **K**′ is not an inertial system. The position vector of a mass point is $\boldsymbol{r}(t)$ with respect to **K** and $\boldsymbol{r}'(t)$ with respect to **K**′, with $\boldsymbol{r}(t) = \boldsymbol{r}'(t)$. The *velocities* are related by

$$\boldsymbol{v}' = \boldsymbol{v} - \boldsymbol{\omega} \times \boldsymbol{r}',$$

where $\boldsymbol{v}'$ refers to **K**′ and $\boldsymbol{v}$ to **K**. Denoting the change per unit time as it is observed from **K**′ by $\mathrm{d}'/\mathrm{d}t$, this means that

$$\frac{\mathrm{d}'}{\mathrm{d}t}\boldsymbol{r} = \frac{\mathrm{d}}{\mathrm{d}t}\boldsymbol{r} - \boldsymbol{\omega} \times \boldsymbol{r} \quad \text{or} \quad \frac{\mathrm{d}}{\mathrm{d}t}\boldsymbol{r} = \frac{\mathrm{d}'}{\mathrm{d}t}\boldsymbol{r} + \boldsymbol{\omega} \times \boldsymbol{r},$$

where

$\mathrm{d}/\mathrm{d}t$: time derivative as observed from **K**,

$\mathrm{d}'/\mathrm{d}t$: time derivative as observed from **K**′.

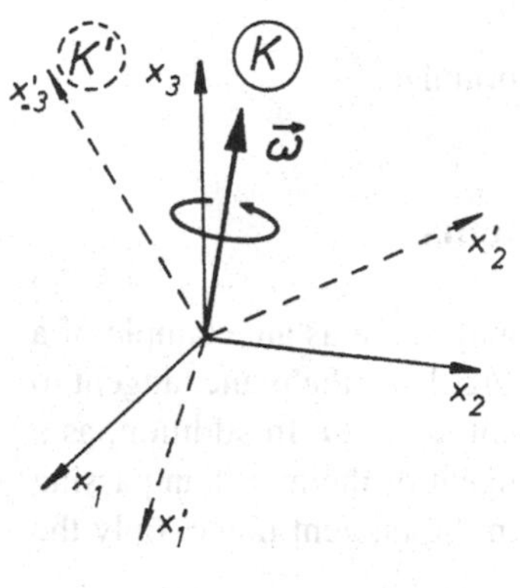

Fig. 1.23. The coordinate system **K**′ rotates about the system **K** with angular velocity $\boldsymbol{\omega}$

The relation between $\mathrm{d}\boldsymbol{r}/\mathrm{d}t$ and $\mathrm{d}'\boldsymbol{r}/\mathrm{d}t$ must be valid for any vector-valued function $\boldsymbol{a}(t)$, viz.

$$\frac{\mathrm{d}}{\mathrm{d}t}\boldsymbol{a} = \frac{\mathrm{d}'}{\mathrm{d}t}\boldsymbol{a} + \boldsymbol{\omega} \times \boldsymbol{a} . \qquad (1.72)$$

Taking $\boldsymbol{\omega}$ to be constant in time, we find that the relationship (1.72) is applied to the velocity $\boldsymbol{a}(t) \equiv \mathrm{d}\boldsymbol{r}/\mathrm{d}t$ as follows:

$$\frac{\mathrm{d}^2}{\mathrm{d}t^2}\boldsymbol{r}(t) = \frac{\mathrm{d}'}{\mathrm{d}t}\left(\frac{\mathrm{d}\boldsymbol{r}}{\mathrm{d}t}\right) + \boldsymbol{\omega} \times \frac{\mathrm{d}\boldsymbol{r}}{\mathrm{d}t} = \frac{\mathrm{d}'}{\mathrm{d}t}\left(\frac{\mathrm{d}'}{\mathrm{d}t}\boldsymbol{r} + \boldsymbol{\omega} \times \boldsymbol{r}\right) + \boldsymbol{\omega} \times \left(\frac{\mathrm{d}'}{\mathrm{d}t}\boldsymbol{r} + \boldsymbol{\omega} \times \boldsymbol{r}\right)$$
$$= \frac{\mathrm{d}^{2\prime}}{\mathrm{d}t^2}\boldsymbol{r} + 2\boldsymbol{\omega} \times \frac{\mathrm{d}'}{\mathrm{d}t}\boldsymbol{r} + \boldsymbol{\omega} \times (\boldsymbol{\omega} \times \boldsymbol{r}) . \qquad (1.73)$$

(If $\boldsymbol{\omega}$ does depend on time, this equation contains one more term, $(\mathrm{d}'\boldsymbol{\omega}/\mathrm{d}t) \times \boldsymbol{r} = (\mathrm{d}\boldsymbol{\omega}/\mathrm{d}t) \times \boldsymbol{r} = \dot{\boldsymbol{\omega}} \times \boldsymbol{r}$.)

Newton's equations are valid in **K** because **K** is inertial; thus

$$m\frac{\mathrm{d}^2}{\mathrm{d}t^2}\boldsymbol{r}(t) = \boldsymbol{F} .$$

Inserting the relation (1.73) between the acceleration $\mathrm{d}^2\boldsymbol{r}/\mathrm{d}t^2$, as seen from **K**, and the acceleration $\mathrm{d}^{2\prime}\boldsymbol{r}/\mathrm{d}t^2$, as seen from **K**$'$, in the equation of motion, one obtains

$$m\frac{\mathrm{d}^{2\prime}}{\mathrm{d}t^2}\boldsymbol{r} = \boldsymbol{F} - 2m\boldsymbol{\omega} \times \frac{\mathrm{d}'}{\mathrm{d}t}\boldsymbol{r} - m\boldsymbol{\omega} \times (\boldsymbol{\omega} \times \boldsymbol{r}) . \qquad (1.74)$$

When observed from **K**$'$, which is not inertial, the mass point is subject not only to the original force $\boldsymbol{F}$ but also to the

Coriolis force $\quad \boldsymbol{C} = -2m\boldsymbol{\omega} \times \boldsymbol{v}' \qquad (1.75)$

and the

centrifugal force $\quad \boldsymbol{Z} = -m\boldsymbol{\omega} \times (\boldsymbol{\omega} \times \boldsymbol{r}) , \qquad (1.76)$

whose directions are easily determined from these formulae.

1.26 Examples of Rotating Reference Systems

Example (i) Any system tied to a point on the earth may serve as an example of a rotating reference frame. Referring to the notation of Fig. 1.24, the plane tangent to the earth at A rotates horizontally about the component $\boldsymbol{\omega}_{\mathrm{v}}$ of $\boldsymbol{\omega}$. In addition, as a whole, it also rotates about the component $\boldsymbol{\omega}_{\mathrm{h}}$ (the tangent of the meridian passing through A). If a mass point moves horizontally, i.e. in the tangent plane, only the

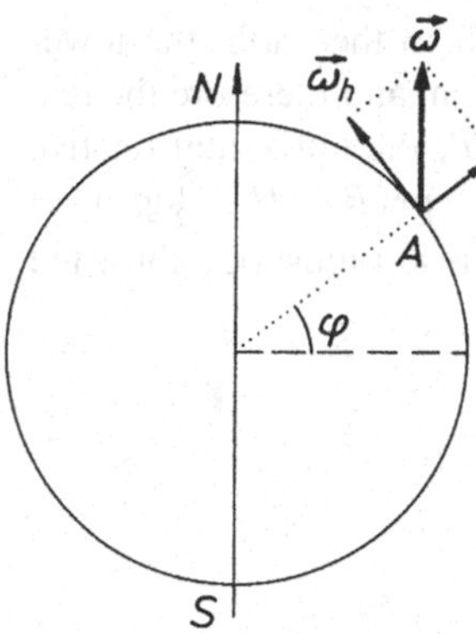

Fig. 1.24. A coordinate system fixed at a point A on the earth's surface rotates about the south–north axis with angular velocity $\omega = \omega_\text{v} + \omega_\text{h}$

component ω_v will be effective in (1.75). Thus, in the *northern* hemisphere the mass point will be deviated to the *right*.

For vertical motion, to a first approximation, only ω_h is effective. In the northern hemisphere this causes an eastward deviation, which can easily be estimated for the example of free fall. For the sake of illustration, we calculate this deviation in two different ways.

(a) *With respect to an inertial system fixed in space.* We assume the mass point m to have a fixed position above point A on the earth's surface. This is sketched in Fig. 1.25, which shows the view looking down on the north pole. The particle's tangent velocity (with reference to **K**!) is $v_\text{T}(R+h) = (R+h)\omega\cos\varphi$. At time $t = 0$ we let it fall freely from the top of a tower of height H. As seen from **K**, m moves horizontally (eastwards) with the constant velocity $v_\text{T}(R+H) = (R+H)\omega\cos\varphi$, while falling vertically with constant acceleration g. Therefore, the height H and the time T needed to reach the ground are related by $H = \frac{1}{2}gT^2$. If at the same time ($t = 0$) the point A at the bottom of the tower left the earth's surface along a tangent, it would move horizontally with a constant velocity $v_\text{T}(R) = R\omega\cos\varphi$. Thus, after a time T, the mass point would hit the ground at a distance

$$\Delta_0 = \big(v_\text{T}(R+H) - v_\text{T}(R)\big)T = H\omega T\cos\varphi\ ,$$

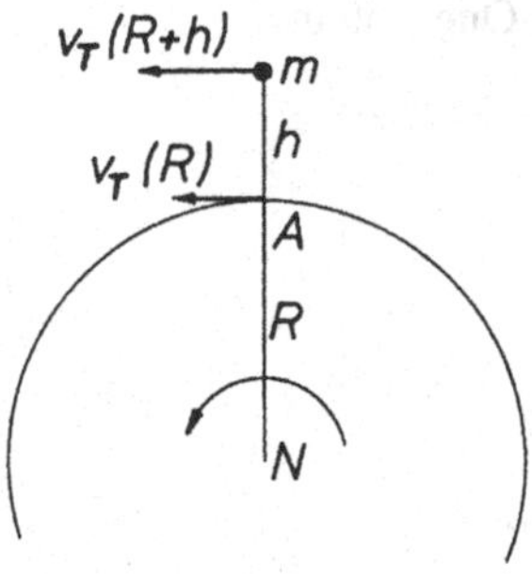

Fig. 1.25. A body falling down vertically is deviated towards the east. Top view of the north pole and the parallel of latitude of A

east of A. In reality, during the time that m needs to fall to the earth, the tower has continued its accelerated motion, in an easterly direction, and therefore the real deviation Δ is smaller than Δ_0. At time t, with $0 \le t \le T$, the horizontal relative velocity of the mass point and the tower is $\left(v_{\mathrm{T}}(R+H) - v_{\mathrm{T}}(R+H-\frac{1}{2}gt^2)\right) = \frac{1}{2}g\omega t^2 \cos\varphi$. This must be integrated from 0 to T and the result must be subtracted from Δ_0. The real deviation is then

$$\begin{aligned}\Delta &\simeq \Delta_0 - \frac{1}{2}g\omega\cos\varphi \int_0^T \mathrm{d}t\, t^2 \\ &= \omega\cos\varphi \int_0^T \mathrm{d}t \left(H - \frac{1}{2}gt^2\right) = \frac{1}{3}g\omega T^3 \cos\varphi \,.\end{aligned}$$

(b) *In the accelerated system moving with the earth.* We start from the equation of motion (1.74). As the empirical constant g is the sum of the gravitational acceleration, directed towards the center of the earth, and the centrifugal acceleration, directed away from it, the centrifugal force (1.76) is already taken into account. (Note that the Coriolis force is linear in ω while the centrifugal force is quadratic in ω. In the range of distances and velocities relevant for terrestial problems both of these are small as compared to the force of attraction by the earth, the centrifugal force being sizeably smaller than the Coriolis force.) Thus, (1.74) reduces to

$$m\frac{\mathrm{d}'^2 \boldsymbol{r}}{\mathrm{d}t^2} = -mg\hat{\boldsymbol{e}}_{\mathrm{v}} - 2m\omega\left(\hat{\boldsymbol{\omega}} \times \frac{\mathrm{d}'\boldsymbol{r}}{\mathrm{d}t}\right) . \tag{1.74$'$}$$

We write the solution in the form $\boldsymbol{r}(t) = \boldsymbol{r}^{(0)}(t) + \omega\boldsymbol{u}(t)$, where $\boldsymbol{r}^{(0)}(t) = (H - \frac{1}{2}gt^2)\hat{\boldsymbol{e}}_{\mathrm{v}}$ is the solution of (1.74′) without the Coriolis force ($\omega = 0$). As $\omega = 2\pi/(1\,\text{day}) = 7.3 \times 10^{-5}\,\mathrm{s}^{-1}$is very small, we determine the function $\boldsymbol{u}(t)$ from (1.74′) approximately by keeping only those terms independent of ω and linear in ω. Inserting the expression for $\boldsymbol{r}(t)$ into (1.74′), we obtain for $\boldsymbol{u}(t)$

$$m\omega\frac{\mathrm{d}'^2}{\mathrm{d}t^2}\boldsymbol{u} \simeq 2mgt\omega\left(\hat{\boldsymbol{\omega}} \times \hat{\boldsymbol{e}}_{\mathrm{v}}\right) .$$

$\hat{\boldsymbol{\omega}}$ is parallel to the earth's axis, $\hat{\boldsymbol{e}}_{\mathrm{v}}$ is vertical. Therefore, $(\hat{\boldsymbol{\omega}} \times \hat{\boldsymbol{e}}_{\mathrm{v}}) = \cos\varphi\hat{\boldsymbol{e}}_{\mathrm{e}}$, where $\hat{\boldsymbol{e}}_{\mathrm{e}}$ is tangent to the earth's surface and points eastwards. One obtains

$$\frac{\mathrm{d}'^2}{\mathrm{d}t^2}\boldsymbol{u} \simeq 2gt\cos\varphi\hat{\boldsymbol{e}}_{\mathrm{e}} ,$$

and, by integrating twice,

$$\boldsymbol{u} \simeq \tfrac{1}{3}gt^3\cos\varphi\hat{\boldsymbol{e}}_{\mathrm{e}} .$$

Thus, the eastward deviation is $\Delta \simeq \frac{1}{3}gT^3\omega\cos\varphi$, as above.

Inserting the relation between T and H, we get

$$\Delta \simeq \frac{2\sqrt{2}}{3}\omega g^{-1/2} H^{3/2} \cos\varphi \simeq 2.189 \times 10^{-5} H^{3/2} \cos\varphi \,.$$

For a numerical example choose $H = 160\,m$, $\varphi = 50°$. This gives $\Delta \simeq 2.8\,\text{cm}$.

Example (ii) Let a mass m be connected to a fixed point O in space and let it rotate with constant angular velocity about that point, as shown in Fig. 1.26. Its kinetic energy is then $T = \frac{1}{2} m R^2 \omega^2$. If we now cut the connection to O, m will leave the circle $(O; R)$ along a tangent with constant velocity $R\omega$. How does the same motion look in a system $\mathbf{K}'$ that rotates synchronously?

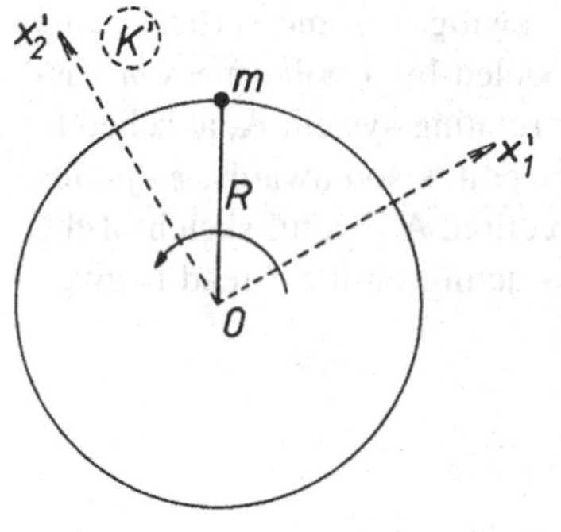

Fig. 1.26. A mass point rotates uniformly about the origin O. $\mathbf{K}'$ is a coordinate system in the plane that rotates synchronously with the particle

From (1.74) one has

$$m \frac{d'^2}{dt^2}\boldsymbol{r} = 2m\omega \frac{d'}{dt}\left(x_2' \boldsymbol{e}_1' - x_1' \boldsymbol{e}_2'\right) + m\omega^2 \boldsymbol{r},$$

or, when written in components,

$$m \frac{d'^2}{dt^2} x_1' = 2m\omega \frac{d'}{dt} x_2' + m\omega^2 x_1' \,, \quad m \frac{d'^2}{dt^2} x_2' = -2m\omega \frac{d'}{dt} x_1' + m\omega^2 x_2' \,.$$

The initial condition at $t = 0$ reads

$$t = 0 \begin{cases} x_1' = R & \dfrac{d'}{dt} x_1' = 0 \,, \\ x_2' = 0 & \dfrac{d'}{dt} x_2' = 0 \,. \end{cases}$$

With respect to $\mathbf{K}$ we would then have

$$x_1(t) = R \,, \quad x_2(t) = R\omega t \,.$$

Therefore, the relationships

$$x_1' = x_1 \cos \omega t + x_2 \sin \omega t \ , \quad x_2' = -x_1 \sin \omega t + x_2 \cos \omega t$$

give us at once the solution of the problem, viz.

$$x_1'(t) = R \cos \omega t + R\omega t \sin \omega t \ , \quad x_2'(t) = -R \sin \omega t + R\omega t \cos \omega t \ .$$

It is instructive to sketch this orbit, as seen from $\mathbf{K}'$, and thereby realize that uniform rectilinear motion looks complicated when observed from a rotating, noninertial system.

Example (iii) A particularly nice example is provided by the Foucault pendulum that the reader might have seen in a laboratory experiment or in a science museum. The model is the following. In a site whose geographical latitude is $0 \le \varphi \le \pi/2$ a mathematical pendulum is suspended in the point with coordinates $(0, 0, l)$ above the ground, and is brought to swing in some vertical plane through that point. Imagine the pendulum to be modeled by a point mass m sustained by a massless thread whose length is l. In the rotating system $\mathbf{K}$, attached to the earth, let the unit vectors be chosen such that $\hat{\boldsymbol{e}}_1$ points southwards, $\hat{\boldsymbol{e}}_2$ points eastwards, while $\hat{\boldsymbol{e}}_3$ denotes the upward vertical direction. A careful sketch of the pendulum and the base vectors shows that the stress acting on the thread is given by

$$\boldsymbol{Z} = Z \left(-\frac{x_1}{l} \hat{\boldsymbol{e}}_1 - \frac{x_2}{l} \hat{\boldsymbol{e}}_2 + \frac{l - x_3}{l} \hat{\boldsymbol{e}}_3 \right) ,$$

where we have normalized the components such that Z is the modulus of this vector field, $Z = |\boldsymbol{Z}|$. Indeed, $l^2 = x_1^2 + x_2^2 + (l - x_3)^2$ so that the sum of the squares of the coefficients in the parentheses is equal to 1. Inserting this expression in the equation of motion (1.75) and denoting, for simplicity, the time derivative $\frac{\mathrm{d}'}{\mathrm{d}t}$ with respect to the rotating system $\mathbf{K}$ by a dot, the equation of motion reads

$$m\ddot{\boldsymbol{r}} = \boldsymbol{Z} + m\boldsymbol{g} - 2m\,(\boldsymbol{\omega} \times \dot{\boldsymbol{r}}) - m\,\boldsymbol{\omega} \times (\boldsymbol{\omega} \times \boldsymbol{r}) \ .$$

For the same reasons as before we neglect the centrifugal force. With the choice of the reference system described above one has

$$\boldsymbol{g} = -g\hat{\boldsymbol{e}}_3 \ , \quad \boldsymbol{\omega} = \omega \begin{pmatrix} -\cos\varphi \\ 0 \\ \sin\varphi \end{pmatrix} \ , \quad \boldsymbol{\omega} \times \dot{\boldsymbol{r}} = \omega \begin{pmatrix} -\dot{x}_2 \sin\varphi \\ \dot{x}_1 \sin\varphi + \dot{x}_3 \cos\varphi \\ -\dot{x}_2 \cos\varphi \end{pmatrix} ,$$

where ω is the modulus of the angular velocity, and φ the geographical latitude.

Writing the equation of motion in terms of its three components one has

$$\begin{aligned}
m\ddot{x}_1 &= -\frac{Z}{l} x_1 + 2m\omega \dot{x}_2 \sin\varphi \ , \\
m\ddot{x}_2 &= -\frac{Z}{l} x_2 - 2m\omega (\dot{x}_1 \sin\varphi + \dot{x}_3 \cos\varphi) \ , \\
m\ddot{x}_3 &= \frac{Z}{l} (l - x_3) - mg + 2m\omega \dot{x}_2 \cos\varphi \ .
\end{aligned}$$

These coupled differential equations are solved most easily in the case of small oscillations. In this approximation set $x_3 \simeq 0$, $\dot{x}_3 \simeq 0$ in the third of these and obtain the modulus of the thread stress from this equation. It is found to be

$$Z = mg - 2m\omega\dot{x}_2 \cos\varphi \,.$$

Next, insert this approximate expression into the first two equations. For consistency with the approximation of small oscillations terms of the type $x_i\dot{x}_k$ must be neglected. In this approximation and introducing the abbreviations

$$\omega_0^2 = \frac{g}{l}\,, \qquad \alpha = \omega \sin\varphi \,,$$

the first two equations become

$$\ddot{x}_1 = -\omega_0^2 x_1 + 2\alpha\dot{x}_2 \,,$$
$$\ddot{x}_2 = -\omega_0^2 x_2 - 2\alpha\dot{x}_1 \,.$$

Solutions of these equations can be constructed by writing them as one complex equation in the variable $z(t) = x_1(t) + \mathrm{i}x_2(t)$,

$$\ddot{z}(t) = -\omega_0^2 z(t) - 2\mathrm{i}\alpha\dot{z}(t) \,.$$

The ansatz $z(t) = C\mathrm{e}^{\mathrm{i}\gamma t}$ yields two solutions for the circular frequency γ, viz.

$$\gamma_1 = -\alpha + \sqrt{\alpha^2 + \omega_0^2}\,, \qquad \gamma_2 = -\alpha - \sqrt{\alpha^2 + \omega_0^2}\,.$$

Below we will study the solutions for these general expressions. The historical experiment performed in 1851 by Foucault in the Panthéon in Paris, however, had parameters such that α was very small as compared to ω_0, $\alpha^2 \ll \omega_0^2$. Indeed, given the latitude of Paris, $\varphi = 48.5^0$, and the parameters of the pendulum chosen by Foucault, $l = 67\,\mathrm{m}$, $m = 28\,\mathrm{kg}$, and, from these, the period $T = 16.4\,\mathrm{s}$, one obtains

$$\omega_0 = \frac{2\pi}{T} = 0.383\,\mathrm{s}^{-1}\,,$$
$$\alpha = \frac{2\pi}{1\ \mathrm{day}} \sin\varphi = \frac{2\pi}{86400} \sin(48.5^0) = 5.45 \times 10^{-5}\,\mathrm{s}^{-1}\,.$$

Therefore $\gamma_{1/2} \simeq -\alpha \pm \omega_0$ and the solutions read

$$z(t) \simeq (c_1 + \mathrm{i}c_2)\mathrm{e}^{-\mathrm{i}(\alpha-\omega_0)t} + (c_3 + \mathrm{i}c_4)\mathrm{e}^{-\mathrm{i}(\alpha+\omega_0)t}\,.$$

It remains to split this function into its real and imaginary parts and to adjust the integration constants to a given initial condition. Suppose the pendulum, at time zero, is elongated along the 1-direction by a distance a and is launched without initial velocity, i.e.

$$x_1(0) = a\,,\ \dot{x}_1(0) = 0\,, \quad x_2(0) = 0\,,\ \dot{x}_2(0) = 0\,,$$

the approximate solution is found to be

$$x_1(t) \simeq a\big[\cos(\alpha t)\cos(\omega_0 t) + (\alpha/\omega_0)\sin(\alpha t)\sin(\omega_0 t)\big] ,$$
$$x_2(t) \simeq a\big[-\sin(\alpha t)\cos(\omega_0 t) + (\alpha/\omega_0)\cos(\alpha t)\sin(\omega_0 t)\big] .$$

As a result the pendulum still swings approximately in a plane. That plane of oscillation rotates very slowly about the local vertical, in a clockwise direction on the northern hemisphere, in a counter-clockwise direction on the southern hemisphere. The mark that the tip of the pendulum would leave on the ground is bent slightly to the right on the northern hemisphere, to the left on the southern hemisphere. For a complete turn of the plane of oscillation it needs the time $24/\sin\varphi$ hours. Rigth on the north pole or on the south pole this time is exactly 24 hours. For the latitude of Paris, it is approximately 32 hours, while at the equator there is no rotation at all.

In order to better illustrate the motion of a Foucault pendulum for small amplitudes let us also consider the case where α is not small as compared to the unperturbed frequency ω_0. For the same initial condition as above, $x_1(0) = a$, $\dot{x}_1(0) = 0$, $x_2(0) = 0 = \dot{x}_2(0)$, the solution now reads

$$x_1(t) = a\big[\cos(\alpha t)\cos(\overline{\omega} t) + (\alpha/\overline{\omega})\sin(\alpha t)\sin(\overline{\omega} t)\big] ,$$
$$x_2(t) = a\big[-\sin(\alpha t)\cos(\overline{\omega} t) + (\alpha/\overline{\omega})\cos(\alpha t)\sin(\overline{\omega} t)\big] ,$$

where $$\overline{\omega} = \sqrt{\omega_0^2 + \alpha^2} .$$

It is useful to calculate also the components of the velocity. One finds

$$\dot{x}_1 = -a\frac{\omega_0^2}{\overline{\omega}}\cos(\alpha t)\sin(\overline{\omega} t) ,$$
$$\dot{x}_2 = a\frac{\omega_0^2}{\overline{\omega}}\sin(\alpha t)\sin(\overline{\omega} t) .$$

The two components of the velocity vanish simultaneously at the times

$$t_n = \frac{n\pi}{\overline{\omega}} = \frac{n}{2}\overline{T} , \quad n = 0, 1, 2, \ldots .$$

This means that at these points of return both components go through zero and change signs, the projection of the pendulum motion on the horizontal plane shows spikes. Figure 1.27 gives a qualitative top view of the motion.

For a quantitative analysis we choose the circular frequency α comparable to $\bar{\omega}$. In the two examples given next these frequencies are chosen relatively rational, $\alpha/\bar{\omega} = 1/4$. (Clearly, this choice is not realistic for the case of the earth and the original Foucault pendulum.) For a rational ratio $\alpha/\bar{\omega} = n/m$ the curve swept out by the tip of the pendulum on the horizontal plane closes. In all other cases it will not close. Figure 1.28 shows the solution given above for the initial condition

$$x_1(0) = 1 , \ \dot{x}_1(0) = 0 , \quad x_2(0) = 0 , \ \dot{x}_2(0) = 0 ,$$

It closes after four oscillations and exhibits the spikes discussed above.

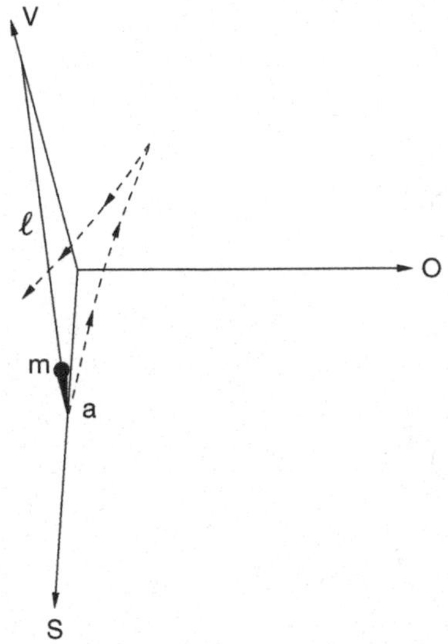

Fig. 1.27. A Foucault pendulum, seen from above, starts at the distance a in the South, without initial velocity. The mark it makes on the horizontal plane is bent to the right and exhibits spikes at the turning points

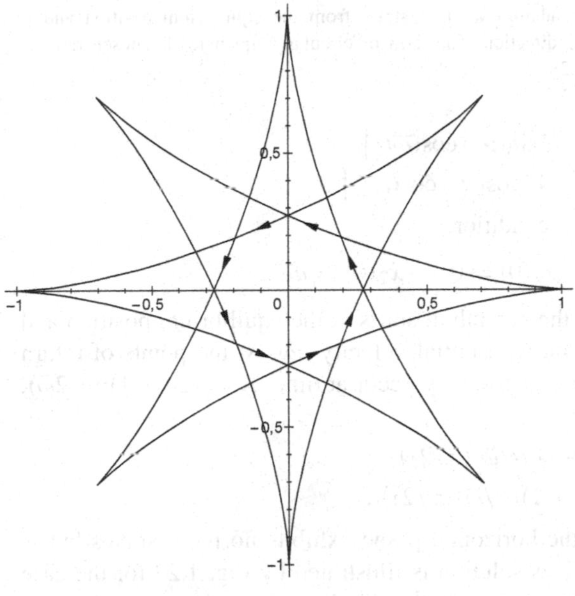

Fig. 1.28. Mark left on the horizontal plane by a pendulum that starts from $x_1(0) = 1$ without initial velocity. The ratio of circular frequencies is chosen rational, $\alpha/\bar{\omega} = 1/4$

Another solution is the following

$$x_1(t) = a \sin(\alpha t) \sin(\overline{\omega} t) ,$$
$$x_2(t) = a \cos(\alpha t) \sin(\overline{\omega} t) .$$

From this one finds the components of the velocity to be

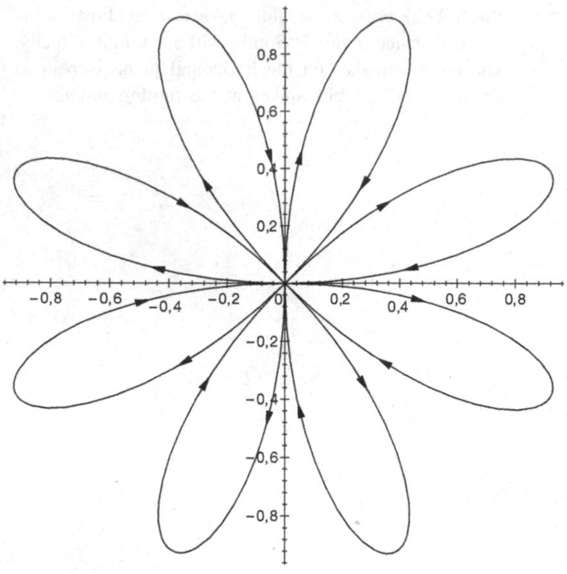

Fig. 1.29. Mark swept out by the pendulum when it starts from the equilibrium position and is kicked with initial velocity $\bar{\omega}$ in the 2-direction. The ratio of circular frequencies is chosen rational and has the same value as in Fig. 1.28

$$\dot{x}_1 = a\big[\alpha\cos(\alpha t)\sin(\overline{\omega}t) + \overline{\omega}\sin(\alpha t)\cos(\overline{\omega}t)\big]\,,$$
$$\dot{x}_2 = a\big[-\alpha\sin(\alpha t)\sin(\overline{\omega}t) + \overline{\omega}\cos(\alpha t)\cos(\overline{\omega}t)\big]\,,$$

which corresponds to the initial condition

$$x_1(0) = 0\,,\quad \dot{x}_1(0) = 0\,,\quad x_2(0) = 0\,,\quad \dot{x}_2(0) = a\overline{\omega}\,.$$

This solution is the one where the pendulum starts at the equilibrium position and is being kicked in the 2-direction with initial velocity $a\bar{\omega}$. At the points of return $x_1^2+x_2^2 = a^2\sin^2(\bar{\omega}t)$ is maximal. Thus, they occur at times $t_n = (2n+1)\pi/(2\bar{\omega})$, and one has

$$\dot{x}_1(t_n) = a\alpha(-)^n\cos\big((2n+1)(\alpha/\overline{\omega})(\pi/2)\big)\,,$$
$$\dot{x}_2(t_n) = -a\alpha(-)^n\sin\big((2n+1)(\alpha/\overline{\omega})(\pi/2)\big)\,.$$

This means that the track on the horizontal plane exhibits no more spikes but is "rounded off" at these points. This solution is illustrated by Fig. 1.29 for the case of the same rational ratio of α and $\bar{\omega}$ as in the previous example.

1.27 Scattering of Two Particles that Interact via a Central Force: Kinematics

In our discussion of central forces acting between two particles we have touched only briefly on the infinite orbits, i.e. those which come from and escape to infinity. In this section and in the two that follow we wish to analyze these scattering orbits in more detail and to study the kinematics and the dynamics of the scattering process. The description of scattering processes is of central importance for physics at the smallest dimensions. In the laboratory one can prepare and identify free, incoming or outgoing states by means of macroscopic particle sources and detectors. That is, one observes the scattering states long *before* and long *after* the scattering process proper, at large distances from the interaction region, but one cannot observe what is happening in the vicinity of the interaction region. The outcome of such scattering processes may therefore be the only, somewhat indirect, source of information on the dynamics at small distances. To quote an example, the scattering of α-particles on atomic nuclei, which Rutherford calculated on the basis of classical mechanics (see Sect. 1.28 (ii) and Sect. 1.29) was instrumental in discovering nuclei and in measuring their sizes.

We consider two particles of masses m_1 and m_2 whose interaction is given by a spherically symmetric potential $U(r)$ (repulsive or attractive). The potential is assumed to tend to zero at infinity at least like $1/r$. In the laboratory the experiment is usually performed in such a way that particle 2 is taken to be at rest (this is the *target*) while particle 1 (the *projectile*) comes from infinity and scatters off particle 2 so that both escape to infinity. This is sketched in Fig. 1.30a. This type of motion looks asymmetric in the two particles because in addition to the relative motion it contains the motion of the center of mass, which moves along with the projectile (to the right in the figure). If one introduces a second frame of reference whose origin is the center of mass, the motion is restricted to the relative motion alone

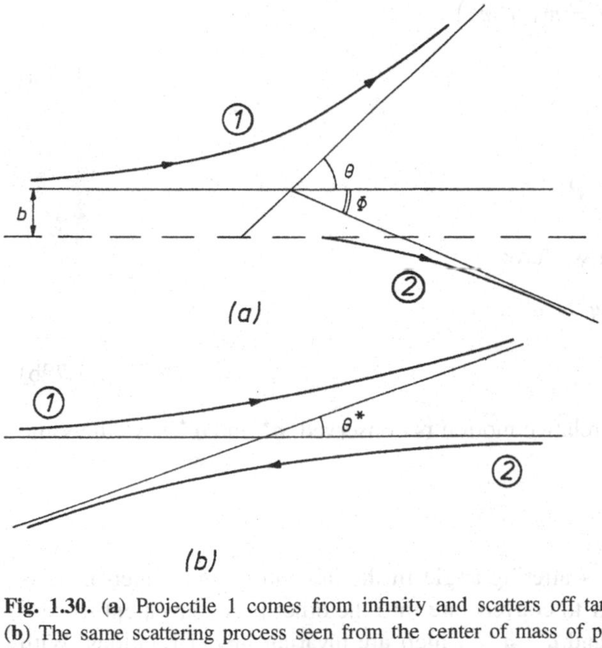

Fig. 1.30. (**a**) Projectile 1 comes from infinity and scatters off target 2, which is initially at rest. (**b**) The same scattering process seen from the center of mass of particles 1 and 2. The asymmetry between projectile and target disappears

(which is the relevant one dynamically) and one obtains the symmetric picture shown in Fig. 1.30b. Both, the laboratory system and the center-of-mass system are inertial systems. We can characterize the two particles by their momenta long before and long after the collision, in either system, as follows:

in the laboratory system:

$\boldsymbol{p}_i$ before, $\boldsymbol{p}_i'$ after the collision, $i = 1, 2$;

in the center-of-mass system:

$\boldsymbol{q}^*$ and $-\boldsymbol{q}^*$ before, $\boldsymbol{q}'^*$ and $-\boldsymbol{q}'^*$ after the collision.

If we deal with an elastic collision, i.e. if the internal state of the particles does not change in the collision, then $\boldsymbol{p}_2 = 0$ and energy conservation together imply that

$$\frac{\boldsymbol{p}_1^2}{2m_1} = \frac{\boldsymbol{p}_1'^2}{2m_1} + \frac{\boldsymbol{p}_2'^2}{2m_2} . \tag{1.77}$$

In addition, momentum conservation gives

$$\boldsymbol{p}_1 = \boldsymbol{p}_1' + \boldsymbol{p}_2' . \tag{1.78}$$

Decomposing in terms of center-of-mass and relative momenta, and making use of the equations obtained in Sect. 1.7.3, one obtains for the initial state

$$\begin{aligned} \boldsymbol{p}_1 &= \frac{m_1}{M}\boldsymbol{P} + \boldsymbol{q}^* , \quad \left(M \stackrel{\text{def}}{=} m_1 + m_2\right) \\ \boldsymbol{p}_2 &= \frac{m_2}{M}\boldsymbol{P} - \boldsymbol{q}^* = 0 ; \end{aligned} \tag{1.79a}$$

that is,

$$\boldsymbol{P} = \frac{M}{m_2}\boldsymbol{q}^* \quad \text{and} \quad \boldsymbol{p}_1 = \boldsymbol{P} .$$

Likewise, after the collision we have

$$\begin{aligned} \boldsymbol{p}_1' &= \frac{m_1}{M}\boldsymbol{P} + \boldsymbol{q}'^* = \frac{m_1}{m_2}\boldsymbol{q}^* + \boldsymbol{q}'^* , \\ \boldsymbol{p}_2' &= \frac{m_2}{M}\boldsymbol{P} - \boldsymbol{q}'^* = \boldsymbol{q}^* - \boldsymbol{q}'^* . \end{aligned} \tag{1.79b}$$

As the kinetic energy of the relative motion is conserved, $\boldsymbol{q}^*$ and $\boldsymbol{q}'^*$ have the same magnitude,

$$|\boldsymbol{q}^*| = |\boldsymbol{q}'^*| \stackrel{\text{def}}{=} q^* .$$

Let θ and θ^* denote the scattering angle in the laboratory and center-of-mass frames, respectively. In order to convert one into the other it is convenient to consider the quantities $\boldsymbol{p}_1 \cdot \boldsymbol{p}_1'$ and $\boldsymbol{q}^* \cdot \boldsymbol{q}'^*$, which are invariant under rotations. With $\boldsymbol{p}_1 = \boldsymbol{q}^* M/m_2$ and $\boldsymbol{p}_1' = \boldsymbol{q}^* m_1/m_2 + \boldsymbol{q}'^*$ one has

$$\boldsymbol{p}_1 \cdot \boldsymbol{p}_1' = \frac{M}{m_2}\left(\frac{m_1}{m_2}\boldsymbol{q}^{*2} + \boldsymbol{q}^* \cdot \boldsymbol{q}'^*\right) = \frac{M}{m_2}q^{*2}\left(\frac{m_1}{m_2} + \cos\theta^*\right).$$

On the other hand,

$$\boldsymbol{p}_1 \cdot \boldsymbol{p}_1' = |\boldsymbol{p}_1||\boldsymbol{p}_1'|\cos\theta = \frac{M}{m_2}q^*\left|\frac{m_1}{m_2}\boldsymbol{q}^* + \boldsymbol{q}'^*\right|\cos\theta$$

$$= \frac{M}{m_2}q^{*2}\sqrt{\left[1 + 2\frac{m_1}{m_2}\cos\theta^* + \left(\frac{m_1}{m_2}\right)^2\right]}\cos\theta\,.$$

From this follows

$$\cos\theta = \left(\frac{m_1}{m_2} + \cos\theta^*\right) \Big/ \sqrt{\left[1 + 2\frac{m_1}{m_2}\cos\theta^* + \left(\frac{m_1}{m_2}\right)^2\right]},$$

or

$$\sin\theta = \sin\theta^* \Big/ \sqrt{\left[1 + 2\frac{m_1}{m_2}\cos\theta^* + \left(\frac{m_1}{m_2}\right)^2\right]},$$

or, finally,

$$\tan\theta = \frac{\sin\theta^*}{(m_1/m_2) + \cos\theta^*}\,. \tag{1.80}$$

In Fig. 1.30a the target particle escapes in the direction characterized by the angle ϕ in the laboratory system. By observing that the triangle $(\boldsymbol{p}_2', \boldsymbol{q}^*, \boldsymbol{q}'^*)$ has two equal sides and that $\boldsymbol{q}^*$ has the same direction as $\boldsymbol{p}_1$ one can easily show that ϕ is related to the scattering angle in the center-of-mass system by

$$\phi = \frac{\pi - \theta^*}{2}\,. \tag{1.81}$$

Several special cases can be read off the formulae (1.79) and (1.80).

(i) If the mass m_1 of the projectile is much smaller than the mass m_2 of the target, $m_1 \ll m_2$, then $\theta^* \simeq \theta$. The difference between the laboratory and center-of-mass frames disappears in the limit of a target that is very heavy compared to the projectile.

(ii) If the masses are equal, $m_1 = m_2$, (1.80) and (1.81) give the relations

$$\theta = \theta^*/2\,, \quad \theta + \phi = \pi/2\,.$$

With respect to the laboratory system the outgoing particles leave in directions perpendicular to each other. In particular, in the case of a *central collision*, $\theta^* = \pi$, and, because of $\boldsymbol{q}'^* = -\boldsymbol{q}^*$,

$$\boldsymbol{p}_1' = 0\,, \quad \boldsymbol{p}_2' = \boldsymbol{p}_1\,.$$

The projectile comes to a complete rest, while the target particle takes over the momentum of the incoming projectile.

1.28 Two-Particle Scattering with a Central Force: Dynamics

Consider a scattering problem in the laboratory system sketched in Fig. 1.31. The projectile (1) comes in from infinity with initial momentum $\boldsymbol{p}_1$, while the target (2) is initially at rest. The initial configuration is characterized by the vector $\boldsymbol{p}_1$ and by a two-dimensional vector $\boldsymbol{b}$, perpendicular to $\boldsymbol{p}_1$, which indicates the azimuthal angle and the distance from the z-axis (as drawn in the figure) of the incident particle. This *impact vector* is directly related to the angular momentum:

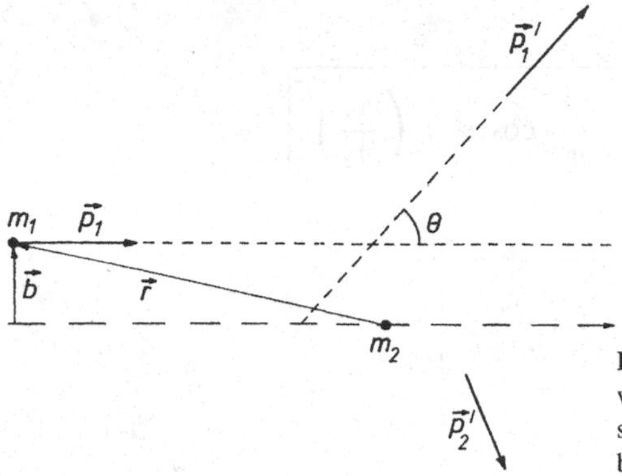

Fig. 1.31. Kinematics of a scattering process with two particles, seen from the laboratory system. The particle with mass m_2 is at rest before the scattering

$$\boldsymbol{l} = \boldsymbol{r} \times \boldsymbol{q}^* = \frac{m_2}{M}\boldsymbol{r} \times \boldsymbol{p}_1 = \frac{m_2}{M}\boldsymbol{b} \times \boldsymbol{p}_1 = \boldsymbol{b} \times \boldsymbol{q}^* \,. \tag{1.82}$$

Its modulus $b = |\boldsymbol{b}|$ is called the *impact parameter* and is given by

$$b = \frac{M}{m_2|\boldsymbol{p}_1|}|\boldsymbol{l}| = \frac{1}{q^*}|\boldsymbol{l}| \,. \tag{1.83}$$

If the interaction is spherically symmetric (as assumed here), or if it is axially symmetric about the z-axis, the direction of $\boldsymbol{b}$ in the plane perpendicular to the z-axis does not matter. Only its modulus, the impact parameter (1.83), is dynamically relevant.

For a given potential $U(r)$ we must determine the angle θ into which particle 1 will be scattered, once its momentum $\boldsymbol{p}_1$ and its relative angular momentum are given. The general analysis presented in Sects. 1.7.1 and 1.24 tells us that we must solve the equivalent problem of the scattering of a fictitious particle of mass $\mu = m_1 m_2/M$, subject to the potential $U(r)$. This is sketched in Fig. 1.32. We have

$$E = \frac{q^{*2}}{2\mu} \,, \quad \boldsymbol{l} = \boldsymbol{b} \times \boldsymbol{q}^* \,. \tag{1.84}$$

Let P be the pericenter, i.e. the point of closest approach. Figure 1.32 shows the scattering process for a repulsive potential and for different values of the impact

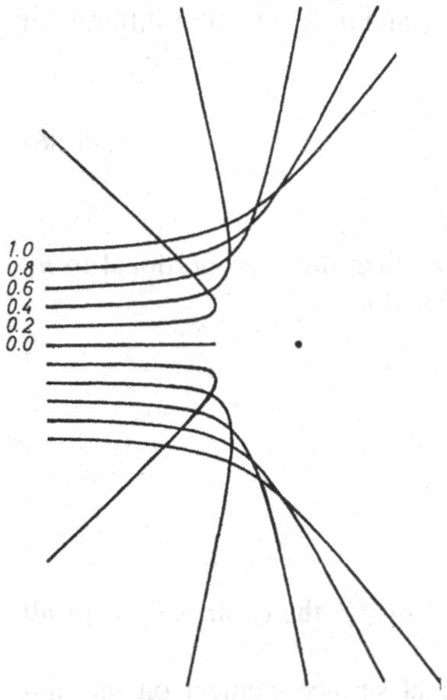

Fig. 1.32. Scattering orbits in the repulsive potential $U(r) = A/r$ (with $A > 0$). The impact parameter is measured in units of the characteristic length $\lambda \stackrel{\text{def}}{=} A/E$, with E the energy of the incoming particle. Cf. Practical Example 1.5

parameter. In Sect. 1.24 we showed that every orbit is symmetric with respect to the straight line joining the force center O and the pericenter P. Therefore, the two asymptotes to the orbit must also be symmetric with respect to OP[7]. Thus, if φ_0 is the angle between OP and the asymptotes, we have

$$\theta^* = |\pi - 2\varphi_0| .$$

The angle φ_0 is obtained from (1.68), making use of the relations (1.84)

$$\begin{aligned}\varphi_0 &= \int_{r_P}^{\infty} \frac{l\,\mathrm{d}r}{r^2\sqrt{2\mu(E - U(r)) - l^2/r^2}} \\ &= \int_{r_P}^{\infty} \frac{b\,\mathrm{d}r}{r^2\sqrt{1 - b^2/r^2 - 2\mu U(r)/q^{*2}}} . \end{aligned} \tag{1.85}$$

For a given $U(r)$, φ_0, and hence the scattering angle θ^* are calculated from this equation as functions of q^* (i.e. of the energy, via (1.84) and of b. However, some care is needed depending on whether or not the connection between b and q^* is unique. There are potentials such as the attractive $1/r^2$ potential where a given scattering angle is reached from two or more different values of the impact parameter. This happens when the orbit revolves about the force center more than once (see Example (iii) below).

A measure for the scattering in the potential $U(r)$ is provided by the *differential cross section* $\mathrm{d}\sigma$. It is defined as follows. Let n_0 be the number of particles incident on the unit area per unit time; $\mathrm{d}n$ is the number of particles per unit time that are

[7] The orbit possesses asymptotes only if the potential tends to zero sufficiently fast at infinity. As we shall learn in the next section, the relatively weak decrease $1/r$ is already somewhat strange.

scattered with scattering angles that lie between θ^* and $\theta^* + d\theta^*$. The differential cross section is then defined by

$$d\sigma \stackrel{\text{def}}{=} \frac{1}{n_0} dn . \tag{1.86}$$

Its physical dimension is $[d\sigma] =$ area.

If the relation between $b(\theta^*)$ and θ^* is unique, then dn is proportional to n_0 and to the area of the annulus with radii b and $b + db$,

$$dn = n_0 2\pi b(\theta^*)\, db ,$$

and therefore

$$d\sigma = 2\pi b(\theta^*)\, db = 2\pi b(\theta^*) \left| \frac{db(\theta^*)}{d\theta^*} \right| d\theta^* .$$

If to a fixed θ^* there correspond several values of $b(\theta^*)$, the contributions of all branches of this function must be added.

It is convenient to refer $d\sigma$ to the infinitesimal surface element on the unit sphere $d\Omega^* = \sin\theta^* d\theta^* d\phi^*$ and to integrate over the azimuth ϕ^*. With $d\omega \equiv 2\pi \sin\theta^* d\theta^*$ we then have

$$d\sigma = \frac{b(\theta^*)}{\sin\theta^*} \left| \frac{db(\theta^*)}{d\theta^*} \right| d\omega . \tag{1.87}$$

We study three instructive examples.

Example (i) Scattering off an ideally reflecting sphere. With the notations of Fig. 1.33

$$b = R \sin\frac{\Delta\alpha}{2} = R \cos\frac{\theta^*}{2} .$$

Here we have used the relationship $\Delta\alpha = \pi - \theta^*$, which follows from the equality of the angle of incidence and the angle of reflection. Thus

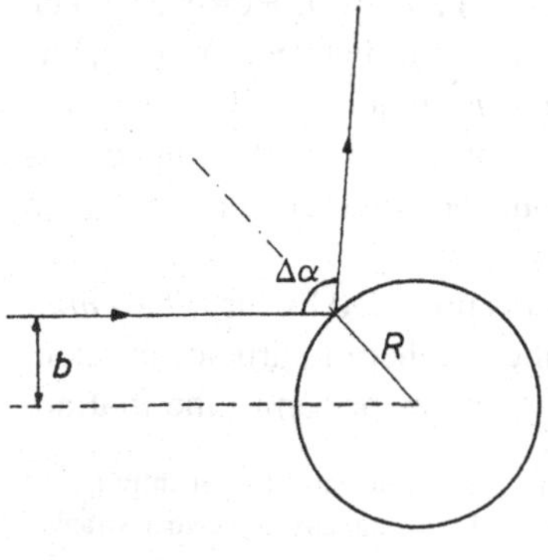

Fig. 1.33. Scattering by an ideally reflecting sphere of radius R

$$\frac{\mathrm{d}b}{\mathrm{d}\theta^*} = -\frac{R}{2}\sin\frac{\theta^*}{2} \quad \text{and} \quad \frac{\mathrm{d}\sigma}{\mathrm{d}\omega} = \frac{R^2}{2}\frac{(\cos\theta^*/2)(\sin\theta^*/2)}{\sin\theta^*} = \frac{R^2}{4} .$$

Integrating over $\mathrm{d}\omega$ we obtain the *total elastic cross section*

$$\sigma_{\mathrm{tot}} = \pi R^2 ,$$

a result that has a simple geometric interpretation: the particle sees the projection of the sphere onto a plane perpendicular to its momentum.

Example (ii) Scattering of particles off nuclei (Rutherford scattering). The potential is $U(r) = \kappa/r$ with $\kappa = q_1 q_2$, where q_1 is the charge of the α-particle (this is a Helium nucleus, which has charge $q_1 = 2e$), while q_2 is the charge of the nucleus that one is studying. Equation (1.85) can be integrated by elementary methods and one finds (making use of a good table of integrals) that

$$\varphi_0 = \arctan\left(\frac{bq^{*2}}{\mu\kappa}\right), \quad \text{or} \tag{1.88}$$

$$\tan\varphi_0 = \frac{bq^{*2}}{\mu\kappa} , \tag{1.88'}$$

from which follows

$$b^2 = \frac{\kappa^2\mu^2}{q^{*4}}\tan^2\varphi_0 = \frac{\kappa^2\mu^2}{q^{*4}}\cot^2\frac{\theta^*}{2} ,$$

and, finally, Rutherford's formula

$$\frac{\mathrm{d}\sigma}{\mathrm{d}\omega} = \left(\frac{\kappa}{4E}\right)^2 \frac{1}{\sin^4(\theta^*/2)} . \tag{1.89}$$

This formula, which is also valid in the context of quantum mechanics, was the key to the discovery of atomic nuclei. It gave the first hint that Coulomb's law is valid at least down to distances of the order of magnitude 10^{-12} cm.

In this example the differential cross section diverges in the forward direction, $\theta^* = 0$, and the total elastic cross section $\sigma_{\mathrm{tot}} = \int \mathrm{d}\omega(\mathrm{d}\sigma/\mathrm{d}\omega)$ is infinite. The reason for this is the slow decrease of the potential at infinity. $U(r) = \kappa/r$ can be felt even at infinity, it is "long ranged". This difficulty arises with all potentials whose range is infinite.

Example (iii) Two-body scattering for an attractive inverse square potential. The potential is $U(r) = -\alpha/r^2$, where α is a positive constant. For positive energy $E > 0$ all orbits are scattering orbits. If $l^2 > 2\mu\alpha$, we have

$$\varphi - \varphi_0 = r_{\mathrm{P}}^{(0)} \int_{r_0}^{r} \frac{\mathrm{d}r'}{r'\sqrt{r'^2 - r_{\mathrm{P}}^2}}$$

with μ the reduced mass, $r_P = \sqrt{(l^2 - 2\mu\alpha)/2\mu E}$ the distance at perihelion, and $r_P^{(0)} = l/\sqrt{2\mu E}$. If the projectile comes in along the x-axis the solution is

$$\varphi(r) = \frac{l}{\sqrt{l^2 - 2\mu\alpha}} \arcsin\left(r_P/r\right) .$$

We verify that for $\alpha = 0$ there is no scattering. In this case

$$\varphi^{(0)}(r) = \arcsin\left(r_P^{(0)}/r\right) ,$$

which means that the projectile moves along a straight line parallel to the x-axis, at a distance $r_P^{(0)}$ from the scattering center. For $\alpha \neq 0$ the azimuth at r_P is

$$\varphi\left(r = r_P\right) = \frac{l}{\sqrt{l^2 - 2\mu\alpha}} \frac{\pi}{2} .$$

Therefore, after the scattering the particle moves in the direction

$$\frac{l}{\sqrt{l^2 - 2\mu\alpha}} \pi .$$

It turns around the force center n times if the condition

$$\frac{l}{\sqrt{l^2 - 2\mu\alpha}} \left(\arcsin\frac{r_P}{\infty} - \arcsin\frac{r_P}{r_P} \right) = \frac{r_P^{(0)}}{r_P} \frac{\pi}{2} > n\pi$$

is fulfilled. Thus, $n = r_P^{(0)}/2r_P$, independently of the energy.

For $l^2 < 2\mu\alpha$ the integral above is (for the same initial condition)

$$\varphi(r) = \frac{r_P^{(0)}}{b} \ln \frac{b + \sqrt{b^2 + r^2}}{r} ,$$

where we have set $b = \sqrt{(2\mu\alpha - l^2)/2\mu E}$. The particle revolves about the force center, along a shrinking spiral. As the radius goes to zero, the angular velocity $\dot{\varphi}$ increases beyond any limit such that the product $\mu r^2 \dot{\varphi} = l$ stays constant (Kepler's second law.)

1.29 Example: Coulomb Scattering of Two Particles with Equal Mass and Charge

It is instructive to study Rutherford scattering in center-of-mass and relative coordinates and thereby derive the individual orbits of the projectile and target particles. We take the masses to be equal, $m_1 = m_2 \equiv m$, and the charges to be equal, $q_1 = q_2 \equiv Q$, for the sake of simplicity. The origin O of the laboratory system

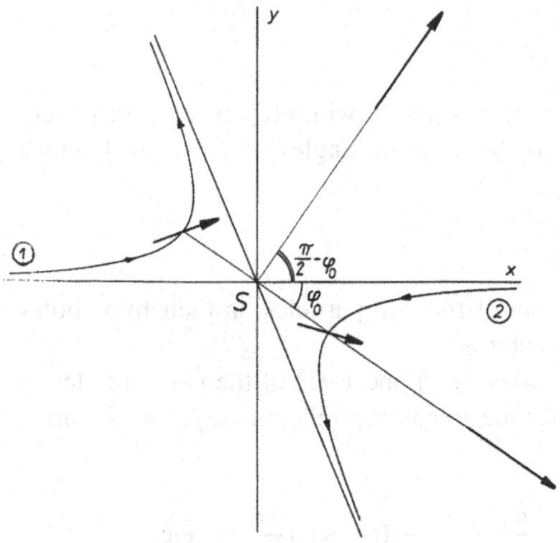

Fig. 1.34. Scattering of two equally charged particles of equal masses, under the action of the Coulomb force. The hyperbola branches are the orbits with respect to the center of mass. The arrows indicate the velocities at pericenter and long after the scattering with respect to the laboratory system

is chosen such that it coincides with the center of mass at the moment of closest approach of the two particles, see Fig. 1.34. Let $\boldsymbol{r}_1$ and $\boldsymbol{r}_2$ be the coordinates of the two particles in the laboratory system and $\boldsymbol{r}_1^*$ and $\boldsymbol{r}_2^*$ their coordinates in the center-of-mass-system. If $\boldsymbol{r}_S$ denotes the center-of-mass and $\boldsymbol{r} = \boldsymbol{r}_1^* - \boldsymbol{r}_2^*$ the relative coordinate, then

$$\boldsymbol{r}_1 = \boldsymbol{r}_S + \tfrac{1}{2}\boldsymbol{r}\,, \quad \boldsymbol{r}_1^* = -\boldsymbol{r}_2^* = \tfrac{1}{2}\boldsymbol{r}\,, \quad \boldsymbol{r}_2 = \boldsymbol{r}_S - \tfrac{1}{2}\boldsymbol{r}\,. \tag{1.90}$$

The total momentum is $\boldsymbol{P} = \boldsymbol{p}_1 = 2\boldsymbol{q}^*$; the total energy decomposes into relative and center-of-mass motion as follows:

$$E = E_r + E_S = \frac{\boldsymbol{q}^{*2}}{2\mu} + \frac{\boldsymbol{P}^2}{2M}\,.$$

As $\mu = m/2$ and $M = 2m$, we have

$$E_r = E_S = \frac{q^{*2}}{m}\,.$$

The orbit of the center of mass S is

$$\boldsymbol{r}_S(t) = \sqrt{\frac{E_r}{m}}\,t\boldsymbol{e}_1\,. \tag{1.91}$$

For the relative motion we have from Sect. 1.7.2

$$r(\varphi) = \frac{p}{\varepsilon\cos(\varphi - \varphi_0) - 1}\,, \quad \varphi \in \left[0,\, 2\varphi_0\right],$$

$$p = \frac{2l^2}{mQ^2}\,, \quad \varepsilon = \sqrt{1 + \frac{4E_r l^2}{mQ^4}}\,, \tag{1.92}$$

and from (1.88′)

$$\cos\varphi_0 = \frac{1}{\varepsilon}\,, \quad \sin\varphi_0 = \frac{\sqrt{\varepsilon^2-1}}{\varepsilon}\,.$$

Thus, in the center-of-mass frame $r_1^* = r_2^* = r(\varphi)/2$, with $r(\varphi)$ from (1.92). Here φ is the orbit parameter, its relation to the azimuth angles of particles 1 and 2 being

$$\varphi_1 = \pi - \varphi\,, \quad \varphi_2 = 2\pi - \varphi\,.$$

This means, in particular, that the motion of the two particles on their hyperbolas (Fig. 1.34) is synchronous in the parameter φ.

It is not difficult to derive the velocities $\boldsymbol{v}_1(t)$ and $\boldsymbol{v}_2(t)$ of the two particles in the laboratory system from (1.90–92). One needs the relation $\mathrm{d}\varphi/\mathrm{d}t = 2l/mr^2$. From this and from

$$\frac{\mathrm{d}x_1}{\mathrm{d}t} = \sqrt{\frac{E_\mathrm{r}}{m}} + \frac{1}{2}\frac{\mathrm{d}}{\mathrm{d}t}(r\cos\varphi_1) = \sqrt{\frac{E_\mathrm{r}}{m}} - \frac{1}{2}\frac{\mathrm{d}}{\mathrm{d}\varphi}(r\cos\varphi)\frac{\mathrm{d}\varphi}{\mathrm{d}t}\,, \quad \text{etc.}$$

one finds the result

$$\frac{\mathrm{d}x_1}{\mathrm{d}t} = 2\sqrt{\frac{E_\mathrm{r}}{m}} - \frac{l}{mp}\sin\varphi = \frac{l}{mp}\left(2\sqrt{\varepsilon^2-1} - \sin\varphi\right)$$
$$\frac{\mathrm{d}y_1}{\mathrm{d}t} = \frac{l}{mp}(1-\cos\varphi)\,. \tag{1.93}$$

For $\boldsymbol{v}_2(t)$ one obtains

$$\frac{\mathrm{d}x_2}{\mathrm{d}t} = \frac{l}{mp}\sin\varphi\,, \quad \frac{\mathrm{d}y_2}{\mathrm{d}t} = -\frac{l}{mp}(1-\cos\varphi)\,. \tag{1.94}$$

Three special cases, two of which are marked with arrows in Fig. 1.34, are read off these formulae.

(i) At the beginning of the motion, $\varphi = 0$:

$$\boldsymbol{v}_1 = \left(2\sqrt{\frac{E_\mathrm{r}}{m}},\, 0\right)\,, \quad \boldsymbol{v}_2 = (0,0)\,.$$

(ii) At the pericenter, $\varphi = \varphi_0$:

$$\boldsymbol{v}_1 = \frac{l}{mp\varepsilon}\left((2\varepsilon-1)\sqrt{\varepsilon^2-1}\,,\ \varepsilon-1\right), \quad \boldsymbol{v}_2 = \frac{l}{mp\varepsilon}\left(\sqrt{\varepsilon^2-1}\,,\ -(\varepsilon-1)\right).$$

(iii) After the scattering, $\varphi = 2\varphi_0$:

$$\boldsymbol{v}_1 = \frac{2l(\varepsilon^2-1)}{mp\varepsilon^2}\left(\sqrt{\varepsilon^2-1},\, 1\right)\,, \quad \boldsymbol{v}_2 = \frac{2l}{mp\varepsilon^2}\left(\sqrt{\varepsilon^2-1},\, -(\varepsilon^2-1\right)\,.$$

Thus, the slope of $\boldsymbol{v}_1$ is $1/\sqrt{\varepsilon^2-1} = 1/\tan\varphi_0$, while the slope of $\boldsymbol{v}_2$ is $-\sqrt{\varepsilon^2-1} = -\tan\varphi_0$.

Of course, it is also possible to give the functions $x_i(\varphi)$ and $y_i(\varphi)$ in closed form, once $t(\varphi)$ is calculated from (1.65):

$$t(\varphi) = \frac{mp^2}{2l} \int_{\varphi_0}^{\varphi} \frac{d\varphi'}{\left(1 - \varepsilon \cos(\varphi' - \varphi_0)\right)^2} . \tag{1.95}$$

(The reader should do this.) Figure 1.35 shows the scattering orbits in the center-of-mass system for the case $\varepsilon = 2/\sqrt{3}$, i.e. for $\varphi_0 = 30°$, in the basis of the dimensionless variables $2x_i/p$ and $2y_i/p$. The same picture shows the positions of the two particles in the *laboratory* system as a function of the dimensionless time variable

$$\tau \stackrel{\text{def}}{=} t \frac{2l}{mp^2} .$$

According to (1.95) this variable is chosen so that the pericenter is reached for $\tau = 0$.

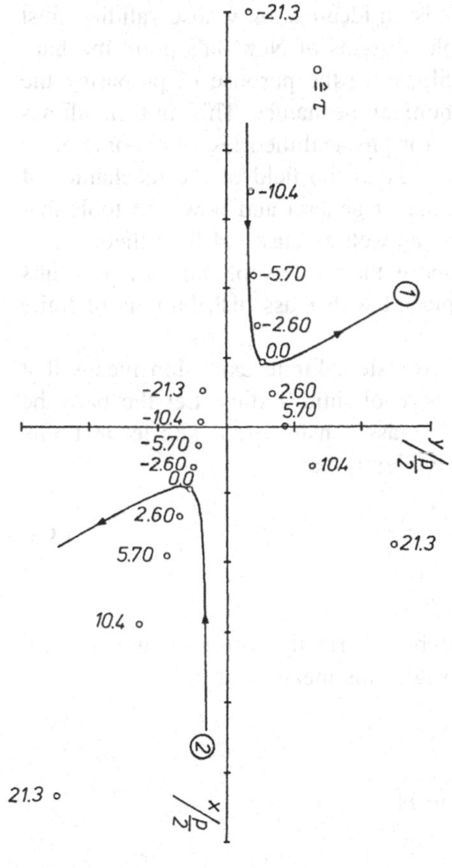

Fig. 1.35. Coulomb scattering of two particles ($m_1 = m_2$, $q_1 = q_2$) with $\varphi_0 = 30°$. The hyperbola branches are the scattering orbits in the center-of-mass system. The open points show the positions of the two particles in the laboratory system at the indicated times

The problem considered here has a peculiar property that one meets in asking where the target (particle 2) was at time $t = -\infty$. The answer is not evident from the figure and one must return to (1.92). With $\mathrm{d}x_2/\mathrm{d}\varphi = r^2 \sin\varphi / 2p$ one finds from this equation that

$$x_2(\varphi_0) - x_2(0) = \frac{p}{2} \int_0^{\varphi_0} \frac{\sin\varphi}{(1 - \cos\varphi - \sqrt{\varepsilon^2 - 1}\sin\varphi)^2} \mathrm{d}\varphi \,.$$

This integral is logarithmically divergent. This means that in the laboratory system particle 2 also came from $x_2 = -\infty$. This somewhat strange result gives a first hint at the peculiar nature of the "long-range" potential $1/r$ that will be met again in quantum mechanics and quantum field theory.

1.30 Mechanical Bodies of Finite Extension

So far we have exclusively considered pointlike mechanical objects, i.e. particles that carry a finite mass but have no finite spatial extension. In its application to macroscopic mechanical bodies this is an idealization whose validity must be checked in every single case. The simple systems of Newton's point mechanics that we studied in this chapter primarily serve the purpose of preparing the ground for a systematic construction of canonical mechanics. This, in turn, allows the development of more general principles for physical theories, after some more abstraction and generalization. One thereby leaves the field of the mechanics of macroscopic bodies proper but develops a set of general and powerful tools that are useful in describing continuous systems as well as classical field theories.

This section contains a few remarks about the validity of our earlier results for those cases where mass points are replaced with mass distributions of finite extension.

Consider a mechanical body of finite extension. Finite extension means that the body can always be enclosed by a sphere of finite radius. Let the body be characterized by a time-independent (rigid) mass density $\varrho(\boldsymbol{x})$, and let m be its total mass. Integrating over all space, one evidently has

$$\int \mathrm{d}^3x \varrho(\boldsymbol{x}) = m \,. \tag{1.96}$$

The dimension of ϱ is mass/(length)^3.

For example, assume the mass density to be spherically symmetric with respect to the center O. Taking this point as the origin this means that

$$\varrho(\boldsymbol{x}) = \varrho(r) \,, \quad r \stackrel{\text{def}}{=} |\boldsymbol{x}| \,.$$

In spherical coordinates the volume element is

$$\mathrm{d}^3x = \sin\theta \, \mathrm{d}\theta \, \mathrm{d}\phi r^2 \, \mathrm{d}r \,.$$

Since ϱ does not depend on θ and ϕ, the integration over these variables can be carried out, so that the condition (1.96) becomes

$$4\pi \int_0^\infty r^2 \, \mathrm{d}r \, \varrho(r) = m \, . \tag{1.97}$$

Equation (1.96) suggests the introduction of a differential mass element

$$\mathrm{d}m \stackrel{\text{def}}{=} \varrho(\boldsymbol{x}) \mathrm{d}^3 x \, . \tag{1.98}$$

In a situation where the resulting differential force $\mathrm{d}\boldsymbol{K}$ is applied to this mass element, it is plausible to generalize the relation (1.8b) between force and acceleration as follows:

$$\ddot{\boldsymbol{x}} \mathrm{d}m = \mathrm{d}\boldsymbol{K} \, . \tag{1.99}$$

(This postulate is due to L. Euler and was published in 1750.) We are now in a position to treat the interaction of two extended celestial bodies. We solve this problem in several steps.

(i) Potential and force field of an extended star. Every mass element situated in $\boldsymbol{x}$ creates a differential potential energy for a pointlike probe of mass m_0 situated in $\boldsymbol{y}$ (inside or outside the mass distribution), given by

$$\mathrm{d}U(\boldsymbol{y}) = -\frac{G \, \mathrm{d}m m_0}{|\boldsymbol{x} - \boldsymbol{y}|} = -G m_0 \frac{\varrho(\boldsymbol{x})}{|\boldsymbol{x} - \boldsymbol{y}|} \mathrm{d}^3 x \, . \tag{1.100}$$

The probe experiences the differential force

$$\mathrm{d}\boldsymbol{K} = -\boldsymbol{\nabla}_y \mathrm{d}U = -\frac{G m_0 \varrho(x)}{|\boldsymbol{x} - \boldsymbol{y}|^2} \frac{\boldsymbol{y} - \boldsymbol{x}}{|\boldsymbol{y} - \boldsymbol{x}|} \mathrm{d}^3 x \, . \tag{1.101}$$

Either formula, (1.100) or (1.101), can be integrated over the entire star. For instance, the total potential energy of the mass m_0 is

$$U(\boldsymbol{y}) = -G m_0 \int \frac{\varrho(\boldsymbol{x})}{|\boldsymbol{x} - \boldsymbol{y}|} \mathrm{d}^3 x \, . \tag{1.102}$$

The vector $\boldsymbol{x}$ scans the mass distribution, while $\boldsymbol{y}$ denotes the point where the potential is to be calculated. The force field that belongs to this potential follows from (1.102), as usual, by taking the gradient with respect to $\boldsymbol{y}$, viz.

$$\boldsymbol{K}(\boldsymbol{y}) = -\boldsymbol{\nabla}_y U(\boldsymbol{y}) \, . \tag{1.103}$$

(ii) Celestial body with spherical symmetry. Let $\varrho(\boldsymbol{x}) = \varrho(s)$, with $s = |\boldsymbol{x}|$, and let $\varrho(s) = 0$ for $s \geq R$. In (1.102) we take the direction of the vector $\boldsymbol{y}$ as the

z-axis. Denoting by $r \stackrel{\text{def}}{=} |\boldsymbol{y}|$ the modulus of $\boldsymbol{y}$ and integrating over the azimuth ϕ, we find that

$$U = -2\pi G m_0 \int_{-1}^{+1} \mathrm{d}z \int_0^\infty s^2 \mathrm{d}s \frac{\varrho(s)}{\sqrt{r^2 + s^2 - 2rsz}}\,, \quad z \stackrel{\text{def}}{=} \cos\theta\,.$$

The integral over z is elementary,

$$\int_{-1}^{+1} \mathrm{d}z (r^2 + s^2 - 2rsz)^{-1/2} = -\frac{1}{rs}[|r-s| - (r+s)] = \begin{cases} 2/r & \text{for} \quad r > s \\ 2/s & \text{for} \quad r < s \end{cases}.$$

One sees that U is spherically symmetric, too, and that it is given by

$$U(r) = -4\pi G m_0 \left(\frac{1}{r} \int_0^r s^2 \mathrm{d}s \varrho(s) + \int_r^\infty s \mathrm{d}s \varrho(s) \right). \tag{1.104}$$

For $r \geq R$ the second integral does not contribute, because $\varrho(s)$ vanishes for $s \geq R$. The first integral extends from O to R and, from (1.97), gives the constant $m/4\pi$. Thus one obtains

$$U(r) = -\frac{G m_0 m}{r} \quad \text{for} \quad r \geq R\,. \tag{1.105}$$

> In the space outside its mass distribution a spherically symmetric star with total mass m creates the same potential as a mass point m placed at its center of symmetry.

It is obvious that this result is of great importance for the application of Kepler's laws to planetary motion.

(iii) Interaction of two celestial bodies of finite extension. If the probe of mass m_0 has a finite extension, too, and is characterized by the mass density $\varrho_0(\boldsymbol{y})$, (1.102) is replaced by the differential potential

$$\mathrm{d}U(\boldsymbol{y}) = -G\varrho_0(\boldsymbol{y})\mathrm{d}^3y \int \frac{\varrho(\boldsymbol{x})}{|\boldsymbol{x} - \boldsymbol{y}|} \mathrm{d}^3x\,.$$

This is the potential energy of the mass element $\varrho(\boldsymbol{y})\mathrm{d}^3y$ in the field of the first star. The total potential energy is obtained from this by integrating over $\boldsymbol{y}$:

$$U = -G \int \mathrm{d}^3x \int \mathrm{d}^3y \frac{\varrho(\boldsymbol{x})\varrho_0(\boldsymbol{y})}{|\boldsymbol{x} - \boldsymbol{y}|}\,. \tag{1.106}$$

If both densities are spherically symmetric, their radii being R and R_0, we obtain again (1.105) whenever the distance of the two centers is larger than $(R + R_0)$.

(iv) Potential of a star with finite extension that is not spherically symmetric. Assume the density $\varrho(\boldsymbol{x})$ still to be finite (that is, $\varrho(\boldsymbol{x}) = 0$ for $|\boldsymbol{x}| \geq R$) but not necessarily spherically symmetric. In calculating the integral (1.102) the following expansion of the inverse distance is particularly useful:

$$\frac{1}{|\boldsymbol{x}-\boldsymbol{y}|} = 4\pi \sum_{l=0}^{\infty} \frac{1}{2l+1} \frac{r_<^l}{r_>^{l+1}} \sum_{\mu=-l}^{l} Y_{l\mu}^*(\hat{x}) Y_{l\mu}(\hat{y}) . \tag{1.107}$$

Here $r_< = |\boldsymbol{x}|$, $r_> = |\boldsymbol{y}|$ if $|\boldsymbol{y}| > |\boldsymbol{x}|$, and correspondingly $r_< = |\boldsymbol{y}|$, $r_> = |\boldsymbol{x}|$ if $|\boldsymbol{y}| < |\boldsymbol{x}|$. The symbols $Y_{l\mu}$ denote well-known special functions, *spherical harmonics*, whose arguments are the polar angles of $\boldsymbol{x}$ and $\boldsymbol{y}$:

$$(\theta_x, \phi_x) \equiv \hat{x} , \quad (\theta_y, \phi_y) \equiv \hat{y} .$$

These functions are *normalized* and *orthogonal* in the following sense:

$$\int_0^\pi \sin\theta \, \mathrm{d}\theta \int_0^{2\pi} \mathrm{d}\phi Y_{l\mu}^*(\theta,\phi) Y_{l'\mu'}(\theta,\phi) = \delta_{ll'}\delta_{\mu\mu'} \tag{1.108}$$

(see e.g. Abramowitz, Stegun 1965). Inserting this expansion in (1.102) and choosing $|\boldsymbol{y}| > R$, one obtains

$$U(\boldsymbol{y}) = -Gm_0 \sum_{l=0}^{\infty} \frac{4\pi}{2l+1} \sum_{\mu=-l}^{+l} \frac{q_{l\mu}}{r^{l+1}} Y_{l\mu}(\boldsymbol{y}) , \tag{1.109}$$

where

$$q_{l\mu} \stackrel{\text{def}}{=} \int \mathrm{d}^3x Y_{l\mu}^*(\hat{x}) s^l \varrho(\boldsymbol{x}) . \tag{1.110}$$

The first spherical harmonic is a constant: $Y_{l=0\,\mu=0} = 1/\sqrt{4\pi}$. If $\varrho(\boldsymbol{x})$ is taken to be spherically symmetric, one obtains

$$\begin{aligned} q_{l\mu} &= \sqrt{4\pi} \int_0^R s^2 \, \mathrm{d}s s^l \varrho(s) \int_0^\pi \sin\theta \, \mathrm{d}\theta \int_0^{2\pi} \mathrm{d}\phi Y_{00} Y_{l\mu}^* \\ &= \sqrt{4\pi} \int_0^R s^2 \, \mathrm{d}s \varrho(s) \delta_{l0}\delta_{\mu 0} = \frac{m}{\sqrt{4\pi}} \delta_{l0}\delta_{\mu 0} , \end{aligned}$$

so that (1.109) leads to the result (1.105), as expected. The coefficients $q_{l\mu}$ are called multipole moments of the density $\varrho(\boldsymbol{x})$. The potentials that they create,

$$U_{l\mu}(y) = -Gm_0 \frac{4\pi q_{l\mu}}{(2l+1)r^{l+1}} Y_{l\mu}^*(\hat{y}) , \tag{1.111}$$

are called multipole potentials. In the case of spherical symmetry only the multipole moment with $l = 0$ is nonzero, while in the absence of this symmetry many or all multipole moments will contribute.

1.31 Time Averages and the Virial Theorem

Let us return to the n-particle system as described by the equations of motion (1.28). We assume that the system is closed and autonomous, i.e. that there are only internal, time-independent forces. We further assume that these are potential forces but not necessarily central forces. For just $n = 3$, general solutions of the equations of motion are known only for certain special situations. Very little is known for more than three particles. Therefore, the following approach is useful because it yields at least some qualitative information.

We suppose that we know the solutions $\boldsymbol{r}_i(t)$, and therefore also the momenta $\boldsymbol{p}_i(t) = m\dot{\boldsymbol{r}}_i(t)$. We then construct the following mapping from phase space onto the real line:

$$v(t) \stackrel{\text{def}}{=} \sum_{i=1}^{n} \boldsymbol{r}_i(t) \cdot \boldsymbol{p}_i(t) . \tag{1.112}$$

This function is called the *virial*. If a specific solution has the property that no particle ever escapes to infinity or takes on an infinitely large momentum, then $v(t)$ remains bounded for all times. Defining time averages as follows:

$$\langle f \rangle \stackrel{\text{def}}{=} \lim_{\Delta\to\infty} \frac{1}{2\Delta} \int_{-\Delta}^{+\Delta} f(t)\mathrm{d}t , \tag{1.113}$$

the average of the time derivative of $v(t)$ is then shown to vanish, viz.

$$\langle \dot{v} \rangle = \lim_{\Delta\to\infty} \frac{1}{2\Delta} \int_{-\Delta}^{+\Delta} \mathrm{d}t \frac{\mathrm{d}v(t)}{\mathrm{d}t} = \lim_{\Delta\to\infty} \frac{v(\Delta) - v(-\Delta)}{2\Delta} = 0 .$$

Since

$$\dot{v}(t) = \sum_{i=1}^{n} m_i \dot{\boldsymbol{r}}_i^2(t) - \sum_{i=1}^{n} \boldsymbol{r}_i(t) \cdot \boldsymbol{\nabla}_i U\big(\boldsymbol{r}_1(t), \ldots, \boldsymbol{r}_n(t)\big) ,$$

we obtain for the time average

$$2\langle T \rangle - \left\langle \sum_{1}^{n} \boldsymbol{r}_i \cdot \boldsymbol{\nabla}_i U \right\rangle = 0 . \tag{1.114}$$

This result is called the *virial theorem*. It takes a particularly simple form when U is a homogeneous function of degree k in its arguments $\boldsymbol{r}_1, \ldots, \boldsymbol{r}_n$. In this case $\sum \boldsymbol{r}_i \cdot \boldsymbol{\nabla}_i U = kU$, so that (1.114) and the principle of energy conservation give

$$2\langle T \rangle - k\langle U \rangle = 0 , \quad \langle T \rangle + \langle U \rangle = E . \tag{1.115}$$

Examples of interest follow.

(i) Two-body systems with harmonic force. Transforming to center-of-mass and relative coordinates, one has

$$v(t) = m_1 \boldsymbol{r}_1 \cdot \dot{\boldsymbol{r}}_1 + m_2 \boldsymbol{r}_2 \cdot \dot{\boldsymbol{r}}_2 = M \boldsymbol{r}_S \cdot \dot{\boldsymbol{r}}_S + \mu \boldsymbol{r} \cdot \dot{\boldsymbol{r}} .$$

The function v remains bounded only if the center of mass is at rest, $\dot{\boldsymbol{r}}_S = 0$. However, the kinetic energy is then equal to the kinetic energy of the relative motion so that (1.115) applies to the latter and to $U(r)$. In this example $U(r) = \alpha r^2$, i.e. $k = 2$. The time averages of the kinetic energy and potential energy of relative motion are the same and are equal to half the energy,

$$\langle T \rangle = \langle U \rangle = \tfrac{1}{2} E .$$

(ii) In the case of the Kepler problem the potential is $U(r) = -\alpha/r$, where r denotes the relative coordinate. Thus $k = -1$. For $E < 0$ (only then is $v(t)$ bounded) one finds for the time averages of kinetic and potential energies of relative motion

$$\langle T \rangle = -E \; ; \quad \langle U \rangle = 2E .$$

Note that this is valid only in $\mathbb{R}^3 \backslash \{0\}$ for the variable $\boldsymbol{r}$. The origin where the force becomes infinite should be excluded. For the two-body system this is guaranteed whenever the relative angular momentum is nonzero.

(iii) For an n-particle system ($n \geq 3$) with gravitational forces some information can also be obtained. We first note that $v(t)$ is the derivative of the function

$$w(t) \stackrel{\text{def}}{=} \sum_{i=1}^{n} \frac{1}{2} m_i \boldsymbol{r}_i^2(t) ,$$

which is bounded, provided that no particle ever escapes to infinity. As one can easily show,

$$\ddot{w}(t) = 2T + U = E + T .$$

Since $T(t)$ is positive at all times, $w(t)$ can be estimated by means of the general solution of the differential equation $\ddot{y}(t) = E$. Indeed

$$w(t) \geq \tfrac{1}{2} E t^2 + \dot{w}(0) t + w(0) .$$

If the total energy is positive, then $\lim_{t \to \pm\infty} w(t) = \infty$, which means that at least one particle will escape to infinity asymptotically (see also Thirring 1992, Sect. 4.5).

Appendix: Practical Examples

1. Kepler Ellipses. Study numerical examples for finite motion of two celestial bodies in their center-of-mass frame (Sect. 1.7.2).

Solution. The relevant equations are found at the end of Sect. 1.7.2. It is convenient to express m_1 and m_2 in terms of the total mass $M = m_1 + m_2$ and to set $M = 1$. The reduced mass is then $\mu = m_1 m_2$. For given masses the form of the orbits is determined by the parameters

$$p = \frac{l^2}{A\mu} \quad \text{and} \quad \varepsilon = \sqrt{1 + \frac{2El^2}{\mu A^2}}\,, \tag{A.1}$$

which in turn are determined by the energy E and the angular momentum. It is easy to calculate and to draw the orbits on a PC. Figure 1.6a shows the example $m_1 = m_2$ with $\varepsilon = 0.5$, $p = 1$, while Fig. 1.6b shows the case $m_1 = m_2/9$ with $\varepsilon = 0.5$, $p = 0.66$. As the origin is the center of mass, the two stars are at opposite positions at any time.

2. Motion of a Double Star. Calculate the two orbital ellipses of the stars of the preceding example pointwise, as a function of time, for a given time interval Δt.

Solution. In Example 1 the figures show $r(\varphi)$ as a function of φ. They do not indicate how the stars move on their orbits as a function of *time*. In order to obtain $r(t)$, one returns to (1.19) and inserts the relative coordinate $r(\varphi)$. Separation of variables yields

$$t_{n+1} - t_n = \frac{\mu p^2}{l} \int_{\phi_n}^{\phi_{n+1}} \frac{\mathrm{d}\varphi}{(1 + \varepsilon \cos\varphi)^2} \tag{A.2}$$

for the orbital points n and $n+1$. (The pericenter has $\phi_{\mathrm{P}} = 0$.) The quantity $\mu p^2/l$ has the dimension of time. Introduce the period from (1.23) and use this as the unit of time,

$$T = 2\pi \frac{\mu^{1/2} a^{3/2}}{A^{1/2}} = \pi \frac{A\mu^{1/2}}{2^{1/2}(-E)^{3/2}}\,.$$

Then

$$\frac{\mu p^2}{l} = \left(1 - \varepsilon^2\right)^{3/2} \frac{T}{2\pi}\,.$$

The integral in (A.2) can be done exactly. Substituting

$$x \stackrel{\mathrm{def}}{=} \sqrt{\frac{1-\varepsilon}{1+\varepsilon}} \tan\frac{\varphi}{2}$$

one has

$$
\begin{aligned}
I &\equiv \int \frac{\mathrm{d}\varphi}{(1+\varepsilon\cos\varphi)^2} = \frac{2}{\sqrt{1-\varepsilon^2}} \int \mathrm{d}x \frac{1+[(1+\varepsilon)/(1-\varepsilon)]x^2}{(1+x^2)^2} \\
&= \frac{2}{\sqrt{1-\varepsilon^2}} \left\{ \int \frac{\mathrm{d}x}{1+x^2} + \frac{2\varepsilon}{1-\varepsilon} \int \frac{x^2\mathrm{d}x}{(1+x^2)^2} \right\} ,
\end{aligned}
$$

whose second term can be integrated by parts. The result is

$$
\begin{aligned}
I = &\frac{2}{(1-\varepsilon^2)^{3/2}} \arctan\left(\sqrt{\frac{1-\varepsilon}{1+\varepsilon}} \tan\frac{\varphi}{2} \right) \\
&- \frac{\varepsilon}{1-\varepsilon^2} \frac{\sin\varphi}{1+\varepsilon\cos\varphi} + C ,
\end{aligned}
\tag{A.3}
$$

so that

$$
\begin{aligned}
\frac{t_{n+1}-t_n}{T} = &\frac{1}{\pi} \left[\arctan\left(\sqrt{\frac{1-\varepsilon}{1+\varepsilon}} \tan\frac{\varphi}{2} \right) \right. \\
&\left. - \frac{1}{2}\varepsilon\sqrt{1-\varepsilon^2} \frac{\sin\varphi}{1+\varepsilon\cos\varphi} \right]_{\phi_n}^{\phi_{n+1}} .
\end{aligned}
\tag{A.4}
$$

One can compute the function $\Delta t(\Delta\phi, \phi)$, for a fixed increment $\Delta\phi$ and mark the corresponding positions on the orbit. Alternatively, one may give a fixed time interval $\Delta t/T$ and determine succeeding orbital positions by solving the implicit equation (A.4) in terms of φ.

3. Precession of Perihelion. (a) For the case of bound orbits in the Kepler problem show that the differential equation for $\varphi = \varphi(r)$ takes the form

$$
\frac{\mathrm{d}\varphi}{\mathrm{d}r} = \frac{1}{r} \sqrt{\frac{r_\mathrm{P} r_\mathrm{A}}{(r-r_\mathrm{P})(r_\mathrm{A}-r)}} ,
\tag{A.5}
$$

where r_P and r_A denote pericenter and apocenter, respectively. Integrate this equation with the boundary condition $\varphi(r = r_\mathrm{P}) = 0$.

(b) The potential is now modified into $U(r) = -A/r + B/r^2$. Determine the solution $\varphi = \varphi(r)$ and discuss the precession of the pericenter after one turn, in comparison with the Kepler case, as a function of $B \lessgtr 0$ where $|B| \ll l^2/2\mu$.

Solutions. (a) For elliptical orbits, $E < 0$, and one has

$$
\frac{\mathrm{d}\varphi}{\mathrm{d}r} = \frac{1}{\sqrt{2\mu(-E)}} \frac{1}{r} \frac{1}{\sqrt{-r^2 - \frac{A}{E}r + \frac{l^2}{2\mu E}}} .
$$

Apocenter and pericenter are given by the roots of the quadratic form $(-r^2 - Ar/E + l^2/2\mu E)$:

$$
r_\mathrm{A/P} = \frac{p}{1 \mp \varepsilon} = -\frac{A}{2E}(1 \pm \varepsilon)
\tag{A.6}
$$

(these are the points where $\mathrm{d}r/\mathrm{d}t = 0$). With

$$r_{\mathrm{P}} r_{\mathrm{A}} = \frac{A^2}{4E^2}(1-\varepsilon^2) = -\frac{l^2}{2\mu E}$$

we obtain (A.5). This equation can be integrated. With the condition $\varphi(r_{\mathrm{P}}) = 0$ one obtains

$$\varphi(r) = \arccos\left[\frac{1}{r_{\mathrm{A}} - r_{\mathrm{P}}}\left(2\frac{r_{\mathrm{A}} r_{\mathrm{P}}}{r} - r_{\mathrm{A}} - r_{\mathrm{P}}\right)\right]. \tag{A.7}$$

As $\varphi(r_{\mathrm{A}}) - \varphi(r_{\mathrm{P}}) = \pi$, one confirms that the pericenter, force center, and apocenter lie on a straight line. Two succeeding pericenter constellations have azimuths differing by 2π, i.e. they coincide. There is no precession of the pericenter.

(b) Let r_{P} and r_{A} be defined as in (A.6). The new apocenter and pericenter positions, in the perturbed potential, are denoted by r'_{A} and r'_{P}, respectively. One has

$$(r - r_{\mathrm{P}})(r_{\mathrm{A}} - r) + \frac{B}{E} = (r - r'_{\mathrm{P}})(r'_{\mathrm{A}} - r),$$

and therefore

$$r'_{\mathrm{P}} r'_{\mathrm{A}} = r_{\mathrm{P}} r_{\mathrm{A}} - \frac{B}{E}. \tag{A.8}$$

Equation (A.5) is modified as follows:

$$\frac{\mathrm{d}\varphi}{\mathrm{d}r} = \frac{1}{r}\sqrt{\frac{r_{\mathrm{P}} r_{\mathrm{A}}}{(r - r'_{\mathrm{P}})(r'_{\mathrm{A}} - r)}} = \sqrt{\frac{r_{\mathrm{P}} r_{\mathrm{A}}}{r'_{\mathrm{P}} r'_{\mathrm{A}}}}\sqrt{\frac{r'_{\mathrm{P}} r'_{\mathrm{A}}}{(r - r'_{\mathrm{P}})(r'_{\mathrm{A}} - r)}}.$$

This equation can be integrated as before under (a):

$$\varphi(r) = \sqrt{\frac{r_{\mathrm{P}} r_{\mathrm{A}}}{r'_{\mathrm{P}} r'_{\mathrm{A}}}} \arccos\left[\frac{1}{r'_{\mathrm{A}} - r'_{\mathrm{P}}}\left(2\frac{r'_{\mathrm{A}} r'_{\mathrm{P}}}{r} - r'_{\mathrm{A}} - r'_{\mathrm{P}}\right)\right]. \tag{A.9}$$

From (A.8) two successive pericenter configurations differ by

$$2\pi\sqrt{\frac{r_{\mathrm{P}} r_{\mathrm{A}}}{r'_{\mathrm{P}} r'_{\mathrm{A}}}} = \frac{2\pi l}{\sqrt{l^2 + 2\mu B}}. \tag{A.10}$$

This difference can be studied numerically, as a function of positive or negative B. Positive B means that the additional potential is repulsive so that, from (A.10), the pericenter will "stay behind". Negative B means additional attraction and causes the pericenter to "advance".

4. Rosettelike Orbits. Study the finite orbits in the attractive potential $U(r) = a/r^{\alpha}$, for some values of the exponent α in the neighborhood of $\alpha = 1$.

Solution. Use as a starting point the system (1.70′–71′) of first-order differential equations, written in dimensionless form:

$$\frac{\mathrm{d}\varrho}{\mathrm{d}\tau} = \pm\sqrt{2b\varrho^{-\alpha} - \varrho^{-2} - 2} \stackrel{\text{def}}{=} f(\varrho)\,, \qquad \frac{\mathrm{d}\varphi}{\mathrm{d}\tau} = \frac{1}{\varrho^2}\,. \tag{A.11}$$

From this calculate the second derivatives:

$$\frac{\mathrm{d}^2\varrho}{\mathrm{d}\tau^2} = \frac{\mathrm{d}}{\mathrm{d}\varrho}\left(\frac{\mathrm{d}\varrho}{\mathrm{d}\tau}\right)\frac{\mathrm{d}\varrho}{\mathrm{d}\tau} = \frac{1}{\varrho^3}\left(1 - b\alpha\varrho^{2-\alpha}\right) \stackrel{\text{def}}{=} g(\varrho)\,, \quad \frac{\mathrm{d}^2\varrho}{\mathrm{d}\tau^2} = -\frac{2}{\varrho^3} f(\varrho)\,.$$

Equation (A.11) can be solved approximately by means of simple Taylor series:

$$\varrho_{n+1} = \varrho_n + hf(\varrho_n) + \tfrac{1}{2}h^2 g(\varrho_n) + O(h^3)\,,$$

$$\varphi_{n+1} = \varphi_n + h\frac{1}{\varrho_n^2} - h^2\frac{1}{\varrho_n^3} f(\varrho_n) + O(h^3)\,, \tag{A.12}$$

for the initial conditions $\tau_0 = 0$, $\varrho(0) = R_0$, $\varphi(0) = 0$. The step size h for the time variable can be taken to be constant. Thus, if one plots the rosette pointwise, one can follow the temporal evolution of the motion. (In Figs. 1.18–22 we have chosen h to be variable, instead, taking $h = h_0\varrho/R_0$, with $h_0 = 0.02$.)

5. Scattering Orbits for a Repulsive Potential. A particle of fixed momentum $\boldsymbol{p}$ is scattered in the field of the potential $U(r) = A/r$, (with $A > 0$). Study the scattering orbits as a function of the impact parameter.

Solution. The orbit is given by

$$r = r(\varphi) = \frac{p}{1 + \varepsilon\cos(\varphi - \varphi_0)} \tag{A.13}$$

with $\varepsilon > 1$. The energy E must be positive. We choose $\varphi_0 = 0$ and introduce the impact parameter $b = l/|\boldsymbol{p}|$ and the quantity $\lambda \stackrel{\text{def}}{=} A/E$ as a characteristic length of the problem. The equation of the hyperbola (A.13) then reads

$$\frac{r(\varphi)}{\lambda} = \frac{2b^2/\lambda^2}{1 + \sqrt{1 + 4b^2/\lambda^2}\cos\varphi}\,. \tag{A.13′}$$

Introducing Cartesian coordinates (see Sect. 1.7.2), we find that (A.13′) becomes

$$\frac{4x^2}{\lambda^2} - \frac{y^2}{b^2} = 1\,.$$

This hyperbola takes on a symmetric position with respect to the coordinate axes, its asymptotes having the slopes $\tan\varphi_0$ and $-\tan\varphi_0$, respectively, where

$$\varphi_0 = \arctan\left(\frac{2b}{\lambda}\right)\,. \tag{A.14}$$

We restrict the discussion to the left-hand branch of the hyperbola. We want the particle always to come in along the same direction, say along the negative x-axis. For a given impact parameter b this is achieved by means of a rotation about the focus on the positive x-axis, viz.

$$u = (x - c)\cos\varphi_0 + y\sin\varphi_0\,,$$
$$v = -(x - c)\sin\varphi_0 + y\cos\varphi_0\,, \tag{A.15}$$

where $c = \sqrt{1 + 4b^2/\lambda^2}/2$ is the distance of the focus from the origin, and $y = \pm b\sqrt{4x^2/\lambda^2 - 1}$. For all b, the particle comes in from $-\infty$ along a direction parallel to the u-axis, with respect to the coordinate system (u, v). Starting from the pericenter ($x_0/\lambda = -\frac{1}{2}$, $y_0 = 0$), let y run upwards and downwards and use (A.15) to calculate the corresponding values of x and y (see Fig. 1.32).

6. Temporal Evolution for Rutherford Scattering. For the example in Sect. 1.29 calculate and plot a few positions of the projectile and target as a function of time, in the laboratory system.

Solution. In the laboratory system the orbits are given, as functions of φ, by (1.90–92). With $\varphi_1 = \pi - \varphi$, $\varphi_2 = 2\pi - \varphi$

$$\boldsymbol{r}_1 = \boldsymbol{r}_S + \frac{1}{2}\boldsymbol{r} = \sqrt{\frac{E_r}{m}}\,t(1,0) + \frac{p}{2}\frac{1}{\varepsilon\cos(\varphi - \varphi_0) - 1}(-\cos\varphi,\ \sin\varphi)\,,$$
$$\boldsymbol{r}_2 = \boldsymbol{r}_S - \frac{1}{2}\boldsymbol{r} = \sqrt{\frac{E_r}{m}}\,t(1,0) - \frac{p}{2}\frac{1}{\varepsilon\cos(\varphi - \varphi_0) - 1}(\cos\varphi,\ -\sin\varphi)\,. \tag{A.16}$$

The integral (1.95) that relates the variables t and φ is calculated as in Example 2. Noting that here $\varepsilon > 1$ and making use of the formulae

$$\arctan x = -\frac{i}{2}\ln\frac{1 + ix}{1 - ix}\,, \qquad \frac{mp}{l}\sqrt{\frac{E_r}{m}} = \sqrt{\varepsilon^2 - 1}\,,$$

we find that

$$\sqrt{\frac{E_r}{m}}\,t(\varphi) = \frac{p}{2}\left(\frac{1}{\varepsilon^2 - 1}\ln\frac{1+u}{1-u} + \frac{\varepsilon}{\sqrt{\varepsilon^2 - 1}}\frac{\sin(\varphi - \varphi_0)}{\varepsilon\cos(\varphi - \varphi_0) - 1}\right)\,, \tag{A.17}$$

where u stands for the expression

$$u \equiv \sqrt{\frac{\varepsilon + 1}{\varepsilon - 1}}\tan\frac{\varphi - \varphi_0}{2}\,.$$

Furthermore, we have

$$\cos\varphi_0 = \frac{1}{\varepsilon}\,, \qquad \sin\varphi_0 = \frac{\sqrt{\varepsilon^2 - 1}}{\varepsilon}\,,$$
$$\tan\frac{\varphi - \varphi_0}{2} = \frac{\sin\varphi - \sin\varphi_0}{\cos\varphi + \cos\varphi_0} = \frac{\varepsilon\sin\varphi - \sqrt{\varepsilon^2 - 1}}{1 + \varepsilon\cos\varphi}\,.$$

Equation (A.17) gives the relation between φ and t. Using dimensionless coordinates $(2x/p, 2y/p)$, one plots points for equidistant values of φ and notes the corresponding value of the dimensionless time variable

$$\tau \stackrel{\text{def}}{=} \frac{2}{p}\sqrt{\frac{E_{\text{r}}}{m}} .$$

Figure 1.35 shows the example $\varepsilon = 0.155$, $\varphi_0 = 30°$. Alternatively, one may choose a fixed time interval with respect to $t(\varphi_0) = 0$ and calculate the corresponding values of φ from (A.17).

2. The Principles of Canonical Mechanics

Canonical mechanics is a central part of general mechanics, where one goes beyond the somewhat narrow framework of Newtonian mechanics with position coordinates in the three-dimensional space, towards a more general formulation of mechanical systems belonging to a much larger class. This is the first step of abstraction, leaving behind ballistics, satellite orbits, inclined planes, and pendulum-clocks; it leads to a new kind of description that turns out to be useful in areas of physics far beyond mechanics. Through d'Alembert's principle we discover the concept of the Lagrangian function and the framework of Lagrangian mechanics that is built onto it. Lagrangian functions are particularly useful for studying the role symmetries and invariances of a given system play in its description. By means of the Legendre transformation we are then led to the Hamiltonian function, which is central to the formulation of canonical mechanics, as developed by Hamilton and Jacobi.

Although these two frameworks of description at first seem artificial and unnecessarily abstract, their use pays in very many respects: the formulation of mechanics over the phase space yields a much deeper insight into its dynamical and geometrical structure. At the same time, this prepares the foundation and formal framework for other physical theories, without which, for example, quantum mechanics cannot be understood and perhaps could not even be formulated.

2.1 Constraints and Generalized Coordinates

2.1.1 Definition of Constraints

Whenever the mass points of a mechanical system cannot move completely independently because they are subject to certain geometrical conditions, we talk about *constraints*. These must be discussed independently because they reduce the number of degrees of freedom and therefore change the equations of motion.

(i) The constraints are said to be *holonomic* (from the Greek: constraints are given by an "entire law") if they can be described by a set of independent equations of the form

$$f_\lambda(\boldsymbol{r}_1, \boldsymbol{r}_2, \ldots, \boldsymbol{r}_n, t) = 0\,; \quad \lambda = 1, 2, \ldots, \Lambda\,. \tag{2.1}$$

F. Scheck, *Mechanics*, Graduate Texts in Physics, 5th ed.,
DOI 10.1007/978-3-642-05370-2_2, © Springer-Verlag Berlin Heidelberg 2010

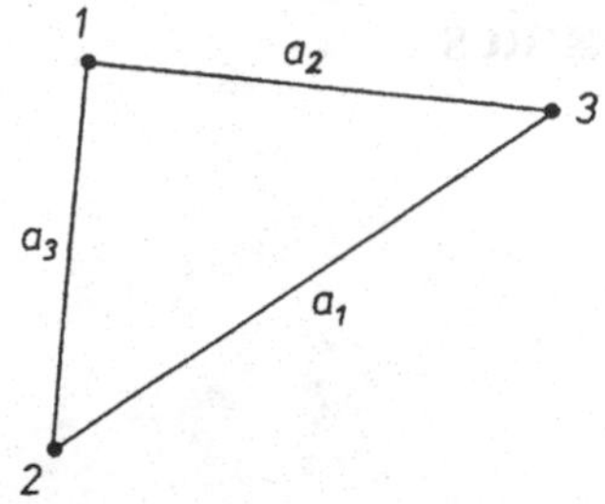

Fig. 2.1. A system of three mass points at constant distances from one another has six degrees of freedom (instead of nine)

Independent means that at any point $(\boldsymbol{r}_1, \ldots, \boldsymbol{r}_n)$ and for all t, the rank of the matrix $\{\partial f_\lambda / \partial \boldsymbol{r}_k\}$ is maximal, i.e. equals Λ. As an example take the three-body system with the condition that all interparticle distances be constant (see Fig. 2.1):

$$f_1 \equiv |\boldsymbol{r}_1 - \boldsymbol{r}_2| - a_3 = 0\,,$$
$$f_2 \equiv |\boldsymbol{r}_2 - \boldsymbol{r}_3| - a_1 = 0\,,$$
$$f_3 \equiv |\boldsymbol{r}_3 - \boldsymbol{r}_1| - a_2 = 0\,.$$

Here $\Lambda = 3$. Without these constraints the number of degrees of freedom would be $f = 3n = 9$. The constraints reduce it to $f = 3n - \Lambda = 6$.

(ii) The constraints are said to be *nonholonomic* if they take the form

$$\sum_{k=1}^{n} \boldsymbol{\omega}_k^i(\boldsymbol{r}_1, \ldots, \boldsymbol{r}_n) \cdot \mathrm{d}\boldsymbol{r}_k = 0\,, \quad i = 1, \ldots, \Lambda \tag{2.2}$$

but cannot be integrated to the form of (2.1). Note that (2.1), by differentiation, gives a condition of type (2.2), viz.

$$\sum_{k=1}^{n} \nabla_k f_\lambda(\boldsymbol{r}_1, \ldots, \boldsymbol{r}_n) \cdot \mathrm{d}\boldsymbol{r}_k = 0\,.$$

This, however, is a complete differential. In contrast, a nonholonomic constraint (2.2) is not integrable and cannot be made so by multiplication with a function, a so-called integrating factor. This class of conditions is the subject of the analysis of *Pfaffian systems*. As we study only holonomic constraints in this book, we do not go into this any further and refer to the mathematics literature for the theory of Pfaffian forms.

(iii) In either case one distinguishes constraints that are (a) *dependent on time* – these are called *rheonomic* ("running law") *constraints*; and (b) *independent of time* – these are called *scleronomic* ("rigid law") *constraints.*

(iv) There are other kinds of constraints, which are expressed in the form of inequalities. Such constraints arise, for instance, when an n-particle system (a gas, for example) is enclosed in a vessel: the particles move freely inside the vessel but cannot penetrate its boundaries. We do not consider such constraints in this book.

2.1.2 Generalized Coordinates

Any set of independent coordinates that take into account the constraints are called generalized coordinates. For example, take a particle moving on the surface of a sphere of radius R around the origin, as sketched in Fig. 2.2. Here the constraint is holonomic and reads $x^2 + y^2 + z^2 = R^2$, so that $f = 3n - 1 = 3 - 1 = 2$. Instead of the *dependent* coordinates $\{x, y, z\}$ one introduces the *independent* coordinates $q_1 = \theta$, $q_2 = \varphi$.

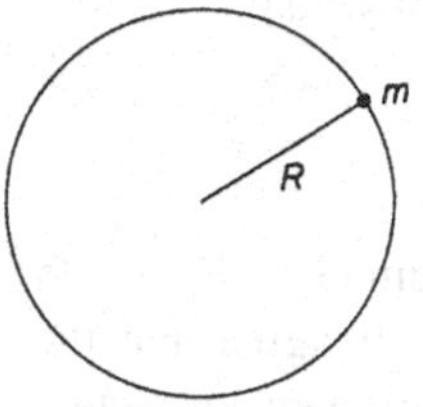

Fig. 2.2. A particle whose motion is restricted to the surface of a sphere has only two degrees of freedom

In general, the set of $3n$ space coordinates of the n-particle system will be replaced by a set of $(3n - \Lambda)$ generalized coordinates, viz.

$$\{\boldsymbol{r}_1, \boldsymbol{r}_2, \ldots \boldsymbol{r}_n\} \to \{q_1, q_2, \ldots, q_f\}\,, \qquad f = 1, 2, \ldots, 3n - \Lambda\,, \tag{2.3}$$

which, in fact, need not have the dimension of length. The aim is now twofold:

(i) determine the number of degrees of freedom f and find f generalized coordinates that take account of the constraints automatically and that are adapted, in an optimal way, to the system one is studying;
(ii) develop simple principles from which the equations of motion are obtained directly, in terms of the generalized coordinates.

We begin by formulating d'Alembert's principle, which is an important auxiliary construction on the way to the goal formulated above.

2.2 D'Alembert's Principle

Consider a system of n mass points with masses $\{m_i\}$ and coordinates $\{\boldsymbol{r}_i\}$, $i = 1, 2, \ldots, n$, subject to the holonomic constraints

$$f_\lambda(\boldsymbol{r}_1, \ldots, \boldsymbol{r}_n, t) = 0\,, \quad \lambda = 1, \ldots, \Lambda\,. \tag{2.4}$$

2.2.1 Definition of Virtual Displacements

A *virtual displacement* $\{\delta\boldsymbol{r}_i\}$ of the system is an arbitrary, infinitesimal change of the coordinates that is compatible with the constraints and the applied forces[1].

[1] Here we make use of this somewhat archaic but very intuitive notion. Geometrically speaking, virtual displacements are described by tangent vectors of the smooth hypersurface in $\mathbb{R}^{3n}$ that is defined by (2.4). D'Alembert's principle can and should be formulated in the geometric framework of Chap. 5.

It is performed at a *fixed* time and therefore has nothing to do with the actual, infinitesimal *motion* $\{\mathrm{d}\boldsymbol{r}_i\}$ of the system during the time change $\mathrm{d}t$ (i.e. the *real* displacement).

Loosely speaking, one may visualize the mechanical system as a half-timbered building that must fit in between the neighboring houses, on a given piece of land (these are the constraints), and that should be stable. In order to test its stability, one shakes the construction a little, without violating the constraints. One imagines the elements of the building to be shifted infinitesimally in all possible, allowed directions, and one observes how the construction responds as a whole.

2.2.2 The Static Case

To begin with, let us assume that the system is in equilibrium, i.e. $\boldsymbol{F}_i = 0$, $i = 1, \ldots, n$, where $\boldsymbol{F}_i$ is the total force applied to particle i. Imagine that the constraint is taken care of by applying an additional force $\boldsymbol{Z}_i$ to every particle i (such forces are called *forces of constraint*). Then

$$\boldsymbol{F}_i = \boldsymbol{K}_i + \boldsymbol{Z}_i \,, \tag{2.5}$$

where $\boldsymbol{Z}_i$ is the force of constraint and $\boldsymbol{K}_i$ the real, dynamic force. Clearly, because all the $\boldsymbol{F}_i$ vanish, the total *virtual work* vanishes:

$$\sum_{i=1}^{n} \boldsymbol{F}_i \cdot \delta\boldsymbol{r}_i = 0 = \sum_{i=1}^{n} [\boldsymbol{K}_i + \boldsymbol{Z}_i] \cdot \delta\boldsymbol{r}_i \,. \tag{2.6}$$

However, since the virtual displacements must be compatible with the constraints, the total work of the forces of constraint alone vanishes, too: $\sum_1^n \boldsymbol{Z}_i \cdot \delta\boldsymbol{r}_i = 0$. Then, from (2.6) we obtain

$$\sum_{i=1}^{n} \boldsymbol{K}_i \cdot \delta\boldsymbol{r}_i = 0 \,. \tag{2.7}$$

In contrast to (2.6), this equation does *not* imply, in general, that the individual terms vanish. This is because the $\delta\boldsymbol{r}_i$ are generally not independent.

2.2.3 The Dynamical Case

If the system is moving, then we have $\boldsymbol{F}_i - \dot{\boldsymbol{p}}_i = 0$ and of course also $\sum_{i=1}^n (\boldsymbol{F}_i - \dot{\boldsymbol{p}}_i) \cdot \delta\boldsymbol{r}_i = 0$. As the total work of the forces of constraint vanishes, $\sum_{i=1}^n (\boldsymbol{Z}_i \cdot \delta\boldsymbol{r}_i) = 0$, we obtain the basic equation expressing *d'Alembert's principle of virtual displacements*:

$$\sum_{i=1}^{n} (\boldsymbol{K}_i - \dot{\boldsymbol{p}}_i) \cdot \delta\boldsymbol{r}_i = 0 \,, \tag{2.8}$$

from which all constraints have disappeared. As in the case of (2.7), the individual terms, in general, do not vanish, because the $\delta \boldsymbol{r}_i$ depend on each other.

Equation (2.8) is the starting point for obtaining the equations of motion for the generalized coordinates. We proceed as follows.

As the conditions (2.4) are independent they can be solved locally for the coordinates $\boldsymbol{r}_i$, i.e.

$$\boldsymbol{r}_i = \boldsymbol{r}_i(q_1, \dots, q_f, t) \,, \quad i = 1, \dots, n \,, \quad f = 3n - \Lambda \,.$$

From these we can deduce the auxiliary formulae

$$\boldsymbol{v}_i \equiv \dot{\boldsymbol{r}}_i = \sum_{k=1}^{f} \frac{\partial \boldsymbol{r}_i}{\partial q_k} \dot{q}_k + \frac{\partial \boldsymbol{r}_i}{\partial t} \,, \tag{2.9}$$

$$\frac{\partial \boldsymbol{v}_i}{\partial \dot{q}_k} = \frac{\partial \boldsymbol{r}_i}{\partial q_k} \,, \tag{2.10}$$

$$\delta \boldsymbol{r}_i = \sum_{k=1}^{f} \frac{\partial \boldsymbol{r}_i}{\partial q_k} \delta q_k \,. \tag{2.11}$$

Note that there is no time derivative in (2.11), because the $\delta \boldsymbol{r}_i$ are virtual displacements, i.e. are made at a fixed time. The first term on the left-hand side of (2.8) can be written as

$$\sum_{i=1}^{n} \boldsymbol{K}_i \cdot \delta \boldsymbol{r}_i = \sum_{k=1}^{f} Q_k \delta q_k \quad \text{with} \quad Q_k \stackrel{\text{def}}{=} \sum_{i=1}^{n} \boldsymbol{K}_i \cdot \frac{\partial \boldsymbol{r}_i}{\partial q_k} \,. \tag{2.12}$$

The quantities Q_k are called *generalized forces* (again, they need not have the dimension of force). The second term also takes the form $\sum_{k=1}^{f} \{\dots\} \delta q_k$, as follows:

$$\sum_{i=1}^{n} \dot{\boldsymbol{p}}_i \cdot \delta \boldsymbol{r}_i = \sum_{i=1}^{n} m_i \ddot{\boldsymbol{r}}_i \cdot \delta \boldsymbol{r}_i = \sum_{i=1}^{n} m_i \sum_{k=1}^{f} \ddot{\boldsymbol{r}}_i \cdot \frac{\partial \boldsymbol{r}_i}{\partial q_k} \delta q_k \,.$$

The scalar products $(\ddot{\boldsymbol{r}}_i \cdot \partial \boldsymbol{r}_i / \partial q_k)$ can be written as

$$\ddot{\boldsymbol{r}}_i \cdot \frac{\partial \boldsymbol{r}_i}{\partial q_k} = \frac{\mathrm{d}}{\mathrm{d}t} \left(\dot{\boldsymbol{r}}_i \cdot \frac{\partial \boldsymbol{r}_i}{\partial q_k} \right) - \dot{\boldsymbol{r}}_i \cdot \frac{\mathrm{d}}{\mathrm{d}t} \frac{\partial \boldsymbol{r}_i}{\partial q_k} \,.$$

Note further that

$$\frac{\mathrm{d}}{\mathrm{d}t} \frac{\partial \boldsymbol{r}_i}{\partial q_k} = \frac{\partial}{\partial q_k} \dot{\boldsymbol{r}}_i = \frac{\partial \boldsymbol{v}_i}{\partial q_k}$$

and, on taking the partial derivative of (2.9) with respect to $\dot{q}_k$, that $\partial \boldsymbol{r}_i / \partial q_k = \partial \boldsymbol{v}_i / \partial \dot{q}_k$.

From these relations one obtains

$$\sum_{i=1}^{n} m_i \ddot{\boldsymbol{r}}_i \cdot \frac{\partial \boldsymbol{r}_i}{\partial q_k} = \sum_{i=1}^{n} \left\{ m_i \frac{\mathrm{d}}{\mathrm{d}t} \left(\boldsymbol{v}_i \cdot \frac{\partial \boldsymbol{v}_i}{\partial \dot{q}_k} \right) - m_i \boldsymbol{v}_i \cdot \frac{\partial \boldsymbol{v}_i}{\partial q_k} \right\} .$$

The two terms in the expression in curly brackets contain the form $\boldsymbol{v} \cdot \partial \boldsymbol{v}/\partial x = (\partial \boldsymbol{v}^2/\partial x)/2$, with $x = q_k$ or $\dot{q}_k$, so that we finally obtain

$$\sum_{i=1}^{n} \dot{\boldsymbol{p}}_i \cdot \delta \boldsymbol{r}_i = \sum_{k=1}^{f} \left\{ \frac{\mathrm{d}}{\mathrm{d}t} \left[\frac{\partial}{\partial \dot{q}_k} \left(\sum_{i=1}^{n} \frac{m_i}{2} \boldsymbol{v}_i^2 \right) \right] - \frac{\partial}{\partial q_k} \left(\sum_{i=1}^{n} \frac{m_i}{2} \boldsymbol{v}_i^2 \right) \right\} \delta q_k . \tag{2.13}$$

Inserting the results (2.12) and (2.13) into (2.8) yields an equation that contains only the quantities δq_k, but not $\delta \boldsymbol{r}_i$. The displacements δq_k are independent (in contrast to the $\delta \boldsymbol{r}_i$, which are not). Therefore, in the equation that we obtain from (2.8) by replacing all $\delta \boldsymbol{r}_i$ by the δq_k, as described above, every term must vanish individually. Thus we obtain the set of equations

$$\frac{\mathrm{d}}{\mathrm{d}t} \left(\frac{\partial T}{\partial \dot{q}_k} \right) - \frac{\partial T}{\partial q_k} = Q_k , \quad k = 1, \ldots, f , \tag{2.14}$$

where $T = \sum_{i=1}^{n} m_i \boldsymbol{v}_i^2/2$ is the kinetic energy. Of course, very much like Q_k, T must be expressed in terms of the variables q_i and $\dot{q}_i$ so that (2.14) really does become a system of differential equations for the $q_k(t)$.

2.3 Lagrange's Equations

Suppose that, in addition, the real forces $\boldsymbol{K}_i$ are potential forces, i.e.

$$\boldsymbol{K}_i = -\nabla_i U . \tag{2.15}$$

In this situation the *generalized* forces Q_k are potential forces, too. Indeed, from (2.12)

$$Q_k = -\sum_{i=1}^{n} \nabla_i U(\boldsymbol{r}_1, \ldots, \boldsymbol{r}_n) \cdot \frac{\partial \boldsymbol{r}_i}{\partial q_k} = -\frac{\partial}{\partial q_k} U(q_1, \ldots, q_f, t) , \tag{2.16}$$

under the assumption that U is transformed to the variables q_k. As U does not depend on the $\dot{q}_k$, T and U can be combined to

$$L(q_k, \dot{q}_k, t) = T(q_k, \dot{q}_k) - U(q_k, t) \tag{2.17}$$

so that (2.14) takes the simple form

$$\boxed{\frac{\mathrm{d}}{\mathrm{d}t} \left(\frac{\partial L}{\partial \dot{q}_k} \right) - \frac{\partial L}{\partial q_k} = 0 .} \tag{2.18}$$

The function $L(q_k, \dot{q}_k, t)$ is called the *Lagrangian function*. Equations (2.18) are called *Lagrange's equations*. They contain the function L (2.17) with

$$\begin{aligned} U(q_1,\dots,q_f,t) &= U(\boldsymbol{r}_1(q_1,\dots,q_f,t),\dots \boldsymbol{r}_n(q_1,\dots,q_f,t))\,, \\ T(q_k,\dot{q}_k) &= \frac{1}{2}\sum_{i=1}^{n} m_i \left(\sum_{k=1}^{f} \frac{\partial \boldsymbol{r}_i}{\partial q_k}\dot{q}_k + \frac{\partial \boldsymbol{r}_i}{\partial t}\right)^2 \\ &\equiv a + \sum_{k=1}^{f} b_k \dot{q}_k + \sum_{k=1}^{f}\sum_{l=1}^{f} c_{kl}\dot{q}_k\dot{q}_l\,, \end{aligned} \tag{2.19}$$

where

$$\begin{aligned} a &= \frac{1}{2}\sum_{i=1}^{n} m_i \left(\frac{\partial \boldsymbol{r}_i}{\partial t}\right)^2 , \\ b_k &= \sum_{i=1}^{n} m_i \frac{\partial \boldsymbol{r}_i}{\partial q_k}\cdot\frac{\partial \boldsymbol{r}_i}{\partial t}\,, \\ c_{kl} &= \frac{1}{2}\sum_{i=1}^{n} m_i \frac{\partial \boldsymbol{r}_i}{\partial q_k}\cdot\frac{\partial \boldsymbol{r}_i}{\partial q_l}\,. \end{aligned} \tag{2.20}$$

The special form $L = T - U$ of the Lagrangian function is called its *natural form*. (For the reasons explained in Sect. 2.11 below, L is not unique.) For scleronomic constraints both a and all b_k vanish. In this case T is a homogeneous function of degree 2 in the variables q_k.

We note that d'Alembert's equations (2.14) are somewhat more general than Lagrange's equations (2.18): the latter follow only if the forces are potential forces. In contrast, the former also hold if the constraints are formulated in a differential form (2.11) that cannot be integrated to holonomic equations.

2.4 Examples of the Use of Lagrange's Equations

We study three elementary examples.

Example (i) A particle of mass m moves on a segment of a sphere in the earth's gravitational field. The dynamical force is $\boldsymbol{K} = (0, 0, -mg)$, the constraint is $|\boldsymbol{r}| = R$, and the generalized coordinates may be chosen to be $q_1 = \theta$ and $q_2 = \varphi$, as shown in Fig. 2.3. The generalized forces are

$$\begin{aligned} Q_1 &= \boldsymbol{K}\cdot\frac{\partial \boldsymbol{r}}{\partial q_1} = -RK_z\sin\theta = Rmg\sin\theta\,, \\ Q_2 &= 0\,. \end{aligned}$$

These are potential forces, $Q_1 = -\partial U/\partial q_1$, $Q_2 = -\partial U/\partial q_2$, with $U(q_1, q_2) = mgR[1 + \cos q_1]$. Furthermore, $T = mR^2[\dot{q}_1^2 + \dot{q}_2^2 \sin^2 q_1]/2$, and therefore

$$L = \tfrac{1}{2}mR^2[\dot{q}_1^2 + \dot{q}_2^2\sin^2 q_1] - mgR[1 + \cos q_1]\,.$$

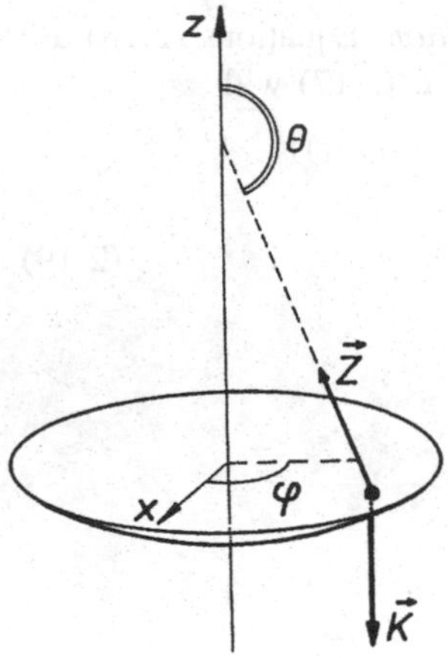

Fig. 2.3. A small ball on a segment of a sphere in the earth's gravitational field. The force of constraint $\boldsymbol{Z}$ is such that it keeps the particle on the given surface. As such it is equivalent to the constraint

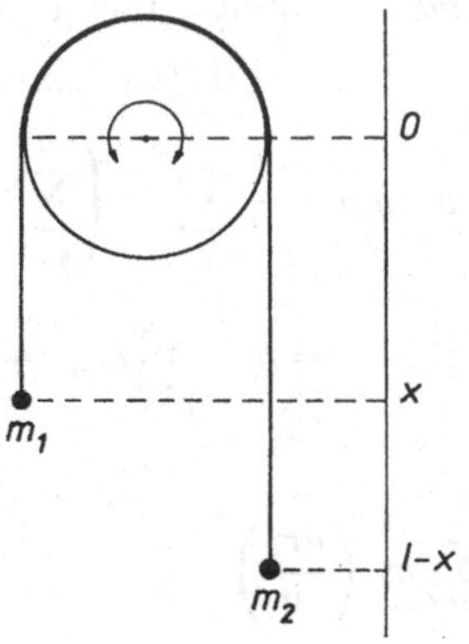

Fig. 2.4. Two pointlike weights m_1 and m_2 are connected by a massless thread, which rests on a wheel. The motion is assumed to be frictionless

We now calculate the derivatives $\partial L/\partial q_i$ and $\mathrm{d}(\partial L/\partial \dot{q}_i)/\mathrm{d}t$:

$$\frac{\partial L}{\partial q_1} = m\, R^2 \dot{q}_2^2 \sin q_1 \cos q_1 + m\, g\, R \sin q_1 \;, \qquad \frac{\partial L}{\partial q_2} = 0 \;,$$
$$\frac{\partial L}{\partial \dot{q}_1} = m\, R^2 \dot{q}_1 \;, \qquad \frac{\partial L}{\partial \dot{q}_2} = m\, R^2 \dot{q}_2 \sin^2 q_1$$

to obtain the equations of motion

$$\ddot{q}_1 - \left[\dot{q}_2^2 \cos q_1 + \frac{g}{R}\right] \sin q_1 = 0 \;, \qquad m\, R^2 \frac{\mathrm{d}}{\mathrm{d}t}(\dot{q}_2 \sin^2 q_1) = 0 \;.$$

Example (ii) Atwood's machine is sketched in Fig. 2.4. The wheel and the thread are assumed to be massless; the wheel rotates without friction. We then have

$$T = \tfrac{1}{2}(m_1 + m_2)\dot{x}^2 \;,$$
$$U = -m_1 g x - m_2 g(l - x) \;,$$
$$L = T - U \;.$$

The derivatives of L are $\partial L/\partial x = (m_1 - m_2)g$, $\partial L/\partial \dot{x} = (m_1 + m_2)\dot{x}$, so that the equation of motion $\mathrm{d}(\partial L/\partial \dot{x})/\mathrm{d}t = \partial L/\partial x$ becomes

$$\ddot{x} = \frac{m_1 - m_2}{m_1 + m_2}\, g \;.$$

It can be integrated at once. If the mass of the wheel cannot be neglected, its rotation will contribute to the kinetic energy T by the amount $T = I(\mathrm{d}\theta/\mathrm{d}t)^2/2$, where I is the relevant moment of inertia and $\mathrm{d}\theta/\mathrm{d}t$ its angular velocity. Let R be the radius of the wheel. The angular velocity is proportional to $\dot{x}$, viz. $R(\mathrm{d}\theta/\mathrm{d}t) = \dot{x}$. Therefore, the kinetic energy is changed to $T = (m_1 + m_2 + I/R^2)\dot{x}^2/2$. (The rotary motion of a rigid body such as this wheel is dealt with in Chap. 3.)

Example (iii) Consider a particle of mass m held by a massless thread and rotating about the point S, as shown in Fig. 2.5. The thread is shortened at a constant rate c per unit time. Let x and y be Cartesian coordinates in the plane of the circle, φ the polar angle in that plane. The generalized coordinate is $q = \varphi$, and we have $x = (R_0 - ct)\cos q$, $y = (R_0 - ct)\sin q$; thus $T = m(\dot{x}^2 + \dot{y}^2)/2 = m[\dot{q}^2(R_0 - ct)^2 + c^2]/2$. In this example T is not a homogeneous function of degree 2 in $\dot{q}$ (the constraint is rheonomic!). The equation of motion now reads $m\dot{q}(R_0 - ct)^2 = \text{const.}$

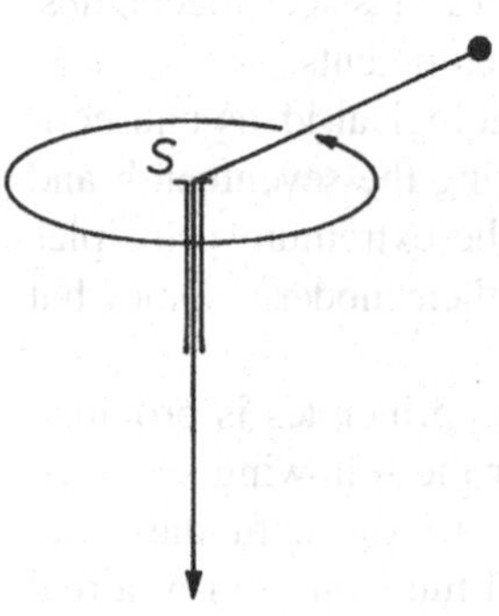

Fig. 2.5. The mass point m rotates about the point S. At the same time, the thread holding the mass point is shortened continuously

2.5 A Digression on Variational Principles

Both conditions (2.7) and (2.8) of d'Alembert's principle, for the static case and dynamic case, respectively, are expressions for an equilibrium: if one "shakes" the mechanical system one is considering, in a way that is compatible with the constraints, the total (virtual) work is equal to zero. In this sense the state of the system is an extremum; the physical state, i.e. the one that is actually realized, has the distinct property, in comparison with all other possible states one might imagine, that it is stable against small changes of the positions (in the static case) or against small changes of the orbits (in the dynamic case). Such an observation is familiar from geometric optics. Indeed, *Fermat's principle* states that in an arbitrary system of mirrors and refracting glasses a light ray always chooses a path that assumes an extreme value. The light's path is either the shortest or the longest between its source and the point where it is detected.

D'Alembert's principle and the experience with Fermat's principle in optics raise the question whether it is possible to define a functional, for a given mechanical system, that bears some analogy to the length of path of a light ray. The actual physical orbit that the system chooses (for given initial condition) would make this function an extremum. Physical orbits would be some kind of geodesic on a manifold determined by the forces; that is, they would usually be the shortest (or the longest) curves connecting initial and final configurations.

There is indeed such a functional for a large class of mechanical systems: the time integral over a Lagrangian function such as (2.17). This is what we wish to develop, step by step, in the following sections.

In fact, in doing so one discovers a gold mine: this extremum, or variational, principle can be generalized to field theories, i.e. systems with an infinite number of degrees of freedom, as well as to quantized and relativistic systems. Today it looks as though *any* theory of fundamental interactions can be derived from a variational principle. Consequently it is rewarding to study this new, initially somewhat abstract, principle and to develop some feeling for its use. This effort pays in that it allows for a deeper understanding of the rich structure of classical mechanics, which in turn serves as a model for many theoretical developments.

One should keep in mind that philosophical and cosmological ideas and concepts were essential to the development of mechanics during the seventeenth and eighteenth centuries. It is not surprising, therefore, that the extremum principles reflect philosophical ideas in a way that can still be felt in their modern, somewhat axiomatic, formulation.

The mathematical basis for the discussion of extremum principles is provided by *variational calculus*. In order to prepare the ground for the following sections, but without going into too much detail, we discuss here a typical, fundamental problem of variational calculus. It consists in finding a real function $y(x)$ of a real variable x such that a given functional $I[y]$ of this function assumes an extreme value. Let

$$I[y] \stackrel{\text{def}}{=} \int_{x_1}^{x_2} \mathrm{d}x\, f(y(x), y'(x), x)\ , \quad y'(x) \equiv \frac{\mathrm{d}}{\mathrm{d}x}\, y(x) \tag{2.21}$$

be a functional of y, with f a given function of y, y' (the derivative of y with respect to x), and the variable x. x_1 and x_2 are arbitrary, but fixed, endpoints. The problem is to determine those functions $y(x)$ which take given values $y_1 = y(x_1)$ and $y_2 = y(x_2)$ at the endpoints and which make the functional $I[y]$ an *extremum*. In other words, one supposes that all possible functions $y(x)$ that assume the given boundary values are inserted into the integral (2.21) and that its numerical value is calculated. What we are looking for are those functions for which this value assumes an extremal value, i.e. is a maximum or a minimum, or, possibly, a saddle point.

As a first step we investigate the quantity

$$I(\alpha) \stackrel{\text{def}}{=} \int_{x_1}^{x_2} f(y(x,\alpha), y'(x,\alpha), x)\mathrm{d}x\ , \tag{2.22}$$

where $y(x, \alpha) = y(x) + \alpha\eta(x)$ with $\eta(x_1) = 0 = \eta(x_2)$. This means that we embed $y(x)$ in a set of comparative curves that fulfill the same boundary conditions

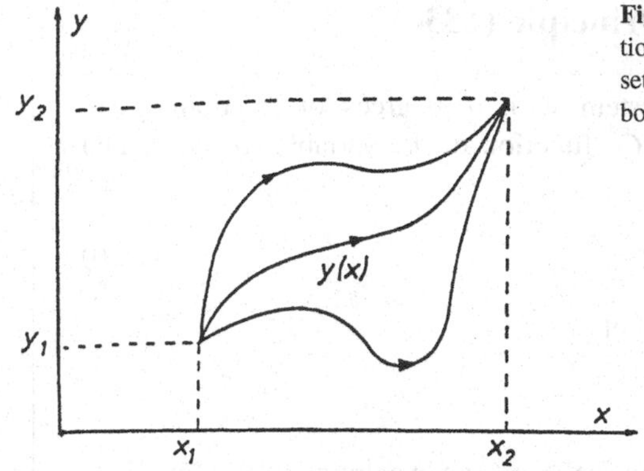

Fig. 2.6. The curve $y(x)$ that makes the functional $I[y]$ an extremum is embedded in a set of comparative curves assuming the same boundary values as $y(x)$

as $y(x)$. Figure 2.6 shows an example. The next step is to calculate the so-called variation of I, that is, the quantity

$$\delta I \stackrel{\text{def}}{=} \frac{\mathrm{d}I}{\mathrm{d}\alpha}\,\mathrm{d}\alpha = \int_{x_1}^{x_2} \mathrm{d}x \left\{ \frac{\partial f}{\partial y}\frac{\mathrm{d}y}{\mathrm{d}\alpha} + \frac{\partial f}{\partial y'}\frac{\mathrm{d}y'}{\mathrm{d}\alpha} \right\} \mathrm{d}\alpha\,.$$

Clearly, $\mathrm{d}y'/\mathrm{d}\alpha = (\mathrm{d}/\mathrm{d}x)(\mathrm{d}y/\mathrm{d}\alpha)$. If the second term is integrated by parts,

$$\int_{x_1}^{x_2} \mathrm{d}x \frac{\partial f}{\partial y'}\frac{\mathrm{d}}{\mathrm{d}x}\left(\frac{\mathrm{d}y}{\mathrm{d}\alpha}\right) = -\int_{x_1}^{x_2} \mathrm{d}x \frac{\mathrm{d}y}{\mathrm{d}\alpha}\frac{\mathrm{d}}{\mathrm{d}x}\left(\frac{\partial f}{\partial y'}\right) + \left.\frac{\partial f}{\partial y'}\frac{\mathrm{d}y}{\mathrm{d}\alpha}\right|_{x_1}^{x_2},$$

the boundary terms do not contribute, because $\mathrm{d}y/\mathrm{d}\alpha = \eta(x)$ vanishes at x_1 and at x_2. Thus

$$\delta I = \int_{x_1}^{x_2} \mathrm{d}x \left\{ \frac{\partial f}{\partial y} - \frac{\mathrm{d}}{\mathrm{d}x}\frac{\partial f}{\partial y'} \right\} \frac{\mathrm{d}y}{\mathrm{d}\alpha}\,\mathrm{d}\alpha\,. \tag{2.23}$$

The expression

$$\frac{\partial f}{\partial y} - \frac{\mathrm{d}}{\mathrm{d}x}\frac{\partial f}{\partial y'} \stackrel{\text{def}}{=} \frac{\delta f}{\delta y} \tag{2.24}$$

is called the *variational derivative* of f by y. It is useful to introduce the notation $(\mathrm{d}y/\mathrm{d}\alpha)\mathrm{d}\alpha \stackrel{\text{def}}{=} \delta y$ and to interpret δy as an infinitesimal variation of the curve $y(x)$. $I(\alpha)$ assumes an extreme value, i.e. $\delta I = 0$. As this must hold true for arbitrary variations δy, the integrand in (2.23) must vanish:

$$\frac{\partial f}{\partial y} - \frac{\mathrm{d}}{\mathrm{d}x}\left(\frac{\partial f}{\partial y'}\right) = 0\,. \tag{2.25}$$

This is *Euler's differential equation of variational calculus*. With a substitution of $L(q, \dot{q}, t)$ for $f(y, y', x)$, a comparison with (2.18) shows that it is identical with Lagrange's equation $\mathrm{d}(\partial L/\partial \dot{q})/\mathrm{d}t - \partial L/\partial q = 0$ (here in one dimension). This surprising result is the starting point for the variational principle proposed by Hamilton.

2.6 Hamilton's Variational Principle (1834)

Postulate. To a mechanical system with f degrees of freedom $\underset{\sim}{q} = \{q_1, q_2, \ldots, q_f\}$ we associate a C^2 function of the variables $\underset{\sim}{q}$ and $\dot{\underset{\sim}{q}}$ and of the time t,

$$L(\underset{\sim}{q}, \dot{\underset{\sim}{q}}, t) , \tag{2.26}$$

called the Lagrangian function. Let

$$\underset{\sim}{\varphi}(t) = (\varphi_1(t), \ldots, \varphi_f(t))$$

in the interval $t_1 \le t \le t_2$ be a physical orbit (i.e. a solution of the equations of motion) that assumes the boundary values $\underset{\sim}{\varphi}(t_1) = \underset{\sim}{a}$ and $\underset{\sim}{\varphi}(t_2) = \underset{\sim}{b}$. This orbit is such that the *action integral*

$$I[\underset{\sim}{q}] \stackrel{\text{def}}{=} \int_{t_1}^{t_2} \mathrm{d}t \; L(\underset{\sim}{q}(t), \dot{\underset{\sim}{q}}(t), t) \tag{2.27}$$

assumes an extreme value[2].

The physical orbit, i.e. the solution of the equations of motion for the specified boundary conditions, is singled out from all other possible orbits that the system might choose and that have the same boundary values by the requirement that the action integral be an extremum. For suitable choices of the boundary values $(t_1, \underset{\sim}{a})$ and $(t_2, \underset{\sim}{b})$ this will be a minimum. However, the example worked out in Exercise 2.18 shows that it can also be a maximum. Saddle point values are possible, too. We return to this question in Sect. 2.36 (ii) below.

2.7 The Euler–Lagrange Equations

A necessary condition for the action integral $I[\underset{\sim}{q}]$ to assume an extreme value, for $\underset{\sim}{q} = \underset{\sim}{\varphi}(t)$, is that $\underset{\sim}{\varphi}(t)$ be the integral curve of the Euler–Lagrange equations

$$\boxed{\frac{\delta L}{\delta q_k} = \frac{\partial L}{\partial q_k} - \frac{\mathrm{d}}{\mathrm{d}t}\left(\frac{\partial L}{\partial \dot{q}_k}\right) = 0 \,, \quad k = 1, \ldots, f \,.} \tag{2.28}$$

The proof of this statement proceeds in analogy to that in Sect. 2.5. Indeed, set $\underset{\sim}{q}(t, \alpha) = \underset{\sim}{\varphi}(t) + \alpha \, \underset{\sim}{\psi}(t)$ with $-1 \le \alpha \le +1$ and $\underset{\sim}{\psi}(t_1) = 0 = \underset{\sim}{\psi}(t_2)$. If I is an extremum for $\underset{\sim}{q} = \underset{\sim}{\varphi}(t)$, then

$$\left.\frac{\mathrm{d}}{\mathrm{d}\alpha} I(\alpha)\right|_{\alpha=0} = 0 \quad \text{with} \quad I(\alpha) \stackrel{\text{def}}{=} \int_{t_1}^{t_2} \mathrm{d}t \; L(\underset{\sim}{q}(t, \alpha), \dot{\underset{\sim}{q}}(t, \alpha), t)$$

2 The name action arises because L has the dimension of energy: the product (energy $\times$ time) is called action, and this is indeed the dimension of the action integral.

and with

$$\frac{\mathrm{d}}{\mathrm{d}\alpha} I(\alpha) = \int_{t_1}^{t_2} \mathrm{d}t \sum_{k=1}^{f} \left\{ \frac{\partial L}{\partial q_k} \frac{\mathrm{d}q_k}{\mathrm{d}\alpha} + \frac{\partial L}{\partial \dot{q}_k} \frac{\mathrm{d}\dot{q}_k}{\mathrm{d}\alpha} \right\} .$$

With regard to the second term of this expression we note that partial integration in the variable t yields the difference of boundary values $K(t_2, \alpha) - K(t_1, \alpha)$ where $K(t, \alpha)$ is the function

$$K(t, \alpha) \stackrel{\text{def}}{=} \sum_{1}^{f} \frac{\partial L}{\partial \dot{q}_k} \frac{\mathrm{d}}{\mathrm{d}\alpha} q_k(t, \alpha) .$$

Now, by assumption, the end points and the values $\varphi(t_2)$ and $\varphi(t_1)$ are held fixed so that

$$\frac{\mathrm{d}}{\mathrm{d}\alpha} q_k(t, \alpha) = \psi_k(t)$$

is zero for $t = t_1$ and for $t = t_2$. Therefore, the boundary values vanish, $K(t_1, \alpha) = 0 = K(t_2, \alpha)$, integration by parts of the second term gives

$$\int_{t_1}^{t_2} \mathrm{d}t \, \frac{\partial L}{\partial \dot{q}_k} \frac{\mathrm{d}\dot{q}_k}{\mathrm{d}\alpha} = - \int_{t_1}^{t_2} \mathrm{d}t \frac{\mathrm{d}}{\mathrm{d}t} \left(\frac{\partial L}{\partial \dot{q}_k} \right) \frac{\mathrm{d}q_k}{\mathrm{d}\alpha} .$$

Inserting this result one obtains

$$\left. \frac{\mathrm{d}I(\alpha)}{\mathrm{d}\alpha} \right|_{\alpha=0} = \int_{t_1}^{t_2} \mathrm{d}t \sum_{k=1}^{f} \left[\frac{\partial L}{\partial q_k} - \frac{\mathrm{d}}{\mathrm{d}t} \left(\frac{\partial L}{\partial \dot{q}_k} \right) \right] \psi_k(t) = 0 .$$

The functions $\psi_k(t)$ are arbitrary and independent. Therefore, the integrand must vanish termwise. Thus, (2.28) is proved. □

Lagrange's equations follow from the variational principle of Hamilton. They are the same as the equations we obtained from d'Alembert's principle in the case where the forces were potential forces. As a result, we obtain f ordinary differential equations of second order in the time variable, f being the number of degrees of freedom of the system under consideration.

2.8 Further Examples of the Use of Lagrange's Equations

The equations of motion (2.28) generalize Newton's second law. We confirm this statement, in a first step, by verifying that in the case of the n-particle system without any constraints these equations take the Newtonian form. The second example goes beyond the framework of "natural" Lagrangian functions and, in fact, puts us on a new and interesting track that we follow up in the subsequent sections.

Example (i) *An n-particle system with potential forces.* As there are no forces of constraint we take as coordinates the position vectors of the particles. For L we choose the natural form

$$L = T - U = \frac{1}{2}\sum_{i=1}^{n} m_i \dot{\boldsymbol{r}}_i^2 - U(\boldsymbol{r}_1, \ldots, \boldsymbol{r}_n, t) ,$$

$$\underline{q} \equiv \{q_1, \ldots, q_{f=3n}\} = \{\boldsymbol{r}_1, \ldots, \boldsymbol{r}_n\} ,$$

$$\frac{\partial L}{\partial q_k} = -\frac{\partial U}{\partial q_k} ; \quad \frac{\mathrm{d}}{\mathrm{d}t}\frac{\partial L}{\partial \dot{q}_k} = m_{i(k)}\ddot{q}_k .$$

The notation $m_{i(k)}$ is meant to indicate that in counting the q_k one must insert the correct mass of the corresponding particle, i.e. m_1 for q_1, q_2, q_3, then m_2 for q_4, q_5, q_6, and so on. Written differently, we obtain $m_i \ddot{\boldsymbol{r}}_i = -\boldsymbol{\nabla}_i U$. Thus, in this case the Euler–Lagrange equations are nothing but the well-known equations of Newton. Therefore, the mechanics that we studied in Chap. 1 can be derived as a special case from the variational principle of Hamilton.

Example (ii) *A charged particle in electric and magnetic fields.* Here we set $\boldsymbol{q} \equiv \underline{q} = \{q_1, q_2, q_3\} = \{x, y, z\}$. The motion of a charged, pointlike particle under the action of time- and space-dependent electric and magnetic fields is described by the equation

$$m\ddot{\boldsymbol{q}} = e\boldsymbol{E}(\boldsymbol{q}, t) + \frac{e}{c}\dot{\boldsymbol{q}}(t) \times \boldsymbol{B}(\boldsymbol{q}, t) , \tag{2.29}$$

where e is its charge. The expression on the right-hand side is the Lorentz force. The electric and magnetic fields may be expressed in terms of scalar and vector potentials as follows:

$$\boldsymbol{E}(\boldsymbol{q}, t) = -\boldsymbol{\nabla}_{\boldsymbol{q}}\Phi(\boldsymbol{q}, t) - \frac{1}{c}\frac{\partial}{\partial t}\boldsymbol{A}(\boldsymbol{q}, t)$$
$$\boldsymbol{B}(\boldsymbol{q}, t) = \boldsymbol{\nabla}_{\boldsymbol{q}} \times \boldsymbol{A}(\boldsymbol{q}, t) , \tag{2.30}$$

where Φ denotes the scalar potential and $\boldsymbol{A}$ denotes the vector potential. The equation of motion (2.29) is obtained, for example, from the following Lagrangian function (whose form we postulate at this point):

$$L(\boldsymbol{q}, \dot{\boldsymbol{q}}, t) = \frac{1}{2}m\dot{\boldsymbol{q}}^2 - e\Phi(\boldsymbol{q}, t) + \frac{e}{c}\dot{\boldsymbol{q}} \cdot \boldsymbol{A}(\boldsymbol{q}, t) . \tag{2.31}$$

Indeed, using the chain rule, one verifies that

$$\frac{\partial L}{\partial q_i} = -e\frac{\partial \Phi}{\partial q_i} + \frac{e}{c}\sum_{k=1}^{3}\dot{q}_k\frac{\partial A_k}{\partial q_i} ,$$

$$\frac{\mathrm{d}}{\mathrm{d}t}\frac{\partial L}{\partial \dot{q}_i} = m\ddot{q}_i + \frac{e}{c}\frac{\mathrm{d}A_i}{\mathrm{d}t} = m\ddot{q}_i + \frac{e}{c}\left[\sum_{k=1}^{3}\dot{q}_k\frac{\partial A_i}{\partial q_k} + \frac{\partial A_i}{\partial t}\right] ,$$

so that from (2.28) there follows the correct equation of motion,

$$m\ddot{q}_i = e\left[-\frac{\partial \Phi}{\partial q_i} - \frac{1}{c}\frac{\partial A_i}{\partial t}\right] + \frac{e}{c}\sum_{k=1}^{3}\dot{q}_k\left[\frac{\partial A_k}{\partial q_i} - \frac{\partial A_i}{\partial q_k}\right]$$
$$= eE_i + \frac{e}{c}(\dot{\boldsymbol{q}} \times \boldsymbol{B})_i .$$

Note that with respect to rotations of the frame of reference the Lagrangian function (2.31) stays invariant, hence is a *scalar* while the equation of motion (2.29) is a *vectorial* equation. Both sides transform like vector fields under rotations. Obviously, scalar, invariant quantities are simpler than quantities that have a specific, but nontrivial transformation behaviour. We will come back to this remark repeatedly in subsequent sections.

2.9 A Remark About Nonuniqueness of the Lagrangian Function

In Example (ii) of Sect. 2.8 the potentials can be chosen differently without changing the observable field strengths (2.30) nor the equations of motion (2.29). Let χ be a scalar, differentiable function of position and time. Replace then the potentials as follows:

$$\begin{aligned} \boldsymbol{A}(\boldsymbol{q},t) &\to \boldsymbol{A}'(\boldsymbol{q},t) = \boldsymbol{A}(\boldsymbol{q},t) + \nabla\chi(\boldsymbol{q},t)\,, \\ \Phi(\boldsymbol{q},t) &\to \Phi'(\boldsymbol{q},t) = \Phi(\boldsymbol{q},t) - \frac{1}{c}\frac{\partial}{\partial t}\chi(\boldsymbol{q},t)\,. \end{aligned} \tag{2.32}$$

The effect of this transformation on the Lagrangian function is the following:

$$\begin{aligned} L'(\boldsymbol{q},\dot{\boldsymbol{q}},t) &\stackrel{\text{def}}{=} \frac{1}{2}m\dot{\boldsymbol{q}}^2 - e\Phi' + \frac{e}{c}\dot{\boldsymbol{q}}\cdot\boldsymbol{A}' \\ &= L(\boldsymbol{q},\dot{\boldsymbol{q}},t) + \frac{e}{c}\left[\frac{\partial\chi}{\partial t} + \dot{\boldsymbol{q}}\cdot\nabla\chi\right] = L(\boldsymbol{q},\dot{\boldsymbol{q}},t) + \frac{\mathrm{d}}{\mathrm{d}t}\left(\frac{e}{c}\chi(\boldsymbol{q},t)\right). \end{aligned}$$

We see that L is modified by the total time derivative of a function of q and t. The potentials are not observable and are therefore not unique. What the example tells us is that the Lagrangian function is not unique either and therefore that it certainly cannot be an observable. L' leads to the same equations of motion as L. The two differ by the total time derivative of a function $M(\underline{q},t)$,

$$L'(\underline{q},\dot{\underline{q}},t) = L(\underline{q},\dot{\underline{q}},t) + \frac{\mathrm{d}}{\mathrm{d}t}M(\underline{q},t) \tag{2.33}$$

(here with $M = e\chi/c$). The statement that L' describes the same physics as L is quite general. As the transformation from L to L' is induced by the gauge transformation (2.32) of the potentials, we shall call transformations of the kind (2.33) *gauge transformations of the Lagrangian function*. The general case is the subject of the next section.

2.10 Gauge Transformations of the Lagrangian Function

Proposition. Let $M(q,t)$ be a C^3 function and let

$$L'(q,\dot{q},t) = L(q,\dot{q},t) + \sum_{k=1}^{f} \frac{\partial M}{\partial q_k}\dot{q}_k + \frac{\partial M}{\partial t}\,.$$

Then $q(t)$ is the integral curve of $\delta L'/\delta q_k = 0$, $k = 1,\ldots,f$, if and only if it is solution of $\delta L/\delta q_k = 0$, $k = 1,\ldots,f$.

Proof. For $k = 1,\ldots,f$ calculate

$$\frac{\delta L'}{\delta q_k} = \frac{\delta L}{\delta q_k} + \left[\frac{\partial}{\partial q_k} - \frac{\mathrm{d}}{\mathrm{d}t}\frac{\partial}{\partial \dot{q}_k}\right]\frac{\mathrm{d}M}{\mathrm{d}t}$$

$$= \frac{\delta L}{\delta q_k} + \frac{\mathrm{d}}{\mathrm{d}t}\left\{\frac{\partial M}{\partial q_k} - \frac{\partial}{\partial \dot{q}_k}\left(\sum_{i=1}^{f}\frac{\partial M}{\partial q_i}\dot{q}_i + \frac{\partial M}{\partial t}\right)\right\} = \frac{\delta L}{\delta q_k}\,.$$

The additional terms that depend on $M(q,t)$ cancel. So, if $\delta L/\delta q_k = 0$, then $\delta L'/\delta q_k = 0$, and vice versa. □

Note that M should not depend on the $\dot{q}_j$. The reason for this becomes clear from the following observation. We could have proved the proposition by means of Hamilton's principle. If we add the term $\mathrm{d}M(q,t)/\mathrm{d}t$ to the integrand of (2.27), we obtain simply the difference $M(q_2,t_2) - M(q_1,t_1)$. As the variation leaves the end points and the initial and final times fixed, this difference gives no contribution to the equations of motion. These equations are therefore the same for L and for L'. It is then clear why M should not depend on $\dot{q}$: if one fixes t_1, t_2 as well as q_1, q_2, one cannot require the derivatives $\dot{q}$ to be fixed at the end points as well. This may also be read off Fig. 2.6.

The harmonic oscillator of Sect. 1.17.1 may serve as an example. The natural form of the Lagrangian is $L = T - U$, i.e.

$$L = \frac{1}{2}\left[\left(\frac{\mathrm{d}z_1}{\mathrm{d}\tau}\right)^2 - z_1^2\right]\,,$$

and leads to the correct equations of motion (1.39). The function

$$L' = \frac{1}{2}\left[\left(\frac{\mathrm{d}z_1}{\mathrm{d}\tau}\right)^2 - z_1^2\right] + z_1\frac{\mathrm{d}z_1}{\mathrm{d}\tau}$$

leads to the same equations because we have added $M = (\mathrm{d}z_1^2/\mathrm{d}\tau)/2$.

Lagrange's equations are even invariant under arbitrary, one-to-one, differentiable transformations of the generalized coordinates. Such transformations are called *diffeomorphisms*: they are defined to be one-to-one maps $f\colon U \to V$ for which both f and its inverse f^{-1} are differentiable. The following proposition deals with transformations of this class.

2.11 Admissible Transformations of the Generalized Coordinates

Proposition. Let $G : q \mapsto Q$ be a diffeomorphism (which should be at least C^2), $g = G^{-1}$ its inverse,

$$Q_i = G_i(q, t) \quad \text{and} \quad q_k = g_k(Q, t) \,, \quad i, k = 1, \ldots, f \,.$$

In particular, one then knows that

$$\det(\partial g_j / \partial Q_k) \neq 0 \,. \tag{2.34}$$

Then the equations $\delta L/\delta q_k = 0$ are equivalent to $\delta \bar{L}/\delta Q_k = 0$, $k = 1, \ldots, f$; i.e. $Q(t)$ is a solution of the Lagrange equations of the transformed Lagrangian function

$$\bar{L} = L \circ g = L\Big(g_1(Q, t), \ldots, g_f(Q, t), \sum_{k=1}^{f} \frac{\partial g_1}{\partial Q_k} \dot{Q}_k + \frac{\partial g_1}{\partial t}, \ldots, \sum_{k=1}^{f} \frac{\partial g_f}{\partial Q_k} \dot{Q}_k + \frac{\partial g_f}{\partial t}, t\Big) \tag{2.35}$$

if and only if $q(t)$ is a solution of the Lagrange equations for $L(q, \dot{q}, t)$.

Proof. Take the variational derivatives of $\bar{L}$ by the Q_k, $\delta \bar{L}/\delta Q_k$, i.e. calculate

$$\begin{aligned}\frac{\mathrm{d}}{\mathrm{d}t}\left(\frac{\partial \bar{L}}{\partial \dot{Q}_k}\right) &= \sum_{l=1}^{f} \frac{\mathrm{d}}{\mathrm{d}t}\left(\frac{\partial L}{\partial \dot{q}_l} \frac{\partial \dot{q}_l}{\partial \dot{Q}_k}\right) = \sum_{l=1}^{f} \frac{\mathrm{d}}{\mathrm{d}t}\left(\frac{\partial L}{\partial \dot{q}_l} \frac{\partial q_l}{\partial Q_k}\right) \\ &= \sum_{l=1}^{f} \left[\frac{\partial L}{\partial \dot{q}_l} \frac{\partial \dot{q}_l}{\partial Q_k} + \frac{\partial q_l}{\partial Q_k} \frac{\mathrm{d}}{\mathrm{d}t} \frac{\partial L}{\partial \dot{q}_l}\right].\end{aligned} \tag{2.36}$$

In the second step we have made use of $\dot{q}_l = \sum_k (\partial g_l / \partial Q_k) \dot{Q}_k + \partial g_l / \partial t$, from which follows $\partial \dot{q}_l / \partial \dot{Q}_k = \partial g_l / \partial Q_k$.

Calculating

$$\frac{\partial \bar{L}}{\partial Q_k} = \sum_{l=1}^{f} \left[\frac{\partial L}{\partial q_l} \frac{\partial q_l}{\partial Q_k} + \frac{\partial L}{\partial \dot{q}_l} \frac{\partial \dot{q}_l}{\partial Q_k}\right]$$

and subtracting (2.36) yields

$$\frac{\delta \bar{L}}{\delta Q_k} = \sum_{l=1}^{f} \frac{\partial g_l}{\partial Q_k} \frac{\delta L}{\delta q_l} \,. \tag{2.37}$$

By assumption the transformation matrix $\{\partial g_l/\partial Q_k\}$ is not singular; cf. (2.34). This proves the proposition. □

Another way of stating this result is this: the variational derivatives $\delta L/\delta q_k$ are *covariant* under diffeomorphic transformations of the generalized coordinates.

It is not correct, therefore, to state that the Lagrangian function is "$T - U$". Although this is a natural form, in those cases where kinetic and potential energies are defined, but it is certainly not the only one that describes a given problem. In general, L is a function of $\underset{\sim}{q}$ and $\dot{\underset{\sim}{q}}$, as well as of time t, and no more. How to construct a Lagrangian function is more a question of the symmetries and invariances of the physical system one wishes to describe. There may well be cases where there is no kinetic energy or no potential energy, in the usual sense, but where a Lagrangian can be found, up to gauge transformations (2.33), which gives the correct equations of motion. This is true, in particular, in applying the variational principle of Hamilton to theories in which fields take over the role of dynamical variables. For such theories, the notion of kinetic and potential parts in the Lagrangian must be generalized anyway, if they are defined at all.

The proposition proved above tells us that with any set of generalized coordinates there is an infinity of other, equivalent sets of variables. Which set is chosen in practice depends on the special features of the system under consideration. For example, a clever choice will be one where as many integrals of the motion as possible will be manifest. We shall say more about this as well as about the geometric meaning of this multiplicity later. For the moment we note that the transformations must be diffeomorphisms. In transforming to new coordinates we wish to conserve the number of degrees of freedom as well as the differential structure of the system. Only then can the physics be independent of the special choice of variables.

2.12 The Hamiltonian Function and Its Relation to the Lagrangian Function *L*

It is easy to convince oneself of the following fact. If the Lagrangian function L has no explicit time dependence then the function

$$\tilde{H}(\underset{\sim}{q}, \dot{\underset{\sim}{q}}) \stackrel{\text{def}}{=} \sum_{k=1}^{f} \dot{q}_k \frac{\partial L}{\partial \dot{q}_k} - L(\underset{\sim}{q}, \dot{\underset{\sim}{q}}) \tag{2.38}$$

is a constant of the motion. Indeed, differentiating with respect to time and making use of the equations $\delta L/\delta q_k = 0$, one has

$$\frac{\mathrm{d}\tilde{H}}{\mathrm{d}t} = \sum_{i=1}^{f} \left[\ddot{q}_i \frac{\partial L}{\partial \dot{q}_i} + \dot{q}_i \frac{\mathrm{d}}{\mathrm{d}t} \frac{\partial L}{\partial \dot{q}_i} - \frac{\partial L}{\partial q_i} \dot{q}_i - \frac{\partial L}{\partial \dot{q}_i} \ddot{q}_i \right] = 0 \, .$$

Take as an example $L = m\dot{\boldsymbol{r}}^2/2 - U(\boldsymbol{r}) \equiv T - U$. Equation (2.38) gives $\tilde{H}(\boldsymbol{r}, \dot{\boldsymbol{r}}) = 2T - (T - U) = T + U = (m\dot{\boldsymbol{r}}^2/2) + U(\boldsymbol{r})$. If we set $m\dot{\boldsymbol{r}} = \boldsymbol{p}$, $\tilde{H}$ goes over into

$H(\boldsymbol{r}, \boldsymbol{p}) = \boldsymbol{p}^2/2m + U(\boldsymbol{r})$. In doing so, we note that the momentum $\boldsymbol{p}$ is given by the partial derivative of L by $\dot{\boldsymbol{x}}$, $\boldsymbol{p} = (\partial L/\partial \dot{x}, \partial L/\partial \dot{y}, \partial L/\partial \dot{z})$. This leads us to the definition in the general case[3]

$$\boxed{p_k \stackrel{\text{def}}{=} \frac{\partial L}{\partial \dot{q}_k}\,,} \tag{2.39}$$

where p_k is called the *momentum canonically conjugate to the coordinate* q_k. One reason for this name is that, for the simple example above, the definition (2.39) leads to the ordinary momentum. Furthermore, the Euler–Lagrange equation

$$\frac{\delta L}{\delta q_k} = \frac{\partial L}{\partial q_k} - \frac{\mathrm{d}}{\mathrm{d}t}\left(\frac{\partial L}{\partial \dot{q}_k}\right) = 0$$

tells us that this momentum is an integral of the motion whenever $\partial L/\partial q_k = 0$. In other words, if L does not depend explicitly on one (or several) of the q_k,

$$L = L(q_1, \ldots, q_{k-1}, q_{k+1}, \ldots q_f, \dot{q}_1, \ldots, \dot{q}_k, \ldots \dot{q}_f, t)\,,$$

then the corresponding, conjugate momentum (momenta) is an (are) integral(s) of the motion, $p_k = \text{const}$. If this is the case, such generalized coordinates q_k are said to be *cyclic coordinates*.

The question arises under which conditions (2.38) can be transformed to the form $H(q, \underset{\sim}{p}, t)$. The answer is provided by what is called *Legendre transformation*, to whose analysis we now turn.

2.13 The Legendre Transformation for the Case of One Variable

Let $f(x)$ be a real, differentiable function (at least C^2). Let $y \stackrel{\text{def}}{=} f(x)$, $z \stackrel{\text{def}}{=} \mathrm{d}f/\mathrm{d}x$ and assume that $\mathrm{d}^2 f/\mathrm{d}x^2 \neq 0$. Then, by the implicit function theorem, $x = g(z)$, the inverse function of $z = \mathrm{d}f(x)/\mathrm{d}x$, exists. The theorem also guarantees the existence of the Legendre transform of f, which is defined as follows:

$$(\mathcal{L}f)(x) \stackrel{\text{def}}{=} x\frac{\mathrm{d}f}{\mathrm{d}x} - f(x) = g(z)z - f(g(z)) \stackrel{\text{def}}{=} \mathcal{L}f(z)\,. \tag{2.40}$$

Thus, as long as $\mathrm{d}^2 f/\mathrm{d}x^2 \neq 0$, $\mathcal{L}f(z)$ is well defined. It is then possible to construct also $\mathcal{L}\mathcal{L}f(z)$, i.e. to apply the Legendre transformation twice. One obtains

$$\frac{\mathrm{d}}{\mathrm{d}z}\mathcal{L}f(z) = g(z) + z\frac{\mathrm{d}g}{\mathrm{d}z} - \frac{\mathrm{d}f}{\mathrm{d}x}\frac{\mathrm{d}x}{\mathrm{d}z} = x + z\frac{\mathrm{d}g}{\mathrm{d}z} - z\frac{\mathrm{d}g}{\mathrm{d}z} = x\,.$$

[3] There are cases where one must take care with the position of indices: q^j (superscript), but $p_j = \partial L/\partial \dot{q}^j$ (subscript). Here we do not have to distinguish between the two positions yet. This will be important, though, in Chaps. 4 and 5.

Its second derivative does not vanish, because

$$\frac{d^2}{dz^2}\mathcal{L}f(z) = \frac{dx}{dz} = \frac{1}{d^2 f/dx^2} \neq 0 \,.$$

Therefore, if we set $\mathcal{L}f(z) = \Phi(z) = xz - f$,

$$\mathcal{L}\mathcal{L}f(z) \equiv \mathcal{L}\Phi = z\frac{d\Phi}{dz} - \Phi(z) = zx - xz + f = f \,.$$

This means that the transformation

$$f \to \mathcal{L}f$$

is one-to-one whenever $d^2 f/dx^2 \neq 0$.

For the sake of illustration we consider two examples.

Example (i) Let $f(x) = mx^2/2$. Then $z = df/dx = mx$ and $d^2 f/dx^2 = m \neq 0$. Thus $x = g(z) = z/m$ and $\mathcal{L}f(z) = (z/m)z - m(z/m)^2/2 = z^2/2m$.

Example (ii) Let $f(x) = x^\alpha/\alpha$. Then $z = x^{\alpha-1}$, $d^2 f/dx^2 = (\alpha - 1)x^{\alpha-2} \neq 0$, provided $\alpha \neq 1$, and, if $\alpha \neq 2$, provided also $x \neq 0$. The inverse is

$$x = g(z) = z^{1/(\alpha-1)}$$

and therefore

$$\mathcal{L}f(z) = z^{1/(\alpha-1)}z - \frac{1}{\alpha}z^{\alpha/(\alpha-1)} = \frac{\alpha-1}{\alpha}z^{\alpha/(\alpha-1)} \equiv \frac{1}{\beta}z^\beta \text{ with } \beta \equiv \frac{\alpha}{\alpha-1} \,.$$

We note the relation $1/\alpha + 1/\beta = 1$. As a result we have

$$f(x) = \frac{1}{\alpha}x^\alpha \underset{\mathcal{L}}{\leftrightarrow} \mathcal{L}f(z) = \frac{1}{\beta}z^\beta \quad \text{with} \quad \frac{1}{\alpha} + \frac{1}{\beta} = 1 \,.$$

If a Lagrangian function is given (here for a system with $f = 1$), the Legendre transform is nothing but the passage to the Hamiltonian function that we sketched in Sect. 2.12. Indeed, if x is taken to be the variable $\dot{q}$ and $f(x)$ the function $L(q, \dot{q}, t)$ of $\dot{q}$, then according to (2.40)

$$\mathcal{L}L(q, \dot{q}, t) = \dot{q}(q, p, t) \cdot p - L(q, \dot{q}(q, p, t), t) = H(q, p, t) \,,$$

where $\dot{q}(q, p, t)$ is the inverse function of

$$p = \frac{\partial L}{\partial \dot{q}}(q, \dot{q}, t) \,.$$

The inverse exists if $\partial^2 L/\partial \dot{q}^2$ is nonzero. If this condition is fulfilled, $\dot{q}$ can be eliminated and is expressed by q, p, and t. In the case studied here, the initial function also depends on other variables such as q and t. Clearly their presence does not affect the Legendre transformation, which concerns the variable $\dot{q}$. (However, in the general case, it will be important to state with respect to which variable the transform is taken.)

With the same condition as above one can apply the Legendre transformation to the Hamiltonian function, replacing p by $p(q, \dot{q}, t)$ and obtaining the Lagrangian function again.

The generalization to more than one degree of freedom is easy but requires a little more writing.

2.14 The Legendre Transformation for the Case of Several Variables

Let the function $F(x_1, \ldots, x_m; u_1, \ldots, u_n)$ be C^2 in all x_k and assume that

$$\det\left(\frac{\partial^2 F}{\partial x_k \partial x_i}\right) \neq 0 \,. \tag{2.41}$$

The equations

$$y_k = \frac{\partial F}{\partial x_k}(x_1, \ldots x_m; u_1, \ldots u_n) \,, \quad k = 1, 2, \ldots, m \tag{2.42}$$

can then be solved locally in terms of the x_i, i.e.

$$x_i = \varphi_i(y_1, \ldots, y_m; u_1, \ldots u_n) \,, \quad i = 1, 2, \ldots, m \,. \tag{2.42}$$

The Legendre transform of F is defined as follows:

$$G(y_1, \ldots, y_m; u_1, \ldots u_n) \equiv \mathcal{L}F = \sum_{k=1}^{m} y_k \varphi_k - F \,. \tag{2.43}$$

We then have

$$\frac{\partial G}{\partial y_k} = \varphi_k \,; \qquad \frac{\partial G}{\partial u_i} = -\frac{\partial F}{\partial u_i} \quad \text{and} \quad \det\left(\frac{\partial^2 G}{\partial y_k \partial y_l}\right) \det\left(\frac{\partial^2 F}{\partial x_i \partial x_j}\right) = 1 \,.$$

As in the one-dimensional case this transformation is then one-to-one. This result can be applied directly to the Lagrangian function if we identify the variables $(x_1 \ldots x_m)$ with $(\dot{q}_1 \ldots \dot{q}_f)$ and $(u_1 \ldots u_n)$ with $(q_1 \ldots q_f, t)$. We start from the function $L(q, \dot{q}, t)$ and define the generalized momenta as in (2.39):

$$p_k \stackrel{\text{def}}{=} \frac{\partial}{\partial \dot{q}_k} L(\underset{\sim}{q}, \underset{\sim}{\dot{q}}, t) \,.$$

These equations can be solved locally and uniquely in terms of the $\dot{q}_k$ precisely if the condition

$$\boxed{\det\left(\frac{\partial^2 L}{\partial \dot{q}_k \partial \dot{q}_i}\right) \neq 0} \tag{2.44}$$

is fulfilled[4]. In this case $\dot{q}_k = \dot{q}_k(\underset{\sim}{q}, \underset{\sim}{p}, t)$ and the Hamiltonian function is given by

$$H(\underset{\sim}{q}, \underset{\sim}{p}, t) = \mathcal{L}L(\underset{\sim}{q}, \underset{\sim}{p}, t) = \sum_{k=1}^{f} p_k \dot{q}_k(\underset{\sim}{q}, \underset{\sim}{p}, t) - L(\underset{\sim}{q}, \dot{\underset{\sim}{q}}(\underset{\sim}{q}, \underset{\sim}{p}, t), t) .$$

With the same condition (2.44), a two-fold application of the Legendre transformation leads back to the original Lagrangian function.

Is it possible to formulate the equations of motion by means of the Hamiltonian instead of the Lagrangian function? The answer follows directly from our equations above, viz.

$$\dot{q}_k = \frac{\partial H}{\partial p_k} , \quad \det\left(\frac{\partial^2 H}{\partial p_j \partial p_i}\right) \neq 0 , \quad \text{and}$$

$$\frac{\partial H}{\partial q_k} = -\frac{\partial L}{\partial q_k} = -\dot{p}_k .$$

We obtain the following system of equations of motion:

$$\boxed{\dot{q}_k = \frac{\partial H}{\partial p_k} ; \quad \dot{p}_k = -\frac{\partial H}{\partial q_k} \quad k = 1, \ldots, f .} \tag{2.45}$$

These equations are called the *canonical equations*. They contain only the Hamiltonian function $H(\underset{\sim}{q}, \underset{\sim}{p}, t)$ and the variables $\underset{\sim}{q}, \underset{\sim}{p}, t$. We note that (2.45) is a system of $2f$ ordinary differential equations of *first* order. They replace the f differential equations of *second* order that we obtained in the Lagrangian formalism. They are completely equivalent to the Euler–Lagrange equations, provided (2.44) holds.

2.15 Canonical Systems

Definition. A mechanical system is said to be *canonical* if it admits a Hamiltonian function such that its equations of motion take the form (2.45).

Proposition. Every Lagrangian system that fulfills the condition (2.44) is canonical. The converse holds also: if $\det(\partial^2 H/\partial p_k \partial p_i) \neq 0$, then every canonical system with f degrees of freedom obeys the Euler–Lagrange equations with $L(\underset{\sim}{q}, \dot{\underset{\sim}{q}}, t)$ given by

$$L(\underset{\sim}{q},\dot{\underset{\sim}{q}},t) = \mathcal{L}H(\underset{\sim}{q},\dot{\underset{\sim}{q}},t) = \sum_{k=1}^{f} \dot{q}_k p_k(\underset{\sim}{q}, \dot{\underset{\sim}{q}}, t) - H(\underset{\sim}{q}, \underset{\sim}{p}(\underset{\sim}{q}, \dot{\underset{\sim}{q}}, t), t) . \tag{2.46}$$

[4] In mechanics the kinetic energy and, hence, the Lagrangian are positive-definite (but not necessarily homogeneous) quadratic functions of the variables $\dot{q}$ In this situation, solving the defining equations for p_k in terms of the q_i yields a unique solution also globally.

Remarks: One might wonder about the specific form (2.40) or (2.43) of the Legendre transformation which when supplemented by the condition (2.44) on the second derivatives, guarantees its bijective, in fact diffeomorphic nature. The following two remarks may be helpful in clarifying matters further.

1. For simplicity, let us write the equations for the case of one degree of freedom, $f = 1$, the generalization to more than one degree of freedom having been clarified in Sect. 2.14. Depending on whether the Lagrangian function or the Hamiltonian function is given, one constructs the hybrid

$$\widetilde{H}(q,\dot{q}) = \dot{q}\frac{\partial L(q,\dot{q})}{\partial \dot{q}} - L(q,\dot{q}) \,, \quad \text{or}$$
$$\widetilde{L}(q,p) = \frac{\partial H(q,p)}{\partial p}\dot{p} - H(q,p) \,,$$

i.e. auxiliary quantities that still depend on the "wrong" variables. If the first or the second of the conditions

$$\frac{\partial^2 L(q,\dot{q})}{\partial \dot{q}\partial \dot{q}} \neq 0 \,, \qquad \frac{\partial^2 H(q,p)}{\partial p \partial p} \neq 0$$

is fulfilled then the equations

$$p = \frac{\partial L}{\partial \dot{q}} \,, \qquad \dot{q} = \frac{\partial H}{\partial p}$$

can be solved for $\dot{q}$ as a function of q and p in the first case, for p as a function of q and $\dot{q}$ in the second, so that the transition from $L(q,\dot{q})$ to $H(q,p)$, or the inverse, from $H(q,p)$ to $L(q,\dot{q})$ becomes possible. An important aspect of Legendre transformation is the obvious symmetry between L and H. The condition on the second derivatives guarantee its uniqueness.

2. The condition on the second derivative tells us that the function to be transformed is either *convex* or *concave*. In this connection it might be useful to consult exercise 2.14 and its solution. In fact, for the Legendre transformation to exist, the weaker condition of *convexity* of the function (or its negative) is the essential requirement, not its differentiability. This weaker form is important for other branches of physics, such as thermodynamics of equilibrium states, or quantum field theory.

2.16 Examples of Canonical Systems

We illustrate the results of the previous sections by two instructive examples.

Example (i) *Motion of a particle in a central field.* As the angular momentum is conserved the motion takes place in a plane perpendicular to $\boldsymbol{l}$. We introduce polar coordinates in that plane and write the Lagrangian function in its natural form. With $x_1 = r\cos\varphi$, $x_2 = r\sin\varphi$, one finds $\boldsymbol{v}^2 = \dot{r}^2 + r^2\dot{\varphi}^2$, and thus

$$L = T - U(r) = \tfrac{1}{2}m(\dot{r}^2 + r^2\dot{\varphi}^2) - U(r). \tag{2.47}$$

Here $q_1 = r$, $q_2 = \varphi$, and $p_1 \equiv p_r = m\dot{r}$, $p_2 \equiv p_\varphi = mr^2\dot{\varphi}$. The determinant of the matrix of second derivatives of L by the $\dot{q}_j$ is

$$\det\left(\frac{\partial^2 L}{\partial\dot{q}_j \partial\dot{q}_i}\right) = \det\begin{pmatrix} m & 0 \\ 0 & mr^2 \end{pmatrix} = m^2 r^2 \neq 0 \quad \text{for} \quad r \neq 0 .$$

The Hamiltonian function can be constructed uniquely and is given by

$$H(\underset{\sim}{q}, \underset{\sim}{p}) = \frac{p_r^2}{2m} + \frac{p_\varphi^2}{2mr^2} + U(r) . \tag{2.48}$$

The canonical equations (2.45) read as follows:

$$\dot{r} = \frac{\partial H}{\partial p_r} = \frac{1}{m} p_r , \quad \dot{\varphi} = \frac{\partial H}{\partial p_\varphi} = \frac{1}{m}\frac{p_\varphi}{r^2} , \tag{2.49a}$$

$$\dot{p}_r = -\frac{\partial H}{\partial r} = \frac{p_\varphi^2}{mr^3} - \frac{\partial U}{\partial r} , \quad \dot{p}_\varphi = 0 . \tag{2.49b}$$

Comparison with Example 1.24 shows that $p_\varphi \equiv l$ is the modulus of angular momentum and is conserved. Indeed, from the expression (2.47) for L, we note that φ is a cyclic coordinate. The first equation (2.49b), when multiplied by p_r and then integrated once, gives (1.62) of Example 1.24. This shows that $H(\underset{\sim}{q}, \underset{\sim}{p})$ is conserved when taken along a solution curve of (2.45).

Example (ii) *A charged particle in electromagnetic fields.* Following the method of Example (ii) of Sect. 2.8 we have

$$L = \frac{1}{2}m\dot{\boldsymbol{q}}^2 - e\Phi(\boldsymbol{q}, t) + \frac{e}{c}\dot{\boldsymbol{q}} \cdot \boldsymbol{A}(\boldsymbol{q}, t) . \tag{2.50}$$

The canonically conjugate momenta are given by

$$p_i = \frac{\partial L}{\partial \dot{q}_i} = m\dot{q}_i + \frac{e}{c}A_i(\boldsymbol{q}, t) .$$

These equations can be solved for $\dot{q}_i$,

$$\dot{q}_i = \frac{1}{m}p_i - \frac{e}{cm}A_i ,$$

so that one obtains

$$H = \sum_{i=1}^{3} \frac{p_i}{m}\left(p_i - \frac{e}{c}A_i\right) - \frac{1}{2m}\sum_{i=1}^{3}\left(p_i - \frac{e}{c}A_i\right)^2$$

$$+ e\Phi - \frac{e}{mc}\sum_{i=1}^{3}\left(p_i - \frac{e}{c}A_i\right)A_i$$

or

$$H(\boldsymbol{q}, \boldsymbol{p}, t) = \frac{1}{2m}\left(\boldsymbol{p} - \frac{e}{c}\boldsymbol{A}(\boldsymbol{q}, t)\right)^2 + e\Phi(\boldsymbol{q}, t) . \tag{2.51}$$

Note the following difference:

$$m\dot{\boldsymbol{q}} = \boldsymbol{p} - \frac{e}{c}\boldsymbol{A}$$

is the *kinematic momentum*,

$$p_i \quad \text{with} \quad p_i = \frac{\partial L}{\partial \dot{q}_i}$$

is the (generalized) momentum *canonically conjugate to* q_i.

2.17 The Variational Principle Applied to the Hamiltonian Function

It is possible to obtain the canonical equations (2.45) directly from Hamilton's variational principle (Sects. 2.5 and 2.6). For this we apply the principle to the following function:

$$F(\underset{\sim}{q}, \underset{\sim}{p}, \dot{\underset{\sim}{q}}, \dot{\underset{\sim}{p}}, t) \stackrel{\text{def}}{=} \sum_{k=1}^{f} p_k \dot{q}_k - H(\underset{\sim}{q}, \underset{\sim}{p}, t) , \tag{2.52}$$

taking the $\underset{\sim}{q}, \underset{\sim}{p}, \dot{\underset{\sim}{q}}, \dot{\underset{\sim}{p}}$ as four sets of independent variables. In the language of Sect. 2.5, $(\underset{\sim}{q}, \underset{\sim}{p})$ corresponds to y, $(\dot{\underset{\sim}{q}}, \dot{\underset{\sim}{p}})$ to y', and t to x. Requiring that

$$\delta \int_{t_1}^{t_2} F \mathrm{d}t = 0 \tag{2.53}$$

and varying the variables q_k and p_k independently, we get the Euler–Lagrange equations $\delta F/\delta q_k = 0$, $\delta F/\delta p_k = 0$. When written out, these are

$$\frac{\mathrm{d}}{\mathrm{d}t}\frac{\partial F}{\partial \dot{q}_k} = \frac{\partial F}{\partial q_k} , \quad \text{or} \quad \dot{p}_k = -\frac{\partial H}{\partial q_k} , \quad \text{and}$$

$$\frac{\mathrm{d}}{\mathrm{d}t}\frac{\partial F}{\partial \dot{p}_k} = \frac{\partial F}{\partial p_k} , \quad \text{or} \quad 0 = \dot{q}_k - \frac{\partial H}{\partial p_k} .$$

Thus, we again obtain the canonical equations (2.45). We shall make use of this result below when discussing canonical transformations.

2.18 Symmetries and Conservation Laws

In Sects. 1.12 and 1.13 we studied the ten classical integrals of the motion of the closed n-particle system, as derived directly from Newton's equations. In this section and in the subsequent ones we wish to discuss these results, as well as generalizations of them, in the framework of Lagrangian functions and the Euler–Lagrange equations.

Here and below we study closed, autonomous systems with f degrees of freedom to which we can ascribe Lagrangian functions $L(\underset{\sim}{q}, \dot{\underset{\sim}{q}})$ without explicit time dependence. Take the natural form for L,

$$L = T(\underset{\sim}{q}, \dot{\underset{\sim}{q}}) - U(\underset{\sim}{q}) , \tag{2.54}$$

where T is a homogeneous function of degree 2 in the $\dot{q}_k$. According to Euler's theorem on homogeneous functions we have

$$\sum_{i=1}^{f} \dot{q}_i \frac{\partial T}{\partial \dot{q}_i} = 2T , \tag{2.55}$$

so that

$$\sum_{i=1}^{f} p_i \dot{q}_i - L = \sum \frac{\partial L}{\partial \dot{q}_i} \dot{q}_i - L = T + U = E .$$

This expression represents the energy of the system. For autonomous systems, E is conserved along any orbit. Indeed, making use of the Euler–Lagrange equations, one finds that

$$\begin{aligned} \frac{\mathrm{d}E}{\mathrm{d}t} &= \frac{\mathrm{d}}{\mathrm{d}t}\Big(\sum p_i \dot{q}_i\Big) - \sum \frac{\partial L}{\partial q_i}\dot{q}_i - \sum \frac{\partial L}{\partial \dot{q}_i}\ddot{q}_i \\ &= \frac{\mathrm{d}}{\mathrm{d}t}\Big(\sum p_i \dot{q}_i\Big) - \frac{\mathrm{d}}{\mathrm{d}t}\Big(\sum \frac{\partial L}{\partial \dot{q}_i}\dot{q}_i\Big) = 0 . \end{aligned}$$

Note that we made use of the Euler-Lagrange equations.

Remarks: In the framework of Lagrangian mechanics a dynamical quantity such as, for instance, the energy which is a candidate for a constant of the motion, at first is a function $E(\underset{\sim}{q}, \dot{\underset{\sim}{q}})$ on velocity space spanned by $\underset{\sim}{q}$ and $\dot{\underset{\sim}{q}}$. Likewise, in the framework of Hamiltonian canonical mechanics it is a function $E(\underset{\sim}{q}, \underset{\sim}{p})$ on phase space that is spanned by $\underset{\sim}{q}$ and $\underset{\sim}{p}$. Of course, such a function on either velocity space or phase space, in general, is not constant. It is constant only – if it represents an integral of the motion – along solutions of the equations of motion. In other terms, its time derivative is equal to zero only if it is evaluated along physical orbits along which $\underset{\sim}{q}$ and $\dot{\underset{\sim}{q}}$, or $\underset{\sim}{q}$ and $\underset{\sim}{p}$, respectively, are related to each other via the equations of motion. For this reason and as discussed in Sect (1.16), this kind of time derivative is called the *orbital derivative,* as a short-hand for time derivative taken along the orbit. The important point to note is that in order to study the variation of a given function along physical orbits we need not know the solutions proper. Knowledge of the differential equations that describe the motion is sufficient for calculating the orbital derivative and to find out, for instance, whether that function is an integral of the system.

Suppose that the mechanical system one is considering is invariant under a class of continuous transformations of the coordinates that can be deformed smoothly into the identical mapping. The system then possesses *integrals of the motion*, i.e. there are dynamical quantities that are constant along orbits of the system. The interesting observation is that it is sufficient to study these transformations in an infinitesimal neighborhood of the identity. This is made explicit in the following theorem by Emmy Noether, which applies to transformations of the space coördinates.

2.19 Noether's Theorem

Let the Lagrangian function $L(\underset{\sim}{q}, \dot{\underset{\sim}{q}})$ describing an autonomous system be invariant under the transformation $\underset{\sim}{q} \mapsto h^s(\underset{\sim}{q})$, where s is a real, continuous parameter and where $h^{s=0}(\underset{\sim}{q}) = \underset{\sim}{q}$ is the identity (see Fig. 2.7). Then there exists an integral of the motion, given by

$$I(\underset{\sim}{q}, \dot{\underset{\sim}{q}}) = \sum_{i=1}^{f} \frac{\partial L}{\partial \dot{q}_i} \frac{\mathrm{d}}{\mathrm{d}s} h^s(q_i) \bigg|_{s=0} . \tag{2.56}$$

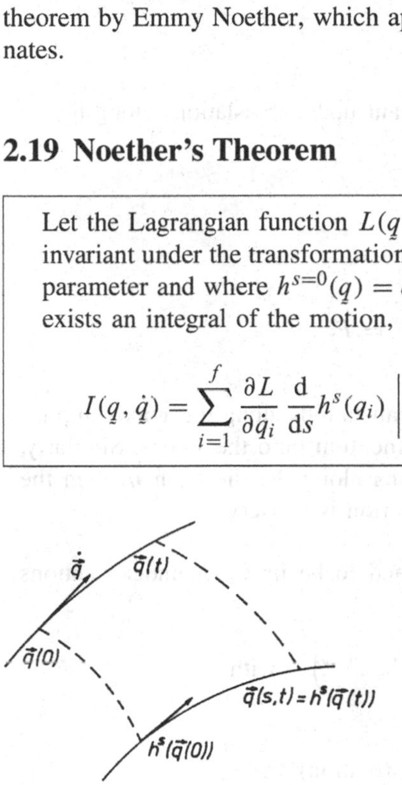

Fig. 2.7. A differentiable, one-parameter transformation of the orbits in the neighborhood of the identical mapping. If it leaves the Lagrangian function invariant, there exists a constant of the motion corresponding to it

Proof. Let $\underset{\sim}{q} = \underset{\sim}{\varphi}(t)$ be a solution of the Euler–Lagrange equations. Then, by assumption, $\underset{\sim}{q}(s, t) = \underset{\sim}{\Phi}(s, t) = h^s(\underset{\sim}{\varphi}(t))$ is a solution too. This means that

$$\frac{\mathrm{d}}{\mathrm{d}t} \frac{\partial L}{\partial \dot{q}_i}(\underset{\sim}{\Phi}(s, t), \dot{\underset{\sim}{\Phi}}(s, t)) = \frac{\partial L}{\partial q_i}(\underset{\sim}{\Phi}(s, t), \dot{\underset{\sim}{\Phi}}(s, t)) , \quad i = 1, \ldots, f . \tag{2.57}$$

Furthermore, L is invariant by assumption, i.e.

$$\frac{\mathrm{d}}{\mathrm{d}s} L(\Phi(s,t), \dot{\Phi}(s,t)) = \sum_{i=1}^{f} \left[\frac{\partial L}{\partial q_i} \frac{\mathrm{d}\Phi_i}{\mathrm{d}s} + \frac{\partial L}{\partial \dot{q}_i} \frac{\mathrm{d}\dot{\Phi}_i}{\mathrm{d}s} \right] = 0 . \tag{2.58}$$

Combining (2.57) and (2.58) we obtain

$$\sum_{i=1}^{f} \left[\frac{\mathrm{d}}{\mathrm{d}t} \left(\frac{\partial L}{\partial \dot{q}_i} \right) \frac{\mathrm{d}\Phi_i}{\mathrm{d}s} + \frac{\partial L}{\partial \dot{q}_i} \frac{\mathrm{d}}{\mathrm{d}t} \left(\frac{\mathrm{d}\Phi_i}{\mathrm{d}s} \right) \right] = 0 = \frac{\mathrm{d}}{\mathrm{d}t} I . \qquad \square$$

We study two examples; let the Lagrangian function have the form

$$L = \frac{1}{2} \sum_{p=1}^{n} m_p \dot{\boldsymbol{r}}_p^2 - U(\boldsymbol{r}_1, \ldots, \boldsymbol{r}_n) .$$

Example (i) Assume that the system is invariant under translations along the x-axis:

$$h^s : \boldsymbol{r}_p \mapsto \boldsymbol{r}_p + s\boldsymbol{e}_x , \quad p = 1, \ldots, n .$$

We then have

$$\frac{\mathrm{d}}{\mathrm{d}s} h^s(\boldsymbol{r}_p) \big|_{s=0} = \boldsymbol{e}_x \quad \text{and} \quad I = \sum_{p=1}^{n} m_p \dot{x}^{(p)} = P_x .$$

The result is the following. Invariance under translations along the x-axis implies conservation of the projection of the total momentum onto the x-axis. Similarly, if the Lagrangian is invariant under translations along the direction $\hat{\boldsymbol{n}}$, then the component of total momentum along that direction is conserved.

Example (ii) The same system is now assumed to be invariant under rotations about the z-axis, cf. Fig. 2.8:

$$\boldsymbol{r}_p = \left(x^{(p)}, y^{(p)}, z^{(p)} \right) \to \boldsymbol{r}'_p = \left(x'^{(p)}, y'^{(p)}, z'^{(p)} \right) \quad \text{with}$$

$$\begin{aligned} x'^{(p)} &= x^{(p)} \cos s + y^{(p)} \sin s , \\ y'^{(p)} &= -x^{(p)} \sin s + y^{(p)} \cos s , \quad \text{(passive rotation)} \\ z'^{(p)} &= z^{(p)} \end{aligned}$$

Here one obtains

$$\frac{\mathrm{d}}{\mathrm{d}s} \boldsymbol{r}'_p \big|_{s=0} = \left(y^{(p)}, -x^{(p)}, 0 \right) = \boldsymbol{r}_p \times \boldsymbol{e}_z$$

and

$$I = \sum_{p=1}^{n} m_p \dot{\boldsymbol{r}}_p \cdot (\boldsymbol{r}_p \times \boldsymbol{e}_z) = \sum \boldsymbol{e}_z \cdot (m_p \dot{\boldsymbol{r}}_p \times \boldsymbol{r}_p) = -l_z .$$

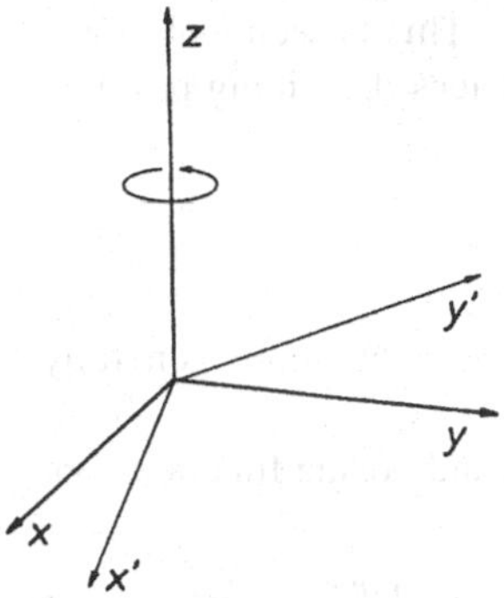

Fig. 2.8. If the Lagrangian function for a mass point is invariant by rotations about the z-axis, the projection of angular momentum onto that axis is conserved

The conserved quantity is found to be the projection of the total angular momentum onto the z-axis. More generally, if L is invariant under rotations about any direction in space (the potential energy must then be spherically symmetric), then the total angular momentum is conserved as a whole.

To which extent there exists an inverse to Noether's theorem, that is to say under which conditions the existence of an integral of the motion implies invariance of the system with respect to a continuous transformation, will be clarified in Sects. 2.34 and 2.41 below.

2.20 The Generator for Infinitesimal Rotations About an Axis

In the two previous examples and, generally, in Noether's theorem h^s is a one-parameter group of diffeomorphisms that have the special property that they can be deformed, in a continuous manner, into the identity. The integral of the motion, I, only depends on the derivative of h^s at $s = 0$, which means that the transformation group is needed only in the neighborhood of the identity. Here we wish to pursue the analysis of Example (ii), Sect. 2.19. This will give a first impression of the importance of continuous groups of transformations in mechanics.

For infinitesimally small s the rotation about the z-axis of Example (ii) can be written as follows:

$$h^s(\boldsymbol{r}) = \left\{ \begin{pmatrix} 1 & 0 & 0 \\ 0 & 1 & 0 \\ 0 & 0 & 1 \end{pmatrix} - s \begin{pmatrix} 0 & -1 & 0 \\ 1 & 0 & 0 \\ 0 & 0 & 0 \end{pmatrix} \right\} \begin{pmatrix} x \\ y \\ z \end{pmatrix} + \mathrm{O}(s^2)$$
$$\equiv (\mathbb{1} - s\mathbf{J}_z)\boldsymbol{r} + \mathrm{O}(s^2) \,. \tag{2.59}$$

The 3×3 matrix $\mathbf{J}_z$ is said to be the *generator for infinitesimal rotations about the z-axis*. In fact, one can show that the rotation about the z-axis by a *finite* angle

$$\boldsymbol{r}' = \begin{pmatrix} \cos\varphi & \sin\varphi & 0 \\ -\sin\varphi & \cos\varphi & 0 \\ 0 & 0 & 1 \end{pmatrix} \boldsymbol{r} \stackrel{\text{def}}{=} \mathbf{R}_z(\varphi)\boldsymbol{r} \tag{2.60}$$

can be constructed from the infinitesimal rotations (2.59). This is seen as follows. For simplicity let us first study the example of 2×2-matrices describing rotations in a plane (the (x, y)-plane, for instance). The matrix

$$\mathbf{M} \stackrel{\text{def}}{=} \begin{pmatrix} 0 & -1 \\ 1 & 0 \end{pmatrix} .$$

has the properties $\mathbf{M}^2 = -\mathbb{1}$, $\mathbf{M}^3 = -\mathbf{M}$, $\mathbf{M}^4 = +\mathbb{1}$, etc. or, more generally, $\mathbf{M}^{2n} = (-1)^n \mathbb{1}$, $\mathbf{M}^{2n+1} = (-1)^n \mathbf{M}$.

Then, from the well-known Taylor series for the sine and cosine functions one has

$$\mathbf{A} \stackrel{\text{def}}{=} \begin{pmatrix} \cos\varphi & \sin\varphi \\ -\sin\varphi & \cos\varphi \end{pmatrix} = \mathbb{1} \sum_{n=0}^{\infty} \frac{(-1)^n}{(2n)!} \varphi^{2n} - \mathbf{M} \sum_{n=0}^{\infty} \frac{(-1)^n}{(2n+1)!} \varphi^{2n+1} ,$$

and, inserting the formulae for the even and odd powers of the matrix $\mathbf{M}$,

$$\mathbf{A} = \sum_{0}^{\infty} \frac{1}{(2n)!} \mathbf{M}^{2n} \varphi^{2n} - \sum_{0}^{\infty} \frac{1}{(2n+1)!} \mathbf{M}^{2n+1} \varphi^{2n+1} = \exp(-\mathbf{M}\varphi) . \tag{2.61}$$

It is then not difficult to convince oneself that the 3×3 matrix $\mathbf{R}_z(\varphi)$ of (2.60) can also be written as an exponential series, as in (2.61), viz.

$$\mathbf{R}_z(\varphi) = \exp(-\mathbf{J}_z \varphi) . \tag{2.62}$$

Indeed, consider the 3×3-matrix defined in (2.59)

$$\mathbf{J}_z = \begin{pmatrix} 0 & -1 & 0 \\ 1 & 0 & 0 \\ 0 & 0 & 0 \end{pmatrix}$$

and verify that its even and odd powers are

$$\mathbf{J}_z^{2n} = (-)^n \begin{pmatrix} 1 & 0 & 0 \\ 0 & 1 & 0 \\ 0 & 0 & 0 \end{pmatrix} , \qquad \mathbf{J}_z^{2n+1} = (-)^n \mathbf{J}_z .$$

Inserting these formulae, one has

$$\mathbf{R}_z(\varphi) = \begin{pmatrix} \cos\varphi & \sin\varphi & 0 \\ -\sin\varphi & \cos\varphi & 0 \\ 0 & 0 & 1 \end{pmatrix} = \begin{pmatrix} 1 & 0 & 0 \\ 0 & 1 & 0 \\ 0 & 0 & 1 \end{pmatrix} + \begin{pmatrix} \cos\varphi - 1 & \sin\varphi & 0 \\ -\sin\varphi & \cos\varphi - 1 & 0 \\ 0 & 0 & 1 \end{pmatrix}$$

$$= \mathbb{1}_{3\times 3} + \sum_{n=1}^{\infty} \frac{1}{(2n)!} \mathbf{J}_z^{2n} \varphi^{2n} - \sum_{n=0}^{\infty} \frac{1}{(2n+1)!} \mathbf{J}_z^{2n+1} \varphi^{2n+1} = \exp(-\mathbf{J}_z \varphi) .$$

The result (2.62) can be understood as follows. In (2.59) take $s = \varphi/n$ with n a positive integer, large compared to 1. Assume then that we perform n such rotations in a series, i.e.

$$\left(\mathbb{1} - \frac{\varphi}{n} \mathbf{J}_z \right)^n ,$$

Finally, let n go to infinity. In this limit use Euler's formula for the exponential, $\lim (1 + x/n)^n = \mathrm{e}^x$, to obtain

$$\lim_{n\to\infty}\left(\mathbb{1}-\frac{\varphi}{n}\mathbf{J}_z\right)^n = \exp(-\mathbf{J}_z\varphi)\ .$$

Clearly, these results can be extended to rotations about any other direction in space.

The appearance of finite-dimensional matrices in the argument of an exponential function is perhaps not familiar to the reader. There is nothing mysterious about such exponentials. They are defined through the power series

$$\exp\{\mathbf{A}\} = 1 + \mathbf{A} + \mathbf{A}^2/2! + \ldots + \mathbf{A}^k/k! + \ldots\ ,$$

where $\mathbf{A}$, like any finite power $\mathbf{A}^k$ in the series, is an $n\times n$ matrix. As the exponential is an entire function (its Taylor expansion converges for any finite value of the argument), there is no problem of convergence of this series.

2.21 More About the Rotation Group

Let $\boldsymbol{x} = (x_1, x_2, x_3)$ be a point on a physical orbit $\boldsymbol{x}(t)$, x_1, x_2 and x_3 being its (Cartesian) coordinates with respect to the frame of reference $\mathbf{K}$. The same point, when described within the frame $\mathbf{K}'$ whose origin coincides with that of $\mathbf{K}$ but which is rotated by the angle φ about the direction $\hat{\boldsymbol{\varphi}}$, is represented by

$$\boldsymbol{x}|_{\mathbf{K}'} = (x'_1, x'_2, x'_3) \quad\text{with}\quad x'_i = \sum_{k=1}^{3} R_{ik}x_k \quad\text{or}\quad \boldsymbol{x}' = \mathbf{R}\boldsymbol{x}\ . \tag{2.63}$$

(This is a *passive* rotation.) By definition, rotations leave the length unchanged. Thus $\boldsymbol{x}'^2 = \boldsymbol{x}^2$, or, when written out more explicitly,

$$(\mathbf{R}\boldsymbol{x})\cdot(\mathbf{R}\boldsymbol{x}) = \boldsymbol{x}\mathbf{R}^{\mathrm{T}}\mathbf{R}\boldsymbol{x} \stackrel{!}{=} \boldsymbol{x}^2\ ,$$

or, in components and even more explicitly,

$$\sum_{i=1}^{3} x'_i x'_i = \sum_{k=1}^{3}\sum_{l=1}^{3}\left(\sum_{i=1}^{3} R_{ik}R_{il}\right) x_k x_l \stackrel{!}{=} \sum_{k=1}^{3}\sum_{l=1}^{3}\delta_{kl}x_k x_l\ .$$

One thus obtains the condition

$$\sum_{i=1}^{3}(\mathbf{R}^{\mathrm{T}})_{ki}R_{il} = \delta_{kl}\ ,$$

i.e. $\mathbf{R}$ must be a real, orthogonal matrix:

$$\mathbf{R}^{\mathrm{T}}\mathbf{R} = \mathbb{1}\ . \tag{2.64}$$

From (2.64) one concludes that $(\det\mathbf{R})^2 = 1$ or $\det\mathbf{R} = \pm 1$.

We restrict the discussion to the rotation matrices with determinant $+1$ and leave aside space inversion (cf. Sect. 1.13). The matrices $\mathbf{R}$ with $\det\mathbf{R} = +1$ form a group, the *special orthogonal group in three real dimensions*

$$\mathrm{SO}(3) = \{\mathbf{R} : \mathbb{R}^3 \to \mathbb{R}^3 \text{ linear} \mid \det\mathbf{R} = +1, \mathbf{R}^{\mathrm{T}}\mathbf{R} = \mathbb{1}\}\ . \tag{2.65}$$

As shown in Sect. 1.13, every such $\mathbf{R}$ depends on 3 real parameters and can be deformed continuously into the identity $\mathbf{R}^0 = \mathbb{1}$. A possible parametrization is the following. Take a vector $\boldsymbol{\varphi}$ whose *direction* $\hat{\boldsymbol{\varphi}} \stackrel{\text{def}}{=} \boldsymbol{\varphi}/\varphi$ defines the axis about which the rotation takes place and whose *modulus* $\varphi = |\boldsymbol{\varphi}|$ defines the angle of rotation, as indicated in Fig. 2.9:

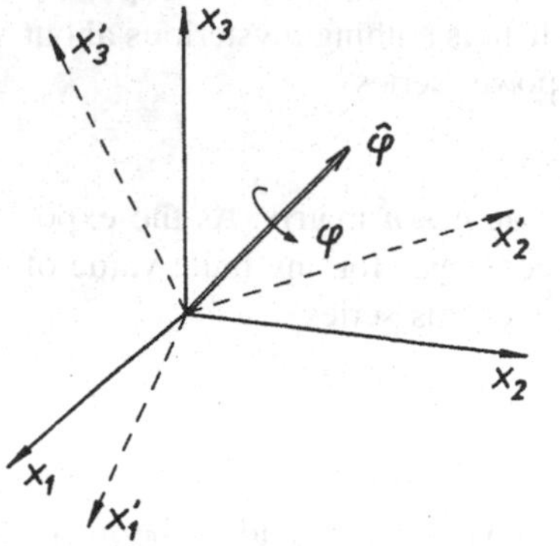

Fig. 2.9. Rotation of the coordinate system about the direction $\hat{\boldsymbol{\varphi}}$ by the (fixed) angle φ

$$\mathbf{R} \equiv \mathbf{R}(\varphi) \quad \text{with} \quad \boldsymbol{\varphi} : \hat{\boldsymbol{\varphi}} = \boldsymbol{\varphi}/\varphi\,, \quad \varphi = |\boldsymbol{\varphi}|\,, \quad 0 \le \varphi \le 2\pi\,. \tag{2.66}$$

(We shall meet other parametrizations in developing the theory of the rigid body in Chap. 3.) The action of $\mathbf{R}(\boldsymbol{\varphi})$ on $\boldsymbol{x}$ can be expressed explicitly in terms of the vectors $\boldsymbol{x}$, $\hat{\boldsymbol{\varphi}} \times \boldsymbol{x}$ and $\hat{\boldsymbol{\varphi}} \times (\hat{\boldsymbol{\varphi}} \times \boldsymbol{x})$, for a passive rotation, by

$$\boldsymbol{x}' = \mathbf{R}(\varphi)\boldsymbol{x} = (\hat{\boldsymbol{\varphi}} \cdot \boldsymbol{x})\hat{\boldsymbol{\varphi}} - \hat{\boldsymbol{\varphi}} \times \boldsymbol{x} \sin\varphi - \hat{\boldsymbol{\varphi}} \times (\hat{\boldsymbol{\varphi}} \times \boldsymbol{x}) \cos\varphi\,. \tag{2.67}$$

This is shown as follows. The vectors $\hat{\boldsymbol{\varphi}}$, $\hat{\boldsymbol{\varphi}} \times \boldsymbol{x}$, and $\hat{\boldsymbol{\varphi}} \times (\hat{\boldsymbol{\varphi}} \times \boldsymbol{x})$ are mutually orthogonal. For example, if the 3-axis is taken along the direction $\hat{\boldsymbol{\varphi}}$, i.e. if $\hat{\boldsymbol{\varphi}} = (0, 0, 1)$, then $\boldsymbol{x} = (x_1, x_2, x_3)$, $\hat{\boldsymbol{\varphi}} \times \boldsymbol{x} = (-x_2, x_1, 0)$, and $\hat{\boldsymbol{\varphi}} \times (\hat{\boldsymbol{\varphi}} \times \boldsymbol{x}) = (-x_1, -x_2, 0)$. With respect to the new coordinate system the same vector has the components

$$x_1' = x_1 \cos\varphi + x_2 \sin\varphi\,, \quad x_2' = -x_1 \sin\varphi + x_2 \cos\varphi\,, \quad x_3' = x_3\,,$$

in agreement with (2.67). One now verifies that (2.67) holds true also when $\hat{\boldsymbol{\varphi}}$ is not along the 3-axis. The first term on the right-hand side of (2.67) contains the information that the projection of $\boldsymbol{x}$ onto $\hat{\boldsymbol{\varphi}}$ stays invariant, while the remaining two terms represent the rotation in the plane perpendicular to $\hat{\boldsymbol{\varphi}}$. Making use of the identity $\boldsymbol{a} \times (\boldsymbol{b} \times \boldsymbol{c}) = \boldsymbol{b}(\boldsymbol{a} \cdot \boldsymbol{c}) - \boldsymbol{c}(\boldsymbol{a} \cdot \boldsymbol{b})$, (2.67) becomes

$$\boldsymbol{x}' = \boldsymbol{x} \cos\varphi - \hat{\boldsymbol{\varphi}} \times \boldsymbol{x} \sin\varphi + (1 - \cos\varphi)(\hat{\boldsymbol{\varphi}} \cdot \boldsymbol{x})\hat{\boldsymbol{\varphi}}\,. \tag{2.68}$$

We now show the following.
(i) $\mathbf{R}(\boldsymbol{\varphi})$ as parametrized in (2.67) belongs to SO(3):

$$\begin{aligned} \boldsymbol{x}'^2 &= (\hat{\boldsymbol{\varphi}} \cdot \boldsymbol{x})^2 + (\hat{\boldsymbol{\varphi}} \times \boldsymbol{x})^2 \sin^2\varphi + (\hat{\boldsymbol{\varphi}} \times (\hat{\boldsymbol{\varphi}} \times \boldsymbol{x}))^2 \cos^2\varphi \\ &= \boldsymbol{x}^2[\cos^2\alpha + \sin^2\alpha \sin^2\varphi + \sin^2\alpha \cos^2\varphi] = \boldsymbol{x}^2\,. \end{aligned}$$

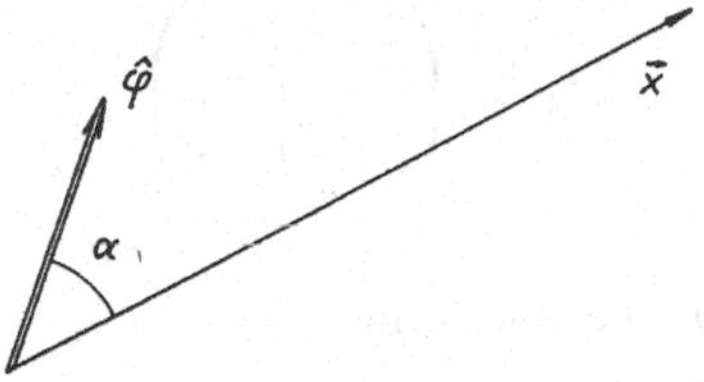

Fig. 2.10. Definition of the angle α between the position vector x and the direction about which the rotation takes place

Here α denotes the angle between the vectors $\hat{\varphi}$ and x (see Fig. 2.10). If $\hat{\varphi}$ and $\hat{\varphi}'$ are parallel, then $\mathsf{R}(\hat{\varphi}')\mathsf{R}(\hat{\varphi}) = \mathsf{R}(\hat{\varphi}+\hat{\varphi}')$. This means that $\mathsf{R}(\hat{\varphi})$ can be deformed continuously into the identity $\mathsf{R}(\mathbf{0}) = \mathbb{1}$ and therefore that $\det \mathsf{R}(\varphi) = +1$.

(ii) Every R of SO(3) can be written in the form (2.67). Consider first those vectors x which remain unchanged under R (up to a factor), $\mathsf{R}x = \lambda x$. This means that

$$\det(\mathsf{R} - \lambda\mathbb{1}) = 0$$

must hold. This is a cubic polynomial with real coefficients. Therefore, it always has at least one real eigenvalue λ, which is $+1$ or -1 because of the condition $(\mathsf{R}x)^2 = x^2$. In the plane perpendicular to the corresponding eigenvector, R must have the form

$$\begin{pmatrix} \cos\Psi & \sin\Psi \\ -\sin\Psi & \cos\Psi \end{pmatrix}$$

in order to fulfill the condition (2.64). Finally, Ψ must be equal to φ because $\det \mathsf{R} = 1$ and because R can be deformed continuously into the identity. Thus, $\mathsf{R}(\varphi)$ must have the decomposition (2.67).

2.22 Infinitesimal Rotations and Their Generators

Assume now that $\varphi \equiv \varepsilon \ll 1$. Then, from (2.67) and (2.68), respectively,

$$x' = (\hat{\varphi}\cdot x)\hat{\varphi} - (\hat{\varphi}\times x)\varepsilon - \hat{\varphi}\times(\hat{\varphi}\times x) + \mathrm{O}(\varepsilon^2) = x - (\hat{\varphi}\times x)\varepsilon + \mathrm{O}(\varepsilon^2). \quad (2.69)$$

Writing this out in components, one obtains

$$\begin{aligned} x' &= x - \varepsilon \begin{pmatrix} 0 & -\hat{\varphi}_3 & \hat{\varphi}_2 \\ \hat{\varphi}_3 & 0 & -\hat{\varphi}_1 \\ -\hat{\varphi}_2 & \hat{\varphi}_1 & 0 \end{pmatrix} x + \mathrm{O}(\varepsilon^2) \\ &= x - \varepsilon \left[\begin{pmatrix} 0 & 0 & 0 \\ 0 & 0 & -1 \\ 0 & 1 & 0 \end{pmatrix} \hat{\varphi}_1 + \begin{pmatrix} 0 & 0 & 1 \\ 0 & 0 & 0 \\ -1 & 0 & 0 \end{pmatrix} \hat{\varphi}_2 \right. \\ &\quad \left. + \begin{pmatrix} 0 & -1 & 0 \\ 1 & 0 & 0 \\ 0 & 0 & 0 \end{pmatrix} \hat{\varphi}_3 \right] x + \mathrm{O}(\varepsilon^2)\,. \end{aligned} \quad (2.70)$$

This is the decomposition of the infinitesimal rotation into rotations about the three directions $\hat{\varphi}_1$, $\hat{\varphi}_2$, and $\hat{\varphi}_3$. Denoting the matrices in this expression by

$$\mathbf{J}_1 \stackrel{\text{def}}{=} \begin{pmatrix} 0 & 0 & 0 \\ 0 & 0 & -1 \\ 0 & 1 & 0 \end{pmatrix} , \quad \mathbf{J}_2 \stackrel{\text{def}}{=} \begin{pmatrix} 0 & 0 & 1 \\ 0 & 0 & 0 \\ -1 & 0 & 0 \end{pmatrix} , \quad \mathbf{J}_3 \stackrel{\text{def}}{=} \begin{pmatrix} 0 & -1 & 0 \\ 1 & 0 & 0 \\ 0 & 0 & 0 \end{pmatrix} , \tag{2.71}$$

and using the abbreviation $\mathbf{J} = (\mathbf{J}_1, \mathbf{J}_2, \mathbf{J}_3)$, (2.70) takes the form

$$\boldsymbol{x}' = [\mathbb{1} - \varepsilon\hat{\boldsymbol{\varphi}} \cdot \mathbf{J}]\boldsymbol{x} + O(\varepsilon^2) . \tag{2.72}$$

Following Sect. 2.20 choose $\varepsilon = \varphi/n$ and apply the same rotation n times. In the limit $n \to \infty$ one obtains

$$\boldsymbol{x}' = \lim_{n\to\infty} \left(\mathbb{1} - \frac{\varphi}{n}\hat{\boldsymbol{\varphi}} \cdot \mathbf{J}\right)^n \boldsymbol{x} = \exp(-\boldsymbol{\varphi} \cdot \mathbf{J})\boldsymbol{x} . \tag{2.73}$$

Thus, the *finite* rotation $\mathbf{R}(\boldsymbol{\varphi})$ is represented by an exponential series in the matrices $\mathbf{J} = (\mathbf{J}_1, \mathbf{J}_2, \mathbf{J}_3)$ and the vector $\boldsymbol{\varphi}$. $\mathbf{J}_k$ is said to be *the generator for infinitesimal rotations about the axis k*.

As before, the first equation of (2.73) can be visualized as n successive rotations by the angle φ/n. In the limit of n going to infinity this becomes an infinite product of infinitesimal rotations. By Gauss' formula this is precisely the exponential indicated in the second equation of (2.73). It is to be understood as the well-known exponential series $\sum_{n=0}^{\infty}(1/n!)\mathbf{A}^n$ in the 3×3 matrix $\mathbf{A} \stackrel{\text{def}}{=} (-\boldsymbol{\varphi} \cdot \mathbf{J})$.

The matrices $\mathbf{R}(\boldsymbol{\varphi})$ form a compact *Lie group* (its parameter space is compact). Its generators (2.71) obey the *Lie algebra* associated with this group. This means that the commutator (or *Lie product*),

$$[\mathbf{J}_i, \mathbf{J}_k] \stackrel{\text{def}}{=} \mathbf{J}_i\mathbf{J}_k - \mathbf{J}_k\mathbf{J}_i ,$$

of any two of them is defined and belongs to the same set $\{\mathbf{J}_k\}$. Indeed, from (2.71) one finds that

$$[\mathbf{J}_1, \mathbf{J}_2] = \mathbf{J}_3 , \quad [\mathbf{J}_1, \mathbf{J}_3] = -\mathbf{J}_2$$

together with four more relations that follow from these by cyclic permutation of the indices. As the Lie product of any two elements of $\{\mathbf{J}_1, \mathbf{J}_2, \mathbf{J}_3\}$ is again an element of this set, one says that the algebra of the $\mathbf{J}_k$ *closes* under the Lie product.

Via (2.72) and (2.73) the generators yield a *local* representation of that part of the rotation group which contains the unit element, the identical mapping. This is not sufficient to reconstruct the *global* structure of this group. We do not reach its component containing matrices with determinant -1. It can happen, therefore, that two groups have the same Lie algebra but are different globally. This is indeed the case for SO(3), which has the same Lie algebra as SU(2), the group of complex 2×2 matrices, which are unitary and have determinant $+1$, *the unitary unimodular group in two complex dimensions*. The elements of the rotation group are differentiable in its parameters (the rotation angles). In this sense it is a differentiable manifold and one may ask questions such as: Is this manifold compact? (The rotation group is.) Is it simply connected? (The rotation group is doubly connected, see Sect. 5.2.3 (iv) and Exercise 3.11.)

2.23 Canonical Transformations

Of course, the choice of a set of generalized coordinates and of the corresponding generalized, canonically conjugate momenta is not unique. For example, Proposition 2.11 taught us that any diffeomorphic mapping of the original coordinates q onto new coordinates Q leaves invariant Lagrange's formalism. The new set describes the same physics by means of a different parametrization. Such transformations are useful, however, whenever one succeeds in making some or all of the new coordinates cyclic. In this case the corresponding generalized momenta are constants of the motion. Following Sect. 2.12, we say that a coordinate Q_k is cyclic if L does not depend explicitly on it,

$$\frac{\partial L}{\partial Q_k} = 0 . \tag{2.74}$$

If this is the case, then also $\partial H/\partial Q_k = 0$, and

$$\dot{P}_k = -\frac{\partial H}{\partial Q_k} = 0 , \tag{2.75}$$

from which we conclude that $P_k = \alpha_k = \text{const}$. The canonical system described by

$$H(Q_1, \ldots, Q_{k-1}, Q_{k+1}, \ldots, Q_f; P_1, \ldots, P_{k-1}, \alpha_k, P_{k+1}, \ldots P_f, t)$$

is reduced to a system with $f-1$ degrees of freedom. For instance, if all Q_k are cyclic, i.e. if

$$H = H(P_1, \ldots, P_f; t) ,$$

the solution of the canonical equations is elementary, because

$$\dot{P}_i = 0 \rightarrow P_i = \alpha_i = \text{const} , \quad i = 1, \ldots, f, \quad \text{and}$$

$$\dot{Q}_i = \left.\frac{\partial H}{\partial P_i}\right|_{P_i=\alpha_i} \stackrel{\text{def}}{=} v_i(t) ,$$

from which the solutions are obtained by integration, viz.

$$Q_i = \int_{t_0}^{t} v_i(t)\mathrm{d}t + \beta_i , \quad i = 1, \ldots, f .$$

The $2f$ parameters $\{\alpha_i, \beta_i\}$ are constants of integration.

This raises a general question: Is it possible to transform the coordinates and momenta in such a way that the canonical structure of the equations of motion is preserved and that some or all coordinates become cyclic? This question leads to the definition of *canonical transformations*.

Diffeomorphic transformations of the variables $\underset{\sim}{q}$ and $\underset{\sim}{p}$ and of the Hamiltonian function $H(\underset{\sim}{q}, \underset{\sim}{p}, t)$, generated by a smooth function of old and new variables in the way described below

$$\{\underset{\sim}{q}, \underset{\sim}{p}\} \to \{\underset{\sim}{Q}, \underset{\sim}{P}\} ,$$

$$H(\underset{\sim}{q}, \underset{\sim}{p}, t) \to \tilde{H}(\underset{\sim}{Q}, \underset{\sim}{P}, t) , \tag{2.76}$$

are said to be canonical if they preserve the structure of the canonical equations (2.45)[5]. Thus, with (2.45) we shall also find that

$$\dot{Q}_i = \frac{\partial \tilde{H}}{\partial P_i} , \quad \dot{P}_i = -\frac{\partial \tilde{H}}{\partial Q_i} . \tag{2.77}$$

In order to satisfy this requirement the variational principle (2.53) of Sect. 2.17 must hold for the system $\{\underset{\sim}{q}, \underset{\sim}{p}, H\}$ as well as for the system $\{\underset{\sim}{Q}, \underset{\sim}{P}, \tilde{H}\}$, viz.

$$\delta \int_{t_1}^{t_2} \left[\sum_1^f p_i \dot{q}_i - H(\underset{\sim}{q}, \underset{\sim}{p}, t) \right] \mathrm{d}t = 0 , \tag{2.78}$$

$$\delta \int_{t_1}^{t_2} \left[\sum_1^f P_i \dot{Q}_i - \tilde{H}(\underset{\sim}{Q}, \underset{\sim}{P}, t) \right] \mathrm{d}t = 0 . \tag{2.79}$$

Proposition 2.10 tells us that this is certainly true if the integrands in (2.78) and (2.79) do not differ by more than the total time derivative of a function M:

$$\sum_{i=1}^f p_i \dot{q}_i - H(\underset{\sim}{q}, \underset{\sim}{p}, t) = \sum_{j=1}^f P_j \dot{Q}_j - \tilde{H}(\underset{\sim}{Q}, \underset{\sim}{P}, t) + \frac{\mathrm{d}}{\mathrm{d}t} M , \tag{2.80}$$

where M depends on old and new variables (but not on their time derivatives) and, possibly, time. There are four ways of choosing M, corresponding to the possible choices of old coordinates/momenta and new coordinates/momenta. These four classes can be obtained from one another by Legendre transformation. They are as follows.

(A) The choice

$$M(\underset{\sim}{q}, \underset{\sim}{Q}, t) \equiv \Phi(\underset{\sim}{q}, \underset{\sim}{Q}, t) . \tag{2.81}$$

In this case we obtain

$$\frac{\mathrm{d}M}{\mathrm{d}t} \equiv \frac{\mathrm{d}\Phi(\underset{\sim}{q}, \underset{\sim}{Q}, t)}{\mathrm{d}t} = \frac{\partial \Phi}{\partial t} + \sum_{j=1}^f \left[\frac{\partial \Phi}{\partial q_j} \dot{q}_j + \frac{\partial \Phi}{\partial Q_j} \dot{Q}_j \right] . \tag{2.82}$$

[5] See the precise definition in Sect. 5.5.4 below

As $\underset{\sim}{q}$ and $\underset{\sim}{Q}$ are independent variables, (2.80) is fulfilled if and only if the following equations hold true:

$$p_i = \frac{\partial \Phi}{\partial q_i}, \quad P_j = -\frac{\partial \Phi}{\partial Q_j}, \quad \tilde{H} = H + \frac{\partial \Phi}{\partial t}. \tag{2.83}$$

The function Φ (and likewise any other function M) is said to be the *generating function of the canonical transformation*. The first equation of (2.83) can be solved for $Q_k(\underset{\sim}{q}, \underset{\sim}{p}, t)$ if

$$\det\left(\frac{\partial^2 \Phi}{\partial q_i \partial Q_j}\right) \neq 0, \tag{2.84a}$$

and the second can be solved for $Q_k(\underset{\sim}{q}, \underset{\sim}{P}, t)$ if

$$\det\left(\frac{\partial^2 \Phi}{\partial Q_i \partial Q_j}\right) \neq 0. \tag{2.84b}$$

(B) The choice

$$M(\underset{\sim}{q}, \underset{\sim}{P}, t) = S(\underset{\sim}{q}, \underset{\sim}{P}, t) - \sum_{k=1}^{f} Q_k(\underset{\sim}{q}, \underset{\sim}{P}, t) P_k. \tag{2.85}$$

This is obtained by taking the Legendre transform of the generating function (2.81) with respect to $\underset{\sim}{Q}$:

$$(\mathcal{L}\Phi)(\underset{\sim}{Q}) = \sum Q_k \frac{\partial \Phi}{\partial Q_k} - \Phi(\underset{\sim}{q}, \underset{\sim}{Q}, t) = -\left[\sum Q_k P_k + \Phi\right].$$

We then have

$$S(\underset{\sim}{q}, \underset{\sim}{P}, t) \stackrel{\text{def}}{=} \sum_{k=1}^{f} Q_k(\underset{\sim}{q}, \underset{\sim}{P}, t) P_k + \Phi(\underset{\sim}{q}, \underset{\sim}{Q}(\underset{\sim}{q}, \underset{\sim}{P}, t), t). \tag{2.86}$$

With the condition (2.84b) Q_k can be solved for $\underset{\sim}{q}$ and $\underset{\sim}{P}$. From (2.83) and (2.86) we conclude that

$$p_i = \frac{\partial S}{\partial q_i}, \quad Q_k = \frac{\partial S}{\partial P_k}, \quad \tilde{H} = H + \frac{\partial S}{\partial t}. \tag{2.87}$$

The same equations are obtained if the generating function (2.85) is inserted into (2.80), taking into account that $\underset{\sim}{q}$ and $\underset{\sim}{P}$ are independent.

(C) The choice

$$M(\underset{\sim}{Q}, \underset{\sim}{p}, t) = U(\underset{\sim}{Q}, \underset{\sim}{p}, t) + \sum_{k=1}^{f} q_k(\underset{\sim}{Q}, \underset{\sim}{p}, t) p_k. \tag{2.88}$$

For this we take the Legendre transform of Φ with respect to $\underset{\sim}{q}$:

$$(\mathcal{L}\Phi)(\underset{\sim}{q}) = \sum q_i \frac{\partial \Phi}{\partial q_i} - \Phi(\underset{\sim}{q}, \underset{\sim}{Q}, t)$$
$$= \sum q_i p_i - \Phi(\underset{\sim}{q}, \underset{\sim}{Q}, t) .$$

We then have

$$U(\underset{\sim}{Q}, \underset{\sim}{p}, t) \stackrel{\text{def}}{=} -\sum_{k=1}^{f} q_k(\underset{\sim}{Q}, \underset{\sim}{p}, t) p_k + \Phi(\underset{\sim}{q}(\underset{\sim}{Q}, \underset{\sim}{p}, t), \underset{\sim}{Q}, t) \tag{2.89}$$

and obtain the equations

$$q_k = -\frac{\partial U}{\partial p_k}, \quad P_k = -\frac{\partial U}{\partial Q_k}, \quad \tilde{H} = H + \frac{\partial U}{\partial t} . \tag{2.90}$$

(D) The choice

$$M(\underset{\sim}{P}, \underset{\sim}{p}, t) = V(\underset{\sim}{P}, \underset{\sim}{p}, t) - \sum_{k=1}^{f} Q_k(\underset{\sim}{q}(\underset{\sim}{P}, \underset{\sim}{p}, t), \underset{\sim}{P}, t) P_k + \sum_{k=1}^{f} q_k(\underset{\sim}{P}, \underset{\sim}{p}, t) p_k .$$

This fourth possibility is obtained from S, for instance, by taking its Legendre transform with respect to $\underset{\sim}{q}$:

$$(\mathcal{L}S)(\underset{\sim}{q}) = \sum q_i \frac{\partial S}{\partial q_i} - S(\underset{\sim}{q}, \underset{\sim}{P}, t) = \sum q_i p_i - S ,$$

so that

$$V(\underset{\sim}{P}, \underset{\sim}{p}, t) \stackrel{\text{def}}{=} -\sum_{k=1}^{f} q_k(\underset{\sim}{P}, \underset{\sim}{p}, t) p_k + S(\underset{\sim}{q}(\underset{\sim}{P}, \underset{\sim}{p}, t), \underset{\sim}{P}, t) .$$

In this case one obtains the equations

$$q_k = -\frac{\partial V}{\partial p_k}, \quad Q_k = \frac{\partial V}{\partial P_k}, \quad \tilde{H} = H + \frac{\partial V}{\partial t} . \tag{2.91}$$

This classification of generating functions for canonical transformations may at first seem rather complicated. When written in this form, the general structure of canonical transformations is not transparent. In reality, the four types (A–D) are closely related and can be treated in a unified way. This is easy to understand if one realizes that generalized coordinates are in no way distinguished over generalized momenta and, in particular, that coordinates can be transformed into momenta and vice versa. In Sects. 2.25 and 2.27 below we shall introduce a unified formulation that clarifies these matters. Before doing this we consider two examples.

2.24 Examples of Canonical Transformations

Example (i) Class B is distinguished from the others by the fact that it contains the identical mapping. In order to see this let

$$S(\underset{\sim}{q}, \underset{\sim}{P}) = \sum_{i=1}^{f} q_i P_i \; . \tag{2.92}$$

We confirm, indeed, from (2.87) that

$$p_i = \frac{\partial S}{\partial q_i} = P_i \; ; \quad Q_j = \frac{\partial S}{\partial P_j} = q_j \; ; \quad \tilde{H} = H \; .$$

Class A, on the other hand, contains that transformation which interchanges the role of coordinates and momenta. Indeed, taking

$$\Phi(\underset{\sim}{q}, \underset{\sim}{Q}) = \sum_{k=1}^{f} q_k Q_k \tag{2.93}$$

we find that (2.83) gives $p_i = Q_i$, $P_k = -q_k$, $\tilde{H}(\underset{\sim}{Q}, \underset{\sim}{P}) = H(-\underset{\sim}{P}, \underset{\sim}{Q})$.

Example (ii) For the harmonic oscillator there is a simple canonical transformation that makes Q cyclic. Start from

$$H(q, p) = \frac{p^2}{2m} + \frac{1}{2} m\omega^2 q^2 \quad (f = 1) \tag{2.94}$$

and apply the canonical transformation generated by

$$\Phi(q, Q) = \tfrac{1}{2} m\omega q^2 \cot Q \; . \tag{2.95}$$

In this case the equations (2.83) are

$$p = \frac{\partial \Phi}{\partial q} = m\omega q \cot Q \; ,$$

$$P = -\frac{\partial \Phi}{\partial Q} = \frac{1}{2} \frac{m\omega q^2}{\sin^2 Q} \; ,$$

or, by solving for q and p,

$$q = \sqrt{\frac{2P}{m\omega}} \sin Q \; , \quad p = \sqrt{2m\omega P} \cos Q \; ,$$

and, finally, $\tilde{H} = \omega P$. Thus, Q is cyclic, and we have

$$\dot{P} = -\frac{\partial \tilde{H}}{\partial Q} = 0 \rightarrow P = \alpha = \text{const} ,$$
$$\dot{Q} = \frac{\partial \tilde{H}}{\partial P} = \omega \rightarrow Q = \omega t + \beta . \tag{2.96}$$

When translated back to the original coordinates this gives the familiar solution

$$q(t) = \sqrt{\frac{2\alpha}{m\omega}} \sin(\omega t + \beta) .$$

As expected, the general solution depends on two integration constants whose interpretation is obvious: α determines the amplitude (it is assumed to be positive) and β the phase of the oscillation.

Whenever the new momentum P is a constant and the new coordinate Q a linear function of time, P is said to be an *action variable*, Q an *angle variable*. We return to action–angle variables below.

2.25 The Structure of the Canonical Equations

First, we consider a system with one degree of freedom, $f = 1$. We assume that it is described by a Hamiltonian function $H(\underset{\sim}{q}, \underset{\sim}{p}, t)$. As in Sect. 1.16 we set

$$\underset{\sim}{x} \stackrel{\text{def}}{=} \begin{pmatrix} q \\ p \end{pmatrix} , \quad \text{or} \quad \underset{\sim}{x} = \begin{pmatrix} x_1 \\ x_2 \end{pmatrix} \quad \text{with} \quad x_1 \equiv q ,\; x_2 \equiv p , \tag{2.97a}$$

as well as[6]

$$H_{,x} \stackrel{\text{def}}{=} \begin{pmatrix} \dfrac{\partial H}{\partial x_1} \\ \dfrac{\partial H}{\partial x_2} \end{pmatrix} \equiv \begin{pmatrix} \dfrac{\partial H}{\partial q} \\ \dfrac{\partial H}{\partial p} \end{pmatrix} \quad \text{and} \quad \mathbf{J} \stackrel{\text{def}}{=} \begin{pmatrix} 0 & 1 \\ -1 & 0 \end{pmatrix} . \tag{2.97b}$$

The canonical equations then take the form

$$-\mathbf{J}\dot{\underset{\sim}{x}} = H_{,x} \tag{2.98}$$

or

$$\boxed{\dot{\underset{\sim}{x}} = \mathbf{J} H_{,x} .} \tag{2.99}$$

The second equation follows from the observation that $\mathbf{J}^{-1} = -\mathbf{J}$. Indeed,

$$\mathbf{J}^2 = -\mathbb{1} \quad \text{and} \quad \mathbf{J}^T = \mathbf{J}^{-1} = -\mathbf{J} . \tag{2.100}$$

[6] The derivative of H by x_k is written as $H_{,x_k}$. More generally, the set of all derivatives of H by $\underset{\sim}{x}$ is abbreviated by $H_{,x}$.

The solutions of (2.99) have the form

$$\underset{\sim}{x}(t, s, \underset{\sim}{y}) = \Phi_{t,s}(\underset{\sim}{y}) \quad \text{with} \quad \Phi_{s,s}(\underset{\sim}{y}) = \underset{\sim}{y} \,, \tag{2.101}$$

where s and $\underset{\sim}{y}$ are the initial time and initial configuration, respectively.

For an arbitrary number f of degrees of freedom we have in a similar way (see also Sect. 1.18)

$$\underset{\sim}{x} \stackrel{\text{def}}{=} \begin{pmatrix} q_1 \\ q_2 \\ \vdots \\ q_f \\ p_1 \\ p_2 \\ \vdots \\ p_f \end{pmatrix}, \quad H_{,x} \stackrel{\text{def}}{=} \begin{pmatrix} \partial H/\partial q_1 \\ \vdots \\ \partial H/\partial q_f \\ \partial H/\partial p_1 \\ \vdots \\ \partial H/\partial p_f \end{pmatrix}, \quad \mathbf{J} \stackrel{\text{def}}{=} \begin{pmatrix} 0_{f\times f} & \mathbb{1}_{f\times f} \\ -\mathbb{1}_{f\times f} & 0_{f\times f} \end{pmatrix}. \tag{2.102}$$

The canonical equations have again the form (2.98) or (2.99) with

$$\mathbf{J} = \begin{pmatrix} 0 & \mathbb{1} \\ -\mathbb{1} & 0 \end{pmatrix}, \tag{2.103}$$

where $\mathbb{1}$ denotes the $f \times f$ unit matrix. Clearly, $\mathbf{J}$ has the same properties (2.100) as for $f = 1$.

2.26 Example: Linear Autonomous Systems in One Dimension

Before proceeding further we consider a simple example: the class of linear, autonomous systems with one degree of freedom. *Linear* means that $\dot{\underset{\sim}{x}} = \mathbf{A}\underset{\sim}{x}$, where $\mathbf{A}$ is a 2×2 matrix. Equation (2.98) now reads

$$-\mathbf{J}\dot{\underset{\sim}{x}} = -\mathbf{J}\mathbf{A}\underset{\sim}{x} = H_{,x} \,. \tag{2.104}$$

This means in turn that H must have the general form

$$H = \tfrac{1}{2}[aq^2 + 2bqp + cp^2] \equiv \tfrac{1}{2}[ax_1^2 + 2bx_1x_2 + cx_2^2] \,. \tag{2.105}$$

Thus

$$\dot{\underset{\sim}{x}} = \mathbf{A}\underset{\sim}{x} = \mathbf{J}H_{,x} = \begin{pmatrix} 0 & 1 \\ -1 & 0 \end{pmatrix} \begin{pmatrix} \partial H/\partial x_1 \\ \partial H/\partial x_2 \end{pmatrix} = \begin{pmatrix} bx_1 + cx_2 \\ -ax_1 - bx_2 \end{pmatrix} \tag{2.105'}$$

and the matrix $\mathbf{A}$ is given by

$$\mathbf{A}=\begin{pmatrix} b & c \\ -a & -b \end{pmatrix} .$$

Note that its trace vanishes, $\mathrm{Tr}\mathbf{A}=0$. It is not difficult to solve (2.105′) directly in matrix form, viz.

$$\underset{\sim}{x}=\exp[(t-s)\mathbf{A}]\underset{\sim}{y}\equiv\Phi_{t-s}(\underset{\sim}{y}) . \tag{2.106}$$

The exponential is calculated by its series expansion. The square of $\mathbf{A}$ is proportional to the unit matrix,

$$\mathbf{A}^2=\begin{pmatrix} b & c \\ -a & -b \end{pmatrix}\begin{pmatrix} b & c \\ -a & -b \end{pmatrix}=(b^2-ac)\begin{pmatrix} 1 & 0 \\ 0 & 1 \end{pmatrix}\equiv-\Delta\mathbb{1} .$$

Therefore, all even powers of $\mathbf{A}$ are multiples of the unit matrix; all odd powers are multiples of $\mathbf{A}$:

$$\mathbf{A}^{2n}=(-1)^n\Delta^n\mathbb{1} , \quad \mathbf{A}^{2n+1}=(-1)^n\Delta^n\mathbf{A} .$$

$\Delta \stackrel{\text{def}}{=} ac-b^2$ is the determinant of $\mathbf{A}$. For the sake of illustration we assume that Δ is positive. We then have (see also Sect. 2.20)

$$\exp\{(t-s)\mathbf{A}\}=\mathbb{1}\cos\left(\sqrt{\Delta}(t-s)\right)+\mathbf{A}\frac{1}{\sqrt{\Delta}}\sin\left(\sqrt{\Delta}(t-s)\right) . \tag{2.107}$$

Thus, the solution (2.106) is obtained as follows, setting $\omega \stackrel{\text{def}}{=} \sqrt{\Delta}=\sqrt{ac-b^2}$:

$$\begin{aligned}\underset{\sim}{x}&=\Phi_{t-s}(\underset{\sim}{y})\equiv\mathbf{P}(t-s)\underset{\sim}{y}\\ &=\begin{pmatrix} \cos\omega(t-s)+\dfrac{b}{\omega}\sin\omega(t-s) & \dfrac{c}{\omega}\sin\omega(t-s) \\ -\dfrac{a}{\omega}\sin\omega(t-s) & \cos\omega(t-s)-\dfrac{b}{\omega}\sin\omega(t-s) \end{pmatrix}\underset{\sim}{y} .\end{aligned}$$

It describes harmonic oscillations. (If, instead, we choose $\Delta<0$, it describes exponential behavior.) The solution is a linear function of the initial configuration, $x^i=\sum_{k=1}^{2}P_{ik}(t-s)y^k$ from which we obtain $\mathrm{d}x^i=\sum_{k=1}^{2}P_{ik}\mathrm{d}y^k$. The volume element $\mathrm{d}x^1\mathrm{d}x^2$ in phase space is invariant if $\det(\partial x^i/\partial y^k)=\det(P_{ik})=1$. This is indeed the case:

$$\det(P_{ik})=\cos^2\omega(t-s)-\left[\frac{b^2}{\omega^2}-\frac{ac}{\omega^2}\right]\sin^2\omega(t-s)=1 .$$

(Recall the remark at the end of Sect. 1.21.1.) This “conservation of phase volume” is sketched in Fig. 2.11 for the case $a=c=1$, $b=0$, i.e. the harmonic oscillator. In fact, this is nothing but the content of Liouville's theorem to which we return below, in a more general context (Sects. 2.29 and 2.30).

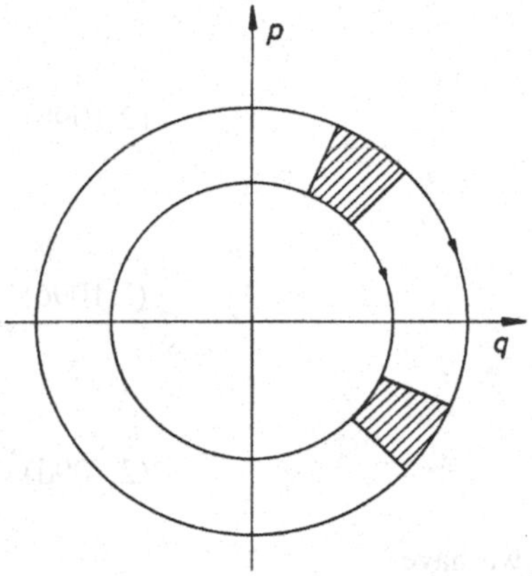

Fig. 2.11. Phase portraits for the harmonic oscillator (units as in Sect. 1.17.1). The hatched area wanders about the origin with constant angular velocity and without changing its shape

2.27 Canonical Transformations in Compact Notation

The $2f$-dimensional phase space of a Hamiltonian system carries an interesting geometrical structure which is encoded in the canonical equations

$$\begin{pmatrix} \dot{\underset{\sim}{q}} \\ \dot{\underset{\sim}{p}} \end{pmatrix} = \begin{pmatrix} \partial H/\partial \underset{\sim}{p} \\ -\partial H/\partial \underset{\sim}{q} \end{pmatrix}$$

and in the canonical transformations that leave these equations form invariant. This structure becomes apparent, for the first time, if the canonical transformations (A–D) and the conditions on their derivatives are formulated in the compact notation of Sect. 2.25.

In Sect. 2.23 (2.84a) we saw that the condition

$$\det\left(\frac{\partial^2 \Phi}{\partial q_i \partial Q_k}\right) \neq 0 \tag{2.108a}$$

had to be imposed on canonical transformations of class A. Only with this condition could the equation $p_i = \partial\Phi/\partial q_i$ be solved for $Q_k(\underset{\sim}{q}, \underset{\sim}{p}, t)$. Similarly, in the other three cases we had the requirements

$$\det\left(\frac{\partial^2 S}{\partial q_i \partial P_k}\right) \neq 0\,, \quad \det\left(\frac{\partial^2 U}{\partial Q_k \partial p_i}\right) \neq 0\,, \quad \det\left(\frac{\partial^2 V}{\partial P_k \partial p_i}\right) \neq 0\,. \tag{2.108b}$$

From (2.83) one reads off the conditions

$$\frac{\partial p_i}{\partial Q_k} = \frac{\partial^2 \Phi}{\partial Q_k \partial q_i} = -\frac{\partial P_k}{\partial q_i}\,. \tag{2.109a}$$

Similarly, from (2.87),

$$\frac{\partial p_i}{\partial P_k} = \frac{\partial^2 S}{\partial P_k \partial q_i} = \frac{\partial Q_k}{\partial q_i} \, . \tag{2.109b}$$

From (2.90) of class C,

$$\frac{\partial q_i}{\partial Q_k} = -\frac{\partial^2 U}{\partial Q_k \partial p_i} = \frac{\partial P_k}{\partial p_i} \, . \tag{2.109c}$$

and from (2.91) of class D,

$$\frac{\partial q_i}{\partial P_k} = -\frac{\partial^2 V}{\partial P_k \partial p_i} = -\frac{\partial Q_k}{\partial p_i} \, . \tag{2.109d}$$

Returning to the compact notation of Sect. 2.25 we have

$$\underset{\sim}{x} \stackrel{\text{def}}{=} \{q_1 \dots q_f;\, p_1 \dots p_f\} \quad \text{and} \quad \underset{\sim}{y} \stackrel{\text{def}}{=} \{Q_1 \dots Q_f;\, P_1 \dots P_f\} \, . \tag{2.110}$$

Equations (2.109a–d) all contain derivatives of the form

$$\frac{\partial x_\alpha}{\partial y_\beta} \stackrel{\text{def}}{=} M_{\alpha\beta} \, , \quad \frac{\partial y_\alpha}{\partial x_\beta} = (\mathbf{M}^{-1})_{\alpha\beta} \, , \quad \alpha, \beta = 1, \dots 2f \, . \tag{2.111}$$

Clearly, one has

$$\sum_{\gamma=1}^{2f} \frac{\partial x_\alpha}{\partial y_\gamma} \frac{\partial y_\gamma}{\partial x_\beta} = \sum_{\gamma=1}^{2f} M_{\alpha\gamma} (\mathbf{M}^{-1})_{\gamma\beta} = \delta_{\alpha\beta} \, .$$

We now show that the diversity of (2.109) can be summarized as follows:

$$M_{\alpha\beta} = \sum_{\mu=1}^{2f} \sum_{\nu=1}^{2f} J_{\alpha\mu} J_{\beta\nu} (\mathbf{M}^{-1})_{\nu\mu} \, . \tag{2.112}$$

Taking account of the relation $\mathbf{J}^{-1} = -\mathbf{J}$ this equation is written alternatively as

$$-\mathbf{JM} = (\mathbf{JM}^{-1})^{\mathrm{T}} \, .$$

We prove it by calculating its two sides separately. The left-hand side is

$$-\mathbf{JM} = -\begin{pmatrix} 0 & \mathbb{1} \\ -\mathbb{1} & 0 \end{pmatrix} \begin{pmatrix} \partial q/\partial Q & \partial q/\partial P \\ \partial p/\partial Q & \partial p/\partial P \end{pmatrix} = \begin{pmatrix} -\partial p/\partial Q & -\partial p/\partial P \\ \partial q/\partial Q & \partial q/\partial P \end{pmatrix} ,$$

while the right-hand side is

$$(\mathbf{JM}^{-1})^{\mathrm{T}} = \begin{pmatrix} \partial P/\partial q & -\partial Q/\partial q \\ \partial P/\partial p & -\partial Q/\partial p \end{pmatrix} .$$

Equations (2.109) tell us that these two matrices are in fact equal. Thus, (2.112) is proved. This equation is rewritten as follows. Writing out the transposition on the right-hand side, we obtain $\mathbf{JM} = -(\mathbf{M}^{-1})^{\mathrm{T}}\mathbf{J}^{\mathrm{T}} = +(\mathbf{M}^{-1})^{\mathrm{T}}\mathbf{J}$. Multiplying this equation with $\mathbf{M}^{\mathrm{T}}$ from the left, we see that $\mathbf{M}$ obeys the equation

$$\boxed{\mathbf{M}^{\mathrm{T}}\mathbf{JM} = \mathbf{J}} \tag{2.113}$$

no matter which type of canonical transformation is being studied.

What is the significance of this equation? According to (2.111) and (2.109a–d), $\mathbf{M}$ is the matrix of second derivatives of generating functions for canonical transformations. The matrices $\mathbf{M}$ obeying (2.113) form a group that imprints a characteristic symmetry on phase space. The matrix $\mathbf{J}$, on the other hand, turns out to be invariant under canonical transformations. For this reason, it plays the role of a metric in phase space. These statements are proved and analyzed in the next section.

2.28 On the Symplectic Structure of Phase Space

The set of all matrices that obey (2.113) form a group, the *real symplectic group* $\mathrm{Sp}_{2f}(\mathbb{R})$ over the space $\mathbb{R}^{2f}$. This is a group that is defined over a space with *even* dimension and that is characterized by a skew-symmetric, invariant bilinear form. As a first step we shall verify that the $\mathbf{M}$ indeed form a group G.

(1) There exists an operation that defines the composition of any two elements $\mathbf{M}_1$ and $\mathbf{M}_2$, $\mathbf{M}_3 = \mathbf{M}_1\mathbf{M}_2$. Obviously, this is matrix multiplication here. $\mathbf{M}_3$ is again an element of G, as one verifies by direct calculation:

$$\mathbf{M}_3^{\mathrm{T}}\mathbf{JM}_3 = (\mathbf{M}_1\mathbf{M}_2)^{\mathrm{T}}\mathbf{J}(\mathbf{M}_1\mathbf{M}_2) = \mathbf{M}_2^{\mathrm{T}}(\mathbf{M}_1^{\mathrm{T}}\mathbf{JM}_1)\mathbf{M}_2 = \mathbf{J}\,.$$

(2) This operation is associative because matrix multiplication has this property.
(3) There is a unit element in G: $\mathbf{E} = \mathbb{1}$. Indeed, $\mathbb{1}^{\mathrm{T}}\mathbf{J}\mathbb{1} = \mathbf{J}$.
(4) For every $\mathbf{M} \in$ G there is an inverse given by

$$\mathbf{M}^{-1} = \mathbf{J}^{-1}\mathbf{M}^{\mathrm{T}}\mathbf{J}\,.$$

This is verified as follows:
(a) Equation (2.113) implies that $(\det \mathbf{M})^2 = 1$, i.e. $\mathbf{M}$ is not singular and has an inverse.
(b) $\mathbf{J}$ also belongs to G since $\mathbf{J}^{\mathrm{T}}\mathbf{JJ} = \mathbf{J}^{-1}\mathbf{JJ} = \mathbf{J}$.
(c) One now confirms that $\mathbf{M}^{-1}\mathbf{M} = \mathbb{1}$:

$$\mathbf{M}^{-1}\mathbf{M} = (\mathbf{J}^{-1}\mathbf{M}^{\mathrm{T}}\mathbf{J})\mathbf{M} = \mathbf{J}^{-1}(\mathbf{M}^{\mathrm{T}}\mathbf{JM}) = \mathbf{J}^{-1}\mathbf{J} = \mathbb{1}$$

and, finally, that $\mathbf{M}^{\mathrm{T}}$ also belongs to G:

$$(\mathbf{M}^{\mathrm{T}})^{\mathrm{T}}\mathbf{JM}^{\mathrm{T}} = (\mathbf{MJ})\mathbf{M}^{\mathrm{T}} = (\mathbf{MJ})(\mathbf{JM}^{-1}\mathbf{J}^{-1}) = (\mathbf{MJ})(\mathbf{J}^{-1}\mathbf{M}^{-1}\mathbf{J}) = \mathbf{J}\,.$$

(In the second step we have taken $\mathbf{M}^{\mathrm{T}}$ from (2.113); in the third step we have used $\mathbf{J}^{-1} = -\mathbf{J}$ twice.)

Thus, we have proved that the matrices $\mathbf{M}$ that fulfill (2.113) form a group. The underlying space is the phase space $\mathbb{R}^{2f}$.

There is a skew-symmetric bilinear form on this space that is invariant under transformations pertaining to $\mathrm{G} = \mathrm{Sp}_{2f}$ and that can be understood as a generalized scalar product of vectors over $\mathbb{R}^{2f}$. For two arbitrary vectors $\underset{\sim}{x}$ and $\underset{\sim}{y}$ we define

$$[\underset{\sim}{x}, \underset{\sim}{y}] \stackrel{\mathrm{def}}{=} \underset{\sim}{x}^{\mathrm{T}} \mathbf{J} \underset{\sim}{y} = \sum_{i,k=1}^{2f} x_i J_{ik} y_k \,. \tag{2.114}$$

One can easily verify that this form is invariant. Let $\mathbf{M} \in \mathrm{Sp}_{2f}$ and let $\underset{\sim}{x}' = \mathbf{M}\underset{\sim}{x}$, $\underset{\sim}{y}' = \mathbf{M}\underset{\sim}{y}$. Then

$$[\underset{\sim}{x}', \underset{\sim}{y}'] = [\mathbf{M}\underset{\sim}{x}, \mathbf{M}\underset{\sim}{y}] = \underset{\sim}{x}^{\mathrm{T}} \mathbf{M}^{\mathrm{T}} \mathbf{J} \mathbf{M} \underset{\sim}{y} = \underset{\sim}{x}^{\mathrm{T}} \mathbf{J} \underset{\sim}{y} = [\underset{\sim}{x}, \underset{\sim}{y}] \,.$$

The bilinear form has the following properties, which are read off (2.114).

(i) It is skew-symmetric:

$$[\underset{\sim}{y}, \underset{\sim}{x}] = -[\underset{\sim}{x}, \underset{\sim}{y}] \,. \tag{2.115a}$$

Proof.

$$[\underset{\sim}{y}, \underset{\sim}{x}] = (\underset{\sim}{x}^{\mathrm{T}} \mathbf{J}^{\mathrm{T}} \underset{\sim}{y})^{\mathrm{T}} = -(\underset{\sim}{x}^{\mathrm{T}} \mathbf{J} \underset{\sim}{y})^{\mathrm{T}} = -\underset{\sim}{x}^{\mathrm{T}} \mathbf{J} \underset{\sim}{y} = -[\underset{\sim}{x}, \underset{\sim}{y}] \,. \tag{2.115b}$$

□

(ii) It is linear in both arguments. For instance,

$$[\underset{\sim}{x}, \lambda_1 \underset{\sim}{y}_1 + \lambda_2 \underset{\sim}{y}_2] = \lambda_1 [\underset{\sim}{x}, \underset{\sim}{y}_1] + \lambda_2 [\underset{\sim}{x}, \underset{\sim}{y}_2] \,. \tag{2.115c}$$

If $[\underset{\sim}{x}, \underset{\sim}{y}] = 0$ for all $\underset{\sim}{y} \in \mathbb{R}^{2f}$, then $\underset{\sim}{x} \equiv 0$. This means that the form (2.114) is not degenerate. Thus, it has all properties that one expects for a scalar product.

The symplectic group Sp_{2f} is the symmetry group of $\mathbb{R}^{2f}$, together with the structure $[\underset{\sim}{x}, \underset{\sim}{y}]$ (2.114), in the same way as $\mathrm{O}(2f)$ is the symmetry group of the same space with the structure of the ordinary scalar product $(\underset{\sim}{x}, \underset{\sim}{y}) = \sum_{k=1}^{2f} x_k y_k$. Note, however, that the symplectic structure (as a nondegenerate form) is only defined for *even* dimension $n = 2f$, while the Euclidean structure $(\underset{\sim}{x}, \underset{\sim}{y})$ is defined for both even and odd dimension and is nondegenerate in either case.

Consider now $2f$ vectors over $\mathbb{R}^{2f}$, $\underset{\sim}{x}^{(1)}, \ldots, \underset{\sim}{x}^{(2f)}$ that are assumed to be linearly independent. Then take the oriented volume of the parallelepiped spanned by these vectors:

$$[\underset{\sim}{x}^{(1)}, \underset{\sim}{x}^{(2)}, \ldots, \underset{\sim}{x}^{(2f)}] \stackrel{\mathrm{def}}{=} \det \begin{pmatrix} x_1^{(1)} & \ldots & x_1^{(2f)} \\ \vdots & & \vdots \\ x_{2f}^{(1)} & \ldots & x_{2f}^{(2f)} \end{pmatrix} . \tag{2.116}$$

Lemma. If $\pi(1), \pi(2), \ldots, \pi(2f)$ denotes the permutation π of the indices $1, 2, \ldots, 2f$ and $\sigma(\pi)$ its signature (i.e. $\sigma = +1$ if it is even, $\sigma = -1$ if it is odd), then

$$[\underset{\sim}{x}^{(1)}, \ldots, \underset{\sim}{x}^{(2f)}] = \frac{(-1)^{[f/2]}}{f!2^f} \times \sum_{\pi} \sigma(\pi)[\underset{\sim}{x}^{\pi(1)}, \underset{\sim}{x}^{\pi(2)}][\underset{\sim}{x}^{\pi(3)}, \underset{\sim}{x}^{\pi(4)}] \ldots$$
$$\ldots [\underset{\sim}{x}^{\pi(2f-1)}, \underset{\sim}{x}^{\pi(2f)}] \,. \tag{2.117}$$

Proof. When written out explicitly, the right-hand side reads

$$\left(\frac{(-1)^{[f/2]}}{f!2^f} \sum_{n_1 \ldots n_{2f}} J_{n_1 n_2} J_{n_3 n_4} \ldots J_{n_{2f-1} n_{2f}} \right) \left(\sum_{\pi} \sigma(\pi) \underset{\sim}{x}_{n_1}^{\pi(1)} \ldots \underset{\sim}{x}_{n_{2f}}^{\pi(2f)} \right) .$$

The second factor of this expression is precisely the determinant (2.116) if $\{n_1, \ldots, n_{2f}\}$ is an even permutation of $\{1, 2, \ldots, 2f\}$: it is minus that determinant if it is an odd permutation. Denoting this permutation by π' and its signature by $\sigma(\pi')$, this last expression is equal to

$$[\underset{\sim}{x}^{(1)}, \ldots, \underset{\sim}{x}^{(2f)}] \left(\frac{(-1)^{[f/2]}}{f!2^f} \sum_{\pi'} J_{\pi'(1)\pi'(2)} \ldots J_{\pi'(2f-1)\pi'(2f)} \sigma(\pi') \right) .$$

We now show that the factor in brackets equals 1, thus proving the lemma. This goes as follows. We know that $J_{i,i+f} = +1$, $J_{j+f,j} = -1$, while all other elements vanish. In calculating $\sum_{\pi} \sigma(\pi) J_{\pi(1)\pi(2)} \ldots J_{\pi(2f-1)\pi(2f)}$ we have the following possibilities,

(a) $J_{i_1,i_1+f} J_{i_2,i_2+f} \ldots J_{i_f,i_f+f}$ with $1 \le i_k \le 2f$, all i_k being different from each other. There are $f!$ such products and they all have the value $+1$ because all of them are obtained from $J_{1,f} \ldots J_{f,2f}$ by exchanging the indices pairwise. The signature $\sigma(\pi)$ is the same for all of them; call it $\sigma(a)$.
(b) Exchange now *one* pair of indices, i.e. $J_{i_1,i_1+f} \ldots J_{i_l+f,i_l} \ldots J_{i_f,i_f+f}$. There are $f \times f!$ products of this type and they all have the value -1. They all have the same signature $\sigma(\pi) \equiv \sigma(b)$ and $\sigma(b) = -\sigma(a)$.
(c) Exchange *two* pairs of indices to obtain $J_{i_1,i_1+f} \ldots J_{i_l+f,i_l} \ldots J_{i_k+f,i_k} \ldots J_{i_f,i_f+f}$. There are $[f(f-1)/2] \times f!$ products of this class and their value is again $+1$. Their signature is $\sigma(c) = \sigma(a)$; etc.

Thus, with the signature factor included, all terms contribute with the same sign and we obtain

$$f![1 + f + f(f-1)/2 + \ldots + 1] = f!2^f \,.$$

It remains to determine $\sigma(a)$. In $J_{1,f+1} J_{2,f+2} \ldots J_{f,2f}$ (which is $+1$) the order of the indices $(1, f+1, 2, f+2, \ldots, f, 2f)$ is obtained from $(1, 2, \ldots f, f+$

$1, \ldots, 2f)$ by $(f-1)+(f-2)+\ldots+1 = f(f-1)/2$ exchanges of neighbors. Thus $\sigma(a) = (-)^{f(f-1)/2}$. As one easily convinces oneself this is the same as $(-1)^{[f/2]}$. □

The lemma serves to prove the following proposition.

Proposition. Every $\mathbf{M}$ pertaining to Sp_{2f} has determinant $+1$:

$$\text{if} \quad \mathbf{M} \in \mathrm{Sp}_{2f} \quad \text{then} \quad \det \mathbf{M} = +1 \; .$$

Proof. By the product formula for determinants we have

$$[\mathbf{M}\underset{\sim}{x}^{(1)}, \ldots \mathbf{M}\underset{\sim}{x}^{(2f)}] = (\det \mathbf{M})[\underset{\sim}{x}^{(1)}, \ldots \underset{\sim}{x}^{(2f)}] \; .$$

As the vectors $\underset{\sim}{x}^{(1)} \ldots \underset{\sim}{x}^{(2f)}$ are linearly independent, their determinant does not vanish. Now, from the lemma (2.117),

$$[\mathbf{M}\underset{\sim}{x}^{(1)}, \ldots \mathbf{M}\underset{\sim}{x}^{(2f)}] = [\underset{\sim}{x}^{(1)}, \ldots, \underset{\sim}{x}^{(2f)}] \; .$$

We conclude that $\det \mathbf{M} = +1$. □

Remarks: In this section we have been talking about vectors on phase space $\mathbb{P}$ while until now $\underset{\sim}{x}$ etc. were points of $\mathbb{P}$. This was justified because we assumed the phase space to be $\mathbb{R}^{2f}$ for which every tangent space can be identified with its base space. If $\mathbb{P}$ is not flat any more, but is a differentiable manifold, our description holds in local coordinate systems (also called charts). This is worked out in more detail in Chap. 5

2.29 Liouville's Theorem

As in Sect. 1.19 we denote the *solutions* of Hamilton's equations by

$$\Phi_{t,s}(x) = (\varphi^1_{t,s}(\underset{\sim}{x}), \ldots, \varphi^{2f}_{t,s}(\underset{\sim}{x})) \; . \tag{2.118}$$

Also as before we call $\Phi_{t,s}(\underset{\sim}{x})$ the *flow* in phase space. Indeed, if $\underset{\sim}{x}$ denotes the initial configuration the system assumes at the initial time s, (2.118) describes how the system flows across phase space and goes over to the configuration $\underset{\sim}{y}$ assumed at time t. The temporal evolution of a canonical system can be visualized as the flow of an incompressible fluid: the flow conserves volume and orientation. Given a set of initial configurations, which, at time s fill a certain oriented domain U_s of phase space, this same ensemble will be found to lie in an oriented phase-space domain U_t, at time t (later or earlier than s), in such a way that U_s and U_t have the same volume $V_s = V_t$ and their orientation is the same. This is the content of Liouville's theorem.

In order to work out its significance we formulate and prove this theorem in two, equivalent ways. The first formulation consists in showing that the matrix

(2.119) of partial derivatives is symplectic. This matrix is precisely the Jacobian of the transformation $\mathrm{d}\underset{\sim}{x} \mapsto \mathrm{d}\underset{\sim}{y} = (\mathbf{D}\underset{\sim}{\Phi})\mathrm{d}\underset{\sim}{x}$. As it is symplectic, it has determinant $+1$. In the second formulation (which is equivalent to the first) we show that the flow has divergence zero, which means that there is no net flow out of U_s nor into U_s.

2.29.1 The Local Form

The matrix of partial derivatives of $\underset{\sim}{\Phi}$ being abbreviated by

$$\mathbf{D}\underset{\sim}{\Phi}_{t,s}(x) \stackrel{\text{def}}{=} \left(\frac{\partial \underset{\sim}{\Phi}^i_{t,s}(\underset{\sim}{x})}{\partial x_k}\right) \tag{2.119}$$

the theorem reads as follows.

> **Liouville's theorem**. Let $\underset{\sim}{\Phi}_{t,s}(\underset{\sim}{x})$ be the flow of the differential equation $-\mathbf{J}\dot{\underset{\sim}{x}} = H_{,x}$. For all $\underset{\sim}{x}, t$ and s for which the flow is defined,
>
> $$\mathbf{D}\underset{\sim}{\Phi}_{t,s}(\underset{\sim}{x}) \in \mathrm{Sp}_{2f}\,. \tag{2.120}$$
>
> The matrix of partial derivatives is symplectic and therefore has determinant $+1$.

Before we proceed to prove this theorem we wish to interpret the consequences of (2.120). The flux $\underset{\sim}{\Phi}_{t,s}(\underset{\sim}{x})$ is a mapping that maps the point $\underset{\sim}{x}$ (assumed by the system at time s) onto the point $\underset{\sim}{x}_t = \underset{\sim}{\Phi}_{t,s}(\underset{\sim}{x})$ (assumed at time t). Suppose we consider neighboring initial conditions filling the volume element $\mathrm{d}x_1 \ldots \mathrm{d}x_{2f}$. The statement (2.120) then tells us that this volume as well as its orientation is conserved under the flow. Indeed, the matrix (2.119) is nothing but the Jacobian of this mapping.

Proof. We have $-\mathbf{J}[\partial\underset{\sim}{\Phi}_{t,s}(\underset{\sim}{x})/\partial t] = H_{,x}(t) \circ \underset{\sim}{\Phi}_{t,s}$. Taking the differential of the equation $-\mathbf{J}\dot{\underset{\sim}{x}} = H_{,x}$ by $\underset{\sim}{x}$, and using the chain rule, we obtain $-\mathbf{J}[\partial\mathbf{D}\underset{\sim}{\Phi}_{t,s}(\underset{\sim}{x})/\partial t] = (\mathbf{D}H_{,x})(\underset{\sim}{\Phi}, t)\mathbf{D}\underset{\sim}{\Phi}_{t,s}(\underset{\sim}{x})$ and finally

$$\frac{\partial}{\partial t}[(\mathbf{D}\underset{\sim}{\Phi}_{t,s}(x))^{\mathrm{T}}\mathbf{J}(\mathbf{D}\underset{\sim}{\Phi}_{t,s}(x))] = -(\mathbf{D}\underset{\sim}{\Phi}_{t,s})^{\mathrm{T}}[\mathbf{D}H_{,x} - (\mathbf{D}H_{,x})^{\mathrm{T}}](\mathbf{D}\underset{\sim}{\Phi}_{t,s}) = 0\,. \tag{2.121}$$

This expression is zero because $\mathbf{D}H_{,x} = (\partial^2 H/\partial x_k \partial x_j)$ is symmetric. Equation (2.120) is obvious for $t = s$. It then follows from (2.121) that it holds for all t. Thus, the theorem is proved. □

The following converse of Liouville's theorem also holds. Let $\underset{\sim}{\Phi}_{t,s}$ be the flow of the differential equation $-\mathbf{J}\dot{\underset{\sim}{x}} = F(\underset{\sim}{x}, t)$ and assume that it fulfills (2.120). Then there exists locally a Hamiltonian function $H(\underset{\sim}{x}, t)$ such that $H_{,x} = F(\underset{\sim}{x}, t)$. This is seen as follows. The equation analogous to (2.121) now says that $\mathbf{D}F - (\mathbf{D}F)^{\mathrm{T}} = 0$, or that the curl of F vanishes: curl $F = 0$. If this is so, F can be written locally as a gradient: $F = H_{,x}$.

2.29.2 The Global Form

The statement of Liouville's theorem can be made more transparent by the example of a set of initial conditions that fill a finite oriented domain U_s whose volume is V_s. At time s we have

$$V_s = \int_{U_s} \mathrm{d}\underset{\sim}{x} \ ,$$

the integral being taken over the domain U_s of phase space. At another time t we have

$$V_t = \int_{U_t} \mathrm{d}\underset{\sim}{y} = \int_{U_s} \mathrm{d}\underset{\sim}{x} \det\left(\frac{\partial \underset{\sim}{y}}{\partial \underset{\sim}{x}}\right) = \int_{U_s} \mathrm{d}\underset{\sim}{x} \det(\mathbf{D}\underset{\sim}{\Phi}_{t,s}) \ ,$$

because in transforming an oriented multiple integral to new variables, the volume element is multiplied with the determinant of the corresponding Jacobi matrix. If we take t in the neighborhood of s we can expand in terms of $(t-s)$:

$$\underset{\sim}{\Phi}_{t,s}(\underset{\sim}{x}) = \underset{\sim}{x} + \underset{\sim}{F}(\underset{\sim}{x},t)\cdot(t-s) + \mathrm{O}((t-s)^2) \ , \quad \text{where}$$

$$\underset{\sim}{F}(\underset{\sim}{x},t) = \mathbf{J}H_{,x} = \left(\frac{\partial H}{\partial \underset{\sim}{p}}, -\frac{\partial H}{\partial \underset{\sim}{q}}\right) .$$

From the definition (2.119) the derivative by $\underset{\sim}{x}$ is

$$\mathbf{D}\underset{\sim}{\Phi}_{t,s}(\underset{\sim}{x}) = \mathbb{1} + \mathbf{D}\underset{\sim}{F}(\underset{\sim}{x},t)\cdot(t-s) + \mathrm{O}((t-s)^2) \ ,$$

or, when written out explicitly,

$$\frac{\partial \Phi^i_{t,s}(\underset{\sim}{x})}{\partial x^k} = \delta_{ik} + \frac{\partial F^i}{\partial x^k}(t-s) + \mathrm{O}((t-s)^2) \ .$$

In taking the determinant, one makes use of the following formula:

$$\det(\mathbb{1} + \mathbf{A}\varepsilon) \equiv \det(\delta_{ik} + A_{ik}\varepsilon) = 1 + \varepsilon \mathrm{Tr}\mathbf{A} + \mathrm{O}(\varepsilon^2) \ ,$$

where $\mathrm{Tr}\mathbf{A} = \sum_i A_{ii}$ denotes the trace of $\mathbf{A}$ and where ε is to be identified with $(t-s)$. We obtain

$$\det(\mathbf{D}\underset{\sim}{\Phi}_{t,s}(\underset{\sim}{x})) = 1 + (t-s)\sum_{i=1}^{2f} \frac{\partial F^i}{\partial x^i} + \mathrm{O}((t-s)^2) \ .$$

The trace $\sum_{i=1}^{2f} \partial F^i / \partial x^i$ is a divergence in the $2f$-dimensional phase space. It is easy to see that it vanishes if $\underset{\sim}{F} = \mathbf{J} H_{,x}$, viz.

$$\operatorname{div} \underset{\sim}{F} \stackrel{\text{def}}{=} \sum_{i=1}^{2f} \frac{\partial F^i}{\partial x^i} = \frac{\partial}{\partial \underset{\sim}{q}} \left(\frac{\partial H}{\partial \underset{\sim}{p}} \right) + \frac{\partial}{\partial \underset{\sim}{p}} \left(-\frac{\partial H}{\partial \underset{\sim}{q}} \right) = 0 \, .$$

This shows that $V_s = V_t$. As long as the flow is defined, the domain U_s of initial conditions can change its position and its shape but not its volume or its orientation.

2.30 Examples for the Use of Liouville's Theorem

Example (i) A particularly simple example is provided by the linear, autonomous system with $f = 1$ that we studied in Sect. 2.26. Here the action of the flow is simply multiplication of the initial configuration $\underset{\sim}{x}$ by the matrix $\mathbf{P}(t-s)$ whose determinant is $+1$. In the special case of the harmonic oscillator, for instance, all phase points move on circles around the origin, with constant and universal angular velocity. A given domain U_s moves around the origin unchanged, like the hand of a clock. This is sketched in Figs. 2.11 and 2.12.

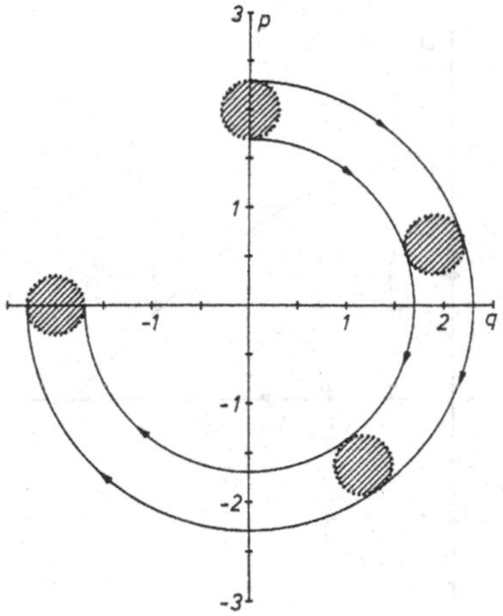

Fig. 2.12. The harmonic oscillator. A circular domain of initial configurations wanders uniformly about the origin. In the units introduced in Sect. 1.17.1 the period is $\tau^0 = 2\pi$. The four positions shown here correspond to the times $\tau = 0,\ 0.2\tau^0,\ 0.4\tau^0$, and $0.75\tau^0$

Example (ii) The example of the mathematical pendulum is somewhat less trivial. We note its equations of motion in the dimensionless form of Sect. 1.17.2,

$$\frac{\mathrm{d}z_1}{\mathrm{d}\tau} = z_2(\tau) \, , \quad \frac{\mathrm{d}z_2}{\mathrm{d}\tau} = -\sin z_1(\tau) \, ,$$

where τ is the dimensionless time variable $\tau = \omega t$, while the reduced energy was defined to be

$$\varepsilon = E/mgl = \tfrac{1}{2}z_2^2 + (1 - \cos z_1) .$$

The quantity ε is constant along every phase portrait (the solution of the canonical equations).

Figure 1.10 in Sect. 1.17.2 shows the phase portraits in phase space (z_1, z_2) (z_1 is the same as q, z_2 is the same as p). For example, a disklike domain of initial configurations U_s behaves under the flow as indicated in Figs. 2.13–15. In dimensionless units, the period of the harmonic oscillator is $\tau^{(0)} = \omega T^{(0)} = 2\pi$. The figures show three positions of the domain U_t into which U_s has moved at the times $k \cdot \tau^{(0)}$ indicated in the captions. As the motion is periodic, one should think of these figures being glued on a cylinder of circumference 2π, such that the lines (π, p) and $(-\pi, p)$ coincide. The deformation of the initial domain is clearly visible. It is particularly noticeable whenever one of the phase points moves along the separatrix (cf. Sect. 1.23 (1.59)). This happens for $\varepsilon = 2$, i.e. for the initial condition ($q = 0, p = 2$), for example. For large, positive time such a point wanders slowly towards the point ($q = \pi, p = 0$). Neighboring points with $\varepsilon > 2$ "swing through", while those with $\varepsilon < 2$ "oscillate", i.e. turn around the origin several times. The figures show very clearly that despite these deformations the original volume and orientation are preserved.

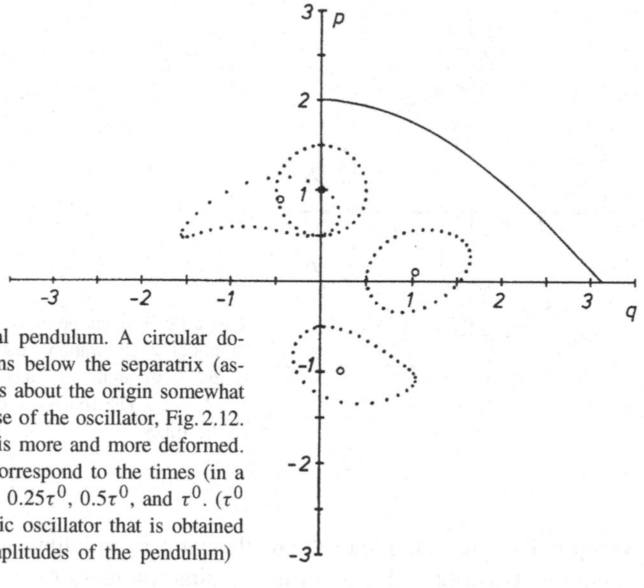

Fig. 2.13. The mathematical pendulum. A circular domain of initial configurations below the separatrix (assumed at time $\tau = 0$) moves about the origin somewhat more slowly than for the case of the oscillator, Fig. 2.12. As the domain proceeds, it is more and more deformed. The positions shown here correspond to the times (in a clockwise direction): $\tau = 0$, $0.25\tau^0$, $0.5\tau^0$, and τ^0. (τ^0 is the period of the harmonic oscillator that is obtained approximately for small amplitudes of the pendulum)

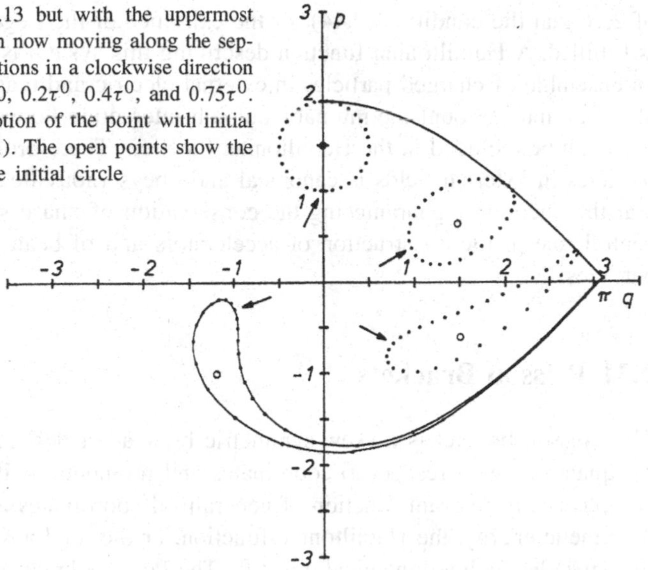

Fig. 2.14. Same as Fig. 2.13 but with the uppermost point of the circular domain now moving along the separatrix. The successive positions in a clockwise direction shown are reached at $\tau = 0$, $0.2\tau^0$, $0.4\tau^0$, and $0.75\tau^0$. The arrows indicate the motion of the point with initial configuration ($q = 0$, $p = 1$). The open points show the motion of the center of the initial circle

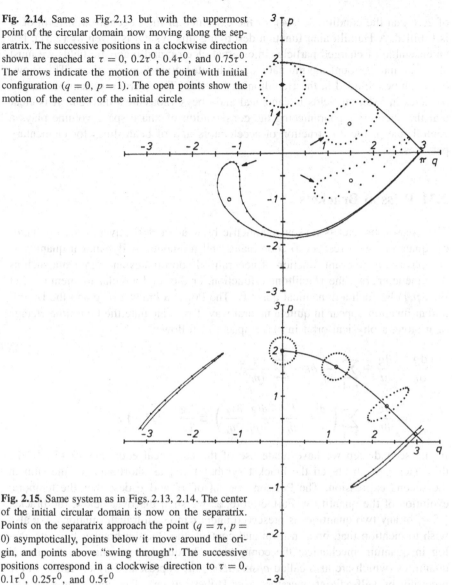

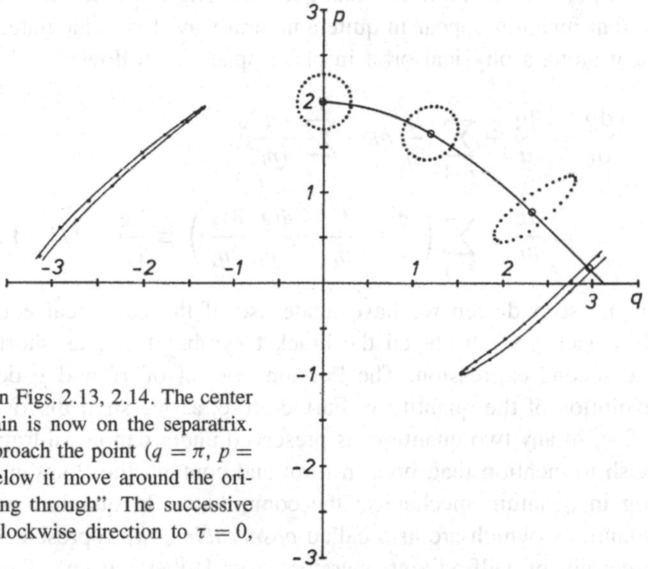

Fig. 2.15. Same system as in Figs. 2.13, 2.14. The center of the initial circular domain is now on the separatrix. Points on the separatrix approach the point ($q = \pi$, $p = 0$) asymptotically, points below it move around the origin, and points above "swing through". The successive positions correspond in a clockwise direction to $\tau = 0$, $0.1\tau^0$, $0.25\tau^0$, and $0.5\tau^0$

Example (iii) Charged particles in external electromagnetic fields obey the equation of motion (2.29):

$$m\ddot{\boldsymbol{r}} = \frac{e}{c}\dot{\boldsymbol{r}} \times \boldsymbol{B} + e\boldsymbol{E}\,.$$

As we saw in the Example (ii) of Sect. 2.8, this equation follows from a Lagrangian function such as the one given in (2.31). We showed in Example (ii) of 2.16 that the condition (2.44) for the existence of the Legendre transform of L is fulfilled. A Hamiltonian function describing this system is given by (2.51). For an ensemble of charged particles in external electric and magnetic fields we must also take into account the mutual Coulomb interaction between them. This, however, can be included in the Hamiltonian function. Therefore, a system of charged particles in external fields is canonical and obeys Liouville's theorem. It is clear that this theorem's guaranteeing the conservation of phase space volume plays a central role in the construction of accelerators and of beam lines for elementary particles.

2.31 Poisson Brackets

The Poisson bracket is a skew-symmetric bilinear of derivatives of two dynamical quantities with respect to coordinates and momenta. A dynamical quantity is any physically relevant function of generalized coordinates and momenta such as the kinetic energy, the Hamiltonian function, or the total angular momentum. Let $g(\underset{\sim}{q}, \underset{\sim}{p}, t)$ be such a dynamical quantity. The Poisson bracket of g and the Hamiltonian function appear in quite a natural way if we calculate the total time change of g along a physical orbit in phase space, as follows:

$$\begin{aligned}\frac{\mathrm{d}g}{\mathrm{d}t} &= \frac{\partial g}{\partial t} + \sum_{i=1}^{f} \frac{\partial g}{\partial q_i}\dot{q}_i + \sum_{i=1}^{f} \frac{\partial g}{\partial p_i}\dot{p}_i \\ &= \frac{\partial g}{\partial t} + \sum_{i=1}^{f} \left(\frac{\partial g}{\partial q_i}\frac{\partial H}{\partial p_i} - \frac{\partial g}{\partial p_i}\frac{\partial H}{\partial q_i} \right) \equiv \frac{\partial g}{\partial t} + \{H, g\} \,.\end{aligned}$$

In the second step we have made use of the canonical equations (2.45). In the third step we introduced the bracket symbol $\{\ ,\ \}$ as shorthand for the sum in the second expression. The Poisson bracket of H and g describes the temporal evolution of the quantity g. Furthermore, as we shall discover below, the bracket $\{f, g\}$ of any two quantities is preserved under canonical transformations. We also wish to mention that, both in form and content, the Poisson bracket finds its analog in quantum mechanics: the commutator. In quantum mechanics, dynamical quantities (which are also called *observables*) are represented by operators (more precisely by self-adjoint operators over Hilbert space). The commutator of two operators contains the information whether or not the corresponding observables can be measured simultaneously. Therefore, the Poisson bracket is not only an important notion of canonical mechanics but also reveals some of its underlying structure and hints at the relationship between classical mechanics and quantum mechanics.

Let $f(\underset{\sim}{x})$ and $g(\underset{\sim}{x})$ be two dynamical quantities, i.e. functions of coordinates and momenta, which are at least C^1. (They may also depend explicitly on time. As this is of no importance for what follows here, we suppress this possible dependence.) Their Poisson bracket $\{f, g\}$ is a scalar product of the type (2.114) and is defined as follows[7]:

$$\{f, g\}(\underset{\sim}{x}) \stackrel{\text{def}}{=} \sum_{i=1}^{f} \left(\frac{\partial f}{\partial p_i} \frac{\partial g}{\partial q_i} - \frac{\partial f}{\partial q_i} \frac{\partial g}{\partial p_i} \right) . \tag{2.122}$$

We also have

$$\begin{aligned} \{f, g\}(\underset{\sim}{x}) &= -[f_{,x}, g_{,x}](\underset{\sim}{x}) \\ &= -\left(\frac{\partial f}{\partial q_i} \cdots \frac{\partial f}{\partial q_f} \frac{\partial f}{\partial p_1} \cdots \frac{\partial f}{\partial p_f} \right) \begin{pmatrix} 0 & 1\!\!1 \\ -1\!\!1 & 0 \end{pmatrix} \begin{pmatrix} \frac{\partial g}{\partial q_1} \\ \vdots \\ \frac{\partial g}{\partial q_f} \\ \frac{\partial g}{\partial p_1} \\ \vdots \\ \frac{\partial g}{\partial p_f} \end{pmatrix} \end{aligned} \tag{2.123}$$

This latter form reveals an important property of the Poisson bracket: it is invariant under canonical transformations. Let Ψ be such a transformation:

$$\underset{\sim}{x} = (q_1, \ldots, q_f, p_1, \ldots, p_f) \longmapsto \underset{\sim}{\Psi}(\underset{\sim}{x}) = (Q_1(x), \ldots, Q_f(x), P_1(x), \ldots, P_f(x)).$$

From Sects. 2.27 and 2.28 we know that $D\underset{\sim}{\Psi}(\underset{\sim}{x}) \in \mathrm{Sp}_{2f}$. It is important to realize that Ψ maps the phase space onto itself, $\Psi : \mathbb{R}^{2f} \to \mathbb{R}^{2f}$, while f and g map the phase space $\mathbb{R}^{2f}$ onto the real numbers $\mathbb{R}$. f and g, in other words, are prescriptions how to form real functions of their arguments, taken from $\mathbb{R}^{2f}$, such as $f = q^2$, $g = (q^2 + p^2)/2$, etc. These prescriptions can be applied to the old variables $(\underset{\sim}{q}, \underset{\sim}{p})$ or, alternatively, to the new ones $(\underset{\sim}{Q}, \underset{\sim}{P})$. We then have the following

Proposition. For all f, g, and $\underset{\sim}{x}$

$$\{f \circ \Psi, g \circ \Psi\}(\underset{\sim}{x}) = \{f, g\} \circ \Psi(\underset{\sim}{x}) , \tag{2.124}$$

provided $\Psi(\underset{\sim}{x})$ is a canonical transformation. In words: if one transforms the quantities f and g to the new variables and then takes their Poisson bracket, the result is the same as that obtained in transforming their original Poisson bracket to the new variables.

7 We define the bracket such that it corresponds to the commutator $[f, g]$ of quantum mechanics, without change of sign.

Proof. Take the derivatives

$$\frac{\partial}{\partial x_i}(f \circ \Psi)(\underset{\sim}{x}) = \sum_{k=1}^{2f} \left. \frac{\partial f}{\partial y_k} \right|_{y=\Psi(x)} \cdot \frac{\partial \Psi_k}{\partial x_i}$$

or, in compact notation

$$(f \circ \Psi)_{,x} = (\mathbf{D}\Psi)^T(\underset{\sim}{x}) \cdot f_{,y}(\Psi(\underset{\sim}{x})) \, .$$

By assumption, Ψ is canonical, i.e. $\mathbf{D}\Psi$ and $(\mathbf{D}\Psi)^T$ are sympletic. Therefore

$$\begin{aligned}[(f \circ \Psi)_{,x}, (g \circ \Psi)_{,x}] &= [(\mathbf{D}\Psi)^T(\underset{\sim}{x}) f_{,y}(\Psi(\underset{\sim}{x})), (\mathbf{D}\Psi)^T(\underset{\sim}{x}) g_{,y}(\Psi(\underset{\sim}{x}))] \\ &= [f_{,y}(\Psi(\underset{\sim}{x})), g_{,y}(\Psi(\underset{\sim}{x}))] = -\{f, g\} \circ \Psi(\underset{\sim}{x}) \, . \quad \square\end{aligned}$$

The proposition has the following corollary.

Corollary. If (2.124) holds identically, or if the weaker condition

$$\{x_i \circ \Psi, x_k \circ \Psi\}(\underset{\sim}{x}) = \{x_i, x_k\} \circ \Psi(\underset{\sim}{x}) \tag{2.125}$$

holds true for all $\underset{\sim}{x}$ and i, k, then $\Psi(\underset{\sim}{x})$ is canonical.

Proof. From the definition (2.122) we have $\{x_i, x_k\} = -J_{ik}$. By assumption, this is invariant under the transformation Ψ, i.e.

$$\{y_m, y_n\}(\underset{\sim}{x}) = -[(\mathbf{D}\Psi)^T(\underset{\sim}{x})\hat{\mathbf{e}}_m \, , \quad (\mathbf{D}\Psi)^T(\underset{\sim}{x})\hat{\mathbf{e}}_n] = -[\hat{\mathbf{e}}_m, \hat{\mathbf{e}}_n] = -J_{mn} \, ,$$

where $\hat{\mathbf{e}}_m$ and $\hat{\mathbf{e}}_n$ are unit vectors in $\mathbb{R}^{2f}$. Thus,

$$(\mathbf{D}\Psi)\mathbf{J}(\mathbf{D}\Psi)^T = \mathbf{J} \, . \qquad \square$$

Note that (2.125), when written in terms of $\underset{\sim}{q}$, $\underset{\sim}{p}$, $\underset{\sim}{Q}$, and $\underset{\sim}{P}$, reads

$$\begin{aligned}\{Q_i, Q_j\}(\underset{\sim}{x}) &= \{q_i, q_j\}(\underset{\sim}{x}) = 0 \, , \\ \{P_i, P_j\}(\underset{\sim}{x}) &= \{p_i, p_j\}(\underset{\sim}{x}) = 0 \, , \\ \{P_i, Q_j\}(\underset{\sim}{x}) &= \{p_i, q_j\}(\underset{\sim}{x}) = \delta_{ij} \, .\end{aligned} \tag{2.126}$$

We close this section with the remark that canonical transformations can be characterized in four equivalent ways. The transformation $\Psi : (\underset{\sim}{q}, \underset{\sim}{p}) \longrightarrow (\underset{\sim}{Q}, \underset{\sim}{P})$ is canonical if

(a) it leaves unchanged the canonical equations (2.45), or
(b) it leaves invariant all Poisson brackets between dynamical quantities f and g, or
(c) it just leaves invariant the set of Poisson brackets (2.126), or
(d) the matrix of its derivatives is symplectic, $\mathbf{D}\Psi \in \mathrm{Sp}_{2f}$.

2.32 Properties of Poisson Brackets

It is possible to write the canonical equations (2.45) by means of Poisson brackets, in a more symmetric form. Indeed, as one may easily verify, they read

$$\dot{q}_k = \{H, q_k\} , \quad \dot{p}_k = \{H, p_k\} \tag{2.127}$$

or in the compact notation of Sect. 2.27,

$$\dot{x}_k = -[H_{,x}, x_{k,x}] = \sum_{i=1}^{2f} J_{ki} \frac{\partial H}{\partial x_i} .$$

Let $g(\underset{\sim}{q}, \underset{\sim}{p}, t)$ be a dynamical quantity, assumed to be at least C^1 in all its variables. As above, we calculate the total derivative of g with respect to time and make use of the canonical equations:

$$\begin{aligned} \frac{\mathrm{d}}{\mathrm{d}t} g(\underset{\sim}{q}, \underset{\sim}{p}, t) &= \frac{\partial g}{\partial t} + \sum_{k=1}^{f} \left(\frac{\partial g}{\partial q_k} \dot{q}_k + \frac{\partial g}{\partial p_k} \dot{p}_k \right) \\ &= \frac{\partial g}{\partial t} + \{H, g\} . \end{aligned} \tag{2.128}$$

This generalizes (2.127) to arbitrary dynamical quantities. If g is an integral of the motion, then

$$\frac{\partial g}{\partial t} + \{H, g\} = 0 , \tag{2.129a}$$

or, if g has no explicit time dependence,

$$\{H, g\} = 0 . \tag{2.129b}$$

Obviously, the Poisson bracket (2.122) has all the properties of the symplectic scalar product (2.115a, c). Besides these, it has the following properties. The bracket of $g(\underset{\sim}{q}, \underset{\sim}{p}, t)$ with q_k is equal to the derivative of g by p_k, while its bracket with p_k is minus its derivative by q_k:

$$\{g, q_k\} = \frac{\partial g}{\partial p_k} , \quad \{g, p_k\} = -\frac{\partial g}{\partial q_k} . \tag{2.130}$$

Furthermore, for any three quantities $u(\underset{\sim}{q}, \underset{\sim}{p}, t)$, $v(\underset{\sim}{q}, \underset{\sim}{p}, t)$, and $w(\underset{\sim}{q}, \underset{\sim}{p}, t)$ that are at least C^2 we can derive the following identity:

$$\text{the } \textit{Jacobi identity} \quad \{u, \{v, w\}\} + \{v, \{w, u\}\} + \{w, \{u, v\}\} = 0 . \tag{2.131}$$

This important identity can be verified either by direct calculation, using the definition (2.122), or by expressing the brackets via (2.123) in terms of the scalar product (2.114), as follows. For the sake of clarity we use the abbreviation $\partial u / \partial x^i \stackrel{\text{def}}{=} u_i$ for partial derivatives and, correspondingly, u_{ik} for the second derivatives $\partial^2 u/\partial x^i \partial x^k$. We then have

$$\{u, \{v, w\}\} = -[u_{,x}, \{v, w\}_{,x}] = +[u_{,x}, [v_{,x}, w_{,x}]_{,x}]$$
$$= \sum_{i=1}^{2f} \sum_{k=1}^{2f} \sum_{m=1}^{2f} \sum_{n=1}^{2f} u_i J_{ik} \partial/\partial x^k (v_m J_{mn} w_n) \,.$$

Thus, the left-hand side of (2.131) is given by

$$\{u, \{v, w\}\} + \{v, \{w, u\}\} + \{w, \{u, v\}\}$$
$$= \sum_{ikmn} u_i J_{ik} J_{mn} (v_{mk} w_n + v_m w_{nk}) + \sum_{ikmn} v_i J_{ik} J_{mn} (w_{mk} u_n + w_m u_{nk})$$
$$+ \sum_{ikmn} w_i J_{ik} J_{mn} (u_{mk} v_n + u_m v_{nk}) \,.$$

The six terms of this sum are pairwise equal and opposite. For example, take the last term on the right-hand side and make the following replacement in the indices: $m \to i$, $i \to n$, $k \to m$, and $n \to k$. As these are summation indices, the value of the term is unchanged. Thus, it becomes $\sum w_n J_{nm} J_{ik} u_i v_{km}$. As $v_{km} = v_{mk}$, but $J_{nm} = -J_{mn}$, it cancels the first term. In a similar fashion one sees that the second and third terms cancel, and similarly the fourth and the fifth.

The Jacobi identity (2.131) is used to demonstrate the following assertion.

Poisson's theorem. The Poisson bracket of two integrals of the motion is again an integral of the motion.

Proof. Let $\{u, v\} = w$. Then, from (2.128)

$$\frac{\mathrm{d}}{\mathrm{d}t} w = \frac{\partial}{\partial t} w + \{H, w\}$$

and by (2.131)

$$\frac{\mathrm{d}}{\mathrm{d}t} w = \frac{\mathrm{d}}{\mathrm{d}t} \{u, v\}$$
$$= \left\{\frac{\partial u}{\partial t}, v\right\} + \left\{u, \frac{\partial v}{\partial t}\right\} - \{u, \{v, H\}\} - \{v, \{H, u\}\}$$
$$= \left\{\frac{\partial u}{\partial t} + \{H, u\}, v\right\} + \left\{u, \frac{\partial v}{\partial t} + \{H, v\}\right\}$$
$$= \left\{\frac{\mathrm{d}u}{\mathrm{d}t}, v\right\} + \left\{u, \frac{\mathrm{d}v}{\mathrm{d}t}\right\} \,. \qquad (2.132)$$

Therefore, if $(\mathrm{d}u/\mathrm{d}t) = 0$ and $(\mathrm{d}v/\mathrm{d}t) = 0$, then also $(\mathrm{d}/\mathrm{d}t)\{u, v\} = 0$. □

Even if u, v are not conserved, (2.132) is an interesting result: the time derivative of the Poisson bracket obeys the product rule.

2.33 Infinitesimal Canonical Transformations

Those canonical transformations which can be deformed continuously into the identity form a particularly important class. In this case one can construct canonical transformations that differ from the identity only infinitesimally. This means that one can study the local action of a canonical transformation – in close analogy to the case of the rotation group we studied in Sect. 2.22.

We start from class B canonical transformations (2.85) (see Sect. 2.23) and from the identical mapping

$$S_{\mathrm{E}} = \sum_{k=1}^{f} q_k P_k : (\underset{\sim}{q}, \underset{\sim}{p}, H) \mapsto (\underset{\sim}{Q} = \underset{\sim}{q}, \underset{\sim}{P} = \underset{\sim}{p}, \tilde{H} = H) \tag{2.133}$$

of Example (i) in Sect. 2.24. Let ε be a parameter, taken to be infinitesimally small, and $\sigma(\underset{\sim}{q}, \underset{\sim}{P})$ a differentiable function of old coordinates and new momenta. (For the moment we only consider transformations without explicit time dependence.) We set

$$S(\underset{\sim}{q}, \underset{\sim}{P}, \varepsilon) = S_{\mathrm{E}} + \varepsilon\sigma(\underset{\sim}{q}, \underset{\sim}{P}) + \mathrm{O}(\varepsilon^2) . \tag{2.134}$$

The function

$$\sigma(\underset{\sim}{q}, \underset{\sim}{P}) = \left.\frac{\partial S}{\partial \varepsilon}\right|_{\varepsilon=0} \tag{2.135}$$

is said to be the *generating function of the infinitesimal transformation* (2.134). From (2.87) we obtain

$$Q_i = \frac{\partial S}{\partial P_i} = q_i + \varepsilon\frac{\partial \sigma}{\partial P_i} + \mathrm{O}(\varepsilon^2) , \tag{2.136a}$$

$$p_j = \frac{\partial S}{\partial q_j} = P_j + \varepsilon\frac{\partial \sigma}{\partial q_j} + \mathrm{O}(\varepsilon^2) . \tag{2.136b}$$

Here the derivatives $\partial\sigma/\partial P_i$, $\partial\sigma/\partial q_i$ depend on the old coordinates and on the new momenta. However, if we remain within the first order in the parameter ε, then, for consistency, all P_j must be replaced by p_j. (P_j differs from p_j by terms of order ε. If we kept it, we would in fact include some, but not all, terms of second order ε^2 in (2.136)). From (2.136) we then have

$$\delta q_i = Q_i - q_i = \frac{\partial \sigma(\underset{\sim}{q}, \underset{\sim}{p})}{\partial p_i}\varepsilon , \tag{2.137a}$$

$$\delta p_j = P_j - p_j = -\frac{\partial \sigma(\underset{\sim}{q}, \underset{\sim}{p})}{\partial q_j}\varepsilon . \tag{2.137b}$$

This can be written in a symmetric form, using (2.130),

$$\delta q_i = \{\sigma(\underset{\sim}{q}, \underset{\sim}{p}), q_i\}\varepsilon \ , \tag{2.138a}$$

$$\delta p_j = \{\sigma(\underset{\sim}{q}, \underset{\sim}{p}), p_j\}\varepsilon \ . \tag{2.138b}$$

The equations (2.138) have the following interpretation. The infinitesimal canonical transformation (2.134) shifts the generalized coordinate (momentum) proportionally to ε and to the Poisson bracket of the generating function (2.135) and that coordinate (or momentum, respectively). A case of special interest is the following. Let

$$S(\underset{\sim}{q}, \underset{\sim}{P}, \varepsilon = \mathrm{d}t) = S_E + H(\underset{\sim}{q}, \underset{\sim}{p})\mathrm{d}t \ . \tag{2.139}$$

With $\mathrm{d}t$ replacing the parameter ε, (2.137), or (2.138), are nothing but thc canonical equations (taking $\delta q_i \equiv \mathrm{d}q_i$, $\delta p_k \equiv \mathrm{d}p_k$):

$$\mathrm{d}q_i = \{H, q_i\}\mathrm{d}t \ , \quad \mathrm{d}p_j = \{H, p_j\}\mathrm{d}t \ . \tag{2.140}$$

Thus, the Hamiltonian function serves to "boost" the system: H is the generating function for the infinitesimal canonical transformation that corresponds to the actual motion $(\mathrm{d}q_i, \mathrm{d}p_j)$ of the system in the time interval $\mathrm{d}t$.

2.34 Integrals of the Motion

We may wish to ask how a given dynamical quantity $f(\underset{\sim}{q}, \underset{\sim}{p}, t)$ behaves under an infinitesimal transformation of the type (2.134). Formal calculation gives us the anwer:

$$\begin{aligned}\delta_\sigma f(\underset{\sim}{q}, \underset{\sim}{p}) &= \sum_{k=1}^{f}\left(\frac{\partial f}{\partial q_k}\delta q_k + \frac{\partial f}{\partial p_k}\delta p_k\right) \\ &= \sum_{k=1}^{f}\left(\frac{\partial f}{\partial q_k}\frac{\partial \sigma}{\partial p_k} - \frac{\partial f}{\partial p_k}\frac{\partial \sigma}{\partial q_k}\right)\varepsilon \\ &= \{\sigma, f\}\varepsilon \ .\end{aligned} \tag{2.141}$$

For example, choosing $\varepsilon = \mathrm{d}t$, $\sigma = H$, (2.141) yields with $\partial f/\partial t = 0$

$$\frac{\mathrm{d}f}{\mathrm{d}t} = \{H, f\} \ , \tag{2.142}$$

i.e. we recover (2.128) for the time change of f. In turn, we may ask how the Hamiltonian function H behaves under an infinitesimal canonical transformation generated by the function $f(\underset{\sim}{q}, \underset{\sim}{p})$. The answer is given in (2.141), viz.

$$\delta_f H = \{f, H\}\varepsilon \ . \tag{2.143}$$

In particular, the vanishing of the bracket $\{f, H\}$ means that H stays invariant under this transformation. If this is indeed the case, then, with $\{f, H\} = -\{H, f\}$ and (2.142), we conclude that f is an integral of the motion. To work out more clearly this reciprocity we write (2.142) in the notation

$$\delta_H f = \{H, f\}\mathrm{d}t \tag{2.142$'$}$$

and compare it with (2.143). One sees that $\delta_f H$ vanishes if and only if $\delta_H f$ vanishes. The infinitesimal canonical transformation generated by $f(\underset{\sim}{q}, \underset{\sim}{p})$ leaves the Hamiltonian function invariant if and only if f is constant along physical orbits.

We note the close analogy to the Noether's theorem for Lagrangian systems (Sect. 2.19). We wish to illustrate this by two examples.

Example (i) Consider an n-particle system described by

$$H = \sum_{i=1}^{n} \frac{\boldsymbol{p}_i^2}{2m_i} + U(\boldsymbol{r}_1, \ldots, \boldsymbol{r}_n) \tag{2.144}$$

that is invariant under *translations in the direction* $\hat{\boldsymbol{a}}$. Thus the canonical transformation

$$S(\boldsymbol{r}_1, \ldots, \boldsymbol{r}_n; \boldsymbol{p}'_1, \ldots, \boldsymbol{p}'_n) = \sum_{i=1}^{n} \boldsymbol{r}_i \cdot \boldsymbol{p}'_i + \boldsymbol{a} \cdot \sum_{i=1}^{n} \boldsymbol{p}'_i \tag{2.145}$$

leaves H invariant, $\boldsymbol{a}$ being a constant vector whose modulus is a and which points in the direction $\hat{\boldsymbol{a}}$. (The unprimed variables $\boldsymbol{r}_i$, $\boldsymbol{p}_i$ are to be identified with the old variables q_k, p_k, while the primed ones are to be identified with Q_k, P_k.) From (2.136) we have

$$Q_k = \frac{\partial S}{\partial P_k} ,$$

which is $\boldsymbol{r}'_i = \boldsymbol{r}_i + \boldsymbol{a}$ here, with $(k = 1, \ldots, f = 3n)$, $(i = 1, \ldots, n)$, and

$$p_k = \frac{\partial S}{\partial q_k} , \quad \text{which is} \quad \boldsymbol{p}_i = \boldsymbol{p}'_i .$$

In fact it is sufficient to choose the modulus of the translation vector $\boldsymbol{a}$ infinitesimally small. The infinitesimal translation is then generated by

$$\sigma = \left.\frac{\partial S}{\partial a}\right|_{a=0} = \hat{\boldsymbol{a}} \cdot \sum_{i=1}^{n} \boldsymbol{p}_i .$$

As H is invariant, we have $\{\sigma, H\} = 0$ and, from (2.142), $\mathrm{d}\sigma/\mathrm{d}t = 0$. We conclude that σ, the projection of total momentum onto the direction $\hat{a}$, is an integral of the motion.

Example (ii) Assume now that the same system (2.144) is invariant under arbitrary rotations of the coordinate system. If we consider an infinitesimal rotation characterized by $\boldsymbol{\varphi} = \varepsilon\hat{\boldsymbol{\varphi}}$, then, from (2.72),

$$\left.\begin{aligned} \boldsymbol{r}'_i &= [\mathbb{1} - (\boldsymbol{\varphi}\cdot\mathbf{J})]\boldsymbol{r}_i + \mathrm{O}(\varepsilon^2) \\ \boldsymbol{p}'_i &= [\mathbb{1} - (\boldsymbol{\varphi}\cdot\mathbf{J})]\boldsymbol{p}_i + \mathrm{O}(\varepsilon^2) \end{aligned}\right\} (i = 1, \dots, n) .$$

The generating function $S(\boldsymbol{r}_i, \boldsymbol{p}'_k)$ is given by

$$S = \sum_{i=1}^{n} \boldsymbol{r}_i \cdot \boldsymbol{p}'_i - \sum_{i=1}^{n} \boldsymbol{p}'_i \cdot (\boldsymbol{\varphi}\cdot\mathbf{J})\boldsymbol{r}_i . \tag{2.146}$$

The notation is as follows: $(\boldsymbol{\varphi}\cdot\mathbf{J})$ is a shorthand for the 3×3 matrix

$$(\boldsymbol{\varphi}\cdot\mathbf{J})_{ab} = \varepsilon[\hat{\varphi}_1(\mathbf{J}_1)_{ab} + \hat{\varphi}_2(\mathbf{J}_2)_{ab} + \hat{\varphi}_3(\mathbf{J}_3)_{ab}] .$$

The second term on the right-hand side of (2.146) contains the scalar product of the vectors $(\boldsymbol{\varphi}\cdot\mathbf{J})\boldsymbol{r}_i$ and $\boldsymbol{p}_i$. First we verify that the generating function (2.146) does indeed describe a rotation. We have

$$Q_k = \frac{\partial S}{\partial P_k} , \quad \text{and thus} \quad \boldsymbol{r}'_i = \boldsymbol{r}_i - (\boldsymbol{\varphi}\cdot\mathbf{J})\boldsymbol{r}_i ,$$

and

$$p_k = \frac{\partial S}{\partial q_k} , \quad \text{and thus} \quad \boldsymbol{p}_i = \boldsymbol{p}'_i - \boldsymbol{p}'_i(\boldsymbol{\varphi}\cdot\mathbf{J}) .$$

J_i being antisymmetric, the second equation becomes $\boldsymbol{p}_i = [\mathbb{1} + (\boldsymbol{\varphi}\cdot\mathbf{J})]\boldsymbol{p}'_i$. If this is multiplied with $[\mathbb{1} - (\boldsymbol{\varphi}\cdot\mathbf{J})]$ from the left we obtain the correct transformation rule $\boldsymbol{p}'_i = [\mathbb{1} - (\boldsymbol{\varphi}\cdot\mathbf{J})]\boldsymbol{p}_i$ up to the terms of second order in ε (which must be omitted, for the sake of consistency). Equation (2.146) now yields

$$\sigma = \left.\frac{\partial S}{\partial \varepsilon}\right|_{\varepsilon=0} = -\sum_{i=1}^{n} \boldsymbol{p}_i(\hat{\boldsymbol{\varphi}}\cdot\mathbf{J})\boldsymbol{r}_i .$$

From (2.69) and (2.72) this can be expressed as the cross product of $\hat{\boldsymbol{\varphi}}$ and $\boldsymbol{r}_i$,

$$(\hat{\boldsymbol{\varphi}}\cdot\mathbf{J})\boldsymbol{r}_i = \hat{\boldsymbol{\varphi}} \times \boldsymbol{r}_i ,$$

so that, finally

$$\boldsymbol{p}_i \cdot (\hat{\boldsymbol{\varphi}}\cdot\mathbf{J})\boldsymbol{r}_i = \boldsymbol{p}_i \cdot (\hat{\boldsymbol{\varphi}} \times \boldsymbol{r}_i) = \hat{\boldsymbol{\varphi}}(\boldsymbol{r}_i \times \boldsymbol{p}_i) = \hat{\boldsymbol{\varphi}} \cdot \boldsymbol{l}_i .$$

The integral of the motion is seen to be the projection of total angular momentum $\boldsymbol{l} = \sum_{i=1}^{n} \boldsymbol{l}^{(i)}$ onto the direction $\hat{\boldsymbol{\varphi}}$. As H was assumed to be invariant for *all* directions, we conclude that the whole vector $\boldsymbol{l} = (l_1, l_2, l_3)$ is conserved and that

$$\{H, l_a\} = 0 , \quad a = 1, 2, 3 . \tag{2.147}$$

2.35 The Hamilton–Jacobi Differential Equation

As we saw in Sect. 2.23, the solution of the equations of motion of a canonical system becomes elementary if we succeed in making all coordinates *cyclic* ones. A special situation where this is obviously the case is met when H, the transformed Hamiltonian function, is zero. The question then is whether one can find a time-dependent, canonical transformation by which H vanishes, viz.

$$\{\underset{\sim}{q}, \underset{\sim}{p}, H(\underset{\sim}{q}, \underset{\sim}{p}, t)\} \underset{S^*(\underset{\sim}{q}, \underset{\sim}{P}, t)}{\longrightarrow} \left\{\underset{\sim}{Q}, \underset{\sim}{P}, \tilde{H} = H + \frac{\partial S^*}{\partial t} = 0\right\} . \tag{2.148}$$

Let us denote this special class of generating functions by $S^*(\underset{\sim}{q}, \underset{\sim}{P}, t)$. For H to vanish we obtain the requirement

$$\boxed{\tilde{H} = H\left(q_i, p_k = \frac{\partial S^*}{\partial q_k}, t\right) + \frac{\partial S^*}{\partial t} = 0 .} \tag{2.149}$$

This equation is called the *differential equation of Hamilton and Jacobi*. It is a partial differential equation, of first order in time, for the unknown function $S^*(\underset{\sim}{q}, \underset{\sim}{\alpha}, t)$, where $\underset{\sim}{\alpha} = (\alpha_1, \dots, \alpha_f)$ are constants. Indeed, as $\tilde{H} = 0$, we have $\dot{P}_k = 0$, so that the new momenta are constants, $P_k = \alpha_k$. Therefore, S^* is a function of the $(f+1)$ *variables* $(q_1, \dots, q_f, t)$ and of the (constant) *parameters* $(\alpha_1, \dots, \alpha_f)$. $S^*(\underset{\sim}{q}, \underset{\sim}{\alpha}, t)$ is called the *action function*.

The new coordinates Q_k are also constants. They are given by

$$Q_k = \frac{\partial S^*(\underset{\sim}{q}, \underset{\sim}{\alpha}, t)}{\partial \alpha_k} = \beta_k . \tag{2.150}$$

Equation (2.150) can be solved for

$$q_k = q_k(\underset{\sim}{\alpha}, \underset{\sim}{\beta}, t) \tag{2.151}$$

precisely if

$$\det\left(\frac{\partial^2 S^*}{\partial \alpha_k \partial q_l}\right) \neq 0 . \tag{2.152}$$

If the function $S^*(\underset{\sim}{q}, \underset{\sim}{\alpha}, t)$ fulfills this condition it is said to be a complete solution of (2.149). In this sense, the partial differential equation (2.149) is equivalent to the system (2.45) of canonical equations. Equation (2.149) is an important topic in the theory of partial differential equations; its detailed discussion is beyond the scope of this book.

If the Hamiltonian function does not depend explicitly on time, $H(\underset{\sim}{q}, \underset{\sim}{p})$ is constant along solution curves and is equal to the energy E. It is then sufficient to study the function

$$S(\underset{\sim}{q},\underset{\sim}{\alpha}) \stackrel{\text{def}}{=} S^*(\underset{\sim}{q},\underset{\sim}{\alpha},t) - Et\,, \tag{2.153}$$

called the *reduced action*. It obeys a time-independent partial-differential equation that follows from (2.149), viz.

$$\boxed{H\left(q_i, \frac{\partial S}{\partial q_k}\right) = E\,.} \tag{2.154}$$

This is known as the *characteristic equation* of Hamilton and Jacobi.

2.36 Examples for the Use of the Hamilton–Jacobi Equation

Example (i) Consider the motion of a free particle for which $H = \boldsymbol{p}^2/2m$. The Hamilton–Jacobi differential equation now reads

$$\frac{1}{2m}(\nabla_{\boldsymbol{r}} S^*(\boldsymbol{r},\boldsymbol{\alpha},t))^2 + \frac{\partial S^*}{\partial t} = 0\,.$$

Its solution is easy to guess. It is

$$S^*(\boldsymbol{r},\boldsymbol{\alpha},t) = \boldsymbol{\alpha}\cdot\boldsymbol{r} - \frac{\boldsymbol{\alpha}^2}{2m}t + c\,.$$

From (2.150) we obtain

$$\boldsymbol{\beta} = \nabla_{\alpha} S^* = \boldsymbol{r} - \frac{\boldsymbol{\alpha}}{m}t\,,$$

which is the expected solution $\boldsymbol{r}(t) = \boldsymbol{\beta} + \boldsymbol{\alpha}t/m$. $\boldsymbol{\beta}$ and $\boldsymbol{\alpha}$ are integration constants; they are seen to represent the initial position and momentum, respectively. The solution as obtained from (2.149) reveals an interesting property. Let $\boldsymbol{r}(t) = (r_1(t), r_2(t), r_3(t))$. Then

$$\dot{r}_i = \frac{1}{m}\frac{\partial S^*}{\partial r_i}\,, \quad i = 1, 2, 3\,.$$

This means that the trajectories $\boldsymbol{r}(t)$ of the particle are everywhere perpendicular to the surfaces $S^*(\boldsymbol{r},\boldsymbol{\alpha},t) = \text{const}$. The relation between these surfaces and the particle's trajectories (which are orthogonal to them) receives a new interpretation in quantum mechanics. A quantum particle does not follow a classical trajectory. It is described by waves whose wave fronts are the analog of the surfaces $S^* = \text{const}$.

Example (ii) Consider the case of the Hamiltonian function $H = \boldsymbol{p}^2/2m + U(\boldsymbol{r})$. In this case we turn directly to the reduced action function (2.153) for which (2.154) reads

$$\frac{1}{2m}(\nabla S)^2 + U(\boldsymbol{r}) = E\,. \tag{2.155}$$

As $E = \boldsymbol{p}^2/2m + U(\boldsymbol{r})$ is constant along solutions, this equation reduces to

$$(\boldsymbol{\nabla} S)^2 = \boldsymbol{p}^2 .$$

Its general solution can be written as an integral

$$S = \int_{\boldsymbol{r}_0}^{\boldsymbol{r}_1} (\boldsymbol{p} \cdot \mathrm{d}\boldsymbol{r}) + S_0 , \tag{2.156}$$

provided the integral is taken along the trajectory with energy E.

Remarks:

(i) The generating function $S^*(\underset{\sim}{q}, \underset{\sim}{\alpha}, t)$ is closely related to the action integral (2.27). We assume that the Hamiltonian function H is such that the Legendre transformation between H and the Lagrangian function L exists. Taking the time derivative of S^* and making use of (2.149), we find

$$\frac{\mathrm{d}S^*}{\mathrm{d}t} = \frac{\partial S^*}{\partial t} + \sum_i \frac{\partial S^*}{\partial \dot{q}_i} \dot{q}_i = \left[-H(\underset{\sim}{q}, \underset{\sim}{p}, t) + \sum_i p_i \dot{q}_i \right]_{\underset{\sim}{p} = \partial S^*/\partial \underset{\sim}{q}}$$

As the variables $\underset{\sim}{p}$ can be eliminated (by the assumption we made), the right-hand side of this equation can be read as the Lagrangian function $L(\underset{\sim}{q}, \dot{\underset{\sim}{q}}, t)$. Integrating over time from t_0 to t, we have

$$S^*(\underset{\sim}{q}(t), \underset{\sim}{\alpha}, t) = \int_{t_0}^{t} \mathrm{d}t' L(\underset{\sim}{q}, \dot{\underset{\sim}{q}}, t') . \tag{2.157}$$

In contrast to the general action integral (2.27), where q and $\dot{q}$ are independent, we must insert the solution curves into L in the integrand on the right-hand side of (2.157). Thus, the action integral, if it is taken for the physical solutions $\underset{\sim}{q}(t)$, is the generating function for canonical transformations that "boost" the system from time t_0 to time t.

(ii) Consider the integral on the right-hand side of (2.157), taken between t_1 and t_2, and evaluated for the physical trajectory $\underset{\sim}{\varphi}(t)$ which goes through the boundary values $\underset{\sim}{a}$ at time t_1, and $\underset{\sim}{b}$ at time t_2. This function is called *Hamilton's principal function*. We assume that the Lagrangian does not depend explicitly on time. The principal function, which we denote by I_0, then depends on the time *difference* $\tau := t_2 - t_1$ only,

$$I_0 \equiv I_0(\underset{\sim}{a}, \underset{\sim}{b}, \tau) = \int_{t_1}^{t_2} \mathrm{d}t\, L\left(\dot{\underset{\sim}{\varphi}}(t), \underset{\sim}{\varphi}(t)\right) .$$

It is instructive to compare this function with the action integral (2.27): In (2.27) $I[\underset{\sim}{q}]$ is a smooth *functional*, $\underset{\sim}{q}$ being an arbitrary smooth function of time that connects the boundary values given there. In contrast to this, I_0 is calculated from a solution $\varphi(t)$ of the equations of motion which goes through the boundary values $(t_1, \underset{\sim}{a})$ and $(t_2, \underset{\sim}{b})$ and, hence, is a smooth *function* of $\underset{\sim}{a}$, $\underset{\sim}{b}$, and $(t_2 - t_1)$.

We consider now a smooth change of the initial and final values of the (generalized) coordinates and of the running time τ. This means that we replace the solution $\varphi(t)$ by another solution $\underset{\sim}{\phi}(s, t)$ which meets the following conditions. $\underset{\sim}{\phi}(s, t)$ is differentiable in the parameter s. For $s \to 0$ it goes over into the original solution, $\underset{\sim}{\phi}(s = 0, t) = \varphi(t)$. During the time $\tau' = \tau + \delta\tau$ it runs from $\underset{\sim}{a}' = \underset{\sim}{a} + \delta\underset{\sim}{a}$ to $\underset{\sim}{b}' = \underset{\sim}{b} + \delta\underset{\sim}{b}$. How does I_0 respond to these smooth changes? The answer is worked out in Exercise 2.30 and is as follows.

Let $\underset{\sim}{p}^a$ and $\underset{\sim}{p}^b$ denote the values of the momenta canonically conjugate to $\underset{\sim}{q}$, at times t_1 and t_2, respectively. One finds

$$\frac{\partial I_0}{\partial \tau} = -E \,, \quad \frac{\partial I_0}{\partial a_i} = -p_i^a \,, \quad \frac{\partial I_0}{\partial b_k} = p_k^b \,,$$

or, written as a variation,

$$\delta I_0 = -E\,\delta\tau - \sum_{i=1}^{f} p_i^a\,\delta a_i + \sum_{k=1}^{f} p_k^b\,\delta b_k \,.$$

The function I_0 and these results can be used to determine the nature of the extremum (2.27): maximum, minimum, or saddle point. For that purpose we consider a set of neighboring physical trajectories which all go through the same initial positions $\underset{\sim}{a}$ but differ in their initial momenta $\underset{\sim}{p}^a$. We follow each one of these trajectories over a fixed time $\tau = t_2 - t_1$ and compare the final positions as functions of the initial momenta $\underset{\sim}{b}(\underset{\sim}{p}^a)$, that is we determine the partial derivatives $M_{ik} = \partial b_i / \partial p_k^a$. The inverse of this matrix M is the matrix of mixed second partial derivatives of I_0 with respect to $\underset{\sim}{a}$ and $\underset{\sim}{b}$,

$$(M^{-1})_{ik} \equiv \frac{\partial p_i^a}{\partial b_k} = -\frac{\partial^2 I_0}{\partial a_i\,\partial b_k} \,.$$

In general, we expect M to have maximal rank. Then its inverse exists and I_0 is a minimum or maximum. However, for certain values of the running time τ, it may happen that one or several of the b_i remain unchanged by variations of the initial momenta. In this case the matrix M has rank smaller than maximal. Such final positions $\underset{\sim}{b}$ for which M becomes singular, together with the corresponding initial positions $\underset{\sim}{a}$, are called *conjugate points*. If, in computing I_0, we happened to choose conjugate points for the boundary values, I_0 is no longer a minimum (or maximum).

A simple example is provided by force-free motion on S_R^2, the surface of the sphere with radius R in $\mathbb{R}^3$. Obviously, the physical orbits are the great circles through the initial position $\underset{\sim}{a}$. If $\underset{\sim}{b}$ is not the antipode of $\underset{\sim}{a}$, then there is a longest and a shortest arc of great circle joining $\underset{\sim}{a}$ and $\underset{\sim}{b}$. If, however, $\underset{\sim}{a}$ and $\underset{\sim}{b}$ are antipodes then all trajectories starting from $\underset{\sim}{a}$ with momenta $\underset{\sim}{p}^a$ which have the same absolute value but different directions, reach $\underset{\sim}{b}$ all at the same time. The point $\underset{\sim}{b}$ is conjugate to $\underset{\sim}{a}$, I_0 is a saddle point of the action integral (2.27).

As a second example, let us study the one-dimensional harmonic oscillator. We use the reduced variables defined in Sect. 1.17.1. With (a, p^a) and (b, p^b) denoting the boundary values in phase space, τ the running time from a to b, the corresponding solution of the equations of motion reads

$$\varphi(t) = \frac{1}{\sin\tau}\left[a\sin(t_2 - t) + b\sin(t - t_1)\right] .$$

This trajectory is periodic. In the units used here the period is $T = 2\pi$. The boundary values of the momentum are

$$p^a = \dot{\varphi}(t_1) = \frac{-a\cos\tau + b}{\sin\tau} , \quad p^b = \dot{\varphi}(t_2) = \frac{-a + b\cos\tau}{\sin\tau} ,$$

while the energy is given by

$$E = \frac{1}{2}\left[\dot{\varphi}^2(t) + \varphi^2(t)\right] = \frac{a^2 + b^2 - 2ab\cos\tau}{2\sin^2\tau}$$
$$= \frac{1}{2}\left[a^2 + (p^a)^2\right] = \frac{1}{2}\left[b^2 + (p^{(b)})^2\right] .$$

In a similar fashion one calculates the Lagrangian $L = \frac{1}{2}\left[\dot{\varphi}^2(t) - \varphi^2(t)\right]$ as well as the function I_0 along the given trajectory

$$I_0(a, b, \tau) = \int_{t_1}^{t_2} \mathrm{d}t\, L(\dot{\varphi}(t), \varphi(t)) = \frac{(a^2 + b^2)\cos\tau - 2ab}{2\sin\tau} .$$

One confirms that, indeed, $\partial I_0/\partial\tau = -E$, $\partial I_0/\partial a = -p^a$, $\partial I_0/\partial b = p^b$. The matrix M, which in this example is one-dimensional, and its inverse are seen to be

$$M = \frac{\partial b}{\partial p^a} = \sin\tau , \quad M^{-1} = -\frac{\partial^2 I_0}{\partial b\,\partial a} = \frac{1}{\sin\tau} .$$

M^{-1} becomes singular at $\tau = \pi$ and at $\tau = 2\pi$, i.e. after half a period $T/2$ and after one full period T, respectively.

Keeping the initial position a fixed, but varying the initial momentum p^a, the final position is given by $b(p^a, \tau) = p^a \sin\tau + a\cos\tau$. Expressed in terms of a, p^a and τ the integral I_0 becomes

$$I_0(a, p^a, \tau) = \frac{1}{2}\sin\tau\left[\left((p^a)^2 - a^2\right)\cos\tau - 2ap^a\sin\tau\right] .$$

It is instructive to plot $b(p^a, \tau)$ as a function of the running time τ, for different values of the initial momentum p^a. As long as $0 \le \tau < \pi$ these curves do not intersect (except for the point a). When $\tau = \pi$ they all meet in $b(p^a, \pi) = -a$, independently of p^a. At this point M^{-1} becomes singular. Thus, the points a and $-a$ are conjugate points. As long as τ stays smaller than π, the action integral I is a minimum. For $\tau = \pi$ all trajectories with given initial position a, but different initial momenta p^a, go through the point $b = -a$ – as required by Hamilton's principle. As $I_0(a, p^a, \tau = \pi)$ is always zero, the extremum of I is a saddle point.

2.37 The Hamilton–Jacobi Equation and Integrable Systems

There are several general methods of solving the Hamilton–Jacobi differential equation (2.149) for situations of practical interest (see e.g. Goldstein 1984). Instead of going into these, we address the general question of the existence of local, or even global, solutions of the canonical equations. We shall discuss the class of completely integrable Hamiltonian systems and give a few examples. The general definition of angle and action variables is then followed by a short description of perturbation theory for quasiperiodic Hamiltonian systems, which is of relevance for celestial mechanics.

2.37.1 Local Rectification of Hamiltonian Systems

Locally the Hamilton–Jacobi equation (2.149) possesses complete solutions, i.e. in a neighborhood of an arbitrary point $\underset{\sim}{x}_0 = (\underset{\sim}{q}_0, \underset{\sim}{p}_0)$ of phase space one can always find a canonical transformation whose generating function $S^*(\underset{\sim}{q}, \underset{\sim}{p}, t)$ obeys the condition (2.152), $\det(\partial^2 S^* / \partial q_i \partial \alpha_k) \neq 0$, and which transforms the Hamiltonian function to $\tilde{H} = 0$. This follows, for example, from the explicit solution (2.157) or a generalization thereof,

$$S^*(\underset{\sim}{q}(t), \underset{\sim}{\alpha}, t) = S_0^*(\underset{\sim}{q}_0) + \int_{(\underset{\sim}{q}_0, t_0)}^{(\underset{\sim}{q}, t)} \mathrm{d}t' \, L(\underset{\sim}{q}, \dot{\underset{\sim}{q}}, t') \,. \tag{2.158}$$

Here $S_0^*(\underset{\sim}{q}_0)$ is a function that represents a given initial condition for S^* such that $\underset{\sim}{p}_0 = \partial S_0^*(\underset{\sim}{q}) / \partial \underset{\sim}{q}|_{\underset{\sim}{q}_0}$. In the second term we have to insert the physical solution that connects $(\underset{\sim}{q}_0, t_0)$ with $(\underset{\sim}{q}, t)$ and is obtained from the Euler–Lagrange equations (2.28). Finally, t and t_0 must be close enough to each other so that physical orbits $\underset{\sim}{q}(t)$, which, at $t = t_0$, pass in a neighborhood of $\underset{\sim}{q}_0$, do not intersect. (Note that we talk here about intersection of the graphs $(t, \underset{\sim}{q}(t))$.) This is the reason the existence of complete solutions is guaranteed only *locally*. Of course, this is no more than a statement about *existence* of solutions for the equations of motion: it says nothing about their construction in practice. To find explicit solutions it may be equally difficult to solve the equations of motion (i.e. either the Euler–Lagrange equations or the canonical equations) or to find complete solutions of the Hamilton–Jacobi differential equation. However, without knowing the solutions explicitly, one can derive fairly general, interesting properties for the case of autonomous systems. We consider an autonomous Hamiltonian system, defined by the Hamiltonian function

$H(\underset{\sim}{q},\underset{\sim}{p})$. H is chosen such that the condition $\det(\partial^2 H/\partial p_i \partial p_k) \neq 0$ is fulfilled, i.e. such that the Legendre transformation exists and is bijective. At first we note that instead of (2.153) we can choose the more general form

$$S^*(\underset{\sim}{q},\underset{\sim}{\alpha},t) = S(\underset{\sim}{q},\underset{\sim}{\alpha}) - \Sigma(\underset{\sim}{\alpha})t\ , \tag{2.159}$$

where $\Sigma(\underset{\sim}{\alpha})$ is an arbitrary differentiable function of the new momenta (which are conserved). Equation (2.154) is then replaced by

$$H\left(\underset{\sim}{q},\frac{\partial S}{\partial \underset{\sim}{q}}\right) = \Sigma(\underset{\sim}{\alpha})\ . \tag{2.160}$$

As we transform to the new coordinates $(\underset{\sim}{Q},\underset{\sim}{P} = \underset{\sim}{\alpha})$, with all Q_j cyclic, (2.160) means that

$$\hat{H}(\underset{\sim}{\alpha}) \equiv H(\underset{\sim}{q}(\underset{\sim}{Q},\underset{\sim}{\alpha}),\underset{\sim}{\alpha}) = \Sigma(\underset{\sim}{\alpha})\ . \tag{2.160$'$}$$

For example, we could choose $\Sigma(\underset{\sim}{\alpha}) = \alpha_f = E$, thus returning to (2.154), with the prescription that $P_f \equiv \alpha_f$ be equal to the energy E. Without restriction of generality we assume that, locally, the derivative $\partial H/\partial p_f$ is not zero (otherwise one must reorder the phase-space variables). The equation $H(q_1 \dots q_f, p_1 \dots p_f) = \Sigma$ can then be solved locally for p_f, viz.

$$p_f = -h(q_1 \dots q_{f-1}, q_f; p_1 \dots p_{f-1}, \Sigma)\ .$$

Taking q_f to be a formal time variable, $\tau \equiv q_f$, the function h can be understood to be the Hamiltonian function of a time-dependent system that has $(f-1)$ degrees of freedom and depends on the constant Σ. Indeed, one can show that the following canonical equations of motion hold true:

$$\frac{dq_i}{d\tau} = \frac{\partial h}{\partial p_i}\ ,\quad \frac{dp_i}{d\tau} = -\frac{\partial h}{\partial q_i}\ ,\quad \text{for}\quad i = 1,2,\dots,f-1\ .$$

To see this, take the derivative of the equation

$$H(q_1 \dots q_{f-1},\tau; p_1 \dots p_{f-1}, -h(q_1 \dots q_{f-1},\tau; p_1 \dots p_{f-1},\Sigma)) = \Sigma$$

with respect to p_i, with $i = 1,2,\dots,f-1$,

$$\frac{\partial H}{\partial p_i} + \frac{\partial H}{\partial p_f}\frac{\partial p_f}{\partial p_i} = 0\ .$$

However, $\partial H/\partial p_i = \dot{q}_i$, $\partial H/\partial p_f = \dot{q}_f$, and $\partial p_f/\partial p_i = -\partial h/\partial p_i$, and hence $dq_i/d\tau = \dot{q}_i/\dot{q}_f = \partial h/\partial p_i$. In a similar fashion, taking the derivative with respect to q_i, one obtains

$$\frac{\partial H}{\partial q_i} + \frac{\partial H}{\partial p_f}\left(-\frac{\partial h}{\partial q_i}\right) = 0\ ,$$

from which the second canonical equation is obtained, with h the Hamiltonian function. The Hamilton–Jacobi differential equation for this formally time-dependent system

$$\frac{\partial S^*}{\partial \tau} + h\left(q_1 \ldots q_{f-1}, \tau; \frac{\partial S^*}{\partial q_1} \ldots \frac{\partial S^*}{\partial q_{f-1}}, \Sigma\right) = 0$$

locally always possesses a complete integral $S^*(q_1 \ldots q_{f-1}, \alpha_1 \ldots \alpha_{f-1}, \Sigma, \tau)$. S^* being a complete solution means that

$$\det\left(\frac{\partial^2 S^*}{\partial q_j \partial \alpha_i}\right) \neq 0 \quad (i, j = 1, 2, \ldots, f-1)$$

Assuming that $\Sigma(\underset{\sim}{\alpha})$ in (2.160) depends explicitly on α_f, one can show that the above condition is fulfilled also for i and j running through 1 to f (hint: take the derivative of (2.160) by α_f). This then proves the following rectification theorem for autonomous Hamiltonian systems.

Rectification Theorem. Let $(\underset{\sim}{q}, \underset{\sim}{p})$ be a point of phase space where not all of the derivatives $\partial H/\partial q_i$ and $\partial H/\partial p_j$ vanish,

$$(\underset{\sim}{q}, \underset{\sim}{p}) = \left(\frac{\partial H}{\partial \underset{\sim}{p}}, -\frac{\partial H}{\partial \underset{\sim}{q}}\right) \neq (0, 0) .$$

(In other words, this point should not be an equilibrium position of the system.) Then the reduced equation (2.160) locally has a complete integral $S(\underset{\sim}{q}, \underset{\sim}{\alpha})$, i.e. condition (2.152) is fulfilled.

The new coordinates Q_i are cyclic and are given by $Q_i = \partial S^*/\partial \alpha_i$. Their time derivatives follow from (2.160′) and the canonical equations. They are

$$\dot{Q}_i = \frac{\partial \hat{H}}{\partial \alpha_i} = \frac{\partial}{\partial \alpha_i} \Sigma(\underset{\sim}{\alpha}) .$$

With the special choice $\Sigma(\underset{\sim}{\alpha}) = \alpha_f = E$, for instance, we obtain

$$\begin{aligned} \dot{Q}_i &= 0 \quad \text{for} \quad i = 1, 2, \ldots, f-1 \\ \dot{Q}_f &= 1 \\ \dot{P}_k &= 0 \quad \text{for} \quad k = 1, 2, \ldots, f \end{aligned} \tag{2.161}$$

and therefore

$$Q_i \equiv \beta_i = \text{const.}, \quad i = 1, 2, \ldots, f-1; \quad Q_f = t - t_0 = \frac{\partial S^*(\underset{\sim}{q}, \underset{\sim}{\alpha})}{\partial \alpha_f},$$

$$P_k \equiv \alpha_k = \text{const.} \quad k = 1, 2, \ldots, f .$$

The significance of this theorem is the following: the flow of an autonomous system can be rectified as shown in Fig. 2.16, in the neighborhood of every point of phase space that is not an equilibrium position. Viewed locally, a transformation of phase space variables smoothes the flux to a uniform, rectilinear flow, (e.g.) parallel to the Q_f-axis. Outside their equilibrium positions all autonomous Hamiltonian systems are locally equivalent[8]. Therefore, interesting properties specific to a given Hamiltonian (or more general) dynamical systems concern the *global* structure of its flow and its equilibrium positions. We shall return to these questions in Chap. 6.

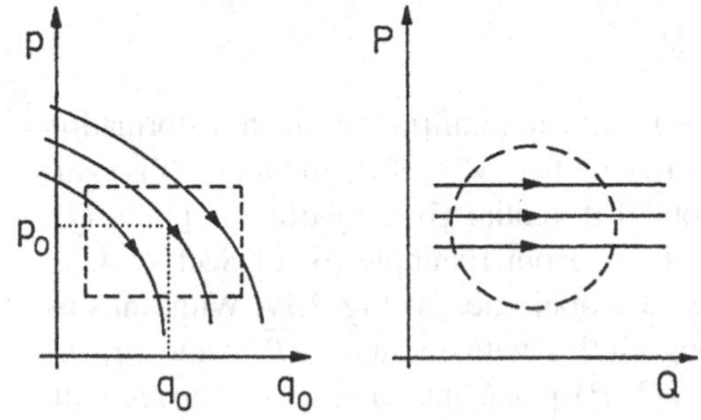

Fig. 2.16. Locally and outside of an equilibrium position a dynamical system can be rectified

Example. *The harmonic oscillator in one dimension.* We shall study the harmonic oscillator using the reduced variables defined in Sect. 1.17.1. For the sake of clarity we write q instead of z_1, p instead of z_2, and t instead of τ. (Thus $q \equiv z_1$ and $p \equiv z_2$ carry the dimension $(\text{energy})^{1/2}$, while t is measured in units of ω^{-1}.) In these units $H = (p^2 + q^2)/2$. Choosing the function on the right-hand side of (2.159) as follows: $\Sigma(\alpha) = P > 0$, the corresponding Hamilton–Jacobi equation (2.160) reads

$$\frac{1}{2}\left(\frac{\partial S}{\partial q}\right)^2 + \frac{1}{2}q^2 = P\,.$$

Its integration is straightforward. Because $\partial S/\partial q = \sqrt{2P - q^2}$,

$$S(q, P) = \int_0^q \sqrt{2P - q'^2}\,dq' \quad \text{with } |q| < \sqrt{2P}\,.$$

We have

$$\frac{\partial^2 S}{\partial q\,\partial P} = \frac{1}{\sqrt{2P - q^2}} \neq 0$$

[8] This is a special case of the more general rectification theorem for general, autonomous, differentiable systems: in the neighborhood of any point $\underline{x}_0$ that is not an equilibrium position (i.e. where $F(\underline{x}_0) \neq 0$), the system $\dot{\underline{x}} = F(\underline{x})$ of first-order differential equations can be transformed to the form $\dot{\underline{z}} = (1, 0, \ldots, 0)$, i.e. $\dot{z}_1 = 1$, $\dot{z}_2 = 0 = \ldots = \dot{z}_f$. For a proof see e.g. Arnol'd (1973).

and

$$p = \frac{\partial S}{\partial q} = \sqrt{2P - q^2} \; ,$$

$$Q = \frac{\partial S}{\partial P} = \int_0^q \frac{1}{\sqrt{2P - q'^2}} dq' = \arcsin \frac{q}{\sqrt{2P}} \; .$$

Because of the arcsin function, Q should be restricted to the interval $(-\frac{\pi}{2}, \frac{\pi}{2})$. However, solving for q and p one obtains

$$q = \sqrt{2P} \sin Q \; , \quad p = \sqrt{2P} \cos Q \; ,$$

so that this restriction can be dropped. It is easy to confirm that the transformation $(q, p) \mapsto (Q, P)$ is canonical, e.g. by verifying that $M = \partial(q, p)/\partial(Q, P)$ is symplectic, or else that $(P\mathrm{d}Q - p\mathrm{d}q)$ is a total differential given by $\mathrm{d}(P \sin Q \cos Q)$. Of course, the result is already known to us from Example (ii) of Sect. 2.24. In the present case the rectification is even a global one, cf. Fig. 2.17. With units as chosen here, the phase point runs along circles with radius $\sqrt{2P}$ in the (q, p)-plane, with angular velocity 1. In the (Q, P)-plane the same point moves with uniform velocity 1 along a straight line parallel to the Q-axis. As the frequency is independent of the amplitude, the velocities on all phase orbits are the same in either representation (this is typical for the harmonic oscillator).

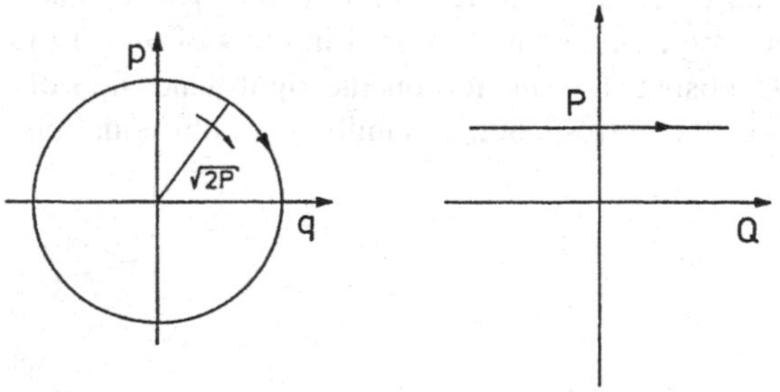

Fig. 2.17. For an oscillator the rectification is global

2.37.2 Integrable Systems

Mechanical systems that can be integrated completely and globally are the exception in the many varieties of dynamical systems. In this section we wish to collect a few general properties and propositions and to give some examples of integrable systems.

The chances of finding complete solutions for a given system, loosely speaking, are the greater the more integrals of the motion are known.

Example (i) Motion of a particle in one dimension, under the influence of a potential $U(q)$ (see Sect. 1.16). The system has one degree of freedom $f = 1$, the dimension of phase space is $\dim \mathbb{P} = 2$, and there is one integral of the motion: that of the energy.

Example (ii) Motion of a particle in three dimensions, with a central potential $U(r)$ (see Sect. 1.24). Here $f = 3$ and $\dim \mathbb{P} = 6$. Integrals of the motion are provided by the energy E, the three components l_i of angular momentum, and, as a consequence, the square of angular momentum $\boldsymbol{l}^2$.

Generally, the dynamical quantities $g_2(q, p), \ldots, g_m(q, p)$ are integrals of the motion if the Poisson brackets of the Hamiltonian function H and g_i vanish, $\{H, g_i\} = 0$, for $i = 2, 3, \ldots, m$. Each one of these functions $g_i(q, p)$ may serve as the generating function for an infinitesimal canonical transformation (cf. Sect. 2.33). By the reciprocity discussed in Sect. 2.33, H is left invariant by this transformation. The question remains, however, in which way the other integrals of the motion transform under the infinitesimal transformation generated by a specific g_i. In Example (ii) above, l_3 generates an infinitesimal rotation about the 3-axis, and we have

$$\{l_3, H\} = 0\,, \quad \{l_3, \boldsymbol{l}^2\} = 0\,, \quad \{l_3, l_1\} = -l_2\,, \quad \{l_3, l_2\} = l_1\,.$$

In other words, while the values of the energy E and the modulus of the angular momentum $l = \sqrt{\boldsymbol{l}^2}$ are invariant, the rotation about the 3-axis changes the values of l_1 and l_2. A solution with fixed values of $(E, \boldsymbol{l}^2, l_3, l_1, l_2)$ becomes a solution with the values $(E, \boldsymbol{l}^2, l_3, l_1' \simeq l_1 - \varepsilon l_2, l_2' \simeq l_2 + \varepsilon l_1)$.

Thus, there are integrals of the motion that "commute" (i.e. whose Poisson bracket $\{g_i, g_k\}$ vanishes), as well as others that do not. These two groups must be distinguished because only the former is relevant for the question of integrability. This leads us to the following.

Definition. The linearly independent dynamical quantities $g_1(q, p) \equiv H(q, p)$, $g_2(q, p), \ldots, g_m(q, p)$ are said to be *in involution* if the Poisson bracket for any pair of them vanishes,

$$\{g_i(q, p), g_k(q, p)\} = 0\,, \quad i, k = 1, 2, \ldots, m\,. \tag{2.162}$$

In Example (ii) above, H, $\boldsymbol{l}^2$, and l_3 (or any other fixed component l_j) are in involution. Let us consider a few more examples.

Example (iii) Among the ten integrals of the motion of the two-body system with central force (cf. Sect. 1.12), the following six are in involution

$$H_{\text{rel}} = \frac{\boldsymbol{p}^2}{2\mu} + U(r)\,, \boldsymbol{P}\,, \boldsymbol{l}^2\,, l_3\,, \tag{2.163}$$

H_{rel} being the energy of the relative motion, $\boldsymbol{P}$ the momentum of the center of mass, and $\boldsymbol{l}$ the relative angular momentum.

Example (iv) (This anticipates Chap. 3). In the case of a force-free rigid body (which has $f = 6$), the kinetic energy $H_{\text{rel}} = \boldsymbol{\omega} \cdot \boldsymbol{L}/2$, the momentum of the center of mass $\boldsymbol{P}$, and $\boldsymbol{L}^2$ and L_3 are in involution (cf. Sect. 3.13).

All quoted examples are globally integrable (in fact, they are integrable by quadratures only). Their striking common feature is that the number of integrals of the motion equals the number, f, of degrees of freedom. For instance, the two-body problem of Example (iii) has $f = 6$ and possesses the six integrals (2.163) in involution. If we consider the three-body system with central forces instead, the number of degrees of freedom is $f = 9$, while the number of integrals of the motion that are in involution remains the same as in the case of two bodies, namely 6. Indeed, the three-body problem is not generally integrable.

Example (v) If, in turn, we manage to integrate a canonical system by means of the Hamilton–Jacobi differential equation (2.149), we obtain the f integrals of the motion (2.150): $Q_k = S^*(\underset{\sim}{q}, \underset{\sim}{\alpha}, t)/\partial\alpha_k$, $k = 1, 2, \ldots, f$, which trivially have the property $\{Q_i, Q_k\} = 0$.

In conclusion, it seems as though the existence of f independent integrals of the motion is sufficient to render the system of $2f$ canonical equations integrable. These matters are clarified by the following theorem of Liouville.

Theorem on Integrable Systems. Let $\{g_1 \equiv H, g_2, \ldots, g_f\}$ be dynamical quantities defined on the $2f$-dimensional phase space $\mathbb{P}$ of an autonomous, canonical system described by the Hamiltonian function H. The $g_i(x)$ are assumed to be in involution,

$$\{g_i, g_k\} = 0\,, \quad i, k = 1, \ldots, f\,, \tag{2.164}$$

and to be independent in the following sense: at each point of the hypersurface

$$S = \{\underset{\sim}{x} \in \mathbb{P} | g_i(\underset{\sim}{x}) = c_i, \quad i = 1, \ldots, f\} \tag{2.165}$$

the differentials $\mathrm{d}g_1, \ldots, \mathrm{d}g_f$ are linearly independent. Then:

(a) S is a smooth hypersurface that stays invariant under the flow corresponding to H. If, in addition, S is compact and connected, then it can be mapped diffeomorphically onto an f-dimensional torus

$$T^f = S^1 \times \ldots \times S^1 \quad (f \text{ factors})\,. \tag{2.165$'$}$$

(Here S^1 is the circle with radius 1, cf. also Sect. 5.2.3, Example (iii) below).

(b) Every S^1 can be described by means of an angle coordinate $\theta_i \in [0, 2\pi)$. The most general motion on S is a quasiperiodic motion, which is a solution of the transformed equations of motion

$$\frac{\mathrm{d}\theta_i}{\mathrm{d}t} = \omega^{(i)}\,, \quad i = 1, \ldots, f\,. \tag{2.166}$$

(c) The canonical equations can be solved by quadratures (i.e. by ordinary integration).

The proof is clearest if one makes use of the elegant tools of Chap. 5. As the reader is probably not yet familiar with them at this point, we skip the proof and refer to Arnol'd (1988, Sect. 49) where it is given in quite some detail. A motion Φ in $\mathbb{P}$ is said to be *quasiperiodic*, with base frequencies $\omega^{(1)}, \ldots, \omega^{(f)}$, if all components of $\Phi(t, s, \underset{\sim}{y})$ are periodic (the periods being $2\pi/\omega^{(i)}$) and if these frequencies are rationally independent, i.e. with $r_i \in \mathbb{Z}$, we have

$$\sum_{i=1}^{f} r_i \omega^{(i)} = 0 \quad \text{only if} \quad r_i = \ldots = r_f = 0\,. \tag{2.167}$$

Let us study two more examples.

Example (vi) Two coupled linear oscillators (cf. Practical Example 2.1). Here $f = 2$, the Hamiltonian function being given by

$$H = \frac{1}{2m}(p_1^2 + p_2^2) + \frac{1}{2}m\omega_0^2(q_1^2 + q_2^2) + \frac{1}{2}m\omega_1^2(q_1 - q_2)^2\,.$$

The following are two integrals of the motion in involution related to H by $g_1 + g_2 = H$:

$$g_1 = \frac{1}{4m}(p_1 + p_2)^2 + \frac{1}{4}m\omega_0^2(q_1 + q_2)^2\,,$$

$$g_2 = \frac{1}{4m}(p_1 - p_2)^2 + \frac{1}{4}m(\omega_0^2 + 2\omega_1^2)(q_1 - q_2)^2\,,$$

This decomposition of H corresponds to the transformation to the two normal-mode oscillations of the system, $z_{1/2} = (q_1 \pm q_2)/\sqrt{2}$, g_1 and g_2 being the energies of these decoupled oscillations. Following Example (ii) of Sect. 2.24, we introduce new canonical coordinates $\{Q_i = \theta_i, P_j \equiv I_j\}$ such that

$$H = g_1 + g_2 = \omega_0 I_1 + \sqrt{\omega_0^2 + 2\omega_1^2}\, I_2\,.$$

Then $\theta_1 = \omega_0 t + \beta_1$, $\theta_2 = \sqrt{\omega_0^2 + 2\omega_1^2}\, t + \beta_2$. For fixed values of I_1 and I_2 the surface S (2.165) is the torus T^2. If the two frequencies $\omega^{(1)} = \omega_0$, $\omega^{(2)} = \sqrt{\omega_0^2 + 2\omega_1^2}$ are rationally dependent, i.e. if $n_1\omega^{(1)} = n_2\omega^{(2)}$ with n_1, n_2 positive integers, then the motion is periodic with period $T = 2\pi/n_1\omega^{(1)} = 2\pi/n_2\omega^{(2)}$. Any orbit on the torus T^2 closes. If, on the contrary, the frequencies are rationally independent, the orbits never close. In this case any orbit is dense on the torus.

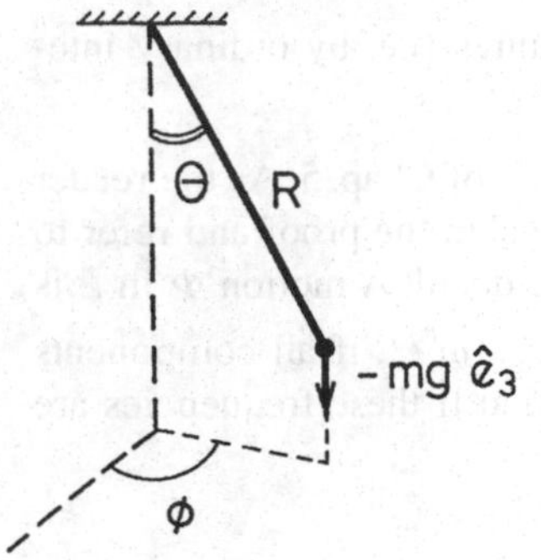

Fig. 2.18. Coordinates used to describe the spherical pendulum

Example (vii) *The spherical mathematical pendulum.* Let R be the length of the pendulum, θ the deviation from the vertical, and ϕ the azimuth in the horizontal plane (see Fig. 2.18). We have

$$H = \frac{p_\theta^2}{2mR^2} + \frac{p_\phi^2}{2mR^2 \sin^2\theta} + mgR(1-\cos\theta) ,$$

where $p_\theta = mR^2 \mathrm{d}\theta/\mathrm{d}t$, $p_\phi = mR^2 \sin^2\theta \mathrm{d}\phi/\mathrm{d}t$. The coordinate ϕ is cyclic. Hence, $p_\phi \equiv l = \text{const}$. There are two integrals of the motion $g_1 = H$, $g_2 = p_\phi$, and we can verify that they are in involution, $\{g_1, g_2\} = 0$. Therefore, according to the theorem above, the system is completely integrable by quadratures. Indeed, taking $q_1 = \theta$, $q_2 = \phi$, $\tau = \omega t$, $p_1 = \mathrm{d}q_1/\mathrm{d}\tau$, and $p_2 = \sin^2 q_1 \mathrm{d}q_2/\mathrm{d}\tau$, and introducing the parameters

$$\varepsilon \equiv \frac{E}{mgR} , \quad \omega^2 \equiv \frac{g}{R} , \quad a^2 \equiv \frac{l^2}{m^2 g R^3} ,$$

we obtain

$$\varepsilon = \frac{1}{2}p_1^2 + \frac{a^2}{2\sin^2 q_1} + (1-\cos q_1) \equiv \frac{1}{2}p_1^2 + U(q_1) .$$

The equations of motion read

$$\frac{\mathrm{d}q_1}{\mathrm{d}\tau} = p_1 = \pm\sqrt{2(\varepsilon - U(q_1))} ,$$

$$\frac{\mathrm{d}p_1}{\mathrm{d}\tau} = \frac{a^2 \cos q_1}{\sin^3 q_1} - \sin q_1 ,$$

$$\frac{\mathrm{d}q_2}{\mathrm{d}\tau} = \frac{a}{\sin^2 q_1} .$$

They are completely integrable. From the first equation we obtain

$$\tau = \int \frac{\mathrm{d}q_1}{\sqrt{2(\varepsilon - U(q_1))}} .$$

Combining the first and the third yields

$$q_2 = a \int \frac{\mathrm{d}q_1}{\sin^2 q_1 \sqrt{2(\varepsilon - U(q_1))}} \, .$$

2.37.3 Angle and Action Variables

Suppose we are given an autonomous Hamiltonian system with (for the moment) $f = 1$ that has periodic solutions for energies E belonging to a certain interval $[E_0, E_1]$. Let Γ_E be a periodic orbit with energy E. Then the period $T(E)$ of the orbit Γ_E is equal to the derivative $\mathrm{d}F(E)/\mathrm{d}E$ of the surface $F(E)$ that is enclosed by this orbit in phase space (see Exercises 2.1 and 2.27),

$$T(E) = \frac{\mathrm{d}}{\mathrm{d}E} \oint_{\Gamma_E} p \, \mathrm{d}q \equiv \frac{\mathrm{d}F(E)}{\mathrm{d}E} \, .$$

The period $T(E)$ being related to the circular frequency by $\omega(E) = 2\pi/T(E)$ we define the quantity

$$I(E) \stackrel{\text{def}}{=} \frac{1}{2\pi} F(E) = \frac{1}{2\pi} \oint_{\Gamma_E} p \, \mathrm{d}q \, . \qquad (2.168)$$

$I(E)$ is called the *action variable*. Except for equilibrium positions, $T(E) = 2\pi \mathrm{d}I(E)/\mathrm{d}E$ is nonzero. Hence, the inverse function $E = E(I)$ exists. Therefore, it is meaningful to construct a canonical transformation $\{q, p\} \to \{\theta, I\}$ such that the transformed Hamiltonian function is just $E(I)$ and I is the new momentum. From (2.154) and (2.87) this means that

$$p = \frac{\partial S(q, I)}{\partial q} \, , \quad \theta = \frac{\partial S(q, I)}{\partial I} \, , \quad H\left(q, \frac{\partial S}{\partial q}\right) = E(I) \, . \qquad (2.169)$$

The new generalized coordinate θ is called the *angle variable*. We then have $I = \text{const} \in \Delta$, where the interval Δ follows from the interval $[E_0, E_1]$ for E. The equation of motion for θ takes the simple form

$$\dot{\theta} = \frac{\partial E(I)}{\partial I} \equiv \omega^{(I)} = \text{const} \, .$$

With the $(Q \equiv \theta, P \equiv I)$ description of phase space, the orbits lie in a strip parallel to the θ axis, whose width is Δ. Each periodic orbit has the representation $(\theta = \omega^{(I)} t + \theta_0, I = \text{const})$, i.e. in the new variables it runs parallel to the abscissa. However, as θ is to be understood modulo 2π, the phase space is bent to form part of a cylinder with radius 1 and height Δ. The periodic orbits lie on the manifold $\Delta \times S^1$ in $\mathbb{P}$.

For a system with more than one degree of freedom, $f > 1$, for which there are f integrals in involution, the angle variables are taken to be the angular coordinates that describe the torus (2.165′). The corresponding action variables are

$$I_k(c_1, \ldots, c_f) = \frac{1}{2\pi} \oint_{\Gamma_k} \sum p_i \mathrm{d}q_i \,,$$

where one integrates over the curve in $\mathbb{P}$ that is the image of ($\theta_i = \text{const for } i \neq k$, $\theta_k \in S^1$). The manifold on which the motion takes place then has the form

$$\Delta_1 \times \ldots \Delta_f \times (S^1)^f = \Delta_1 \times \ldots \times \Delta_f \times T^f \,. \tag{2.170}$$

Example (vi) illustrates the case $f = 2$ for two decoupled oscillators. In Example (vii) the quantities ε (energy) and a (azimuthal angular momentum) are constants of the motion, and we have

$$I_1(\varepsilon, a) = \frac{1}{2\pi} \oint p_1 \, \mathrm{d}q_1 = \frac{1}{2\pi} \oint \sqrt{2(\varepsilon - U(q_1))} \, \mathrm{d}q_1 \,,$$

$$I_2(\varepsilon, a) = \frac{1}{2\pi} \oint p_2 \, \mathrm{d}q_2 = a \,.$$

Solving the first equation for ε, $\varepsilon = \varepsilon(I_1, a)$, we obtain the frequency for the motion in θ (the deviation from the vertical),

$$\dot{\theta} = \frac{\partial \varepsilon(I_1, a)}{\partial I_1} \equiv \omega_1 \,.$$

2.38 Perturbing Quasiperiodic Hamiltonian Systems

The theory of perturbations of integrable quasiperiodic Hamiltonian systems is obviously fundamental for celestial mechanics and for Hamiltonian dynamics in general. This is an important and extensive branch of mathematics that we cannot deal with in detail for lack of space. We can only sketch the basic questions addressed in perturbation theory and must refer to the literature for a more adequate account.

Consider an autonomous, integrable system for which there is a set of action-angle variables. Let the system be described by $H_0(I)$. We now add to it a small Hamiltonian perturbation so that the Hamiltonian function of the perturbed system reads

$$H(\underset{\sim}{\theta}, \underset{\sim}{I}, \mu) = H_0(\underset{\sim}{I}) + \mu H_1(\underset{\sim}{\theta}, \underset{\sim}{I}, \mu) \,. \tag{2.171}$$

Here H_1 is assumed to be 2π-periodic in the angle variables $\underset{\sim}{\theta}$, while μ is a real parameter that controls the strength of the perturbation.

To quote an example, let us consider the *restricted three-body problem*, which is defined as follows. Two mass points P_1 and P_2 whose masses are m_1 and m_2, respectively, move on circular orbits about their center of mass, under the action of gravitation. A third mass point P is added that moves in the orbit plane of P_1 and P_2, and whose mass is negligible compared to m_1 and m_2 so that it does not perturb the motion of the original two-body system. The problem consists in finding the motion of P. Obviously, this is a model for the motion of the moon in the field of the sun and the earth, of the motion of an asteroid with respect to the system of the sun and Jupiter (the heaviest of the planets in our planetary system), or of the motion of satellites in the neighborhood of the earth and the moon.

Thus, the general problem is defined as follows:

(a) $H(\underset{\sim}{\theta}, \underset{\sim}{I}, \mu)$ is a real analytic function of $\underset{\sim}{\theta} \in T^f$, of $\underset{\sim}{I} \in \Delta_1 \times \ldots \times \Delta_f$, as in (2.170), and of $\mu \in I \subset \mathbb{R}$, where the interval I includes the origin.

(b) H is periodic in the variables θ_i, i.e.

$$H(\underset{\sim}{\theta} + 2\pi \hat{\boldsymbol{e}}_i, \underset{\sim}{I}, \mu) = H(\underset{\sim}{\theta}, \underset{\sim}{I}, \mu) , \quad i = 1, 2, \ldots, f ,$$

where $\hat{\boldsymbol{e}}_i$ is the ith unit vector.

(c) For $\mu = 0$ the problem has a form that is integrable directly and completely. The condition $\det(\partial^2 H / \partial I_k \partial I_j) \neq 0$ holds. The unperturbed solutions read

$$\begin{aligned} \theta_i^{(0)}(t) &= \frac{\partial H_0(\underset{\sim}{I})}{\partial I_i} t + \beta_i^{(0)} , \\ I_i^{(0)} &= \alpha_i^{(0)} , \quad i = 1, 2, \ldots, f , \quad \text{with} \quad \alpha_i^{(0)} \in \Delta_i . \end{aligned} \tag{2.172}$$

The aim of perturbation theory is to construct solutions of the perturbed system for small values of μ. We assumed H to be real and analytic in μ. Therefore, any solution (2.172) can be continued in any *finite* time interval I_t and for small values of μ with, say, $|\mu| < \mu_0(I_t)$, where μ_0 is suitably chosen and is a function of the interval I_t. Unfortunately, the question that is of real physical interest is much more difficult: it is the question whether there exist solutions of the perturbed system that are defined for *all* times. Only if one succeeds in constructing such solutions is there a chance to decide, for instance, whether the periodic motion of our planetary system is stable at large time scales. In fact, this question still has no final answer[9].

Perturbation theory makes use of two basic ideas. The first is to do a systematic expansion in terms of the parameter μ and to solve the equations generated in this way, order by order. Let

$$\begin{aligned} \theta_k &= \theta_k^{(0)} + \mu\theta_k^{(1)} + \mu^2\theta_k^{(2)} + \ldots ; \quad \theta_k^{(0)} = \omega_k t + \beta_k , \\ I_k &= I_k^{(0)} + \mu I_k^{(1)} + \mu^2 I_k^{(2)} + \ldots ; \quad I_k^{(0)} = \alpha_k , \end{aligned} \tag{2.173}$$

and then insert these expansions in the canonical equations,

$$\dot{\theta}_k = \{H, \theta_k\} , \quad \dot{I}_k = \{H, I_k\} ,$$

and compare terms of the same order μ^n. For instance, at first order μ^1 one finds

[9] There is evidence, from numerical studies, that the motion of the planet Pluto is chaotic, i.e. that it is intrinsically unstable over large time scales (G.J. Sussman and J. Wisdom, Science **241**(1988) 433). Because Pluto couples to the other planets, though weakly, this irregular behavior eventually spreads to the whole system.

$$\dot{\theta}_k^{(1)} = \{H_1, \theta_k^{(0)}\} \simeq \frac{\partial H_1(\underset{\sim}{\theta}^{(0)}, \underset{\sim}{I}^{(0)})}{\partial I_k^{(0)}} ,$$
$$\dot{I}_k^{(1)} = \{H_1, I_k^{(0)}\} \simeq -\frac{\partial H_1(\underset{\sim}{\theta}^{(0)}, \underset{\sim}{I}^{(0)})}{\partial \theta^{(0)}} . \tag{2.174}$$

We have to insert the unperturbed solutions $\underset{\sim}{\theta}^{(0)}$ and $\underset{\sim}{I}^{(0)}$ on the right-hand side, for consistency, because otherwise there would appear terms of higher order in μ. As H_1 was assumed to be periodic, it can be written as a Fourier series,

$$H_1(\underset{\sim}{\theta}^{(0)}, \underset{\sim}{I}^{(0)}) = \sum_{m_1 \dots m_f} C_{m_1 \dots m_f}(\underset{\sim}{\alpha}) \exp\left\{ \mathrm{i} \sum_{k=1}^{f} m_k \theta_k^{(0)} \right\}$$
$$= \sum C_{m_1 \dots m_f}(\underset{\sim}{\alpha}) \exp\left\{ \mathrm{i} \sum m_k(\omega_k t + \beta_k) \right\} .$$

Equations (2.174) can then be integrated. The solutions contain terms whose time dependence is given by

$$\frac{1}{\sum m_k \omega_k} \exp\left\{ \mathrm{i} \sum m_k \omega_k t \right\} .$$

Such terms will remain small, for small perturbations, unless their denominator vanishes. If, in turn, $\sum m_k \omega_k = 0$, $\theta^{(1)}$ and $I^{(1)}$ will grow linearly in time. This kind of perturbation is said to be a *secular perturbation*.

The simplest case is the one where the frequencies ω_k are rationally independent, cf. (2.167). The time average of a continuous function F over the quasiperiodic flow $\theta^{(0)}(t) = \omega t + \beta$ is equal to the space average of F on the torus T^f[10],

$$\lim \frac{1}{T} \int_0^T F(\underset{\sim}{\theta}(t))\, \mathrm{d}t = \frac{1}{(2\pi)^f} \int_{T^f} \mathrm{d}\theta_1 \dots \mathrm{d}\theta_f F(\underset{\sim}{\theta}) \stackrel{\text{def}}{=} \langle F \rangle . \tag{2.175}$$

Taking account of the secular term alone, one then obtains from (2.174) the approximate equations

$$\dot{\theta}_k^{(1)} = \frac{\partial}{\partial I_k^{(0)}} \langle H_1 \rangle , \quad \dot{I}_k^{(1)} = 0 . \tag{2.176}$$

The second idea is to transform the initial system (2.171) by means of successive canonical transformations in such a way that the transformed Hamiltonian function $\tilde{H}$ depends only on the action variables $\underset{\sim}{I}$, up to terms of increasingly high order in the parameter μ. This program requires a detailed discussion and needs advanced and refined tools of analysis. Here we can only quote the main result, which is relevant for questions of stability of Hamiltonian systems.

---

[10] Equation (2.175) holds for functions $f_k = \exp\{\mathrm{i} \sum k_i \theta_i(t)\}$ with $\theta_i(t) = \omega_i t + \beta_i$, where it gives, in fact, $\langle f_k \rangle = 0$, except for $k_1 = \dots = k_f = 0$. Any continuous F can be approximated by a finite linear combination $F = \sum C_k f_k$.

2.39 Autonomous, Nondegenerate Hamiltonian Systems in the Neighborhood of Integrable Systems

The manifold on which the motions of an autonomous integrable system $H_0(\underset{\sim}{I})$ take place is the one given in (2.170). We assume that the frequencies $\{\omega_i\}$ are rationally independent (see (2.167)). For fixed values of the action variables $I_k = \alpha_k$ every solution curve runs around the torus T^f and covers it densely. One says that the quasiperiodic motion is *ergodic*. After a sufficiently long time the orbit returns to an arbitrarily small neighborhood of its starting point but does not close. This situation is decribed by the term *nonresonant torus*[11].

We now add a small Hamiltonian perturbation to this system so that it is described by

$$H(\underset{\sim}{\theta}, \underset{\sim}{I}, \mu) = H_0(\underset{\sim}{I}) + \mu H_1(\underset{\sim}{\theta}, \underset{\sim}{I}, \mu) . \tag{2.177}$$

The question then is in which sense this system is stable. Does the perturbation modify only slightly the manifold of motions of the system $H_0(\underset{\sim}{I})$, or does it destroy it completely?

The most important result that to a large extent answers this question is provided by a theorem of Kolmogorov, Arnol'd and Moser that we wish to state here without proof in admittedly somewhat qualitative terms.

Theorem (KAM). If the frequencies of an integrable, Hamiltonian system H_0 are rationally independent and if, in addition, these frequencies are sufficiently irrational, then, for small values of μ, the perturbed system $H = H_0 + \mu H_1$ has solutions that are predominantly quasiperiodic, too, and that differ only slightly from those of the unperturbed system H_0. Most of the nonresonant tori of H_0 are deformed, but only slightly. Thus, the perturbed system possesses nonresonant tori as well, on which the orbits are dense.

Here, sufficiently irrational means the following. A single frequency is sufficiently irrational if there are positive real numbers γ and α such that

$$\left|\omega - \frac{n}{m}\right| \geq \gamma m^{-\alpha} \tag{2.178a}$$

for all integers m and n. Similarly, f rationally independent frequencies are sufficiently irrational if there are positive constants γ and α such that

$$\left|\sum r_i \omega_i\right| \geq \gamma |\boldsymbol{r}|^{-\alpha} , \quad r_i \in \mathbb{Z} . \tag{2.178b}$$

It is instructive to study the special case of systems with two degrees of freedom, $f = 2$, because they exhibit many interesting properties that can be analyzed in

[11] If, in turn, the frequencies are rationally dependent, the tori are said to be *resonant tori*, cf. Example (vi) of Sect. 2.37.2. In this case the motion is quasiperiodic with a number of frequencies that is smaller than f.

detail (see e.g. Guckenheimer, Holmes 1986, Sect. 4.8). The general case is treated, e.g., by Thirring (1989) and Rüssmann (1979).

The KAM theorem was a decisive step forward in our understanding of the dynamics of quasiperiodic, Hamiltonian systems. It yields good results on long-term stability, although with certain, and somewhat restrictive, assumptions. Therefore, the qualitative behavior of only a restricted class of systems can be derived from it. An example is provided by the restricted three-body problem sketched above (Rüssmann 1979). Unfortunately, our planetary system falls outside the range of applicability of the theorem. Also, the theorem says nothing about what happens when the frequencies $\{\omega_i\}$ are not rationally independent, i.e. when there are resonances. We shall return to this question in Sect. 6.5.

2.40 Examples. The Averaging Principle

2.40.1 The Anharmonic Oscillator

Consider a perturbed oscillator in one dimension, the perturbation being proportional to the fourth power of the coordinate. The Hamiltonian function is

$$H = \frac{p^2}{2m} + \frac{1}{2} m\omega_0^2 q^2 + \mu q^4 \,, \tag{2.179}$$

or, in the notation of (2.177),

$$H_0 = \frac{p^2}{2m} + \frac{1}{2} m\omega_0^2 q^2 \,, \quad H_1 = q^4 \,.$$

In the absence of the anharmonic perturbation, the energy $E^{(0)}$ of a periodic orbit is related to the maximal amplitude $q_{\max}$ by $(q_{\max})^2 = 2E^{(0)}/m\omega_0^2$. We take

$$x \stackrel{\text{def}}{=} \frac{q}{q_{\max}} \quad \text{and} \quad \varepsilon \stackrel{\text{def}}{=} \mu \frac{4E^{(0)}}{m^2\omega_0^4} \,,$$

so that the potential energy becomes

$$U(q) = \frac{1}{2} m\omega_0^2 q^2 + \mu q^4 = E^{(0)}(1 + \varepsilon x^2)x^2 \,.$$

We study this system using two different approaches.

(i) If we want the perturbed oscillation to have the same maximal amplitude $q_{\max}$, i.e. $x_{\max} = 1$, the energy must be chosen to be $E = E^{(0)}(1+\varepsilon)$. The aim is to compute the period of the perturbed solution to order ε. From (1.55) we have

$$\begin{aligned} T &= \sqrt{\frac{2m}{E}} \int_{-q_{\max}}^{q_{\max}} \mathrm{d}q \left(1 - \frac{m\omega_0^2}{2E} q^2 + \frac{\mu}{E} q^4\right)^{-1/2} \\ &= \frac{2}{\omega_0\sqrt{1+\varepsilon}} \int_{-1}^{+1} \mathrm{d}x \left(1 - \frac{x^2}{1+\varepsilon} - \frac{\varepsilon}{1+\varepsilon} x^4\right)^{-1/2} \,. \end{aligned}$$

In the neighborhood of $\varepsilon = 0$ one has

$$T(\varepsilon = 0) = \frac{2}{\omega_0} \int_{-1}^{+1} \frac{dx}{\sqrt{1 - x^2}} = \frac{2\pi}{\omega_0} ,$$

$$\left. \frac{dT}{d\varepsilon} \right|_{\varepsilon=0} = -\frac{1}{\omega_0} \left\{ \int_{-1}^{+1} \frac{dx}{\sqrt{1 - x^2}} + \int_{-1}^{+1} \frac{x^2 dx}{\sqrt{1 - x^2}} \right\} = -\frac{3}{4} \frac{2\pi}{\omega_0} .$$

Thus, the perturbed solution with the same maximal amplitude has the frequency $\omega = \omega_0(1 + 3\varepsilon/4) + O(\varepsilon^2)$. It reads

$$q(t) \simeq q_{max} \sin\left((1 + 3\varepsilon/4)\omega_0 t + \varphi_0\right) . \qquad (2.180)$$

Comparing this with the unperturbed solution $q^{(0)}(t) = q_{max} \sin(\omega_0 t + \varphi_0)$, we see that $q(t)$ is in opposite phase to $q^{(0)}(t)$ after the time $\Delta = 4\pi/(3\varepsilon\omega_0)$. Thus, with increasing time, the perturbed solution moves far away from the unperturbed one.

(ii) Let us analyze the same system but this time making use of the methods of Sect. 2.38. The action variable (2.168) of the unperturbed oscillator is given by

$$\begin{aligned} I^{(0)} &= \frac{1}{2\pi} 2 \int_{-q_{max}}^{q_{max}} dq \sqrt{2m(E^{(0)} - m\omega_0^2 q^2/2)} \\ &= \frac{2E^{(0)}}{\pi\omega_0} \int_{-1}^{+1} dx \sqrt{1 - x^2} = \frac{E^{(0)}}{\omega_0} , \end{aligned}$$

and therefore we have $H_0(I^{(0)}) = E^{(0)} = I^{(0)}\omega_0$. The angle variable $\theta^{(0)}$ was determined in the example of Sect. 2.37.1 (cf. also Sect. 2.24, Example (ii)). We have

$$q^{(0)}(t) = q_{max} \sin\theta^{(0)}$$

with

$$q_{max} = \sqrt{\frac{2I^{(0)}}{m\omega_0}} \quad \text{and} \quad \theta^{(0)} = \omega_0 t + \varphi_0 .$$

Inserting this into the perturbation yields

$$H_1(\theta^{(0)}, I^{(0)}) = \frac{4I^{(0)2}}{m^2\omega_0^2} \sin^4\theta^{(0)} .$$

We now calculate the average of $\sin^4\theta^{(0)}$ over the torus $T^1 = S^1$:

$$\int_0^{2\pi} \sin^4\theta^{(0)} d\theta^{(0)} = \frac{3}{8} 2\pi = \frac{3\pi}{4} .$$

The average of H_1 (2.175) is then $\langle H_1 \rangle = 3I^{(0)2}/2m^2\omega_0^2$. Inserting this into (2.176) we get

$$\dot{\theta}^{(1)}(t) = \frac{\partial}{\partial I^{(0)}} \langle H_1 \rangle = \frac{3 I^{(0)}}{m^2 \omega_0^2} \,, \qquad \dot{I}^{(1)}(t) = 0 \,. \tag{2.181}$$

To first order in the parameter μ, which measures the strength of the perturbation, we obtain according to (2.173)

$$\frac{1}{t} \theta(t) \simeq \omega_0 + \frac{3\mu I^{(0)}}{m^2 \omega_0^2} = \omega_0 \left(1 + \frac{3}{4} \varepsilon \right) ,$$
$$I(T) \simeq I^{(0)} \,, \tag{2.182}$$

with ε as defined above. Clearly (2.182) is precisely our earlier result (2.180): the frequency increases a little, but the action variable stays constant.

2.40.2 Averaging of Perturbations

The result (2.176) for the motion in first-order perturbation theory contains the average of H_1 over the torus (2.175). This average is the same as the time average (if the frequencies are rationally independent). This is a special case of a more general situation that may be described as follows. For the sake of simplicity we consider the case $f = 1$. The unperturbed system has the period $T_0 = 2\pi/\omega_0$. Take t to be a time large compared to T_0, but still small compared to $\Delta \simeq T_0/\mu$, where μ again measures the strength of the perturbation. It is instructive to consider the example of Sect. 2.40.1, where $\Delta = 4\pi/(3\varepsilon\omega_0)$ is the time after which the perturbed system is completely out of phase. Taking, for example, the solution (2.180) with $\varphi_0 = 0$, we have

$$q(t) \simeq \sin\left(\frac{2\pi}{T_0} t\right) \cos\left(\frac{2\pi}{\Delta} t\right) + \cos\left(\frac{2\pi}{T_0} t\right) \sin\left(\frac{2\pi}{\Delta} t\right) ,$$

which, for $T_0 < t \ll \Delta$, is approximately

$$q(t) \simeq \sin\left(\frac{2\pi}{T_0} t\right) + \frac{2\pi}{\Delta} t \cos\left(\frac{2\pi}{T_0} t\right) = \sin\left(\frac{2\pi}{T_0} t\right) + \frac{3}{2} \varepsilon \omega_0 t \cos\left(\frac{2\pi}{T_0} t\right) .$$

Thus, the unperturbed solution is modified by a small term that is the product of a term proportional to t and of $\cos(2\pi t/T_0)$, the latter being of comparatively rapid oscillation. During the same time the action variable does not change, or changes only to second order in the perturbation.

More generally, if the equations $\dot{\theta}^{(0)} = \theta^{(0)}(I^{(0)})$, $\dot{I}^{(0)} = 0$ are subject to a perturbation such that the perturbed equations of motion read

$$\dot{\theta} = \theta^{(0)}(I^{(0)}) + \mu f(\theta, I) \,,$$
$$\dot{I} = \mu g(\theta, I) \,, \tag{2.183}$$

where f and g are periodic functions in θ, then the change of the action variable over time t will be approximately

$$\delta I \simeq \mu t \left\{ \frac{1}{t} \int_0^t g(\theta^{(0)}(t'), I^{(0)}) \mathrm{d}t' \right\} .$$

As $t > T_0$, the term in curly brackets is approximately the time average, taken over the unperturbed motion. Here, this is equal to the average over the torus T^1. Therefore, one expects the average behavior of $I(t)$ to be described by the differential equation

$$\dot{\bar{I}} = \mu \langle g \rangle = \frac{\mu}{2\pi} \int_0^{2\pi} \mathrm{d}\theta g(\theta, I) . \tag{2.184}$$

Returning to the special case of a Hamiltonian perturbation,

$$H = H_0(I) + \mu H_1(\theta, I) ,$$

the second equation (2.183) reads

$$\dot{I} = -\mu \frac{\partial H_1}{\partial \theta} .$$

As H_1 was assumed to be periodic, the average of $\partial H_1/\partial\theta$ over the torus vanishes and we obtain the averaged equation (2.184),

$$\dot{\bar{I}} = 0 ,$$

in agreement with the results of perturbation theory. These results tell us that the action variable does not change over time intervals of the order of t, with $T_0 < t < \Delta$. Dynamical quantities that have this property are said to be *adiabatic invariants*. The characteristic time interval that enters the definition of such invariants is t, with $t < \Delta \simeq T_0/\mu$. Therefore, it is meaningful to make the replacement $\mu = \eta t$ in the perturbed system $H(\underset{\sim}{\theta}, \underset{\sim}{I}, \mu)$. For times $0 \leq t < 1/\eta$ the system changes slowly, or adiabatically. A dynamical quantity $F(\underset{\sim}{\theta}, \underset{\sim}{I}, \mu) : \mathbb{P} \to \mathbb{R}$ is called *adiabatic invariant* if for every positive constant c there is an η_0 such that for $\eta < \eta_0$ and $0 < t < 1/\eta$

$$|F(\underset{\sim}{\theta}(t), \underset{\sim}{I}(t), \eta t) - F(\underset{\sim}{\theta}(0), \underset{\sim}{I}(0), 0)| < c \tag{2.185}$$

(see e.g. Arnol'd 1978, 1983).

Note that the perturbation on the right-hand side of (2.183) need not be Hamiltonian. Thus, we can also study more general dynamical systems of the form

$$\dot{\underset{\sim}{x}} = \mu \underset{\sim}{f}(\underset{\sim}{x}, t, \mu) , \tag{2.186}$$

where $\underset{\sim}{x} \in \mathbb{P}$, $0 \leq \mu \ll 1$, and where $\underset{\sim}{f}$ is periodic with period T_0 in the time variable t. Defining

$$\langle \underset{\sim}{f} \rangle(\underset{\sim}{x}) \stackrel{\text{def}}{=} \frac{1}{T_0} \int_0^{T_0} \underset{\sim}{f}(\underset{\sim}{x}, t, 0) \mathrm{d}t ,$$

we may decompose $\underset{\sim}{f}$ into its average and an oscillatory part,

$$\underset{\sim}{f} = \langle \underset{\sim}{f} \rangle (x) + \underset{\sim}{g}(\underset{\sim}{x}, t, \mu) \, .$$

Substituting

$$\underset{\sim}{x} = \underset{\sim}{y} + \mu \underset{\sim}{S}(\underset{\sim}{y}, t, \mu)$$

and taking the differential with respect to t gives

$$\begin{aligned} \frac{\partial y_k}{\partial t} + \mu \sum_i \frac{\partial S_k}{\partial y_i} \frac{\partial y_i}{\partial t} &= \sum_i \left(\delta_{ik} + \mu \frac{\partial S_k}{\partial y_i} \right) \frac{\partial y_i}{\partial t} = \dot{x}_k - \mu \frac{\partial S_k}{\partial t} \\ &= \mu \langle f_k \rangle (\underset{\sim}{x}) + \mu g_k(\underset{\sim}{x}, t, \mu) - \mu \frac{\partial S_k}{\partial t} \, . \end{aligned}$$

If S is chosen such that $\partial S_k / \partial t = g_k(\underset{\sim}{y}, t, 0)$, and if terms of higher than first order in μ are neglected, then (2.186) becomes the average, autonomous system

$$\dot{\underset{\sim}{y}} = \mu \langle \underset{\sim}{f} \rangle (\underset{\sim}{y}) \, , \tag{2.187}$$

(see Guckenheimer, Holmes 1986, Sect. 4.1). Let us return once more to the example of Sect. 2.40.1. In the first approach we had asked for that solution of the perturbed system which had the same maximal amplitude as the unperturbed one. Now we have learnt to "switch on" the perturbation, in a time-dependent fashion, by letting $\mu = \eta t$ increase slowly (adiabatically) from $t = 0$ to t. Our result tells us that the adiabatically increasing perturbation deforms the solution with energy $E^{(0)}$ and amplitude $q_{\max}$ smoothly into the perturbed solution with energy $E = E^{(0)}(1 + \varepsilon)$ that has the same amplitude as the unperturbed one.

A final word of caution is in order. The effects of small perturbations are by no means always smooth and adiabatic – in contrast to what the simple examples above seem to suggest. For example, if the time dependence of the perturbation is in resonance with one of the frequencies of the unperturbed system, then even a small perturbing term will upset the system dramatically.

2.41 Generalized Theorem of Noether

In the original form of Noether's theorem, Sect. 2.19, the Lagrangian function was assumed to be strictly invariant under continuous transformations containing the identity. Invariance of $L(q_1, \ldots, q_f, \dot{q}_1, \ldots, \dot{q}_f)$ with respect to one-parameter symmetry transformations of the variables q_i implied that the equations of motion were covariant, that is were form invariant under such transformations. This is one of the reasons why the notion of Lagrangian function is of central importance: in many situations it is far simpler to construct invariants rather than covariant differential equations. The theorem in its strict form was illustrated by the closed n-particle system with central forces. It was shown that

- its invariance under space translations yielded conservation of total momentum,
- its invariance under rotations in $\mathbb{R}^3$, yielded conservation of total angular momentum.
- By extending the definition of independent generalized coordinates it was also possible to demonstrate the relationship between invariance under translations in time and conservation of energy, s. Exercise 2.17 and its solution.

Throughout this section, for simplicity, we drop the "under-tilda" on points of velocity space or phase space and write q for $(q_1, \dots, q_f)$, $\dot{q}$ for $(\dot{q}_1, \dots, \dot{q}_f)$, and p for $(p_1, \dots, p_f)$.

In this section we discuss further versions of Noether's theorem which generalize the previous case in two respects. First we recall that covariance of the equations of motion is also guaranteed if the Lagrangian function is not strictly invariant but is modified by an additive time differential of a function of q_i and t,

$$L(q, \dot{q}, t) \mapsto L'(q, \dot{q}, t) = L(q, \dot{q}, t) + \frac{\mathrm{d}}{\mathrm{d}t} M(t, q) . \tag{2.188}$$

The function M should be a smooth function (or, at least, a C^2 function) of the coordinates q_i and possibly time but otherwise is arbitrary. As an example consider the Lagrangian function of n freely moving particles,

$$L(\boldsymbol{x}_1, \dots, \boldsymbol{x}_n, \dot{\boldsymbol{x}}_1, \dots, \dot{\boldsymbol{x}}_n) = T_{\mathrm{kin}} = \frac{1}{2} \sum_{i=1}^{n} m_i \dot{\boldsymbol{x}}_i^2 .$$

Obviously, L is invariant under arbitrary Galilei transformations, the corresponding Euler-Lagrange equations are covariant, i.e. if one of the two following equations holds then also the other holds,

$$\frac{\mathrm{d}^2 \boldsymbol{x}_i(t)}{\mathrm{d}t^2} = 0 \Longleftrightarrow \frac{\mathrm{d}^2 \boldsymbol{x}'_i(t')}{\mathrm{d}t'^2} = 0 , \quad i = 1, 2, \dots, n .$$

Thus, a general Galilei transformation (1.32) (barring time reversal and space reflections)

$$t \mapsto t' = t + s , \quad s \in \mathbb{R} , \tag{2.189a}$$
$$\boldsymbol{x} \mapsto \boldsymbol{x}' = \mathbf{R}\boldsymbol{x} + \boldsymbol{w}t + \boldsymbol{a} , \qquad \mathbf{R} \in \mathrm{SO}(3) , \ \boldsymbol{w}, \boldsymbol{a} \text{ real}, \tag{2.189b}$$

does not change the form of the equations of motion. However, it does change the Lagrangian function, viz.

$$\begin{aligned} L'(\boldsymbol{x}'_i, \dot{\boldsymbol{x}}') &= L(\boldsymbol{x}_i, \dot{\boldsymbol{x}}_i) + \sum_{i=1}^{n} m_i \left(\mathbf{R}\dot{\boldsymbol{x}}_i\right) \cdot \boldsymbol{w} + \frac{1}{2} \sum_{i=1}^{n} m_i \boldsymbol{w}^2 \\ &= L(\boldsymbol{x}_i, \dot{\boldsymbol{x}}_i) + \sum_{i=1}^{n} m_i \dot{\boldsymbol{x}}_i \cdot \left(\mathbf{R}^{-1}\boldsymbol{w}\right) + \frac{1}{2} \sum_{i=1}^{n} m_i \boldsymbol{w}^2 . \end{aligned}$$

The new function L' differs from L by the time differential of the function

$$M(\boldsymbol{x}_1, \dots, \boldsymbol{x}_n, t) = \sum_{i=1}^{n} m_i \boldsymbol{x}_i \cdot \left(\mathbf{R}^{-1}\boldsymbol{w}\right) + \tfrac{1}{2}t\Big(\sum_{i=1}^{n} m_i\Big)\boldsymbol{w}^2 \; .$$

This is seen to be a gauge transformation in the sense of Sect. 2.10 which leaves the equations of motion unchanged. The example shows that Noether's theorem can be extended to cases where the Lagrangian function is modified by the gauge terms introduced in Sect. 2.10.

A further generalization of the theorem consists in admitting gauge functions in (2.188) which depend on the variables t, q_i, as well as on $\dot{q}_i$ provided a supplementary condition is introduced which guarantees that any new acceleration terms $\ddot{q}_i$ caused by the symmetry transformation vanish identically[12].

Given a mechanical system with f degrees of freedom, to which one can associate a Lagrangian function $L(q, \dot{q}, t)$ and coordinates $(t, q_1, \dots, q_f, \dot{q}_1, \dots, \dot{q}_f)$ on $\mathbb{R}_t \times \mathbb{P}$ (direct product of time axis and phase space), consider transformations of the coordinates

$$t' = g(t, q, \dot{q}, s) \; , \tag{2.190a}$$

$$q'^{i} = h^i(t, q, \dot{q}, s) \; . \tag{2.190b}$$

The functions g and h^i should be (at least) twice differentiable in their $2f + 2$ arguments. The real parameter s varies within an interval that includes zero, and for $s = 0$ (2.190a) and (2.190b) are the identity transformations

$$g(t, q, \dot{q}, s = 0) = t \; , \qquad h^i(t, q, \dot{q}, s = 0) = q^i \; , \quad i = 1, 2, \dots, f \; .$$

As for the case of strict invariance of the Lagrangian function only the neighbourhood of $s = 0$ matters for our purposes. This means that g and h^i may be expanded up to first order in s,

$$\delta t := t' - t = \left.\frac{\partial g}{\partial s}\right|_{s=0} s + \mathcal{O}(s^2) \equiv \tau(t, q, \dot{q})s + \mathcal{O}(s^2) \; , \tag{2.191a}$$

$$\delta q^i := q'^{i} - q^i = \left.\frac{\partial h^i}{\partial s}\right|_{s=0} s + \mathcal{O}(s^2) \equiv \kappa^i(t, q, \dot{q})s + \mathcal{O}(s^2) \; , \tag{2.191b}$$

terms of order s^2 and higher being neglected. The first derivatives defined in (2.191a) and in (2.191b),

$$\tau(t, q, \dot{q}) = \left.\frac{\partial g(t, q, \dot{q}, s)}{\partial s}\right|_{s=0} \; , \quad \kappa^i(t, q, \dot{q}) = \left.\frac{\partial h^i(t, q, \dot{q}, s)}{\partial s}\right|_{s=0} \; ,$$

are the generators for infinitesimal transformations g and h^i.

An arbitrary smooth curve $t \to q(t)$ is mapped by g and h^i to a curve $t' \to q'(t')$. To first order in s, their time derivatives fulfill the relation

$$\frac{\mathrm{d}q'^{i}}{\mathrm{d}t'} = \frac{\mathrm{d}q'^{i}}{\mathrm{d}t}\frac{\mathrm{d}t}{\mathrm{d}t'} = \frac{\dot{q}^i + s\dot{\kappa}^i}{1 + s\dot{\tau}} = \dot{q}^i + s(\dot{\kappa}^i - \dot{q}^i\dot{\tau}) \; . \tag{2.192}$$

[12] W. Sarlet, F. Cantrijn; SIAM Review **23** (1981) 467.

The action functional on which Hamilton's principle rests, stays invariant, up to gauge terms, if there is a function $M(t, q, \dot{q})$ such that

$$\int_{t_1'}^{t_2'} \mathrm{d}t' \, L\Big(q'(t'), \frac{\mathrm{d}}{\mathrm{d}t'}q'(t'), t'\Big)$$

$$= \int_{t_1}^{t_2} \mathrm{d}t \, L\Big(q(t), \frac{\mathrm{d}}{\mathrm{d}t}q(t), t\Big) + s \int_{t_1}^{t_2} \mathrm{d}t \, \frac{\mathrm{d}M(t, q, \dot{q})}{\mathrm{d}t} + \mathcal{O}(s^2)$$

for every smooth curve $t \to q(t)$. The integral on the left-hand side can be transformed to an integral over t from t_1 to t_2,

$$\int_{t_1'}^{t_2'} \mathrm{d}t' \, \cdots = \int_{t_1}^{t_2} \mathrm{d}t \Big(\frac{\mathrm{d}t'}{\mathrm{d}t}\Big) \, \cdots \, .$$

Then, for every smooth curve we must have

$$L\Big(q'(t'), \frac{\mathrm{d}}{\mathrm{d}t'}q'(t'), t'\Big)\frac{\mathrm{d}t'}{\mathrm{d}t}$$

$$= L\Big(q(t), \frac{\mathrm{d}}{\mathrm{d}t}q(t), t\Big) + s\frac{\mathrm{d}M(t, q, \dot{q})}{\mathrm{d}t} \, , \tag{2.193a}$$

this being an identity in the variables t, q, and $\dot{q}$. To first order in s and with $\frac{\mathrm{d}t'}{\mathrm{d}t} = 1 + s\dot{\tau}$ this equation yields

$$\frac{\partial L}{\partial t}\delta t + \sum_i \frac{\partial L}{\partial q^i}\delta q^i + \sum_i \frac{\partial L}{\partial \dot{q}^i}\delta \dot{q}^i + sL(t, q, \dot{q})\dot{\tau} = s\frac{\mathrm{d}M(t, q, \dot{q})}{\mathrm{d}t} \, . \tag{2.193b}$$

What we have to do next is to insert (2.192) and to calculate the total time derivatives $\dot{\tau}$, $\dot{\kappa}^i$, and $\dot{M}(t, q, \dot{q})$, obtaining

$$\begin{aligned}
\delta \dot{q}^i &= s(\dot{\kappa}^i - \dot{q}^i\dot{\tau}) \, , \\
\dot{\tau} &= \frac{\partial \tau}{\partial t} + \sum_i \frac{\partial \tau}{\partial q^i}\dot{q}^i + \sum_i \frac{\partial \tau}{\partial \dot{q}^i}\ddot{q}^i \, , \\
\dot{\kappa}^i &= \frac{\partial \kappa^i}{\partial t} + \sum_k \frac{\partial \kappa^i}{\partial q^k}\dot{q}^k + \sum_k \frac{\partial \kappa^i}{\partial \dot{q}^k}\ddot{q}^k \, , \\
\dot{M} &= \frac{\partial M}{\partial t} + \sum_i \frac{\partial M}{\partial q^i}\dot{q}^i + \sum_i \frac{\partial M}{\partial \dot{q}^i}\ddot{q}^i \, .
\end{aligned}$$

Collecting all terms in (2.193b) which are proportional to s, inserting the auxiliary formulae just given, comparison of coefficients in (2.193b) gives the somewhat lengthy expression

$$\frac{\partial L}{\partial t}\tau+\sum_i \frac{\partial L}{\partial q^i}\kappa^i$$
$$+\sum_i \frac{\partial L}{\partial \dot{q}^i}\Big\{\Big(\frac{\partial \kappa^i}{\partial t}+\sum_j \frac{\partial \kappa^i}{\partial q^j}\dot{q}^j+\sum_j \frac{\partial \kappa^i}{\partial \dot{q}^j}\ddot{q}^j\Big)$$
$$-\dot{q}^i\Big(\frac{\partial \tau}{\partial t}+\sum_j \frac{\partial \tau}{\partial q^j}\dot{q}^j+\sum_j \frac{\partial \tau}{\partial \dot{q}^j}\ddot{q}^j\Big)\Big\}$$
$$+L(t,q,\dot{q})\Big(\frac{\partial \tau}{\partial t}+\sum_j \frac{\partial \tau}{\partial q^j}\dot{q}^j+\sum_j \frac{\partial \tau}{\partial \dot{q}^j}\ddot{q}^j\Big)$$
$$=\Big(\frac{\partial M}{\partial t}+\sum_i \frac{\partial M}{\partial q^i}\dot{q}^i+\sum_i \frac{\partial M}{\partial \dot{q}^i}\ddot{q}^i\Big)\,. \qquad (2.193c)$$

At this point one imposes the condition on the terms containing accelerations $\ddot{q}^j$ that was formulated above. This yields a first set of f equations. Indeed, collecting all such terms for every value of the index j, one obtains

$$L(t,q,\dot{q})\frac{\partial \tau}{\partial \dot{q}^j}+\sum_i \frac{\partial L}{\partial \dot{q}^i}\Big(\frac{\partial \kappa^i}{\partial \dot{q}^j}-\dot{q}^i\frac{\partial \tau}{\partial \dot{q}^j}\Big)=\frac{\partial M}{\partial \dot{q}^j}\,, \qquad (2.194a)$$
$$j=1,\dots,f\,.$$

If these equations are fulfilled the lengthy equation (2.193c) reduces to one further equation that will be important for identifying integrals of the motion. It reads

$$\frac{\partial L}{\partial t}\tau+\sum_i \frac{\partial L}{\partial q^i}\kappa^i+\sum_i \frac{\partial L}{\partial \dot{q}^i}\Big\{\frac{\partial \kappa^i}{\partial t}+\sum_j \frac{\partial \kappa^i}{\partial q^j}\dot{q}^j-\dot{q}^i\Big(\frac{\partial \tau}{\partial t}+\sum_j \frac{\partial \tau}{\partial q^j}\dot{q}^j\Big)\Big\}$$
$$+L(t,q,\dot{q})\Big(\frac{\partial \tau}{\partial t}+\sum_j \frac{\partial \tau}{\partial q^j}\dot{q}^j\Big)=\frac{\partial M}{\partial t}+\sum_i \frac{\partial M}{\partial q^i}\dot{q}^i\,. \qquad (2.194b)$$

Thus, one obtains in total $(f+1)$ equations which hold for arbitrary smooth curves $t\to q(t)$. These equations simplify when $q(t)=\varphi(t)$ is a solution of the Euler-Lagrange equations for L,

$$\frac{\mathrm{d}}{\mathrm{d}t}\frac{\partial L}{\partial \dot{q}^i}-\frac{\partial L}{\partial q^i}=0\,,\quad i=1,2,\dots,f\,,\qquad q(t)=\varphi(t)\,.$$

The strategy aiming at uncovering integrals of the motion is the following: Write eq. (2.193c), as far as possible, as a sum of terms which contain only *total* time derivatives, and make use of the equations of motion, to replace where ever this is necessary, $\frac{\partial L}{\partial q^i}$ by $\frac{\mathrm{d}}{\mathrm{d}t}(\frac{\partial L}{\partial \dot{q}^i})$. Note that those expressions in (2.193c) that are contained in round brackets are already in the form of total time derivatives. Only the first two terms on the left-hand side still contain partial derivatives. Repeating (2.193c), inserting the equations of motion in the second term, it becomes

$$\frac{\partial L}{\partial t}\tau+\sum_i\left(\frac{\mathrm{d}}{\mathrm{d}t}\frac{\partial L}{\partial \dot{q}^i}\right)\kappa^i+\sum_i\frac{\partial L}{\partial \dot{q}^i}\frac{\mathrm{d}\kappa^i}{\mathrm{d}t}$$
$$-\sum_i\frac{\partial L}{\partial \dot{q}^i}\dot{q}^i\frac{\mathrm{d}\tau}{\mathrm{d}t}+L\frac{\mathrm{d}\tau}{\mathrm{d}t}-\frac{\mathrm{d}M}{\mathrm{d}t}=0\,. \tag{2.195a}$$

The sum of the second and third terms of this equation is a total differential. Collecting the first and the fourth terms, and making use once more of the equations of motion, one has

$$\frac{\partial L}{\partial t}\tau-\sum_i\frac{\partial L}{\partial \dot{q}^i}\dot{q}^i\frac{\mathrm{d}\tau}{\mathrm{d}t}$$
$$=\frac{\mathrm{d}L}{\mathrm{d}t}\tau-\sum_i\left(\frac{\mathrm{d}}{\mathrm{d}t}\frac{\partial L}{\partial \dot{q}^i}\right)\dot{q}^i\tau-\sum_i\frac{\partial L}{\partial \dot{q}^i}\left(\frac{\mathrm{d}}{\mathrm{d}t}\dot{q}^i\right)\tau-\sum_i\frac{\partial L}{\partial \dot{q}^i}\dot{q}^i\frac{\mathrm{d}\tau}{\mathrm{d}t}$$
$$=\frac{\mathrm{d}L}{\mathrm{d}t}\tau-\frac{\mathrm{d}}{\mathrm{d}t}\sum_i\left(\frac{\partial L}{\partial \dot{q}^i}\dot{q}^i\tau\right). \tag{2.195b}$$

Inserting this in (2.195a) transforms this equation into one that contains indeed only total derivatives with respect to t,

$$\frac{\mathrm{d}}{\mathrm{d}t}(L\tau)+\frac{\mathrm{d}}{\mathrm{d}t}\sum_i\left[\frac{\partial L}{\partial \dot{q}^i}\left(\kappa^i-\dot{q}^i\tau\right)\right]-\frac{\mathrm{d}}{\mathrm{d}t}M=0\,. \tag{2.195c}$$

This shows that the dynamical quantity $I:\mathbb{R}_t\times TQ\to\mathbb{R}$,

$$\boxed{\begin{aligned}I(t,q,\dot{q})&=L(t,q,\dot{q})\tau(t,q,\dot{q})\\&\quad+\sum_i\frac{\partial L}{\partial \dot{q}^i}\left[\kappa^i(t,q,\dot{q})-\dot{q}^i\tau(t,q,\dot{q})\right]-M(t,q,\dot{q})\end{aligned}} \tag{2.196}$$

is constant when taken along solutions $q(t)=\varphi(t)$ of the equations of motions. We call I, eq. (2.196), the *Noether invariant.*

The following examples serve the purpose of illustrating the nature of the Noether invariant and its relationship to the symmetries of the mechanical system described by the Lagrangian function.

Example (i) If the generating function τ as well as the gauge function M vanish identically,

$$\tau(t,q,\dot{q})\equiv 0\,,\quad M(t,q,\dot{q})\equiv 0\,, \tag{2.197}$$

then one is back to the case of strict invariance, Sect. 2.19. The invariant (2.196) then is identical with the expression (2.56) for which several examples were given in Sect. 2.19.

Example (ii) Consider the closed n-particle system described by the Lagrangian function

$$L = \frac{1}{2}\sum_{k=1}^{n} m_k \dot{\boldsymbol{x}}^{(k)2} - U(\boldsymbol{x}^{(1)}, \ldots, \boldsymbol{x}^{(k)}) . \tag{2.198}$$

Obviously, this function is invariant under translations in time. By choosing, accordingly,

$$\tau(t, q, \dot{q}) = -1 , \quad \kappa^i(t, q, \dot{q}) = 0 , \quad M(t, q, \dot{q}) = 0 \tag{2.199}$$

the Noether invariant (2.196) is found to be

$$I = -L + \sum_i \dot{q}^i \frac{\partial L}{\partial \dot{q}^i} = -(T_{\text{kin}} - U) + 2T_{\text{kin}} = T_{\text{kin}} + U = E . \tag{2.200}$$

Thus, invariance under time translations implies conservation of the total energy.

Example (iii) For the same system (2.198) choose the generating functions and the gauge function as follows,

$$\tau(t, q, \dot{q}) = 0 , \quad \kappa^{(k)1}(t, q, \dot{q}) = t , \quad M(t, q, \dot{q}) = \sum_{k=1}^{n} m_k x^{(k)1} , \tag{2.201}$$

with k numbering the particles from 1 to n. The number of degrees of freedom being $f = 3n$ the functions κ^i are numbered by that index k and the three cartesian directions. Inserting (2.201) into (2.196) the Noether invariant is found to be

$$I = t \sum_{k=1}^{n} m_k \dot{x}^{(k)1} - \sum_{k=1}^{n} m_k x^{(k)1} . \tag{2.202}$$

This is seen to be the 1-component of the linear combination

$$t M \boldsymbol{v}_S - M \boldsymbol{r}_S(t) = t \boldsymbol{P} - M \boldsymbol{r}_S(t) , \qquad (M = \sum_{k=1}^{n} m_k)$$

of the center-of-mass's momentum $\boldsymbol{P}$ and of its orbit $\boldsymbol{r}_S(t)$ and is equal to the 1-component of $M\boldsymbol{r}(0)$. This is the center-of-mass principle obtained earlier in Sect. 1.12.

Remarks:

1. There are more examples for the use of the generalized version of Noether's theorem which apply to specific forms of the interaction. For instance, in the case of the Kepler problem with its characteristic $1/r$-potential, one can derive the conservation of the Hermann-Bernoulli-Laplace vector (usually called Lenz-Runge vector) (see also exercise 2.22 and its solution). This example is worked out in Boccaletti and Pucacco (1998).

2. The theorem of E. Noether has a converse in the following sense. Taking the derivative of the function $I(q, \dot{q}, t)$, eq. (2.196), with respect to $\dot{q}^j$ and using the equations (2.194a) and (2.194b) one sees that

$$\frac{\partial I}{\partial \dot{q}^j} = \sum_k \frac{\partial^2 L}{\partial \dot{q}^j \partial \dot{q}^k} \left(\kappa^k - \dot{q}^k \tau \right) .$$

The matrix of second, mixed partial derivatives that multiplies the right hand side,

$$\mathbf{A} = \left\{ A_{jk} \right\} := \left\{ \frac{\partial^2 L}{\partial \dot{q}^j \partial \dot{q}^k} \right\}$$

is well-known from the Legendre transformation from L to H. Assume its determinant to be different from zero,

$$D = \det \mathbf{A} \neq 0 ,$$

(which is the condition for the Legendre transformation to exist!) so that $\mathbf{A}$ possesses an inverse. Denoting the entries of the inverse by

$$\mathbf{A}^{-1} = \left\{ A^{kl} \right\} , \quad \text{i.e.} \sum_k A_{jk} A^{kl} = \delta^l_j ,$$

the initial equation can be solved for κ^k,

$$\kappa^k(t, q, \dot{q}) = \sum_l A^{kl} \frac{\partial I}{\partial \dot{q}^l} + \dot{q}^k \tau(t, q, \dot{q}) . \tag{2.203a}$$

Inserting this expression in (2.196) and solving for τ one obtains

$$\tau(t, q, \dot{q}) = \frac{1}{L} \left\{ I(t, q, \dot{q}) + M(t, q, \dot{q}) - \sum_l A^{kl} \frac{\partial I}{\partial \dot{q}^l} \frac{\partial L}{\partial \dot{q}^k} \right\} . \tag{2.203b}$$

Thus, to every integral of the motion $I(q, \dot{q}, t)$ of the dynamical system described by the Lagrangian function $L(q, \dot{q}, t)$ there correspond the infinitesimal transformations (2.203a) and (2.203b). For all solutions $t \to \varphi(t)$ of the equations of motion these generating functions leave Hamilton's action integral invariant.

Note, however, that $M(q, \dot{q}, t)$, to a large extent, is an arbitrary function and that, as a consequence, the function $\tau(q, \dot{q}, t)$ ist not unique. For a given integral of the motion there are infinitely many symmetry transformations.

3. There is a corollary to the statement given in the previous remark. Given an integral $I(q, \dot{q}, t) = I^{(0)}(q, \dot{q}, t)$ of the motion for the mechanical system described by $L(q, \dot{q}, t)$, an integral that corresponds to the transformation generated by

$$\tau^{(0)} \equiv 0 , \quad \kappa^i = \kappa^{(0)i}(t, q, \dot{q}) , \quad \text{with} \quad M = M^{(0)}(t, q, \dot{q}) .$$

Then the following transformations

$$\tau = \tau(t, q, \dot{q}) , \quad \kappa^i = \kappa^{(0)i}(t, q, \dot{q}) + \tau(t, q, \dot{q}) \dot{q}^i ,$$

together with the choice

$$M = M^{(0)}(t, q, \dot{q}) + L(t, q, \dot{q}) \tau(t, q, \dot{q})$$

lead to the same integral of the motion. The verification of this is left as an exercise.

Appendix: Practical Examples

1. Small Oscillations. Let a Lagrangian system be described in terms of f generalized coordinates $\{q_i\}$, each of which can oscillate around an equilibrium position q_i^0. The potential energy $U(q_1, \ldots, q_f)$ having an absolute minimum U_0 at $(q_1^0, \ldots, q_f^0)$ one may visualize this system as a lattice defined by the equilibrium positions $(q_1^0, \ldots, q_f^0)$, the edges of which can oscillate around this configuration. The limit of *small oscillations* is realized if the potential energy can be approximated by a quadratic form in the neighborhood of its minimum, viz.

$$U(q_1, \ldots, q_f) \simeq \frac{1}{2} \sum_{i,k=1}^{f} u_{ik}(q_i - q_i^0)(q_k - q_k^0) . \tag{A.1}$$

Note that for the mathematical pendulum (which has $f = 1$) this is identical with the limit of small deviations from the vertical, i.e. the limit of harmonic oscillation. For $f > 1$ this is a system of coupled harmonic oscillators.

Derive the equations of motion and find the normal modes of this system.

Solution. It is clear that only the symmetric part of the coefficients u_{ik} is dynamically relevant, $a_{ik} \stackrel{\text{def}}{=} (u_{ik} + u_{ki})/2$. As U has a minimum, the matrix

$$\mathbf{A} = \{a_{ik}\}$$

is not only real and symmetric but also positive. This means that all its eigenvalues are real and positive-semidefinite. It is useful to replace the variables q_i by the deviations from equilibrium, $z_i = q_i - q_i^0$. The kinetic energy is a quadratic form of the time derivatives of q_i or, equivalently of z_i, with symmetric coefficients:

$$T = \frac{1}{2} \sum_{i,k} t_{ik} \dot{z}_i \dot{z}_k .$$

The matrix $\{t_{ik}\}$ is not singular and is positive as well. Therefore, one can choose the natural form for the Lagrangian function

$$L = \frac{1}{2} \sum_{i,k} (t_{ik} \dot{z}_i \dot{z}_k - a_{ik} z_i z_k) , \tag{A.2}$$

from which follows the system of coupled equations

$$\sum_{k=1}^{f} t_{ik} \ddot{z}_k + \sum_{j=1}^{f} a_{ij} z_j = 0 , \quad i = 1, \ldots, f . \tag{A.3}$$

For $f = 1$ this is the equation of the harmonic oscillator. This suggests solving the general case by means of the substitution

$$z_i = a_i \, \mathrm{e}^{\mathrm{i}\Omega t} .$$

The complex form is chosen in order to simplify the calculations. In the end we shall have to take the real part of the eigenmodes. Inserting this expression for z into the equations of motion (A.3) yields the following system of coupled linear equations:

$$\sum_{j=1}^{f}(-\Omega^2 t_{ij} + a_{ij})a_j = 0 \,. \tag{A.4}$$

This has a nontrivial solutions if and only if the determinant of its coefficient vanishes,

$$\det(a_{ij} - \Omega^2 t_{ij}) \stackrel{!}{=} 0 \,. \tag{A.5}$$

This equation has f positive-semidefinite solutions

$$\Omega_l^2 \,, \quad l = 1, \ldots, f \,,$$

which are said to be the eigenfrequencies of the system.

As an example we consider two identical harmonic oscillators (frequency ω_0) that are coupled by means of a harmonic spring. The spring is not active when both oscillators are at rest (or, more generally, whenever the difference of their positions is the same as at rest). It is not difficult to guess the eigenfrequencies of this system: (i) the two oscillators swing in phase, the spring remains inactive; (ii) the oscillators swing in opposite phase. Let us verify this behavior within the general analysis. We have

$$T = \tfrac{1}{2}m(\dot{z}_1^2 + \dot{z}_2^2) \,,$$
$$U = \tfrac{1}{2}m\omega_0^2(z_1^2 + z_2^2) + \tfrac{1}{2}m\omega_1^2(z_1 - z_2)^2 \,.$$

Taking out the common factor m, the system (A.4) reads

$$\begin{pmatrix} (\omega_0^2 + \omega_1^2) - \Omega^2 & -\omega_1^2 \\ -\omega_1^2 & (\omega_0^2 - \omega_1^2) - \Omega^2 \end{pmatrix} \begin{pmatrix} a_1 \\ a_2 \end{pmatrix} = 0 \,. \tag{A.4$'$}$$

The condition (A.5) yields a quadratic equation whose solutions are

$$\Omega_1^2 = \omega_0^2 \,, \quad \Omega_2^2 = \omega_0^2 + 2\omega_1^2 \,.$$

Inserting these, one by one, into the system of equations (A.4′), one finds

$$\text{for} \quad \Omega_1 = \omega_0 \,, \quad a_2^{(1)} = a_1^{(1)} \,,$$
$$\text{for} \quad \Omega_2 = \sqrt{\omega_0^2 + 2\omega_1^2} \,, \quad a_2^{(2)} = -a_1^{(2)} \,.$$

(The normalization is free. We choose $a_1^{(i)} = 1/\sqrt{2}$, $i = 1, 2$). Thus, we indeed obtain the expected solutions. The linear combinations above, i.e.

$$Q_1 \stackrel{\text{def}}{=} \sqrt{m} \sum_i a_i^{(1)} z_i = (z_1 + z_2)\sqrt{m/2}\,,$$

$$Q_2 \stackrel{\text{def}}{=} \sqrt{m} \sum_i a_i^{(2)} z_i = (z_1 - z_2)\sqrt{m/2}$$

decouple the system completely. The Lagrangian function becomes

$$L = \frac{1}{2} \sum_{l=1}^{2} (\dot{Q}_l^2 - \Omega_l^2 Q_l^2)\,.$$

It describes two independent linear oscillators. The new variables Q_i are said to be *normal coordinates* of the system. They are defined by the eigenvectors of the matrix $(a_{ij} - \Omega_l^2 t_{ij})$ and correspond to the eigenvalues Ω_l^2.

In the general case ($f > 2$) one proceeds in an analogous fashion. Determine the frequencies from (A.5) and insert them, one by one, into (A.4). Solve this system and determine the eigenvectors $(a_1^{(l)}, \ldots, a_f^{(l)})$ (up to normalization) that pertain to the eigenvalues Ω_l^2.

If all eigenvalues are different, the eigenvectors are uniquely determined up to normalization. We write (A.4) for two different eigenvalues,

$$\sum_j (-\Omega_q^2 t_{ij} + a_{ij}) a_j^{(q)} = 0\,, \tag{A.6a}$$

$$\sum_i (-\Omega_p^2 t_{ij} + a_{ij}) a_i^{(p)} = 0\,, \tag{A.6b}$$

and multiply the first equation by $a_i^{(p)}$ from the left, the second by $a_j^{(q)}$ from the left. We sum the first over i and the second over j and take their difference. Both t_{ij} and a_{ij} are symmetric. Therefore, we obtain

$$(\Omega_p^2 - \Omega_q^2) \sum_{i,j} a_i^{(p)} t_{ij} a_j^{(q)} = 0\,.$$

As $\Omega_p^2 \neq \Omega_q^2$, the double sum must vanish if $p \neq q$. For $p = q$, we can normalize the eigenvectors such that the double sum gives 1. We conclude that

$$\sum_{i,j} a_i^{(p)} t_{ij} a_j^{(q)} = \delta_{pq}\,.$$

Equation (A.6a) and the result above can be combined to obtain

$$\sum_{i,j} a_i^{(p)} a_{ij} a_j^{(q)} = \Omega_p^2 \sum a_i^{(p)} t_{ij} a_j^{(q)} = \Omega_p^2 \delta_{pq}\,.$$

This result tells us that the matrices t_{ij} and a_{ij} are diagonalized simultaneously. We then set

$$z_i = \sum_p a_i^{(p)} Q_p \tag{A.7}$$

and insert this into the Lagrangian function to obtain

$$L = \frac{1}{2} \sum_{p=1}^{f} (\dot{Q}_p^2 - \Omega_p^2 Q_p^2) . \tag{A.8}$$

Thus, we have achieved the transformation to normal coordinates.

If some of the frequencies are degenerate, the corresponding eigenvectors are no longer uniquely determined. It is always possible, however, to choose s linearly independent vectors in the subspace that belongs to $\Omega_{r_1} = \Omega_{r_2} = \ldots = \Omega_{r_s}$ (s denotes the degree of degeneracy). This construction is given in courses on linear algebra.

One can go further and try several examples on a PC: a linear chain of n oscillators with harmonic couplings, a planar lattice of mass points joined by harmonic springs, etc., for which the matrices t_{ik} and a_{ik} are easily constructed. If one has at one's disposal routines for matrix calculations, it is not difficult to find the eigenfrequencies and the normal coordinates.

2. The Planar Mathematical Pendulum and Liouville's Theorem. Work out (numerically) Example (ii) of Sect. 2.30 and illustrate it with some figures.

Solution. We follow the notation of Sect. 1.17.2, i.e. we take $z_1 = \varphi$ as the generalized coordinate and $z_2 = \dot{\varphi}/\omega$ as the generalized momentum, where $\omega = \sqrt{g/l}$ is the frequency of the corresponding harmonic oscillator and $\tau = \omega t$. Thus, time is measured in units of $(\omega)^{-1}$, The energy is measured in units of mgl, i.e.

$$\varepsilon = \frac{E}{mgl} = \frac{1}{2} z_2^2 + (1 - \cos z_1) . \tag{A.9}$$

ε is positive-semidefinite. $\varepsilon < 2$ pertains to the oscillating solutions, $\varepsilon = 2$ is the separatrix, and $\varepsilon > 2$ pertains to the rotating solutions. The equations of motion (1.40) yield the second-order differential equation for z_1

$$\frac{d^2 z_1}{d\tau^2} = -\sin z_1(\tau) . \tag{A.10}$$

First, one verifies that z_1 and z_2 are indeed conjugate variables, provided one uses τ as time variable. In order to see this start from the dimensionless Lagrangian function

$$\lambda \stackrel{\text{def}}{=} \frac{L}{mgl} = \frac{1}{2\omega^2} \left(\frac{d\varphi}{dt} \right)^2 - (1 - \cos\varphi) = \frac{1}{2} \left(\frac{dz_1}{d\tau} \right)^2 - (1 - \cos z_1)$$

and take its derivative with respect to $\dot{z}_1 = (dz_1/d\tau)$. This gives $z_2 = (dz_1/d\tau) = (d\varphi/dt)/\omega$, as expected.

For drawing the phase portraits, Fig. 1.10, it is sufficient to plot z_2 as a function of z_1, as obtained from (A.9). This is not sufficient, however, if we wish to follow the motion along the phase curves, as a function of time. As we wish to study the time evolution of an ensemble of initial conditions, we must integrate the differential equation (A.10). This integration can be done numerically, e.g. by means of a Runge–Kutta procedure (cf. Abramowitz, Stegun 1965, Sect. 25.5.22). Equation (A.9) has the form $y'' = -\sin y$. Let h be the step size and y_n and y'_n the values of the function and its derivative respectively at τ_n. Their values at $\tau_{n+1} = \tau_n + h$ are obtained by the following series of steps. Let

$$
\begin{aligned}
k_1 &= -h \sin y_n \,, \\
k_2 &= -h \sin\left(y_n + \frac{h}{2} y'_n + \frac{h}{8} k_1 \right) \,, \\
k_3 &= -h \sin\left(y_n + h y'_n + \frac{h}{2} k_2 \right) \,.
\end{aligned}
\tag{A.11}
$$

Then

$$
\begin{aligned}
y_{n+1} &= y_n + h[y'_n + \tfrac{1}{6}(k_1 + 2k_2)] + O(h^4) \,, \\
y'_{n+1} &= y'_n + \tfrac{1}{6}k_1 + \tfrac{2}{3}k_2 + \tfrac{1}{6}k_3 + O(h^4) \,.
\end{aligned}
\tag{A.12}
$$

Note that y is our z_1 while y' is z_2 and that the two are related by (A.9) to the reduced energy ε. Equations (A.12) are easy to implement on a computer. Choose an initial configuration ($y_0 = z_1(0)$, $y'_0 = z_2(0)$), take $h = \pi/30$, for example, and run the program until the time variable has reached a given endpoint τ. Using the dimensionless variable τ, the harmonic oscillator (corresponding to small oscillations of the pendulum) has the period $T^{(0)} = 2\pi$. It is convenient, therefore, to choose the end point to be $T^{(0)}$ or fractions thereof. This shows very clearly the retardation of the pendulum motion as compared to the oscillator: points on pendulum phase portraits with $0 < \varepsilon \ll 2$ move almost as fast as points on the oscillator portrait; the closer ε approaches 2 from below, the more they are retarded compared to the oscillator. Points on the separatrix ($\varepsilon = 2$) that start from, say, ($z_1 = 0, z_2 = 2$) can never move beyond the first quadrant of the (z_1, z_2)-plane. They approach the point $(\pi, 0)$ asymptotically, as τ goes to infinity.

In the examples shown in Figs. 2.13–15 we study the flow of an initial ensemble of 32 points on a circle with radius $r = 0.5$ and the center of that circle, for the time intervals indicated in the figures. This allows one to follow the motion of each individual point. As an example, in Fig. 2.14 we have marked with arrows the consecutive positions of the point that started from the configuration (0, 1).

Of course, one may try other shapes for the initial ensemble (instead of the circle) and follow its flow through phase space. A good test of the program is to replace the right-hand side of (A.10) with $-z_1$. This should give the picture shown in Fig. 2.12.

3. The Mechanics of Rigid Bodies

The theory of rigid bodies is a particularly important part of general mechanics. Firstly, next to the spherically symmetric mass distributions that we studied in Sect. 1.30, the top is the simplest example of a body with finite extension. Secondly, its dynamics is a particularly beautiful model case to which one can apply the general principles of canonical mechanics and where one can study the consequences of the various space symmetries in an especially transparent manner. Thirdly, its equations of motion (Euler's equations) provide an interesting example of *nonlinear* dynamics. Fourthly, the description of the rigid body leads again to the compact Lie group SO(3) that we studied in connection with the invariance of equations of motion with respect to rotations. The configuration space of a nondegenerate top is the direct product of the three-dimensional space $\mathbb{R}^3$ and of the group SO(3), in the following sense. The momentary configuration of a rigid body is determined if we know (i) the position of its center of mass, and (ii) the orientation of the body relative to a given inertial system. The center of mass is described by a position vector $\boldsymbol{r}_S(t)$ in $\mathbb{R}^3$, the orientation is described by three, time-dependent angles which span the parameter manifold of SO(3).

Finally, there are special cases of the theory of rigid bodies which can be integrated analytically, or can be analyzed by geometrical means. Thus, one meets further nontrivial examples of integrable systems.

3.1 Definition of Rigid Body

A rigid body can be visualized in two ways:

(A) A system of n mass points, with masses $m_1, \ldots, m_n$, which are joined by *rigid* links, is a rigid body. Figure 3.1 shows the example $n = 4$.

(B) A body with a given continuous mass distribution $\varrho(\boldsymbol{r})$ whose shape does not change, is also a rigid body. The hatched volume shown in Fig. 3.2 is an example.

In case (A) the total mass is given by

$$M = \sum_{i=1}^{n} m_i \tag{3.1}$$

F. Scheck, *Mechanics*, Graduate Texts in Physics, 5th ed.,
DOI 10.1007/978-3-642-05370-2_3, © Springer-Verlag Berlin Heidelberg 2010

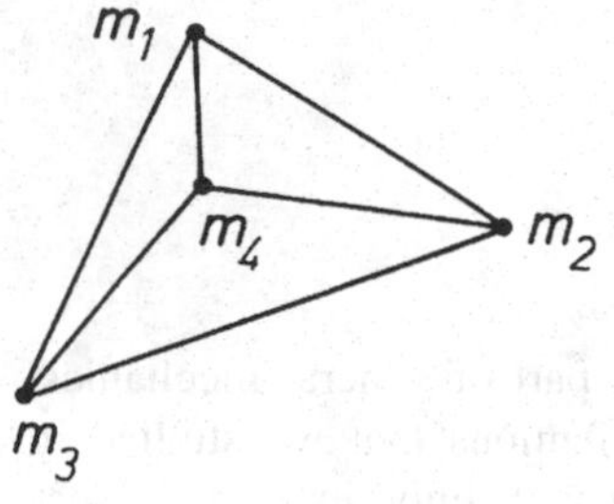

Fig. 3.1. A finite number of mass points whose distances are fixed at all times form a rigid body. The figure shows the example $n = 4$

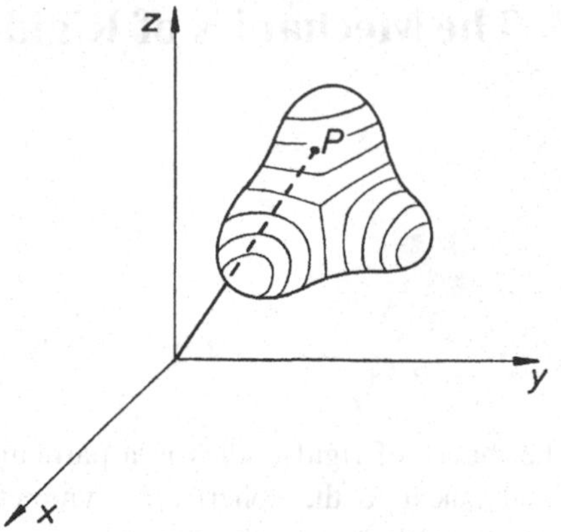

Fig. 3.2. A rigid body consisting of a fixed, invariable mass distribution

while for case (B) it is

$$M = \int d^3r \varrho(r) , \tag{3.2}$$

(cf. Sect. 1.30).

The two definitions lead to the same type of mechanical system. This observation depends in an essential way on the assumption that the body has no internal degrees of freedom whatsoever. If, to the contrary, the shape of the distribution $\varrho(\boldsymbol{r})$ of case (B) is allowed to change in the course of time, there will be internal forces. One expects the dynamics of an extended object with continuous mass distribution to be quite different from that of the system shown in Fig. 3.1 when that object is not rigid. This is the subject of the mechanics of continua, not dealt with here.

It is useful to introduce two classes of coordinate system for the description of rigid bodies and their motion:

(i) a coordinate system **K** that is fixed in space and is assumed to be an inertial system;
(ii) an intrinsic (or body-fixed) coordinate system $\overline{\mathbf{K}}$ which is fixed in the body and therefore follows its motion actively.

Figure 3.3 shows examples of these two types of reference system. The inertial system **K** (which we may also call the observer's or "laboratory" system) is useful for a simple description of the motion. The intrinsic, body-fixed system $\overline{\mathbf{K}}$ in general is *not* an inertial system because its origin follows the motion of the body as a whole and, hence, may be an accelerated frame. It is useful because, with respect to this system, the mass distribution and all static properties derived from it are described in the most simple way. Take for example the mass density. If looked at from $\overline{\mathbf{K}}$, $\bar{\varrho}(\boldsymbol{r})$ is a given function, fixed once and for ever, irrespective of the motion of the body. With respect to **K**, on the other hand, it is a time-dependent

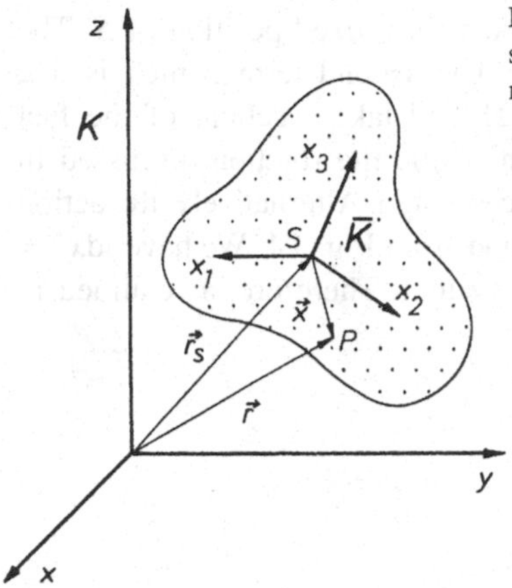

Fig. 3.3. Coordinate system **K** fixed in space and intrinsic system $\overline{\mathbf{K}}$, which is fixed in the body and follows its motion

function $\varrho(\boldsymbol{r}, t)$ that depends on how the body moves in space. (For an example see Exercise 3.9.)

The origin S of $\overline{\mathbf{K}}$ is an arbitrary but fixed point in the body; (it will often be useful to choose the center of mass for S). Let $\boldsymbol{r}_S(t)$ be the position vector of S with respect to the inertial system **K.** Another point P of the body has position vector $\boldsymbol{r}(t)$ with respect to **K**, and $\boldsymbol{x}$ with respect to $\overline{\mathbf{K}}$. As it describes P relative to S, $\boldsymbol{x}$ is independent of time, by construction.

The number of degrees of freedom of a rigid body can be read off Fig. 3.3. Its position in space is completely determined by the following data: the position $\boldsymbol{r}_S(t)$ of S and the orientation of the intrinsic system $\overline{\mathbf{K}}$ with respect to another system centered on S whose axes are parallel to those of **K**. For this we need six quantities: the three components of $\boldsymbol{r}_S$, as well as three angles that fix the relative orientation of $\overline{\mathbf{K}}$. Therefore, *a nondegenerate rigid body has six degrees of freedom.* (The degenerate case of the rod is an exception. The rod is a rigid body whose mass points all lie on a line. It has only five degrees of freedom.) It is essential to distinguish carefully the (space-fixed) inertial system **K** from the (body-fixed) system $\overline{\mathbf{K}}$. Once one has understood the difference between these two reference systems and the role they play in the description of the rigid body, the theory of the top becomes simple and clear.

3.2 Infinitesimal Displacement of a Rigid Body

If we shift and rotate the rigid body infinitesimally, a point P of the body is displaced as follows:

$$\mathrm{d}\boldsymbol{r} = \mathrm{d}\boldsymbol{r}_S + \mathrm{d}\boldsymbol{\varphi} \times \boldsymbol{x} \,, \tag{3.3}$$

where we have used the notation of Fig. 3.3. The displacement $\mathrm{d}\boldsymbol{r}_S$ of the point S is the parallel shift of the body as a whole. The direction $\hat{n} = \mathrm{d}\boldsymbol{\varphi}/|\mathrm{d}\boldsymbol{\varphi}|$ and the angle $|d\varphi|$ characterize the rotation of the body, for a fixed position of S. The translational part of (3.3) is immediately clear. The second term, which is due to the rotation, follows from (2.68) of Sect. 2.21 and takes account of the fact that here we are dealing with an *active* rotation, while the rotation discussed in Sect. 2.21 was a *passive* one – hence the difference in sign. Alternatively, the action of this infinitesimal rotation can also be understood from Fig. 3.4. We have $|d\boldsymbol{x}| = |\boldsymbol{x}| \cdot |d\boldsymbol{\varphi}| \sin\alpha$, $(\hat{\boldsymbol{n}}, \boldsymbol{x}, d\boldsymbol{x})$ forming a right-hand system. Therefore, as claimed in (3.3), $d\boldsymbol{x} = d\boldsymbol{\varphi} \times \boldsymbol{x}$.

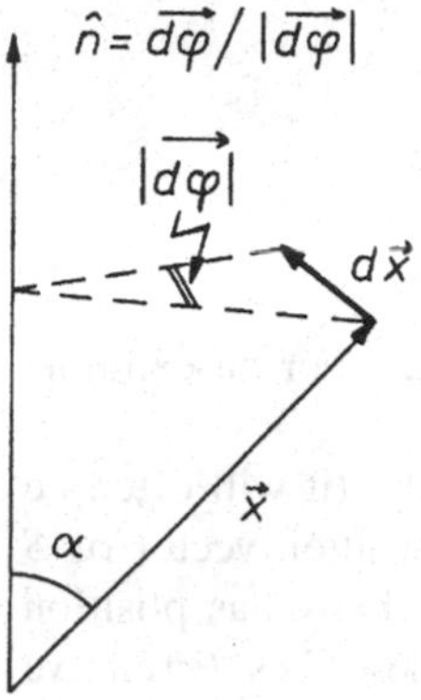

Fig. 3.4. Drawing of the action of a small rotation of the rigid body and from which relation (3.3) can be read off

From (3.3) follows an important relation between the velocities of the points P and S,

$$\boldsymbol{v} \stackrel{\text{def}}{=} \frac{d\boldsymbol{r}}{dt} \quad \text{and} \quad \boldsymbol{V} \stackrel{\text{def}}{=} \frac{d\boldsymbol{r}_S}{dt}, \tag{3.4a}$$

respectively, and the angular velocity

$$\boldsymbol{\omega} \stackrel{\text{def}}{=} \frac{d\boldsymbol{\varphi}}{dt}. \tag{3.4b}$$

It reads

$$\boldsymbol{v} = \boldsymbol{V} + \boldsymbol{\omega} \times \boldsymbol{x}. \tag{3.5}$$

Thus, the velocity of P is the sum of the *translation velocity* of the body as a whole and of a term linear in the *angular velocity* $\boldsymbol{\omega}$. We now show that this angular velocity is universal in the sense that it characterizes the rotational motion of the body but does not depend on the choice of S, the origin of $\overline{\mathbf{K}}$. In order to see this, choose another point S' with coordinate $\boldsymbol{r}'_S = \boldsymbol{r}_S + \boldsymbol{a}$. The relation (3.5) also applies to this choice,

$$\boldsymbol{v} = \boldsymbol{V}' + \boldsymbol{\omega}' \times \boldsymbol{x}'.$$

On the other hand we have $\boldsymbol{r} = \boldsymbol{r}'_S + \boldsymbol{x}' = \boldsymbol{r}_S + \boldsymbol{a} + \boldsymbol{x}'$, and hence $\boldsymbol{x} = \boldsymbol{x}' + \boldsymbol{a}$ and $\boldsymbol{v} = \boldsymbol{V} + \boldsymbol{\omega} \times \boldsymbol{a} + \boldsymbol{\omega} \times \boldsymbol{x}'$. These two expressions for the same velocity hold for any $\boldsymbol{x}$ or $\boldsymbol{x}'$. From this we conclude that

$$V' = V + \omega \times a \,, \tag{3.6a}$$

$$\omega' = \omega \,. \tag{3.6b}$$

This shows the universality of the angular velocity.

3.3 Kinetic Energy and the Inertia Tensor

From now on we place S, the origin of the intrinsic system $\overline{\mathbf{K}}$, at the center of mass of the body. (Exceptions to this will be mentioned explicitly.) With the definition (1.29) of the center of mass, this implies in case (A) that

$$\sum_{i=1}^{n} m_i x^{(i)} = 0 \tag{3.7a}$$

and in case (B) that

$$\int \mathrm{d}^3 x\, x \varrho(x) = 0 \,. \tag{3.7b}$$

We calculate the kinetic energy for both cases (A) and (B), for the sake of illustration.

(i) In the discrete model of the rigid body and making use of (3.5) we find

$$\begin{aligned} T &= \frac{1}{2}\sum_{i=1}^{n} m_i v^{(i)2} = \frac{1}{2}\sum m_i \left(V + \omega \times x^{(i)}\right)^2 \\ &= \frac{1}{2}\left(\sum m_i\right) V^2 + V \cdot \sum m_i \left(\omega \times x^{(i)}\right) + \frac{1}{2}\sum m_i \left(\omega \times x^{(i)}\right)^2 . \end{aligned} \tag{3.8}$$

In the second term of this expression one may use the identity

$$V \cdot \left(\omega \times x^{(i)}\right) = x^{(i)} \cdot (V \times \omega)$$

to obtain

$$V \cdot \sum m_i \left(\omega \times x^{(i)}\right) = (V \times \omega) \cdot \sum m_i x^{(i)} = 0 \,.$$

This term vanishes because of the condition (3.7a).

The third term on the right-hand side of (3.8) contains the square of the vector $\omega \times x^{(i)}$. Omitting for a moment the particle index, we can transform this as follows:

$$\begin{aligned} (\omega \times x)^2 &= \omega^2 x^2 \sin^2 \alpha = \omega^2 x^2 \left(1 - \cos^2 \alpha\right) \\ &= \omega^2 x^2 - (\omega \cdot x)^2 = \sum_{\mu=1}^{3}\sum_{\nu=1}^{3} \omega^\mu \left(x^2 \delta_{\mu\nu} - x_\mu x_\nu\right) \omega^\nu . \end{aligned}$$

The decomposition of this last expression in Cartesian coordinates serves the purpose of separating the coordinates $x^{(i)}$ from the components of the angular velocity $\boldsymbol{\omega}$. The former scan the rigid body, while the latter are universal and hence independent of the body. Inserting these auxiliary results into (3.8) we obtain a simple form for the kinetic energy,

$$T = \frac{1}{2}MV^2 + \frac{1}{2}\sum_{\mu=1}^{3}\sum_{\nu=1}^{3}\omega^{\mu} J_{\mu\nu}\omega^{\nu} \tag{3.9}$$

where we have set

$$\boxed{J_{\mu\nu} \stackrel{\text{def}}{=} \sum_{i=1}^{n} m_i \left[\boldsymbol{x}^{(i)2}\delta_{\mu\nu} - x_{\mu}^{(i)}x_{\nu}^{(i)}\right] .} \tag{3.10a}$$

(ii) The calculation is completely analogous for the continuous model of the rigid body,

$$\begin{aligned} T &= \frac{1}{2}\int d^3x\varrho(\boldsymbol{x})(\boldsymbol{V} + \boldsymbol{\omega}\times\boldsymbol{x})^2 \\ &= \frac{1}{2}\boldsymbol{V}^2\int d^3x\varrho(\boldsymbol{x}) + (\boldsymbol{V}\times\boldsymbol{\omega})\int d^3x\varrho(\boldsymbol{x})\boldsymbol{x} \\ &\quad + \frac{1}{2}\int d^3x\varrho(\boldsymbol{x})\omega^{\mu}\left[\boldsymbol{x}^2\delta_{\mu\nu} - x_{\mu}x_{\nu}\right]\omega^{\nu} . \end{aligned}$$

The integral in the first term is the total mass. The second term vanishes because of the condition (3.7b). Thus, the kinetic energy takes the same form (3.9), with $J_{\mu\nu}$ now given by[1]

$$\boxed{J_{\mu\nu} \stackrel{\text{def}}{=} \int d^3x\varrho(\boldsymbol{x})\left[\boldsymbol{x}^2\delta_{\mu\nu} - x_{\mu}x_{\nu}\right] .} \tag{3.10b}$$

As a result, the kinetic energy of a rigid body (3.9) has the general decomposition

$$T = T_{\text{trans}} + T_{\text{rot}} , \tag{3.11}$$

whose first term is the translational kinetic energy

$$T_{\text{trans}} = \tfrac{1}{2}MV^2 \tag{3.12}$$

and whose second term is the rotational kinetic energy

$$\boxed{T_{\text{rot}} = \tfrac{1}{2}\boldsymbol{\omega}\mathbf{J}\boldsymbol{\omega} .} \tag{3.13}$$

$\mathbf{J} = \{J_{\mu\nu}\}$ is a tensor of rank two, i.e. it transforms under rotations as follows. If

$$x_{\mu} \rightarrow x'_{\mu} = \sum_{\nu=1}^{3} R_{\mu\nu}x_{\nu}$$

[1] In general, $J_{\mu\nu}$ depends on time, whenever the $\boldsymbol{x}_i$ refer to a fixed reference frame in space and when the body rotates with respect to that frame (see Sect. 3.11).

with $\mathbf{R} \in \mathrm{SO}(3)$, then

$$J_{\mu\nu} \rightarrow J'_{\mu\nu} = \sum_{\lambda=1}^{3} \sum_{\varrho=1}^{3} R_{\mu\lambda} R_{\nu\varrho} J_{\lambda\varrho} \, . \tag{3.14}$$

Being completely determined by the mass distribution, this tensor is characteristic for the rigid body. It is called the *inertia tensor.* This name reflects its formal similarity to the inertial mass (which is a scalar, though).

The tensor $\mathbf{J}$ is defined over a three-dimensional Euclidean vector space V. Generally speaking, second-rank tensors are bilinear forms over V. Inertia tensors belong to the subset of real, symmetric, (and as we shall see below) positive tensors over V. We shall not go into the precise mathematical definitions here. What is important for what follows is the transformation behavior (3.14); that is, omitting indices $\mathbf{J}' = \mathbf{R}\mathbf{J}\,\mathbf{R}^{\mathrm{T}}$.

3.4 Properties of the Inertia Tensor

In this and the two subsequent sections we study the inertia tensor as a static property of the rigid body. This means we assume the body to be at rest or, equivalently, make use of a coordinate system that is rigidly linked to the body and follows its motion. The inertia tensor contains an invariant term that is already diagonal,

$$\int \mathrm{d}^3 x \varrho(\boldsymbol{x}) \boldsymbol{x}^2 \delta_{\mu\nu} \, ,$$

and a term that depends on the specific choice of the intrinsic reference system,

$$- \int \mathrm{d}^3 x \varrho(\boldsymbol{x}) x_\mu x_\nu \, ,$$

and that in general is not diagonal. The following properties of the inertia tensor can be derived from its definition (3.10).

(i) $\mathbf{J}$ is linear and therefore additive in the mass density $\varrho(\boldsymbol{x})$. This means that the inertia tensor of a body obtained by joining two rigid bodies equals the sum of the inertia tensors of its components. Quantities that have this additive property are also said to be *extensive*.

(ii) $\mathbf{J}$ is represented by a real, symmetric matrix that reads explicitly

$$\mathbf{J} = \int \mathrm{d}^3 x \varrho(\boldsymbol{x}) \begin{pmatrix} x_2^2 + x_3^2 & -x_1 x_2 & -x_1 x_3 \\ -x_2 x_1 & x_3^2 + x_1^2 & -x_2 x_3 \\ -x_3 x_1 & -x_3 x_2 & x_1^2 + x_2^2 \end{pmatrix} \, . \tag{3.15}$$

Every real and symmetric matrix can be brought to diagonal form by means of an orthogonal transformation $\mathbf{R}_0 \in \mathrm{SO}(3)$

$$\mathbf{R}_0\mathbf{J}\mathbf{R}_0^{-1} = \overset{\circ}{\mathbf{J}} = \begin{pmatrix} I_1 & 0 & 0 \\ 0 & I_2 & 0 \\ 0 & 0 & I_3 \end{pmatrix} , \qquad (3.16)$$

where I_1, $I - 2$, and I_3 are the eigenvalues of $\mathbf{J}$. Thus, by a suitable choice of the body-fixed system of reference the inertia tensor becomes diagonal. Reference systems that have this property are again orthogonal systems and are said to be *principal-axes systems.* Of course, the same representation (3.15) holds also in a system of principal axes. As the inertia tensor $\mathbf{J}$ is then diagonal, its off-diagonal entries vanish and it reads

$$\overset{\circ}{\mathbf{J}} = \int \mathrm{d}^3 y \varrho(\mathbf{y}) \begin{pmatrix} y_2^2 + y_3^2 & 0 & 0 \\ 0 & y_3^2 + y_1^2 & 0 \\ 0 & 0 & y_1^2 + y_2^2 \end{pmatrix} . \qquad (3.17)$$

This formula is useful in calculating the moments of inertia. But, more generally and without doing such a calculation, it allows to derive the following general properties and inequalities for the eigenvalues:

$$I_i \geq 0 , \quad i = 1, 2, 3 , \qquad (3.18a)$$

$$I_1 + I_2 \geq I_3 , \quad I_2 + I_3 \geq I_1 , \quad I_3 + I_1 \geq I_2 . \qquad (3.18b)$$

Thus, the matrix $\mathbf{J}$ is indeed positive. Its eigenvalues I_i are called *(principal) moments of inertia.*

Diagonalization of the inertia tensor is a typical eigenvalue problem of linear algebra. The problem is to find those directions $\hat{\boldsymbol{\omega}}^{(i)}$, $i = 1, 2, 3$, for which

$$\mathbf{J}\hat{\boldsymbol{\omega}}^{(i)} = I_i \hat{\boldsymbol{\omega}}^{(i)} . \qquad (3.19)$$

This linear system of equations has a nontrivial solution provided its determinant vanishes,

$$\det\big(\mathbf{J} - I_i \mathbb{1}\big) = 0 . \qquad (3.20)$$

Equation (3.20) is a cubic equation for the unknown I_i. According to (3.17) and (3.18a) it has three real, positive semidefinite solutions. The eigenvector $\hat{\boldsymbol{\omega}}^{(k)}$ that belongs to the eigenvalue I_k is obtained from (3.19), which is to be solved three times, for $k = 1$, 2, and 3. The matrix $\mathbf{R}_0$ in (3.16) is then given by

$$\mathbf{R}_0 = \begin{pmatrix} \hat{\omega}_1^{(1)} & \hat{\omega}_2^{(1)} & \hat{\omega}_3^{(1)} \\ \hat{\omega}_1^{(2)} & \hat{\omega}_2^{(2)} & \hat{\omega}_3^{(2)} \\ \hat{\omega}_1^{(3)} & \hat{\omega}_2^{(3)} & \hat{\omega}_3^{(3)} \end{pmatrix} . \qquad (3.21)$$

It is not difficult to show that two eigenvectors $\hat{\boldsymbol{\omega}}^{(i)}$, $\hat{\boldsymbol{\omega}}^{(k)}$, which belong to distinct eigenvalues I_i and I_k, respectively, are orthogonal. For this take the difference $\hat{\boldsymbol{\omega}}^{(i)}\mathbf{J}\hat{\boldsymbol{\omega}}^{(k)} - \hat{\boldsymbol{\omega}}^{(k)}\mathbf{J}\hat{\boldsymbol{\omega}}^{(i)}$. With (3.19) this becomes

$$\hat{\boldsymbol{\omega}}^{(i)}\mathbf{J}\hat{\boldsymbol{\omega}}^{(k)} - \hat{\boldsymbol{\omega}}^{(k)}\mathbf{J}\hat{\boldsymbol{\omega}}^{(i)} = \big(I_k - I_i\big)\big(\hat{\boldsymbol{\omega}}^{(k)} \cdot \hat{\boldsymbol{\omega}}^{(i)}\big) .$$

The left-hand side vanishes because $\mathbf{J}$ is symmetric. Therefore, if $I_k \neq I_i$, then

$$\hat{\omega}^{(k)} \cdot \hat{\omega}^{(i)} = 0 . \tag{3.22}$$

It may happen that two (or more) eigenvalues are equal, $I_i = I_k$, in which case we cannot prove the above orthogonality. However, as the system (3.19) is linear, any linear combination of $\hat{\omega}^{(i)}$ and of $\hat{\omega}^{(k)}$, say $\hat{\omega}^{(i)} \cos\alpha + \hat{\omega}^{(k)} \sin\alpha$, is also an eigenvector of $\mathbf{J}$, with eigenvalue $I_i = I_k$. It is then clear that we can always choose, by hand, two orthogonal linear combinations. The degeneracy just tells us that there is no preferred choice of principal axes. We illustrate this by means of the following model. Suppose the inertia tensor, after diagonalization, has the form

$$\mathring{\mathbf{J}} = \begin{pmatrix} A & 0 & 0 \\ 0 & A & 0 \\ 0 & 0 & B \end{pmatrix}$$

with $A \neq B$. Any further rotation about the 3-axis has the form

$$\mathbf{R} = \begin{pmatrix} \cos\theta & \sin\theta & 0 \\ -\sin\theta & \cos\theta & 0 \\ 0 & 0 & 1 \end{pmatrix}$$

and leaves $\mathring{\mathbf{J}}$ invariant. Thus any direction in the $(1,2)$-plane is a principal axis, too, and corresponds to the moment of inertia A. In this plane we choose two orthogonal axes. Because $B \neq A$ the third principal axis is perpendicular to these.

(iii) The inertia tensor and specifically its eigenvalues (the moments of inertia) are static properties of the body, very much like its mass. As we shall see below, the angular momentum and the kinetic energy are proportional to I_k when the body rotates about the corresponding eigenvector $\hat{\omega}^{(k)}$.

A body whose moments of inertia are all different, $I_1 \neq I_2 \neq I_3$, is said to be an *asymmetric, or triaxial, top.* If two of the moments are equal, $I_1 = I_2 \neq I_3$, we talk about the *symmetric top.* If all three moments are equal, $I_1 = I_2 = I_3$, we call it *a spherical top*[2].

(iv) If the rigid body has a certain amount of symmetry in shape and mass distribution, the determination of its center of mass and its principal axes is a lot easier. For instance, we have the following proposition:

> **Proposition:** If the shape and mass distribution of a rigid body is symmetric with respect to reflection in a plane (see Fig. 3.5), its center of mass and two of its principal axes lie in that plane. The third principal axis is perpendicular to it.

Proof. As a first trial choose an orthogonal frame of reference whose 1- and 2-axes are in the plane and whose 3-axis is perpendicular to it. For symmetry reasons, to any mass element with positive x_3 there corresponds an equal mass element with

[2] This does not necessarily mean that the rigid body has a spherical shape.

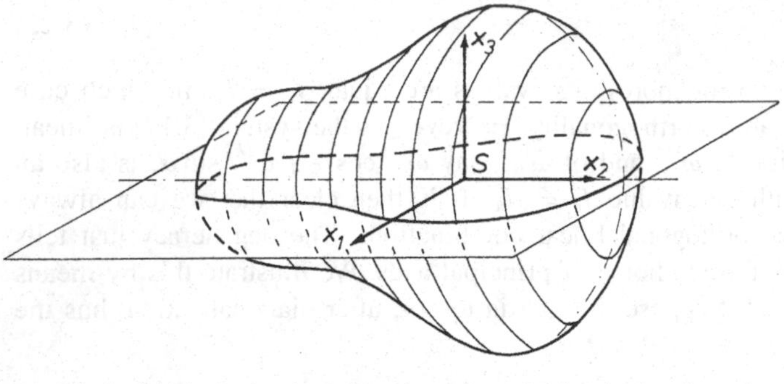

Fig. 3.5. A rigid body that is symmetric under reflection in the plane shown in the figure

negative x_3. Therefore, $\int d^3x x_3 \varrho(\boldsymbol{x}) = 0$. Comparison with (3.7b) shows that the first part of the proposition is true: S lies in the plane of symmetry. Suppose now S is found and the system (x_1, x_2, x_3) is centered in S. In the expression (3.15) for $\mathbf{J}$ the following integrals vanish:

$$\int d^3x \varrho(\boldsymbol{x}) x_1 x_3 = 0 , \quad \int d^3\varrho(\boldsymbol{x}) x_2 x_3 = 0 .$$

This is so because, for fixed x_1 (or x_2, respectively), the positive values of x_3 and the corresponding negative values $-x_3$ give equal and opposite contributions. What remains is

$$\mathbf{J} = \begin{pmatrix} J_{11} & J_{12} & 0 \\ J_{12} & J_{22} & 0 \\ 0 & 0 & I_3 \end{pmatrix} .$$

However, this matrix is diagonalizable by a rotation *in* the plane of symmetry (i.e. one about the 3-axis). This proves the second part of the proposition. □

Similar arguments apply to the case when the body possesses *axial symmetry*, i.e. if it is symmetric under rotations about a certain axis. In this case the center of mass S lies on the symmetry axis and that axis is a principal axis. The remaining two are perpendicular to it. The corresponding moments of inertia being degenerate, they must be chosen by hand in the plane through S that is perpendicular to the symmetry axis.

Remark: In calculations involving the inertia tensor the following symbolic notations can be very useful.

Let any vector or vector field $\boldsymbol{a}$ over $\mathbb{R}^3$ be written as $|\boldsymbol{a}\rangle$. Its dual which when acting on any other vector (field) $|\boldsymbol{c}\rangle$ is denoted $\langle\boldsymbol{a}|$ and so looks like a kind of mirror image of $|\boldsymbol{a}\rangle$. With this notation an expression such as $\langle\boldsymbol{a}|\boldsymbol{c}\rangle$ is nothing but the ordinary scalar product $\boldsymbol{a} \cdot \boldsymbol{c}$. On the other hand, an object such as $|\boldsymbol{b}\rangle\langle\boldsymbol{a}|$ is a tensor which acts on other vectors $\boldsymbol{c}$ by $|\boldsymbol{b}\rangle\langle\boldsymbol{a}|\boldsymbol{c}\rangle = (\boldsymbol{a} \cdot \boldsymbol{c})\boldsymbol{b}$, thus yielding new vectors parallel to $\boldsymbol{b}$. This is to say that $|\boldsymbol{a}\rangle = (a_1, a_2, a_3)^T$ is a *column* vector while its dual $\langle \boldsymbol{a}| = (a_1, a_2, a_3)$ is a *row* vector. Applying standard rules of matrix calculus, one has

$$\langle \boldsymbol{b}|\boldsymbol{a}\rangle = \begin{pmatrix} b_1 & b_2 & b_3 \end{pmatrix} \begin{pmatrix} a_1 \\ a_2 \\ a_3 \end{pmatrix} = \sum_{k=1}^{3} b_k a_k = \boldsymbol{b} \cdot \boldsymbol{a} ,$$

$$|\boldsymbol{b}\rangle\langle \boldsymbol{a}| = \begin{pmatrix} b_1 \\ b_2 \\ b_3 \end{pmatrix} \begin{pmatrix} a_1 & a_2 & a_3 \end{pmatrix} = \begin{pmatrix} b_1 a_1 & b_1 a_2 & b_1 a_3 \\ b_2 a_1 & b_2 a_2 & b_2 a_3 \\ b_3 a_1 & b_3 a_2 & b_3 a_3 \end{pmatrix} .$$

For example the definition (3.10b) written in this notation, becomes

$$\mathbf{J} = \int \mathrm{d}^3 x \, \varrho(\boldsymbol{x}) \big[\langle \boldsymbol{x}|\boldsymbol{x}\rangle \mathbb{1}_3 - |\boldsymbol{x}\rangle\langle \boldsymbol{x}| \big] ,$$

where $\mathbb{1}_3$ is the 3×3 unit matrix. This notation emphasizes the fact that $\mathbf{J}$ is an object that acts on vectors (or vector fields) and yields as the result another vector (or vector field).

In fact, this notation is the same as Dirac's "ket" and "bra" notation that the reader will encounter in quantum theory. In the older literature on vector analysis the tensor $|\boldsymbol{b}\rangle\langle \boldsymbol{a}|$ was called dyadic product of $\boldsymbol{b}$ and $\boldsymbol{a}$.

3.5 Steiner's Theorem

Let $\mathbf{J}$ be the inertia tensor as calculated according to (3.15) in a body-fixed system $\overline{\mathbf{K}}$ with origin S, the center of mass. Let $\overline{\mathbf{K}}'$ be another body-fixed system which is obtained by shifting $\overline{\mathbf{K}}$ by a given translation vector $\boldsymbol{a}$, as shown in Fig. 3.6. Let $\mathbf{J}'$ be the inertia tensor as calculated in the second system,

$$J'_{\mu\nu} = \int \mathrm{d}^3 x' \varrho(\boldsymbol{x}') \left[\boldsymbol{x}'^2 \delta_{\mu\nu} - x'_\mu x'_\nu \right]$$

with $\boldsymbol{x}' = \boldsymbol{x} + \boldsymbol{a}$. Then $\mathbf{J}'$ and $\mathbf{J}$ are related by

$$J'_{\mu\nu} = J_{\mu\nu} + M \left[\boldsymbol{a}^2 \delta_{\mu\nu} - a_\mu a_\nu \right] . \tag{3.23}$$

In the compact "bracket" notation introduced above, it reads $\mathbf{J}' = \mathbf{J} + \mathbf{J}_a$, with $\mathbf{J}_a = M\,[\langle \boldsymbol{a}|\boldsymbol{a}\rangle \mathbb{1}_3 - |\boldsymbol{a}\rangle\langle \boldsymbol{a}|]$.

The proof is not difficult. Insert $\boldsymbol{x}' = \boldsymbol{x} + \boldsymbol{a}$ into the first equation and take account of the fact that all integrals with integrands that are linear in $\boldsymbol{x}$ vanish because of the center-of-mass condition (3.7b). In Fig. 3.6 $\overline{\mathbf{K}}'$ has axes parallel to those of $\overline{\mathbf{K}}$.

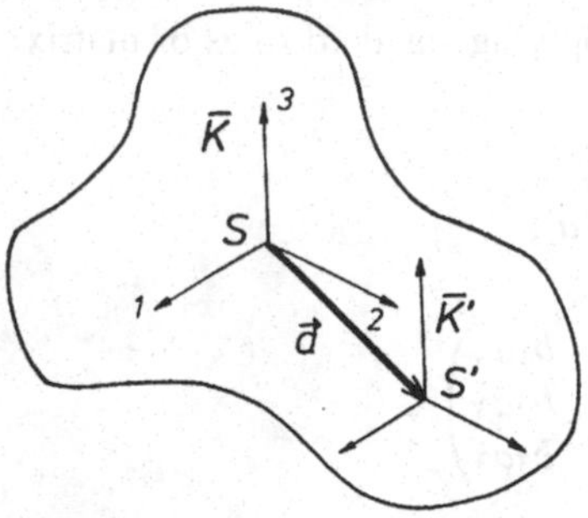

Fig. 3.6. The system $\overline{\mathbf{K}}$ is attached to the center of mass S. One wishes to determine the inertia tensor with respect to another body-fixed system $\overline{\mathbf{K}}'$, which is centered on the point S'

If $\overline{\mathbf{K}}'$ is rotated from $\overline{\mathbf{K}}$ by the rotation $\mathbf{R}$, in addition to the shift, (3.23) generalizes to

$$J'_{\mu\nu} = \sum_{\sigma,\tau=1}^{3} R_{\mu\sigma} R_{\nu\tau} \left(J_{\sigma\tau} + M \left[\boldsymbol{a}^2 \delta_{\sigma\tau} - a_\sigma a_\tau \right] \right) , \quad \text{or} \tag{3.24}$$

$$\mathbf{J}' = \mathbf{R}(\mathbf{J} + \mathbf{J}_a)\mathbf{R}^{-1} .$$

The content of this formula is the following. First, $\overline{\mathbf{K}}'$ is rotated by $\mathbf{R}^{-1}$ to a position where its axes are parallel to those of $\overline{\mathbf{K}}$. At this point, Steiner's theorem is applied, in the form of (3.23). Finally, the rotation is undone by applying $\mathbf{R}$.

3.6 Examples of the Use of Steiner's Theorem

Example (i) For a ball of radius R and with spherically symmetric mass distribution $\varrho(\boldsymbol{x}) = \varrho(r)$ and for any system attached to its center, the inertia tensor is diagonal. In addition, the three moments of inertia are equal, $I_1 = I_2 = I_3 \equiv I$. Adding them up and using (3.17), we find that

$$3I = 2 \int \varrho(r) r^2 \mathrm{d}^3 x = 8\pi \int_0^R \varrho(r) r^4 \mathrm{d}r$$

and therefore

$$I = \frac{8\pi}{3} \int_0^R \varrho(r) r^4 \mathrm{d}r .$$

We also have the relation

$$M = 4\pi \int_0^R \varrho(r) r^2 \mathrm{d}r$$

for the total mass of the ball. If, furthermore, its mass distribution is *homogeneous*, then

$$\varrho(r) = \frac{3M}{4\pi R^3}, \quad \text{for} \quad r \le R, \quad \text{and} \quad I = \frac{2}{5}MR^2 .$$

Example (ii) Consider a body composed of two identical, homogeneous balls of radius R which are soldered at their point of contact T. This point is the center of mass and, obviously, the (primed) axes drawn in Fig. 3.7 are principal axes. We make use of the additivity of the inertia tensor and apply Steiner's theorem. The individual ball carries half the total mass. Hence its moment of inertia is $I_0 = MR^2/5$. In a system centered in T whose 1- and 3-axes are tangent, *one* ball would have the moments of inertia, by Steiner's theorem,

$$I_1' = I_3' = I_0 + \frac{M}{2}R^2 ; \quad I_2' = I_0 .$$

The same axes are principal axes for the system of two balls and we have

$$I_1 = I_3 = 2\left(I_0 + \frac{M}{2}R^2\right) = \frac{7}{5}MR^2 ,$$

$$I_2 = 2I_0 = \frac{2}{5}MR^2 .$$

Example (iii) The homogeneous children's top of Fig. 3.8 is another example for Steiner's theorem, in its form (3.23), because its point of support O does not coincide with the center of mass S. The mass density is homogeneous. It is not difficult to show that the center of mass is at a distance $3h/4$ from O on the symmetry axis. The inertia tensor is diagonal in the unprimed system (centered in S) as well as in the primed system (centered in O). The volume is $V = \pi R^2 h/3$, and the density is $\varrho = 3M/\pi R^2 h$. Using cylindrical coordinates,

$$x_1' = r\cos\varphi , \quad x_2' = r\sin\varphi , \quad x_3' = z ,$$

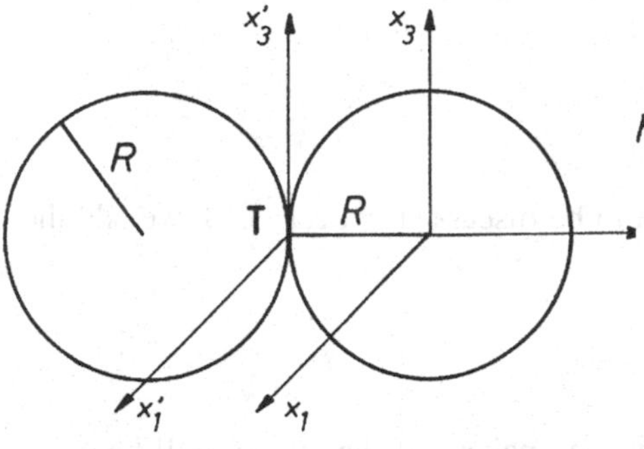

Fig. 3.7. A rigid body consisting of two identical balls that are tangent to each other. The primed axes are principal axes

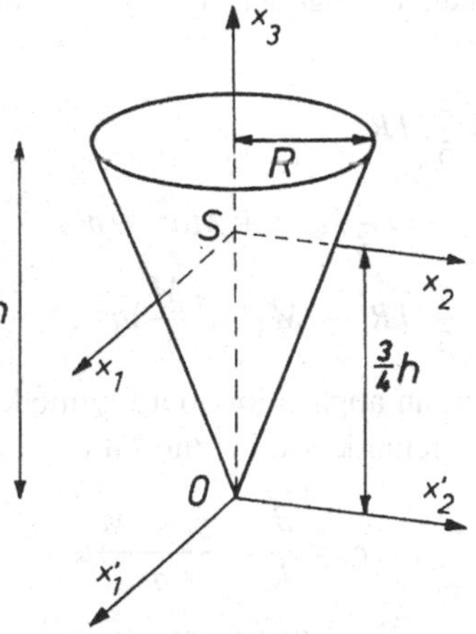

Fig. 3.8. The children's top is an example of the application of Steiner's theorem

the moments of inertia are easily calculated within the primed system. One finds that

$$I_1' = I_2' = \varrho \int \mathrm{d}^3x' \left(x_2'^2 + x_3'^2\right) = \tfrac{3}{5}M\left(\tfrac{1}{4}R^2 + h^2\right)$$
$$I_3' = \varrho \int \mathrm{d}^3x' \left(x_1'^2 + x_2'^2\right) = \tfrac{3}{10}MR^2 .$$

The moments of inertia in the unprimed system are obtained from Steiner's theorem, viz.

$$I_1 = I_2 = I_1' - Ma^2 , \quad \text{with} \quad a = \tfrac{3}{4}h ,$$

and thus

$$I_1 = I_2 = \tfrac{3}{20}M\left(R^2 + \tfrac{1}{4}h^2\right) \quad \text{and} \quad I_3 = I_3' = \tfrac{3}{10}MR^2 .$$

Example (iv) Inside an originally homogeneous ball of mass M and radius R a pointlike mass m is placed at a distance d from the ball's center, $0 < d < R$. The inertia tensor is an *extensive* quantity, hence the inertia tensors of the ball and of the point mass add. Let a be the distance of the ball's center to the center of mass S, and b be the distance from the point mass to S. With these notations also shown in Fig. 3.9a and making use of the center of mass condition $mb - Ma = 0$ one finds

$$a = md/(m + M) , \quad b = Md/(m + M) = (M/m)a .$$

A sytem of principal axes is obvious: Let the line joining the ball's center and the point mass be the 3-axis (symmetry axis), then choose two orthogonal directions in the plane orthogonal to the 3-axis through the center. Using Steiner's theorem one has

$$I_3 = \frac{2}{5}MR^2 ,$$
$$I_1 = I_2 = \frac{2}{5}MR^2 + Ma^2 + mb^2$$
$$= \frac{2}{5}MR^2 + M\left(1 + \frac{M}{m}\right)a^2 .$$

In view of an application to a toy model to be discussed in Sect. 3.18, we add the following remark. Define the ratios

$$\alpha = \frac{a}{R} , \quad \delta = \frac{d}{R} = \frac{m + M}{m}\alpha .$$

A condition that will be of relevance for the analysis of that model will be

$$(1 - \alpha)I_3 < I_1 = I_2 < (1 + \alpha)I_3 . \qquad (*)$$

In the example worked out here this condition reads

$$-\frac{2}{5} < \frac{m+M}{m}\alpha < \frac{2}{5}$$

which says, when expressed in terms of δ, that this parameter should be less than 2/5. Note that it is the upper limit in $(*)$ which gives this bound.

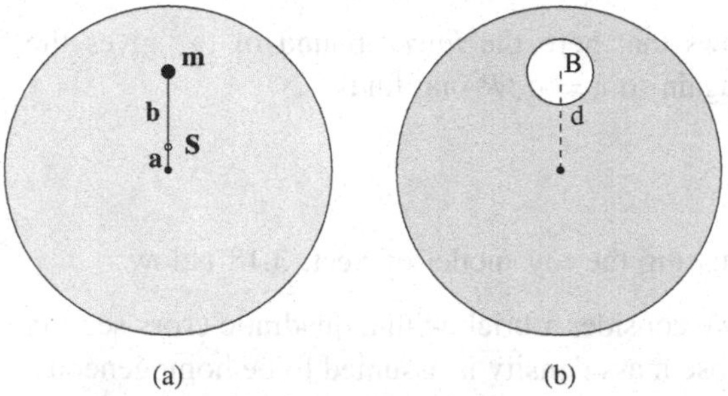

Fig. 3.9. (**a**): A point mass is added to a ball of homogeneous mass density thus changing the original spherical top into a symmetric top (**b**): In the same homegeneous ball a hole is cut out that makes the top a symmetric but no more a spherical one

Example (**v**) Consider the same ball (M, R) as in the previous example but suppose that this time a small hollow sphere is cut out of it whose center B is at a distance d from the ball's center and whose radius is r, cf. Fig. 3.9b. Referring to that figure the center of mass now lies *below* the center of the ball. The mass of the ball which is cut out is

$$m = \left(\frac{r}{R}\right)^3 M .$$

As in example (iv) let a and b denote the distances from the ball's center to the center of mass S, and from the center of the hollow sphere to S, respectively. Then $d = b - a$, and the center of mass condition reads

$$Ma + (-m)b = 0 .$$

(Remember that the mass and the inertia tensor are extensive quantities. The minus sign stems from the fact that one has taken away the mass of the hole!) Choosing principal axes like in the previous case the moments of inertia are

$$I_3 = \frac{2}{5}MR^2 - \frac{2}{5}mr^2 ,$$

$$I_1 = I_2 = \frac{2}{5}MR^2 + Ma^2 - \frac{2}{5}mr^2 - mb^2 .$$

It is interesting to follow up the inequality $(*)$ also in this example. Inserting the formulae for the moments of inertia it reads

$$-\alpha\frac{2}{5}\left(MR^2 - mr^2\right) < Ma^2 - mb^2 < \alpha\frac{2}{5}\left(MR^2 - mr^2\right) .$$

The middle part is seen to be negative,

$$Ma^2 - mb^2 = -Ma^2 \frac{M-m}{m} ,$$

so that the inequality should be multiplied by (-1). This yields

$$\frac{2}{5}\left(1 - \frac{mr^2}{MR^2}\right) > \frac{M-m}{m}\alpha > -\frac{2}{5}\left(1 - \frac{mr^2}{MR^2}\right) .$$

Comparison to example (iv) shows that here the *lower* bound of $(*)$ gives the essential restriction. Converting again to $\delta = d/R$ one finds

$$\delta < 2/5\left(1 - r^5/R^5\right) .$$

This result will be useful in discussing the toy model of Sect. 3.18 below.

Example (vi) As a last example we consider a brick with a quadratic cross section (side length a_1) and height a_3 whose mass density is assumed to be homogeneous. If one chooses the coordinate system $\overline{\mathbf{K}}$ shown in Fig. 3.10, the inertia tensor is already diagonal. With $\varrho_0 = M/a_1a_2a_3$ one finds $I_1 = M(a_2^2 + a_3^2)/12$, cyclic in $1, 2, 3$. The aim is now to compute the inertia tensor in the body-fixed system $\overline{\mathbf{K}}'$, whose 3-axis lies along one of the main diagonals of the brick. As $a_1 = a_2$, we find $I_1 = I_2$. Therefore, as a first step, one can rotate the 1-axis about the initial x_3-axis by an arbitrary amount. For example, one can choose it along a diagonal of the cross section, without changing the inertia tensor (which is diagonal). In

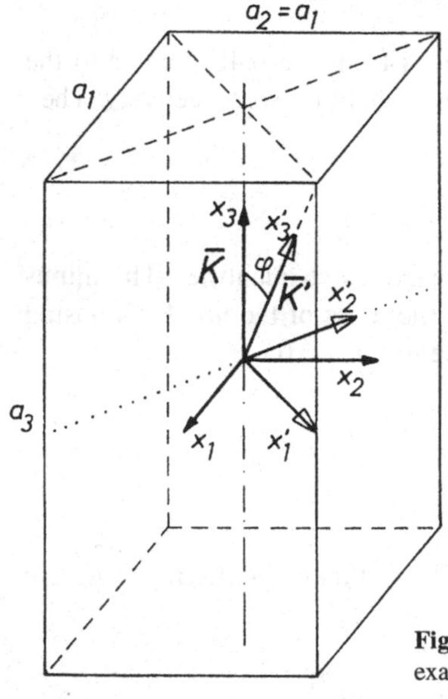

Fig. 3.10. A brick with homogeneous mass density, as an example of a symmetric rigid body

a second step, $\overline{\mathrm{K}}'$ is reached by a rotation about the x_2'-axis by the angle $\phi = \arctan(a_1\sqrt{2}/a_3)$

$$\bar{\mathrm{K}} \xrightarrow[\mathbf{R}_\phi]{} \bar{\mathrm{K}}' \,, \quad \mathbf{R}_\phi = \begin{pmatrix} \cos\phi & 0 & -\sin\phi \\ 0 & 1 & 0 \\ \sin\phi & 0 & \cos\phi \end{pmatrix} .$$

According to (3.24) the relation between the inertia tensors is $\mathbf{J}' = \mathbf{R}\mathbf{J}\mathbf{R}^{\mathrm{T}}$. $\mathbf{J}'$ is not diagonal. One finds that

$$J_{11}' = I_1\cos^2\phi + I_3\sin^2\phi = \frac{M}{12}\frac{4a_1^4 + a_1^2a_3^2 + a_3^4}{2a_1^2 + a_3^2} ,$$

$$J_{22}' = I_2 = \frac{M}{12}\left(a_1^2 + a_3^2\right) ,$$

$$J_{33}' = I_1\sin^2\phi + I_3\cos^2\phi = \frac{M}{12}\frac{2a_1^4 + 4a_1^2a_3^2}{2a_1^2 + a_3^2} ,$$

$$J_{12}' = 0 = J_{21}' = J_{23}' = J_{32}' ,$$

$$J_{13}' = J_{31}' = (I_1 - I_3)\sin\phi\cos\phi = \frac{M}{12}\frac{(a_3^2 - a_1^2)a_1a_3\sqrt{2}}{2a_1^2 + a_3^2} .$$

The x_2'-axis is a principal axis; the x_1'- and x_3'-axes are not, with one exception: if the body is a cube, i.e. if $a_1 = a_3$, J_{13}' and J_{31}' vanish. Thus, for a homogeneous cube, any orthogonal system attached to its center of gravity is a system of principal axes. For equal (and homogeneous) mass densities a cube of height a behaves like a ball with radius $R = a\sqrt[5]{5/16\pi} \simeq 0.630a$. In turn, if we require the moments of inertia to be equal, for a cube and a ball of the same mass M, we must have $R = a\sqrt{5}/2\sqrt{3} \simeq 0.645a$.

3.7 Angular Momentum of a Rigid Body

The angular momentum of a rigid body can be decomposed into the angular momentum of its center of mass and the relative (internal) angular momentum. This follows from the general analysis of the mechanical systems we studied in Sects. 1.8–1.12. As we learnt there, the relative angular momentum is independent of the choice of the laboratory system and therefore is the dynamically relevant quantity.

The relative angular momentum of a rigid body, i.e. the angular momentum with respect to its center of mass, is given by

$$\boldsymbol{L} = \sum_{i=1}^{n} m_i \boldsymbol{r}_i \times \dot{\boldsymbol{r}}_i \tag{3.25a}$$

if we choose to describe the body by the discrete model (A). For case (B) it is, likewise,

$$L = \int d^3x\varrho(\boldsymbol{x})\boldsymbol{x} \times \dot{\boldsymbol{x}} \,. \tag{3.25b}$$

From (3.5) we have $\dot{\boldsymbol{x}} = \boldsymbol{\omega} \times \boldsymbol{x}$. Adopting the continuous version (B) from now on, this becomes

$$\boldsymbol{L} = \int d^3x\varrho(\boldsymbol{x})\boldsymbol{x} \times (\boldsymbol{\omega} \times \boldsymbol{x}) = \int d^3x\varrho(\boldsymbol{x})\left[\boldsymbol{x}^2\boldsymbol{\omega} - (\boldsymbol{x} \cdot \boldsymbol{\omega})\boldsymbol{x}\right] .$$

The last expression on the right-hand side is just the product of the inertia tensor and the angular velocity $\boldsymbol{\omega}$, viz.

$$\boxed{\boldsymbol{L} = \mathbf{J}\boldsymbol{\omega}} \,. \tag{3.26}$$

Indeed, writing this in components and making use of (3.10b), we have

$$L_\mu = \sum_{\nu=1}^{3} \int d^3x\varrho(\boldsymbol{x})\left[\boldsymbol{x}^2\delta_{\mu\nu} - x_\mu x_\nu\right]\omega_\nu \,. \tag{3.26'}$$

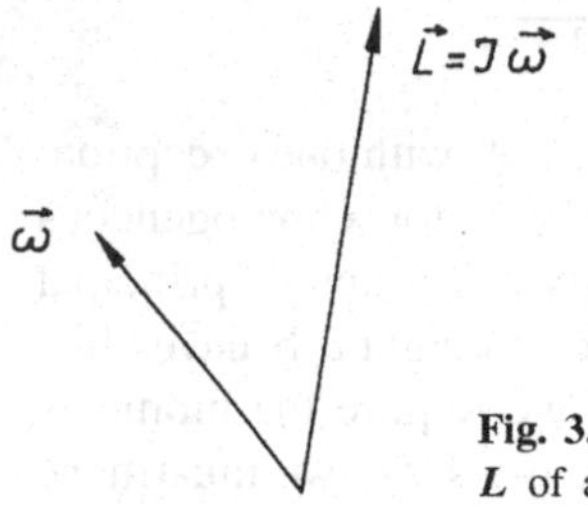

Fig. 3.11. The momentary angular velocity $\boldsymbol{\omega}$ and the angular momentum $\boldsymbol{L}$ of a rigid body, in general, do not point in the same direction

The relation (3.26) tells us that the angular momentum is obtained by applying the inertia tensor to the angular velocity. We note that $\boldsymbol{L}$ does not point in the same direction as $\boldsymbol{\omega}$, cf. Fig. 3.11, unless $\boldsymbol{\omega}$ is one of the eigenvectors of the inertia tensor. In this case

$$\boldsymbol{L} = I_i\boldsymbol{\omega} \,, \quad \left(\boldsymbol{\omega} \| \boldsymbol{\omega}^{(i)}\right) . \tag{3.27}$$

Thereby, the eigenvalue problem (3.19) receives a further physical interpretation: it defines those directions of the angular velocity $\boldsymbol{\omega}$ for which the angular momentum $\boldsymbol{L}$ is parallel to $\boldsymbol{\omega}$. In this case, if $\boldsymbol{L}$ is conserved (i.e. fixed in space), the top rotates about this direction with constant angular velocity.

The expression (3.13) for the rotational energy can be rewritten by means of relation (3.26):

$$T_{\text{rot}} = \tfrac{1}{2}\,\boldsymbol{\omega} \cdot \boldsymbol{L} \tag{3.28}$$

i.e. $2T_{\text{rot}}$ is equal to the projection of $\boldsymbol{\omega}$ onto $\boldsymbol{L}$. If $\boldsymbol{\omega}$ points along one of the principal axes, (3.28) becomes, by (3.27),

$$T_{\rm rot} = \tfrac{1}{2} I_i \omega^2 \,, \quad \left(\boldsymbol{\omega} \| \boldsymbol{\omega}^{(i)}\right) . \tag{3.29}$$

This expression for $T_{\rm rot}$ shows very clearly the analogy to the kinetic energy of the translational motion, (3.12).

To conclude let us write the relationship between angular momentum and angular velocity by means of the "bracket" notation,

$$|\boldsymbol{L}\rangle = \int \mathrm{d}^3x \left[\langle \boldsymbol{x}|\boldsymbol{x}\rangle \mathbb{1}_3 - |\boldsymbol{x}\rangle\langle \boldsymbol{x}| \right] |\boldsymbol{\omega}\rangle = \mathbf{J}|\boldsymbol{\omega}\rangle \,.$$

This formula shows very clearly the action of the 3×3-matrix $\mathbf{J}$ on the column vector $|\boldsymbol{\omega}\rangle$ and may be more transparent than the expression (3.26′) in terms of coordinates.

3.8 Force-Free Motion of Rigid Bodies

If there are no external forces, the center of mass moves uniformly along a straight line (Sect. 1.9). The angular momentum $\boldsymbol{L}$ is conserved (Sects. 1.10–11),

$$\frac{\mathrm{d}}{\mathrm{d}t} \boldsymbol{L} = 0 \,. \tag{3.30}$$

Similarly the kinetic energy of the rotational motion is conserved,

$$\frac{\mathrm{d}}{\mathrm{d}t} T_{\rm rot} = \frac{1}{2} \frac{\mathrm{d}}{\mathrm{d}t} (\boldsymbol{\omega} \mathbf{J} \boldsymbol{\omega}) = \frac{1}{2} \frac{\mathrm{d}}{\mathrm{d}t} (\boldsymbol{\omega} \cdot \boldsymbol{L}) = 0 \,. \tag{3.31}$$

(This follows from conservation of the total energy (Sect. 1.11) and of the total momentum. The kinetic energy of translational motion is then conserved separately.) We study three special cases.

(i) *The spherical top.* The inertia tensor is diagonal, its eigenvalues are degenerate, $I_1 = I_2 = I_3 \equiv I$. We have $\boldsymbol{L} = I_i \boldsymbol{\omega}$. As $\boldsymbol{L}$ is constant, this implies that $\boldsymbol{\omega}$ is constant too,

$$\boldsymbol{L} = \text{const} \Rightarrow \boldsymbol{\omega} = \frac{1}{I} \boldsymbol{L} = \text{const} \,.$$

The top rotates uniformly about a fixed axis.

(ii) *The rigid rod.* This is a degenerate top. It is a linear, i.e. one-dimensional, rigid body for which

$$I_1 = I_2 \equiv I \,,$$
$$I_3 = 0 \,,$$

where the moments of inertia refer to the axes shown in Fig. 3.12. As it has no mass outside the 3-axis, the rod cannot rotate about that axis. From (3.27) we have $L_1 = I\omega_1$, $L_2 = I\omega_2$, $L_3 = 0$. Therefore, leaving aside the center-of-mass

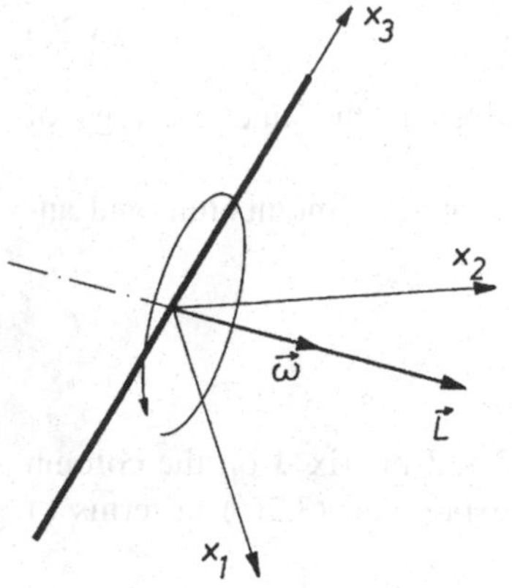

Fig. 3.12. The rigid rod as an example of a degenerate rigid body

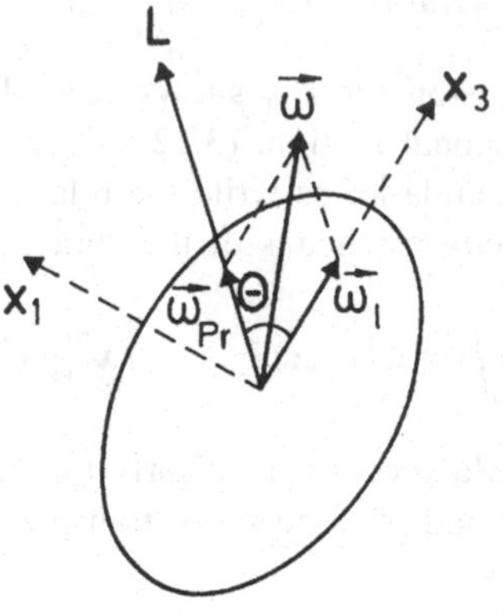

Fig. 3.13. Example of a symmetric body; $I_1 = I_2 \neq I_3$

motion, force-free motion of the rod can only be uniform rotation about any axis perpendicular to the 3-axis.

(iii) *The (nondegenerate) symmetric top.* This is an important special case and we shall analyze its motion in some detail, here and below, using different approaches. Taking the 3-axis along the symmetry axis, we have $I_1 = I_2 \neq I_3$. Suppose $\boldsymbol{L}$, the angular momentum, is given. We choose the 1-axis in the plane spanned by $\boldsymbol{L}$ and the 3-axis. The 2-axis being perpendicular to that plane, we have $L_2 = 0$ and hence $\omega_2 = 0$. In other words, $\boldsymbol{\omega}$ is also in the (1,3)-plane, as shown in Fig. 3.13. It is then easy to analyze the motion of the symmetric top for the case of no external forces. The velocity $\dot{\boldsymbol{x}} = \boldsymbol{\omega} \times \boldsymbol{x}$ of all points *on* the symmetry axis is perpendicular to the (1,3)-plane (it points "backwards" in the figure). Therefore, the symmetry axis rotates uniformly about $\boldsymbol{L}$, which is a fixed vector in space. This part of the motion is called *regular precession.* It is convenient to write $\boldsymbol{\omega}$ as the sum of components along $\boldsymbol{L}$ and along the 3-axis,

$$\boldsymbol{\omega} = \boldsymbol{\omega}_l + \boldsymbol{\omega}_{\mathrm{pr}} . \tag{3.32}$$

Clearly, the longitudinal component $\boldsymbol{\omega}_l$ is irrelevant for the precession. The component $\boldsymbol{\omega}_{\mathrm{pr}}$ is easily calculated from Fig. 3.14a. With $\omega_{\mathrm{pr}} = |\boldsymbol{\omega}_{\mathrm{pr}}|$ and $\omega_1 = \omega_{\mathrm{pr}} \sin\theta$, as well as $\omega_1 = L_1/I_1$ and $L_1 = |\boldsymbol{L}| \sin\theta$, one obtains

$$\omega_{\mathrm{pr}} = \frac{|\boldsymbol{L}|}{I_1} . \tag{3.33}$$

Because the symmetry axis (i.e. the 3-axis) precesses about $\boldsymbol{L}$ (which is fixed in space) and because at all times $\boldsymbol{L}$, $\boldsymbol{\omega}$, and the 3-axis lie in a plane, the angular velocity $\boldsymbol{\omega}$ also precesses about $\boldsymbol{L}$. In other words, $\boldsymbol{\omega}$ and the symmetry axis rotate uniformly and synchronously about the angular momentum $\boldsymbol{L}$, as shown in Fig. 3.14b. The cone traced out by $\boldsymbol{\omega}$ is called the *space cone*, while the one traced out by the symmetry axis is called the *nutation cone.*

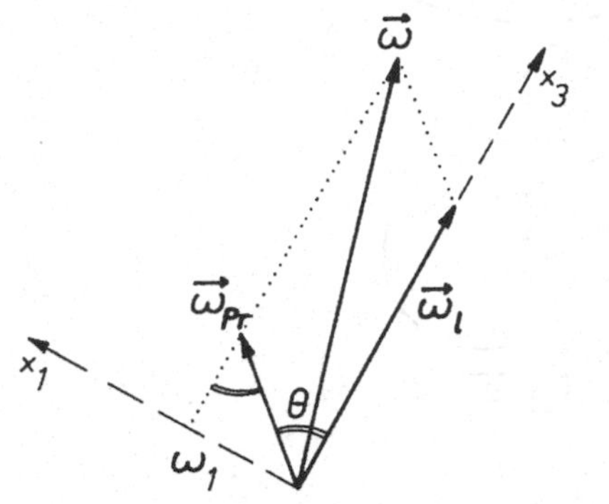

Fig. 3.14a. The angular velocity $\boldsymbol{\omega}$ is written as the sum of its component ω_l along the symmetry axis and its component ω_{pr} along the angular momentum

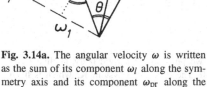

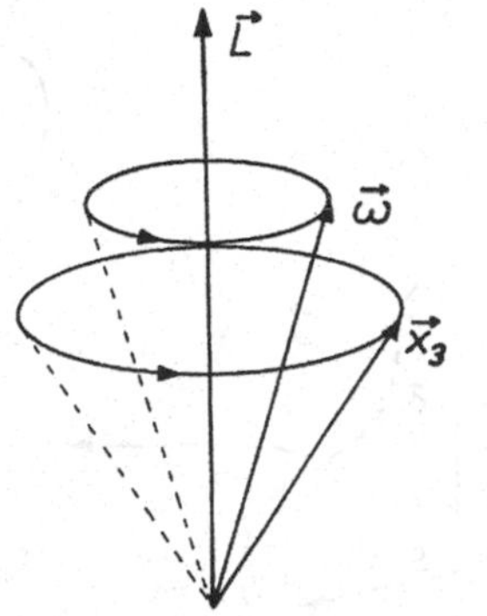

Fig. 3.14b. The symmetry axis (x_3) of the symmetric top and the momentary angular velocity precess uniformly about the angular momentum, which is fixed in space

In addition to its precession as a whole, the body also rotates uniformly about its symmetry axis. The angular velocity for this part of the motion is

$$\omega_3 = \frac{L_3}{I_3} = \frac{|\boldsymbol{L}|\cos\theta}{I_3} . \tag{3.34}$$

Note that the analysis given above describes the motion of the symmetric rigid body as it is seen by an observer in the space-fixed laboratory system, i.e. the system where $\boldsymbol{L}$ is constant. It is instructive to ask how the same motion appears to an observer fixed in the body for whom the 3-axis is constant. We shall return to this question in Sect. 3.13 below.

3.9 Another Parametrization of Rotations: The Euler Angles

Our aim is to derive the equations of motion for the rigid body. As we stressed in the introduction (Sect. 3.1), it is essential to identify the various reference systems that are needed for the description of the rigid body and its motion and to distinguish them clearly, at any point of the discussion. We shall proceed as follows. At time $t = 0$, let the body have the position shown in Fig. 3.15. Its system of principal axes (below, we use the abbreviation PA for 'principal axes') $\overline{\mathbf{K}}$, at $t = 0$, then assumes the position shown in the left-hand part of the figure. We make a copy of this system, call it $\mathbf{K}$, keep that copy fixed, and use this as the inertial system of reference. Thus, at $t = 0$, the body-fixed system and the inertial system coincide. At a later time t let the body have the position shown in the right-hand part of Fig. 3.15. Its center-of-mass, by the action of the external forces, has moved along the trajectory drawn in the figure (if there are no external forces, its motion is uniform and along a straight line). In addition, the body as a whole is rotated away from its original orientation.

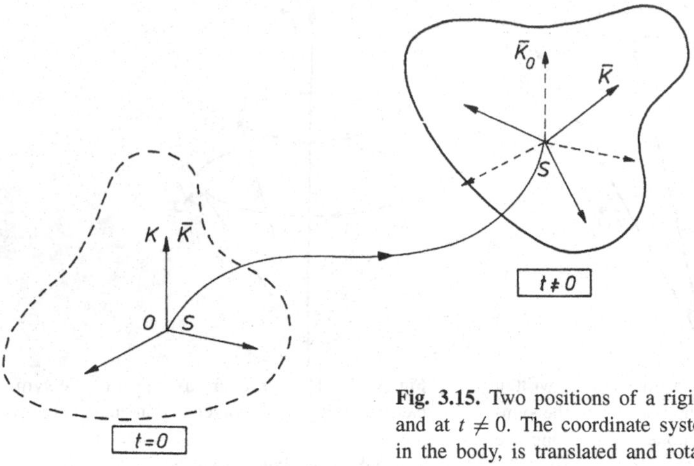

Fig. 3.15. Two positions of a rigid body, at time $t = 0$ and at $t \neq 0$. The coordinate system $\overline{\mathbf{K}}$, which is fixed in the body, is translated and rotated

Choose now one more reference system, denoted $\overline{\mathbf{K}}_0$, that is attached to S, its axes being *parallel*, at all times, to the axes of the inertial system $\mathbf{K}$. The actual position of the rigid body at time t is then completely determined once we know the position $\boldsymbol{r}_S(t)$ of the center of mass and the relative position of the PA system with respect to the auxiliary system $\overline{\mathbf{K}}_0$. The first part, the knowledge of $\boldsymbol{r}_S(t)$, is nothing but the separation of center-of-mass motion that we studied earlier, in a more general context. Therefore, the problem of describing the motion of a rigid body is reduced to the description of its motion relative to a reference system centered in S, the center of mass, and whose axes have fixed directions in space.

The relative rotation from $\overline{\mathbf{K}}_0$ to $\overline{\mathbf{K}}$ can be parametrized in different ways. We may adopt the parametrization that we studied in Sect. 2.22, i.e. write the rotation matrix in the form $\mathbf{R}(\boldsymbol{\varphi}(t))$, where the vector $\boldsymbol{\varphi}$ is now a function of time. We shall do so in Sects. 3.12 and 3.13 below.

An alternative, and equivalent, parametrization is the one in terms of *Eulerian angles*. It is useful, for example, when describing rigid bodies in the framework of canonical mechanics, and we shall use it below, in Sects. 3.15–3.16. It is defined as follows. Write the general rotation $\mathbf{R}(t) \in \mathrm{SO}(3)$ as a product of three successive rotations in the way sketched in Fig. 3.16,

$$\mathbf{R}(t) = \mathbf{R}_3(\gamma)\mathbf{R}_\eta(\beta)\mathbf{R}_{3_0}(\alpha) \,. \qquad (3.35)$$

The coordinate system is rotated first about the initial 3-axis by an angle α. In a second step it is rotated about the intermediate 2-axis by an angle β, and lastly it is rotated about the new (and final) 3-axis by an angle γ.

With this choice, the general motion of a rigid body is described by six functions of time, $\{\boldsymbol{r}_S(t),\, \alpha(t),\, \beta(t),\, \gamma(t)\}$, in accordance with the fact that it has six degrees of freedom. Both parametrizations, i.e. by means of

$$\left\{\boldsymbol{r}_S(t),\, \mathbf{R}(\boldsymbol{\varphi}(t))\right\} \quad \text{with} \quad \boldsymbol{\varphi}(t) = \left\{\varphi_1(t),\, \varphi_2(t),\, \varphi_3(t)\right\} , \qquad (3.36)$$

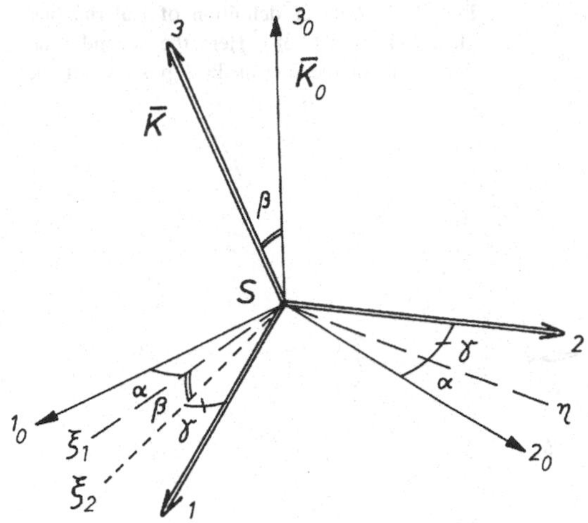

Fig. 3.16. Definition of Eulerian angles as in (3.35). The second rotation is about the intermediate position of the 2-axis

which is the one developed in Sect. 2.21 (2.67), and the one just described, i.e. by means of

$$\{\boldsymbol{r}_S(t),\ \theta_i(t)\} \quad \text{with} \quad \theta_1(t) \equiv \alpha(t),\ \theta_2(t) \equiv \beta(t),\ \theta_3(t) \equiv \gamma(t)\ , \tag{3.37}$$

are useful and will be used below. We remark, further, that the definition of the Eulerian angles described above is the one used in quantum mechanics.

3.10 Definition of Eulerian Angles

Traditionally, the dynamics of rigid bodies makes use of a somewhat different definition of Eulerian angles. This definition is distinguished from the previous one by the choice of the axis for the second rotation in (3.35). Instead of (the intermediate position of) the 2-axis η, the coordinate frame is rotated about the intermediate position of the 1-axis ξ,

$$\mathbf{R}(t) = \mathbf{R}_3(\Psi)\mathbf{R}_\xi(\theta)\mathbf{R}_{3_0}(\Phi)\ . \tag{3.38}$$

Figure 3.17 illustrates this choice of successive rotations. For the sake of clarity, we have suppressed the two intermediate positions of the 2-axis. The transformation from one definition to the other can be read off Figs. 3.16 and 3.17, which were drawn such that $\overline{\mathbf{K}}_0$ and $\overline{\mathbf{K}}$ have the same relative position. It is sufficient to exchange 1- and 2-axes in these figures as follows:

Fig. 3.16		Fig. 3.17
(2_0-axis)	$\longrightarrow$	(1_0-axis)
(1_0-axis)	$\longrightarrow$	$-$(2_0-axis)

keeping the 3-axes the same. This comparison yields the relations

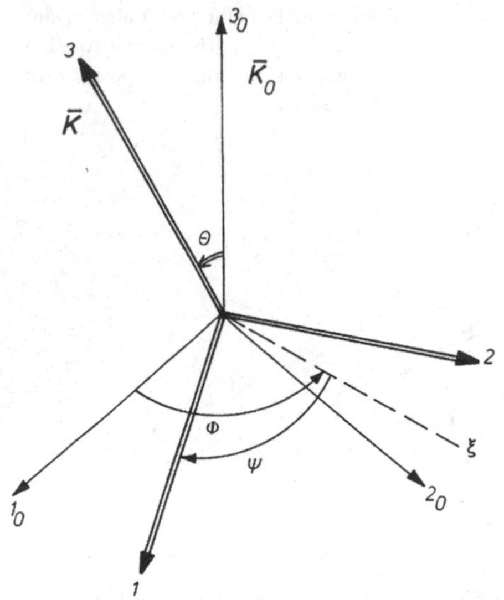

Fig. 3.17. Another definition of Eulerian angles, following (3.38). Here the second rotation is about the intermediate position of the 1-axis

$$\Phi = \alpha + \tfrac{\pi}{2}(\mathrm{mod}\, 2\pi)\ , \quad \theta = \beta\ , \quad \Psi = \gamma - \tfrac{\pi}{2}(\mathrm{mod}\, 2\pi)\ . \tag{3.39}$$

It is easy to convince oneself that the intervals of definition for the Eulerian angles

$$0 \le \alpha \le 2\pi\ , \quad 0 \le \beta \le \pi\ , \quad \text{and} \quad 0 \le \gamma \le 2\pi \tag{3.40}$$

allow us to describe every rotation from $\overline{\mathbf{K}}_0$ to $\overline{\mathbf{K}}$. If one chooses intervals for Φ, θ, and Ψ,

$$0 \le \Phi \le 2\pi\ , \quad 0 \le \theta \le \pi\ , \quad \text{and} \quad 0 \le \Psi \le 2\pi\ , \tag{3.40$'$}$$

it is clear that the additive terms 2π in (3.39), must be adjusted so as not to leave these intervals (see the Appendix on some mathematical notions).

3.11 Equations of Motion of Rigid Bodies

When the rigid body is represented by a finite number of mass points (with fixed links between them), the formulation of center-of-mass motion and relative motion follows directly from the principles of Sects. 1.9 and 1.10. If one chooses a representation in terms of a continuous mass distribution this is not true a priori. Strictly speaking, we leave here the field of mechanics of (finitely many) particles. Indeed, the principle of center-of-mass motion follows only with Euler's generalization (1.99), Sect. 1.30, of (1.8b). Similarly, for finitely many point particles the principle of angular momentum is a consequence of the equations of motion (1.28). Here it follows only if the additional assumption is made that the stress tensor (to be defined in the mechanics of continua) is symmetric. Alternatively, one may introduce this principle as an independent law. It seems that this postulate goes back to L. Euler (1775).

Let $\boldsymbol{P} = M\boldsymbol{V}$ denote the total momentum, with $\boldsymbol{V} = \dot{\boldsymbol{r}}_S(t)$, and let $\boldsymbol{F}$ be the resultant of the external forces. The principle of center-of-mass motion reads

$$\frac{\mathrm{d}}{\mathrm{d}t}\boldsymbol{P} = \boldsymbol{F}\,, \quad \text{where} \quad \boldsymbol{F} = \sum_{i=1}^{n} \boldsymbol{F}^{(i)}\,. \tag{3.41}$$

If, in addition, $\boldsymbol{F}$ is a potential force, $\boldsymbol{F} = -\boldsymbol{\nabla} U(\boldsymbol{r}_S)$, we can define a Lagrangian,

$$L = \tfrac{1}{2}M\dot{\boldsymbol{r}}_S^2 + T_{\text{rot}} - U\big(\boldsymbol{r}_S\big)\,. \tag{3.42}$$

Here, it is important to keep in mind the system of reference with respect to which one writes down the rotational kinetic energy. Choosing $\overline{\mathbf{K}}_0$ (the system attached to S and parallel to the inertial system fixed in space), we have

$$T_{\text{rot}} = \tfrac{1}{2}\boldsymbol{\omega}(t)\tilde{\mathbf{J}}(t)\boldsymbol{\omega}(t)\,. \tag{3.43a}$$

Note that the inertia tensor depends on time. This is so because the body rotates with respect to $\overline{\mathbf{K}}_0$. Clearly, T_{rot} is an invariant form. When expressed with respect to the PA system (or any other *body-fixed* system), it is

$$T_{\text{rot}} = \tfrac{1}{2}\bar{\boldsymbol{\omega}}(t)\mathbf{J}\bar{\boldsymbol{\omega}}(t)\,. \tag{3.43b}$$

$\mathbf{J}$ is now constant; if we choose the PA system, it is diagonal, $J_{mn} = I_m\delta_{mn}$. In (3.43a) the angular velocity $\boldsymbol{\omega}(t)$ is seen from $\overline{\mathbf{K}}_0$, and hence from the laboratory, while in (3.43b) it refers to a system fixed in the body. As we learnt in studying the free motion of the symmetric top, the time evolution of the angular velocity looks different in a frame with axes of fixed direction in space than in a frame fixed in the body.

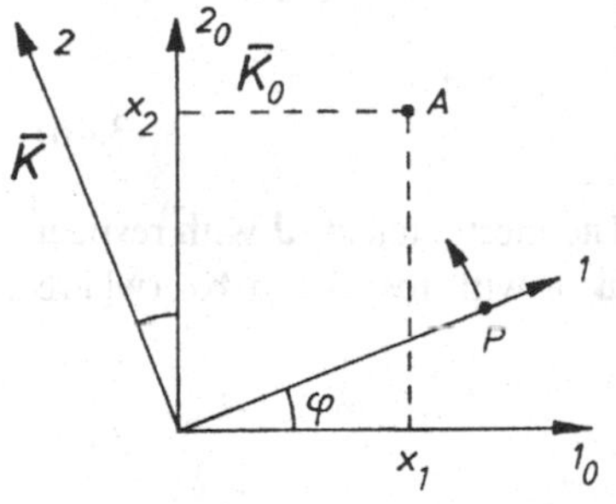

Fig. 3.18. Rotation of the body-fixed system about the 3-axis

In order to clarify the situation with the two kinds of system of reference, we consider first the simplified case of a rotation about the 3-axis. Here we have to study only the transformation behavior in the (1,2)-plane. Consider first a given point A, with fixed coordinates (x_1, x_2, x_3) with respect to $\overline{\mathbf{K}}_0$. If described with respect to $\overline{\mathbf{K}}$, cf. Fig. 3.18, the same point has the coordinates

$$\begin{aligned}\bar{x}_1 &= x_1\cos\varphi + x_2\sin\varphi\,,\\ \bar{x}_2 &= -x_1\sin\varphi + x_2\cos\varphi\,,\\ \bar{x}_3 &= x_3\,.\end{aligned} \tag{3.44a}$$

These equations express the passive form of rotation that we studied in Sects. 2.21 and 2.22. Take now a point P, fixed on the 1-axis of $\overline{\mathbf{K}}$, and assume that this latter system rotates uniformly with respect to $\overline{\mathbf{K}}_0$. We then have

$$\varphi \equiv \varphi(t) = \omega t \; ; \quad P : \bar{x}_1 = a \; , \quad \bar{x}_2 = 0 = \bar{x}_3 \; ;$$

and, by inverting the formulae (3.44a),

$$\begin{aligned} x_1 &= \bar{x}_1 \cos\omega t - \bar{x}_2 \sin\omega t \; , \\ x_2 &= \bar{x}_1 \sin\omega t + \bar{x}_2 \cos\omega t \; , \\ x_3 &= \bar{x}_3 \; , \end{aligned} \tag{3.44b}$$

so that the point P moves according to (P: $x_1 = a\cos\omega t$, $x_2 = a\sin\omega t$, $x_3 = 0$). This is the active form of rotation. Turning now to an arbitrary rotation, we replace (3.44a) by

$$\mathbf{R}(\varphi) = \exp\left(-\sum_{i=1}^{3} \varphi_i \mathbf{J}_i\right) \; , \tag{3.45a}$$

where $\mathbf{J} = \{\mathbf{J}_1, \mathbf{J}_2, \mathbf{J}_3\}$ are the generators for infinitesimal rotations about the corresponding axes (cf. (2.73) in Sect. 2.22). Equation (3.44b) is the inverse of (3.44a). Hence, in the general case,

$$(\mathbf{R}(\varphi))^{-1} = (\mathbf{R}(\varphi))^{\mathrm{T}} = \mathbf{R}(-\varphi) = \exp\left(\sum_{i=1}^{3} \varphi_i \mathbf{J}_i\right) \; . \tag{3.45b}$$

The vectors $\boldsymbol{\omega}$ (angular velocity) and $\boldsymbol{L}$ (angular momentum) are physical quantities. They obey the (passive) transformation law

$$\bar{\boldsymbol{\omega}} = \mathbf{R}\boldsymbol{\omega} \; , \quad \bar{\boldsymbol{L}} = \mathbf{R}\boldsymbol{L} \; , \tag{3.46}$$

$\boldsymbol{\omega}$ and $\boldsymbol{L}$ referring to $\overline{\mathbf{K}}_0$, $\bar{\boldsymbol{\omega}}$ and $\bar{\boldsymbol{L}}$ referring to $\overline{\mathbf{K}}$. The inertia tensor $\mathbf{J}$ with respect to $\overline{\mathbf{K}}$ (where it is constant) and the same tensor taken with respect to $\overline{\mathbf{K}}_0$ (where it depends on time) are related by

$$\mathbf{J} = \mathbf{R}\tilde{\mathbf{J}}\mathbf{R}^{\mathrm{T}} \; , \quad \tilde{\mathbf{J}}(t) = \mathbf{R}^{\mathrm{T}}(t)\mathbf{J}\mathbf{R}(t) \; . \tag{3.47}$$

This follows from the proposition (3.24), with $\boldsymbol{a} = 0$. Clearly, T_{rot} is a scalar and hence is invariant. Indeed, inserting (3.46) into (3.43b), we obtain (3.43a),

$$T_{\mathrm{rot}} = \tfrac{1}{2}(\mathbf{R}\boldsymbol{\omega})\mathbf{J}(\mathbf{R}\boldsymbol{\omega}) = \tfrac{1}{2}\boldsymbol{\omega}\left(\mathbf{R}^{\mathrm{T}}\mathbf{J}\mathbf{R}\right)\boldsymbol{\omega} = \tfrac{1}{2}\boldsymbol{\omega}\tilde{\mathbf{J}}\boldsymbol{\omega} \; .$$

The equation of motion describing the rotation is obtained from the principle of angular momentum. With reference to the system $\overline{\mathbf{K}}_0$, it tells us that the time change of the angular momentum equals the resultant external torque,

$$\frac{\mathrm{d}}{\mathrm{d}t}\boldsymbol{L} = \boldsymbol{D} \; . \tag{3.48}$$

Thus, adopting the discrete model (A) for the rigid body, we have

$$L = \sum_{i=1}^{n} m_i \boldsymbol{x}^{(i)} \times \dot{\boldsymbol{x}}^{(i)} , \tag{3.49}$$

$$D = \sum_{i=1}^{n} \boldsymbol{x}^{(i)} \times \boldsymbol{F}^{(i)} . \tag{3.50}$$

In summary, the equations of motion (3.41) and (3.48) have the general form

$$M\ddot{\boldsymbol{r}}_S(t) = \boldsymbol{F}\big(\boldsymbol{r}_S, \dot{\boldsymbol{r}}_S, \theta_i, \dot{\theta}_i, t\big) , \tag{3.51}$$

$$\dot{\boldsymbol{L}} = \boldsymbol{D}\big(\boldsymbol{r}_S, \dot{\boldsymbol{r}}_S, \theta_i, \dot{\theta}_i, t\big) , \tag{3.52}$$

where (3.51) refers to the inertial system of reference and (3.52) refers to the system $\overline{\mathbf{K}}_0$, which is centered in S and has its axes parallel to those of the inertial system.

3.12 Euler's Equations of Motion

In this section we apply the equation of motion (3.52) to the rigid body and, in particular, work out its specific form for this case. Inverting the second equation of (3.46) we have

$$\boldsymbol{L} = \mathbf{R}^{\mathrm{T}}(t)\bar{\boldsymbol{L}} = \tilde{\mathbf{J}}(t)\boldsymbol{\omega}(t) .$$

Differentiating with respect to time, we obtain

$$\dot{\boldsymbol{L}} = \dot{\tilde{\mathbf{J}}}\boldsymbol{\omega} + \tilde{\mathbf{J}}\dot{\boldsymbol{\omega}}$$

and, by means of (3.47), also

$$\dot{\tilde{\mathbf{J}}}(t) = \frac{\mathrm{d}}{\mathrm{d}t}\big[\mathbf{R}^{\mathrm{T}}(t)\mathbf{J}\mathbf{R}(t)\big] = \dot{\mathbf{R}}^{\mathrm{T}}\mathbf{J}\mathbf{R} + \mathbf{R}^{\mathrm{T}}\mathbf{J}\dot{\mathbf{R}} .$$

If we again replace $\mathbf{J}$ by $\tilde{\mathbf{J}}$ in this last expression, this becomes

$$\dot{\tilde{\mathbf{J}}}(t) = \big(\dot{\mathbf{R}}^{\mathrm{T}}\mathbf{R}\big)\tilde{\mathbf{J}} + \tilde{\mathbf{J}}\big(\mathbf{R}^{\mathrm{T}}\dot{\mathbf{R}}\big) .$$

We now study the specific combination of the rotation matrix and the time derivative of its transpose that appears in the time derivative of $\mathbf{J}(t)$. Let us define

$$\boldsymbol{\Omega}(t) \stackrel{\mathrm{def}}{=} \dot{\mathbf{R}}^{\mathrm{T}}(t)\mathbf{R}(t) \equiv \dot{\mathbf{R}}^{-1}(t)\mathbf{R}(t) . \tag{3.53}$$

The transpose of this matrix, $\boldsymbol{\Omega}^{\mathrm{T}} = \mathbf{R}^{\mathrm{T}}\dot{\mathbf{R}}$, is equal to $-\boldsymbol{\Omega}$. This follows by taking the time derivative of the orthogonality condition $\mathbf{R}^{\mathrm{T}}\mathbf{R} = \mathbb{1}$, whereby

$$\dot{\mathbf{R}}^{\mathrm{T}}\mathbf{R}+\mathbf{R}^{\mathrm{T}}\dot{\mathbf{R}}=0\,,$$

and hence

$$\boldsymbol{\Omega}+\boldsymbol{\Omega}^{\mathrm{T}}=0\,.$$

Thus, we obtain

$$\dot{\tilde{\mathbf{J}}}=\boldsymbol{\Omega}\tilde{\mathbf{J}}+\tilde{\mathbf{J}}\boldsymbol{\Omega}^{\mathrm{T}}=\boldsymbol{\Omega}\tilde{\mathbf{J}}-\tilde{\mathbf{J}}\boldsymbol{\Omega}=[\boldsymbol{\Omega},\,\tilde{\mathbf{J}}]\,,\tag{3.54}$$

where $[\,,]$ denotes the commutator, $[A,\,B]\stackrel{\mathrm{def}}{=}AB-BA$. In order to compute the action of $\boldsymbol{\Omega}$ on an arbitrary vector, one must first calculate the time derivative of the rotation matrix (3.45a). The exponential is to be understood as a shorthand for its series expansion, cf. Sect. 2.20. Differentiating termwise and assuming $\dot{\boldsymbol{\varphi}}$ to be parallel to $\hat{\boldsymbol{\varphi}}$, one obtains

$$\frac{\mathrm{d}}{\mathrm{d}t}\mathbf{R}(\varphi(t))=-\left[\sum_{i=1}^{3}\dot{\varphi}_i(t)\mathbf{J}_i\right]\mathbf{R}(\varphi(t))=-\left[\sum_{i=1}^{3}\omega_i\mathbf{J}_i\right]\mathbf{R}\,.\tag{3.55}$$

From Sect. 2.22 we know that the action of $(\sum_{i=1}^{3}\omega_i\mathbf{J})$ on an arbitrary vector $\boldsymbol{b}$ of $\mathbb{R}^3$ can be expressed by means of the cross product, viz.

$$\left(\sum_{i=1}^{3}\omega_i\mathbf{J}_i\right)\boldsymbol{b}=\boldsymbol{\omega}\times\boldsymbol{b}\,.$$

Obviously, an analogous formula applies to the inverse of $\mathbf{R}$,

$$\frac{\mathrm{d}}{\mathrm{d}t}\mathbf{R}^{\mathrm{T}}(\varphi(t))=\left(\sum_{i=1}^{3}\omega_i\mathbf{J}_i\right)\mathbf{R}^{\mathrm{T}}(\varphi(t))\,.$$

Therefore, taking $\boldsymbol{b}=\mathbf{R}^{\mathrm{T}}\boldsymbol{a}$, we have

$$\dot{\mathbf{R}}^{\mathrm{T}}(t)\boldsymbol{a}=\boldsymbol{\omega}\times\left(\mathbf{R}^{\mathrm{T}}\boldsymbol{a}\right)\,,\tag{3.56a}$$

and from this

$$\boldsymbol{\Omega}(t)\boldsymbol{a}=\dot{\mathbf{R}}^{\mathrm{T}}\mathbf{R}\boldsymbol{a}=\boldsymbol{\omega}\times\left(\mathbf{R}^{\mathrm{T}}\mathbf{R}\boldsymbol{a}\right)=\boldsymbol{\omega}\times\boldsymbol{a}\,.\tag{3.56b}$$

This gives us

$$\begin{aligned}\dot{\boldsymbol{L}}&=\dot{\tilde{\mathbf{J}}}\boldsymbol{\omega}+\tilde{\mathbf{J}}\dot{\boldsymbol{\omega}}=(\boldsymbol{\Omega}\tilde{\mathbf{J}}-\tilde{\mathbf{J}}\boldsymbol{\Omega})\boldsymbol{\omega}+\tilde{\mathbf{J}}\dot{\boldsymbol{\omega}}\\&=\boldsymbol{\Omega}\tilde{\mathbf{J}}\boldsymbol{\omega}+\tilde{\mathbf{J}}\dot{\boldsymbol{\omega}}=\boldsymbol{\omega}\times(\tilde{\mathbf{J}}\boldsymbol{\omega})+\tilde{\mathbf{J}}\dot{\boldsymbol{\omega}}=\boldsymbol{\omega}\times\boldsymbol{L}+\tilde{\mathbf{J}}\dot{\boldsymbol{\omega}}\,,\end{aligned}\tag{3.57}$$

where we have used the equation $\boldsymbol{\Omega}\boldsymbol{\omega}=0$, which follows from (3.56b). It remains to compute $\dot{\boldsymbol{\omega}}$,

$$\dot{\boldsymbol{\omega}}=\frac{\mathrm{d}}{\mathrm{d}t}\left(\mathbf{R}^{\mathrm{T}}\bar{\boldsymbol{\omega}}\right)=\mathbf{R}^{\mathrm{T}}\dot{\bar{\boldsymbol{\omega}}}+\dot{\mathbf{R}}^{\mathrm{T}}\bar{\boldsymbol{\omega}}\,.$$

The second term vanishes because, by (3.56a),

$$\dot{\mathbf{R}}^{\mathrm{T}}\bar{\omega} = \omega \times \left(\mathbf{R}^{\mathrm{T}}\bar{\omega}\right) = \omega \times \omega = 0\,.$$

Thus, $\dot{\omega} = \mathbf{R}^{\mathrm{T}}\dot{\bar{\omega}}$. Inserting (3.57) into the equation of motion (3.52), we obtain

$$\dot{\boldsymbol{L}} = \omega \times \boldsymbol{L} + \tilde{\mathbf{J}}\mathbf{R}^{\mathrm{T}}\dot{\bar{\omega}} = \boldsymbol{D}\,.$$

This form of the equations of motion has the drawback that it contains both quantities referring to a system of reference with space-fixed directions and quantities referring to a body-fixed system. However, it is not difficult to convert them completely to the system fixed in the body: multiply these equations with $\mathbf{R}$ from the left and note that $\mathbf{R}(\omega \times \boldsymbol{L}) = \mathbf{R}\omega \times \mathbf{R}\boldsymbol{L} = \bar{\omega} \times \bar{\boldsymbol{L}}$. In this way we obtain *Euler's equations* in their final form,

$$\boxed{\mathbf{J}\dot{\bar{\omega}} + \bar{\omega} \times \bar{\boldsymbol{L}} = \bar{\boldsymbol{D}}}\,. \qquad (3.58)$$

All quantities now refer to the body-fixed system $\overline{\mathbf{K}}$. In particular, $\mathbf{J}$ is the (constant) inertia tensor as computed in Sect. 3.4 above. If the intrinsic system $\overline{\mathbf{K}}$ is chosen to be a PA system, $\mathbf{J}$ is diagonal. Finally, we note that $\bar{\boldsymbol{L}} = \mathbf{J}\bar{\omega}$. This shows that the equations of motion (3.58) for the unknown functions $\bar{\omega}(t)$ are nonlinear.

Remark: Because of its antisymmetry, the action of the matrix $\boldsymbol{\Omega}$ on any vector $\boldsymbol{a}$ is always the one given in (3.56b), with ω to be calculated from the rotation $\mathbf{R}(t) = \exp\{-\mathbf{S}\}$, with $\mathbf{S} = \sum \varphi_i(t)\mathbf{J}_i$. We have

$$\begin{aligned}\boldsymbol{\Omega} = \dot{\mathbf{R}}^{\mathrm{T}}(t)\mathbf{R}(t) &= \left(\frac{\mathrm{d}}{\mathrm{d}t}\mathrm{e}^{\mathbf{S}}\right)\mathrm{e}^{-\mathbf{S}} \\ &= \left(\dot{\mathbf{S}} + \frac{1}{2}\dot{\mathbf{S}}\mathbf{S} + \frac{1}{2}\mathbf{S}\dot{\mathbf{S}} + \ldots\right)(\mathbb{1} - \mathbf{S} + \ldots) \\ &= \dot{\mathbf{S}} + \frac{1}{2}[\mathbf{S}, \dot{\mathbf{S}}] + \mathcal{O}(\varphi^2)\,.\end{aligned}$$

Making use of the commutators $[\mathbf{J}_i, \mathbf{J}_j] = \sum_k \varepsilon_{ijk}\mathbf{J}_k$ one derives the identity

$$[\mathbf{S}, \dot{\mathbf{S}}] \equiv [\mathbf{S}(\varphi), \dot{\mathbf{S}}(\varphi)] = \mathbf{S}(\varphi \times \dot{\varphi})\,,$$

from which one computes ω. If we make the assumption (as we did in Sect. 2.22 and also in this section above) that $\varphi = \varphi\hat{\boldsymbol{n}}$ and $\dot{\varphi}$ have the same direction, then $\mathbf{S}$ commutes with $\dot{\mathbf{S}}$ so that ω and $\dot{\varphi}$ coincide.

One may, of course, also consider a situation where both the modulus and the direction of φ change with time. In this case φ and $\dot{\varphi}$ are no longer parallel, and $\mathbf{S}$ and $\dot{\mathbf{S}}$ no longer commute. It then follows from (3.56b), from $\mathbf{S}\boldsymbol{a} = \varphi \times \boldsymbol{a}$, and from the calculation above that

$$\omega = \dot{\varphi} + \frac{1}{2}\varphi \times \dot{\varphi} + \mathcal{O}(\varphi^2)\,.$$

In very much the same way one shows that

$$\overline{\omega} = \dot{\varphi} - \frac{1}{2}\varphi \times \dot{\varphi} + \mathcal{O}(\varphi^2) .$$

In deriving Euler's equations the difference to the situation where $\dot{\varphi}$ and φ are taken to be parallel is irrelevant because we may always add a *constant* rotation such that the modulus φ of φ is small, and $\omega \approx \dot{\varphi}$ (see also Sect. 5.7.4).

3.13 Euler's Equations Applied to a Force-Free Top

As a first illustration of Euler's equations we study the force-free motion of rigid bodies. If no external forces are present, the center of mass, by (3.51), moves with constant velocity along a straight line. The right-hand side of Euler's equations (3.58) vanishes, $\boldsymbol{D} = 0$. If $\overline{\mathbf{K}}$ is chosen to be the PA system, then $J_{ik} = I_i \delta_{ik}$ and $\bar{L}_i = I_i \bar{\omega}_i$, so that (3.58) reads

$$I_i \dot{\bar{\omega}}_i + (\bar{\omega} \times \bar{\boldsymbol{L}})_i = 0 .$$

More explicitly, because $(\bar{\omega} \times \bar{\boldsymbol{L}})_1 = I_3 \bar{\omega}_2 \bar{\omega}_3 - I_2 \bar{\omega}_3 \bar{\omega}_2 = (I_3 - I_2) \bar{\omega}_2 \bar{\omega}_3$ (with cyclic permutation of the indices), the equations of motion read

$$\begin{aligned} I_1 \dot{\bar{\omega}}_1 &= \big(I_2 - I_3\big) \bar{\omega}_2 \bar{\omega}_3 , \\ I_2 \dot{\bar{\omega}}_2 &= \big(I_3 - I_1\big) \bar{\omega}_3 \bar{\omega}_1 , \\ I_3 \dot{\bar{\omega}}_3 &= \big(I_1 - I_2\big) \bar{\omega}_1 \bar{\omega}_2 . \end{aligned} \tag{3.59}$$

(i) *The asymmetric or triaxial top.* Here $I_1 \neq I_2 \neq I_3 \neq I_1$. The equations (3.59) being nonlinear, their solution in the general case is certainly not obvious. Yet, as we shall see below, their solution can be reduced to quadratures by making use of the conservation of energy and angular momentum. Before turning to this analysis, we discuss a qualitative feature of its motion that can be read off (3.59). Without loss of generality we assume the ordering

$$I_1 < I_2 < I_3 . \tag{3.60}$$

Indeed, the principle axes can always be chosen and numbered in such a way that the 1-axis is the axis of the smallest moment of inertia, the 3-axis that of the largest. The right-hand sides of the first and third equations of (3.59) then have negative coefficients, while the right-hand side of the second equation has a positive coefficient. Thus, the stability behavior of a rotation about the 2-axis (the one with the intermediate moment of inertia) will be different, under the effect of a small perturbation, from that of rotations about the 1- or 3-axes. Indeed, in the latter cases the rotation is found to be stable, while in the former it is unstable, (see Sect. 6.2.5).

We now set $x(t) \stackrel{\text{def}}{=} \bar{\omega}_3(t)$ and make use of the two conservation laws that hold for free motion:

$$2T_{\text{rot}} = \sum_{i=1}^{3} I_i \bar{\omega}_i^2 = \text{const} , \tag{3.61}$$

$$\boldsymbol{L}^2 = \sum_{i=1}^{3} \left(I_i \bar{\omega}_i\right)^2 = \text{const} . \tag{3.62}$$

Taking the combinations

$$\boldsymbol{L}^2 - 2T_{\text{rot}} I_1 = I_2(I_2 - I_1)\bar{\omega}_2^2 + I_3(I_3 - I_1)x^2 ,$$
$$\boldsymbol{L}^2 - 2T_{\text{rot}} I_2 = -I_1(I_2 - I_1)\bar{\omega}_1^2 + I_3(I_3 - I_2)x^2 ,$$

we deduce the following equations:

$$\bar{\omega}_1^2 = -\frac{1}{I_1(I_2 - I_1)}\left[\boldsymbol{L}^2 - 2T_{\text{rot}} I_2 - I_3(I_3 - I_2)x^2\right] \equiv -\alpha_0 + \alpha_2 x^2 ,$$
$$\bar{\omega}_2^2 = \frac{1}{I_2(I_2 - I_1)}\left[\boldsymbol{L}^2 - 2T_{\text{rot}} I_1 - I_3(I_3 - I_1)x^2\right] \equiv \beta_0 - \beta_2 x^2 .$$

With the convention (3.60), all differences of moments of inertia are written so as to make the coefficients α_0, α_2, β_0, β_2 positive. Inserting these auxiliary relations into the third equation of (3.59) yields the differential equation

$$I_3 \dot{x}(t) = (I_1 - I_2)\sqrt{\left(\beta_0 - \beta_2 x^2\right)\left(-\alpha_0 + \alpha_2 x^2\right)} \tag{3.63}$$

for $x(t)$. It can be solved by separation of variables and hence by ordinary integration (quadrature). Clearly, $\bar{\omega}_1(t)$ and $\bar{\omega}_2(t)$ obey analogous differential equations that are obtained from (3.63) by cyclic permutation.

(ii) *The symmetric top.* Without loss of generality we assume

$$I_1 = I_2 \neq I_3 \quad \text{and} \quad I_1 \neq 0 , \; I_3 \neq 0 . \tag{3.64}$$

The solution of the equations (3.59) is elementary in this case. First, we note that

$$I_3 \dot{\bar{\omega}}_3 = 0 , \quad \text{i.e.} \quad \bar{\omega}_3 = \text{const} .$$

Introducing the notation

$$\omega_0 \stackrel{\text{def}}{=} \bar{\omega}_3 \frac{I_3 - I_1}{I_1} (= \text{const}) \tag{3.65}$$

we see that the first two equations of (3.59) become

$$\dot{\bar{\omega}}_1 = -\omega_0 \bar{\omega}_2 \quad \text{and} \quad \dot{\bar{\omega}}_2 = \omega_0 \bar{\omega}_1 ,$$

their solutions being

$$\bar{\omega}_1(t) = \omega_\perp \cos(\omega_0 t + \tau) , \quad \bar{\omega}_2(t) = \omega_\perp \sin(\omega_0 t + \tau) . \tag{3.66}$$

Here $\omega_\perp$ and τ are integration constants that are chosen at will. ω_0, in turn, is already fixed by the choice of the integration constant $\bar{\omega}_3$ in (3.65). As a result, one obtains

$$\bar{\boldsymbol{\omega}} = \left(\omega_\perp \cos(\omega_0 t + \tau)\,,\ \omega_\perp \sin(\omega_0 t + \tau)\,,\ \bar{\omega}_3\right)$$

and $\bar{\boldsymbol{\omega}}^2 = \omega_\perp^2 + \bar{\omega}_3^2$. The vector $\bar{\boldsymbol{\omega}}$ has constant length: it rotates uniformly about the 3-axis of the PA system. This is the symmetry axis of the top.

As to the angular momentum with respect to the intrinsic system, one has

$$\begin{aligned}
\bar{L}_1 &= I_1 \omega_\perp \cos(\omega_0 t + \tau)\,,\\
\bar{L}_2 &= I_1 \omega_\perp \sin(\omega_0 t + \tau)\,,\\
\bar{L}_3 &= I_3 \bar{\omega}_3\,,\\
\bar{\boldsymbol{L}}^2 &= I_1^2 \omega_\perp^2 + I_3^2 \bar{\omega}_3\,.
\end{aligned} \tag{3.67}$$

This shows that $\bar{\boldsymbol{L}}$ rotates uniformly about the symmetry axis, too. Furthermore, at any time the symmetry axis $\hat{\bar{\boldsymbol{f}}}$, $\bar{\boldsymbol{\omega}}$ and $\bar{\boldsymbol{L}}$ lie in one plane.

It is not difficult to work out the relation of the constants of integration $\omega_\perp$, $\bar{\omega}_3$ (or ω_0) to the integrals of the motion that are characteristic for motion without external forces: the kinetic energy T_{rot} and the modulus of angular momentum. One has

$$2T_{\text{rot}} = \sum_{i=1}^{3} I_i \bar{\omega}_i^2 = I_1 \omega_\perp^2 + I_3 \bar{\omega}_3^2 = I_1 \left[\omega_\perp^2 + \frac{I_1 I_3}{(I_3 - I_1)^2}\omega_0^2\right]\,,$$

$$\boldsymbol{L}^2 = \bar{\boldsymbol{L}}^2 = I_1^2 \omega_\perp^2 + I_3^2 \bar{\omega}_3^2 = I_1^2 \left[\omega_\perp^2 + \frac{I_3^2}{(I_3 - I_1)^2}\omega_0^2\right]\,,$$

from which one obtains

$$\omega_\perp^2 = \frac{1}{I_1(I_1 - I_3)}\left[\boldsymbol{L}^2 - 2I_3 T_{\text{rot}}\right]\,, \tag{3.68}$$

$$\omega_0^2 = \frac{I_1 - I_3}{I_1^2 I_3}\left[2I_1 T_{\text{rot}} - \boldsymbol{L}^2\right]\,. \tag{3.69}$$

Finally, one may wish to translate these results to a description of the same motion with respect to the system $\overline{\mathbf{K}}_0$ of Fig. 3.15. Because there are no external forces, this is an inertial system. Denoting the symmetry axis of the top by $\hat{\bar{\boldsymbol{f}}}$ (this is the 3-axis of $\overline{\mathbf{K}}$), the same unit vector, with respect to $\mathbf{K}_0$, depends on time and is given by

$$\hat{\boldsymbol{f}}(t) = \mathbf{R}^{\mathrm{T}}(t)\hat{\bar{\boldsymbol{f}}}\,.$$

Since $\bar{\boldsymbol{L}}$, $\bar{\boldsymbol{\omega}}$, and $\hat{\bar{\boldsymbol{f}}}$ are always in a plane, so are $\boldsymbol{L} = \mathbf{R}^{\mathrm{T}}\bar{\boldsymbol{L}}$, $\boldsymbol{\omega} = \mathbf{R}^{\mathrm{T}}\bar{\boldsymbol{\omega}}, \hat{\boldsymbol{f}}$. Being conserved, the angular momentum $\boldsymbol{L}$ is a constant vector in space, while $\boldsymbol{\omega}$ and $\hat{\boldsymbol{f}}$ rotate uniformly and synchronously about this direction.

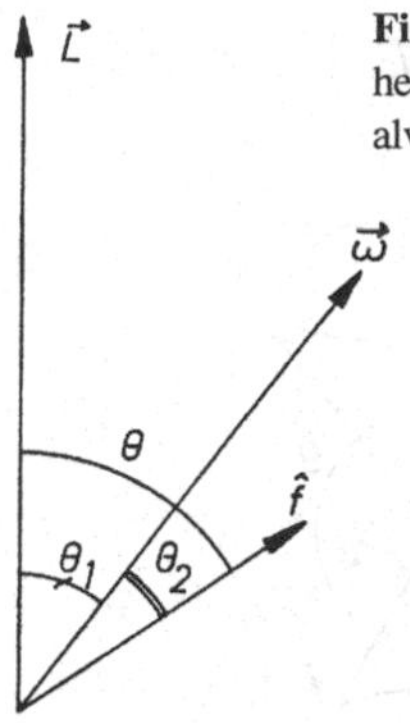

Fig. 3.19. For a free symmetric top the angular momentum $\boldsymbol{L}$ is conserved and hence fixed in space. The axis of symmetry $\hat{\boldsymbol{f}}$, the angular velocity $\boldsymbol{\omega}$, and $\boldsymbol{L}$ always lie in a plane. $\hat{\boldsymbol{f}}$ and $\boldsymbol{\omega}$ perform a uniform precession about $\boldsymbol{L}$

As shown in Fig. 3.19, we call θ_1, θ_2 the angles between $\boldsymbol{L}$ and $\boldsymbol{\omega}$ and between $\boldsymbol{\omega}$ and $\hat{\boldsymbol{f}}$, respectively, and let

$$\theta \stackrel{\text{def}}{=} \theta_1 + \theta_2 \,.$$

We show that $\cos\theta$ and $\cos\theta_2$ must always have the same sign. This will help us to find the possible types of motion. We have

$$2T_{\text{rot}} = \boldsymbol{L}\cdot\boldsymbol{\omega} \quad \text{and} \quad \cos\theta_1 = \frac{2T_{\text{rot}}}{|\boldsymbol{L}||\boldsymbol{\omega}|} \,.$$

As T_{rot} is conserved and positive, $\cos\theta_1$ is constant and positive. Thus

$$-\tfrac{\pi}{2} \le \theta_1 \le \tfrac{\pi}{2} \,.$$

Furthermore, making use of the invariance of the scalar product, we have

$$\begin{aligned} \boldsymbol{\omega}\cdot\hat{\boldsymbol{f}} &= \bar{\boldsymbol{\omega}}\cdot\hat{\bar{\boldsymbol{f}}} = |\boldsymbol{\omega}|\cos\theta_2 = \bar{\omega}_3 = \bar{L}_3/I_3 \\ &= \bar{\boldsymbol{L}}\cdot\hat{\bar{\boldsymbol{f}}}/I_3 = \boldsymbol{L}\cdot\hat{\boldsymbol{f}}/I_3 = |\boldsymbol{L}|\cos\theta/I_3 \,. \end{aligned}$$

It follows, indeed, that $\cos\theta$ and $\cos\theta_2$ have the same sign, at any moment of the motion. As a consequence, there can be only two types of motion, one where this sign is positive (Fig. 3.20a) and one where the sign is negative (Fig. 3.20b), θ_1 being constrained as shown above, $-\frac{\pi}{2} \le \theta_1 \le \frac{\pi}{2}$. Figure 3.20 shows the situation for $I_3 > I_1 = I_2$, i.e. for a body that is elongated like an egg or a cigar. If $I_3 < I_1 = I_2$, i.e. for a body that has the shape of a disc or a pancake, the angular momentum lies between $\boldsymbol{\omega}$ and the symmetry axis of the top. Finally, we can write down one more relation between the angles θ and θ_2. Take the 1-axis of the intrinsic system in the plane spanned by $\boldsymbol{L}$ and $\hat{\boldsymbol{f}}$ (as in Sect. 3.8 (iii)). Then we have

$$\bar{L}_1/\bar{L}_3 = \tan\theta = I_1\bar{\omega}_1/(I_3\bar{\omega}_3) = \big(I_1/I_3\big)\tan\theta_2 \,.$$

(iii) *A practical example: the Earth.* To a good approximation the Earth can be regarded as a slightly flattened, disklike, symmetric top. Its symmetry axis is defined by the geographic poles. Because its axis of rotation is slightly inclined, by

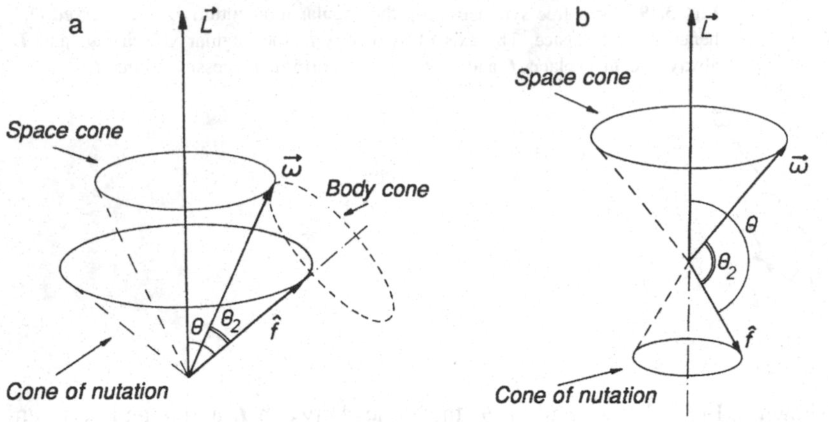

Fig. 3.20. The two types of motion of the axis of symmetry $\hat{f}$ and the angular velocity $\boldsymbol{\omega}$ about the angular momentum $\boldsymbol{L}$

about 0.2″, with respect to the symmetry axis, it performs a precession motion. Neglecting the external forces acting on the Earth, we can estimate the period of this precession as follows. We have

$$I_1 = I_2 < I_3 \quad \text{with} \quad (I_3 - I_1)/I_1 \simeq 1/300 \,. \tag{3.70}$$

The frequency of precession is given by (3.65). Thus, the period is

$$T = \frac{2\pi}{\omega_0} = \frac{2\pi I_1}{(I_3 - I_1)\bar{\omega}_3} \,.$$

Inserting $2\pi/\bar{\omega}_3 = 1$ day and the ratio (3.70) one gets $T \simeq 300$ days. Experimentally, one finds a period of about 430 days and an amplitude of a few meters. The deviation of the measured period from the estimate is probably due to the fact that the Earth is not really rigid.

In fact, the Earth is not free and is subject to external forces and torques exerted on it by the Sun and the Moon. The precession estimated above is superimposed upon a much longer precession with a mean period of 25 800 years (the so-called Platonic year). However, the fact that the free precession estimated above is so much faster than this extremely slow gyroscopic precession justifies the assumption of force-free motion on which we based our estimate.

3.14 The Motion of a Free Top and Geometric Constructions

The essential features of the motion of a free, asymmetric rigid body can be understood qualitatively, without actually solving the equations (3.63), by means of the following constructions. The first of these refers to a reference system fixed in space; the second refers to the intrinsic PA system and both make use of the conservation laws for energy and angular momentum.

(i) *Poinsot's construction (with respect to a space-fixed system).* The conservation law (3.61) can be written in two equivalent ways in terms of quantities in the reference system fixed in space,

$$2T_{\rm rot} = \boldsymbol{\omega}(t) \cdot \boldsymbol{L} = \boldsymbol{\omega}(t)\tilde{\mathbf{J}}(t)\boldsymbol{\omega}(t) = \text{const} . \tag{3.71}$$

As $\boldsymbol{L}$ is fixed, (3.71) tells us that the projection of $\boldsymbol{\omega}(t)$ onto $\boldsymbol{L}$ is constant. Thus, the tip of the vector $\boldsymbol{\omega}(t)$ always lies in a plane perpendicular to $\boldsymbol{L}$. This plane is said to be the *invariant plane.* The second equality in (3.71) tells us, on the other hand, that the tip of $\boldsymbol{\omega}(t)$ must also lie on an ellipsoid, whose position in space changes with time, viz.

$$\sum_{i,k=1}^{3} \tilde{J}_{ik}(t)\omega_i(t)\omega_k(t) = 2T_{\rm rot} .$$

These two surfaces are shown in Fig. 3.21. As we also know that

$$2T_{\rm rot} = \sum_{i=1}^{3} I_i\bar{\omega}_i^2 ,$$

the principal diameters a_i of this ellipsoid are given by $a_i = \sqrt{2T_{\rm rot}/I_i}$. For fixed energy, the ellipsoid has a fixed shape,

$$\sum_{i=1}^{3} \frac{\bar{\omega}_i^2}{2T_{\rm rot}/I_i} = 1 . \tag{3.72}$$

When looked at from the laboratory system, however, the ellipsoid moves as a whole. To understand this motion, we note the relation

$$\frac{\partial T_{\rm rot}}{\partial \omega_i} = \frac{1}{2}\frac{\partial}{\partial \omega_i}\left(\sum_{k,l=1}^{3} \omega_k \tilde{J}_{kl}\omega_l\right) = \sum_{m=1}^{3} \tilde{J}_{im}\omega_m = L_i ,$$

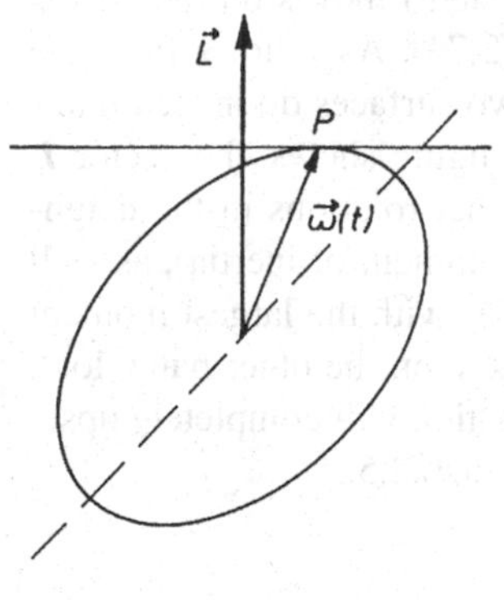

Fig. 3.21. The tip of $\omega(t)$ wanders on the invariant plane and on a time-dependent ellipsoid tangent to that plane

which tells us that $\boldsymbol{L} = \boldsymbol{\nabla}_\omega T_{\mathrm{rot}}$. Thus, at any moment, $\boldsymbol{L}$ is perpendicular to the tangent plane to the ellipsoid at the point P. In other words, the invariant plane is tangent to the ellipsoid. The momentary axis of rotation is just $\boldsymbol{\omega}(t)$. Therefore, the motion of the ellipsoid is such that it *rolls* over the invariant plane without gliding. In the course of the motion, the point P traces out two curves, one on the invariant plane and one on the ellipsoid. These curves are somewhat complicated in the general case. In the case of a symmetric body with, say, $I_1 = I_2 > I_3$, they are seen to be circles.

(ii) *General construction within a principal-axes system.* Using a PA system, we have $\bar{L}_i = I_i \bar{\omega}_i$, so that the conserved quantities (3.61–3.62) can be written as follows:

$$2T_{\mathrm{rot}} = \sum_{i=1}^{3} \frac{\bar{L}_i^2}{I_i}\,, \tag{3.73}$$

$$\boldsymbol{L}^2 = \sum_{i=1}^{3} \bar{L}_i^2\,. \tag{3.74}$$

The first of these, when read as an equation for the variables $\bar{L}_1$, $\bar{L}_2$, $\bar{L}_3$, describes an ellipsoid with principal axes

$$a_i = \sqrt{2T_{\mathrm{rot}} I_i}\,, \quad i = 1,\, 2,\, 3\,. \tag{3.75}$$

With the convention (3.60) they obey the inequalities $a_1 < a_2 < a_3$. From (3.60) one also notes that

$$2T_{\mathrm{rot}} I_1 \le \bar{\boldsymbol{L}}^2 = \boldsymbol{L}^2 \le 2T_{\mathrm{rot}} I_3\,. \tag{3.76}$$

The second equation, (3.74) is a sphere with radius

$$R = \sqrt{\boldsymbol{L}^2} \quad \text{and} \quad a_1 \le R \le a_3\,. \tag{3.77}$$

Taking both equations together, we conclude that the extremity of $\bar{\boldsymbol{L}}$ (this is the angular momentum as seen from the body-fixed PA system) moves on the curves of intersection of the ellipsoid (3.73) and the sphere (3.74). As follows from the inequalities (3.76), or equivalently from (3.77), these two surfaces do indeed intersect. This yields the picture shown in Fig. 3.22. As the figure shows, the vector $\bar{\boldsymbol{L}}$ performs periodic motions in all cases. One also sees that rotations in the neighborhood of the 1-axis (principal axis with the smallest moment of inertia), as well rotations in the neighborhood of the 3-axis (principal axis with the largest moment of inertia) are stable. Rotations with $\bar{\boldsymbol{L}}$ close to the 2-axis, on the other hand, look unstable. One is led to suspect that even a small perturbation will completely upset the motion. That this is indeed so will be shown in Sect. 6.2.5.

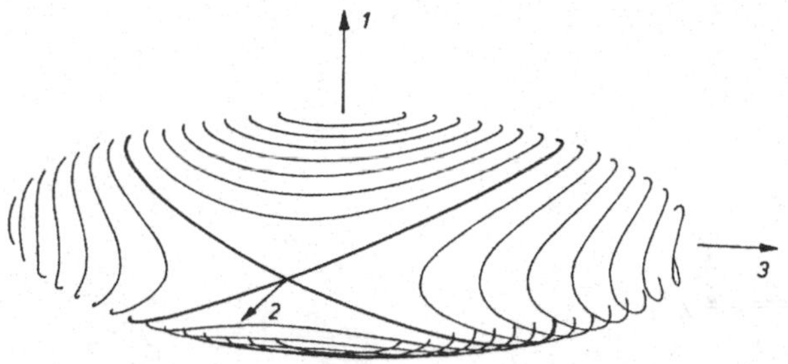

Fig. 3.22. The angular momentum $\bar{L}$, as seen from a reference system fixed in the body, moves along the curves of intersection of the spheres (3.74) and of the ellipsoids (3.73)

3.15 The Rigid Body in the Framework of Canonical Mechanics

The aim of this section is to derive once more the equations of motion of rigid bodies, this time by means of a Lagrangian function that is expressed in terms of Eulerian angles. In a second step we wish to find the generalized momenta that are canonically conjugate to these variables. Finally, via a Legendre transformation, we wish to construct a Hamiltonian function for the rigid body.

(i) *Angular velocity and Eulerian angles.* In a first step we must calculate the components of the angular velocity $\bar{\omega}$ with respect to a PA system, following (3.35), and express them in terms of Eulerian angles as defined in Sect. 3.10. A simple, geometric way of doing this is to start from Fig. 3.16 or 3.23. To the three time-dependent rotations in (3.35) there correspond the angular velocities $\boldsymbol{\omega}_\alpha$, $\boldsymbol{\omega}_\beta$, and $\boldsymbol{\omega}_\gamma$. Here, $\boldsymbol{\omega}_\alpha$ points along the 3_0-axis, $\boldsymbol{\omega}_\beta$ along the axis $S\eta$, and $\boldsymbol{\omega}_\gamma$ along the 3-axis. If 1, 2, and 3 denote the principal axes, as before, and if $(\boldsymbol{\omega}_\alpha)_i$ denotes the component of $\boldsymbol{\omega}_\alpha$ along the axis i, the following decompositions are obtained from Fig. 3.23:

$$(\boldsymbol{\omega}_\beta)_1 = \dot{\beta}\sin\gamma\,, \quad (\boldsymbol{\omega}_\beta)_2 = \dot{\beta}\cos\gamma\,, \quad (\boldsymbol{\omega}_\beta)_3 = 0\,, \tag{3.78}$$

$$(\boldsymbol{\omega}_\alpha)_3 = \dot{\alpha}\cos\beta\,, \quad (\boldsymbol{\omega}_\alpha)_{\xi_2} = -\dot{\alpha}\sin\beta\,, \tag{3.79a}$$

from which follows

$$(\boldsymbol{\omega}_\alpha)_1 = -\dot{\alpha}\sin\beta\cos\gamma\,, \quad (\boldsymbol{\omega}_\alpha)_2 = \dot{\alpha}\sin\beta\sin\gamma\,, \tag{3.79b}$$

and finally

$$(\boldsymbol{\omega}_\gamma)_1 = 0\,, \quad (\boldsymbol{\omega}_\gamma)_2 = 0\,, \quad (\boldsymbol{\omega}_\gamma)_3 = \dot{\gamma}\,. \tag{3.80}$$

Thus, the angular velocity $\bar{\omega} = \boldsymbol{\omega}_\alpha + \boldsymbol{\omega}_\beta + \boldsymbol{\omega}_\gamma$ is given by $\bar{\omega} = (\bar{\omega}_1, \bar{\omega}_2, \bar{\omega}_3)^T$ with

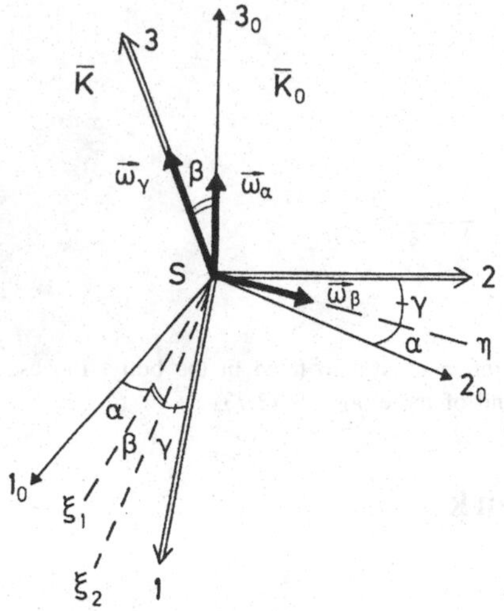

Fig. 3.23. Construction that helps to express ω by the time derivatives of the Eulerian angles. Definition as in Fig. 3.16

$$\begin{aligned} \bar{\omega}_1 &= \dot{\beta} \sin\gamma - \dot{\alpha} \sin\beta \cos\gamma \; , \\ \bar{\omega}_2 &= \dot{\beta} \cos\gamma + \dot{\alpha} \sin\beta \sin\gamma \; , \\ \bar{\omega}_3 &= \dot{\alpha} \cos\beta + \dot{\gamma} \; . \end{aligned} \tag{3.81}$$

It is easy to translate these results to the definition of Eulerian angles as given in Sect. 3.10. The transformation rules (3.39) tell us that in (3.81) $\cos\gamma$ must be replaced by $-\sin\Psi$ and $\sin\gamma$ by $\cos\Psi$, giving

$$\begin{aligned} \bar{\omega}_1 &= \dot{\theta} \cos\Psi + \dot{\Phi} \sin\theta \sin\Psi \; , \\ \bar{\omega}_2 &= -\dot{\theta} \sin\Psi + \dot{\Phi} \sin\theta \cos\Psi \; , \\ \bar{\omega}_3 &= \dot{\Phi} \cos\theta + \dot{\Psi} \; . \end{aligned} \tag{3.82}$$

The functions $\bar{\omega}_i(t)$ obey the system of differential equations (3.58), Euler's equations. Once they are known, by inverting (3.82) and solving for $\dot{\Phi}$, $\dot{\theta}$, and $\dot{\Psi}$, one obtains the following system of coupled differential equations

$$\begin{aligned} \dot{\Phi} &= \left[\bar{\omega}_1 \sin\Psi + \bar{\omega}_2 \cos\Psi\right] / \sin\theta \; , \\ \dot{\theta} &= \bar{\omega}_1 \cos\Psi - \bar{\omega}_2 \sin\Psi \; , \\ \dot{\Psi} &= \bar{\omega}_3 - \left[\bar{\omega}_1 \sin\Psi + \bar{\omega}_2 \cos\Psi\right] \cot\theta \; . \end{aligned} \tag{3.83}$$

The solutions of this system $\{\Phi(t), \theta(t), \Psi(t)\}$ describe the actual motion completely.

Making use of (3.82) we can construct a Lagrangian function in terms of Eulerian angles. Its natural form is

$$L = T - U \; , \tag{3.84}$$

where the kinetic energy is given by

$$T \equiv T_{\text{rot}} = \frac{1}{2}\sum_{i=1}^{3} I_i \bar{\omega}_i^2 = \frac{1}{2} I_1 (\dot{\theta}\cos\Psi + \dot{\Phi}\sin\theta\sin\Psi)^2$$
$$+\frac{1}{2} I_2 (-\dot{\theta}\sin\Psi + \dot{\Phi}\sin\theta\cos\Psi)^2 + \frac{1}{2} I_3 (\dot{\Psi} + \dot{\Phi}\cos\theta)^2 \,. \quad (3.85)$$

Note that we use the second definition of Eulerian angles (see Sect. 3.10 and Fig. 3.17) and that we assume that the center-of-mass motion is already separated off.

The first test is to verify that L, as given by (3.84), yields Euler's equations in the form (3.59) when there are no external forces ($U = 0$). We calculate

$$\frac{\partial L}{\partial \dot{\Psi}} = \frac{\partial T}{\partial \bar{\omega}_3}\frac{\partial \bar{\omega}_3}{\partial \dot{\Psi}} = I_3 \bar{\omega}_3 \,,$$
$$\frac{\partial L}{\partial \Psi} = \frac{\partial T}{\partial \bar{\omega}_1}\frac{\partial \bar{\omega}_1}{\partial \Psi} + \frac{\partial T}{\partial \bar{\omega}_2}\frac{\partial \bar{\omega}_2}{\partial \Psi} = \left(I_1 - I_2\right)\bar{\omega}_1\bar{\omega}_2 \,.$$

Indeed, the Euler–Lagrange equation $(\mathrm{d}/\mathrm{d}t)(\partial L/\partial\dot{\Psi}) = \partial L/\partial\Psi$ is identical with the third of equations (3.59). The remaining two follow by cyclic permutation.

(ii) *Canonical momenta and the Hamiltonian function.* The momenta canonically conjugate to the Eulerian angles are found by taking the partial derivatives of L with respect to $\dot{\Phi}$, $\dot{\theta}$, and $\dot{\Psi}$. The momentum p_Ψ is the easiest to determine:

$$p_\Psi \stackrel{\text{def}}{=} \frac{\partial L}{\partial \dot{\Psi}} = I_3(\dot{\Psi} + \dot{\Phi}\cos\theta) = \bar{L}_3 = \boldsymbol{L}\cdot\hat{\boldsymbol{e}}_3$$
$$= L_1 \sin\theta\sin\Phi - L_2\sin\theta\cos\Phi + L_3\cos\theta \,. \quad (3.86)$$

In the last step, $\hat{\boldsymbol{e}}_3$, the unit vector along the 3-axis, is written in components with respect to $\overline{\mathbf{K}}_0$ (whose axes are fixed in space). The momentum p_Φ is a little more complicated to calculate,

$$p_\Phi \stackrel{\text{def}}{=} \frac{\partial L}{\partial \dot{\Phi}} = \sum_{i=1}^{3} \frac{\partial T}{\partial \bar{\omega}_i}\frac{\partial \bar{\omega}_i}{\partial \dot{\Phi}} = I_1\bar{\omega}_1 \sin\theta\sin\Psi + I_2\bar{\omega}_2\sin\theta\cos\Psi + I_3\bar{\omega}_3\cos\theta$$
$$= \boldsymbol{L}\cdot\hat{\boldsymbol{e}}_{3_0} = L_3 \,. \quad (3.87)$$

Here, we made use of the equation $\bar{L}_i = I_i\bar{\omega}_i$ and of the fact that ($\sin\theta\sin\Psi$, $\sin\theta\cos\Psi$, $\cos\theta$) is the decomposition of the unit vector $\hat{\boldsymbol{e}}_{3_0}$ along the principal axes. Finally, the third generalized momentum is given by

$$p_\theta \stackrel{\text{def}}{=} \frac{\partial L}{\partial \dot{\theta}} = \bar{L}_1\cos\Psi - \bar{L}_2\sin\Psi = \boldsymbol{L}\cdot\hat{\boldsymbol{e}}_\xi \,, \quad (3.88)$$

where $\hat{\boldsymbol{e}}_\xi$ is the unit vector along the line $S\xi$ of Fig. 3.17. One verifies that

$$\det\left(\frac{\partial^2 T}{\partial\dot{\theta}_i\,\partial\dot{\theta}_k}\right) \neq 0 \,,$$

which means that (3.86–3.88) can be solved for the $\bar{\omega}_i$, or, equivalently, for the $\bar{L}_i$. After a little algebra one finds

$$\begin{aligned}\bar{L}_1 &= \frac{1}{\sin\theta}\big(p_\Phi - p_\Psi \cos\theta\big)\sin\Psi + p_\theta \cos\Psi\ , \\ \bar{L}_2 &= \frac{1}{\sin\theta}\big(p_\Phi - p_\Psi \cos\theta\big)\cos\Psi - p_\theta \sin\Psi\ , \\ \bar{L}_3 &= p_\Psi\ .\end{aligned} \tag{3.89}$$

With $T = (\sum \bar{L}_i^2/I_i)/2$ this allows us to construct the Hamiltonian function. One obtains the expression

$$\begin{aligned}H = &\frac{1}{2\sin^2\theta}\big(p_\Phi - p_\Psi\cos\theta\big)^2\left(\frac{\sin^2\Psi}{I_1} + \frac{\cos^2\Psi}{I_2}\right) \\ &+\frac{1}{2}p_\theta^2\left(\frac{\cos^2\Psi}{I_1} + \frac{\sin^2\Psi}{I_2}\right) \\ &+\frac{\sin\Psi\cos\Psi}{2\sin\theta}p_\theta\big(p_\Phi - p_\Psi\cos\theta\big)\left(\frac{1}{I_1} - \frac{1}{I_2}\right) + \frac{1}{2I_3}p_\Psi^2 + U\ .\end{aligned} \tag{3.90}$$

We note that p_Φ is the projection of the angular momentum onto the space-fixed 3_0 axis, while p_Ψ is its projection onto the body-fixed 3-axis. If the potential energy U does not depend on Φ, this variable is cyclic, so that p_Φ is constant, as expected. The expression (3.90) simplifies considerably in the case of a symmetric top for which we can again take $I_1 = I_2$, without loss of generality. If U vanishes, or does not depend on Ψ, then Ψ is also cyclic and p_Ψ is conserved as well.

(iii) *Some Poisson brackets.* If the Eulerian angles are denoted generically by $\{\Theta_i(t)\}$, the Poisson brackets over the phase space (with coordinates Θ_i and p_{Θ_i}) are given by

$$\{f, g\}(\Theta_i, p_{\Theta_i}) = \sum_{i=1}^{3}\left(\frac{\partial f}{\partial p_{\Theta_i}}\frac{\partial g}{\partial \Theta_i} - \frac{\partial f}{\partial \Theta_i}\frac{\partial g}{\partial p_{\Theta_i}}\right)\ . \tag{3.91}$$

The components of the angular momentum with respect to the systems $\overline{\mathbf{K}}$ and $\overline{\mathbf{K}}_0$ have interesting Poisson brackets, both within each system and between them. One finds

$$\big\{L_1, L_2\big\} = -L_3 \quad \text{(cyclic)}\ , \tag{3.92}$$

$$\big\{\bar{L}_1, \bar{L}_2\big\} = +\bar{L}_3 \quad \text{(cyclic)}\ , \tag{3.93}$$

$$\big\{L_i, \bar{L}_j\big\} = 0 \quad \text{for all } i \text{ and } j\ . \tag{3.94}$$

Note the remarkable signs in (3.92) and (3.93). Finally, one verifies that the brackets of the kinetic energy with all L_i, as well as with all $\bar{L}_i$, vanish,

$$\big\{L_i, T\big\} = 0 = \big\{\bar{L}_i, T\big\}\ , \quad i = 1, 2, 3\ . \tag{3.95}$$

3.16 Example: The Symmetric Children's Top in a Gravitational Field

The point of support O does not coincide with the center of mass S, their distance being

$$OS = l \,.$$

Therefore, if I_1 ($= I_2$) is the moment of inertia for rotations about an axis through S that is perpendicular to the symmetry axis (the 3-axis in Fig. 3.24), then Steiner's theorem, Sect. 3.5, tells us that

$$I_1' = I_2' = I_1 + Ml^2$$

is the relevant moment of inertia for rotations about an axis through O that is also perpendicular to the symmetry axis. I_1 and I_1' were calculated in Sect. 3.6 (iii) above. Since $I_1' = I_2'$, the first two terms in T_{rot} (3.85) simplify, so that the Lagrangian function for the spinning top in the earth's gravitational field is given by

$$L = \tfrac{1}{2}\left(I_1 + Ml^2\right)\left(\dot{\theta}^2 + \dot{\Phi}^2 \sin^2\theta\right) + \tfrac{1}{2}I_3(\dot{\Psi} + \dot{\Phi}\cos\theta)^2 - Mgl\cos\theta \,. \quad (3.96)$$

The variables Φ and Ψ are cyclic, the momenta conjugate to them are conserved,

$$p_\Psi = \bar{L}_3 = I_3(\dot{\Psi} + \dot{\Phi}\cos\theta) = \text{const} \,, \quad (3.97\text{a})$$

$$p_\Phi = L_3 = \left(I_1' \sin^2\theta + I_3\cos^2\theta\right)\dot{\Phi} + I_3\dot{\Psi}\cos\theta = \text{const} \,. \quad (3.97\text{b})$$

As long as we neglect frictional forces, the energy is also conserved,

$$E = \tfrac{1}{2}I_1'\left(\dot{\theta}^2 + \dot{\Phi}^2\sin^2\theta\right) + \tfrac{1}{2}I_3(\dot{\Psi} + \dot{\Phi}\cos\theta)^2 + Mgl\cos\theta = \text{const} \,. \quad (3.98)$$

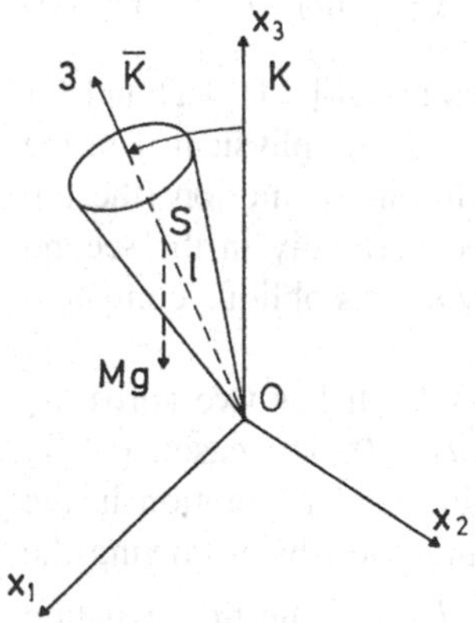

Fig. 3.24. The symmetric children's top in a gravitational field

From (3.97) we can isolate $\dot{\Phi}$ and $\dot{\Psi}$, viz.

$$\dot{\Phi} = \frac{L_3 - \bar{L}_3 \cos\theta}{I_1' \sin^2\theta} , \quad \dot{\Psi} = \frac{\bar{L}_3}{I_3} - \dot{\Phi}\cos\theta . \tag{3.99}$$

Inserting these expressions into (3.98) we obtain an equation of motion that contains only the variable $\theta(t)$. With the new abbreviations

$$E' \stackrel{\text{def}}{=} E - \frac{\bar{L}_3^2}{2I_3} - Mgl , \tag{3.100}$$

$$U_{\text{eff}}(\theta) \stackrel{\text{def}}{=} \frac{(L_3 - \bar{L}_3 \cos\theta)^2}{2I_1' \sin^2\theta} - Mgl(1-\cos\theta) , \tag{3.101}$$

(3.98) becomes the effective equation

$$E' = \tfrac{1}{2} I_1' \dot{\theta}^2 + U_{\text{eff}}(\theta) = \text{const} , \tag{3.102}$$

to which one can apply the methods that we developed in the first chapter. Here, we shall restrict the discussion to a qualitative analysis.

From the positivity of the kinetic energy, the physically admissible domain of variation of the angle is determined by the condition $E' \geq U_{\text{eff}}(\theta)$. Whenever L_3 differs from $\bar{L}_3$, U_{eff} tends to plus infinity both for $\theta \to 0$ and for $\theta \to \pi$. Let

$$u(t) \stackrel{\text{def}}{=} \cos\theta(t) \tag{3.103}$$

and therefore $\dot{\theta}^2 = \dot{u}^2/(1-u^2)$. Equation (3.102) is then equivalent to the following differential equation for $u(t)$:

$$\dot{u}^2 = f(u) , \tag{3.104}$$

where

$$f(u) \stackrel{\text{def}}{=} (1-u^2)\left[(2E'/I_1') + 2Mgl(1-u)/I_1'\right] - \left(L_3 - \bar{L}_3 u\right)^2 / I_1'^2 . \tag{3.105}$$

Only those values of $u(t)$ are physical which lie in the interval $[-1, +1]$ and for which $f(u) \geq 0$. The boundaries $u = 1$ or $u = -1$ can only be physical if in the expression (3.105) $L_3 = \bar{L}_3$ or $L_3 = -\bar{L}_3$. Both conditions of motion (the top standing vertically in the first case and being suspended vertically in the second case) are called *sleeping*. In all other cases the symmetry axis is oblique compared to the vertical.

The function $f(u)$ has the behavior shown in Fig. 3.25. It has two zeros, u_1 and u_2, in the interval $[-1, +1]$. For $u_1 \leq u \leq u_2$, $f(u) \geq 0$. The case $u_1 = u_2$ is possible but arises only for very special initial conditions. The motion in the general case $u_1 < u_2$ can be described qualitatively quite well by following the motion of the symmetry axis on a sphere. Setting $u_0 \stackrel{\text{def}}{=} L_3/\bar{L}_3$, the first equation

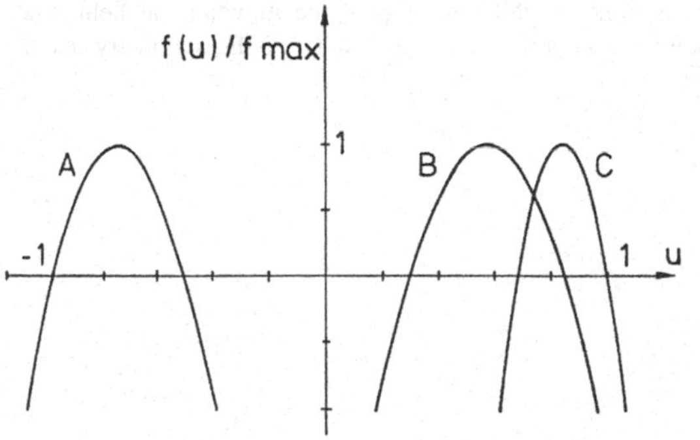

Fig. 3.25. Graph of the function $f(u)$ (3.105) with $u = \cos\theta(t)$. See also Practical Example 1 of the Appendix

(3.99) gives

$$\dot{\Phi} = \frac{\bar{L}_3}{I_1'} \frac{u_0 - u}{1 - u^2} \ . \tag{3.106}$$

Thus, whenever $u_1 \neq u_2$, the extremity of $\hat{f}$ moves on the unit sphere between the parallels of latitude defined by

$$\theta_i = \arccos u_i \ , \quad i = 1, 2 \ .$$

Depending on the position of u_0 relative to u_1 and u_2, we must distinguish three cases.

(i) $u_0 > u_2$ (or $u_0 < u_1$). From (3.106) we see that $\dot{\Phi}$ always has the same sign. Therefore, the motion looks like the one sketched in Fig. 3.26a.
(ii) $u_1 < u_0 < u_2$. In this case $\dot{\Phi}$ has different signs at the upper and lower parallels. The motion of the symmetry axis $\hat{f}$ looks as sketched in Fig. 3.26b.
(iii) $u_0 = u_1$ or $u_0 = u_2$. Here $\dot{\Phi}$ vanishes at the lower or upper parallel, respectively. In the second case, for example, the motion of $\hat{f}$ is the one sketched in Fig. 3.26c.

The motion of the extremity of $\hat{f}$ on the sphere is called *nutation.*

3.17 More About the Spinning Top

The analysis of the previous section can be pushed a little further. For example, one may ask under which condition the rotation about the vertical is stable. This is indeed the aim when one plays with a children's top: one wants to have it spin, if possible vertically, and for as long as possible. In particular, one wishes to know to what extent friction at the point of support disturbs the game.

(i) *Vertical rotation (standing top).* For $\theta = 0$ we have $L_3 = \bar{L}_3$. From (3.101) one finds that $U_{\text{eff}}(0) = 0$ and therefore $E' = 0$ or $E = \bar{L}_3^2/(2I_3) + Mgl$. The

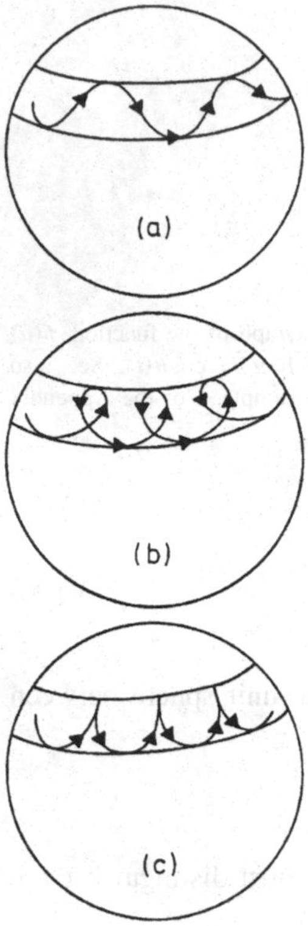

Fig. 3.26. Symmetric children's top in the gravitational field. The figure shows the nutation of the extremity of the symmetry axis $\hat{f}$

rotation is stable only if $U_{\text{eff}}(\theta)$ has a minimum at $\theta = 0$. In the neighborhood of $\theta = 0$ we have

$$U_{\text{eff}} \simeq \left[\bar{L}_3^2/8I_1' - Mgl/2\right]\theta^2 .$$

The second derivative of U_{eff} is positive only if $\bar{L}_3^2 > 4MglI_1'$ or

$$\bar{\omega}_3^2 > 4MglI_1'/I_3^2 . \tag{3.107}$$

(ii) *Including friction.* The motion in the presence of frictional forces can be described qualitatively as follows. Consider a top in an oblique position with $p_\Psi = \bar{L}_3 > p_\Phi = L_3$. The action of friction results in slowing down p_Ψ continuously, while leaving p_Φ practically unchanged until the two are equal, $p_\Psi = p_\Phi$. At this moment the top spins vertically. From then on both p_Φ and p_Ψ decrease synchronously. The top remains vertical until the lower limit of the stability condition (3.107) is reached. For $\bar{\omega}_3$ below that limit the motion is unstable. Even a small perturbation will cause the top to rock and eventually to topple over.

3.18 Spherical Top with Friction: The "Tippe Top"

The tippe top is a symmetric, almost spherical rigid body whose moments of inertia $I_1 = I_2$ and $I_3 \neq I_1$ fulfill a certain inequality, see (3.112) below. It differs from a homogeneous ball essentially only in that its center of mass does not coincide with its geometrical center. If one lets this top spin on a horizontal plane in the earth's gravitational field and includes friction between the top and the plane of support, it behaves in an astounding way. Initially it spins about the symmetry axis of its equilibrium position such that its center of mass is *below* the center of symmetry. The angular momentum points in an almost vertical direction and, hence, is nearly perpendicular to the plane. By the action of gliding friction, however, the top quickly inverts its position so that, in a second stage, it rotates in an "upside-down" position before eventually coming to rest again. After this rapid inversion the angular momentum is again vertical. This means that the sense of rotation with respect to a body-fixed system has changed during inversion. As the center of mass is lifted in the gravitational field, the rotational energy and therefore the angular momentum have decreased during inversion. When the top has reached its upside-down position, it continues spinning while its center of mass is at rest with respect to the laboratory system. During this stage only rotational friction is at work. As this frictional force is small, the top remains in the inverted position for a long time before it slows down and returns to the state of no motion.

Although this toy was apparently already known at the end of the nineteenth century, it was thought for a long time that its strange behavior was too complicated to be understood analytically and that it could only be simulated by numerically solving Euler's equations. This, as we shall see, is not true. Indeed, as was shown recently, the salient features of this top can be described by means of the analytic tools of this chapter and a satisfactory and transparent prediction of its strange behavior is possible. This is the reason why I wish to add it to the traditional list of examples in the theory of spinning rigid bodies.

The analysis is done in two steps: In a first step we prove by a simple geometric argument that in the presence of gliding friction on the plane of support, a specific linear combination of L_3, i.e. the projection of the angular momentum onto the vertical, and of $\overline{L}_3$, its projection onto the top's symmetry axis, is a constant of the motion. On the basis of this conservation law and of the inequality for the moments of inertia, see (3.112) below, one shows that the inverted (spinning) position is energetically favorable compared to the upright position.

In a second step one writes the equations of motion in a specific set of variables which is particularly well adapted to the problem and one analyzes the dynamical behavior as a function of time and the stability or instability of the solutions.

We make the following assumptions: Let the top be a sphere whose mass distribution is inhomogeneous in such a way that the center of mass S does not coincide with its geometrical center Z. The mass distribution is axially symmetric, but not spherically symmetric, so that the moments of inertia referring to the axes perpendicular to the symmetry axis are equal, but differ from the third, $I_1 = I_2 \neq I_3$. By a suitable choice of the unit of length, the radius of the sphere is $R = 1$. The center

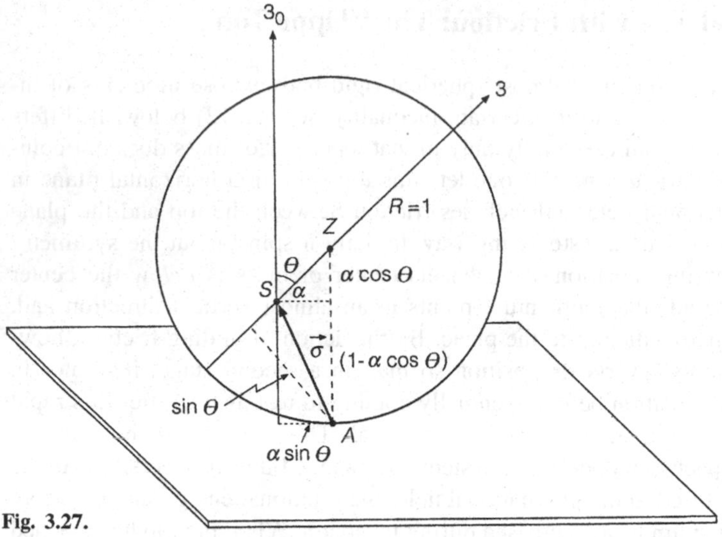

Fig. 3.27.

of mass is situated at a distance α from the center Z, with $0 \leq \alpha < 1$, as sketched in Fig. 3.27. There are three types of frictional force that act on the instantaneous point of support A in the plane: *rolling friction* which is active whenever the top rolls over the plane without gliding; *rotational friction* which acts when the top is spinning about a vertical axis about a fixed point in the plane; and *gliding friction* which acts whenever the top glides over the plane of support. We assume that the support is such that the force due to gliding friction is much larger than those due to the other two kinds of friction. Indeed, it turns out that it is the gliding friction that is responsible for the inversion of the spinning top. Finally, for the sake of simplicity, we assume that during the initial phase in which we are interested the rotational energy T_{rot} is much larger than the gravitational energy $U = mg(1 - \cos\theta)$.

3.18.1 Conservation Law and Energy Considerations

The instantaneous velocity of the point of support A is the sum of the center of mass's horizontal velocity $(\dot{s}_1, \dot{s}_2)$ (i.e. the component parallel to the plane) and of the relative velocities which stem from changes of the Euler angles. From Fig. 3.27 we deduce that a change of Ψ, i.e. a rotation about the body-fixed 3-axis, causes a linear velocity of A in the plane whose magnitude is $v_\Psi = \dot{\Psi} \sin\theta$, while a change of the angle Φ, i.e. a rotation about the space-fixed 3_0-axis, causes a velocity whose magnitude is $v_\Phi = \dot{\Phi}\, \alpha \sin\theta$. Both act along the same direction in the plane, say $\hat{\boldsymbol{n}}$. In contrast, a change in the angle θ gives rise to a velocity with magnitude $v_\theta = \dot{\theta}(1 - \alpha\cos\theta)$ in the direction $\hat{\boldsymbol{t}}$ perpendicular to $\hat{\boldsymbol{n}}$. Although it is not difficult to identify these directions in Fig. 3.27 it is sufficient for our discussion to know that the velocities related to $\dot{\Psi}$ and to $\dot{\Phi}$ have the same direction, while that due to $\dot{\theta}$ is perpendicular to this direction.

The effect of friction is described phenomenologically as in (6.28) below by introducing dissipative terms R_Φ, R_Ψ such that

$$\dot{p}_\Phi = -R_\Phi \,, \qquad \dot{p}_\Psi = -R_\Psi \,, \tag{3.108}$$

(and an analogous equation for $\dot{p}_\theta$). As we know the canonical momenta p_Φ and p_Ψ are the projections L_3 and $\overline{L}_3$, respectively, of the angular momentum onto the 3_0-axis and the 3-axis, respectively. Therefore, R_Φ and R_Ψ are external torques, equal to the cross product of the distance to the corresponding axis of rotation and the frictional force. As the force is the same in both of them, independently of its detailed functional dependence on the velocity, and as these torques are parallel, their ratio is equal to the ratio of the distances,

$$R_\Phi / R_\Psi = \alpha \sin\theta / \sin\theta = \alpha \,. \tag{3.109}$$

As a consequence, while both $p_\Phi \equiv L_3$ and $p_\Psi \equiv \overline{L}_3$ decrease with time, the specific linear combination $\dot{L}_3 - \alpha\dot{\overline{L}}_3 = 0$ vanishes. This yields the integral of the motion

$$\lambda := L_3 - \alpha\overline{L}_3 = \text{constant}. \tag{3.110}$$

We note in passing that this conservation law, which provides the key to an understanding of the tippe top, has an amusing history that can be traced back[3] to 1872! In fact, the quantity λ is the projection of the angular momentum $\boldsymbol{L}$ onto the vector $\boldsymbol{\sigma} = \overrightarrow{AS}$. Indeed, from (3.86) and (3.87) we have

$$\lambda = \boldsymbol{L} \cdot \left(\hat{e}_{3_0} - \alpha\hat{\eta}\right) = \boldsymbol{L} \cdot \boldsymbol{\sigma} \,, \text{ where } \hat{\eta} = \mathbf{R}^{-1}(t)\hat{e}_{\bar{3}} \,.$$

Suppose the top is launched such that the conserved quantity (3.110) is large in the following sense

$$\lambda \gg \sqrt{mgI_1} \,, \tag{3.111}$$

with m the mass of the top. Suppose, furthermore, that the mass distribution is chosen so that the moments of inertia fulfill the inequalities

$$(1-\alpha)I_3 < I_1 < (1+\alpha)I_3 \,. \tag{3.112}$$

The first of these means that the gravitational energy can be neglected in comparison to the rotational energy; the second assumption (3.112) implies that the top has lower energy when it rotates in a completely inverted position (S above Z) than when it rotates in its normal, upright position.

When the top has stopped gliding, its center of mass having come to rest, but rotates about a vertical axis in a (quasi) stationary state, we can conclude that

---

[3] St. Ebenfeld, F. Scheck: Ann. Phys. (New York) **243**, 195 (1995). Note that the vector $\boldsymbol{\sigma}$ equals $-\boldsymbol{a}$ in this reference and that the choice of convention for the rotation $\mathbf{R}(t)$, while consistent with earlier sections of this chapter, is the inverse of the one employed there.

$$\dot{s}_1 = \dot{s}_2 = 0\,, \quad \dot{\theta} = 0\,, \quad \dot{\Psi} + \alpha\dot{\Phi} = 0\,; \tag{3.113}$$

with $s(t)$ denoting the trajectory of the center of mass.

Inserting $I_1 = I_2 \neq I_3$ into the expression (3.85) for the kinetic energy of rotation one finds

$$T_{rot} = \frac{1}{2} I_1 \left(\dot{\theta}^2 + \dot{\Phi}^2 \sin^2\theta \right) + \frac{1}{2} I_3 \left(\dot{\Psi} + \dot{\Phi}\cos\theta \right)^2 .$$

From this follow the generalized momenta

$$L_3 \equiv p_\Phi = \frac{\partial L}{\partial \dot{\Phi}} = \dot{\Phi}\left(I_1 \sin^2\theta + I_3 \cos^2\theta \right) + I_3 \dot{\Psi}\cos\theta\,, \tag{3.114a}$$

$$\overline{L}_3 \equiv p_\Psi = \frac{\partial L}{\partial \dot{\Psi}} = I_3 \left(\dot{\Psi} + \dot{\Phi}\cos\theta \right) . \tag{3.114b}$$

Inserting here the second and third of conditions (3.113), the rotational energy becomes

$$T_{\text{rot}} = \frac{1}{2} F(z)\dot{\Phi}^2 \quad \text{with } F(z \equiv \cos\theta) = I_1(1 - z^2) + I_3(z - \alpha)^2 .$$

The third of conditions (3.113), when inserted in (3.114a) and (3.114b), allows one to re-express the constant of the motion λ in terms of the same function, viz

$$\lambda = \dot{\Phi}\left(I_1 \sin^2\theta + I_3(\cos\theta - \alpha)^2 \right) = \dot{\Phi} F(z)\,,$$

so that the kinetic energy can be written in terms of λ and the function F,

$$T_{\text{rot}} = \frac{\lambda^2}{2F(z)} . \tag{3.115}$$

The rotational energy assumes its smallest value when $F(z)$ takes its largest value. With the assumption (3.112) this happens, in the physical range of θ, for $z = -1$, i.e. $\theta = \pi$. As the function $F(z)$ increases monotonically in the interval $[1, -1]$, the top's rotation in the completely inverted position is favored energetically over rotation in the upright position.

3.18.2 Equations of Motion and Solutions with Constant Energy

Assuming that the forces due to rotational and rolling friction can be neglected, the possible asymptotic states of the spinning top are clearly those in which gliding friction has ceased to be active. These asymptotic states have constant energy. Except for the trivial state of rest, they can only be one of the following: Rotation in the upright or in the completely inverted position, or rotation about a nonvertical direction, changing with time, whereby the top rolls over the plane without gliding. Let us call the former two *rotational*, and the latter *tumbling* motion.

The simple energy consideration of Sect. 3.18.1 leaves unanswered a number of important questions. Given the moments of inertia I_1 and I_3, which of the allowed asymptotic states are *stable*? In the case where an asymptotic state is stable, which initial conditions (i.e. when launching the top) will develop into that state under the action of gliding friction? Finally, in what way will simple criteria such as (3.112) be modified when the gravitational force is taken into account?

In fact, these questions touch upon the field of qualitative dynamics, which is the subject of Chap. 6 (cf. in particular, the notion of Liapunov stability). A complete analysis of this dynamical problem can be found in the reference just given in footnote 3, an article that should be accessible after having studied Chap. 6. Here we confine ourselves to constructing the equations of motion in a form that is well adapted to the problem, and to report the most important results of the analysis.

As described in Sect. 3.9 it is useful to introduce three frames of reference: the space-fixed inertial system **K**, the system $\overline{\mathbf{K}}_0$ which is centered on the center of mass and whose axes are parallel, at all times, to those of **K**, and a body-fixed system $\overline{\mathbf{K}}$ whose 3-axis is the symmetry axis of the top. In writing down the inertia tensor we make use of the following symbolic notation: We write any vector $\boldsymbol{a}$ as $|\boldsymbol{a}\rangle$ and the object which is dual to it as $\langle\boldsymbol{a}|$. An expression of the form $\langle\boldsymbol{b}|\boldsymbol{a}\rangle$ is then just another way of writing the scalar product $\boldsymbol{b}\cdot\boldsymbol{a}$, while $|\boldsymbol{b}\rangle\langle\boldsymbol{a}|$ is a tensor which, when applied to a third vector $\boldsymbol{c}$ yields a vector again, viz.

$$|\boldsymbol{b}\rangle\langle\boldsymbol{a}|\boldsymbol{c}\rangle = (\boldsymbol{a}\cdot\boldsymbol{c})\,\boldsymbol{b}\,.$$

In this notation the inertia tensor with respect to the system $\overline{\mathbf{K}}$ reads

$$\mathbf{J} = I_1\left\{\mathbb{1} + \frac{I_3 - I_1}{I_1}|\hat{e}_{\bar{3}}\rangle\langle\hat{e}_{\bar{3}}|\right\}\,,$$

while its inverse reads[4]

$$\mathbf{J}^{-1} = \frac{1}{I_1}\left\{\mathbb{1} - \frac{I_3 - I_1}{I_3}|\hat{e}_{\bar{3}}\rangle\langle\hat{e}_{\bar{3}}|\right\}\,.$$

Hence, in the frame of reference $\overline{\mathbf{K}}_0$ it has the form

$$\widetilde{\mathbf{J}}(t) = I_1\left\{\mathbb{1} + \frac{I_3 - I_1}{I_1}|\hat{\eta}\rangle\langle\hat{\eta}|\right\}\,, \tag{3.116a}$$

where $\hat{\eta}$ is the representation of the unit vector $\hat{e}_{\bar{3}}$ with respect to $\overline{\mathbf{K}}_0$, i.e.

$$\hat{\eta} = \mathbf{R}^{-1}(t)\hat{e}_{\bar{3}} = \mathbf{R}^T(t)\hat{e}_{\bar{3}}\,.$$

An analogous formula holds for its inverse

$$\widetilde{\mathbf{J}}^{-1}(t) = \frac{1}{I_1}\left\{\mathbb{1} - \frac{I_3 - I_1}{I_3}|\hat{\eta}\rangle\langle\hat{\eta}|\right\}\,. \tag{3.116b}$$

---

[4] In order to become familiar with this notation and calculus the reader should verify that $\langle\hat{e}_{\bar{i}}|\mathbf{J}|\hat{e}_{\bar{k}}\rangle = \text{diag}\,(I_1, I_1, I_3)$, and that $\mathbf{J}^{-1}$ is indeed the inverse of $\mathbf{J}$.

The angular velocity $\boldsymbol{\omega}$ may be taken from (3.56b). Alternatively, making use of (3.116b), it may be expressed in terms of the angular momentum $\boldsymbol{L} = \widetilde{\mathbf{J}} \cdot \boldsymbol{\omega}$:

$$\boldsymbol{\omega}(t) = \frac{1}{I_1}\left\{\boldsymbol{L}(t) - \frac{I_3 - I_1}{I_3}|\hat{\boldsymbol{\eta}}\rangle\langle\hat{\boldsymbol{\eta}}|\boldsymbol{L}\rangle\right\}. \tag{3.117}$$

The time derivative of $\hat{\boldsymbol{\eta}}$ follows from (3.56a), the time derivative of $\boldsymbol{L}$ is given by the external torque $\boldsymbol{N}$ (with respect to the system $\overline{\mathbf{K}}_0$), and the acceleration $\ddot{\boldsymbol{s}}(t)$ of the center of mass is given by the resulting external force $\boldsymbol{F}$ (in the system $\mathbf{K}$). Therefore, the equations of motion are

$$\frac{\mathrm{d}}{\mathrm{d}t}\hat{\boldsymbol{\eta}} = \boldsymbol{\omega} \times \hat{\boldsymbol{\eta}} = \frac{1}{I_1}\boldsymbol{L} \times \hat{\boldsymbol{\eta}}, \tag{3.118a}$$

$$\frac{\mathrm{d}}{\mathrm{d}t}\boldsymbol{L} = \boldsymbol{N}(\hat{\boldsymbol{\eta}}, \boldsymbol{L}, \dot{\boldsymbol{s}}), \tag{3.118b}$$

$$m\ddot{\boldsymbol{s}} = \boldsymbol{F}(\hat{\boldsymbol{\eta}}, \boldsymbol{L}, \dot{\boldsymbol{s}}). \tag{3.118c}$$

(We recall that $\boldsymbol{s}(t)$ is the trajectory of the center of mass and $\dot{\boldsymbol{s}}$ its velocity in the space-fixed system $\mathbf{K}$.)

If we demand that the top remain on the plane at all times (no bouncing), then the 3-component s_3 of the center of mass coordinate is not an independent variable. Indeed, the condition is that both the 3-coordinate of the point A and the 3-component of its velocity $\boldsymbol{v} = \dot{\boldsymbol{s}} - \boldsymbol{\omega} \times \boldsymbol{\sigma}$ are zero at all times. One easily shows that this implies the condition

$$\dot{s}_3 + \frac{\alpha}{I_1}\langle\hat{\boldsymbol{e}}_{\bar{3}_0}|\boldsymbol{L} \times \hat{\boldsymbol{\eta}}\rangle = 0, \tag{3.119}$$

which in turn expresses $\dot{s}_3$ in terms of $\hat{\boldsymbol{\eta}}$ and $\boldsymbol{L}$. The third equation of motion (3.118c) must be replaced with

$$m\ddot{s}_{1,2} = \mathrm{Pr}_{1,2}\,\boldsymbol{F},$$

where the right-hand side denotes the projection of the external force $\boldsymbol{F}$ onto a horizontal plane parallel to the plane of support.

The external force $\boldsymbol{F}$ acting on the center of mass S is the sum of the gravitational force $\boldsymbol{F}_{\mathrm{g}} = -mg\hat{\boldsymbol{e}}_{\bar{3}_0}$, the normal force $\boldsymbol{F}_{\mathrm{n}} = g_n\hat{\boldsymbol{e}}_{\bar{3}_0}$, and the frictional force $\boldsymbol{F}_{\mathrm{fr}} = -g_{\mathrm{fr}}\hat{\boldsymbol{v}}$. In contrast, the point A, being supported by the plane, experiences only the normal force and the frictional force, $\boldsymbol{F}^{(A)} = \boldsymbol{F}_{\mathrm{n}} + \boldsymbol{F}_{\mathrm{fr}}$, so that the external torque is given by

$$\boldsymbol{N} = -\boldsymbol{\sigma} \times \boldsymbol{F}^{(A)} = (\alpha\hat{\boldsymbol{\eta}} - \hat{\boldsymbol{e}}_{\bar{3}_0}) \times (g_{\mathrm{n}}\hat{\boldsymbol{e}}_{\bar{3}_0} - g_{\mathrm{fr}}\hat{\boldsymbol{v}}).$$

This leads immediately to the final form of the equations of motion

$$\frac{\mathrm{d}}{\mathrm{d}t}\hat{\boldsymbol{\eta}} = \boldsymbol{\omega} \times \hat{\boldsymbol{\eta}} = \frac{1}{I_1}\boldsymbol{L} \times \hat{\boldsymbol{\eta}}, \tag{3.120a}$$

$$\frac{\mathrm{d}}{\mathrm{d}t}\boldsymbol{L} = (\alpha\hat{\boldsymbol{\eta}} - \hat{\boldsymbol{e}}_{\bar{3}_0}) \times (g_{\mathrm{n}}\hat{\boldsymbol{e}}_{\bar{3}_0} - g_{\mathrm{fr}}\hat{\boldsymbol{v}}), \tag{3.120b}$$

$$m\ddot{s}_{1,2} = -g_{\mathrm{fr}}\hat{\boldsymbol{v}}. \tag{3.120c}$$

The coefficient g_n in the normal force follows from the equation $\ddot{s}_3 = -g + g_n/m$ if one calculates the left-hand side by means of (3.119). For this one must take the orbital derivative of (3.119) which means replacing the time derivatives of $\boldsymbol{L}$ and of $\hat{\boldsymbol{\eta}}$ by (3.120b) and (3.120a), respectively. The result reads

$$g_n = \frac{mgI_1\left[1 + \alpha(\eta_3 \boldsymbol{L}^2 - L_3\overline{L}_3)/(gI_1^2)\right]}{I_1 + m\alpha^2(1-\eta_3^2) + m\alpha\mu\left[(\eta_3 - \alpha)\hat{\boldsymbol{e}}_{\bar{3}_0} - (1-\alpha\eta_3)\hat{\boldsymbol{\eta}}\right]\cdot\hat{\boldsymbol{v}}} \,. \tag{3.121}$$

Here $\eta_3 = \hat{\boldsymbol{\eta}}\cdot\hat{\boldsymbol{e}}_{\bar{3}_0}$ is the projection onto the vertical. Regarding the frictional force we have assumed $g_{fr} = \mu g_n$, with μ a (positive) coefficient of friction.

Equations (3.120a–c) provide a good starting point for a complete analysis of the tippe top. One the one hand they are useful for studying analytic properties of the various types of solutions; on the other hand they may be used for a numerical treatment of specific solutions (cf. practical example 2 below). Here we report on some of the results, taken from the work quoted in footnote 3, and refer to that article for further details.

(i) Conservation law: It is easy to verify that the conservation law (3.110) also follows from the equations of motion (3.120a). The orbital derivative of λ (i.e. the time derivative taken along orbits of the system by making use of the equations of motion) is given by

$$\frac{d\lambda}{dt} = \frac{d\boldsymbol{L}}{dt}\cdot\boldsymbol{\sigma} + \boldsymbol{L}\cdot\frac{d\boldsymbol{\sigma}}{dt} \,.$$

The second term vanishes because, on account of (3.120a), $d\boldsymbol{\sigma}/dt = -\alpha d\hat{\boldsymbol{\eta}}/dt$ is perpendicular to $\boldsymbol{L}$. The first term vanishes because the torque $\boldsymbol{N}$ is perpendicular to $\boldsymbol{\sigma}$.

(ii) Asymptotic states: The asymptotic states with constant energy obey the equations of motion (3.120a) with $\hat{\boldsymbol{v}} = 0$, the second and the third equation being replaced by

$$\frac{d\boldsymbol{L}}{dt} = \alpha g_n \hat{\boldsymbol{\eta}} \times \hat{\boldsymbol{e}}_{\bar{3}_0} \,, \quad m\ddot{s}_{1,2} = 0 \,,$$

while (3.121) simplifies to

$$g_n = mg\frac{1 + \alpha(\eta_3 \boldsymbol{L}^2 - L_3\overline{L}_3)/(gI_1^2)}{1 + m\alpha^2(1-\eta_3^2)/I_1} \,.$$

The solutions of constant energy have the following general properties:

(a) The projections L_3 and $\overline{L}_3$ of the angular momentum $\boldsymbol{L}$ onto the vertical and the symmetry axes, respectively, are conserved.
(b) The square of the angular momentum $\boldsymbol{L}^2$ as well as the projection $\hat{\boldsymbol{\eta}}\cdot\hat{\boldsymbol{e}}_{\bar{3}_0} = \eta_3$ of $\hat{\boldsymbol{\eta}}$ onto the vertical are conserved.
(c) At all times the vectors $\hat{\boldsymbol{e}}_{\bar{3}_0}$, $\hat{\boldsymbol{\eta}}$, and $\boldsymbol{L}$ lie in a plane.
(d) The center of mass remains fixed in space, $\dot{\boldsymbol{s}} = 0$.

The types of motion that have constant energy have either $\eta_3 = +1$ (rotation in the upright position), or $\eta_3 = -1$ (rotation with complete inversion), or, possibly, $-1 < \eta_3 < +1$ if these are allowed. The latter are tumbling motions whereby the top simultaneously rotates in an oblique (time dependent) orientation and rolls over the plane without gliding. Whether or not tumbling motion is possible depends on the choice of the moments of inertia.

(iii) When does the spinning top turn upside-down? The general and complete answer to the question of which asymptotic state is reached from a given initial condition would occupy too much space. Here, we restrict ourselves to an example which corroborates the results of Sect. 3.18.1. For a given value of the constant of the motion (3.110) we define the following auxiliary quantities

$$A := I_3(1-\alpha) - I_1 + \frac{mga I_3^2}{\lambda^2}(1-\alpha)^4 ,$$

$$B := I_3(1+\alpha) - I_1 - \frac{mga I_3^2}{\lambda^2}(1+\alpha)^4 .$$

A detailed analysis of orbital stability (a so-called Liapunov analysis) for this example yields the following results: If $A > 0$ the state with $\eta_3 = +1$ is asymptotically stable and the top will rotate in the upright position. If, however, $A < 0$ this state is unstable. Furthermore, if $B > 0$ then a state with $\eta_3 = -1$ is asymptotically stable; if $B < 0$ then it is unstable. Whenever λ is sufficiently large, cf. (3.111), the third terms in A and B can be neglected. The two conditions $A < 0$ and $B > 0$, taken together, then yield the inequalities (3.112). In this situation rotation in the upright position is unstable, whereas rotation in a completely inverted position is stable. No matter how the top is launched initially, it will always turn upside-down. This is the genuine "tippe top". In the examples (iv) and (v) of Sect. 3.6 two simple models for such a top were described.

The other possible cases can be found in the reference quoted above. Here is what one finds in case the initial rotation is chosen sufficiently fast (i.e. if λ is large in the sense of (3.111)):

(a) For $I_1 < I_3(1-\alpha)$ both, rotation in the upright position and rotation in the inverted position, are stable. There also exists a tumbling motion (with constant energy) but it is unstable. This top could be called *indifferent* because, depending on the initial condition, it can tend either to the upright or to the inverted position.

(b) For $I_1 > I_3(1+\alpha)$ the two vertical positions are unstable. There is exactly one state of tumbling motion (i.e. rotating and rolling without gliding) which is asymptotically stable. Every initial condition will quickly lead to it.

Appendix: Practical Examples

1. Symmetric Top in a Gravitational Field. Study quantitatively the motion of a symmetric spinning top in the earth's gravitational field (a qualitative description is given in Sect. 3.16).

Solution. It is convenient to introduce dimensionless variables as follows. For the energy E' (3.100) take

$$\varepsilon \stackrel{\text{def}}{=} E'/Mgl\,. \tag{A.1}$$

Instead of the projections L_3 and $\bar{L}_3$ introduce

$$\lambda \stackrel{\text{def}}{=} \frac{L_3}{\sqrt{I_1' Mgl}}\,, \quad \bar{\lambda} \stackrel{\text{def}}{=} \frac{\bar{L}_3}{\sqrt{I_1' Mgl}}\,. \tag{A.2}$$

The function $f(u)$ on the right-hand side of (3.104) is replaced with the dimensionless function

$$\varphi(u) \stackrel{\text{def}}{=} \frac{I_1'}{Mgl} f(u) = 2\left(1-u^2\right)(\varepsilon+1-u) - (\lambda - \bar{\lambda}u)^2\,. \tag{A.3}$$

As one may easily verify, the ratio I_1'/Mgl has dimension (time)^2. Thus, $\omega \stackrel{\text{def}}{=} \sqrt{Mgl/I_1'}$ is a frequency. Finally, using the dimensionless time variable $\tau \stackrel{\text{def}}{=} \omega t$, (3.104) becomes

$$\left(\frac{\mathrm{d}u}{\mathrm{d}\tau}\right)^2 = \varphi(u)\,. \tag{A.4}$$

Vertical rotation is stable if $\bar{L}_3^2 > 4MglI_1'$, i.e. if $\bar{\lambda} > 2$. The top is vertical if $\lambda = \bar{\lambda}$. With $u \to 1$, the critical energy, with regard to stability, is then $\varepsilon_{\text{crit.}}(\lambda = \bar{\lambda}) = 0$. For the suspended top we have $\lambda = -\bar{\lambda}$, $u \to 1$, and the critical energy is

$$\varepsilon_{\text{crit.}}(\lambda = -\bar{\lambda}) = -2\,.$$

The equations of motion now read

$$\left(\frac{\mathrm{d}u}{\mathrm{d}\tau}\right)^2 = \varphi(u) \quad \text{or} \quad \left(\frac{\mathrm{d}\theta}{\mathrm{d}\tau}\right)^2 = \frac{\varphi(u)}{1-u^2} \tag{A.5}$$

and

$$\frac{\mathrm{d}\Phi}{\mathrm{d}\tau} = \bar{\lambda}\frac{u_0 - u}{1-u^2}\,, \quad \text{with} \quad u_0 = \frac{L_3}{\bar{L}_3} = \frac{\lambda}{\bar{\lambda}}\,. \tag{A.6}$$

Curve A of Fig. 3.25 corresponds to the case of a suspended top, i.e. $\lambda = -\bar{\lambda}$ and $u_0 = -1$. We have chosen $\varepsilon = 0$, $\lambda = 3.0$. Curve C corresponds to the vertical top, and we have chosen $\varepsilon = 2$, $\lambda = \bar{\lambda} = 5$. Curve B, finally, describes an intermediate situation. Here we have taken $\varepsilon = 2$, $\lambda = 4$, $\bar{\lambda} = 6$.

The differential equations (A.5) and (A.6) can be integrated numerically, e.g. by means of the Runge–Kutta procedure described in Practical Example 2.2. For this purpose let

$$y = \begin{pmatrix} \dfrac{d\theta}{d\tau} \\ \dfrac{d\Phi}{d\tau} \end{pmatrix}$$

and read (A.11) and (A.12) of the Appendix to Chapter 2 as equations with two components each. This allows us to represent the motion of the axis of symmetry in terms of angular coordinates (θ, Φ) in the strip between the two parallels defined by u_1 and u_2. It requires a little more effort to transcribe the results onto the unit sphere and to represent them, by a suitable projection, as in Fig. 3.26.

2. The Tippe Top. Under the assumption that the coefficient of gliding friction is proportional to g_n, the coefficient that appears in the normal force, numerically integrate the equations of motion (3.120a–c) with the three possible choices for the moments of inertia.

Solution. The assumption is that $g_{fr} = \mu g_n$. Let $v = \|\boldsymbol{v}\|$ be the modulus of the velocity. In order to avoid the discontinuity at $\boldsymbol{v} = 0$ on the right-hand side of (3.120c) one can replace $\hat{\boldsymbol{v}}$ by

$$\hat{\boldsymbol{v}} \longmapsto \tanh(M\|\boldsymbol{v}\|)\frac{\boldsymbol{v}}{\|\boldsymbol{v}\|},$$

where M is a large postive number. Indeed, the factor $\tanh(M\|\boldsymbol{v}\|)$ vanishes at zero and tends quickly, yet in a continuous fashion, to 1. It is useful to introduce appropriate units of length, mass, and time such that $R = 1$, $m = 1$ and $g = 1$. Furthermore, the coefficient of friction μ should be chosen sufficiently large, say $\mu = 0.75$, so that the numerical solutions quickly reach the asymptotic state(s). Compare your results with the examples given by Ebenfeld and Scheck (1995) footnote 3.

4. Relativistic Mechanics

Mechanics, as we studied it in the first three chapters, is based on two fundamental principles. On the one hand one makes use of simple functions such as the Lagrangian function and of functionals such as the action integral whose properties are clear and easy to grasp. In general, Lagrangian and Hamiltonian functions do not represent quantities that are directly measurable. However, they allow us to derive the equations of motion in a general and simple way. Also, they exhibit the specific symmetries of a given dynamical system more clearly than the equations of motion themselves, whose form and transformation properties are usually complicated.

On the other hand, one assumes a very special structure for the space-time manifold that supports mechanical motion. In the cases discussed up until now the equations of motion were assumed to be form-invariant with regard to general Galilei transformations (Sect. 1.13; see also the discussion in Sect. 1.14). This implied, in particular, that Lagrangian functions, kinetic and potential energies, had to be invariant under these transformations.

While the first "building principle" is valid far beyond nonrelativistic point mechanics (provided one is prepared to generalize it to some extent, if necessary), the validity of the principle of Galilei invariance of kinematics and dynamics is far more restricted. True, celestial mechanics as well as the mechanics that we encounter in daily life when playing billiards, riding a bicycle, working with a block-and-tackle, etc., is described by the Galilei-invariant theory of gravitation to a very high accuracy. However, this is not true, in general, for microscopic objects such as elementary particles, and it is never true for nonmechanical theories such as Maxwell's theory of electromagnetic phenomena. Without actually having to give up the general, *formal* framework altogether, one must replace the principle of Galilei invariance by the more general principle of Lorentz or Poincaré invariance. While in a hypothetical Galilei-invariant world, particles can have arbitrarily large velocities, Poincaré transformations contain an upper limit for physical velocities: the (universal) speed of light. Galilei-invariant dynamics then appears as a limiting case, applicable whenever velocities are small compared to the speed of light.

In this chapter we learn why the velocity of light plays such a special role, in what way the Lorentz transformations follow from the universality of the speed of light, and how to derive the main properties of these transformations. Today, basing our conclusions on a great amount of experience and increasingly precise

F. Scheck, *Mechanics*, Graduate Texts in Physics, 5th ed.,
DOI 10.1007/978-3-642-05370-2_4,

experimental information, we believe that *any* physical theory is (at least locally) Lorentz invariant[1]. Therefore, in studying the special theory of relativity, within the example of mechanics, we meet another pillar on which physics rests and whose importance stretches far beyond classical mechanics.

4.1 Failures of Nonrelativistic Mechanics

We wish to demonstrate, by means of three examples, why Galilei invariant mechanics cannot be universally valid.

(i) *Universality of the speed of light.* Experiment tells us that the speed of light, with respect to inertial systems of reference, is a universal constant. Its value is

$$c = 2.997\,924\,58 \times 10^8\,\mathrm{ms}^{-1}\,. \tag{4.1}$$

Our arguments of Sect. 1.14 show clearly that in Galilei-invariant mechanics a universal velocity and, in particular, an upper limit for velocities cannot exist. This is so because any process with characteristic velocity $\boldsymbol{v}$, with respect to an inertial system of reference $\mathbf{K}_1$, can be observed from another inertial system $\mathbf{K}_2$, moving with constant velocity $\boldsymbol{w}$ relative to $\mathbf{K}_1$. With respect to $\mathbf{K}_2$, the process then has the velocity

$$\boldsymbol{v}' = \boldsymbol{v} + \boldsymbol{w}\,, \tag{4.2}$$

in other words, velocities add linearly and therefore can be made arbitrarily large.

(ii) *Particles without mass carry energy and momentum.* In nonrelativistic mechanics the kinetic energy and momentum of a free particle are related by

$$E = T = \frac{1}{2m}\boldsymbol{p}^2\,. \tag{4.3}$$

In nature there are particles whose mass vanishes. For instance, the photon (or light quantum), which is the carrier of electromagnetic interactions, is a particle whose mass vanishes. Nevertheless, a photon carries energy and momentum (as proved by the photoelectric effect, for example), even though relation (4.3) is meaningless in this case: neither is the energy E infinite when $|\boldsymbol{p}|$ is finite nor does the momentum vanish when E has a finite value. In the simplest situation a photon is characterized by a circular frequency ω and a wavelenght λ that are related by $\omega\lambda = 2\pi c$. If the energy E_γ of the photon is proportional to ω and if its momentum is inversely proportional to λ, then (4.3) is replaced with a relation of the form

$$T_\gamma \equiv E_\gamma = \alpha|\boldsymbol{p}|c\,, \tag{4.4}$$

[1] Space inversion **P** and time reversal **T** are excepted because there are interactions in nature that are Lorentz invariant but not invariant under **P** and under **T**.

where the index γ is meant to refer to a photon and where α is a dimensonless number (it will be found to be equal to 1 below). Furthermore, the photon has only kinetic energy, hence E_γ (total energy) = T_γ (kinetic energy).

Further, there are even processes where a *massive* particle decays into several *massless* particles so that its mass is completely converted into kinetic energy. For example, an electrically neutral π meson decays spontaneously into two photons:

$$\pi^0 \text{ (massive) } \rightarrow \gamma + \gamma \text{ (massless) },$$

where $m(\pi^0) = 2.4 \times 10^{-28}$ kg. If the π^0 is at rest before the decay, the momenta of the two photons are found to add up to zero,

$$\boldsymbol{p}_\gamma^{(1)} + \boldsymbol{p}_\gamma^{(2)} = 0 ,$$

while the sum of their energies is equal to $m(\pi^0)$ times the square of the speed of light,

$$T_\gamma^{(1)} + T_\gamma^{(2)} = c(|\boldsymbol{p}_\gamma^{(1)}| + |\boldsymbol{p}_\gamma^{(2)}|) = m(\pi^0)c^2 .$$

Apparently, a massive particle has a finite nonvanishing energy, even when it is at rest:

$$E(\boldsymbol{p} = 0) = mc^2 , \tag{4.5}$$

This energy is said to be its *rest energy*. Its total energy, at finite momentum, is then

$$E(\boldsymbol{p}) = mc^2 + T(\boldsymbol{p}) , \tag{4.6}$$

where $T(\boldsymbol{p})$, at least for small velocities $|\boldsymbol{p}|/m \ll c$, is given by (4.3), while for massles particles ($m = 0$) it is given by (4.4) with $\alpha = 1$.

Of course, one is curious to know how these two statements can be reconciled. As we shall soon learn, the answer is provided by the relativistic energy-momentum relation

$$E(\boldsymbol{p}) = \sqrt{(mc^2)^2 + \boldsymbol{p}^2 c^2} , \tag{4.7}$$

which is generally valid for a free particle of any mass and which contains both (4.3) and (4.4), with $\alpha = 1$. If this is so the kinetic energy is given by

$$T(\boldsymbol{p}) = E(\boldsymbol{p}) - mc^2 = \sqrt{(mc^2)^2 + \boldsymbol{p}^2 c^2} - mc^2 . \tag{4.8}$$

Indeed, for $m = 0$ this gives $T = E = |\boldsymbol{p}|c$, while for $m \neq 0$ and for small momenta $|\boldsymbol{p}|/m \ll c$

$$T(\boldsymbol{p}) \simeq mc^2 \left\{ 1 + \frac{1}{2} \frac{\boldsymbol{p}^2 c^2}{(mc^2)^2} - 1 \right\} = \frac{\boldsymbol{p}^2}{2m} , \tag{4.9}$$

which is independent of the speed of light c!

(iii) *Radioactive decay of moving particles*. There are elementary particles that are unstable but decay relatively "slowly" (quantum mechanics teaches us that this is realized when their lifetime is very much larger than Planck's constant divided by the rest energy, $\tau \gg h/2\pi mc^2$). Their decay can then be studied under various experimental conditions. As an example take the *muon* μ, which is a kind of heavy, and unstable, electron. Its mass is about 207 times larger than the mass of the electron[2],

$$m(\mu)c^2 = 206.77m(\mathrm{e})c^2 . \tag{4.10}$$

The muon decays spontaneously into an electron and two neutrinos (nearly massless particles that have only weak interactions),

$$\mu \to \mathrm{e} + \nu_1 + \nu_2 . \tag{4.11}$$

If one stops a large number of muons in the laboratory and measures their lifetime, one gets[3]

$$\tau^{(0)}(\mu) = (2.197019 \pm 0.000021) \times 10^{-6}\,\mathrm{s} . \tag{4.12}$$

If one performs the same measurement on a beam of muons that move at constant velocity $\boldsymbol{v}$ in the laboratory, one gets

$$\tau^{(v)}(\mu) = \gamma\tau^{(0)}(\mu) , \quad \text{where} \quad \gamma = E/mc^2 = (1 - \boldsymbol{v}^2/c^2)^{-1/2} . \tag{4.13}$$

For example, a measurement at $\gamma = 29.33$ gave the value

$$\tau^{(v)}(\mu) = 64.39 \times 10^{-6}\,\mathrm{s} \simeq 29.3\tau^{(0)}(\mu) .$$

This is an astounding effect: the instability of a muon is an internal property of the muon and has nothing to do with its state of motion. Its mean lifetime is something like a clock built into the muon. Experiment tells us that this clock ticks more slowly when the clock and the observer who reads it are in relative motion than when they are at rest. Relation (4.13) even tells us that the lifetime, as measured by an observer at rest, tends to infinity when the velocity $|\boldsymbol{v}|$ approaches the speed of light.

If, instead, we had applied Galilei-invariant kinematics to this problem, the lifetime in motion would be the same as at rest. Again, there is no contradiction with the relativistic relationship (4.13) because $\gamma \simeq 1 + \boldsymbol{v}^2/2c^2$. For $|\boldsymbol{v}| \ll c$ the nonrelativistic situation is realized.

---

[2] These results as well as references to the original literature are to be found in the Review of Particle Properties, Physics Letters B**592** (2004) 1 and (on the web) http://pdg.lbl.gov.

[3] J. Bailey et al., Nucl. Phys. **B 150** (1979) 1.

4.2 Constancy of the Speed of Light

The starting point and essential basis of the special theory of relativity is the following experimental observation that we formulate in terms of a postulate:

> **Postulate I.** In vacuum, light propagates, with respect to any inertial system and in all directions, with the universal velocity c (4.1). This velocity is a constant of nature.

As the value of the speed of light c is fixed at 299 792 458 ms^{-1} and as there are extremely precise methods for measuring frequencies, and hence time, the meter is defined by the distance that a light ray traverses in the fraction 1/299 792 458 of a second. (This replaces the standard meter, i.e. the measuring rod that is deposited in Paris.)

The postulate is in clear contradiction to the Galilei invariance studied in Sect. 1.13. In the nonrelativistic limit, two arbitrary inertial systems are related by the transformation law (see (1.32))

$$\begin{aligned} \boldsymbol{x}' &= \mathbf{R}\boldsymbol{x} + \boldsymbol{w}t + \boldsymbol{a}\,, \\ t' &= \lambda t + s\,, \quad (\lambda = \pm 1)\,, \end{aligned} \tag{4.14}$$

according to which the velocities of a given process, measured with respect to two different intertial systems, are related by $\boldsymbol{v}' = \boldsymbol{v} + \boldsymbol{w}$. If Postulate I is correct, (4.14) must be replaced with another relation, which must be such that it leaves the velocity of light invariant from one inertial frame to another and that (4.14) holds whenever $|\boldsymbol{v}| \ll c$ holds.

In order to grasp the consequences of this postulate more precisely, imagine the following experiment of principle. We are given two inertial systems **K** and **K**$'$. Let a light source at position $\boldsymbol{x}_A$ emit a signal at time t_A, position and time coordinates referring to **K**. In vacuum, this signal propagates in all directions with constant velocity c and hence lies on a sphere with its center at $\boldsymbol{x}_A$. If we measure this signal at a later time $t_B > t_A$, at a point $\boldsymbol{x}_B$ in space, then obviously $|\boldsymbol{x}_B - \boldsymbol{x}_A| = c(t_B - t_A)$, or, if we take the squares,

$$(\boldsymbol{x}_B - \boldsymbol{x}_A)^2 - c^2(t_B - t_A)^2 = 0\,. \tag{4.15}$$

Points with coordinates $(\boldsymbol{x}, t)$ for which one indicates the three spatial coordinates as well as the time at which something happens at $\boldsymbol{x}$ (emission or detection of a signal, for instance) are called *world points* or *events*. Accordingly, the propagation of a signal described by a parametrized curve $(\boldsymbol{x}(t), t)$ is called a *world line*.

Suppose the world points $(\boldsymbol{x}_A, t_A)$, $(\boldsymbol{x}_B, t_B)$ have the coordinates $(\boldsymbol{x}'_A, t'_A)$ and $(\boldsymbol{x}'_B, t'_B)$, respectively, with regard to the system **K**$'$. Postulate I implies that these points must be connected by the same relation (4.15), i.e.

$$(\boldsymbol{x}'_B - \boldsymbol{x}'_A)^2 - c^2(t'_B - t'_A)^2 = 0$$

with the same, universal constant c. In other words the special form

$$z^2 - (z^0)^2 = 0 , \tag{4.16}$$

relating the spatial distance $|z| = |\boldsymbol{x}_B - \boldsymbol{x}_A|$ of two world points A and B to the difference of their time coordinates $z^0 = c(t_B - t_A)$, must be invariant under all transformations that map inertial systems onto inertial systems. In fact, we confirm immediately that there are indeed subgroups of the Galilei group that leave this form invariant. These are

(i) translations $t' = t + s$ and $\boldsymbol{x}' = \boldsymbol{x} + \boldsymbol{a}$, and
(ii) rotations $t' = t$ and $\boldsymbol{x}' = \mathbf{R}\boldsymbol{x}$.

This is not true, however, for special Galilei transformations, i.e. in the case where the two inertial systems move relative to each other. In this case (4.14) reads $t' = t$, $\boldsymbol{x}' = \boldsymbol{x} + \boldsymbol{w}t$, so that $(\boldsymbol{x}'_A - \boldsymbol{x}'_B)^2 = (\boldsymbol{x}_A - \boldsymbol{x}_B + \boldsymbol{w}(t_A - t_B))^2$, which is evidently not equal to $(\boldsymbol{x}_A - \boldsymbol{x}_B)^2$. What is the most general transformation

$$(t, \boldsymbol{x}) \underset{\Lambda}{\rightarrow} (t', \boldsymbol{x}') \tag{4.17}$$

that replaces (4.14) and is such that the invariance of the form (4.16) is guaranteed?

4.3 The Lorentz Transformations

In order to unify the notation let us introduce the following definitions:

$$x^0 \stackrel{\text{def}}{=} ct ,$$
$$(x^1, x^2, x^3) \stackrel{\text{def}}{=} \boldsymbol{x} .$$

It is customary to denote indices referring to *space* components only by *Latin* letters $i, j, k, \ldots$. If one refers to *space* and *time* components, without distinction, one uses *Greek* letters $\mu, \nu, \varrho, \ldots$ instead. Thus

x^μ : $\mu = 0, 1, 2, 3$ denotes the world point $(x^0 = ct, x^1, x^2, x^3)$, and
x^i : $i = 1, 2, 3$ denotes its spatial components.

One also writes x for a world point and $\boldsymbol{x}$ for its spatial part so that

$$x^\mu = (x^0, \boldsymbol{x}) .$$

Using this notation (4.15) reads

$$(x_B^0 - x_A^0)^2 - (\boldsymbol{x}_B - \boldsymbol{x}_A)^2 = 0 .$$

This form bears some analogy to the squared norm of a vector in n-dimensional Euclidean space $\mathbb{R}^n$, which is written in various ways:

$$x_{\mathrm{E}}^2 = \sum_{i=1}^{n} (x^i)^2 = \sum_{i=1}^{n} \sum_{k=1}^{n} x^i \delta_{ik} x^k = (x, x)_{\mathrm{E}} . \tag{4.18}$$

(The index E stands for Euclidean.) The Kronecker symbol δ_{ik} is a metric tensor here. As such it is invariant under rotations in $\mathbb{R}^n$, i.e.

$$\mathbf{R}^T \delta \mathbf{R} = \delta \, .$$

A well-known example is provided by $\mathbb{R}^3$, the three-dimensional Euclidean space with the metric

$$\delta_{ik} = \begin{pmatrix} 1 & 0 & 0 \\ 0 & 1 & 0 \\ 0 & 0 & 1 \end{pmatrix} .$$

In four space-time dimensions, following the analogy with the example above, we introduce the following metric tensor:

$$g_{\mu\nu} = g^{\mu\nu} = \begin{pmatrix} 1 & 0 & 0 & 0 \\ 0 & -1 & 0 & 0 \\ 0 & 0 & -1 & 0 \\ 0 & 0 & 0 & -1 \end{pmatrix} . \tag{4.19}$$

This enables us to write the invariant form (4.15) as follows:

$$\sum_{\mu=0}^{3} \sum_{\nu=0}^{3} (x_B^\mu - x_A^\mu) g_{\mu\nu} (x_B^\nu - x_A^\nu) = 0 \, . \tag{4.20}$$

Before we move on, we wish to stress that the position of (Greek) indices matters: one must distinguish *upper* (or *contravariant*) indices from *lower* (or *covariant*) indices. For instance, we have

$$x^\mu = (x^0, \boldsymbol{x}) \, , \tag{4.21a}$$

(by definition), but

$$x_\lambda \stackrel{\text{def}}{=} \sum_{\mu=0}^{3} g_{\lambda\mu} x^\mu = (x^0, -\boldsymbol{x}) \, . \tag{4.21b}$$

For example, the generalized scalar product that appears in (4.20) can be written in several ways, viz.

$$(z, z) = (z^0)^2 - z^2 = \sum_{\mu\nu} z^\mu g_{\mu\nu} z^\nu = \sum_\mu z^\mu z_\mu = \sum_\nu z_\nu z^\nu \, . \tag{4.22}$$

Note that the indices to be summed always appear in pairs, one being an upper index and one a lower index. As one can sum only a covariant and a contravariant index, it is useful to introduce *Einstein's summation convention*, which says that expressions such as $A_\alpha B^\alpha$ should be understood to be

$$\sum_{\alpha=0}^{3} A_\alpha B^\alpha \, .$$

Remarks: The *bra* and *ket* notation that we used in Chap. 3 is very useful in the present context, too. A point x of $\mathbb{R}^4$, or likewise a tangent vector $a = (a_0, \boldsymbol{a})^T$, is represented by a four-component column,

$$|x\rangle = \left(x^0, \boldsymbol{x}\right)^T , \quad |a\rangle = \left(a^0, \boldsymbol{a}\right)^T .$$

Objects which are dual to them are written as row vectors but contain the minus sign that follows from the metric tensor $\mathbf{g} = \text{diag}\,(1, -\mathbf{1})$,

$$\langle y| = \left(y^0, -\boldsymbol{y}\right) , \quad \langle b| = \left(b^0, -\boldsymbol{b}\right) .$$

Taking scalar product in the sense of multiplying a 1×3-matrix and a 3×1-matrix yields the correct answers

$$\langle y|x\rangle = y^0 x^0 - \boldsymbol{y}\cdot\boldsymbol{x} ,$$
$$\langle b|a\rangle = b^0 a^0 - \boldsymbol{b}\cdot\boldsymbol{a} ,$$

for the Lorentz invariants which can be formed out of them. The "bra-ket" notation emphasizes that $\langle b|$ is the dual object that acts on $|a\rangle$, very much in the spirit of linear algebra.

The metric tensor defined in (4.19) has the following properties:

(i) It is invariant under the transformations (4.17).
(ii) It fulfills the relations $g_{\alpha\beta}g^{\beta\gamma} = \delta_\alpha^\gamma$, where δ_α^γ is the Kronecker symbol, and $g_{\alpha\beta} = g_{\alpha\mu}g^{\mu\nu}g_{\nu\beta} = g^{\alpha\beta}$.
(iii) Its determinant is $\det\mathbf{g} = -1$.
(iv) Its inverse and its transpose are $\mathbf{g}^{-1} = \mathbf{g} = \mathbf{g}^{\mathrm{T}}$.

The problem posed in (4.17) consists in constructing the most general affine transformation

$$x^\mu \underset{(\boldsymbol{\Lambda},a)}{\longrightarrow} x'^\mu : x'^\mu = \Lambda^\mu_{\ \sigma} x^\sigma + a^\mu \tag{4.23}$$

that guarantees the invariance of the form (4.15). Any such transformation maps inertial frames onto inertial frames because any uniform motion along a straight line is transformed into a state of motion of the same type.

Inserting the general form (4.23) into the form (4.15) or (4.20), and in either system of reference **K** or **K**$'$:

$$(x_B^\mu - x_A^\mu)g_{\mu\nu}(x_B^\nu - x_A^\nu) = 0 = (x_B'^\alpha - x_A'^\alpha)g_{\alpha\beta}(x_B'^\beta - x_A'^\beta) ,$$

we note that the translational part cancels out. As to the homogeneous part $\boldsymbol{\Lambda}$, which is a 4×4 matrix, we obtain the condition

$$\Lambda^\sigma_{\ \mu} g_{\sigma\tau} \Lambda^\tau_{\ \nu} \stackrel{!}{=} \alpha g_{\mu\nu} \tag{4.24}$$

where α is a real positive number that remains undetermined for the moment. In fact, if we decide to write x as a shorthand notation for the contravariant vector x^μ and $\boldsymbol{\Lambda}$ instead of $\Lambda^\mu_{\ \nu}$,

$$x \equiv \{x^\mu\}, \quad \boldsymbol{\Lambda} \equiv \{\Lambda^\mu_{\ \nu}\},$$

then (4.23) and (4.24) can be written in the compact form

$$x' = \boldsymbol{\Lambda} x + a, \tag{4.23'}$$

$$\boldsymbol{\Lambda}^{\mathrm{T}} \mathbf{g} \boldsymbol{\Lambda} = \alpha \mathbf{g}. \tag{4.24'}$$

Here x is a column vector and $\boldsymbol{\Lambda}$ is a 4×4 matrix, and we use the standard rules for matrix multiplication. For example, let us determine the inverse $\boldsymbol{\Lambda}^{-1}$ of $\boldsymbol{\Lambda}$, anticipating that $\alpha = 1$. It is obtained from (4.24′) by multiplying this equation with $\mathbf{g}^{-1} = \mathbf{g}$ from the left:

$$\boldsymbol{\Lambda}^{-1} = \mathbf{g} \boldsymbol{\Lambda}^{\mathrm{T}} \mathbf{g}. \tag{4.25}$$

Writing this out in components, we have

$$(\boldsymbol{\Lambda}^{-1})^\alpha_{\ \beta} = g^{\alpha\mu} \Lambda^\nu_{\ \mu}\, g_{\nu\beta} \left(\stackrel{\text{def}}{=} \Lambda_\beta^{\ \alpha}\right),$$

a matrix that is sometimes also denoted by $\Lambda_\beta^{\ \alpha}$.

Because $\boldsymbol{g}$ is not singular, (4.24) implies that $\boldsymbol{\Lambda}$ is not singular. Indeed, from (4.24),

$$(\det \boldsymbol{\Lambda})^2 = \alpha^4.$$

What do we know about the real number α from experience in physics? To answer this question let us consider two world points (or events) A and O whose difference $z \stackrel{\text{def}}{=} x_A - x_O$ does not necessarily fulfill (4.15) or (4.16). Defining their generalized distance to be $d \stackrel{\text{def}}{=} (z^0)^2 - (\boldsymbol{z})^2$, we calculate this distance with respect to the inertial system $\mathbf{K}'$:

$$d' \stackrel{\text{def}}{=} (z'^0)^2 - (\boldsymbol{z}')^2 = \alpha[(z^0) - (\boldsymbol{z})^2] = \alpha d.$$

Taking, for example, rotations in $\mathbb{R}^3$ that certainly fulfill (4.15), we see that this means that the spatial distance $\sqrt{\boldsymbol{z}^2}$, as measured from the second system of reference, appears stretched or compressed by the factor $\sqrt{\alpha}$. More generally, any spatial distance and any time interval are changed by the factor $\sqrt{\alpha}$, when measured with respect to $\mathbf{K}'$, compared to their value with respect to $\mathbf{K}$. This means either that any dynamics and any equation of motion that depend on spatial distances and on time differences differ in a measurable way in different frames of reference or that the laws of nature are invariant under scale transformations $x^\mu \to x'^\mu = \sqrt{\alpha} x^\mu$.

The first possibility is in contradiction with the Galilei invariance of mechanics. This invariance, which is well confirmed by experiment, must hold in the limit of small velocities. The second possibility contradicts our experience, too: the laws governing the forces of nature, as far as they are known to us, contain parameters with dimension and are by no means invariant under scale transformations of spatial and/or time differences. In fact, this is the main reason we choose the transformation (4.23) (which is still to be determined) to be an *affine* transformation. In conclusion, experience in physics suggests we take the constant α to be equal to 1,

$$\alpha = 1 . \tag{4.26}$$

Another way of formulating this conclusion is by the following:

Postulate II. The most general affine transformation $x \mapsto x' = \Lambda x + a$, $y \mapsto y' = \Lambda y + a$ must leave invariant the generalized distance $z^2 = (z^0)^2 - \boldsymbol{z}^2$ (where $z = y - x$), independent of whether z^2 is zero or not.

This postulate, which is based on experience, can be obtained in still another way. Our starting point was the notion of inertial frame of reference, with respect to which free motion (i.e. motion without external forces) proceeds along a straight line and with constant velocity. In other words, such a frame has the special property that dynamics, i.e. the equations of motion, take a particularly simple fom. The class of all inertial frames is the class of reference frames with respect to which the equations of motion have the same form. By definition and by construction the transformation $x' = \Lambda x + a$ (4.23) maps inertial frames onto inertial frames. As the dynamics is characterized by quantities and parameters with dimensions and as it is certainly not scale invariant, since, furthermore, Postulate I must hold true, transformations (4.23) must leave the squared norm $z^2 = z^\mu g_{\mu\nu} z^\nu$ invariant. Postulate II already contains some empirical information: very much as in nonrelativistic mechanics, lengths and times are relevant, as well as the units that are used to measure them and that are compared at different world points. The following postulate is more general and much stronger than this.

Postulate of Special Relativity. The laws of nature are invariant under the group of transformations (Λ, a).

This postulate contains Postulate II. It goes far beyond it, however, because it says that *all* physical theories, not only mechanics, are invariant under the transformations (Λ, a). Clearly, this is a very strong statement, which reaches far beyond mechanics. It holds true, indeed, also in the physics of elementary particles (space reflection and time reversal being excepted), at spatial dimensions of the order of 10^{-15} m and below. In fact, special relativity belongs to those theoretical foundations of physics whose validity is best established.

According of Postulate II the generalized distance of two world points x and y is invariant, with respect to transformations (4.23), even when it is nonzero:

$$\boxed{(y - x)^2 = (y^\alpha - x^\alpha) g_{\alpha\beta} (y^\beta - x^\beta) = \text{invariant} .}$$

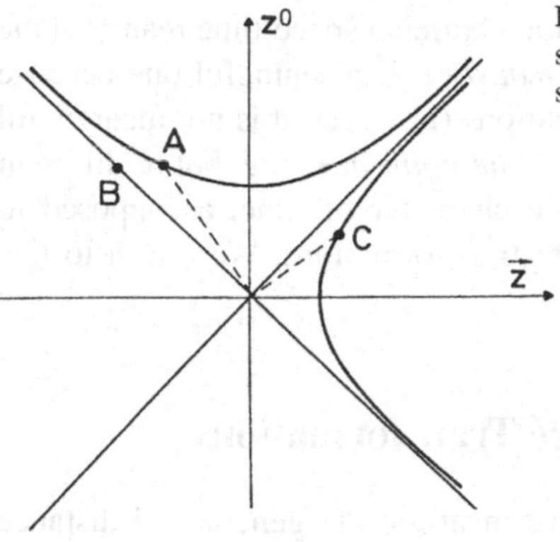

Fig. 4.1. Schematic representation of four-dimensional space–time. z^0 is the time axis, $\boldsymbol{z}$ symbolizes the three space directions

Note that this quantity can be positive, negative, or zero. This can be visualized by plotting the vector $z = y - x$ in such a way that the spatial part $\boldsymbol{z}$ is represented, symbolically, by one axis (the abscissa in Fig. 4.1), while the time component z^0 is represented by a second axis, perpendicular to the first (the ordinate in the figure). The surfaces $z^2 = \text{const}$ are axially symmetric hyperboloids, or if z^2 vanishes, a double cone, embedded in the space-time continuum. The double cone that is tangent, at infinity, to the hyperboloids, is called the *light cone*. Vectors on this cone are said to be *lightlike*; vectors for which $z^2 > 0$, i.e. $(z^0)^2 > \boldsymbol{z}^2$, are said to be *timelike*; vectors for which $z^2 < 0$, i.e. $(z^0)^2 < \boldsymbol{z}^2$, are said to be *spacelike*. (These definitions are important because they are independent of the signature of the metric tensor g. We have chosen the signature $(+, -, -, -)$ but we could have chosen $(-, +, +, +)$ as well.) In Fig. 4.1 the point A is timelike, B is lightlike, and C is spacelike. Considering the action of the transformation $(\boldsymbol{\Lambda}, a)$, we see that the translation $(\mathbb{1}, a)$ has no effect, since a cancels out in the difference $y - x$. The homogeneous part $(\boldsymbol{\Lambda}, 0)$ shifts the points A and C on their respective paraboloids shown in the figure, while it shifts B on the light cone. We give here typical examples for the three cases:

$$\text{timelike vector} \quad (z^0, \boldsymbol{0}) \qquad \text{with} \quad z^0 = \sqrt{z^2}\,, \tag{4.27a}$$

$$\text{spacelike vector} \quad (0, z^1, 0, 0) \qquad \text{with} \quad z^1 = \sqrt{-z^2}\,, \tag{4.27b}$$

$$\text{lightlike vector} \quad (1, 1, 0, 0) \tag{4.27c}$$

These three cases can be taken to be the *normal forms* for timelike, spacelike, and lightlike vectors, respectively. Indeed, every timelike vector can be mapped, by Lorentz transformations, onto the special form (4.27a). Similarly, every spacelike vector can be mapped onto (4.27b), and every lightlike vector can be transformed into (4.27c). This will be shown below in Sect. 4.5.2.

The world points x and y lie in a four-dimensional affine space. Fixing an origin (by choosing a coordinate system, for example) makes this the vector space $\mathbb{R}^4$. The differences $(y - x)$ of world points are elements of this vector space. If we endow this space with the metric structure $g_{\mu\nu}$ (4.19) we obtain what is called the flat *Minkowski space-time manifold* M^4. This manifold is different, in an essential way, from Galilei space–time. In the Galileian space-time manifold the statement that two events took place *simultaneously* is a meaningful one because simultaneity is preserved by Galilei transformations (however, it is not meaningful to claim that two events had happened at the *same point in space*, but at different times.) Absolute simultaneity i.e. the absolute character of time, as opposed to space, no longer holds with regard to Lorentz transformations. We return to this question in more detail in Sect. 4.7.

4.4 Analysis of Lorentz and Poincaré Transformations

By definition, the transformations $(\boldsymbol{\Lambda}, a)$ leave invariant the generalized distance $(x-y)^2 = (x^0 - y^0)^2 - (\boldsymbol{x} - \boldsymbol{y})^2$ of two world points. They form a group, the *inhomogeneous Lorentz group* (iL), or *Poincaré group*. Before turning to their detailed analysis we verify that these transformations indeed form a group.

1. The composition of two Poincaré transformations is again a Poincaré transformation:

$$(\boldsymbol{\Lambda}_2, a_2)(\boldsymbol{\Lambda}_1, a_1) = (\boldsymbol{\Lambda}_2\boldsymbol{\Lambda}_1, \boldsymbol{\Lambda}_2 a_1 + a_2) .$$

   The homogeneous parts are formed by matrix multiplication, the translational part is obtained by applying $\boldsymbol{\Lambda}_2$ to a_1 and adding a_2. It is easy to verify that the product $(\boldsymbol{\Lambda}_2\boldsymbol{\Lambda}_1)$ obeys (4.24) with $\alpha = 1$.
2. The composition of more than two transformations is associative:

$$(\boldsymbol{\Lambda}_3, a_3)[(\boldsymbol{\Lambda}_2, a_2)(\boldsymbol{\Lambda}_1, a_1)] = [(\boldsymbol{\Lambda}_3, a_3)(\boldsymbol{\Lambda}_2, a_2)](\boldsymbol{\Lambda}_1, a_1) ,$$

   because both the homogeneous part $\boldsymbol{\Lambda}_3\boldsymbol{\Lambda}_2\boldsymbol{\Lambda}_1$ and the translational part $\boldsymbol{\Lambda}_3\boldsymbol{\Lambda}_2 a_1 + \boldsymbol{\Lambda}_3 a_2 + a_3$ of this product are associative.
3. There exists a unit element, the identical transformation, which is given by $\mathsf{E} = (\boldsymbol{\Lambda} = \mathbb{1}, a = 0)$.
4. As $\mathbf{g}$ is not singular, by (4.24), every transformation $(\boldsymbol{\Lambda}, a)$ has an inverse. It is not difficult to verify that the inverse is given by $(\boldsymbol{\Lambda}, a)^{-1} = (\boldsymbol{\Lambda}^{-1}, -\boldsymbol{\Lambda}^{-1}a)$.

By taking the translational part to be zero, we see that the matrices $\boldsymbol{\Lambda}$ form a group by themselves. This group is said to be the *homogeneous Lorentz group* (L). The specific properties of the homogeneous Lorentz group follow from (4.24) (with $\alpha = 1$). They are:

1. $(\det\boldsymbol{\Lambda})^2 = 1$. Because $\boldsymbol{\Lambda}$ is real, this implies that either $\det\boldsymbol{\Lambda} = +1$ or $\det\boldsymbol{\Lambda} = -1$. The transformations with determinant +1 are called *proper Lorentz transformations*.
2. $(\Lambda^0{}_0)^2 \geq 1$. Hence, either $\Lambda^0{}_0 \geq +1$ or $\Lambda^0{}_0 \leq -1$. This inequality is obtained from (4.24) by taking the special values $\mu = \nu = 0$, viz.

$$\Lambda^\sigma{}_0 g_{\sigma\tau} \Lambda^\tau{}_0 = (\Lambda^0{}_0)^2 - \sum_{i=1}^{3} (\Lambda^i{}_0)^2 = 1 , \quad \text{or} \quad (\Lambda^0{}_0)^2 = 1 + \sum_{i=1}^{3} (\Lambda^i{}_0)^2 .$$

Transformations with $\Lambda^0_{\ 0} \geq +1$ are said to be *orthochronous*. They yield a "forward" mapping of time, in contrast to the transformations with $\Lambda^0_{\ 0} \leq -1$, which relate future and past.

Thus, there are four types of homogeneous Lorentz transformations, which are denoted as follows: $L_+^\uparrow, L_+^\downarrow, L_-^\uparrow, L_-^\downarrow$. The index + or − refers to the property $\det\Lambda = +1$ and $\det\Lambda = -1$, respectively; the arrow pointing upwards means $\Lambda^0_{\ 0} \geq +1$, while the arrow pointing downwards means $\Lambda^0_{\ 0} \leq -1$. Special examples for the four types are the following.
(i) The identity belongs to the branch $L_+^\uparrow$:

$$\mathbf{E} = \begin{pmatrix} 1 & 0 & 0 & 0 \\ 0 & 1 & 0 & 0 \\ 0 & 0 & 1 & 0 \\ 0 & 0 & 0 & 1 \end{pmatrix} \in L_+^\uparrow \,. \tag{4.28}$$

(ii) Reflection of the space axes (*parity*) belongs to the branch $L_-^\uparrow$:

$$\mathbf{P} = \begin{pmatrix} 1 & & & \\ & -1 & & \\ & & -1 & \\ & & & -1 \end{pmatrix} \in L_-^\uparrow \,. \tag{4.29}$$

(iii) Reversal of the time direction (*time reversal*) belongs to $L_-^\downarrow$:

$$\mathbf{T} = \begin{pmatrix} -1 & & & \\ & 1 & & \\ & & 1 & \\ & & & 1 \end{pmatrix} \in L_-^\downarrow \,. \tag{4.30}$$

(iv) The product **PT** of time reversal and space reflection belongs to $L_+^\downarrow$:

$$\mathbf{PT} = \begin{pmatrix} -1 & & & \\ & -1 & & \\ & & -1 & \\ & & & -1 \end{pmatrix} \in L_+^\downarrow \,. \tag{4.31}$$

At this point, we wish to make a few remarks relevant to what follows. The four discrete transformations (4.28–31) themselves form what is called *Klein's group*,

$$\{\mathbf{E}, \mathbf{P}, \mathbf{T}, \mathbf{PT}\} \,. \tag{4.32}$$

Indeed, one can easily verify that the product of any two of them is an element of the group.

It is also clear that two arbitrary transformations belonging to *different* branches cannot be made to coincide by continuous deformation. Indeed, as long as $\boldsymbol{\Lambda}$ is real, transformations with determinant $+1$ and those with determinant -1 are separated discontinuously from each other. (Likewise, transformations with $\Lambda^0_{\ 0} \geq +1$ and with $\Lambda^0_{\ 0} \leq -1$ cannot be related by continuity). However, for given $\boldsymbol{\Lambda} \in L^{\uparrow}_{+}$, we note that the product $\boldsymbol{\Lambda}\mathbf{P}$ is in $L^{\uparrow}_{-}$, the product $\boldsymbol{\Lambda}\mathbf{T}$ is in $L^{\downarrow}_{-}$, and the product $\boldsymbol{\Lambda}(\mathbf{PT})$ is in $L^{\downarrow}_{+}$. Thus, if we know the transformations belonging to $L^{\uparrow}_{+}$ (the proper, orthochronous Lorentz transformations), those pertaining to the other branches can be generated from them by multiplication with $\mathbf{P}, \mathbf{T}$, and $(\mathbf{PT})$. These relations are summarized in Table 4.1.

Table 4.1. The four disjoint branches of the homogeneous Lorentz group

$L^{\uparrow}_{+}(\det \boldsymbol{\Lambda} = 1, \Lambda^0_{\ 0} \geq 1)$ *Examples*: E, rotations, special Lorentz transformations	$L^{\downarrow}_{+}(\det \boldsymbol{\Lambda} = 1, \Lambda^0_{\ 0} \leq -1)$ *Examples*: $\mathbf{PT}$, as well as all $\boldsymbol{\Lambda}(\mathbf{PT})$ with $\boldsymbol{\Lambda} \in L^{\uparrow}_{+}$
$L^{\uparrow}_{-}(\det \boldsymbol{\Lambda} = -1, \Lambda^0_{\ 0} \geq 1)$ *Examples*: $\mathbf{P}$, as well as all $\boldsymbol{\Lambda}\mathbf{P}$ with $\boldsymbol{\Lambda} \in L^{\uparrow}_{+}$	$L^{\downarrow}_{-}(\det \boldsymbol{\Lambda} = -1, \Lambda^0_{\ 0} \leq -1)$ *Examples*: $\mathbf{T}$, as well as all $\boldsymbol{\Lambda}\mathbf{T}$ with $\boldsymbol{\Lambda} \in L^{\uparrow}_{+}$

Finally, we conclude that the branch $L^{\uparrow}_{+}$ is a subgroup of the homogeneous Lorentz group. Indeed, the composition of two transformations of $L^{\uparrow}_{+}$ is again element of $L^{\uparrow}_{+}$. Furthermore, it contains the unit element as well as the inverse of any of its elements. This subgroup $L^{\uparrow}_{+}$ is called the *proper, orthochronous Lorentz group*. (In contrast to $L^{\uparrow}_{+}$, the remaining three branches are not subgroups.)

4.4.1 Rotations and Special Lorentz Tranformations ("Boosts")

The rotations in three-dimensional space, well-known to us from Sect. 2.22, leave the spatial distance $|\boldsymbol{x} - \boldsymbol{y}|$ invariant. As they do not change the time component of any four-vector, the transformations

$$\boldsymbol{\Lambda}(\mathbf{R}) \equiv \mathcal{R} \stackrel{\text{def}}{=} \begin{pmatrix} 1 & 0 & 0 & 0 \\ 0 & & & \\ \cdot & & \mathbf{R} & \\ 0 & & & \end{pmatrix} \tag{4.33}$$

with $\mathbf{R} \in \mathrm{SO}(3)$ leave invariant the form $(z^0)^2 - (\boldsymbol{z})^2$. Thus, they are Lorentz transformations. Now, obviously $\mathcal{R}^0_{\ 0} = +1$, and $\det\mathcal{R} = \det\mathbf{R} = +1$, so they belong to the branch $L^{\uparrow}_{+}$. Thus, extending the rotations in three-dimensional space by adding a 1 in the time–time component, and zeros in the time–space and the space–time components, as shown in (4.33), we obtain a subgroup of $L^{\uparrow}_{+}$.

We now turn to the relativistic generalization of the special Galilei transformations

$$\boldsymbol{x}' = \boldsymbol{x} - \boldsymbol{v}t\ , \quad t' = t\ . \tag{4.34}$$

Their relativistic counterparts are called *special Lorentz transformations*, or *boosts*. They are obtained as follows.

As we know, boosts describe the situation where two inertial systems of reference **K** and **K**′ move relative to each other with constant velocity $\boldsymbol{v}$. Figure 4.2 shows the example of uniform motion along the spatial 1-axis, $\boldsymbol{v} = v\hat{\boldsymbol{e}}_1$. The space components that are transverse to the 1-axis are certainly not changed, i.e.

$$z'^2 = z^2\,, \quad z'^3 = z^3\,.$$

Regarding the remaining components of the four-vector z, this implies that the form $(z^0)^2 - (z^1)^2$ must be invariant.

$$(z^0)^2 - (z^1)^2 = (z^0 + z^1)(z^0 - z^1) = \text{ invariant}\,.$$

Thus, we must have

$$z'^0 + z'^1 = f(v)(z^0 + z^1)\,, \quad (z'^0 - z'^1) = \frac{1}{f(v)}(z^0 - z^1)$$

with the conditions $f(v) > 0$ and $\lim_{v\to 0} f(v) = 1$. Furthermore, the origin O′ of **K**′ moves with velocity $\boldsymbol{v}$, relative to **K**. Thus, the primed and unprimed 1-component of O′ are, respectively,

$$z'^1 = \frac{1}{2}\left(f - \frac{1}{f}\right)z^0 + \frac{1}{2}\left(f + \frac{1}{f}\right)z^1 = 0\,, \quad z^1 = \frac{v}{c}z^0\,,$$

from which follow

$$(f^2 - 1) + \frac{v}{c}(f^2 + 1) = 0\,,$$

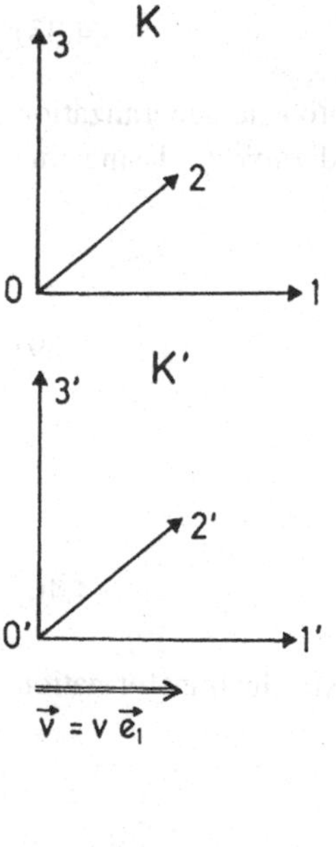

Fig. 4.2. **K** and **K**′ are inertial systems that move at a constant velocity relative to each other. At $t = 0$ or, $\boldsymbol{v} = 0$, the two systems coincide

and, finally,

$$f(v) = \sqrt{\frac{1-(v/c)}{1+(v/c)}} \,. \tag{4.35}$$

At this point let us introduce the following (universally accepted) abbreviations:

$$\beta \stackrel{\text{def}}{=} \frac{|\boldsymbol{v}|}{c}\,, \quad \gamma \stackrel{\text{def}}{=} \frac{1}{\sqrt{1-\beta^2}}\,. \tag{4.36}$$

Using this notation, z^1 and z^0 are seen to transform as follows:

$$\begin{pmatrix} z'^0 \\ z'^1 \end{pmatrix} = \frac{1}{2}\begin{pmatrix} f+1/f & f-1/f \\ f-1/f & f+1/f \end{pmatrix}\begin{pmatrix} z^0 \\ z^1 \end{pmatrix} = \begin{pmatrix} \gamma & -\gamma\beta \\ -\gamma\beta & \gamma \end{pmatrix}\begin{pmatrix} z^0 \\ z^1 \end{pmatrix}\,,$$

where $0 \le \beta \le 1$, $\gamma \ge 1$. Including the 2- and 3-components, the special Lorentz transformation (boost) that we are out to construct reads

$$\boldsymbol{\Lambda}(-\boldsymbol{v}) \equiv \mathbf{L}(-\boldsymbol{v})_{\boldsymbol{v}=v\hat{\boldsymbol{e}}_1} = \begin{pmatrix} \gamma & -\gamma\beta & 0 & 0 \\ -\gamma\beta & \gamma & 0 & 0 \\ 0 & 0 & 1 & 0 \\ 0 & 0 & 0 & 1 \end{pmatrix}\,. \tag{4.37}$$

It has the properties $L^0_{\ 0} \ge +1$ and $\det\mathbf{L} = +1$, and therefore it belongs to $L_+^\uparrow$.

Without loss of generality we could have parametrized the function $f(\boldsymbol{v})$ (4.35) by

$$f(\boldsymbol{v}) = \exp(-\lambda(\boldsymbol{v}))\,. \tag{4.38}$$

As we shall show below (Sect. 4.6), the parameter λ is a relativistic generalization of the (modulus of the) velocity. For this reason it is called *rapidity*. Using this parametrization, the transformation (4.37) takes the form

$$\mathbf{L}(-v\hat{\boldsymbol{e}}_1) = \begin{pmatrix} \cosh\lambda & -\sinh\lambda & 0 & 0 \\ -\sinh\lambda & \cosh\lambda & 0 & 0 \\ 0 & 0 & 1 & 0 \\ 0 & 0 & 0 & 1 \end{pmatrix}\,, \tag{4.39}$$

where λ and $|\boldsymbol{v}|$ are related by

$$\tanh\lambda = \frac{|\boldsymbol{v}|}{c} = \beta\,. \tag{4.40}$$

If the velocitiy $\boldsymbol{v}$ does not point along the direction of the 1-axis, the transformation (4.37) takes the form

$$\mathbf{L}(-\boldsymbol{v}) = \begin{pmatrix} \gamma & -\gamma \dfrac{v^k}{c} \\ -\gamma \dfrac{v^i}{c} & \delta^{ik} + \dfrac{\gamma^2}{1+\gamma} \dfrac{v^i v^k}{c^2} \end{pmatrix} . \tag{4.41}$$

This more general expression is derived by means of the following steps. The matrix (4.37) that describes the case $\boldsymbol{v} = v\hat{\boldsymbol{e}}_1$ is symmetric. It transforms the time and the 1-coordinates in a nontrivial way but leaves unchanged the directions perpendicular to $\boldsymbol{v}$. In particular, we have

$$z'^0 = \gamma[z^0 - \beta z^1] = \gamma \left[z^0 - \frac{1}{c} \boldsymbol{v} \cdot \boldsymbol{z} \right] ,$$

$$z'^1 = \gamma[-\beta z^0 + z^1] = \gamma \left[-\frac{v^1}{c} z^0 + z^1 \right] .$$

If $\boldsymbol{v}$ has an arbitrary direction in space, one could, of course, rotate the coordinate system by means of a rotation $\mathcal{R}$ in such a way that the new 1-axis points along $\boldsymbol{v}$. The boost $\mathbf{L}$ would then have precisely the form of (4.37). Finally, one could undo the rotation. As $\mathbf{L}$ is symmetric, so is the product $\mathcal{R}^{-1}\mathbf{L}\mathcal{R}$. Without calculating this rotation explicitly, we can use the following form for the boost that we wish to construct:

$$\mathbf{L}(-\boldsymbol{v}) = \begin{pmatrix} \gamma & -\gamma \dfrac{v^k}{c} \\ -\gamma \dfrac{v^i}{c} & T^{ik} \end{pmatrix} \quad \text{with} \quad T^{ik} = T^{ki} .$$

For vanishing velocity T^{ik} becomes the unit matrix. Therefore, we can write T^{ik} as follows:

$$T^{ik} = \delta^{ik} + a \frac{v^i v^k}{c^2} .$$

We determine the coefficient a by making use of our knowledge of the coordinates of O′, the origin of $\mathbf{K}'$, in either system of reference. With respect to $\mathbf{K}'$ we have

$$z'^i = -\gamma \frac{1}{c} v^i z^0 + \sum_{k=1}^{3} T^{ik} z^k = 0 .$$

As seen from $\mathbf{K}$, O′ moves at a constant velocity $\boldsymbol{v}$, i.e. $z^k = v^k z^0/c$. From these equations follows the requirement

$$\sum_k T^{ik} \frac{v^k}{c} = \frac{1}{c} \left[1 + a \frac{\boldsymbol{v}^2}{c^2} \right] v^i \stackrel{!}{=} \gamma \frac{v^i}{c} ,$$

or $1 + a\beta^2 = \gamma$, and, finally, $a = \gamma^2/(\gamma+1)$. This completes the construction of (4.41).

4.4.2 Interpretation of Special Lorentz Transformations

First, we verify that the transformation (4.41) becomes the special Galilei transformation (4.34) whenever the velocity is small compared to c. We develop matrix (4.41) in terms of $\beta = |\boldsymbol{v}|/c$, up to first order, viz.

$$\mathsf{L}(-\boldsymbol{v}) = \begin{pmatrix} 1 & -v^k/c \\ -v^i/c & \delta^{ik} \end{pmatrix} + \mathrm{O}(\beta^2) \,.$$

Neglecting the terms of order $\mathrm{O}(\beta^2)$, we indeed obtain $t' = t$ and $z = -vt + z$. Thus, the transformation rule (4.34) holds approximately for $(\boldsymbol{v}/c)^2 \ll 1$. This is an excellent approximation for the planets of our solar system. For example, the earth's orbital velocity is about $30\,\mathrm{km\,s^{-1}}$, and therefore $(\boldsymbol{v}/c)^2 \simeq 10^{-8}$. Elementary particles, on the other hand, can be accelerated to velocities very close to the speed of light. In this case transformation (4.41) is very different from (4.34).

An instructive way of visualizing the special Lorentz transformation (4.41) is to think of $\mathbf{K}'$ as being fixed in an elementary particle that moves at a constant velocity $\boldsymbol{v}$ with respect to the inertial system $\mathbf{K}$. $\mathbf{K}'$ is then said to be the *rest system* of the particle, while $\mathbf{K}$ could be the laboratory system.

Transformation (4.41) describes the transition between laboratory and rest system; it "boosts" the particle from its state of rest to the state with velocity $\boldsymbol{v}$. To make this clear, we anticipate a little by defining the following four-vector:

$$\omega \stackrel{\text{def}}{=} (\gamma c, \gamma \boldsymbol{v})^T \,. \tag{4.42}$$

(A more detailed reasoning will be given in Sect. 4.8 below.) The generalized, squared norm of this vector is $\omega^2 = \gamma^2 c^2 (1 - \boldsymbol{v}^2/c^2) = c^2$. If we apply the matrix (4.41) to ω, we obtain

$$\mathsf{L}(-\boldsymbol{v})\omega = \omega^{(0)} \tag{4.43}$$

with $\omega^{(0)} = (c, \mathbf{0})$. Thus, this vector must be related to the relativistic generalization of velocity or of momentum. We see that $L(-\boldsymbol{v})$ transforms something moving with velocity $\boldsymbol{v}$ to something at rest (velocity 0), hence the minus sign in the definition above. We shall say more about the interpretation of ω later and return to the analysis of the Lorentz transformations.

4.5 Decomposition of Lorentz Transformations into Their Components

4.5.1 Proposition on Orthochronous, Proper Lorentz Transformations

The structure of the homogeneous, proper, orthochronous Lorentz group $L_+^\uparrow$ is clarified by the following theorem.

Decomposition Theorem. Every transformation $\boldsymbol{\Lambda}$ of $L_+^\uparrow$ can be written, in a unique way, as the product of a rotation and a special Lorentz transformation following the rotation:

$$\boldsymbol{\Lambda} = \mathbf{L}(\boldsymbol{v})\mathcal{R} \quad \text{with} \quad \mathcal{R} = \begin{pmatrix} 1 & 0 \\ 0 & \mathbf{R} \end{pmatrix}, \quad \mathbf{R} \in \mathrm{SO}(3) . \tag{4.44}$$

The parameters of the two transformations are given by the following expressions:

$$v^i/c = \Lambda^i{}_0/\Lambda^0{}_0 , \tag{4.45}$$

$$R^{ik} = \Lambda^i{}_k - \frac{1}{1+\Lambda^0{}_0}\Lambda^i{}_0\Lambda^0{}_k . \tag{4.46}$$

Proof. As a first step one verifies that the velocity defined by (4.45) is an admissible velocity, i.e. that it does not exceed the speed of light. This follows from (4.24) (with $\alpha = 1$):

$$\boldsymbol{\Lambda}^{\mathrm{T}}\mathbf{g}\boldsymbol{\Lambda} = \mathbf{g} \tag{4.47}$$

$$\text{or} \quad \Lambda^\mu{}_\sigma g_{\mu\nu}\Lambda^\nu{}_\tau = g_{\sigma\tau} . \tag{4.47$'$}$$

Choosing $\sigma = \tau = 0$, then $\sigma = i$, $\tau = k$, and then $\sigma = 0$, $\tau = i$, we find that (4.47) yields the following equations, respectively,

$$(\Lambda^0{}_0)^2 - \sum_{i=1}^{3}(\Lambda^i{}_0)^2 = 1 , \tag{4.48a}$$

$$\Lambda^0{}_i\Lambda^0{}_k - \sum_{j=1}^{3}\Lambda^j{}_i\Lambda^j{}_k = -\delta_{ik} , \tag{4.48b}$$

$$\Lambda^0{}_0\Lambda^0{}_i - \sum_{j=1}^{3}\Lambda^j{}_0\Lambda^j{}_i = 0 . \tag{4.48c}$$

Now, from (4.48a) we indeed find that

$$\frac{v^2}{c^2} = \frac{\sum (\Lambda^i{}_0)^2}{(\Lambda^0_0)^2} = \frac{(\Lambda^0_0)^2 - 1}{(\Lambda^0_0)^2} \leq 1 \,.$$

Comparison with the general expression (4.41) for a boost then gives

$$L^0_0(v) = \Lambda^0_0 \,; \quad L^0_i(v) = L^i{}_0(v) = \Lambda^i{}_0 \,,$$
$$L^i{}_k(v) = \delta^{ik} + \frac{1}{1+\Lambda^0_0} \Lambda^i{}_0 \Lambda^k_0 \,. \tag{4.49}$$

As a second step we define

$$\mathcal{R} \stackrel{\text{def}}{=} \mathbf{L}^{-1}(v)\Lambda = \mathbf{L}(-v)\Lambda \tag{4.50}$$

and show that $\mathcal{R}$ is a rotation. This follows by means of (4.48a) and (4.48c) and by doing the multiplication on the right-hand side of (4.50), viz.

$$\mathcal{R}^0_0 = (\Lambda^0_0)^2 - \sum_i (\Lambda^i{}_0)^2 = 1 \,,$$

$$\mathcal{R}^0_i = \Lambda^0_0 \Lambda^0_i - \sum_j \Lambda^j_0 \Lambda^j_i = 0 \,.$$

At the same time we calculate the space–space components of the rotation,

$$\mathcal{R}^i{}_k = \Lambda^i{}_k - \Lambda^i{}_0 \Lambda^0_k + \frac{1}{1+\Lambda^0_0} \Lambda^i{}_0 \sum_j \Lambda^j_0 \Lambda^j_k \,.$$

Inserting (4.48c) in the right-hand side yields assertion (4.46).

As a third and last step it remains to show that the decomposition (4.44) is unique. For this purpose assume that there are two different velocities v and $\bar{v}$, as well as two different rotations $\mathbf{R}$ and $\bar{\mathbf{R}}$ of SO(3), such that

$$\Lambda = \mathbf{L}(v)\mathcal{R} = L(\bar{v})\bar{\mathcal{R}}$$

holds true. From this we would conclude that

$$\mathbf{L}(-v)\Lambda\mathcal{R}^{-1} = \mathbb{1} = \mathbf{L}(-v)\mathbf{L}(\bar{v})\bar{\mathcal{R}}\mathcal{R}^{-1} \,.$$

Taking the time–time component of this expression, for example, we would obtain

$$1 = \sum_{\nu=0}^{3} L^0{}_\nu(-v) L^\nu{}_0(\bar{v}) = \left[1 - \frac{1}{c^2} v \cdot \bar{v}\right] / \sqrt{(1 - v^2/c^2)(1 - \bar{v}^2/c^2)} \,.$$

This equation can be correct only if $v = \bar{v}$. If this is so then also $\mathcal{R} = \bar{\mathcal{R}}$. Thus the theorem is proved. □

4.5.2 Corollary of the Decomposition Theorem and Some Consequences

Note the order of the factors of the decomposition (4.44): the rotation $\mathcal{R}$ is applied first and is followed by the boost $\mathbf{L}(\boldsymbol{v})$. One could prove the decomposition of $\boldsymbol{\Lambda} \in L_+^\uparrow$, with a different order of its factors, as well, viz.

$$\boldsymbol{\Lambda} = \mathcal{R}\mathbf{L}(\boldsymbol{w}) \quad \text{with} \quad \mathcal{R} = \begin{pmatrix} 1 & 0 \\ 0 & \mathbf{R} \end{pmatrix} , \qquad \mathbf{R} \in \mathrm{SO}(3) , \tag{4.51}$$

where the vector $\boldsymbol{w}$ is given by

$$\frac{w^i}{c} \stackrel{\text{def}}{=} \frac{\Lambda^0_{\ i}}{\Lambda^0_{\ 0}} \tag{4.52}$$

and where $\mathbf{R}$ is the same rotation as in (4.46). The proof starts from the relation

$$\boldsymbol{\Lambda}\mathbf{g}\boldsymbol{\Lambda}^{\mathrm{T}} = \mathbf{g} , \tag{4.53}$$

$$\Lambda^\sigma_{\ \mu} g^{\mu\nu} \Lambda^\tau_{\ \nu} = g^{\sigma\tau} , \tag{4.53'}$$

which is the analog of (4.47) and which says no more than that if $\boldsymbol{\Lambda}$ belongs to $L_+^\uparrow$, then its inverse $\boldsymbol{\Lambda}^{-1} = \mathbf{g}\boldsymbol{\Lambda}^{\mathrm{T}}\mathbf{g}$ also belongs to $L_+^\uparrow$. Otherwise the steps of the proof are the same as in Sect. 4.5.1. One verifies, by direct calculation, that $\boldsymbol{v} = \mathbf{R}\boldsymbol{w}$. This is not surprising because, by comparing (4.44) and (4.51), we find

$$\mathbf{L}(\boldsymbol{v}) = \mathcal{R}\mathbf{L}(\boldsymbol{w})\mathcal{R}^{-1} = \mathbf{L}(\mathbf{R}\boldsymbol{w}) . \tag{4.54}$$

The decomposition theorem has several important consequences.

(i) The decomposition is useful in proving that every timelike four vector can be mapped to the normal form (4.27a), every spacelike vector to the normal form (4.27b), and every lightlike vector to the form (4.27c). We choose the example of a timelike vector, $z = (z^0, \boldsymbol{z})$ with $z^2 = (z^0)^2 - (\boldsymbol{z})^2 > 0$. By a rotation it assumes the form $(z^0, z^1, 0, 0)$. If z^0 is negative, apply $\mathbf{PT}$ to z so that z^0 becomes positive and hence $z^0 > |z^1|$. As one verifies by explicit calculation, the boost along the 1-axis, with the parameter λ as obtained from

$$\mathrm{e}^\lambda = \sqrt{(z^0 - z^1)/(z^0 + z^1)} ,$$

takes the vector to the form of (4.27a).

(ii) The group $L_+^\uparrow$ is a Lie group and contains the rotation group SO(3) as a subgroup. The decomposition theorem tells us that $L_+^\uparrow$ depends on six real parameters: the three angles of the rotation and the three components of the velocity. Thus, its Lie algebra is made up of six generators. More precisely, to the real angles characterizing the rotations there correspond the *directions* of the boosts and the rapidity parameter λ. This parameter has its value in the interval $[0, \infty]$. While the manifold of the rotation angles is compact, that of λ is not. Indeed, the Lorentz group is found to be noncompact. Therefore, its structure and its representations are not simple and must be studied separately. This is beyond the scope of this book.

(iii) It is not difficult, though, to construct the six generators of $L_+^\uparrow$. We already know the generators for rotations, see Sect. 2.22. Adding the time–time and space–time components they are

$$\mathbf{J}_i = \left(\begin{array}{c|ccc} 0 & 0 & 0 & 0 \\ \hline 0 & & & \\ 0 & & (\mathbf{J}_i) & \\ 0 & & & \end{array}\right) , \tag{4.55}$$

where $(\mathbf{J}_i)$ are the 3×3 matrices given in (2.71). The generators for infinitesimal boosts are derived in an analogous manner. The example of a special Lorentz transformation along the 1-axis (4.39) contains the submatrix

$$\mathbf{A} \stackrel{\text{def}}{=} \begin{pmatrix} \cosh\lambda & \sinh\lambda \\ \sinh\lambda & \cosh\lambda \end{pmatrix} = 1\!\!1 \sum_{n=0}^{\infty} \frac{\lambda^{2n}}{(2n)!} + \mathbf{K} \sum_{n=0}^{\infty} \frac{\lambda^{2n+1}}{(2n+1)!}$$

$$\text{with} \quad \mathbf{K} = \begin{pmatrix} 0 & 1 \\ 1 & 0 \end{pmatrix} .$$

The latter matrix (it is the Pauli matrix $\sigma^{(1)}$) has the following properties:

$$\mathbf{K}^{2n} = 1\!\!1 , \quad \mathbf{K}^{2n+1} = \mathbf{K} .$$

Therefore, we have

$$\mathbf{A} = \sum_{n=0}^{\infty} \left\{ \frac{\lambda^{2n}}{(2n)!} \mathbf{K}^{2n} + \frac{\lambda^{2n+1}}{(2n+1)!} \mathbf{K}^{2n+1} \right\} = \exp(\lambda \mathbf{K}) .$$

Alternatively, writing this exponential series by means of Gauss's formula for the exponential,

$$\mathbf{A} = \lim_{k\to\infty} \left(1\!\!1 + \frac{\lambda}{k} \mathbf{K} \right)^k ,$$

we see that the *finite* boost is generated by successive application of very many *infinitesimal* ones. From this argument we deduce the generator for infinitesimal boosts along the 1-axis:

$$\mathbf{K}_1 = \begin{pmatrix} 0 & 1 & 0 & 0 \\ 1 & 0 & 0 & 0 \\ 0 & 0 & 0 & 0 \\ 0 & 0 & 0 & 0 \end{pmatrix} . \tag{4.56}$$

It is then easy to guess the analogous expressions for the generators $\mathbf{K}_2$ and $\mathbf{K}_3$ for infinitesimal boosts along the 2-axis and the 3-axis, respectively:

$$\mathbf{K}_2 = \begin{pmatrix} 0 & 0 & 1 & 0 \\ 0 & 0 & 0 & 0 \\ 1 & 0 & 0 & 0 \\ 0 & 0 & 0 & 0 \end{pmatrix}, \quad \mathbf{K}_3 = \begin{pmatrix} 0 & 0 & 0 & 1 \\ 0 & 0 & 0 & 0 \\ 0 & 0 & 0 & 0 \\ 1 & 0 & 0 & 0 \end{pmatrix}. \tag{4.57}$$

By the decomposition theorem every Λ of $L_+^\uparrow$ can be written as follows:

$$\Lambda = \exp(-\boldsymbol{\varphi} \cdot \mathbf{J}) \exp(\lambda \hat{\boldsymbol{w}} \cdot \mathbf{K}) , \tag{4.58}$$

where $\mathbf{J} = (\mathbf{J}_1, \mathbf{J}_2, \mathbf{J}_3)$, $\mathbf{K} = (\mathbf{K}_1, \mathbf{K}_2, \mathbf{K}_3)$, and $\lambda = \operatorname{arctanh} |\boldsymbol{w}|/c$.

(iv) It is instructive to compute the commutators of the matrices $\mathbf{J}_i$ and $\mathbf{K}_k$ as given by (4.56), (4.57), and (2.71). One finds that

$$[\mathbf{J}_1, \mathbf{J}_2] \equiv \mathbf{J}_1\mathbf{J}_2 - \mathbf{J}_2\mathbf{J}_1 = \mathbf{J}_3 , \tag{4.59a}$$
$$[\mathbf{J}_1, \mathbf{K}_1] = 0 , \tag{4.59b}$$
$$[\mathbf{J}_1, \mathbf{K}_2] = \mathbf{K}_3 \quad [\mathbf{K}_1, \mathbf{J}_2] = \mathbf{K}_3 , \tag{4.59c}$$
$$[\mathbf{K}_1, \mathbf{K}_2] = -\mathbf{J}_3 . \tag{4.59d}$$

All other commutators are obtained from these by cyclic permutation of the indices.

One can visualize the meaning of relations (4.59a–d) to some extent by recalling that the $\mathbf{J}_i$ and $\mathbf{K}_k$ generate infinitesimal transformations. For instance, (4.59a) tells us that two infinitesimal rotations by the angle ε_1 about the 1-axis and by the angle ε_2 about the 2-axis, when inverted in different order, give a net rotation about the 3-axis, by the angle $\varepsilon_1\varepsilon_2$,

$$\mathbf{R}^{-1}(0, \varepsilon_2, 0)\, \mathbf{R}^{-1}(\varepsilon_1, 0, 0)\, \mathbf{R}(0, \varepsilon_2, 0)\, \mathbf{R}(\varepsilon_1, 0, 0) = \mathbf{R}(0, 0, \varepsilon_1 \cdot \varepsilon_2) + O(\varepsilon_i^3)$$

(The reader should work this out).

Equation (4.59b) states that a boost along a given direction is unchanged by a rotation about the same direction. Equation (4.59c) expresses the fact that the three matrices $(\mathbf{K}_1, \mathbf{K}_2, \mathbf{K}_2)$ transform under rotations like an ordinary vector in $\mathbb{R}^3$ (hence the notation in (4.58)).

The commutation relation (4.59d) is the most interesting. If one applies a boost along the 1-axis, followed by a boost along the 2-axis, and then inverts these transformations in the "wrong" order, there results a pure rotation about the 3-axis. In order to see this clearly, let us consider

$$\mathbf{L}_1 \simeq \mathbb{1} + \lambda_1\mathbf{K}_1 + \tfrac{1}{2}\lambda_1^2\mathbf{K}_1^2 , \quad \mathbf{L}_2 \simeq \mathbb{1} + \lambda_2\mathbf{K}_2 + \tfrac{1}{2}\lambda_2^2\mathbf{K}_2^2$$

with $\lambda_i \ll 1$. To second order in the λ_i we then obtain

$$\mathbf{L}_2^{-1}\mathbf{L}_1^{-1}\mathbf{L}_2\mathbf{L}_1 \simeq \mathbb{1} - \lambda_1\lambda_2[\mathbf{K}_1, \mathbf{K}_2] = \mathbb{1} + \lambda_1\lambda_2\mathbf{J}_3 . \tag{4.60}$$

Here is an example that illustrates this result. An elementary particle, say the electron, carries an intrinsic angular momentum, called spin. Let this particle have the momentum $\boldsymbol{p}_0 = 0$. The series of Lorentz transformations described above eventually bring the momentum back to the value $\boldsymbol{p}_0 = 0$. However, the spin is rotated a little about the 3-axis. This observation is the basis of the so-called Thomas precession, which is discussed in treatises on special relativity and which has a number of interesting applications.

4.6 Addition of Relativistic Velocities

The special Galilei transformation (4.34), or the special Lorentz transformation (4.41), relates the inertial systems $\mathbf{K}_0$ and $\mathbf{K}'$, the parameter $\boldsymbol{v}$ being the relative velocity of the two systems of reference. For example, one may think of $\mathbf{K}'$ as being fixed in a particle that moves with velocity $\boldsymbol{v}$ relative to an observer who is placed at the origin of $\mathbf{K}_0$. We assume that the absolute value of this velocity is smaller than the velocity of light, c. Of course, the system of reference $\mathbf{K}_0$ can be replaced with any other one, $\mathbf{K}_1$, moving with constant velocity $\boldsymbol{w}$ relative to $\mathbf{K}_0$ ($|\boldsymbol{w}|$ being assumed smaller than c, too). What, then, is the special transformation (the boost) that describes the motion of $\mathbf{K}'$, the particle's rest system, as observed from $\mathbf{K}_1$?

In the case of the Galilei transformation (4.34) the answer is obvious: $\mathbf{K}'$ moves relative to $\mathbf{K}_1$ with the constant velocity $\boldsymbol{u} = \boldsymbol{v} + \boldsymbol{w}$. In particular, if $\boldsymbol{v}$ and $\boldsymbol{w}$ are parallel and if both $|\boldsymbol{v}|$ and $|\boldsymbol{w}|$ exceed $c/2$, then the magnitude of $\boldsymbol{u}$ exceeds c.

In the relativistic case the law of addition for velocities is different. Without restriction of generality let us take $\boldsymbol{v}$ along the 1-axis. Let λ be the corresponding rapidity parameter,

$$\tanh\lambda = \frac{|\boldsymbol{v}|}{c} \equiv \frac{v}{c}\,, \quad \text{or} \quad e^{\lambda} = \sqrt{\frac{1+v/c}{1-v/c}}\,,$$

so that the transformation between $\mathbf{K}_0$ and $\mathbf{K}'$ reads

$$\mathsf{L}(\boldsymbol{v} = v\hat{\boldsymbol{e}}_1) = \begin{pmatrix} \cosh\lambda & \sinh\lambda & 0 & 0 \\ \sinh\lambda & \cosh\lambda & 0 & 0 \\ 0 & 0 & 1 & 0 \\ 0 & 0 & 0 & 1 \end{pmatrix}. \tag{4.61}$$

A case of special interest is certainly the one where $\boldsymbol{w}$ is parallel to $\boldsymbol{v}$ and points in the same direction, i.e. the one where one boosts twice along the same direction. $\mathsf{L}(\boldsymbol{w} = w\hat{\boldsymbol{e}}_1)$ has the form (4.61), with λ being replaced by the parameter μ, which fulfills

$$\tanh\mu = \frac{w}{c}\,, \quad \text{or} \quad e^{\mu} = \sqrt{\frac{1+w/c}{1-w/c}}\,.$$

The product $\mathbf{L}(w\hat{e}_1)\mathbf{L}(v\hat{e}_1)$ is again a special Lorentz transformation along the 1-direction. Making use of the addition theorems for hyperbolic functions one finds

$$\mathbf{L}(w\hat{e}_1)\mathbf{L}(v\hat{e}_1) = \begin{pmatrix} \cosh(\lambda+\mu) & \sinh(\lambda+\mu) & 0 & 0 \\ \sinh(\lambda+\mu) & \cosh(\lambda+\mu) & 0 & 0 \\ 0 & 0 & 1 & 0 \\ 0 & 0 & 0 & 1 \end{pmatrix} \equiv \mathbf{L}(u\hat{e}_1) .$$

From this follows the relation

$$e^{\lambda+\mu} = \sqrt{\frac{1+u/c}{1-u/c}} = \sqrt{\frac{(1+v/c)(1+w/c)}{(1-v/c)(1-w/c)}} ,$$

which, in turn, yields the rule for addition of (parallel) velocities, viz.

$$\frac{u}{c} = \frac{v/c+w/c}{1+vw/c^2} . \tag{4.62}$$

This formula has two interesting properties.

(i) If both velocities v and w are small compared to the speed of light, then

$$u = v + w + O(vw/c^2) . \tag{4.63}$$

Thus, (4.62) reduces to the nonrelativistic addition rule, as expected. The first relativistic corrections are of order $1/c^2$.

(ii) As long as v and w are both smaller than c, this holds also for u. If one of them is equal to c, the other one being still smaller than c, or, if both are equal to c, then u is equal to c. In no case does u ever exceed c.

When $\boldsymbol{v}$ and $\boldsymbol{w}$ do not point in the same direction matters become a little more complicated, but the conclusion remains unchanged. As an example, let us consider a boost along the 1-axis, followed by a boost along the 2-axis. This time we choose the form (4.37), or (4.41), noting that the parameters γ and β are related by

$$\gamma_i = 1/\sqrt{1-\beta_i^2} \quad \text{or} \quad \beta_i\gamma_i = \sqrt{\gamma_i^2 - 1} , \quad i = 1, 2 . \tag{4.64}$$

Multiplying the matrices $\mathbf{L}(v_2\hat{e}_2)$ and $\mathbf{L}(v_1\hat{e}_1)$ one finds that

$$\boldsymbol{\Lambda} \equiv \mathbf{L}(v_2\hat{e}_2)\mathbf{L}(v_1\hat{e}_1) = \begin{pmatrix} \gamma_1\gamma_2 & \gamma_1\gamma_2\beta_1 & \gamma_2\beta_2 & 0 \\ \gamma_1\beta_1 & \gamma_1 & 0 & 0 \\ \gamma_1\gamma_2\beta_2 & \gamma_1\gamma_2\beta_1\beta_2 & \gamma_2 & 0 \\ 0 & 0 & 0 & 1 \end{pmatrix} . \tag{4.65}$$

This transformation is neither a boost (because it is not symmetric) nor a pure rotation (because $\Lambda^0{}_0$ is not 1). Being the product of two boosts it is an element of $L_+^\uparrow$. Therefore, it must be a product of the two kinds of transformations, one boost and one rotation. The decomposition theorem (Sect. 4.5.1) in the form of (4.44), when applied to $\boldsymbol{\Lambda}$, gives

$$\Lambda = \mathsf{L}(\boldsymbol{u})\mathcal{R}(\varphi)$$

with $u^i/c = \Lambda^i{}_0/\Lambda^0{}_0 = (\beta_1/\gamma_2, \beta_2, 0)$, while the equations (4.46) for the rotation give, making use of (4.64),

$$R^{11} = R^{22} = (\gamma_1 + \gamma_2)/(1 + \gamma_1\gamma_2)\,, \quad R^{33} = 1\,,$$
$$R^{12} = -R^{21} = -\sqrt{(\gamma_1^2 - 1)(\gamma_2^2 - 1)}/(1 + \gamma_1\gamma_2)\,,\ R^{13} = R^{23} = 0 = R^{31} = R^{32}.$$

Thus, the rotation is about the 3-axis, $\hat{\boldsymbol{\varphi}} = \hat{\boldsymbol{e}}_3$, the angle being

$$\begin{aligned}\varphi &= -\arctan\sqrt{(\gamma_1^2 - 1)(\gamma_2^2 - 1)}/(\gamma_1 + \gamma_2)\\ &= -\arctan\left[\beta_1\beta_2/\left(\sqrt{1-\beta_1^2} + \sqrt{1-\beta_2^2}\right)\right]\,. \end{aligned} \tag{4.66a}$$

For the velocity $\boldsymbol{u}$ one finds

$$\left(\frac{\boldsymbol{u}}{c}\right)^2 = \beta_1^2 + \beta_2^2 - \beta_1^2\beta_2^2 = \frac{\gamma_1^2\gamma_2^2 - 1}{\gamma_1^2\gamma_2^2}\,, \tag{4.66b}$$

so that, indeed, $|\boldsymbol{u}| \le c$. We note that if $\boldsymbol{v}$ and $\boldsymbol{w}$ have arbitrary relative directions the parameter γ pertaining to u is equal to the product of γ_1, γ_2, and $(1+\boldsymbol{v}\cdot\boldsymbol{w}/c^2)$. Whenever both γ_i are larger than 1, then γ is also larger than or equal to 1, and hence the parameter β pertaining to $\boldsymbol{u}$ is smaller than 1. In other words, $|\boldsymbol{u}|$ never exceeds c.

These somewhat complicated relationships simplify considerably when all velocities are small compared to the speed of light. In Sect. 4.4.2 we already checked that the nonrelativistic limit of a special Lorentz transformation $\mathsf{L}(\boldsymbol{v})$ yields precisely the corresponding special Galilei transformation. If in (4.65) both v_1 and v_2 are small compared to c, we obtain

$$\boldsymbol{u} \simeq v_1\hat{\boldsymbol{e}}_1 + v_2\hat{\boldsymbol{e}}_2\,,$$

$$\varphi = -\arctan\left[\frac{v_1 v_2}{c^2}\left(1 + \mathrm{O}\left(\frac{v_i^2}{c^2}\right)\right)\right] \simeq 0\,.$$

The two velocities add like vectors; the rotation about the 3-axis is the identity. The induced rotation in (4.66a) is a purely relativistic phenomenon. Locally, i.e. when expressed infinitesimally, it is due to the commutator (4.59d) that we discussed in Sect. 4.5.2 (iv).

4.7 Galilean and Lorentzian Space–Time Manifolds

While translations (in space and in time) and rotations (in space) are the same within the Galilei and Lorentz groups, the special transformations are different, in an essential way, in the two cases. As a consequence, the space–time manifolds equipped with the Galilei group as the invariance group, or alternatively the Lorentz group, inherit a very different structure. This is what we wish to show in this section.

We start with the example of a special (or boost) transformation with velocity $w = \beta c$ along the 1-axis, understood to be a passive transformation. In the case of the Galilei group it reads, setting $x^0 = ct$,

$$\begin{aligned} x'^0 &= x^0 \,, & x'^2 &= x^2 \,, \\ x'^1 &= x^1 - \beta x^0 \,, & x'^3 &= x^3 \,. \end{aligned} \tag{4.67}$$

(Of course, (4.67) is independent of the speed of light. c is introduced here in view of the comparison with the relativistic case.) In the case of the Lorentz group it reads

$$\begin{aligned} x'^0 &= \gamma[x^0 - \beta x^1] \,, & x'^2 &= x^2 \,, \\ x'^1 &= \gamma[-\beta x^0 + x^1] \,, & x'^3 &= x^3 \,. \end{aligned} \tag{4.68}$$

The coordinates x^μ refer to the inertial system $\mathbf{K}$; the coordinates x'^μ refer to $\mathbf{K}'$, which moves, relative to $\mathbf{K}$, with the velocity $\boldsymbol{w} = \beta c \hat{\boldsymbol{e}}_1$. Suppose we are given three mass points A, B, C, to which no forces are applied and whose coordinates at time $t = 0$ are $\boldsymbol{x}^{(A)} = (0,0,0)$, $\boldsymbol{x}^{(B)} = \boldsymbol{x}^{(C)} = (\Delta,0,0)$, with respect to the system $\mathbf{K}$. A is assumed to be at rest; B moves with the velocity $\boldsymbol{v} = 0.1\, c\hat{\boldsymbol{e}}_1$; C moves with the velocity $\boldsymbol{w} = \beta c \hat{\boldsymbol{e}}_1$ in the same direction as B. We choose $\beta = 1/\sqrt{3} \simeq 0.58$. All three of them move uniformly along straight lines in the (x^1, t)-plane. After time $t = \Delta/c$, for example, they have reached the positions A_1, B_1, C_1, respectively, indicated in Figs. 4.3a and b. If one follows the same motions by placing an observer in the system of reference $\mathbf{K}'$, then in a Lorentz invariant world the picture will be very different from the one in a Galilei invariant world.

(i) According to the nonrelativistic equations (4.67), the positions of the three mass points with respect to $\mathbf{K}'$ and at $t' = 0$ coincide with those with respect to $\mathbf{K}$. After the time $t = \Delta/c$ they have reached the positions A_1', B_1', C_1', respectively, shown in Fig. 4.3a. The figure shows very clearly that time plays a special role, compared to space. Events that are simultaneous with respect to $\mathbf{K}$ are observed by an observer in $\mathbf{K}'$ at the same times, too. As was explained in Sect. 1.14 (ii) it is not possible to compare spatial positions of points at *different* times without knowing the relation (4.67) between the two systems (e.g. comparing A_0 with A_1, $A_0' = A_0$ with A_1'). However, the comparison of positions taken on at *equal* times, is independent of the system of reference one has chosen, and therefore it is physically meaningful. To give an example, if an observer in $\mathbf{K}$ and another observer in $\mathbf{K}'$ measure the spatial positions of A and C at time $t = t' = 0$, as well as at any other time $t = t'$, they will find that A and C move uniformly along straight lines and that the difference of their velocities is $\boldsymbol{w} = \beta c \hat{\boldsymbol{e}}_1$.

(ii) If the two systems of reference are related by the Lorentz transformation (4.68), instead of the Galilei transformation (4.67), the observer in $\mathbf{K}'$ sees the orbits $A_0'A_1'$, $B_0'B_1'$, $C_0'C_1'$ as shown in Fig. 4.3b. This figure leads to two important

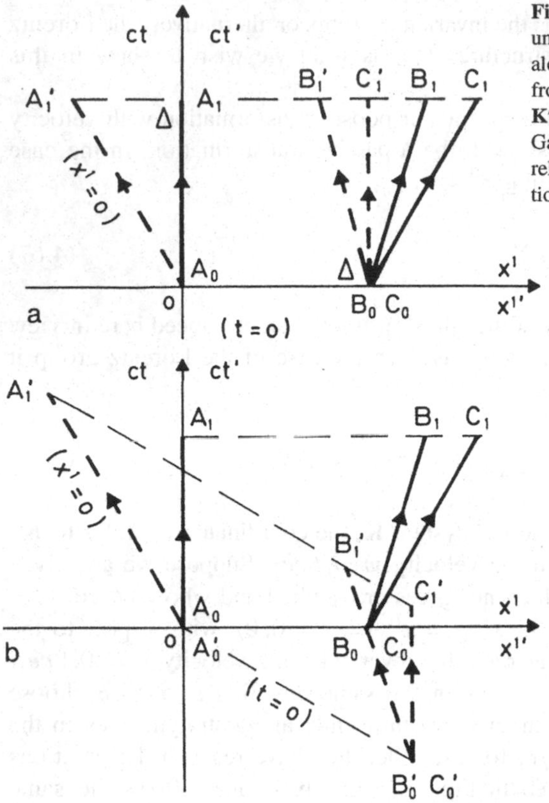

Fig. 4.3. Three mass points moving uniformly, but with different velocities, along straight lines. They are observed from two different inertial systems **K** and **K**′. **(a)** **K** and **K**′ are related by a special Galilei transformation. **(b)** **K** and **K**′ are related by a special Lorentz transformation

observations. Firstly, simultaneity of events is now dependent on the system of reference. The events A_0 and $B_0 = C_0$, which are simultaneous with respect to **K**, lie on the straight line $x'^0 = -\beta x'^1$, when observed from **K**′, and hence occur at different times. (Similarly, the events A_1, B_1 and C_1 are simultaneous with respect to **K**. In **K**′ they fall onto the straight line $x'^0 = -\beta x'^1 + \gamma(1-\beta^2)$.) Secondly, Fig. 4.3b shows a new symmetry between x^0 and x^1, which is not present in the corresponding nonrelativistic figure (4.3a). The images of the lines $t = 0$ and $x^1 = 0$, in **K**′, are symmetric with respect to the bisector of the first quadrant. (As we assigned the coordinates $(\Delta, 0)$ to B_0, $(0, \Delta)$ to A_1, their images B'_0 and A'_1, respectively, have symmetric positions with respect to the same straight line, too.)

More generally, what can we say about the structure of Galilean space–time and of Minkowskian space–time? Both are smooth manifolds with the topology of $\mathbb{R}^4$. The choice of a coordinate system is usually made with regard to the local physical processes one wishes to describe and may be understood as the choice of a "chart" taken from an "atlas" that describes the manifold. (These notions are given precise definitions and interpretations in Chap. 5.)

(i) *Galilei invariant space–time*. In a world where physics is invariant under Galilei transformations, time has an absolute nature: the statement that two events take place at the same time is independent of their spatial distance and of the coordinate system one has chosen. Call P_{G} the (four-dimensional) Galilean space–time; $M = \mathbb{R}_t$ the (one-dimensional) time manifold. Suppose first that we choose an arbitrary coordinate system **K** with respect to which the orbits of physical particles are described by world lines $(t, \boldsymbol{x}(t))$. Consider the projection

$$\pi : P_{\mathrm{G}} \to M : (t, \boldsymbol{x}) \mapsto t \,, \tag{4.69}$$

which assigns its time coordinate t to every point of the world line $(t, \boldsymbol{x}) \in P_{\mathrm{G}}$. Keeping t fixed, the projection π in (4.69) collects all $\boldsymbol{x}$ that are simultaneous. If $\boldsymbol{x}'$ and t' are the images of these $\boldsymbol{x}$ and the fixed t, respectively, under a general Galilei transformation

$$t' = t + s \,, \quad \boldsymbol{x}' = \mathbf{R}\boldsymbol{x} + \boldsymbol{w}t + \boldsymbol{a} \,, \tag{4.70}$$

then the projection defined in (4.69) again collects all simultaneous events,

$$\pi : (t', \boldsymbol{x}') \mapsto t' \,.$$

Thus, the projection has a well-defined meaning, independent of the specific coordinate system one chooses. Consider now an interval I of $\mathbb{R}_t$ that contains the time t. The preimage of I with respect to π has the structure (time interval) $\times$ (three-dimensional affine space),

$$\pi^{-1} : I \to \pi^{-1}(I) \in P_{\mathrm{G}} \,, \quad \text{isomorphic to } I \times \mathbb{E}^3 \,. \tag{4.71}$$

In the terminology of differential geometry this statement and the properties that π has mean that P_{G} is an *affine fibre bundle* over the base manifold $M = \mathbb{R}_t$, with typical fibre $\mathbb{E}^3$. (We do not give the precise definitions here.)

The world line in (4.69) refers to a specific (though arbitrary) observer's system **K**, the observer taking his own position as a point of reference. This corresponds to the statement that one always compares two (or more) physical events in P_{G}. The projection (4.69) asks for events, say A and B, which are simultaneous, i.e. for which $t_A = t_B$. This suggests defining the projection in a truly coordinate-free manner as follows. Let $x_A = (t_A, \boldsymbol{x}_A)$ and $x_B = (t_B, \boldsymbol{x}_B)$ be points of P_{G}. The projection declares all those points to be equivalent, $x_A \sim x_B$, for which $t_A = t_B$.

What else can we say about the structure of P_{G}? If in (4.70) we exlude the special transformations, by taking $\boldsymbol{w} = 0$, then there would exist a canonical projection onto three-dimensional space that would be the same for any choice of the coordinate system. In this case P_{G} would have the global product structure $\mathbb{R}_t \times \mathbb{E}^3$. A fibre bundle that has this global product structure is said to be trivial. However, if we admit the special transformations ($\boldsymbol{w} \neq 0$) in (4.70), then our example discussed earlier and the more general case illustrated by Fig. 4.4 show that the realization of the projection is not independent of the system of reference one has chosen. Although the bundle

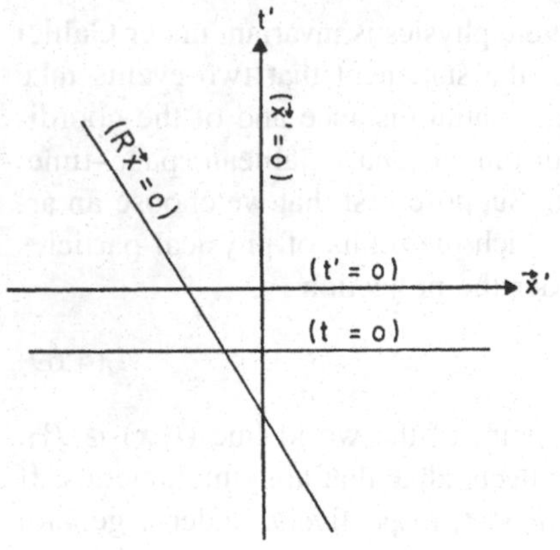

Fig. 4.4. In Galilean space–time P_G, time has an absolute character. However, the projection onto the spatial part of P_G depends on the system of reference one chooses

$$P_G(\pi : P_G \to M = \mathbb{R}_t, \text{Fibre } F = \mathbb{E}^3) \tag{4.72}$$

has the *local* structure $\mathbb{R}_t \times \mathbb{E}^3$, it is not trivial in the sense defined above.
(ii) *Lorentz invariant space–time*. The example given by (4.68) and Fig. 4.3b shows clearly that the space–time endowed with the Lorentz transformations does not have the bundle structure of Galilean space–time. Neither the projection onto the time axis nor that onto three-dimensional space can be defined in a canonical way, i.e. independently of a coordinate system. On the contrary, space and time now appear as truly equivalent, Lorentz transformations mixing space and time in a symmetric way.

Not only *spatial* distances but also *time differences* now depend on the inertial system one chooses. (There is a correlation between spatial and time distances, though, because $(\boldsymbol{x}_{(1)} - \boldsymbol{x}_{(2)})^2 - (x^0_{(1)} - x^0_{(2)})^2$ must be invariant.) As a consequence, moving scales look shorter, while moving clocks tick more slowly. These are new and important phenomena to which we now turn.

4.8 Orbital Curves and Proper Time

The example illustrated by Fig. 4.3b reveals a surprising, and at first somewhat strange, property: a given process of physical motion takes different times, from its beginning to its end, if it is observed from different systems of reference. In order to get rid of this dependence on the system of reference, it is helpful to think of the moving objects A, B, and C as being equipped with their own clocks and, if they are extended objects, with their own measuring scales. This is useful because then we can compare their intrinsic data with the data in other systems of reference. In particular, if the motion is uniform and along a straight line, the comoving systems of reference are inertial and the comparison becomes particularly simple.

In the case of arbitrary, accelerated motion, the best approach is to describe the orbit curve in a geometrical, invariant manner, by means of a Lorentz-invariant orbital parameter. In other words one writes the world line of a mass point in the form $x(\tau)$, where τ is the arc length of this world line. τ is an orbital parameter that is independent of any system of reference. Of course, instead of the dimension *length*, we could give it the dimension *time*, by multiplication with $1/c$. The function $x(\tau)$ describes the spatial and temporal evolution of the motion, in a geometrically invariant way. If τ is given the dimension of time, by multiplication with an appropriate constant with dimensions, one can understand τ to be the time shown by a clock that is taken along in the motion. For this reason, τ is called the *proper time*.

Note, however, that the world line $x(\tau)$ cannot be completely arbitrary. The particle can only move at velocities that do not exceed the speed of light. This is equivalent to the requirement that there must exist a momentary rest system at any point of the orbit. If we choose an arbitrary inertial system of reference, $x(\tau)$ has the representation $x(\tau) = (x^0(\tau), \boldsymbol{x}(\tau))$. Given $x(\tau)$ the velocity vector $\dot{x} = (\dot{x}^0, \dot{\boldsymbol{x}})^T$ is defined by

$$\dot{x}^\mu \stackrel{\text{def}}{=} \frac{\mathrm{d}}{\mathrm{d}\tau} x^\mu(\tau) \,. \tag{4.73}$$

In order to satisfy the requirement stated above, this vector must always be *timelike* (or lightlike), i.e. $(\dot{x}^0)^2 \geq \dot{\boldsymbol{x}}^2$. If this is fulfilled, then the following statement also holds true: if $\dot{x}^0 = \mathrm{d}x^0/\mathrm{d}\tau > 0$ holds in one point of the orbit, then this holds everywhere along the whole orbit. (Figure 4.5 shows an example of a physically possible orbital curve in space–time.) Finally, one can parametrize the orbital parameter τ in such a way that the (invariant) norm of the vector (4.73) always has the value c:

$$\dot{x}^2 \equiv \dot{x}^\mu g_{\mu\nu} \dot{x}^\nu = c^2 \,. \tag{4.74}$$

For a given value of the parameter $\tau = \tau_0$, $\dot{x}$ at the world point $(\tau_0, \boldsymbol{x}(\tau_0))$ can be brought to the form $\dot{x} = (c, 0, 0, 0)$ by means of a Lorentz transformation. Thus,

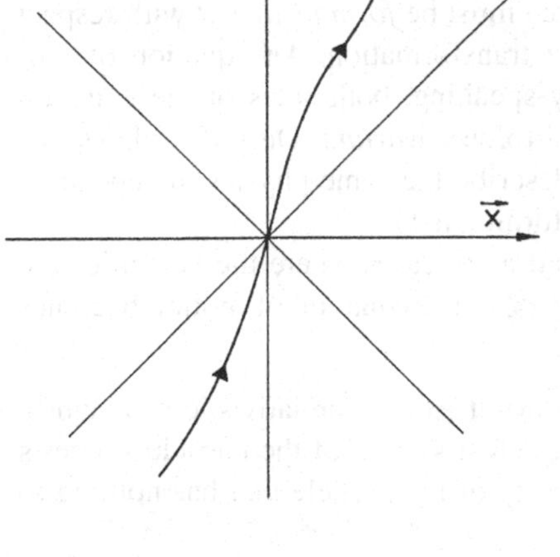

Fig. 4.5. Example of a physically allowed world line. At any point of the orbit the velocity vector is timelike or lightlike (i.e. in the diagram its slope is greater than or equal to 45°)

this transformation leads to the momentary rest system of the particle, and we have

$$\frac{dx^0}{d\tau} = c\,, \quad \text{i.e.} \quad d\tau = \frac{1}{c}dx^0 = dt \quad (\text{at } \tau = \tau_0)\,. \tag{4.75}$$

Note that $\dot{x}$ is precisely the vector ω of (4.42) and that the transformation to the rest system is precisely the one given in (4.43).

The result (4.75) can be interpreted in the following way. If the particle carries a clock along its orbit, this clock measures the proper time τ. Read as a geometrical variable, τ is proportional to the length of arc, $s \stackrel{\text{def}}{=} c\tau$. This is so because the invariant (squared) *line element* ds^2 is given by

$$ds^2 = c^2 d\tau^2 = dx^\mu g_{\mu\nu} dx^\nu = c^2(dt)^2 - (d\boldsymbol{x})^2\,. \tag{4.76}$$

This expression emphasizes again the role of $g_{\mu\nu}$ as the metric tensor.

4.9 Relativistic Dynamics

4.9.1 Newton's Equation

Let **K** be an inertial system of reference and let a particle move with (momentary) velocity $\boldsymbol{v}$ relative to **K**. Further, let $\mathbf{K}_0$ be the rest system of the particle, the axes of $\mathbf{K}_0$ being chosen parallel to those of **K**. The relation between the two systems is then given by the special Lorentz transformation (4.41), with velocity $\boldsymbol{v}$, as indicated in the following:

$$\mathbf{K}_0 \underset{\mathsf{L}(\boldsymbol{v})}{\overset{\mathsf{L}(-\boldsymbol{v})}{\leftrightarrows}} \mathbf{K}\,. \tag{4.77}$$

In trying to generalize Newton's second law (1.8) to relativistic dynamics, we must take care of two conditions.

(i) The postulated relation between the generalized acceleration $d^2x(\tau)/d\tau^2$ and the relativistic analog of the applied force must be *form invariant* with respect to every proper, orthochronous Lorentz transformation. An equation of motion that is form invariant (i.e., loosely speaking, both sides of the equation transform in the same way), is also said to be *covariant*. Only if it obeys this condition will the equation of motion describe the same physics, independent of the reference system in which it is formulated.

(ii) In the rest system of the particle, as well as in cases where the velocities are small compared to the speed of light, $|\boldsymbol{v}| \ll c$, the equation of motion becomes Newton's equation (1.8).

Let m be the mass of the particle as one knows it from nonrelativistic mechanics. The observation that this quantity refers to the rest system of the particle suggests that we should regard it as an intrinsic property of the particle that has nothing to do with its momentary state of motion. For this reason, this quantity is said to be the *rest mass* of the particle. In the case of elementary particles the rest mass is one of the fundamental properties characteristic of the particle. For example, the electron has the rest mass

$$m_{\mathrm{e}} = (9.109\,382\,15 \pm 0.000\,00045) \times 10^{-31}\,\mathrm{kg}\,,$$

while the muon, which otherwise has all the properties of the electron, is characterized by its rest mass being heavier, viz.

$$m_{\mu} \simeq 206.77\, m_{\mathrm{e}}\,.$$

Very much like proper time τ, the rest mass m is a Lorentz scalar. Therefore, with the following form for the generalized equation of motion:

$$m\frac{\mathrm{d}^2}{\mathrm{d}\tau^2}x^{\mu}(\tau) = f^{\mu}\,, \tag{4.78}$$

the left-hand side is a four-vector under Lorentz transformations. Condition (i) states that f^{μ} must be a four-vector as well. If this is so, we can write down the equation of motion (4.78) in the rest system $\mathbf{K}_0$, where we can make use of the second condition (ii). With respect to $\mathbf{K}_0$ and by (4.75), $\mathrm{d}\tau = \mathrm{d}t$. Hence the left-hand side of (4.78) reads

$$m\frac{\mathrm{d}^2}{\mathrm{d}\tau^2}x^{\mu}(\tau)\bigg|_{\mathbf{K}_0} = m\left(\frac{\mathrm{d}}{\mathrm{d}t}c,\ \frac{\mathrm{d}^2}{\mathrm{d}t^2}\boldsymbol{x}\right) = m(0, \ddot{\boldsymbol{x}})\,.$$

Condition (ii) imposes the requirement

$$f^{\mu}|_{\mathbf{K}_0} = (0, \boldsymbol{K})\,,$$

where $\boldsymbol{K}$ is the Newtonian force. We calculate f^{μ} with respect to the inertial system $\mathbf{K}$, as indicated in (4.77):

$$f^{\mu}|_{\mathbf{K}} = \sum_{\nu=0}^{3} L^{\mu}_{\ \nu}(\boldsymbol{v})\, f^{\nu}|_{\mathbf{K}_0}\,. \tag{4.79}$$

Writing this out in space and time components, we have

$$\begin{aligned} \boldsymbol{f} &= \boldsymbol{K} + \frac{\gamma^2}{1+\gamma}\frac{1}{c^2}(\boldsymbol{v}\cdot\boldsymbol{K})\boldsymbol{v}\,, \\ f^0 &= \gamma\frac{1}{c}(\boldsymbol{v}\cdot\boldsymbol{K}) = \frac{1}{c}(\boldsymbol{v}\cdot\boldsymbol{f})\,, \end{aligned} \tag{4.80}$$

where we have used the relationship $\beta^2 = (\gamma^2 - 1)/\gamma^2$. Thus, the covariant force f^{μ} is nothing but the Newtonian force $(0, \boldsymbol{K})$, boosted from the rest system to $\mathbf{K}$.

4.9.2 The Energy–Momentum Vector

The equation of motion (4.78) obtained above suggests defining the following relativistic analog of the momentum $\boldsymbol{p}$:

$$p^\mu \stackrel{\text{def}}{=} m\frac{\mathrm{d}}{\mathrm{d}\tau}x^\mu(\tau) . \tag{4.81}$$

When evaluated in the rest system this takes the form

$$p^\mu|_{\mathbf{K}_0} = (mc, \mathbf{0}) .$$

If it is boosted to the system **K**, as in (4.79), it becomes

$$p^\mu|_{\mathbf{K}} = (\gamma mc, \gamma m\boldsymbol{v}) . \tag{4.82}$$

The same result can be obtained in an alternative way. From (4.76) we see that $\mathrm{d}\tau$ along an orbit is given by

$$\mathrm{d}\tau = \sqrt{(\mathrm{d}t)^2 - (\mathrm{d}\boldsymbol{x})^2/c^2} = \sqrt{1-\beta^2}\mathrm{d}t = \mathrm{d}t/\gamma .$$

Equation (4.81), on the other hand, when evaluated in **K**, gives

$$p^0 = m\gamma\frac{\mathrm{d}}{\mathrm{d}t}(ct) = mc\gamma , \tag{4.82a}$$

$$\boldsymbol{p} = m\gamma\frac{\mathrm{d}}{\mathrm{d}t}\boldsymbol{x} = m\gamma\boldsymbol{v} . \tag{4.82b}$$

The Lorentz scalar parameter m is the rest mass of the particle. It takes over the role of the well-known mass parameter of nonrelativistic mechanics whenever the particle is at rest or moves at small velocities. Note that the nonrelativistic relation $\boldsymbol{p} = m\boldsymbol{v}$ is replaced by (4.82b), i.e. the mass is replaced by the product of the rest mass m and γ. For this reason the product

$$m(\boldsymbol{v}) \stackrel{\text{def}}{=} m\gamma = \frac{1}{\sqrt{1-\boldsymbol{v}^2/c^2}}m$$

is sometimes interpreted as the moving, velocity-dependent mass. It is equal to the rest mass for $\boldsymbol{v} = 0$ but tends to (plus) infinity when $|\boldsymbol{v}|$ approaches the speed of light, c. As stated in Sect. 1.4 it is advisable to avoid this interpretation.

The time component of the four-vector p^μ, when multiplied with c, has the dimension of energy. Therefore, we write

$$\boxed{p^\mu = \left(\frac{1}{c}E, \boldsymbol{p}\right) \quad \text{with} \quad E = \gamma mc^2 , \quad \boldsymbol{p} = \gamma m\boldsymbol{v} .} \tag{4.83}$$

This four-vector is said to be the *energy–momentum vector*. Clearly, its squared norm is invariant under Lorentz transformations. It is found to have the value

$$p^2 \equiv (p, p) = (p^0)^2 - \boldsymbol{p}^2 = \frac{1}{c^2}E^2 - \boldsymbol{p}^2 = m^2c^2 .$$

This last equation yields the important relativistic relationship

$$\boxed{E = \sqrt{\boldsymbol{p}^2c^2 + (mc^2)^2}} \tag{4.84}$$

between the energy E and the momentum $\boldsymbol{p}$ of a free particle. This is the relativistic generalization of the energy–momentum relation we anticipated in (4.7). If $\boldsymbol{p} = 0$, then $E = mc^2$. The quantity mc^2 is called the *rest energy* of the particle with mass m. Thus, E always contains this contribution, even when the momentum vanishes. Consequently, the *kinetic* energy must be defined as follows:

$$T \stackrel{\text{def}}{=} E - mc^2 . \tag{4.85}$$

The first test, of course, is to verify that the well-known relation $T = \boldsymbol{p}^2/2m$ is obtained from (4.85) for small velocities. Indeed, for $\beta \ll 1$,

$$T \simeq \frac{\boldsymbol{p}^2}{2m}\left(1 - \frac{\boldsymbol{p}^2}{4m^2c^2}\right) = T_{\text{nonrel}} - \frac{(\boldsymbol{p}^2)^2}{8m^3c^2} .$$

Clearly, only a complete dynamical theory can answer the questions raised in Sect. 4.1. Nevertheless, the relativistic equation opens up possibilities that were not accessible in nonrelativistic mechanics, and that we wish at least to sketch. Any theory of interactions between particles that is invariant under Lorentz transformations contains the equation (4.84) for free particles. The following consequences can be deduced from this relation between energy and momentum.

(i) Even a particle at rest has energy, $E(\boldsymbol{v} = 0) = mc^2$, proportional to its mass. This is the key to understanding why a massive elementary particle can decay into other particles such that its rest energy is converted, partially or entirely, into *kinetic* energy of the decay products. For example, in the spontaneous decay of a positively charged pion into a positively charged muon and a neutrino,

$$\pi^+(m_\pi = 273.13\, m_e) \to \mu^+(m_\mu = 206.77\, m_e) + \nu(m_\nu \simeq 0) ,$$

about one fourth of its rest mass, namely $((m_\pi - m_\mu)/m_\pi)m_\pi c^2$, is found in the form of kinetic energy of the μ^+ and the ν. This is calculated as follows. Let $(E_q/c, \boldsymbol{q})$, $(E_p/c, \boldsymbol{p})$, and $(E_k/c, \boldsymbol{k})$ denote the four-momenta of the pion, the muon, and the neutrino, respectively. The pion being at rest before the decay (cf. Fig. 4.6) we have

$$q^\mu = \left(\frac{E_q}{c}, \boldsymbol{q}\right) = (m_\pi c, \boldsymbol{0}) , \quad E_p = \sqrt{(m_\mu c^2)^2 + \boldsymbol{p}^2c^2} ; \quad E_k = |\boldsymbol{k}|c .$$

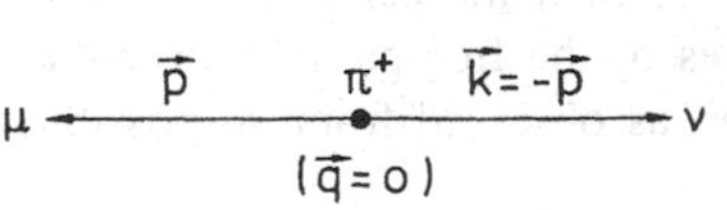

Fig. 4.6. A positively charged pion at rest decays into a (positively charged) muon and an (electrically neutral) neutrino

By conservation of energy and momentum

$$q^{\mu} = p^{\mu} + k^{\mu} , \quad \boldsymbol{k} = -\boldsymbol{p} , \quad \text{and}$$

$$m_{\pi}c^2 = \sqrt{(m_{\mu}c^2)^2 + \boldsymbol{p}^2c^2} + |\boldsymbol{p}|c .$$

This allows us to compute the absolute value of the momentum $\boldsymbol{p}$ or $\boldsymbol{k}$, viz.

$$|\boldsymbol{p}| = |\boldsymbol{k}| = \frac{m_{\pi}^2 - m_{\mu}^2}{2m_{\pi}} c = 58.30\, m_{\mathrm{e}}c .$$

Therefore, the kinetic energy of the neutrino is $T^{(\nu)} = E_{\boldsymbol{k}} = 58.30\, m_{\mathrm{e}}c^2$, while that of the muon is

$$T^{(\mu)} = E_{\boldsymbol{p}} - m_{\mu}c^2 = 8.06\, m_{\mathrm{e}}c^2 .$$

Thus, $T^{(\mu)} + T^{(\nu)} = 66.36\, m_{\mathrm{e}}c^2 \simeq 0.243\, m_{\pi}c^2$, as asserted above. The lion's share of this kinetic energy is carried away by the neutrino, in spite of the fact that muon and neutrino have equal and opposite momenta. On the other hand, the muon shares the major part of the total energy, namely $E_{\boldsymbol{p}} = 214.8\, m_{\mathrm{e}}c^2$, because it is massive.

(ii) In contrast to nonrelativistic mechanics, the transition to vanishing rest mass poses no problems. For $m = 0$ we have $E = |\boldsymbol{p}|c$ and $p^{\mu} = (|\boldsymbol{p}|, \boldsymbol{p})$. A particle without mass nevertheless carries both energy and momentum. Its velocity always has magnitude c, cf. (4.82), no matter how small $\boldsymbol{p}$ is. However, it does not have a rest system. There is no causal way of following the particle and of "catching it up" because the boosts diverge for $|\boldsymbol{v}| \to c$.

We already know an example of massless elementary particles: the photons. Photons correspond to the elementary excitations of the radiation field. As they are massless, one is led to conjecture that the theory of the electromagnetic radiation field cannot be based on nonrelativistic mechanics. Rather, this theory (which is the subject of electrodynamics) must be formulated within a framework that contains the speed of light as a natural limit for velocities. Indeed, Maxwell's theory of electromagnetic phenomena is invariant under Lorentz transformations. Neutrinos some of which have nonvanishing though very small masses, can often be treated as being massless.

We now summarize the findings of this section. The state of a free particle of rest mass m is characterized by the energy–momentum four-vector $p^{\mu} = (E/c, \boldsymbol{p})$, whose norm is invariant and for which

$$p^2 = \frac{1}{c^2}E^2 - \boldsymbol{p}^2 = m^2c^2 .$$

We note that this four-vector is always timelike, or lightlike if $m = 0$. If, as shown in Fig. 4.7, we plot the time component p^0 as the ordinate, the space components (symbolically) as the abscissa, p^{μ} is found to lie on a hyperboloid. As the energy E must be positive, only the upper part of this hyperboloid is relevant. The surface obtained in this way is said to be the *mass shell* of the particle with mass m. It describes either all physically possible states of the free particle, or, alternatively, a fixed state with energy–momentum p^{μ} as observed from all possible inertial systems of references.

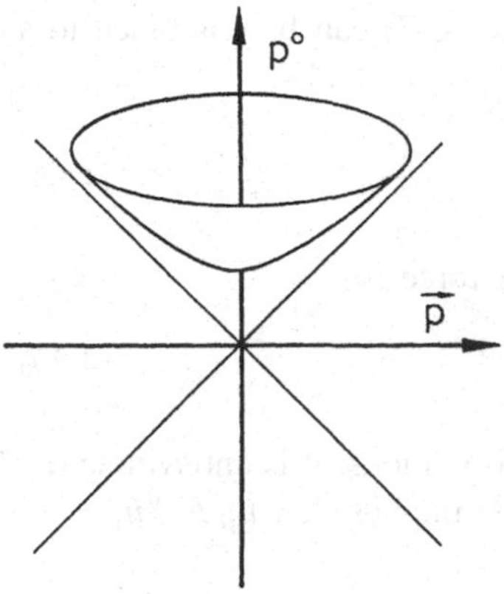

Fig. 4.7. Schematic representation of energy and momentum of a particle. The points ($p^0 = E/c, \boldsymbol{p}$) lie on the upper half of the hyperboloid

4.9.3 The Lorentz Force

A charged particle traversing external electric and magnetic fields at velocity $\boldsymbol{v}$ experiences the Lorentz force (2.29) or (1.49) that we discussed in the context of nonrelativistic dynamics. Here we wish to derive the corresponding relativistic equation of motion (4.78) in a covariant formulation.

With respect to an inertial system of reference, where $\dot{x} = (\gamma c, \gamma \boldsymbol{v})^T$ and $\mathrm{d}/\mathrm{d}\tau = \gamma \mathrm{d}/\mathrm{d}t$, the *spatial* part of (4.78) reads

$$\gamma \frac{\mathrm{d}}{\mathrm{d}t}\boldsymbol{p} = m\gamma \frac{\mathrm{d}}{\mathrm{d}t}(\gamma \boldsymbol{v}) = \gamma e \left(\boldsymbol{E} + \frac{1}{c}\boldsymbol{v} \times \boldsymbol{B} \right) . \tag{4.86}$$

First we show that its *time* component follows from (4.86) and is given by

$$m\gamma \frac{\mathrm{d}}{\mathrm{d}t}(\gamma c) = \gamma \frac{e}{c} \boldsymbol{E} \cdot \boldsymbol{v} . \tag{4.87}$$

This is seen as follows. Calculating the scalar product of (4.86) and $\boldsymbol{v}/c$, its right-hand side becomes $\gamma e/c \boldsymbol{E} \cdot \boldsymbol{v}$. Thus, one obtains

$$\begin{aligned}\gamma \frac{e}{c} \boldsymbol{E} \cdot \boldsymbol{v} &= m\gamma \frac{\boldsymbol{v}}{c} \frac{\mathrm{d}}{\mathrm{d}t}(\gamma \boldsymbol{v}) = mc\gamma \boldsymbol{\beta} \frac{\mathrm{d}}{\mathrm{d}t}(\gamma \boldsymbol{\beta}) \\ &= mc \frac{1}{2} \frac{\mathrm{d}}{\mathrm{d}t}(\gamma \boldsymbol{\beta})^2 = mc \frac{1}{2} \frac{\mathrm{d}}{\mathrm{d}t}(\gamma^2 \beta^2) ,\end{aligned}$$

where we set $\boldsymbol{\beta} = \boldsymbol{v}/c$. As

$$\gamma^2 \beta^2 = \frac{\beta^2}{1 - \beta^2} = \gamma^2 - 1 ,$$

we find

$$\gamma \frac{e}{c} \boldsymbol{E} \cdot \boldsymbol{v} = mc \frac{1}{2} \frac{\mathrm{d}}{\mathrm{d}t} \gamma^2 = mc\gamma \frac{\mathrm{d}}{\mathrm{d}t} \gamma$$

which proves (4.87). Next we show that (4.86) and (4.87) can be combined to a covariant equation of motion, with $u \equiv \dot{x}$:

$$m \frac{\mathrm{d}}{\mathrm{d}\tau} u^\mu = \frac{e}{c} F^{\mu\nu} u_\nu \,. \tag{4.88}$$

This means that the relativistic form of the Lorentz force is

$$K^\mu = \frac{e}{c} F^{\mu\nu} u_\nu \,. \tag{4.89}$$

Here, $F^{\mu\nu}$ is a tensor with respect to Lorentz transformations. It is antisymmetric, $F^{\nu\mu} = -F^{\mu\nu}$, because, with $u_\mu u^\mu = \text{const.}$, (4.88) implies that $u_\mu F^{\mu\nu} u_\nu = 0$. In an arbitrary inertial system it is given by

$$F^{\mu\nu} = \begin{pmatrix} 0 & -E^1 & -E^2 & -E^3 \\ E^1 & 0 & -B^3 & B^2 \\ E^2 & B^3 & 0 & -B^1 \\ E^3 & -B^2 & B^1 & 0 \end{pmatrix} . \tag{4.90}$$

The requirement that $F^{\mu\nu}$ yield the Lorentz force fixes this tensor uniquely. To prove this, we note that u_ν (with a lower index) is $u_\nu = g_{\nu\sigma} u^\sigma = (\gamma c, -\gamma \boldsymbol{v})$ and work out the multiplication on the right-hand side of (4.89). This indeed gives (4.86) and (4.87).

The relativistic Lorentz force has a form that differs from the Newtonian force of Sect. 4.9.1. It is not generated by "boosting" a Newtonian, velocity-independent force but is the result of applying the tensor (4.90) to the velocity u^μ. This tensor, which is antisymmetric, is said to be the *tensor of field strenghts*. Its time–space and space–time components are the components of the electric field,

$$F^{i0} = -F^{0i} = E^i \,, \tag{4.91a}$$

while its space–space components contain the magnetic field according to

$$F^{21} = -F^{12} = B^3 \quad \text{(and cyclic permutations).} \tag{4.91b}$$

The covariant form (4.88) of the equation of motion for a charged particle in electric and magnetic fields shows that these fields cannot be the space components of four-vectors. Instead, they are components of a tensor over Minkowski space M^4, as indicated in (4.90) or (4.91). This means, in particular, that electric and magnetic fields are transformed into each other by special Lorentz transformations. For example, a charged particle that is at rest with respect to an observer generates a static (i.e. time-independent), spherically symmetric electric field. If, on the other hand, the particle and the observer move at a constant velocity $\boldsymbol{v}$ relative to each other, the observer will measure both electric and magnetic fields. (See e.g. Jackson 1998, Sect. 11.10.)

4.10 Time Dilatation and Scale Contraction

Suppose we are given a clock that ticks at regular and fixed time intervals Δt and that we wish to read from different inertial systems. This idea is meaningful because precise measurements of time are done by measuring atomic or molecular frequencies and comparing them with reference frequencies. Such frequencies are internal properties of the atomic or molecular system one is using and do not depend on the state of motion of the system.

For an observer who sees the clock at rest with respect to his inertial system, two consecutive ticks are separated by the space–time interval $\{\mathrm{d}\boldsymbol{x} = 0, \mathrm{d}t = \Delta t\}$. Using this data, he calculates the invariant interval of proper time with the result

$$\mathrm{d}\tau = \sqrt{(\mathrm{d}t)^2 - (\mathrm{d}\boldsymbol{x})^2/c^2} = \Delta t \ .$$

Another observer who moves with constant velocity relative to the first observer, and therefore also relative to the clock, sees that consecutive ticks are separated by the space–time interval $\{\Delta t', \Delta \boldsymbol{x}' = \boldsymbol{v}\Delta t'\}$. From his data he calculates the invariant interval of proper time to be

$$\mathrm{d}\tau' = \sqrt{(\Delta t')^2 - (\Delta \boldsymbol{x}')^2/c^2} = \sqrt{1-\beta^2}\,\Delta t' \ .$$

As proper time is Lorentz invariant, we have $\mathrm{d}\tau' = \mathrm{d}\tau$. This means that the second observer (for whom the clock is in motion) sees the clock tick with a longer period, given by

$$\Delta t' = \frac{\Delta t}{\sqrt{1-\beta^2}} = \gamma \Delta t \ . \qquad (4.92)$$

This is the important phenomenon of *time dilatation*: for an observer who sees the clock in motion, it is slower than at rest, i.e. it ticks at time intervals that are dilated by a factor γ. In Example (iii) of Sect. 4.1 we discussed a situation where time dilatation was actually observed. The experiment quoted there confirms the effect predicted in (4.92) with the following accuracy: the difference $\Delta t - \Delta t'/\gamma$ is zero within the experimental error bar

$$\frac{\tau^{(0)}(\mu) - \tau^{(v)}(\mu)/\gamma}{\tau^{(0)}(\mu)} = (0.2 \pm 0.9) \times 10^{-3} \ .$$

Another, closely related effect of special Lorentz transformations is the *scale or Fitzgerald–Lorentz contraction* that we now discuss. It is somewhat more difficult to describe than time dilatation because the determination of the length of a scale requires, strictly speaking, the measurement of two space points at the same time. As such points are separated by a spacelike distance, this cannot be a causal, and hence physical, measurement. A way out of this problem would be to let two scales of equal length move towards each other and to compare their positions at the moment they overlap. Alternatively, we may use the following simple argument.

Suppose there are two landmarks at the space points

$$\boldsymbol{x}^{(A)} = (0,0,0) \quad \text{and} \quad \boldsymbol{x}^{(B)} = (L_0,0,0) \ ,$$

where the coordinates refer to the inertial system $\mathbf{K}_0$. Because we want to measure their spatial distance, we ask an observer to make a journey from A to B, as shown in Fig. 4.8, with constant velocity $\boldsymbol{v} = (v,0,0)$, and, of course, $v < c$. As seen from $\mathbf{K}_0$ he departs from A at time $t = 0$ and reaches the landmark B at time $t = T_0$, B having moved to C during this time in our space–time diagram (Fig. 4.8). In the case of Galilei transformations, i.e. for nonrelativistic motion, we would conclude that the distance is

$$L_0 \equiv |\boldsymbol{x}^{(B)} - \boldsymbol{x}^{(A)}| = vT_0 \ .$$

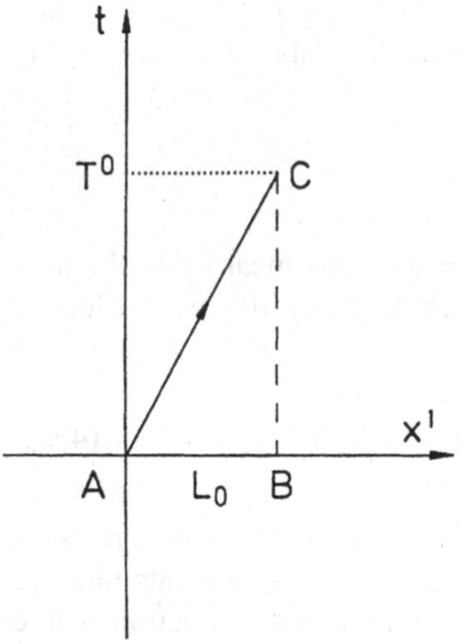

Fig. 4.8. An observer traveling at constant velocity determines the distance from A to B. He finds $L = L_0/\gamma$

In the relativistic, Lorentz-invariant world we find a different result. When the traveler reaches C, his own, comoving clock shows the time $T = T_0/\gamma$, with $\gamma = (1 - v^2/c^2)^{-1/2}$. Thus, he concludes that the length separating A and B is

$$L = vT = vT_0/\gamma = L_0/\gamma \ . \tag{4.93}$$

In other words, the scale AB that moves relative to the traveling observer (with velocity $-\boldsymbol{v}$) appears to him contracted by the factor $1/\gamma$. This is the phenomenon of *scale contraction*, or *Fitzgerald–Lorentz contraction*.

One easily understands that scales oriented along the 2-axis or the 3-axis, or any other direction in the (2,3)-plane, remain unchanged and do not appear contracted. Therefore, the phenomenon of scale contraction means, more precisely, that an extended body that moves relative to an inertial system appears contracted in the direction of its velocity $\boldsymbol{v}$ only. The spatial dimensions perpendicular to $\boldsymbol{v}$ remain unmodified.

The book by Ellis and Williams (1994) contains an elementary but well illustrated discussion of time dilatation and scale contraction as well as the apparent paradoxes of special relativity. Although it was written for laymen, as Ruth Williams told me, it seems to me that this book is not only entertaining but also useful for the reader who wishes to get a better feeling for time and space in special relativity.

4.11 More About the Motion of Free Particles

By definition, the state of motion of a free particle is characterized by its relativistic energy–momentum vector (4.83) being on its mass shell,

$$p^2 = E^2/c^2 - \boldsymbol{p}^2 = m^2c^2 \, . \tag{4.94}$$

We wish to describe this relativistic motion without external forces by means of the methods of canonical mechanics. As we are dealing with free motion in a flat space, the solutions of Hamilton's variational principle will be just straight lines in the space–time continuum. Therefore, we assume the action integral (2.27) to be given by the path integral between two points A and B in space–time, where A and B are timelike relative to each other:

$$I[x] = K \int_A^B \mathrm{d}s \, , \quad \text{with} \quad (x^{(B)} - x^{(A)})^2 > 0 \, . \tag{4.95}$$

As we showed in Sect. 2.36, the action integral is closely related to the generating function S^*, which satisfies the equation of Hamilton and Jacobi. Assuming the solutions to be inserted in (4.95), we have

$$S^* = K \int_A^B \mathrm{d}s \, . \tag{4.96}$$

Here the quantity K is a constant whose dimension is easy to determine: the action has the dimension (energy $\times$ time) and s has the dimension (length). Therefore, K must have the dimension (energy/velocity), or, equivalently, (mass $\times$ velocity). On the other hand, I or S^* must be Lorentz invariant. The only invariant parameters, but those with dimension, are the rest mass of the particle and the velocity of light. Thus, up to a sign, K is the product mc. In fact, as we show below, the correct choice is $K = -mc$.

With respect to an arbitrary, but fixed, inertial system we have $\mathrm{d}s = c\,\mathrm{d}\tau = \sqrt{1 - \boldsymbol{v}^2/c^2}\,c\,\mathrm{d}t$, with $\boldsymbol{v} = \mathrm{d}\boldsymbol{x}/\mathrm{d}t$. Thus,

$$I = -mc^2 \int_{t^{(A)}}^{t^{(B)}} \sqrt{1 - \boldsymbol{v}^2/c^2}\mathrm{d}t \equiv \int_{t^{(A)}}^{t^{(B)}} L\,\mathrm{d}t \, .$$

This yields the (natural form of) the Lagrangian function whose Euler–Lagrange equations describe relativistic free motion. Expanding this Lagrangian function in terms of v/c, we find the expected nonrelativistic form

$$L = -mc^2\sqrt{1 - v^2/c^2} \simeq -mc^2 + \tfrac{1}{2}mv^2 \,, \tag{4.97}$$

to which the term $-mc^2$ is added. The form (4.97) for the Lagrangian function is not quite satisfactory because it refers to a fixed inertial system and therefore is not manifestly invariant. The reason for this is that we introduced a time coordinate. The time variable, being the time component of a four-vector, is not invariant. If instead we introduce some other, Lorentz-*invariant* parameter τ (we give it the dimension of time), then (4.95) reads

$$I = -mc \int_{t(A)}^{t(B)} \mathrm{d}\tau \sqrt{\frac{\mathrm{d}x^\alpha}{\mathrm{d}\tau}\frac{\mathrm{d}x_\alpha}{\mathrm{d}\tau}} \,, \tag{4.98}$$

so that the invariant Lagrangian function reads

$$L_{\text{inv}} = -mc\sqrt{\frac{\mathrm{d}x^\alpha}{\mathrm{d}\tau}\frac{\mathrm{d}x_\alpha}{\mathrm{d}\tau}} = -mc\sqrt{\dot{x}^2} \,, \tag{4.99}$$

where $\dot{x}^\alpha = \mathrm{d}x^\alpha/\mathrm{d}\tau$. One realizes again the $\dot{x}^2$ must be positive, i.e. that $\dot{x}$ must be timelike. The Euler–Lagrange equations that follow from the action (4.98) are

$$\frac{\partial L_{\text{inv}}}{\partial x^\alpha} - \frac{\mathrm{d}}{\mathrm{d}\tau}\frac{\partial L_{\text{inv}}}{\partial \dot{x}^\alpha} = 0 \,,$$

and hence

$$\frac{\mathrm{d}}{\mathrm{d}\tau}\frac{mc\dot{x}_\alpha}{\sqrt{\dot{x}^2}} = 0 \,.$$

Here the momentum canonically conjugate to x^α is

$$p_\alpha = \frac{\partial L_{\text{inv}}}{\partial \dot{x}^\alpha} = -mc\frac{\dot{x}_\alpha}{\sqrt{\dot{x}^2}} \,. \tag{4.100}$$

It satisfies the constraint

$$p^2 - m^2c^2 = 0 \,. \tag{4.101}$$

If we now attempt to construct the Hamiltonian function, following the rules of Chap. 2, we find that

$$H = \dot{x}^\alpha p_\alpha - L_{\text{inv}} = mc\left[-\dot{x}^2/\sqrt{\dot{x}^2} + \sqrt{\dot{x}^2}\right] = 0 \,.$$

The essential reason the Hamiltonian function vanishes is that the description of the motion as given here contains a redundant degree of freedom, namely the time coordinate of $\dot{x}$. The dynamics is contained in the constraint (4.101). One also realizes that the Legendre transformation from L_{inv} to H cannot be performed: the condition for this transformation to exist,

$$\det\left(\frac{\partial^2 L_{\text{inv}}}{\partial \dot{x}^\beta \partial \dot{x}^\alpha}\right) \neq 0 \, ,$$

is not fulfilled. Indeed, calculating the matrix of second derivatives, one obtains

$$\frac{\partial^2 L_{\text{inv}}}{\partial \dot{x}^\beta \partial \dot{x}^\alpha} = -\frac{mc}{(\dot{x}^2)^{3/2}}[\dot{x}^2 g_{\alpha\beta} - \dot{x}_\alpha \dot{x}_\beta] \, .$$

The following argument shows that the determinant of this matrix vanishes. Define

$$A_{\alpha\beta} \stackrel{\text{def}}{=} \dot{x}^2 g_{\alpha\beta} - \dot{x}_\alpha \dot{x}_\beta \, .$$

The homogeneous system of linear equations $A_{\alpha\beta}u^\beta = 0$ has a nontrivial solution precisely if $\det \mathbf{A} = 0$. Therefore, if we can find a nonvanishing $u^\beta \neq (0, 0, 0, 0)$ that is solution of this system, then the determinant of $\mathbf{A}$ vanishes. There is indeed such a solution, namely $u^\beta = c\dot{x}^\beta$, because for any $\dot{x}^\beta \neq 0$

$$A_{\alpha\beta}\dot{x}^\beta = \dot{x}^2 \dot{x}_\alpha - \dot{x}^2 \dot{x}_\alpha = 0 \, .$$

For the first time we meet here a Lagrangian system that is not equivalent to a Hamiltonian system, in a canonical way. In fact, this is an example for a Lagrangian (or Hamiltonian) system with constraints whose analysis must be discussed separately.

In the example discussed above, one could proceed as follows. At first one ignores the constraint (4.101) but introduces it into the Hamiltonian function by means of a so-called Lagrangian multiplier. With H as given above, we take

$$H' = H + \lambda \Psi(p) \, , \quad \text{with} \quad \Psi(p) \stackrel{\text{def}}{=} p^2 - m^2c^2 \, ;$$

λ denoting the multiplier. The coordinates and momenta satisfy the canonical Poisson brackets

$$\{x^\alpha, x_\beta\} = 0 = \{p^\alpha, p_\beta\} \, ; \quad \{p^\alpha, x_\beta\} = \delta^\alpha_\beta \, .$$

The canonical equations read

$$\begin{aligned}
\dot{x}^\alpha &= \{H', x^\alpha\} = \{\lambda, x^\alpha\}\Psi(p) + \lambda\{\Psi(p), x^\alpha\} \\
&= \lambda\{p^2 - m^2c^2, x^\alpha\} = \lambda\{p^2, x^\alpha\} = 2\lambda p^\alpha \, , \\
\dot{p}^\alpha &= \{H', p^\alpha\} = \{\lambda, p^\alpha\}\Psi(p) = 0 \, ,
\end{aligned}$$

where we made use of the constraint $\Psi(p) = 0$. With p^α and $\dot{x}^\alpha$ being related by (4.100), we deduce $\lambda = -\sqrt{\dot{x}^2}/2mc$. The equation of the motion is the same as above, $\dot{p}_\alpha = 0$.

4.12 The Conformal Group

In Sect. 4.3 we argued that the laws of nature that apply to massive particles always involve quantities with dimensions and therefore cannot be scale invariant. As a consequence, the transformation law (4.23) must hold with condition (4.24) and the choice $\alpha = 1$. In a world in which there is only radiation, this restriction does not apply because radiation fields are mediated by massless particles (quanta). Therefore, it is interesting to ask about the most general transformations that guarantee the invariance of the form

$$z^2 = 0 \quad \text{with} \quad z = x_A - x_B \quad \text{and} \quad x_A, x_B \in M^4 \,.$$

The Poincaré transformations that we had constructed for the case $\alpha = 1$ certainly belong to this class. As we learnt in Sect. 4.4, the Poincaré transformations form a group that has 10 parameters. If only the invariance of $z^2 = 0$ is required, then there are two more classes of transformations. These are the dilatations

$$x'^{\mu} = \lambda x^{\mu} \quad \text{with} \quad \lambda \in \mathbb{R} \,,$$

which depend on one parameter and which form a subgroup by themselves. Obviously, they are linear.

One can show that there is still another class of (nonlinear) transformations that leave the light cone invariant. They read (see Exercise 4.15)

$$x^{\mu} \to x'^{\mu} = \frac{x^{\mu} + x^2 c^{\mu}}{1 + 2(c \cdot x) + c^2 x^2} \,. \tag{4.102}$$

They depend on four real parameters and are said to be *special conformal transformations*. They form a subgroup, too: the unit is given by $c^{\mu} = 0$; the composition of two transformations of the type (4.102) is again of the same type, because

$$x'^{\mu} = \frac{x^{\mu} + x^2 c^{\mu}}{\sigma(c, x)} \quad \text{with} \quad \sigma(c, x) \stackrel{\text{def}}{=} 1 + 2(c \cdot x) + c^2 x^2 \,,$$

$$x'^2 = \frac{x^2}{\sigma(c, x)} \,,$$

and

$$x''^{\mu} = \frac{x'^{\mu} + x'^2 d^{\mu}}{\sigma(d, x')} = \frac{x^{\mu} + x^2 (c^{\mu} + d^{\mu})}{\sigma(c + d, x)} \,.$$

Finally, the inverse of (4.102) is given by the choice $d^{\mu} = -c^{\mu}$. Thus, one discovers the *conformal group* over Minkowski space M^4. This group has

$$10 + 1 + 4 = 15$$

parameters. It plays an important role in field theories that do not contain any massive particle.

5. Geometric Aspects of Mechanics

In many respects, mechanics carries geometrical structures. This could be felt very clearly at various places in the first four chapters. The most important examples are the structures of the space–time continua that support the dynamics of nonrelativistic and relativistic mechanics, respectively. The formulation of Lagrangian mechanics over the space of generalized coordinates and their time derivatives, as well as of Hamilton–Jacobi canonical mechanics over the phase space, reveals strong geometrical features of these manifolds. (Recall, for instance, the symplectic structure of phase space and Liouville's theorem.) To what extent mechanics is of geometric nature is illustrated by the fact that, historically, it gave important impulses to the development of differential geometry. In turn, the modern formulation of differential geometry and of some related mathematical disciplines provided the necessary tools for the treatment of problems in qualitative mechanics that are the topic of present-day research. This provides another impressive example of cross-fertilization of pure mathematics and theoretical physics.

In this chapter we show that canonical mechanics quite naturally leads to a description in terms of differential geometric notions. We develop some of the elements of differential geometry and formulate mechanics by means of this language. For lack of space, however, this chapter cannot cover all aspects of the mathematical foundations of mechanics. Instead, it offers an introduction with the primary aim of motivating the necessity of the geometric language and of developing the elements up to a point from where the transition to the mathematical literature on mechanics (see the list of references) should be relatively smooth. This may help to reduce the disparity between texts written in a more physics-oriented language and the modern mathematical literature and thus to encourage the beginner who has to bridge the gap between the two. At the same time this provides a starting point for catching up with recent research developments in modern mechanics.

As a final remark, we note that studying the geometric structure of mechanics, over the last decades, has become important far beyond this discipline. Indeed, we know today that all fundamental interactions of nature carry strong geometric features. Once again, mechanics is the door to, and basis of, all of theoretical physics. In studying these geometric aspects of the fundamental interactions, we will, at times, turn back to mechanics where many of the essential building blocks are developed in a concrete and well understood framework.

F. Scheck, *Mechanics*, Graduate Texts in Physics, 5th ed.,
DOI 10.1007/978-3-642-05370-2_5,

5.1 Manifolds of Generalized Coordinates

In Sect. 2.11 we showed that every diffeomorphic mapping of coordinates $\{\underset{\sim}{q}\}$ onto new coordinates $\{\underset{\sim}{q}'\}$

$$G : \{\underset{\sim}{q}\} \mapsto \{\underset{\sim}{q}'\} : q_i = g_i(\underset{\sim}{q}', t), \dot{\underset{\sim}{q}} = \sum_{k=1}^{f} \frac{\partial g_i}{\partial q'_k} \dot{q}'_k + \frac{\partial g_i}{\partial t} \tag{5.1}$$

leaves the equations of motion form invariant. This means, except for purely practical aspects, any choice of a set of generalized coordinates $\{\underset{\sim}{q}\}$ is as good as any other that is related to the first in a one-to-one and differentiable manner. The physical system one wishes to describe is independent of the specific choice one makes, or, more loosely speaking, "the physics is the same", no matter which coordinates one employs. It is obvious that the transformation must be *uniquely invertible*, or one-to-one, as one should not loose information in either direction. The number of independent degrees of freedom must be the same. Similarly, it is meaningful to require the mapping to be *differentiable* because we do not want to destroy or to change the differential structure of the equations of motion.

Any such choice of coordinates provides a possible, specific realization of the mechanical system. Of course, from a practical point of view, there are appropriate and inappropriate choices, in the sense that the coordinates may be optimally adapted to the problem because they contain as many cyclic coordinates as possible, or, on the contrary, may be such that they inhibit the solution of the equations of motion. This comment concerns the actual *solution* of the equations of motion but not the *structure* of the coordinate manifold into which the mechanical system is embedded.

In mechanics a set of f generalized coordinates arises by constraining an initial set of degrees of freedom by a number of independent, holonomic constraints. For instance, the coordinates of a system of N particles that are initially elements of an $\mathbb{R}^{3N}$ are constrained by $\Lambda = 3N - f$ equations, in such a way that the f independent, generalized coordinates, in general, are not elements of an $\mathbb{R}^f$. Let us recall two examples for the sake of illustration.

(i) The plane mathematical pendulum that we studied in Sects. 1.17.2 and 2.30, Example (ii) has one degree of freedom. The natural choice for a generalized coordinate is the angle measuring the deviation from the vertical, $q \equiv \varphi$. As this coordinate takes values in the interval $[-\pi, +\pi]$, with $q = \pi$ and $q = -\pi$ to be identified, it is an element of the unit circle S^1. The coordinate manifold is the S^1, independent of how we choose q. (For instance, if we choose the arc $q = s = l\varphi$, s is defined on the circle with radius l. This circle is topologically equivalent to S^1.)

(ii) The coordinate manifold of the rigid body (Chap. 3) provides another example. Three of the six generalized coordinates describe the unconstrained motion of the center of mass and are therefore elements of a space $\mathbb{R}^3$. The remaining three describe the spatial orientation of the top with respect to a system of reference whose axes have fixed directions in space. They are angles and belong to the manifold of the rotation group SO(3). As we learnt earlier, this manifold can be parametrized in different ways: for instance, by the direction about which the rotation takes place and by the angle of rotation $(\hat{\boldsymbol{n}}, \varphi)$, or, alternatively, by three Eulerian angles $(\theta_1, \theta_2, \theta_3)$, using one or the other of the definitions given in Sects. 3.9 and 3.10. We shall analyze the structure of this manifold in more detail below, in Sect. 5.2.3. Already at this point it seems plausible that it will turn out to be rather different from a three-dimensional Euclidean space and that we shall need further tools of geometry for its description.

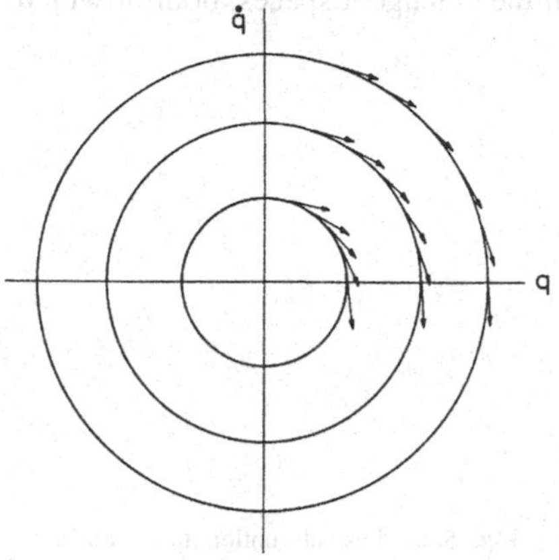

Fig. 5.1. Velocity field in the space of coordinates and their time derivatives for the one-dimensional harmonic oscillator

Actual *solutions* of the equations of motion $\underline{q}(t, t_0, \underline{q}_0) = \Phi_{t,t_0}(\underline{q}_0)$ (cf. Sect. 1.20) are curves in the manifold Q of coordinates. In this sense Q is the physical space that carries the real motion. However, in order to set up the equations of motion and to construct their solutions, we also need the time derivatives $\mathrm{d}\underline{q}/\mathrm{d}t \equiv \dot{\underline{q}}$ of the coordinates as well als Lagrangian functions $L(\underline{q}, \dot{\underline{q}}, t)$ over the space M of the $\underline{q}$ and the $\dot{\underline{q}}$. The Lagrangian function is to be inserted into the action integral $I[\underline{q}]$, functional of $\underline{q}(t)$, from which differential equations of second order in time follow via Hamilton's variational principle (or some other extremum principle). For example, for $f = 1$ the physical solutions can be constructed piecewise if one knows the velocity field. Figure 5.1 shows the example of the harmonic oscillator and its velocity field (cf. also Sect. 1.17.1). More generally, this means that we shall have to study vector fields over M, and hence the tangent spaces $T_x Q$ of the manifold Q, for all elements x of Q.

A similar remark applies to the case where, instead of the variables $(\underline{q}, \dot{\underline{q}})$, we wish to make use of the phase-space variables $(\underline{q}, \underline{p})$. We recall that p was defined to be the partial derivative of the Lagrangian function by $\dot{q}$,

$$p_i \stackrel{\text{def}}{=} \frac{\partial L}{\partial \dot{q}^i} \,. \tag{5.2}$$

L being a (scalar) function on the space of the $\underline{q}$ and the $\dot{\underline{q}}$ (it maps this space onto the real number), i.e. on the union of tangent spaces $T_x Q$, definition (5.2) leads to the corresponding *dual* spaces $T_x^* Q$, the so-called *cotangent spaces*.

These remarks suggest detaching the mechanical system one is considering from a specific choice of generalized coordinates $\{q\}$ and to choose a more abstract formulation by defining and describing the manifold Q of physical motions in a *coordinate-free* language. The choice of sets of coordinates $\{q\}$ or $\{q'\}$ is equivalent to describing Q in terms of *local coordinates*, or, as one also says, in terms of *charts*. Furthermore, one is led to study various geometric objects living on the manifold Q, as well as on its tangent spaces $T_x Q$ and cotangent spaces $T_x^* Q$. Examples are Lagrangian functions that are defined on the tangent spaces and Hamiltonian functions that are defined on the cotangent spaces, both of which give real numbers.

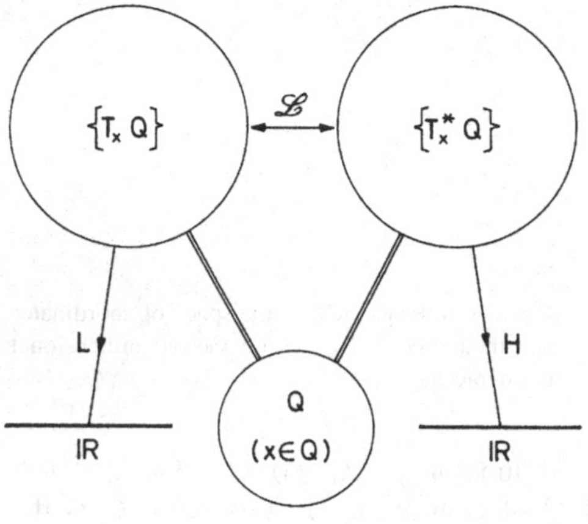

Fig. 5.2. Physical motion takes place in the coordinate manifold Q. The Lagrangian function and the Hamiltonian function are defined on the tangent and cotangent spaces, respectively

Figure 5.2 shows a first sketch of these interrelationships. As we shall learn below, there are many more geometric objects on manifolds other than functions (which are mappings to the reals). An example that we met earlier is vector fields such as the velocity field of a flow in phase space. In order to awake the reader's curiosity, we just remind him of the Poisson brackets, defined on $T_x^* Q$, and of the volume form that appears in Liouville's theorem.

An example of a smooth manifold, well known from linear algebra and from analysis, is provided by the n-dimensional Euclidean space $\mathbb{R}^n$. However, Euclidean spaces are not sufficient to describe general and nontrivial mechanical systems, as is demonstrated by the examples of the coordinate manifolds of the plane pendulum and of the rigid body. As we shall see, the union of all tangent spaces

$$TQ \stackrel{\text{def}}{=} \{T_x Q | x \in Q\} \tag{5.3}$$

and the union of all cotangent spaces

$$T^*Q \stackrel{\text{def}}{=} \{T_x^* Q | x \in Q\} \tag{5.4}$$

are smooth manifolds. The former is said to be the *tangent bundle*, the latter the *cotangent bundle*. Suppose then that we are given a conservative mechanical system, or a system with some symmetry. The set of all solutions lie on hypersurfaces in $2f$-dimensional space that belong to fixed values of the energy or are characterized by the conserved quantities pertaining to the symmetry. In general, these hypersurfaces are smooth manifolds, too, but cannot always be embedded in $\mathbb{R}^{2f}$. Thus, we must learn to describe such physical manifolds by mapping them, at least locally, onto Euclidean spaces of the same dimension. Or, when expressed in a more pictorial way, whatever happens on the manifold M is projected onto a set of charts, each of which represents a local neighborhood of M. If one knows how to join neighboring charts and if one has at one's disposal a complete set of charts, then one obtains a true image of the whole manifold, however complicated it may look globally.

The following sections (5.2–4) serve to define and discuss the notions sketched above and to illustrate them by means of a number of examples. From Sect. 5.5 on we return to mechanics by formulating it in terms of a geometric language, preparing the ground for new insights and results. In what follows (Sects. 5.2–5.4) we shall use the following notation:

Q denotes the manifold of generalized coordinates; its dimension is equal to f, the number of degrees of freedom of the mechanical system one is considering.

M denotes a general smooth (and finite dimensional) manifold of dimension $\dim M = n$.

5.2 Differentiable Manifolds

5.2.1 The Euclidean Space $\mathbb{R}^n$

The definition of a differentiable manifold relates directly to our knowledge of the n-dimensional Euclidean space $\mathbb{R}^n$. This space is a *topological* space. This means that it can be covered by means of a set of open neighborhoods that fulfills some quite natural conditions. For any two distinct points of $\mathbb{R}^n$ one can define neighborhoods of these points that do not overlap: one says the $\mathbb{R}^n$ is a *Hausdorff space*. Furthermore, one can always find a collection B of open sets such that every open subset of $\mathbb{R}^n$ is represented as the union of elements of B. Such a collection B is said to be a basis. It is even possible to construct a countable set of neighborhoods $\{U_i\}$ of any point p of $\mathbb{R}^n$ such that for *any* neighborhood U of p there is an i for which U_i is contained in U. These $\{U_i\}$ can also be made a basis, in the sense defined above: thus the space $\mathbb{R}^n$ certainly has a *countable basis*. All this is summarized by the statement that $\mathbb{R}^n$ is a topological, Hausdorff space with a countable basis.

It is precisely these requirements that are incorporated in the definition of a manifold. Even if they look somewhat complicated at first sight, these properties are very natural in all important branches of mechanics. Therefore, as a physicist one has a tendency to take them for granted and to assume tacitly that the spaces and manifolds of mechanics have these properties. The reader who wishes to define matters very precisely from the start is consequently advised to consult, for example, the mathematical literature quoted in the Appendix and to study the elements of topology and set theory.

The space $\mathbb{R}^n$ has more structure than that. It is an n-dimensional real vector space on which there exists a natural inner product and hence a norm. If $p = (p_1, p_2, \ldots, p_n)$ and $q = (q_1, q_2, \ldots, q_n)$ are two elements of $\mathbb{R}^n$, the inner product and the norm are defined by

$$p \cdot q \stackrel{\text{def}}{=} \sum_{i=1}^{n} p_i q_i \quad \text{and} \quad |p| \stackrel{\text{def}}{=} \sqrt{p \cdot p}\,, \tag{5.5}$$

respectively. Thus $\mathbb{R}^n$ is a metric space. The distance function

$$d(p, q) \stackrel{\text{def}}{=} |p - q| \tag{5.6}$$

following from (5.5) has all properties that a metric should have: it is nondegenerate, i.e. $d(p, q)$ vanishes if and only if $p = q$; it is symmetric $d(q, p) = d(p, q)$; and it obeys Schwarz' inequality

$$d(p, r) \le d(p, q) + d(q, r)\,.$$

Finally, we know that on $\mathbb{R}^n$ one can define *smooth* functions,

$$f : U \subset \mathbb{R}^n \to \mathbb{R}\,,$$

which map open subsets U of $\mathbb{R}^n$ onto the real numbers. The smoothness, or C^∞ property, of a function f means that at every point $u \in U$ all mixed partial derivatives of f exist and are continuous. As an example consider the function f^i which associates to every element $p \in \mathbb{R}^n$ its ith coordinate p_i, as shown in Fig. 5.3,

$$f^i : \mathbb{R}^n \to \mathbb{R} : p = (p_1, \ldots, p_i, \ldots, p_n) \mapsto p_i\,, \quad i = 1, 2, \ldots, n\,. \tag{5.7}$$

These functions $f^i(p) = p_i$ are said to be the *natural coordinate functions* of $\mathbb{R}^n$.

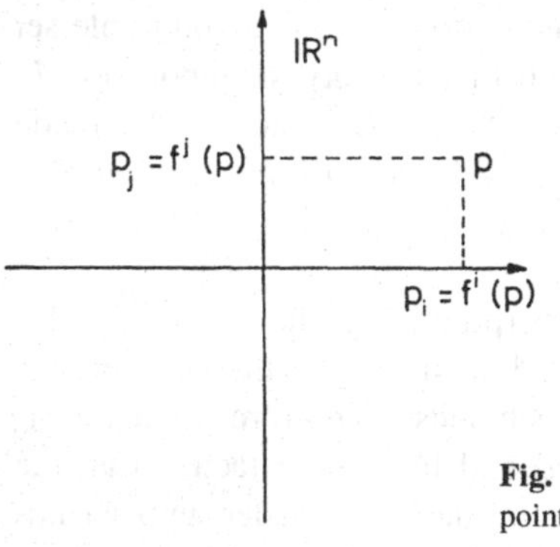

Fig. 5.3. The coordinate functions f^i and f^j assign to each point p of $\mathbb{R}^n$ its coordinates p^i and p^j, respectively

5.2.2 Smooth or Differentiable Manifolds

Physical manifolds like the ones we sketched in Sect. 5.1 often are not Euclidean spaces but topological spaces (Hausdorff with countable basis) that carry differentiable structures. Qualitatively speaking, they resemble Euclidean spaces *locally*, i.e. open subsets of them can be mapped onto Euclidean spaces and these "patches" can be joined like the charts of an atlas.

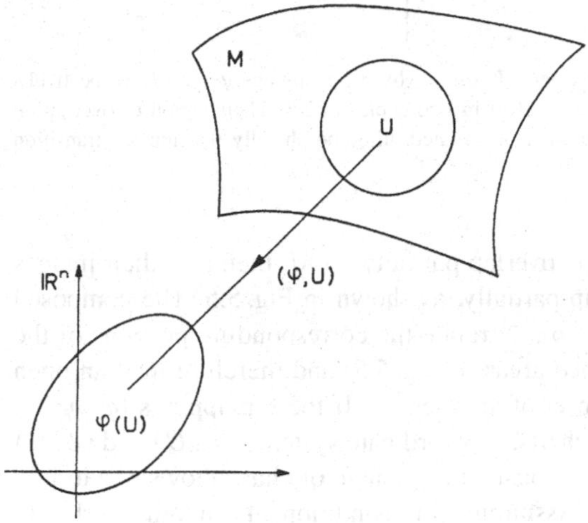

Fig. 5.4. The chart mapping φ maps an open domain U of the manifold M homeomorphically onto a domain $\varphi(U)$ of $\mathbb{R}^n$, where $n = \dim M$

Let M be such a topological space and let its dimension be $\dim M = n$. By definition, a *chart* or *local coordinate system* on M is a homeomorphism,

$$\varphi : U \subset M \to \varphi(U) \subset \mathbb{R}^n , \tag{5.8}$$

of an open set U of M onto an open set $\varphi(U)$ of $\mathbb{R}^n$, in the way sketched in Fig. 5.4. Indeed, applying the mapping (5.8) followed by the coordinate functions (5.7) yields a coordinate representation in $\mathbb{R}^n$

$$x^i = f^i \circ \varphi \quad \text{or} \quad \varphi(p) = (x^1(p), \dots, x^n(p)) \in \mathbb{R}^n \tag{5.9}$$

for every point $p \in U \subset M$. This provides the possibility of defining a diversity of geometrical objects on $U \subset M$ (i.e. *locally* on the manifold M), such as curves, vector fields, etc. Note, however, that this will not be enough, in general: since these objects are to represent physical quantities, one wishes to study them, if possible, on the *whole* of M. Furthermore, relationships between physical quantities must be independent of the choice of local coordinate systems (one says that the physical equations are covariant). This leads rather naturally to the following construction.

Cover the manifold M by means of open subsets $U, V, W, \dots$, such that every point $p \in M$ is contained in at least one of them. For every subset $U, V, \dots$ choose a homeomorphism φ, ψ, ..., respectively, such that U is mapped onto $\varphi(U)$ in $\mathbb{R}^n$,

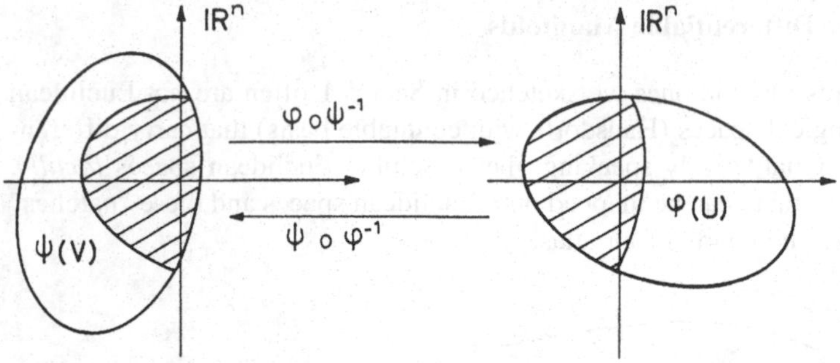

Fig. 5.5. Two overlapping, open domains U and V on M, by the mappings φ and ψ, respectively, are mapped onto the open domains $\varphi(U)$ and $\psi(V)$ in two copies of $\mathbb{R}^n$. Their region of overlap on M is mapped onto the hatched areas. The latter are related diffeomorphically through the transition mappings $(\varphi \circ \psi^{-1})$ or $(\psi \circ \varphi^{-1})$

V onto $\psi(V)$ in $\mathbb{R}^n$, etc. If U and V overlap partially on M, then also their images $\varphi(U)$ and $\psi(V)$ in $\mathbb{R}^n$ will overlap partially, as shown in Fig. 5.5. The composed mapping $\varphi \circ \psi^{-1}$ and its inverse $\psi \circ \varphi^{-1}$ relate the corresponding portions of the images $\varphi(U)$ and $\psi(V)$ (the hatched areas in Fig. 5.5) and therefore map an open subset of one $\mathbb{R}^n$ onto an open subset of another $\mathbb{R}^n$. If these mappings $(\varphi \circ \psi^{-1})$ and $(\psi \circ \varphi^{-1})$ are smooth, the two charts, or coordinate systems, (φ, U) and (ψ, V) are said to have *smooth overlap*. Obviously, this change of chart allows one to join U and V like two patches of M. Assuming this condition of smooth overlap to be trivially true, in the case where U and V do not overlap at all, provides the possibility of describing the entire manifold M by means of an *atlas* of charts.

An atlas is a collection of charts on the manifold M such that

A1. Every point of M is contained in the domain of at least one chart.

A2. Any pair of two charts overlap smoothly (in the sense defined above).

Before we go on let us ask what we have gained so far. Given such an atlas, we can differentiate geometric objects defined on M. This is done in the following way. One projects the objects onto the charts of the atlas and differentiates their images, which are now contained in spaces $\mathbb{R}^n$, using the well-known rules of analysis. As all charts of the atlas are related diffeomorphically, this procedure extends to the whole of M. In this sense an atlas defines a differentiable structure on the manifold M. In other words, with an atlas at hand, it is possible to introduce a mathematically consistent calculus on the manifold M.

There remains a technical difficulty, which, however, can be resolved easily. With the definition given above, it may happen that two formally different atlases yield the same calculus on M. In order to eliminate this possibility one adds the following to definitions A1 and A2:

A3. Each chart that has smooth overlap with all other charts shall be contained in the atlas.

In this case the atlas is said to be *complete* (or *maximal*). It is denoted by $\mathcal{A}$. This completes the framework we need for the description of physical relationships and physical laws on spaces that are not Euclidean $\mathbb{R}^n$ spaces. The objects, defined on the manifold M, can be visualized by mapping them onto charts. In this way, they can be subject to a consistent calculus as we know it from analysis in $\mathbb{R}^n$.

In summary, the topological structure is given by the definition of the manifold M, equipped with an atlas; the differential structure on M is fixed by giving a complete, differentiable atlas $\mathcal{A}$ of charts on M. Thus, a smooth, or differentiable, manifold is defined by the pair $(M, \mathcal{A})$. We remark, in passing, that there are manifolds on which there exist different differentiable structures that are not equivalent.

5.2.3 Examples of Smooth Manifolds

Let us consider a few examples of differentiable manifolds of relevance for mechanics.

(i) The space $\mathbb{R}^n$ is a differentiable manifold. The coordinate functions $(f^1, f^2, \ldots, f^n)$ induce the identical mapping

$$\mathrm{id} : \mathbb{R}^n \to \mathbb{R}^n$$

of $\mathbb{R}^n$ onto itself. Therefore they yield an atlas on $\mathbb{R}^n$ that contains a single chart. To make it a complete atlas, we must add the set ϑ of all charts on $\mathbb{R}^n$ compatible with the identity id. These are the diffeomorphisms $\Phi : U \to \Phi(U) \subset \mathbb{R}^n$ on $\mathbb{R}^n$. The differentiable structure obtained in this way is said to be *canonical*.

(ii) A sphere of radius R in $\mathbb{R}^3$. Consider the sphere

$$S^2_R \stackrel{\text{def}}{=} \{\boldsymbol{x} = (x^1, x^2, x^3) \in \mathbb{R}^3 | \boldsymbol{x}^2 = (x^1)^2 + (x^2)^2 + (x^3)^2 = R^2\} .$$

We may (but need not) think of it as being embedded in a space $\mathbb{R}^3$. An atlas that describes this two-dimensional smooth manifold in spaces $\mathbb{R}^2$ must contain at least two charts. Here we wish to construct an example for them. Call the points

$$N = (0, 0, R) , \quad S = (0, 0, -R)$$

the north pole and south pole, respectively. On the sphere S^2_R define the open subsets

$$U : S^2_R - \{N\} \quad \text{and} \quad V \stackrel{\text{def}}{=} S^2_R - \{S\} .$$

Define the mappings $\varphi : U \to \mathbb{R}^2$, $\psi : V \to \mathbb{R}^2$ onto the charts as follows: φ projects the domain U from the north pole onto the plane $x^3 = 0$ through the equator, while ψ projects the domain V from the south pole onto the same plane (more precisely, a copy thereof), cf. Fig. 5.6. If $\boldsymbol{p} = (x^1, x^2, x^3)$ is a point of U on the sphere, its projection is given by

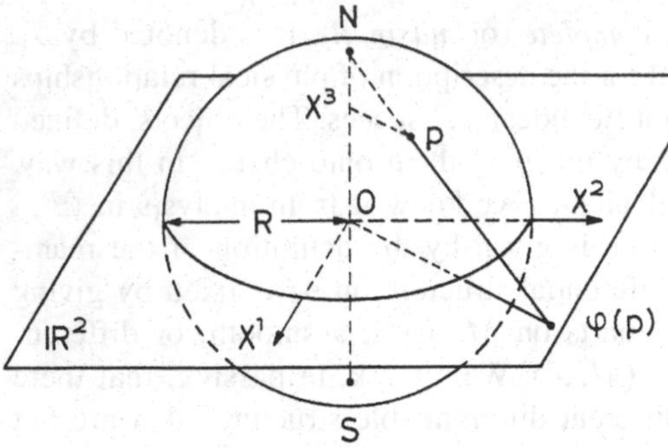

Fig. 5.6. One needs at least two charts for the description of the surface of a sphere. Here these charts are obtained by stereographic projection from the north and south poles, respectively

$$\varphi(\boldsymbol{p}) = \frac{R}{R - x^3}(x^1, x^2) .$$

Taking the same point to be an element of the domain V, we see that its projection onto $\mathbb{R}^2$ is given by

$$\Psi(\boldsymbol{p}) = \frac{R}{R + x^3}(x^1, x^2) .$$

Let us then verify that $(\psi \circ \varphi^{-1})$ is a diffeomorphism on the intersection of the domains U and V. We have

$$\varphi(U \cap V) = \mathbb{R}^2 - \{0\} = \Psi(U \cap V) .$$

Let $\boldsymbol{y} = (y^1, y^2)$ be a point on the plane through the equator without origin $\boldsymbol{y} \in \mathbb{R}^2 - \{0\}$. Its pre-image on the manifold is

$$\boldsymbol{p} = \varphi^{-1}(\boldsymbol{y}) = (x^1 = \lambda y^1, x^2 = \lambda y^2, x^3) ,$$

where $\lambda = (R - x^3)/R$, x^3 being obtained from the condition $\lambda^2 \boldsymbol{u}^2 + (x^3)^2 = R^2$, and where we have set $\boldsymbol{u}^2 = (y^1)^2 + (y^2)^2$. From this one finds

$$x^3 = \frac{\boldsymbol{u}^2 - R^2}{\boldsymbol{u}^2 + R^2} R$$

and $\lambda = 2R^2/(\boldsymbol{u}^2 + R^2)$, from which one obtains, in turn,

$$\boldsymbol{p} = \varphi^{-1}(y) = \frac{1}{\boldsymbol{u}^2 + R^2}(2R^2 y^1, 2R^2 y^2, R(\boldsymbol{u}^2 - R^2)) .$$

Applying the mapping ψ to this point and taking account of the relation $R/(R + x^3) = (\boldsymbol{u}^2 + R^2)/2\boldsymbol{u}^2$, we find on $\mathbb{R}^2 - \{0\}$

$$\Psi \circ \varphi^{-1}(y) = \frac{R^2}{\boldsymbol{u}^2}(y^1, y^2) .$$

Clearly, this is a diffeomorphism from $\mathbb{R}^2 - \{0\}$ onto $\mathbb{R}^2 - \{0\}$. The origin, which is the projection of the south pole by the first mapping, and is the projection of the north pole by the second, must be excluded. Hence the necessity of two charts.

(iii) The torus T^m. m-dimensional tori are the natural manifolds of integrable mechanical systems (see Sect. 2.37.2). The torus T^m is defined as the product of m copies of the unit circle,

$$T^m = S^1 \times S^1 \times \ldots \times S^1 \quad (m \text{ factors}) .$$

For $m = 2$, for instance, it has the shape of the inner tube of a bicycle. The first S^1 goes around the tube, the second describes its cross section. The torus T^2 is also homeomorphic to the space obtained from the square $\{x, y | 0 \le x \le 1, 0 \le y \le 1\}$ by pairwise identification of the points $(0, y)$ and $(1, y)$, and $(x, 0)$ and $(x, 1)$. An atlas for T^2 is provided, for example, by three charts defined as follows:

$$\varphi_k^{-1}(\alpha_k, \beta_k) = (\mathrm{e}^{\mathrm{i}\alpha_k}, \mathrm{e}^{\mathrm{i}\beta_k}) \in T^2 , \quad k = 1, 2, 3 ,$$

where $\alpha_1, \beta_1 \in (0, 2\pi)$, $\alpha_2, \beta_2 \in (-\pi, +\pi)$, $\alpha_3, \beta_3 \in (-\pi/2, 3\pi/2)$.

Readers are invited to make a sketch of the torus and to convince themselves thereby that T^2 is indeed covered completely by the charts given above.

(iv) The parameter manifold of the rotation group SO(3), which is the essential part of the physical coordinate manifold of the rigid body, is a differentiable manifold. Here we wish to describe it in somewhat more detail. For this purpose let us first consider the group SU(2) of unitary (complex) 2×2 matrices $\mathbf{U}$ with determinant 1:

$$\{\mathbf{U} \text{ complex } 2 \times 2 \text{ matrices } | \mathbf{U}^\dagger \mathbf{U} = \mathbb{1} , \quad \det \mathbf{U} = 1\} .$$

These matrices form a group, the *unitary unimodular group in two complex dimensions*. $\mathbf{U}^\dagger$ denotes the complex conjugate of the transposed matrix, $(\mathbf{U}^\dagger)_{pq} = (U_{qp})^*$. It is not difficult to convince oneself that any such matrix can be written as follows:

$$\mathbf{U} = \begin{pmatrix} a & b \\ -b^* & a^* \end{pmatrix} \quad \text{provided} \quad |a|^2 + |b|^2 = 1 .$$

With the complex numbers a and b written as $a = x^1 + \mathrm{i}x^2$ and $b = x^3 + \mathrm{i}x^4$, the condition $\det \mathbf{U} = 1$ becomes

$$(x^1)^2 + (x^2)^2 + (x^3)^2 + (x^4)^2 = 1 .$$

If the x^i are interpreted as coordinates in a space $\mathbb{R}^4$, this condition describes the unit sphere S^3 embedded in that space. Let us parametrize the coordinates by means of angles u, v and w, as follows:

$$\begin{aligned} x^1 &= \cos u \; \cos v \\ x^2 &= \cos u \; \sin v \qquad u \in [0, \pi/2] \\ x^3 &= \sin u \; \cos w \qquad v, w \in [0, 2\pi) \\ x^4 &= \sin u \; \sin w \end{aligned}$$

such as to fulfill the condition on their squares automatically. Clearly, the sphere S^3 is a smooth manifold. Every closed curve on it can be contracted to a point, so it is singly connected. We now wish to work out its relation to SO(3).

For this purpose we return to the representation of rotation matrices $\mathbf{R} \in \mathrm{SO}(3)$ by means of Eulerian angles, as defined in (3.35) of Sect. 3.9. Inserting the expressions (2.71) for the generators and multiplying the three matrices in (3.35), one obtains

$$\mathbf{R}(\alpha, \beta, \gamma) = \begin{pmatrix} \cos\gamma\cos\beta\cos\alpha & \cos\gamma\cos\beta\sin\alpha & -\cos\gamma\sin\beta \\ -\sin\gamma\sin\alpha & +\sin\gamma\cos\alpha & \\ & & \\ -\sin\gamma\cos\beta\cos\alpha & -\sin\gamma\cos\beta\sin\alpha & \sin\gamma\sin\beta \\ -\cos\gamma\sin\alpha & +\cos\gamma\cos\alpha & \\ & & \\ \sin\beta\cos\alpha & \sin\beta\sin\alpha & \cos\beta \end{pmatrix} .$$

In the next step we define the following map from S^3 onto SO(3):

$$f : S^3 \to \mathrm{SO}(3)$$

by

$$\begin{cases} \gamma = v + w \ (\mathrm{mod}\, 2\pi) \\ \beta = 2u \\ \alpha = v - w \ (\mathrm{mod}\, 2\pi) \end{cases}$$

As α and γ take values in $[0, 2\pi)$ and β takes values in $[0, \pi]$, the mapping is surjective. We note the following relations between matrix elements of $\mathbf{R}$ and the angles u, v, w:

$$R_{33} = \cos(2u) \,,$$
$$R_{31} = \sqrt{1 - R_{33}^2}\cos(v - w) \,, \qquad R_{13} = -\sqrt{1 - R_{33}^2}\cos(v + w) \,,$$
$$R_{32} = \sqrt{1 - R_{33}^2}\sin(v - w) \,, \qquad R_{23} = \sqrt{1 - R_{33}^2}\sin(v + w) \,.$$

(The ramaining entries, not shown here, are easily derived.)

Consider a point $\boldsymbol{x} \in S^3$, $\boldsymbol{x}(u, v, w)$ and its antipodal point $\boldsymbol{x}' = -\boldsymbol{x}$, which is obtained by the choice of parameters $u' = u$, $v' = v + \pi (\mathrm{mod}\, 2\pi)$, $w' = w + \pi (\mathrm{mod}\, 2\pi)$. These two points have the same image in SO(3) because $\gamma' = v' + w' = v + w + 2\pi (\mathrm{mod}\, 2\pi) = \gamma + 2\pi (\mathrm{mod}\, 2\pi)$; similarly, $\alpha' = \alpha + 2\pi (\mathrm{mod}\, 2\pi)$, while $\beta' = \beta$. Thus, the manifold of SO(3) is the image of S^3, but $\boldsymbol{x}$ and $-\boldsymbol{x}$ are mapped onto the same element of SO(3). In other words, the manifold of the rotation group is S^3 with antipodal points identified. If opposite points on the sphere are to be identified then there are two distinct classes of closed curves: (i) those which return to the same point and which can be contracted to a point, and (ii) those which start in $\boldsymbol{x}$ and end in $-\boldsymbol{x}$ and which cannot be contracted to a point. This is equivalent to saying that the manifold of SO(3) is doubly connected.

As a side remark we point out that we have touched here on a close relationship between the groups SU(2) and SO(3) that will turn out to be important in describing intrinsic angular momentum (spin) in quantum mechanics. The manifold of the former is the (singly connected) unit sphere S^3.

5.3 Geometrical Objects on Manifolds

Next, let us introduce various geometrical objects that are defined on smooth manifolds and are of relevance for mechanics. There are many examples: *functions* such as the Lagrangian and Hamiltonian functions, *curves* on manifolds such as solution curves of equations of motion, *vector fields* such as the velocity field of a given dynamical system, *forms* such as the volume form that appears in Liouville's theorem, and many more.

We start with a rather general notion: mappings from a smooth manifold M with atlas $\mathcal{A}$ onto another manifold N with atlas $\mathcal{B}$ (where N may be identical with M):

$$F : (M, \mathcal{A}) \to (N, \mathcal{B}) \,. \tag{5.10}$$

The point p, which is contained in an open subset U of M, is mapped onto the point $F(p)$ in N, which, of course, is contained in the image $F(U)$ of U.

Let m and n be the dimensions of M and N, respectively. Assume that (φ, U) is a chart from the atlas $\mathcal{A}$, and (ψ, V) a chart from $\mathcal{B}$ such that $F(U)$ is contained in V. The following composition is then a mapping between the Euclidean spaces $\mathbb{R}^m$ and $\mathbb{R}^n$:

$$\psi \circ F \circ \varphi^{-1} : \varphi(U) \subset \mathbb{R}^m \to \psi(V) \subset \mathbb{R}^n \,. \tag{5.11}$$

At this level it is meaningful to ask the question whether this mapping is continuous or even differentiable. This suggests the following definition: the mapping F (5.10) is said to be *smooth*, or differentiable, if the mapping (5.11) has this property for every point $p \in U \subset M$, every chart $(\varphi, U) \in \mathcal{A}$, and every chart $(\psi, V) \in \mathcal{B}$, the image $F(U)$ being contained in V.

As we shall soon see, we have already met mappings of the kind (5.10) on several occasions in earlier chapters, although we did not formulate them in this compact and general manner. This may be clearer if we notice the following special cases of (5.10). (i) The manifold from which F starts is the one-dimensional Euclidean space $(\mathbb{R}, \vartheta)$, e.g. the time axis $\mathbb{R}_t$. The chart mapping φ is then simply the identity on $\mathbb{R}$. In this case the mapping F (5.10) is a smooth curve on the manifold $(N, \mathcal{B})$, e.g. physical orbits. (ii) The manifold to which F leads is $\mathbb{R}$, i.e. now the chart mapping ψ is the identity. In this case F is a smooth function on M, an example being provided by the Lagrangian function. (iii) Initial and final manifolds are identical. This is the case, for example, for F being a diffeomorphism of M.

5.3.1 Functions and Curves on Manifolds

A *smooth function* on a manifold M is a mapping from M to the real numbers,

$$f : M \to \mathbb{R} : p \in M \mapsto f(p) \in \mathbb{R} , \tag{5.12}$$

which is differentiable, in the sense defined above.

An example is provided by the Hamiltonian function H, which assigns a real number to each point of phase space $\mathbb{P}$, assmuming H to be independent of time. If H has an explicit time dependence, it assigns a real number to each point of $\mathbb{P} \times \mathbb{R}_t$, the direct product of phase space and time axis. As another example consider the charts introduced in Sect. 5.2.2. The mapping $x^i = f^i \circ \varphi$ of (5.9), with the function f^i as defined in (5.7), is a function on M. To each point $p \in U \subset M$ it assigns its ith coordinate in the chart (φ, U).

The set of all smooth functions on M is denoted by $\mathcal{F}(M)$.

In Euclidean space $\mathbb{R}^n$ the notion of a *smooth curve* $\gamma(\tau)$ is a familiar one. When understood as a mapping, it leads from an open interval I of the real axis $\mathbb{R}$ (this can be the time axis $\mathbb{R}_t$, for instance) to the $\mathbb{R}^n$,

$$\gamma : I \subset \mathbb{R} \to \mathbb{R}^n : \tau \in I \mapsto \gamma(\tau) \in \mathbb{R}^n . \tag{5.13a}$$

Here, the interval may start at $-\infty$ and/or may end at $+\infty$. If $\{e_i\}$ is a basis of $\mathbb{R}^n$, then $\gamma(\tau)$ has the decomposition

$$\gamma(\tau) = \sum_{i=1}^{n} \gamma^i(\tau) e_i . \tag{5.13b}$$

On an arbitrary smooth manifold N smooth curves are defined following the general case (5.10), by considering their image in local charts as in (5.11),

$$\gamma : I \subset \mathbb{R} \to N : \tau \in I \mapsto \gamma(\tau) \in N . \tag{5.14}$$

Let (ψ, V) be a chart on N. For the portion of the curve contained in V, the composition $\psi \circ \gamma$ is a smooth curve in $\mathbb{R}^n$ (take (5.11) with $\varphi = \text{id}$). As N is equipped with a complete atlas, we can follow the curve everywhere on N, by following it from one chart to the next.

We wish to add two remarks concerning curves and functions that are important for the sequel. For the sake of simplicity we return to the simpler case (5.13a) of curves on Euclidean space $\mathbb{R}^n$.

(i) Smooth curves are often obtained as solutions of first-order differential equations. Let τ_0 be contained in the interval I, and let $p_0 = \gamma(\tau_0) \in \mathbb{R}^n$ be the point on the curve reached at "time" τ_0. If we take the derivative

$$\dot{\gamma}(\tau) = \frac{\mathrm{d}\gamma(\tau)}{\mathrm{d}\tau} ,$$

then

$$\dot{\gamma}(\tau_0) = \sum_{i=1}^{n} \dot{\gamma}^i(\tau_0) e_i \stackrel{\text{def}}{=} v_{p_0}$$

is the vector tangent to the curve in p_0. Now, suppose we draw all tangent vectors v_p in all points $p \in \gamma(\tau)$ of the curve. Clearly, this reminds us of the stepwise construction of solutions of mechanical equations of motion. However, we need more than that: the tangent vectors must be known in *all* points of an open domain in $\mathbb{R}^n$, not just along one curve $\gamma(\tau)$. Furthermore, the *field* of vectors obtained in this way must be smooth everywhere where it is defined, not just along the curve. $\gamma(\tau)$ is then one representative of a set of solutions of the first-order differential equation

$$\dot{\alpha}(\tau) = v_{\alpha(\tau)} \, . \tag{5.15}$$

As an example, consider a mechanical system with one degree of freedom: the one-dimensional harmonic oscillator. From Sect. 1.17.1, let

$$\underset{\sim}{x} = \begin{pmatrix} q \\ p \end{pmatrix}, \quad H = \frac{1}{2}(p^2 + q^2) \, .$$

The equation of motion reads $\dot{\underset{\sim}{x}} = \mathbf{J} H_{,x} \equiv \underset{\sim}{X}_{\mathrm{H}}$, with

$$\underset{\sim}{X}_{\mathrm{H}} = \begin{pmatrix} \partial H / \partial p \\ -\partial H / \partial q \end{pmatrix} = \begin{pmatrix} p \\ -q \end{pmatrix} .$$

$\underset{\sim}{x}$ is a point in the two-dimensional manifold $N = \mathbb{R}^2$; the vector field $\underset{\sim}{X}_{\mathrm{H}}$ is said to be the *Hamiltonian vector field*. The solutions of the differential equation $\dot{\underset{\sim}{x}} = \underset{\sim}{X}_{\mathrm{H}}$ (5.15)

$$\underset{\sim}{x}(\tau) \equiv \Phi_{\tau-\tau_0}(\underset{\sim}{x}_0) = \begin{pmatrix} q_0 \cos(\tau - \tau_0) + p_0 \sin(\tau - \tau_0) \\ -q_0 \sin(\tau - \tau_0) + p_0 \cos(\tau - \tau_0) \end{pmatrix}$$

are curves on N, each of which is fixed by the initial condition

$$\underset{\sim}{x}(\tau_0) = \underset{\sim}{x}_0 = \begin{pmatrix} q_0 \\ p_0 \end{pmatrix} .$$

(ii) Let y be an arbitrary, but fixed, point of $\mathbb{R}^n$. We consider the set $T_y\mathbb{R}^n$ of all tangent vectors at the point (i.e. vectors that are tangent to all possible curves going through y), as shown in Fig. 5.7. As one can add these vectors and can multiply them with real numbers, they form a real vector space. (In fact, one can show that this vector space $T_y\mathbb{R}^n$ is isomorphic to $\mathbb{R}^n$, the manifold that we consider. Therefore, in this case, we are justified in drawing the vectors v in the same space as the curves themselves, see Fig. 5.7.)

Consider a smooth function $f(x)$ on $\mathbb{R}^n$ (or on some neighborhood of the point y) and a vector $v = \sum v^i e_i$ of $T_y\mathbb{R}^n$, and take the derivative of $f(x)$ at the point y, in the direction of the tangent vector v. This is given by

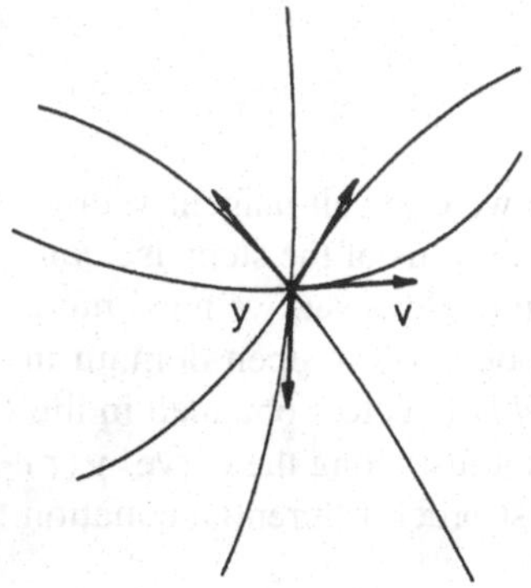

Fig. 5.7. The vectors tangent to all possible smooth curves through a given point y of $\mathbb{R}^n$ span a vector space, the tangent space $T_y\mathbb{R}^n$

$$v(f) \stackrel{\text{def}}{=} \sum_{i=1}^{n} v^i \left.\frac{\partial f}{\partial x^i}\right|_{x=y} . \tag{5.16}$$

This directional derivative assigns to each function $f(x) \in \mathcal{F}(\mathbb{R}^n)$ a real number given by (5.16),

$$v : \mathcal{F}(\mathbb{R}^n) \to \mathbb{R} : f \mapsto v(f) .$$

This derivative has the following properties: if $f(x)$ and $g(x)$ are two smooth functions on $\mathbb{R}^n$, and a and b two real numbers, then

$$\text{V1.} \quad v(af + bg) = av(f) + bv(g) \quad (\textit{R-linearity}) , \tag{5.17}$$

$$\text{V2.} \quad v(fg) = v(f)g(y) + f(y)v(g) \quad (\textit{Leibniz' rule}) . \tag{5.18}$$

5.3.2 Tangent Vectors on a Smooth Manifold

Thinking of a smooth, two-dimensional manifold M as a surface embedded in $\mathbb{R}^3$, we can see that the tangent vectors at the point y of M are contained in the plane through y tangent to M. This tangent space T_yM is the Euclidean space $\mathbb{R}^2$. This is true more generally. Let M be an n-dimensional hypersurface embedded in $\mathbb{R}^{n+1}$. T_yM is a vector space of dimension n, isomorphic to $\mathbb{R}^n$. Any element of T_yM can be used to form a directional derivative of functions on M. These derivatives have properties V1 and V2.

In the case of an arbitrary, abstractly defined, smooth manifold, it is precisely these properties which are used in the *definition of tangent vectors*: a tangent vector v in the point $p \in M$ is a real-valued function

$$v : \mathcal{F}(M) \to \mathbb{R} \tag{5.19}$$

that has properties V1 and V2, i.e.

$$v(af + bg) = av(f) + bv(g) , \tag{V1}$$

$$v(fg) = v(f)g(p) + f(p)v(g) , \tag{V2}$$

where $f, g \in \mathcal{F}(M)$ and $a, b \in \mathbb{R}$. The second property, in particular, shows that v acts like a derivative. This is what we expect from the concrete example of Euclidean space $\mathbb{R}^n$. The space T_pM of all tangent vectors in $p \in M$ is a vector space over $\mathbb{R}$, provided addition of vectors and multiplication with real numbers are defined as usual, viz.

$$\begin{aligned}(v_1 + v_2)(f) &= v_1(f) + v_2(f) \,, \\ (av)(f) &= av(f) \,,\end{aligned} \tag{5.20}$$

for all functions f on M and all real numbers a. This vector space has the same dimension as M.

In general, one cannot take a partial derivative of a function $g \in \mathcal{F}(M)$ on M itself. However, this is possible for the image of g in local charts. Let (φ, U) be a chart, $p \in U$ a point on M, and g a smooth function on M. The derivative of $g \circ \varphi^{-1}$ with respect to the natural coordinate function f^i (5.7), which is taken at the image $\varphi(p)$ in $\mathbb{R}^n$, is well defined. It is

$$\partial_i \Big|_p (g) \equiv \frac{\partial g}{\partial x^i}\Big|_p \stackrel{\text{def}}{=} \frac{\partial(g \circ \varphi^{-1})}{\partial f^i}(\varphi(p)) \,. \tag{5.21}$$

The functions

$$\partial_i \Big|_p \equiv \frac{\partial}{\partial x^i}\Big|_p : \mathcal{F}(M) \to \mathbb{R} : g \mapsto \frac{\partial g}{\partial x^i}\Big|_p \,, \quad i = 1, 2, \ldots, n \,, \tag{5.22}$$

have properties V1 and V2 and hence are tangent vectors to M, at the point $p \in U \subset M$.

The objects defined in (5.22) are useful in two respects. Firstly, they are used to define partial derivatives of smooth functions g on M, by projecting g onto a Euclidean space by means of local charts. Secondly, one can show that the vectors

$$\partial_1|_p, \partial_2|_p, \ldots, \partial_n|_p$$

form a basis of the tangent space T_pM (see e.g. O'Neill 1983), so that any vector of T_pM has the representation

$$v = \sum_{i=1}^{n} v(x^i)\partial_i|_p \tag{5.23}$$

in local charts, x^i being the coordinates defined in (5.9).

We now summarize our findings. A vector space T_pM is pinned to each point p of a smooth, but otherwise arbitrary, manifold M. It has the same dimension as M and its elements are the tangent vectors to M at the point p. If (φ, U) is a chart on M that contains p, the vectors $\partial_i|_p$, $i = 1, \ldots, n$, defined in (5.22), form a basis of T_pM, i.e. they are linearly independent and any vector v of T_pM can be represented as a linear combination of them.

5.3.3 The Tangent Bundle of a Manifold

All points $p, q, r, \ldots$ of a smooth manifold M possess their own tangent spaces T_pM, T_qM, $T_rM, \ldots$ Although these spaces all have the same dimension they are different from each other. For this reason one usually draws them, symbolically, as shown in Fig. 5.8, in such a way that they do not intersect. (Had they been like tangents to M, they would seem to intersect.)

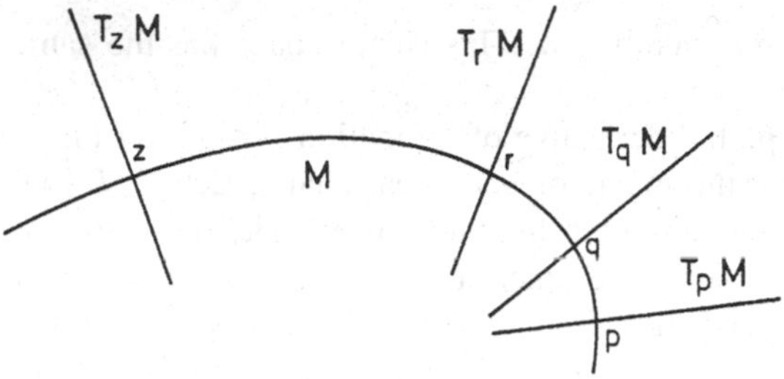

Fig. 5.8. The set of all tangent spaces at the points $p, q, \ldots$ of the manifold M, is the tangent bundle TM of M

One can show, without too much difficulty, that the (disjoint) union of all tangent spaces

$$TM \stackrel{\text{def}}{=} \bigcup_{p \in M} T_pM \tag{5.24}$$

is again a smooth manifold. This manifold TM is said to be the *tangent bundle*, M being the *base space* and the tangent spaces T_pM being the *fibres*. If M has dimension $\dim M = n$, the tangent bundle has dimension

$$\dim TM = 2n \ .$$

Figure 5.8 exhibits symbolically this fibre structure of TM. Very much like the basis itself, the manifold TM is described by means of local charts and by means of a complete atlas of charts. In fact, the differentiable structure on M induces in a natural way a differentiable structure on TM. Without going into the precise definitions at this point, qualitatively we may say this: each chart (φ, U) is a differentiable mapping from a neighborhood U of M onto the $\mathbb{R}^n$. Consider then $TU \stackrel{\text{def}}{=} \cup_{p \in U} T_pM$, i.e. the open subset TU of TM, which is defined once U is given. The mapping φ from U to $\mathbb{R}^n$ induces a mapping of the tangent vectors in p onto tangent vectors in $\varphi(p)$, the image of p;

$$T\varphi \stackrel{\text{def}}{=} TU \to \varphi(U) \times \mathbb{R}^n \ .$$

This mapping is linear and it has all the properties of a chart (we do not show this here, but refer to Sect. 5.4.1 below, which gives the definition of the tangent mapping). As a result, each chart (φ, U) from the atlas for M induces a chart $(T\varphi, TU)$ for TM. This chart is said to be the bundle chart associated with (φ, U).

A point of TM is characterized by two entries

$$(p, v) \quad \text{with} \quad p \in M \quad \text{and} \quad v \in T_pM\,,$$

i.e. by the base point p of the fibre T_pM and by the vector v, an element of this vector space. Furthermore, there is a natural *projection* from TM to the base space M,

$$\pi : TM \to M : (p, v) \mapsto p\,, \quad p \in M, v \in T_pM\,. \tag{5.25}$$

To each element in the fibre T_pM it assigns its base point p.

Lagrangian mechanics provides a particularly beautiful example for the concept of the tangent bundle. Let Q be the manifold of physical motions of a mechanical system and let u be a point of Q, which is represented by coordinates $\{\underset{\sim}{q}\}$ in local charts. Consider all possible smooth curves $\gamma(\tau)$ going through this point, with the orbit parameter always being chosen such that $u = \gamma(0)$. The tangent vectors $v_u = \dot{\gamma}(0)$, which appear as $\{\dot{\underset{\sim}{q}}\}$ in charts, span the vector space T_uQ. The Lagrangian function of an autonomous system is defined locally as a function $L(\underset{\sim}{q}, \dot{\underset{\sim}{q}})$, where $\underset{\sim}{q}$ is an arbitrary point in the physical manifold Q, while $\dot{\underset{\sim}{q}}$ is the set of all tangent vectors at that point, both being written in local charts of TQ. It is then clear that the Lagrangian function is a function on the tangent bundle, as anticipated in Fig. 5.2,

$$L : TQ \to \mathbb{R}\,.$$

It is defined in points (p, v) of the tangent bundle TQ, i.e. locally it is a function of the generalized coordinates $\underset{\sim}{q}$ and the velocities $\dot{\underset{\sim}{q}}$. It is the postulate of Hamilton's principle that determines the physical orbits $\underset{\sim}{q}(t) = \underset{\sim}{\Phi}(t)$ via differential equations obtained by means of the Lagrangian function. We return to this in Sect. 5.5.

As a final remark in this subsection, we point out that TM *locally* has the product structure $M \times \mathbb{R}^n$. However, its *global* structure can be more complicated.

5.3.4 Vector Fields on Smooth Manifolds

Vector fields of the kind sketched in Fig. 5.9 are met everywhere in physics. For a physicist they are examples of an intuitively familiar concept: flow fields of a liquid, velocity fields of swarms of particles, force fields, or more specifically within canonical mechanics, Hamiltonian vector fields. In the preceding two sections we considered all possible tangent vectors $v_p \in T_pM$ in a point p of M. The concept of a vector *field* concerns something else: it is a prescription that assigns to each point p of M precisely *one* tangent vector V_p taken from T_pM[1]. For example, given the stationary flow of a liquid in a vessel, the flow velocity at each point inside the vessel is uniquely determined. At the same time it is an element of the

[1] In what follows we shall often call V_p, i.e. the restriction of the vector field to T_pM, a *representative of the vector field*.

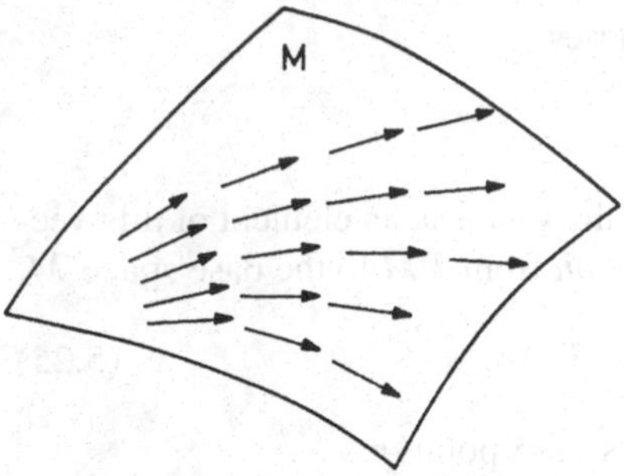

Fig. 5.9. Sketch of a smooth vector field on the manifold M

tangent space that belongs to this point. In other words, at every point the flow field chooses a specific vector from the vector space pertaining to that point.

These general considerations are cast into a precise form by the following definition.

VF1. A *vector field* V on the smooth manifold M is a function that assigns to every point p of M a specific tangent vector V_p taken from the vector space T_pM:

$$V : M \to TM : p \in M \mapsto V_p \in T_pM \,. \tag{5.26}$$

According to (5.19) tangent vectors are applied to smooth functions on M and yield their generalized directional derivatives. In a similar fashion, vector fields act on smooth functions,

$$V : \mathcal{F}(M) \to \mathcal{F}(M) \,,$$

by the following rule: at every point $p \in M$ the representative V_p of the vector field V is applied to the function, viz.

$$(Vf)(p) \stackrel{\text{def}}{=} V_p(f) \,, \quad f \in \mathcal{F}(M) \,. \tag{5.27}$$

This rule allows us to define smoothness of vector fields, as follows.

VF2. The vector field V is said to be smooth if Vf is smooth, for all smooth functions f on M.

The vector field V leads from M to TM by assigning to each $p \in M$ the element (p, V_p) of TM. Applying the projection π, as defined in (5.25), to this element yields the identity on M. Any such mapping

$$\sigma : M \to TM$$

that has the property $\pi \circ \sigma = \mathrm{id}_M$ is said to be a *section in* TM. Hence, smooth vector fields are differentiable sections.

In a chart (φ, U), i.e. in local coordinates, a vector field can be represented locally by means of coordinate vector fields, or *base fields*. For every point p of an open neighborhood $U \subset M$, the base field $\partial_i|_p$, according to (5.22), is defined as a vector field on U:

$$\partial_i : U \to TU : p \in U \mapsto \partial_i|_p \ . \tag{5.28}$$

As the functions $(g \circ \varphi^{-1})$, which appear in (5.21), are differentiable, it is clear that ∂_i is a smooth vector field on U. As in Sect. 5.2.2 let us denote the chart mapping by $\varphi(p) = (x^1(p), \ldots, x^n(p))$. Any smooth vector field V defined on $U \subset M$ has the local representation

$$V = \sum_{i=1}^{n} (V x^i) \partial_i \tag{5.29}$$

on U. Finally, by joining together these local representations on the domains of charts $U, V, \ldots$ of a complete atlas, we obtain a patchwise representation of the vector field that extends over the manifold as a whole. The base fields on two contiguous, overlapping domains U and V of the charts (φ, U) and (ψ, V), respectively, are related as follows. Returning to (5.21) and making use of the chain rule, one has

$$\frac{\partial(g \circ \varphi^{-1})}{\partial f^i} = \sum_{k=1}^{n} \frac{\partial(g \circ \psi^{-1})}{\partial f^k} \frac{\partial(\psi^k \circ \varphi^{-1})}{\partial f^i} \ .$$

Denoting the derivatives (5.21) by $\partial_i^\varphi\big|_p$ and $\partial_i^\psi\big|_p$, i.e. by indicating the chart mapping φ or ψ as a superscript, we find in the overlap of U and V

$$\partial_i^\varphi\big|_p(g) = \sum_{k=1}^{n} \partial_k^\psi\big|_p(g) \frac{\partial(\psi^k \circ \varphi^{-1})}{\partial f^i} \ . \tag{5.30}$$

The matrix appearing on the right-hand side is the Jacobi matrix $\mathbf{J}_{\psi \circ \varphi^{-1}}$ of the transition mapping $(\psi \circ \varphi^{-1})$.

The set of all smooth vector fields on M is usually denoted by $\mathcal{X}(M)$ or $\mathcal{V}(M)$. We already know an example from Sect.5.3.1: the Hamiltonian vector field on a two-dimensional phase space. If $x^1 = q$ and $x^2 = p$ denote local coordinates, then

$$X_{\mathrm{H}} = \frac{\partial H}{\partial p} \partial_1 - \frac{\partial H}{\partial q} \partial_2 \ ,$$

so that $v^i = X_{\mathrm{H}}(x^i)$ gives the vector field of that example.

According to (5.19) a tangent vector v of T_pM assigns to each smooth function f a real number. In the case of a vector *field* this statement applies to every point p of M, cf. (5.27). When we consider this equation as a function of p, we see that the action of the field V on the function f yields another smooth function on M,

$$V \in \mathcal{V}(M) : f \in \mathcal{F}(M) \to Vf \in \mathcal{F}(M)$$
$$: f(p) \mapsto V_p(f) \,. \tag{5.31}$$

This action of V on functions in $\mathcal{F}(M)$ has the properties V1 and V2 of Sect. 5.3.2, i.e. V acts on f like a derivative. Therefore, vector fields can equivalently be understood as derivatives on the set $\mathcal{F}(M)$ of smooth functions on M[2].

Starting from this interpretation of vector fields one can define the *commutator* of two vector fields X and Y of $\mathcal{V}(M)$,

$$Z = [X, Y] \stackrel{\text{def}}{=} XY - YX \,. \tag{5.32}$$

X or Y, when applied to smooth functions, yield again smooth functions. Therefore, as $X(Yf)$ and $Y(Xf)$ are functions, the action of the commutator on f is given by

$$Zf = X(Yf) - Y(Xf) \,.$$

One verifies by explicit calculation that Z fulfills V1 and V2, and, in particular, that $Z(fg) = (Zf)g + f(Zg)$. In doing this calculation one notices that it is important to take the commutator in (5.32). Indeed, the mixed terms $(Xf)(Yg)$ and $(Yf)(Xg)$ only cancel by taking the difference $(XY - YX)$. As a result, the commutator is again a derivative for smooth functions on M, or, equivalently, the commutator is a smooth vector field on M. (This is not true for the products XY and YX.) For each point $p \in M$ (5.32) defines a tangent vector Z_p in T_pM given by $Z_p(f) = X_p(Yf) - Y_p(Xf)$.

The commutator of the base fields in the domain of a given chart vanishes, $[\partial_i, \partial_k] = 0$. This is an expression of the well-known fact that the mixed, second partial derivatives of smooth functions are symmetric. Without going into the details, we close this subsection with the remark that $[X, Y]$ can also be interpreted as the so-called *Lie derivative* of the vector field Y by the vector field X. What this means can be described in a qualitative manner as follows: a vector field X defines a flow, through the collection of solutions of the differential equation $\dot{\alpha}(\tau) = X_{\alpha(\tau)}$, as in (5.15). One can ask the question, given certain differential-geometric objects such as functions, vector fields, etc., how these objects change along the flow of X. In other words, one takes their derivative along the flow of a given vector field X. This special type of derivative is said to be the Lie derivative; in the general case it is denoted by L_X. If acting on vector fields, it is $L_X = [X, Y]$. It has the following property: $L_{[X,Y]} = [L_X, L_Y]$, to which we shall return in Sect. 5.5.5.

---

[2] The precise statement is this: the real vector space of $\mathbb{R}$-linear derivations on $\mathcal{F}(M)$ is isomorphic to the real vector space $\mathcal{V}(M)$.

5.3.5 Exterior Forms

Let γ be a smooth curve on the manifold M,

$$\gamma = \{\gamma^1, \dots, \gamma^n\} : I \subset \mathbb{R} \to M ,$$

that goes through the point $p \in M$ such that $p = \gamma\,(\tau = 0)$. Let f be a smooth function on M. The directional derivative of this function in p, along the tangent vector $v_p = \dot{\gamma}(0)$, is given by

$$\mathrm{d} f_p(v_p) = \frac{\mathrm{d}}{\mathrm{d}\tau} f(\gamma(\tau))\Big|_{\tau=0} . \tag{5.33}$$

This provides an example for a differentiable mapping of the tangent space T_pM onto the real numbers. Indeed,

$$\mathrm{d} f_p : T_pM \to \mathbb{R}$$

assigns to every v_p the real number $\mathrm{d} f(\gamma(\tau))/\mathrm{d}\tau|_{\tau=0}$. This mapping is linear. As is well known from linear algebra, the linear mappings from T_pM to $\mathbb{R}$ span the vector space dual to T_pM. This vector space is denoted by T_p^*M and is said to be the *cotangent space* (cotangent to M) at the point p. The disjoint union of the cotangent spaces over all points p of M,

$$\bigcup_{p\in M} T_p^*M \stackrel{\text{def}}{=} T^*M , \tag{5.34}$$

finally, is called the *cotangent bundle*, in analogy to the tangent bundle (5.24). Let us denote the elements of T_p^*M by ω_p. Of course, in the example (5.33) we may take the point p to be running along a curve $\gamma(\tau)$, or, more generally, if there is a set of curves that cover the whole manifold, we may take it to be wandering everywhere on M. This generates something like a "field" of directional derivatives everywhere on M that is linear and differentiable. Such a geometric object, which is, in a way, dual to the vector fields defined previously, is said to be a *differential form of degree 1*, or simply a *one-form*. Its precise definition goes as follows.

DF1. A one-form is a function

$$\omega : M \to T^*M : p \mapsto \omega_p \in T_p^*M \tag{5.35}$$

that assigns to every point $p \in M$ an element ω_p in the cotangent space T_p^*M. Here, the form ω_p is a linear mapping of the tangent space T_pM onto the reals, i.e. $\omega_p(v_p)$ is a real number.

Since ω acts on tangent vectors v_p at every point p, we can apply this one-form to smooth vector *fields* X: the result $\omega(X)$ is then a real function whose value in p is given by $\omega_p(X_p)$. Therefore, definition DF1 can be supplemented by a criterion that tests whether the function obtained in this way is differentiable, viz. the following.

DF2. The one-form ω is said to be *smooth* if the function $\omega(X)$ is smooth for any vector field $X \in \mathcal{V}(M)$.

The set of all smooth one-forms over M is often denoted by $\mathcal{X}^*(M)$, the notation stressing the fact that it consists of objects that are dual to the vector fields, denoted by $\mathcal{X}(M)$ or $\mathcal{V}(M)$.

An example of a smooth differential form of degree 1 is provided by the differential of a smooth function on M,

$$\mathrm{d}f : TM \to \mathbb{R}\,, \tag{5.36}$$

which is defined such that $(\mathrm{d}f)(X) = X(f)$. For instance, consider the chart mapping (5.9),

$$\varphi(p) = \left(x^1(p), \ldots, x^n(p)\right)\,,$$

where the $x^i(p)$ are smooth functions on M. The differential (5.36) of x^i in the neighborhood $U \subset M$, for which the chart is valid, is

$$\mathrm{d}x^i : TU \to \mathbb{R}\,.$$

Let $v = (v(x^1), \ldots, v(x^n))$ be a tangent vector taken from the tangent space T_pM at a point p of M. Applying the one-form $\mathrm{d}x^i$ to v yields a real number that is just the component $v(x^i)$ of the tangent vector,

$$\mathrm{d}x^i(v) = v(x^i)\,.$$

This is easily understood if one recalls the representation (5.23) of v in a local chart and if one calculates the action of $\mathrm{d}x^i$ on the base vector $\partial_j|_p$ (5.22). One finds, indeed, that

$$\mathrm{d}x^i(\partial_{j|p}) = \left.\frac{\partial}{\partial x^j}\right|_p \mathrm{d}x^i = \delta^i_j\,.$$

With this result in mind one readily understands that the one-forms dx^i form a basis of the cotangent space T^*_pM at each point p of M. The basis $\{\mathrm{d}x^i|_p\}$ of T^*_pM is the dual of the basis $\{\partial_i|_p\}$ of T_pM. The one-forms $\mathrm{d}x^1, \ldots, \mathrm{d}x^n$ are said to be *base differential forms of degree 1* on U. This means, in particular, that any smooth one-form can be written locally as

$$\omega = \sum_{i=1}^{n} \omega(\partial_i)\mathrm{d}x^i\,. \tag{5.37}$$

Here $\omega(\partial_i)$ at each point p is the real number obtained when applying the one-form ω onto the base field ∂_i; see DF1 and DF2. The representation is valid on the domain U of a given chart. As the manifold M can be covered by means of the charts of a complete atlas and as neighboring charts are joined together diffeomorphically, one can continue the representation (5.37) patchwise on the charts (φ, U), (ψ, V), etc. everywhere on M.

As an example, consider the total differential of a smooth function g on M, where M is a smooth manifold described by a complete atlas of, say, two charts. On the domain U of the first chart (φ, U) we have $\mathrm{d}g(\partial_i) = \partial g/\partial x^i$, and hence $\mathrm{d}g = \sum_{i=1}^n (\partial g/\partial x^i)\mathrm{d}x^i$. Similarly, on the domain V of the second chart (ψ, V), $\mathrm{d}g(\partial_i) = \partial g/\partial y^i$ and $\mathrm{d}g = \sum_{i=1}^n (\partial g/\partial y^i)\mathrm{d}y^i$. On the overlap of U and V either of the two local representations is valid. The base fields on U and those on V are related by the Jacobi matrix, cf. (5.30), while the base forms are related by the inverse of that matrix.

Let us summarize the dual concepts of vector fields and one-forms. As indicated in VF1 the vector field X chooses one specific tangent vector X_p from each tangent space T_pM at the point p of M. This representative X_p acts on smooth functions in a differentiable manner, according to the rules V1 and V2. The base fields $\{\partial_i\}$ are special vector fields that are defined locally, i.e. chartwise. The one-form ω, on the other hand, assigns to each point p a specific element ω_p from the cotangent space T_p^*M. Thus, ω_p is a linear mapping acting on elements X_p of T_pM. As a whole, $\omega(X)$ is a smooth function of the base point p. The set of differentials $\mathrm{d}x^i$ are special cases of one-forms in the domains of local charts. They can be continued all over the manifold M, by going from one chart to the next. The set $\{\mathrm{d}x^i|_p\}$ is a basis of the cotangent space T_p^*M that is dual to the basis $\{\partial_i|_p\}$ of the tangent space T_pM.

5.4 Calculus on Manifolds

In this last of the preparatory sections we show how to generate new geometrical objects from those studied in Sect. 5.3 and how to do calculations with them. We introduce the *exterior product* of forms, which generalizes the vector product in $\mathbb{R}^3$, as well as the *exterior derivative*, which provides a systematic generalization of the notions gradient, curl, and divergence, familiar from calculus in the space $\mathbb{R}^3$. We also briefly discuss integral curves of vector fields and, thereby, return to some of the results of Chap. 1. In this context, the central concepts are again those of smooth mapping of a manifold onto another (or itself) and the linear transformations of the tangent and cotangent spaces induced by the mapping.

5.4.1 Differentiable Mappings of Manifolds

In Sect. 5.3 we defined smooth mappings

$$F : (M, \mathcal{A}) \to (N, \mathcal{B}) \tag{5.38}$$

from the manifold M with differentiable structure $\mathcal{A}$ onto the manifold N with differentiable structure $\mathcal{B}$. Differentiability was defined by means of charts and in Euclidean spaces, as indicated in (5.11). It is not difficult to work out the transformation behavior of geometrical objects on M, under the mapping (5.38). For *functions* this is easy. Let f be a smooth function on the target manifold N,

$$f : N \to \mathbb{R} : q \in N \mapsto f(q) \in \mathbb{R} .$$

If q is the image of the point $p \in M$ by the mapping F, i.e. $q = F(p)$, then the composition $(f \circ F)$ is a smooth function on the starting manifold M. It is said to be the *pull-back* of the function f, i.e. the function f on N is "pulled back" to the manifold M, where it becomes $(f \circ F)$. This pull-back by the mapping F is denoted by F^*,

$$F^* f = f \circ F : p \in M \mapsto f(F(p)) \in \mathbb{R} . \tag{5.39}$$

Thus, any smooth function that is given on N can be carried over to M. The converse, i.e. the *push-forward* of a function from the starting manifold M to the target manifold N is possible only if F is invertible and if F^{-1} is smooth as well. For example, this is the case if F is a diffeomorphism.

By (5.38), *vector fields* on M are mapped onto vector fields on N. This is seen as follows. Vector fields act on functions, as described in Sect. 5.3.4. Let X be a vector field on M, X_p its representative in T_pM, the tangent space in $p \in M$, and g a smooth function on the target manifold N. As the composition $(g \circ F)$ is a smooth function on M, we can apply X_p to it, $X_p(g \circ F)$. If this is understood as an assignment

$$(X_F)_q : g \in \mathcal{F}(N) \mapsto X_p(g \circ F) \in \mathbb{R} ,$$

then $(X_F)_q$ is seen to be a tangent vector at the point $q = F(p)$ on the target manifold. For this to be true, conditions V1 and V2 must be fulfilled. V1 being obvious, we only have to verify the Leibniz rule V2. For two functions f and g on N the following equation holds at the points $p \in M$ and $q = F(p) \in N$, respectively,

$$\begin{aligned} X_F(fg) &= X((f \circ F)(g \circ F)) \\ &= X(f \circ F)g(F(p)) + f(F(p))X(g \circ F) \\ &= X_F(f)g(q) + f(q)X_F(g) . \end{aligned}$$

This shows that $(X_F)_q$ is indeed a tangent vector belonging to T_qN. In this way, the differentiable mapping F induces a linear mapping of the tangent spaces, which is said to be the *differential mapping* $\mathrm{d}F$ corresponding to F. The mapping

$$\mathrm{d}F : TM \to TN : X \mapsto X_F \tag{5.40a}$$

is defined at every point p and its image q by[3]

[3] Below we shall also use the notation TF, instead of $\mathrm{d}F$, a notation which is customary in the mathematical literature.

$$\mathrm{d}F_p : T_pM \to T_qN : X_p \mapsto (X_F)_q \,, \quad q = F(p) \,. \tag{5.40b}$$

Its action on functions $f \in \mathcal{F}(N)$ is

$$\mathrm{d}F_p(X)(f) = X(f \circ F), X \in \mathcal{V}(M), f \in \mathcal{F}(N) \,.$$

In Fig. 5.10 we illustrate the mapping F and the induced mapping $\mathrm{d}F$. As a matter of exception, we have drawn the tangent spaces in p and in the image point $q = F(p)$ as genuine tangent planes to M and N, respectively. We note that if F is a diffeomorphism, in particular, then the corresponding differential mapping is a linear isomorphism of the tangent spaces. (For an example see Sect. 6.2.2.)

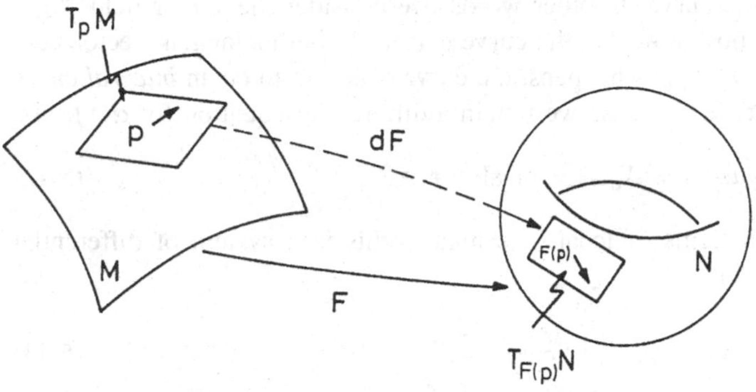

Fig. 5.10. The smooth mapping F from M to N induces a linear mapping $\mathrm{d}F$ (or TF) of the tangent space T_pM onto the tangent space T_qN in the image $q = F(p)$ of p

Given the transformation behavior of vector fields, we can deduce the transformation behavior of *exterior differential forms* as follows. Let $\omega \in \mathcal{X}^*(N)$ be a one-form on N. As we learnt earlier, it acts on vector fields defined on N. As the latter are related to vector fields on M via the mapping (5.40a), one can "pull back" the form ω on N to the starting manifold M. The *pull-back* of the form ω, by the mapping F, is denoted by $F^*\omega$. It is defined by

$$(F^*\omega)(X) = \omega(\mathrm{d}F(X)) \,, \quad X \in \mathcal{V}(M) \,. \tag{5.41}$$

Thus, on the manifold M the form $F^*\omega$ acts on X and yields a real function on M whose value in $p \in M$ is given by the value of the function $\omega(\mathrm{d}F(X))$ in $q = F(p)$.

5.4.2 Integral Curves of Vector Fields

In Sects. 1.16, 1.18–20, we studied the set of solutions of systems of first-order differential equations, for all possible initial conditions. In canonical mechanics, these equations are the equations of Hamilton and Jacobi, in which case the right-hand side of (1.41) contains the Hamiltonian vector field. Smooth vector fields and integral curves of vector fields are geometrical concepts that occur in many areas of physics. We start by defining the tangent field of a curve α on a manifold M. The curve α maps an interval I of the real τ-axis $\mathbb{R}_\tau$ onto M. The tangent vector field on $\mathbb{R}_\tau$ is simply given by the derivative $\mathrm{d}/\mathrm{d}\tau$. From (5.40a), the linear mapping $\mathrm{d}\alpha$ maps this field onto the tangent vectors to the curve α on M. This generates the vector field

$$\dot{\alpha} \stackrel{\text{def}}{=} \mathrm{d}\alpha \circ \frac{\mathrm{d}}{\mathrm{d}\tau},$$

tangent to the curve $\alpha : I \to M$. On the other hand, for an arbitrary smooth vector field X on M, we may consider its representatives in the tangent spaces $T_{\alpha(\tau)}M$ over the points on the curve. In other words, we consider the vector field $X_{\alpha(\tau)}$ along the curve. Suppose now that the curve α is such that its tangent vector field $\dot{\alpha}$ coincides with $X_{\alpha(\tau)}$. If this happens, the curve α is said to be an *integral curve* of the vector field X. In this case we obtain a differential equation for $\alpha(\tau)$, viz.

$$\dot{\alpha} = X \circ \alpha \quad \text{or} \quad \dot{\alpha}(\tau) = X_{\alpha(\tau)} \quad \text{for all} \quad \tau \in I\,. \tag{5.42}$$

When written out in terms of local coordinates, this is a system of differential equations of first order,

$$\frac{\mathrm{d}}{\mathrm{d}\tau}(x^i \circ \alpha) = X^i(x^1 \circ \alpha, \ldots, x^n \circ \alpha)\,, \tag{5.43}$$

which is of the type studied in Chap. 1; cf. (1.41). (Note, however, that the right-hand side of (5.43) does not depend explicitly on τ. This means that the flow of this system is always stationary.) In particular, the theorem of Sect. 1.19 on the existence and uniqueness of solutions is applicable to the system (5.43).

Let us consider an example: the Hamiltonian vector field for a system with one degree of freedom, i.e. on a two-dimensional phase space (see also Sects. 5.3.1 and 5.3.4),

$$X_H = \frac{\partial H}{\partial p}\partial_q - \frac{\partial H}{\partial q}\partial_p\,.$$

The curve $\{x^1 \circ \alpha, x^2 \circ \alpha\}(\tau) = \{q(\tau), p(\tau)\}$ is an integral curve of X_H if and only if the equations

$$\frac{\mathrm{d}q}{\mathrm{d}\tau} = \frac{\partial H}{\partial p} \quad \text{and} \quad \frac{\mathrm{d}p}{\mathrm{d}\tau} = -\frac{\partial H}{\partial q}$$

are fulfilled. If the phase space is the Euclidean space $\mathbb{R}^2$, its representation in terms of charts is trivial (it is the identity on $\mathbb{R}^2$) and we can simply write $\alpha(\tau) = \{q(\tau), p(\tau)\}$.

The theorem of Sect. 1.19 guarantees that for each p on M there is precisely one integral curve α for which p is the initial point (or initial configuration, as we said there) $p = \alpha(0)$. Clearly, one will try to continue that curve on M as far as this is possible. By this procedure one obtains the *maximal* integral curve α_p through p. The theorem of Sect. 1.19 tells us that it is uniquely determined. One says that the vector field X is *complete* if everyone of its maximal integral curves is defined on the entire real axis $\mathbb{R}$.

For a complete vector field the set of all maximal integral curves

$$\Phi(p,\tau) \stackrel{\text{def}}{=} \alpha_p(\tau)$$

yields the *flow* of the vector field. If one keeps the time parameter τ fixed, then $\Phi(p,\tau)$ gives the position of the orbit point in M to which p has moved under the action of the flow, for every point p on M. If in turn one keeps p fixed and varies τ, the flow yields the maximal integral curve going through p. We return to this in Sect. 6.2.1.

In Chap. 1 we studied examples of flows of complete vector fields. The flows of Hamiltonian vector fields have the specific property of preserving volume and orientation. As such, they can be compared to the flow of a frictionless, incompressible fluid.

5.4.3 Exterior Product of One-Forms

We start with two simple examples of forms on the manifold $M = \mathbb{R}^3$. Let $\boldsymbol{K} = (K^1, K^2, K^3)$ be a force field and let $\boldsymbol{v}$ be the velocity field of a given physical motion in $\mathbb{R}^3$. The work per unit time is given by the scalar product $\boldsymbol{K}\cdot\boldsymbol{v}$. This can be written as the action of the one-form $\omega_{\boldsymbol{K}} \stackrel{\text{def}}{=} \sum_{i=1}^{3} K^i \mathrm{d}x^i$ onto the tangent vector $\boldsymbol{v}$, viz.

$$\omega_{\boldsymbol{K}}(\boldsymbol{v}) = \sum K^i v^i = \boldsymbol{K}\cdot\boldsymbol{v}\,.$$

In the second example let $\boldsymbol{v}$ be the velocity field of a flow in the oriented space $\mathbb{R}^3$. We wish to study the flux across some smooth surface in $\mathbb{R}^3$. Consider two tangent vectors $\boldsymbol{t}$ and $\boldsymbol{s}$ at the point $\boldsymbol{x}$ of this surface. The flux (including its sign) across the parallelogram spanned by $\boldsymbol{t}$ and $\boldsymbol{s}$ is given by the scalar product of $\boldsymbol{v}$ with the cross product $\boldsymbol{t}\times\boldsymbol{s}$,

$$\Phi_{\boldsymbol{v}}(\boldsymbol{t},\boldsymbol{s}) = v^1(t^2s^3 - t^3s^2) + v^2(t^3s^1 - t^1s^3) + v^3(t^1s^2 - t^2s^1)\,.$$

This quantity can be understood as an exterior form that acts on *two* tangent vectors. It has the following properties: the form $\Phi_{\boldsymbol{v}}$ is linear in both of its arguments. Furthermore, it is skew-symmetric because, as we interchange $\boldsymbol{t}$ and $\boldsymbol{s}$, the parallelogram changes its orientation, and the flux changes sign. A form with these properties is said to be an *exterior two-form* on $\mathbb{R}^3$.

Two-forms can be obtained from two exterior one-forms, for instance by defining a product of forms that is bilinear and skew-symmetric. This product is called the *exterior product*. It is defined as follows. The exterior product of two base forms $\mathrm{d}x^i$ and $\mathrm{d}x^k$ is denoted by $\mathrm{d}x^i \wedge \mathrm{d}x^k$. It is defined by its action on two arbitrary tangent vectors s and t belonging to T_pM,

$$(\mathrm{d}x^i \wedge \mathrm{d}x^k)(s, t) = s^i t^k - s^k t^i \ . \tag{5.44}$$

The symbol $\wedge$ denotes the "wedge" product. As any one-form can be written as a linear combination of base one-forms, the exterior product of two one-forms ω and θ, in each point p of a manifold M, is given by

$$\begin{aligned}(\omega \wedge \theta)_p(v, w) &= \omega_p(v)\theta_p(w) - \omega_p(w)\theta_p(v) \\ &= \det\begin{pmatrix}\omega_p(v) & \omega_p(w) \\ \theta_p(v) & \theta_p(w)\end{pmatrix} .\end{aligned} \tag{5.45}$$

Here v and w are elements of T_pM. To each point p the one-forms ω and θ assign the elements ω_p and θ_p of T_p^*M, respectively. The exterior product $\omega \wedge \theta$ is defined at each point p, according to (5.45), and hence everywhere on M.

In much the same way as the coordinate one-forms $\mathrm{d}x^i$ serve as a basis for all one-forms, every two-form $\overset{2}{\omega}$ can be represented by a linear combination of base two-forms $\mathrm{d}x^i \wedge \mathrm{d}x^k$ (with $i < k$),

$$\overset{2}{\omega} = \sum_{i<k=1}^{n} \omega_{ik} \mathrm{d}x^i \wedge \mathrm{d}x^k \ , \tag{5.46}$$

the restriction $i < k$ taking account of the relation $\mathrm{d}x^k \wedge \mathrm{d}x^i = -\mathrm{d}x^i \wedge \mathrm{d}x^k$. The coefficients in (5.46) are obtained from the action of $\overset{2}{\omega}$ onto the corresponding base vector fields,

$$\omega_{ik} = \overset{2}{\omega}(\partial_i, \partial_k) \ . \tag{5.47}$$

The exterior product can be extended to three-forms, four-forms, and forms of higher degree. For example, the k-fold exterior product is given by

$$(\omega_1 \wedge \omega_2 \wedge \ldots \wedge \omega_k)(v^{(1)}, v^{(2)}, \ldots, v^{(k)}) = \det(\omega_i(v^{(j)})) \ . \tag{5.48}$$

It is linear in its k arguments and it is totally antisymmetric. Any k-form can be expressed as a linear combination of base k-forms

$$\mathrm{d}x^{i_1} \wedge \mathrm{d}x^{i_2} \wedge \ldots \wedge \mathrm{d}x^{i_k} \ , \quad \text{with} \quad i_1 < i_2 < \ldots < i_k \ . \tag{5.49}$$

There are $\binom{n}{k}$ such base forms. In particular, if $k = 1$ or $k = n$, there is precisely one such base form. On the other hand, for $k > n$, at least two one-forms in (5.49) must be equal. By the antisymmetry of the base forms (5.49), any form of degree higher than n vanishes. Thus, the highest degree a form on an n-dimensional manifold M can have is $k = n$. For $k = n$ the form (5.49) is proportional to the oriented volume element of an n-dimensional vector space.

The examples show that the exterior product is a generalization of the vector product in $\mathbb{R}^3$. In a certain sense, it is even simpler than that because multiple products such as (5.48) or (5.49) pose no problems of where to put parantheses. The exterior product is associative.

5.4.4 The Exterior Derivative

In the preceding paragraph it was shown that one can generate two-forms as well as forms of higher degree by taking exterior products of one-forms. Here we shall learn that there is another possibility of obtaining smooth forms of higher degree: by means of the *exterior derivative*, or *Cartan derivative*.

Let us first summarize, in the form of a definition, what the preceding section taught us about smooth differential forms of degree k.

DF3. A k-form is a function

$$\overset{k}{\omega} : M \to (T^*M)^k : p \mapsto \overset{k}{\omega}_p \,, \tag{5.50}$$

that assigns to each point $p \in M$ an element of $(T_p^*M)^k$, the k-fold direct product of the cotangent space. $\overset{k}{\omega}_p$ is a multilinear, skew-symmetric mapping from $(T_pM)^k$ onto the real numbers, i.e. it acts on k vector fields

$$\overset{k}{\omega}_p(X_1, \ldots, X_k) \in \mathbb{R} \tag{5.51}$$

and is antisymmetric in all k arguments.

The real number (5.51) is a function of the base point p. Therefore, in analogy to DF2 of Sect. 5.3.5, one defines smoothness for exterior forms as follows.

DF4. The k-form $\overset{k}{\omega}$ is said to be *smooth* if the function $\overset{k}{\omega}(X_1, \ldots, X_k)$ is differentiable, for all sets of smooth vector fields $X_i \in \mathcal{V}(M)$. Locally (i.e. in charts) any such k-form can be written, in a unique way, as a linear combination of the base forms (5.49),

$$\overset{k}{\omega} = \sum_{i_1 < i_2 < \ldots < i_k} \omega_{i_1 \ldots i_k} \mathrm{d}x^{i_1} \wedge \ldots \wedge \mathrm{d}x^{i_k} \,. \tag{5.52}$$

The coefficients are given by the action of $\overset{k}{\omega}$ onto the corresponding base vector fields $\partial_{i_1}, \ldots, \partial_{i_k}$.

Functions on M can be understood as forms of degree zero. As we showed in Sect. 5.3.5, the well-known total derivative converts a function into a one-form. Indeed, in a local representation we had

$$\mathrm{d}g = \sum_{i=1}^{n} \frac{\partial g}{\partial x^i} \mathrm{d}x^i \,, \tag{5.53}$$

where $\partial g/\partial x^i$ are the partial derivatives, i.e. the result of the action of the one-form $\mathrm{d}g$ onto the base fields ∂_i, while the $\mathrm{d}x^i$ are the base one-forms.

The Cartan or exterior derivative generalizes this step to smooth forms of arbitrary degree. It maps smooth k-forms onto $(k+1)$-forms, this mapping being linear,

$$\mathrm{d} : \overset{k}{\omega} \rightarrow \overset{k+1}{\omega} \; . \tag{5.54}$$

It is defined uniquely and has the following properties:

CD1. For functions g on M, $\mathrm{d}g$ is the usual total derivative.
CD2. The action of d on the exterior (or wedge) product of two forms of degree k and l is

$$\mathrm{d}(\overset{k}{\omega} \wedge \overset{l}{\omega}) = (\mathrm{d}\,\overset{k}{\omega}) \wedge \overset{l}{\omega} + (-)^k \overset{k}{\omega} \wedge (\mathrm{d}\,\overset{l}{\omega}) \; .$$

CD3. The form $\overset{k}{\omega}$ being represented locally as in (5.52), the action of the exterior derivative on this form is

$$\mathrm{d}\,\overset{k}{\omega} = \sum_{i_1 < \ldots < i_k} \mathrm{d}\omega_{i_1 \ldots i_k}(x^1, \ldots, x^n) \wedge \mathrm{d}x^{i_1} \wedge \ldots \wedge \mathrm{d}x^{i_k} \; .$$

Here, $\mathrm{d}\omega_{i_1 \ldots i_k}(x^1, \ldots, x^n)$ is the total differential and is expressed in terms of base one-forms, as in (5.53).

This exterior derivative is a local and linear operator. Property CD2 can also be described by saying that d is an *antiderivation* (with respect to the exterior product $\wedge$), in the sense that it obeys the Leibniz rule CD2 with extra signs that depend on the degree of the first form. A remarkable property of the exterior derivative is that the composition of d with itself gives zero,

$$\mathrm{d} \circ \mathrm{d} = 0 \; . \tag{5.55}$$

We prove this assertion for the case of smooth functions $g \in \mathcal{F}(M)$. We have $\mathrm{d}g = \sum_{i=1}^{n} (\partial g/\partial x^i)\mathrm{d}x^i$ and, according to CD3,

$$\begin{aligned}(\mathrm{d} \circ \mathrm{d})g = \mathrm{d}(\mathrm{d}g) &= \sum_i \mathrm{d}(\partial g/\partial x^i) \wedge \mathrm{d}x^i \\ &= \sum_i \left(\sum_{k<i} + \sum_{k>i} \right) \frac{\partial^2 g}{\partial x^k \partial x^i} \mathrm{d}x^k \wedge \mathrm{d}x^i \; .\end{aligned}$$

If we exchange $\mathrm{d}x^k$ and $\mathrm{d}x^i$ in the second sum in the brackets on the right-hand side, and if we relabel the indices by exchanging k and i, we obtain, using the antisymmetry of the wedge product,

$$(\mathrm{d} \circ \mathrm{d})g = \sum_{k<i} \left(\frac{\partial^2 g}{\partial x^k \partial x^i} - \frac{\partial^2 g}{\partial x^i \partial x^k} \right) \mathrm{d}x^k \wedge \mathrm{d}x^i = 0 \, .$$

This vanishes because the second, mixed partial derivatives of smooth functions are equal. The fact that (5.55) holds for any k-form follows from this result and from the product rule CD2.

5.4.5 Exterior Derivative and Vectors in $\mathbb{R}^3$

To illustrate the general and somewhat abstract definitions of the preceding sections, we consider the manifold $M = \mathbb{R}^3$, i.e. the three-dimensional Euclidean space of physics. For a smooth function $f(\boldsymbol{x})$ the exterior derivative gives

$$\mathrm{d}f = \sum_{i=1}^{3} (\partial f / \partial x^i) \mathrm{d}x^i \, .$$

This is the well-known total differential of f. When applied to the base field ∂_k, it gives

$$\mathrm{d}f(\partial_k) = \partial f / \partial x^k \, .$$

This generates the triple $\{\partial f/\partial x^1, \partial f/\partial x^2, \partial f/\partial x^3\} = \nabla f$, which represents the gradient of f in $\mathbb{R}^3$.

The exterior product of two forms $\overset{k}{\omega}$ and $\overset{l}{\omega}$ is an exterior form of degree $(k+l)$. Functions have to be understood as zero-forms. Thus, the exterior product of two functions f and g is the ordinary product. In this case, rule CD2 is nothing but the product rule for differentiation:

$$\nabla(fg) = (\nabla f)g + f(\nabla g) \, .$$

Consider now the one-form

$$\overset{1}{\omega}_a = \sum_{i=1}^{3} a_i(\boldsymbol{x}) \mathrm{d}x^i \, . \tag{5.56}$$

Its exterior derivative is

$$\mathrm{d}\overset{1}{\omega}_a = \left(-\frac{\partial a_1}{\partial x^2} + \frac{\partial a_2}{\partial x^1} \right) \mathrm{d}x^1 \wedge \mathrm{d}x^2 + \left(-\frac{\partial a_1}{\partial x^3} + \frac{\partial a_3}{\partial x^1} \right) \mathrm{d}x^1 \wedge \mathrm{d}x^3$$
$$+ \left(-\frac{\partial a_2}{\partial x^3} + \frac{\partial a_3}{\partial x^2} \right) \mathrm{d}x^2 \wedge \mathrm{d}x^3 \, . \tag{5.57}$$

If $\{a_1(\boldsymbol{x}), a_2(\boldsymbol{x}), a_3(\boldsymbol{x})\}$ are understood to be the components of a vector field $\boldsymbol{a}(\boldsymbol{x})$, the coefficients of the two-form $d\overset{1}{\omega}_a$ are seen to be the coefficients of the curl of

$\boldsymbol{a}(\boldsymbol{x})$. These identifications are specific for the dimension 3 of the space $M = \mathbb{R}^3$ and do not hold in general.

The three-dimensional Euclidean space admits a metric (see Sect.5.2.1). Furthermore, it is orientable because three linearly independent vectors define an oriented volume of the parallelepiped they span. Therefore, if $(\hat{\boldsymbol{e}}_1, \hat{\boldsymbol{e}}_2, \hat{\boldsymbol{e}}_3)$ is a set of orthonormal vectors in the tangent space $T_x\mathbb{R}^3$, we can assign to each k-form ω an $(n-k)$-form, i.e. a $(3-k)$-form, denoted $*\omega$, through the definition

$$(*\omega)(\boldsymbol{e}_{k+1}, \ldots, \boldsymbol{e}_3) \stackrel{\text{def}}{=} \omega(\boldsymbol{e}_1, \ldots, \boldsymbol{e}_k) , \quad 0 \le k \le n = 3 . \tag{5.58}$$

This assignment is said to be the *Hodge star* operation. In $\mathbb{R}^3$ it assigns to every three-form a zero-form (a function), to every two-form a one-form, and vice versa. For example, we obtain

$$\begin{aligned}
&*\mathrm{d}x^1 = \mathrm{d}x^2 \wedge \mathrm{d}x^3 && \text{(cyclic permutations)} && \text{(two-form)},\\
&*\mathrm{d}x^2 \wedge \mathrm{d}x^3 = \mathrm{d}x^1 = *(*\mathrm{d}x^1) && \text{(cyclic permutations)} && \text{(one-form)},\\
&*\mathrm{d}x^1 \wedge \mathrm{d}x^2 \wedge \mathrm{d}x^3 = 1 && && \text{(zero-form)}.
\end{aligned}$$

Assigning the one-form (5.56) to the vector field $\boldsymbol{a}(\boldsymbol{x})$, its exterior derivative is given by (5.57). Applying the star operation to this two-form yields the one-form

$$\stackrel{1}{\omega}_{\boldsymbol{b}} \stackrel{\text{def}}{=} * \,\mathrm{d}\stackrel{1}{\omega}_{\boldsymbol{a}} \equiv \sum_{i=1}^{3} b_i(\boldsymbol{x})\mathrm{d}x^1 = \left(\frac{\partial a_2}{\partial x^1} - \frac{\partial a_1}{\partial x^2}\right)\mathrm{d}x^3 + \text{cyclic permutations} ,$$

where we have set $b_1 = \partial a_3/\partial x^2 - \partial a_2/\partial x^3$ (and cyclic permutations). Thus, we obtain again a form of the type (5.56) whose coefficients are the components of curl $\boldsymbol{a}(\boldsymbol{x})$. This result is due to the dimension of the space $\mathbb{R}^3$: the star operation turns a two-form into a one-form, and vice versa. The space of one-forms has dimension $\binom{n}{1}$, the space of two-forms has dimension $\binom{n}{2}$. For $n = 3$ we have $\binom{n}{1} = \binom{n}{2} = 3$, i.e. these dimensions are equal and the two spaces are isomorphic. On the basis of this observation let us work out the relation between the exterior product of Sect. 5.4.3 and the vector product in $\mathbb{R}^3$. For two vectors $\boldsymbol{a}$ and $\boldsymbol{b}$ construct the one-forms $\omega_{\boldsymbol{a}}$ and $\omega_{\boldsymbol{b}}$, respectively, following the pattern of (5.56). Take their exterior product and apply the star operation to it. This gives the one-form

$$\begin{aligned}
*(\stackrel{1}{\omega}_{\boldsymbol{a}} \wedge \stackrel{1}{\omega}_{\boldsymbol{b}}) &= (a_1b_2 - a_2b_1)^*(\mathrm{d}x^1 \wedge \mathrm{d}x^2) + \text{(cyclic permutations)}\\
&= (a_1b_2 - a_2b_1)\mathrm{d}x^3 + \text{(cyclic permutations)}\\
&= \stackrel{1}{\omega}_{\boldsymbol{a}\times\boldsymbol{b}} \, .
\end{aligned} \tag{5.59}$$

This formula explains in which sense the $\wedge$-product generalizes the ordinary vector product.

Finally, to a given vector field $\boldsymbol{a}(\boldsymbol{x})$ we can also associate the following two-form:

$$\stackrel{2}{\omega}_{\boldsymbol{a}} \stackrel{\text{def}}{=} a_1\mathrm{d}x^2 \wedge \mathrm{d}x^3 + \text{(cyclic permutations)}. \tag{5.60}$$

Taking its exterior derivative, we obtain a three-form whose coefficient is the divergence of $\boldsymbol{a}$,

$$\mathrm{d}\overset{2}{\omega}_{\boldsymbol{a}} = \left(\frac{\partial a_1}{\partial x^1} + \frac{\partial a_2}{\partial x^2} + \frac{\partial a_3}{\partial x^3}\right) \mathrm{d}x^1 \wedge \mathrm{d}x^2 \wedge \mathrm{d}x^3 \,. \tag{5.61}$$

Of course, the star operation can be applied to the two expressions (5.60) and (5.61), giving the results

$$*\overset{2}{\omega}_{\boldsymbol{a}} = \overset{1}{\omega}_{\boldsymbol{a}} \quad \text{and} \quad *\,(\mathrm{d}\overset{2}{\omega}_{\boldsymbol{a}}) = \operatorname{div} \boldsymbol{a} \,.$$

The dimension $n = 3$ is essential if one wishes to interpret the vector product $\boldsymbol{a} \times \boldsymbol{b}$ as another vector. This isomorphism does not hold in dimensions other than 3. Note, however, that the cross product $\boldsymbol{a} \times \boldsymbol{b}$ in $\mathbb{R}^3$ is a vector of a different nature than $\boldsymbol{a}$ or $\boldsymbol{b}$. For example, $\boldsymbol{a} \equiv \boldsymbol{r}$ (position vector) and $\boldsymbol{b} \equiv \boldsymbol{p}$ (momentum vector) are *odd* with respect to space reflection, while their vector product $\boldsymbol{l} = \boldsymbol{r} \times \boldsymbol{p}$ (angular momentum vector) is *even*. A vector that is even under space reflection is said to be an *axial vector*.

A final remark: one may be surprised that the one-form (5.56) can be used to describe a vector field, even though vector fields have the coordinate representation $\sum a^i(x)\partial_i$. The reason for this is that $\mathbb{R}^3$ admits a metric that acts on vector fields: $g(v, w)$ with $g(\partial_i, \partial_k) = g_{ik}$. Interpreting the metric $g(v, w)$ as a mapping from w to v shows that it generates an isomorphism between $\mathcal{X}^*(M)$ and $\mathcal{X}(M)$.

5.5 Hamilton–Jacobi and Lagrangian Mechanics

In Sects. 5.1 and 5.3.3 we described qualitatively the manifolds of generalized coordinates as well as their tangent and cotangent bundles on which the Lagrangian function and the Hamiltonian function are respectively defined (cf. Fig. 5.2). In this section we examine these relations in more detailed and precise terms. We study geometric objects that live on the manifolds sketched in Fig. 5.2 and most of which are already known to us from Chap. 2. In particular, we define and study the so-called canonical two-form on phase space, which describes the symplectic structure of phase space (cf. Sect. 2.28), as well as all consequences following from this structure (such as Liouville's theorem, Poisson brackets, etc.). We study the Hamiltonian vector fields, (i.e. the canonical equations in a geometric language), and the geometric formulation of Lagrangian mechanics, as well as the relation between these two descriptions.

5.5.1 Coordinate Manifold Q, Velocity Space TQ, and Phase Space T^*Q

In Sect. 5.3.3 we remarked that Lagrangian functions $L(q, \dot{q}, t)$ are *functions* on the tangent bundle TQ of the coordinate manifold Q, i.e. $L \in \mathcal{F}(TQ)$,

$$L : TQ \to \mathbb{R} \,. \tag{5.62}$$

In writing this down we have used a local coordinate expression. Indeed, $\{q\} = \{q^1, \ldots, q^f\}$ represents the point $u \in Q$ in a chart, $f = \dim Q$ being the number of degrees of freedom, while $\{\dot{q}\} = \{\dot{q}^1, \ldots, \dot{q}^f\}$ gives the local components of an arbitrary tangent vector $v_u = \sum \dot{q}^i \partial_i \in T_u Q$. One should not be confused by the notation: the $\{\dot{q}\}$ are the tangent vectors to all possible curves $\gamma(t)$ passing through $u \in Q$. Only if we are given the *solutions* $q = \Phi(t, t_0, q_0)$ of the equations of motion (which follow from the Lagrangian function) do their tangent vectors generate the velocity field corresponding to real physical motion.

According to (5.62) L is to be understood really as a *function* on the manifold TQ. It is not a mapping of the kind studied in Sect.5.3.5, which assigns to each tangent vector, an element of TQ, a real number (in other words, it is not a one-form). Let us analyze this in a little more detail. First, we confirm that TQ, the tangent bundle of the smooth manifold Q, is again a smooth manifold of dimension $\dim TQ = 2\dim Q$. Therefore, it is possible to define smooth functions on TQ. (The general prescription is this. Let M be a smooth manifold of dimension m. With (φ, U), a local chart of $(M, \mathcal{A})$ belonging to the complete atlas $\mathcal{A}$, we construct the corresponding differential, or tangent, mapping $T\varphi$, following the definition (5.40a). With $U \subset M$, $T\varphi$ maps the domain $TU = U \times T_u M$, $u \in U$, of TM onto $\varphi(U) \times \mathbb{R}^m$. One then shows that $T\mathcal{A} = \{(T\varphi, TU)\}$ is a complete atlas for the manifold TM.)

In the simplest case a Lagrangian function has the local form (the so-called natural form)

$$L = T_{\text{kin}}(q, \dot{q}) - V(q) , \tag{5.63}$$

where V is a potential, while T_{kin} is the kinetic energy whose general form could be

$$T_{\text{kin}} = \frac{1}{2} \sum_{i,k=1}^{f} \dot{q}^i g_{ik}(q) \dot{q}^k \tag{5.64}$$

Here, the tensor $g_{ik}(q)$ is the matrix representation of a metric and may depend on the base point q. For a single particle in $\mathbb{R}^3$ we have $g_{ik} = \delta_{ik}$, with $i, k = 1, 2, 3$. Of course, a potential that does not depend on velocities, say $V(u)$, is initially defined to be a function on Q. However, from Sect.5.4.1, it can easily be transported to TQ. Indeed, if $\pi : TQ \to Q$ is the natural projection (5.25), then the pull-back of the function $V(u)$

$$\pi^* V = V \circ \pi$$

is a function on TQ. The action of $\pi^* V$ on elements v_u of $T_u Q$ is very simple: π projects onto the base point u, i.e. just cuts out the vector component of v_u.

The kinetic energy (5.64), in turn, is defined on TQ from the start, in a nontrivial way. To understand this better, we first give a precise definition of the *metric*. So far we have dealt with the set of smooth vector fields $\mathcal{X}(M)$ and with the set of smooth one-forms $\mathfrak{X}^*(M)$, cf. Sects. 5.3.4–5.3.5. The former are also called *contravariant* tensors of rank 1 and one may write equivalently

$$\mathfrak{X}(M) \equiv \mathcal{T}_0^1(M) . \tag{5.65a}$$

The latter are also said to be *covariant* tensors of rank 1 and one writes correspondingly

$$\mathfrak{X}^*(M) = \mathcal{T}_1^0(M) . \tag{5.65b}$$

We have further considered geometric objects that can be understood to be tensors of higher rank. For example, the two-forms we generated by taking the exterior product of two one-forms, $\overset{2}{\omega} = \overset{1}{\omega}_a \wedge \overset{1}{\omega}_b$, are smooth, bilinear mappings from the product $TM \times TM$ to $\mathbb{R}$. Therefore, they are contravariant tensors of rank 2 that, in addition, are antisymmetric. In general, a tensor T_s^r with r *contra*variant indices and s *co*variant indices is defined to be a multilinear mapping of r copies of T^*M times s copies of TM onto the real numbers, viz.

$$(T_s^r)_p : (T_p^*M)^r (T_pM)^s \to \mathbb{R} . \tag{5.66}$$

A *tensor field* of type $\binom{r}{s}$ assigns to each point $p \in M$ a tensor (5.66), in much the same way as the vector fields (5.26) and the one-forms (5.35) did, both of which are special cases of this general definition. The *set of all smooth tensor fields* of type is denoted by $\mathcal{T}_s^r(M)$.

Here we wish to define the metric, which is another special tensor field. Loosely speaking, a metric serves to define the norm of vectors and the scalar product of vectors (thereby specifying, in particular, orthogonality of vectors). Furthermore, by means of the metric tensor a vector (which is a contravariant rank-1 tensor) is turned into a covariant object (i.e. a covariant tensor of rank 1). In either case, the metric acts on vectors, i.e. on elements of the tangent space. Keeping this in mind, the following definition will be plausible.

ME. *Definition of metric.* A metric on a smooth manifold M is a tensor field g from $\mathcal{T}_2^0(M)$ (the smooth covariant tensor fields of rank 2), whose representative at every point p of M is symmetric and nondegenerate. This means that
(i) $g_p(v_p, w_p) = g_p(w_p, v_p)$ for all $v_p, w_p \in T_pM$ and at each point $p \in M$, and
(ii) if $g_p(v_p, w_p) = 0$ for a fixed $v_p \in T_pM$, but all $w_p \in T_pM$, then $v_p = 0$, at every point $p \in M$.

We can treat the metric as a mapping. In analogy to (5.26) and (5.35) we have

$$g \in \mathcal{T}_2^0(M) : M \to T^*M \times T^*M : p \mapsto g_p , \quad \text{where} \tag{5.67a}$$

$$g_p : T_pM \times T_pM \to \mathbb{R} : v, w \mapsto g_p(v, w) . \tag{5.67b}$$

Locally, i.e. in local charts, the metric can be applied to base fields, yielding the so-called metric tensor

$$g_p(\partial_i, \partial_k) \stackrel{\text{def}}{=} g_{ik}(p) . \tag{5.68}$$

The requirements ME(i) and ME(ii) then imply that (i) $g_{ik}(p) = g_{ki}(p)$, and (ii) the matrix $\{g_{ik}(p)\}$ is nonsingular. Its inverse is denoted by g^{ik}. Using the decomposition (5.29) of vector fields in terms of base fields, we have

$$g_p(v, w) = \sum_{i,k=1}^{n} v^i g_{ik}(p) w^k , \tag{5.69}$$

where v^i and w^k are the components of v_p and w_p, respectively, in a local representation of T_pM. The same statement can be phrased differently: locally the metric tensor can be written as a linear combination of tensor products of base one-forms as follows[4]:

$$g = \sum_{i,k} g_{ik}(p) \mathrm{d}x^i \otimes \mathrm{d}x^k . \tag{5.70}$$

Equipped with this knowledge we readily understand the structure of the form (5.64) of the kinetic energy, which is a function on TQ. Let $v_u \in T_uQ$ be represented locally by $v_u = \sum \dot{q}^i \partial_i$. Then, obviously, $T_{\text{kin}} = g_u(v_u, v_u)$. In fact, we may say much more than that. If g_p, (5.67a), is applied to only one vector field, a mapping from TM to T^*M is obtained,

$$g_p : TM \to T^*M : w \mapsto g_p(\bullet, w) \quad \text{(dot denotes vacancy)} .$$

In other words, $g_p(\bullet, w)$ is a one-form and $g_p(\bullet, w) \stackrel{\text{def}}{=} \omega_w$, which, upon application to a vector $v \in T_pM$, yields the real number $g_p(v, w)$. Thus, the metric assigns to each vector field $X \in \mathcal{X}(M)$ the smooth one form $g(\bullet, X) \in \mathcal{X}^*(M)$, and vice versa. This is precisely what happens when one introduces (in charts) the generalized momenta $p_i = \partial L / \partial \dot{q}^i$, which are canonically conjugate to the q^i. Using (5.63) and (5.64) one obtains

$$p_i = \frac{\partial T}{\partial \dot{q}^i} = \sum_k g_{ik}(p) \dot{q}^k \equiv g_p\left(\bullet, \sum \dot{q}^k \partial_k\right) . \tag{5.71}$$

The transition from the variables $\{q^i, \dot{q}^j\}$ to the variables $\{q^i, p_j\}$ that we studied in Chap. 2 in reality means that one goes over from a description of mechanics on the tangent bundle TQ to a description on the cotangent bundle T^*Q. If there exists a metric on Q then there is the isomorphism sketched above, which allows

[4] Using well-known techniques of linear algebra one can show that at each point $p \in M$ one can find a basis such that g_{ik} is diagonal, i.e. $g = \sum_{i=1}^{n} \varepsilon_i \mathrm{d}x^i \otimes \mathrm{d}x^i$, with $\varepsilon_i = \pm 1$. If all ε_i are equal to $+1$, the metric is said to be Riemannian. In all other cases it is said to be semi-Riemannian.

one to identify the two pictures. In general, however, this canonical identification is not guaranteed. In any case, whether or not a metric exists, TQ and T^*Q are two different spaces. Therefore, the transition from the Lagrangian formulation of mechanics to the Hamiltonian formulation is more than a simple change of variables. Very much like Q and TQ, the cotangent bundle T^*Q is a smooth manifold. In mechanics T^*Q is the phase space. In local charts it is described by coordinates $\{q^i, p_k\}$, where p_k has the character of a one-form, see (5.71). The Lagrangian function is defined on TQ, the Hamiltonian function on T^*Q (cf. Fig. 5.2). The two representations of mechanics are related by the Legendre transformation $\mathcal{L}$, as explained in Chap. 2.

The general case (without assuming a metric on Q) is treated by Abraham and Marsden (1981): mechanics on TQ and its formulation on T^*Q are related by means of the so-called fibre derivative. We cannot go into this more general treatment without introducing further mathematical tools. We point out, however, that the restricted case discussed above exhibits all essential features.

5.5.2 The Canonical One-Form on Phase Space

The Hamiltonian function is defined on the manifold $M \stackrel{\text{def}}{=} T^*Q$, which plays a central role in mechanics. Figure 5.11 shows in more detail the manifolds Q, TQ, T^*Q, and, in addition, the tangent bundle TM of the phase space. We shall return briefly to Lagrangian mechanis (on TQ) in Sect. 5.6 below. Here, our goal is to work out more clearly the geometric-symplectic structure of mechanics in phase space, well known to us from Chap. 2, and to understand it from a higher level. One possible approach is provided by what is called the canonical one-form θ_0 on phase space,

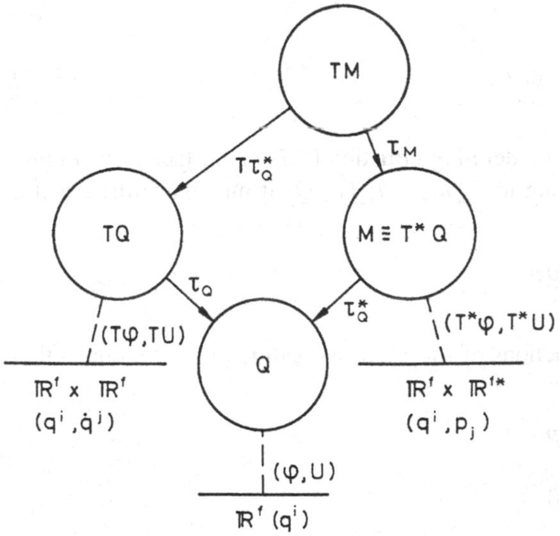

Fig. 5.11. The cotangent bundle $M \stackrel{\text{def}}{=} T^*Q$ is the phase space. M being a smooth manifold itself, it possesses a tangent bundle $TM = T(T^*Q)$. τ_Q and τ_Q^* are the canonical projections from TQ and T^*Q to Q, respectively, while τ_M is the projection from TM to M. TM and TQ, in turn, are related by the tangent mapping corresponding to τ_Q^*

$$\theta_0 : M \to T^*M : m \in M \mapsto (\theta_0)_m \in T^*_m M \,, \tag{5.72}$$

which is defined as follows. Let α be an arbitrary smooth one-form on the coordinate manifold Q:

$$\alpha : Q \to T^*Q : u \in Q \mapsto \alpha_u \in T^*_u Q \,. \tag{5.73}$$

The form θ_0 is to be defined on $M = T^*Q$. As α provides a mapping from Q to M, we can use it to pull back θ_0 from M to Q. This yields the one-form $(\alpha^*\theta_0)$, which lives on the base manifold Q. With this remark in mind we state the following definition.

C1F. The *canonical one-form* θ_0 is the unique form on $M = T^*Q$ whose pull-back onto Q by means of an arbitrary one-form α (5.73) yields precisely this α. Expressed in a formula, the canonical one-form θ_0 fulfills

$$(\alpha^*\theta_0) = \alpha \quad \text{for all} \quad \alpha \in \mathcal{X}^*(Q) \,. \tag{5.74}$$

This requirement fixes θ_0 uniquely.

Remark: In view of its specific role as defined by the rule (5.74) the one-form θ_0 is also called the *tautological form*.

As shown in Fig. 5.11, a chart (φ, U) of the domain $U \subset Q$ induces a chart $(T\varphi, TU)$ for $TU \subset TQ$, as well as a chart $(T^*\varphi, T^*U)$ for $T^*U \subset M = T^*Q$. A point $u \in U$ has the image $\{q^i\} = \{\varphi^i(u)\}$, which belongs to the neighborhood $U' = \varphi(U)$ in $\mathbb{R}^f$. A tangent vector $v_u \in T_u Q$, with base point u, has the image $\{q^i = \varphi^i(u), v^i = T\varphi^i(v) \equiv \dot{q}^i\}$ in $U' \times \mathbb{R}^f$. Similarly, each one-form $\omega_u \in T^*_u Q$ has the image $\{q^i, \alpha_i \equiv p_i\}$ in $U' \times (\mathbb{R}^f)^*$. Thus, the local representation of α_u (5.37) reads

$$\alpha_u = \sum_{j=1}^{f} \alpha_j(\underset{\sim}{q})\mathrm{d}q^j \equiv \sum_{j=1}^{f} p_j \mathrm{d}q^j \,. \tag{5.75}$$

When expressed in local form, the defining equation (5.74) is in fact very simple: $(\theta_0)_m$ being a one-form belonging to $T^*_m M = T^*_m(T^*Q)$ it must have the general local form

$$(\theta_0)_m = \sum_i \sigma_i \mathrm{d}q^i + \sum_k \tau^k \mathrm{d}p_k \,,$$

where σ_i and τ^k are smooth functions of $(\underset{\sim}{q}, \underset{\sim}{p})$. The condition (5.74) requires that these functions be

$$\sigma_i(\underset{\sim}{q}, \underset{\sim}{\alpha}(\underset{\sim}{q})) = \sigma_i(\underset{\sim}{q}, \underset{\sim}{p}) \stackrel{!}{=} p_i \,,$$

$$\tau^k(\underset{\sim}{q}, \underset{\sim}{\alpha}(\underset{\sim}{q})) = \tau^k(\underset{\sim}{q}, \underset{\sim}{p}) \stackrel{!}{=} 0 \,.$$

Thus, the local form of the canonical one-form is the same as α_u (5.75)

$$\boxed{(\theta_0)_m = \sum_{i=1}^{f} p_i \mathrm{d}q^i} \qquad m \in M = T^*Q\,. \tag{5.76}$$

Note, however, that θ_0 is defined on the phase space $M = T^*Q$, i.e. that $(\theta_0)_m$ is an element of $T^*_m M$, in contrast to the arbitrary one-form α, which lives "one storey below".

Remark. The canonical one-form is the key to the geometric formulation of mechanics on phase space. Starting from the definition given above and making use of Fig. 5.11, one can work out the following pattern. Let u be a fixed point on the base manifold Q, v_u a tangent vector from $T_u Q$, and α_u a one-form from $T^*_u Q$. Then $r \stackrel{\text{def}}{=} \alpha_u(v_u)$ is a real number. Using the definition (5.74) we can write it alternatively as $r = (\alpha^*\theta_0)_u(v_u)$. Now, as α maps the basis Q onto T^*Q, the corresponding tangent mapping $T\alpha$ maps TQ onto TM, the base point u being fixed. Let $w_u \in T_{\alpha_u}M$ be the preimage of α_u by the projection τ_M, i.e. $w_u = \tau_M^{-1}(\alpha_u)$. Then we have $w_u = T\alpha(v_u)$, while the same real number r is also given by

$$r = (\theta_0)_{m=\alpha_u}(w_u) = \alpha_u \circ T\tau^*_Q(w_u)\,.$$

This last equation can be used to define θ_0, (Abraham, Marsden 1981, Sect. 3.2.10). With this alternative but equivalent definition the derivation of the local form (5.76) is a bit more tedious.

One can understand that θ_0 is indeed unique by noting that condition C1F is to be fulfilled for all α_u. These forms span the space $T^*_u Q$ completely. As the v_u are arbitrary, too, their preimages w_u span the complete space $T_{\alpha_u}M$.

Loosely speaking, C1F is a prescription that says that arbitrary one-forms on Q should be interpreted as a specific one-form on T^*Q. It is canonical and characteristic for the cotangent bundle insofar as one-forms live on T^*Q and are pulled *back* by mappings (in contrast to vector fields, which are mapped "forward"). The local representation (5.76) is sufficient because one can always join together the charts of a complete atlas and describe θ_0 in this way, on the whole of $M = T^*Q$. Of course, the definition given in (5.74), or the one described briefly in the remark above, are completely free of coordinates.

Let $F = \psi \circ \varphi^{-1}$ be the transition mapping from the chart (φ, U) to the chart (ψ, V). In the overlap of the images of U and of V, F maps the point $\{q\} = \varphi(u)$ to the point $\{Q\} = \psi(u)$. This is the same point in $\mathbb{R}^f$, but it is expressed in terms of different coordinates. A tangent vector $v_u \in T_u Q$ whose coordinate image is $\{\dot{q}\}$ in the first case and $\{\dot{Q}\}$ in the second is transformed by means of the tangent mapping TF, while one-forms are pulled back according to (5.41). As to the canonical one-form, we note that it keeps its local form (5.76). Indeed, we have

$$p_i = \sum_{k=1}^{f} \frac{\partial Q^k}{\partial q^i} P_k\,,$$

and therefore

$$\sum_i p_i \mathrm{d}q^i = \sum_i \sum_k P_k \frac{\partial Q^k}{\partial q^i} \mathrm{d}q^i = \sum_k P_k \mathrm{d}Q^k \,. \tag{5.77}$$

This result is obvious because the definition (5.74) fixes θ_0 on the whole of T^*Q and because the local form (5.76) holds in each chart. The following assertion is somewhat less obvious.

Proposition. Let $F : Q \to Q$ be a diffeomorphism on the base manifold Q. With α a one-form on Q the pull-back of $\alpha_u \in T_u^*Q$ is then defined in either direction, so that F induces a diffeomorphism $T^*F : T^*Q \to T^*Q$. The pull-back of the canonical one-form is given by

$$(T^*F)^*\theta_0 = \theta_0 \,. \tag{5.78}$$

In this sense it is invariant.

Abraham and Marsden (1981, Theorem 3.2.12) provide a proof that does not make use of coordinates. In coordinates, the proof, in essence, follows from the calculation done in (5.77).

5.5.3 The Canonical, Symplectic Two-Form on M

The canonical two-form is defined to be (minus) the exterior derivative of the canonical one-form θ_0 of C1F (5.74), viz.

$$\textbf{C2F.}\quad \omega_0 \stackrel{\text{def}}{=} -\mathrm{d}\theta_0 \,. \tag{5.79}$$

This is a closed form, $\mathrm{d}\omega_0 = -\mathrm{d} \circ \mathrm{d}\theta_0 = 0$. Its representation in local coordinates follows from the local form (5.76) of θ_0. It reads

$$(\omega_0)_m = \sum_{i=1}^{f} \mathrm{d}q^i \wedge \mathrm{d}p_i \,, \quad m \in M \,. \tag{5.80}$$

This form is of special importance because it exhibits the symplectic structure of phase space. This will be clear from the following observations and propositions.

As a two-form on M, ω_0 is a bilinear mapping from $TM \times TM$ to the real numbers. It acts on pairs $(w^{(a)}, w^{(b)})$ of vector fields on M, i.e. $(\omega_0)_m$ is applied to pairs $(w_m^{(a)}, w_m^{(b)})$ of tangent vectors from T_mM, where $w_m^{(a)}$ and $w_m^{(b)}$ are the representatives in T_mM of $w^{(a)}$ and $w^{(b)}$, respectively. In charts any such vector field has the form

$$w = \sum_{i=1}^{f} w^i \frac{\partial}{\partial q^i} + \sum_{k=1}^{f} \bar{w}_k \frac{\partial}{\partial p_k} \,, \tag{5.81}$$

so that

$$(\omega_0)_m(w_m^{(a)}, w_m^{(b)}) = \sum_{i=1}^{f}(w^{(a)i}\bar{w}_i^{(b)} - \bar{w}_i^{(a)}w^{(b)i}) . \tag{5.82}$$

If we agree on ordering coordinates such that $\eta^i = \mathrm{d}q^i$, with $i = 1, \ldots, f$, form the first set of base forms, and $\eta^{i+f} = \mathrm{d}p_i$, $i = 1, \ldots, f$, form the second set, and if we write (ω_0) in the general form

$$(\omega_0)_m = \sum_{i,k} \omega_{ik}\eta^i \wedge \eta^k , \tag{5.82$'$}$$

it is easy to see that its coefficients ω_{ik} are given by

$$\omega_{ik} = \begin{pmatrix} 0_{f\times f} & 1\!\!1_{f\times f} \\ -1\!\!1_{f\times f} & 0_{f\times f} \end{pmatrix} .$$

This matrix is nothing but the matrix $\mathbf{J}$ of (2.102). As $\mathbf{J}$ is regular, one sees that $(\omega_0)_m$ is *nondegenerate and skew-symmetric*. As this holds at each point $m \in M$, the canonical two-form ω_0 is nondegenerate and skew-symmetric on the whole of M. Thus, the form ω_0 must be closely related to the canonical equations (2.99). Before we turn to this relationship we wish to point out an interesting property of the cotangent bundle $M = T^*Q$.

Taking the k-fold exterior products of $(\omega_0)_m$ with itself yields forms of degree $2k$. For example, for $k = 2$ and $k = 3$, respectively,

$$\begin{aligned} (\omega_0)_m \wedge (\omega_0)_m &= \sum_{i_1,i_2=1}^{f} \mathrm{d}q^{i_1} \wedge \mathrm{d}p_{i_1} \wedge \mathrm{d}q^{i_2} \wedge \mathrm{d}p_{i_2} \\ &= -2! \sum_{i_1<i_2} \mathrm{d}q^{i_1} \wedge \mathrm{d}q^{i_2} \wedge \mathrm{d}p_{i_1} \wedge \mathrm{d}p_{i_2} \\ (\omega_0)_m \wedge (\omega_0)_m \wedge (\omega_0)_m &= -3! \sum_{i_1<i_2<i_3} \mathrm{d}q^{i_1} \wedge \mathrm{d}q^{i_2} \wedge \mathrm{d}q^{i_3} \wedge \mathrm{d}p_{i_1} \wedge \mathrm{d}p_{i_2} \wedge \mathrm{d}p_{i_3} . \end{aligned}$$

The form of highest degree that can be constructed in this way has degree $2f$. It reads

$$\underbrace{(\omega_0)_m \wedge \ldots \wedge (\omega_0)_m}_{f\text{-fold}} = f!(-)^{[f/2]}\mathrm{d}q^1 \wedge \mathrm{d}q^2 \wedge \ldots \wedge \mathrm{d}q^f \wedge \mathrm{d}p_1 \wedge \ldots \wedge \mathrm{d}p_f , \tag{5.83a}$$

where $[f/2]$ is the largest integer smaller than or equal to $f/2$. This f-fold product generates the oriented *volume form*

$$\Omega \stackrel{\text{def}}{=} \frac{(-1)^{[f/2]}}{f!}\omega_0 \wedge \ldots \wedge \omega_0 \quad (f \text{ factors}) \tag{5.83b}$$

on T^*Q, whose value in the point m is proportional to the expression (5.83a). This is an important result. On the cotangent bundle of a smooth manifold there always exist the canonical forms θ_0 and ω_0 and thus also the volume form (5.83a). The cotangent bundle of a manifold Q is always orientable, even if its base manifold Q is not. At the same time, we have established the basis for Liouville's theorem. Only the result (5.83a) enables us to talk about flows on phase space that preserve volume and orientation. As a consequence, the specific properties of phase space that we studied by means of the canonical equation (2.99), in the more "pedestrian" approach of Chap. 2, rest on an underlying, deeper geometric structure. The following subsection is devoted to a short discussion of this structure. (As this is a digression, the reader may wish to skip it on a first reading and move on directly to Sect. 5.5.5.)

5.5.4 Symplectic Two-Form and Darboux's Theorem

Very much like the metric on a Riemannian or semi-Riemannian manifold the canonical two-form is a covariant tensor of rank 2 on the manifold M. Like the metric it is nondegenerate. While the metric pertains to the set of symmetric tensors, ω_0 belongs to the set of antisymmetric forms of degree two.

Let M be a smooth manifold of dimension $\dim M = n$, and let ω be a covariant tensor (a general one, at first),

$$\omega \in \mathcal{T}_2^0(M) : M \to T^*M \times T^*M : p \mapsto (\omega)_p \, . \tag{5.84}$$

ω is said to be *nondegenerate* if $(\omega)_p$ has this property at every point $p \in M$. T_pM is a vector space of dimension n, $T_pM \times T_pM$ has dimension $2n$, and $(\omega)_p$ maps $T_pM \times T_pM$ onto the real numbers.

One proves the following assertions.

(a) If $(\omega)_p$ is *symmetric* and nondegenerate, i.e. if the matrix $\omega_{ik} = (\omega)_p(\partial_i, \partial_k)$ is regular, then there is an ordered basis of T_pM and an ordered basis of T_p^*M, dual to the former, such that this matrix is diagonal, its eigenvalues being $\varepsilon_i = \pm 1$ (cf Footnote 4 to (5.70)).

(b) If $(\omega)_p$ is antisymmetric and if the matrix $\{\omega_{ik}\}$ has rank r, then r is an even integer and there is an ordered basis of T_pM and its dual in T_p^*M such that

$$(\omega)_p = \sum_{i=1}^{r/2} \mathrm{d}x^i \wedge \mathrm{d}x^{i+r/2} \, ,$$

i.e. such that the matrix $\{\omega_{ik}\}$ has the form

$$\{\omega_{ik}\} = \begin{pmatrix} 0 & \mathbb{1} & 0 \\ -\mathbb{1} & 0 & 0 \\ 0 & 0 & 0 \end{pmatrix}$$

with $\mathbb{1}$ being the unit matrix of dimension $r/2$. In this case ω can be nondegenerate only if the dimension n of M is *even*; the rank r is then equal to $r = 2n$.

This latter assertion is followed up by the following proposition.

Proposition. Let ω be an antisymmetric two-form on the manifold M (following the pattern of (5.84)). The form ω is nondegenerate if and only if M has *even* dimension, $n = 2k$, and if the k-fold exterior product $\omega \wedge \ldots \wedge \omega$ is a volume form on M.

Expressed differently, this says that if there exists a nondegenerate, skew-symmetric two-form on M, then M is orientable. If this is true, (5.83b) provides an oriented volume form, viz.

$$\Omega_\omega = \frac{(-1)^{[k]}}{(2k)!}\omega \wedge \ldots \wedge \omega \quad (k\text{-fold}) \tag{5.85}$$

with k given by $\dim M = n = 2k$.

The relation to the symplectic group that we studied in Sect. 2.28 becomes clear by way of the following definitions and assertions.

SYF. Every nondegenerate, skew-symmetric two-form σ on a vector space V of even dimension $n = 2k$ is called a *symplectic form*. In the case treated above, we had $\sigma \equiv (\omega)_p$ and $V \equiv T_pM$.

SYV. The pair (V, σ) is said to be a *symplectic vector space* if $\dim V = 2k$ and if σ has the property SYF.

SYT. *Symplectic transformations* are defined to be transformations between vector spaces that preserve the symplectic structure SYV, i.e. if (V, σ) and (W, τ) are symplectic vector spaces, then

$$F : V \to W$$

is symplectic precisely if the pull-back of τ onto V equals σ, $F^*\tau = \sigma$.

The vector spaces V and W need not have the same dimension. However, if they do have the same dimension $n = 2k$, F preserves the oriented volume. This is seen by showing that $F^*\Omega_\tau = \Omega_\sigma$, where Ω_τ and Ω_σ are the standard n-forms (5.85) on W and V, respectively. Symplectic transformations have the following property. The symplectic mappings of a symplectic vector space (V, σ) onto itself,

$$F : (V, \sigma) \to (V, \sigma)\,, \quad F^*\sigma = \sigma\,,$$

form the symplectic group $Sp_{2f}(\mathbb{R})$. In order to show this, let us choose that basis $\{e^i\}$ of V for which σ has the canonical form

$$\{\sigma_{ik}\} = \begin{pmatrix} 0 & \mathbb{1} \\ -\mathbb{1} & 0 \end{pmatrix} \equiv \mathbf{J}\,.$$

In this basis the transformation F is represented by the matrix $\{F^i_k\}$, i.e. $e'^i = \sum_{k=1}^n F^i_k e^k$. The condition $F^*\sigma = \sigma$ says that $\sigma(e'^i, e'^j)$ must be equal to

$\sigma(e^i, e^j)$, i.e. that $\mathbf{F}^T\mathbf{J}\mathbf{F} = \mathbf{J}$. This is precisely (2.113) and tells us that the matrix $\mathbf{F}$ pertains to $Sp_{2f}(\mathbb{R})$.

Note that the definitions and assertions given above apply to the representative $(\omega)_p$ of ω over the base point $p \in M$. They are extended to ω, and thus to the whole of M, by means of the following theorem.

Darboux's Theorem. Let ω be a nondegenerate two-form on the manifold M whose dimension is therefore even, $\dim M = n = 2k$. The form ω is closed, i.e. $d\omega = 0$, precisely if for each point $p \in M$ there exists a chart (φ, U) such that $\varphi(p) = 0$ and such that in every point $p' \in U \subset M$ with

$$\varphi(p') = (x^1(p'), \ldots, x^k(p'), \ldots x^{2k}(p'))$$

ω admits the local representation

$$\omega = \sum_{i=1}^{k} dx^i \wedge dx^{i+k} \qquad (5.86)$$

on the neighborhood U.

For the proof of this theorem, as well as of the other assertions of this section, we refer to Abraham and Marsden (1981).

We close this digression with some definitions and remarks that serve the purpose of generalizing definitions SYF, SYV, and SYT to arbitrary manifolds.

S1. A *symplectic form* on a manifold M of even dimension $\dim M = n = 2k$ is a nondegenerate, skew-symmetric, closed two-form ω,

$$d\omega = 0\,. \qquad (5.87)$$

S2. A pair (M, ω), with ω having property S1, is said to be a *symplectic manifold*.

S3. Those charts where (5.86) holds true (whose existence is guaranteed by Darboux's theorem) are said to be *symplectic charts*. Their local coordinates are called *canonical coordinates*.

S4. A smooth mapping F that relates two symplectic manifolds (M, σ) and (N, τ) is said to be symplectic if $F^*\tau = \sigma$. The symplectic mappings are the canonical transformations of mechanics if the starting and the target manifolds are identical.

These notions belong to what is called *symplectic geometry*. As far as mechanics is concerned, the importance of symplectic geometry should be clear from our discussion. In fact, it seems to be relevant for many more parts of physics and therefore leads directly into modern research. In this connection we refer the reader to Guillemin and Sternberg (1986).

5.5.5 The Canonical Equations

In Chap.2, Sect. 2.25, we showed that the canonical equations (2.45) could be written in the form (2.99), viz.

$$\dot{\underset{\sim}{x}} = \mathbf{J} H_{,x} \equiv (X_{\mathrm{H}})_x \tag{5.88}$$

Here, $\underset{\sim}{x}$ is a point in phase space, while $H_{,x}$ and $\mathbf{J}$ are defined as in (2.102). We realize that (5.88) is a local representation in charts. As indicated by the subscript x, the Hamiltonian vector field on the right-hand side of (5.88) (cf. the definition in Sect.5.3.1) is a coordinate expression in charts. On the basis of the results obtained in Sect. 5.5.3 it is clear that the canonical two-form will serve the purpose of formulating the canonical equations of motion in a coordinate-free manner, i.e. directly on T^*Q, the cotangent bundle of the coordinate manifold Q.

Let $M = T^*Q$, as before. Vector fields on M assign to each $p \in M$ an element of the tangent space T_pM at that point,

$$X \in \mathcal{X}(M) : M \to TM : p \mapsto X_p \,.$$

In charts X_p has the local form (5.81). Equation (5.88) defines the Hamiltonian vector fields in charts, i.e. componentwise. Thus, in the notation of (5.88),

$$(X_{\mathrm{H}})^i = \frac{\partial H}{\partial p_i}\,, \quad \overline{(X_{\mathrm{H}})_k} = -\frac{\partial H}{\partial q^k}\,. \tag{5.89}$$

These partial derivatives of H also appear in the exterior derivative $\mathrm{d}H$. As H is a function on M, its exterior derivative is equal to the total differential. When expressed locally, we have

$$\mathrm{d}H = \sum_{i=1}^{f} \frac{\partial H}{\partial q^i}\,\mathrm{d}q^i + \sum_{j=1}^{f} \frac{\partial H}{\partial p_j}\,\mathrm{d}p_j \,. \tag{5.90}$$

As we know, the Hamiltonian vector field is

$$\begin{pmatrix} (X_{\mathrm{H}})^i \\ (X_{\mathrm{H}})_k \end{pmatrix} = \begin{pmatrix} 0 & \mathbb{1} \\ -\mathbb{1} & 0 \end{pmatrix} \begin{pmatrix} \dfrac{\partial H}{\partial q^i} \\[2ex] \dfrac{\partial H}{\partial p_j} \end{pmatrix},$$

or $(X_{\mathrm{H}})_x = \mathbf{J}(\mathrm{d}H)_x$, where the subscript x is meant to indicate that we still compare coordinate expressions.

As $\mathbf{J}^{-1} = -\mathbf{J}$, we can also write $-\mathbf{J}(X_{\mathrm{H}})_x = (\mathrm{d}H)_x$. From this we can abstract the coordinate-free definition of the Hamiltonian vector field as follows. $\mathbf{J}$ is nothing but the local matrix representation (5.82′) of the canonical two-form ω_0. Such a two-form ω acts on pairs of vector fields. In analogy to the case of the metric, one may instead take ω to act on only *one* vector field, e.g. $\omega(V, \bullet)$

with the dot denoting a vacancy (it stands for the missing second argument). As such it maps the tangent bundle TM onto $\mathbb{R}$, i.e. it operates like an exterior form of degree 1. With this remark in mind, the following definition becomes readily understandable.

> **HVF.** Let (M, ω) be a symplectic manifold, i.e. $\dim M = 2f$ is even and ω has properties S1. The Hamiltonian function H is assumed to be given as a smooth function on $M = T^*Q$. The *Hamiltonian vector field* X_H is defined through the condition
>
> $$\boxed{\omega(X_H, \bullet) = \mathrm{d}H\,.} \qquad (5.91)$$
>
> The triple (M, ω, X_H) is said to be a *Hamiltonian system*.

With $Y \in \mathcal{X}(M)$ an arbitrary vector field on M, we have from (5.91)

$$\omega(X_H, Y) = \mathrm{d}H(Y)\,.$$

As ω is nondegenerate, this equation fixes X_H uniquely. Indeed, if there were two different vector fields X_H and X'_H for the same function H, then $\omega(X_H - X'_H, Y) = 0$ for all Y. This is possible only if $X_H - X'_H$ vanishes identically. On the other hand, $\mathrm{d}H(Y)$ cannot be zero for all Y, unless $H = 0$. Hence, for each H there is a unique X_H. In local coordinates the defining equation (5.91) yields precisely the expressions (5.89). This is verified by direct calculation,

$$\omega_p(X_H, \bullet) = \sum_{i=1}^{f} (X_H)^i \mathrm{d}p_i - \sum_{k=1}^{f} \overline{(X_H)_k}\, \mathrm{d}q^k\,.$$

Comparing with $\mathrm{d}H$ (5.90) yields (5.89). The definition (5.91) is independent of coordinates, however, and it is not restricted to the case of finite dimension.

The integral curves of the vector field X_H, i.e. the solutions of the differential equation

$$\dot{\gamma}(t) = (X_H)_{\gamma(t)}\,, \qquad (5.92)$$

describe the possible physical motions of the system defined by the Hamiltonian function H. When expressed in local coordinates, (5.92) becomes (5.88) and hence the local form of the canonical equations of motion (2.45).

If H has no explicit time dependence and if $\gamma(t)$ is a solution of (5.92), then

$$\frac{\mathrm{d}}{\mathrm{d}t} H(\gamma(t)) = \mathrm{d}H(\dot{\gamma}) = \mathrm{d}H(X_H(\gamma(t))) = \omega(X_H(\gamma), X_H(\gamma)) = 0\,.$$

This is the well-known fact that H is constant along solutions of the equations of motion.

It is not difficult to formulate once more Liouville's theorem, Sect. 2.29, using the tools and results developed so far. When phrased in geometric terms it reads as follows.

> **Liouville's Theorem.** Let (M, ω, X_H) be a Hamiltonian system, i.e. let the nondegenerate, closed two-form ω and the Hamiltonian vector field X_H be given on a manifold with even dimension. Denote by Φ_t the flow of the vector field X_H (this is the set of all integral curves corresponding to all possible initial conditions). For all t the flow Φ_t is symplectic, i.e. $\Phi_t^* \omega = \omega$. As a consequence, the oriented volume Ω_ω (5.85) is conserved.

In Sect. 2.29 we proved this theorem in two equivalent ways. The proof in terms of geometry is instructive in several respects. The reader who wishes to skip it, on a first reading, should move on immediately to Sect. 5.5.6. The proof makes use of the *Lie derivative* and of the fact that the symplectic form ω is closed. The Lie derivative L_X, which refers to a smooth vector field X, is obtained from the following geometric picture. The vector field X defines (at least locally on M) the flow Φ_τ, i.e. the set of all solutions of the differential equation (5.42). Consider an arbitrary differentiable geometric object T on M such as a function, another vector field, a k-form or an $\binom{r}{s}$-tensor field. We ask the question in which way the object T changes differentially, along the lines of the flow Φ_τ of the vector field X. For a function the answer is very simple. At the point $p \in M$ this is just the directional derivative

$$\mathrm{d} f_p(X_p) \stackrel{\text{def}}{=} (L_X f)_p \,,$$

described in (5.33). The same derivative can also be written as

$$\left.\frac{\mathrm{d}}{\mathrm{d}\tau} f(\Phi_\tau(p))\right|_{\tau=0} = \left.\frac{\mathrm{d}}{\mathrm{d}\tau} \Phi_\tau^* f(p)\right|_{\tau=0} ,$$

where $\Phi_{\tau=0}(p) = p$ and where the right-hand side is to be understood as in (5.39). If T is another vector field $T \equiv Y$, its Lie derivative is given by the commutator $[X, Y] \stackrel{\text{def}}{=} L_X Y$, as explained in Sect. 5.3.4. (One may define L_X to be a differential operator on the smooth tensor fields on M, with the condition that it operate on functions and on vector fields as described above, see Abraham and Marsden (1981). The following definition is equivalent to this.)

By the existence and uniqueness theorem for differential equations of the type (5.42) the flow Φ_τ of X is a (local) diffeomorphism of M. Therefore, the geometric object T can be transported forward or backward along that flow (cf. Sect. 5.4.1). In particular, it can be differentiated along the flux lines of Φ_τ[5].

[5] For this reason V.I. Arnol'd (1978) calls the Lie derivative the fisherman's derivative. The fisherman sees only the river in front of him. He sees all kinds of objects floating by on the river and takes their differential along the lines of the river's flow.

Consider now the special case $T \equiv \alpha$ being an exterior k-form on M, X a vector field, and Φ_τ its (local) flow. According to what we said above, the Lie derivative fulfills the identity

$$\frac{\mathrm{d}}{\mathrm{d}\tau}\Phi_\tau^*\alpha = \Phi_\tau^* L_X\alpha \,. \tag{5.93}$$

The Lie derivative L_X, at the point $q = \Phi_\tau(p)$, pulled back to the point p, is the derivative with respect to the orbit parameter τ of the pull-back of the form α. Like α, $L_X\alpha$ is a k-form. Functions are to be read as zero-forms for which $L_X f = \mathrm{d}f(X)$. One can show that the Lie derivative can be expressed by means of the exterior derivative. If the vector field X is inserted in the position of the first argument of the form α, then $\alpha(X, \bullet(k-1)\bullet)$ is a $(k-1)$-form (positions 2 to k are vacant). Taking the exterior derivative of the latter yields again a k-form, $\mathrm{d}(\alpha(X, \bullet(k-1)\bullet))$. If, in turn, we differentiate α first we obtain the $(k+1)$-form $\mathrm{d}\alpha$. Inserting X into this $(k+1)$-form leads again to a k-form, namely $(\mathrm{d}\alpha)(X, \bullet(k)\bullet)$.[6] We then have

$$L_X\alpha = (\mathrm{d}\alpha)(X, \bullet(k)\bullet) + \mathrm{d}(\alpha(X, \bullet(k-1)\bullet)) \,. \tag{5.94}$$

(The proof goes by induction, see e.g. Abraham and Marsden (1981).) With the identities (5.93) and (5.94) Liouville's theorem follows immediately. Inserting the symplectic form ω, as well as the Hamiltonian vector field X_H, we obtain

$$\begin{aligned}\frac{\mathrm{d}}{\mathrm{d}t}\Phi_t^*\omega &= \Phi_t^* L_{X_\mathrm{H}}\omega \\ &= \Phi_t^*[(\mathrm{d}\omega)(X_\mathrm{H}, \bullet, \bullet) + \mathrm{d}(\omega(X_\mathrm{H}, \bullet))] \,.\end{aligned}$$

The first term vanishes because ω is closed. The second vanishes, too, because $\mathrm{d}(\omega(X_\mathrm{H}, \bullet)) = \mathrm{d} \circ \mathrm{d}H = 0$, by the definition (5.91). Finally, as $\Phi_{t=0}$ is the identity, we obtain $\Phi_t^*\omega = \omega$, for all t for which the flow is defined. This proves the theorem.

5.5.6 The Poisson Bracket

An essential ingredient in the proof of Liouville's theorem is the fact that the symplectic two-form ω is closed. In this section we establish (once more) the relationship between this form and the Poisson bracket, with the aim of understanding better the significance of $\mathrm{d}\omega = 0$. (In Sect.2.32 we showed that the Poisson bracket of two dynamical quantities is identical to the symplectic, skew-symmetric scalar product of their derivatives, hence the comment "once more".)

The dynamical quantities f and g that are to be inserted in the Poisson bracket (2.122) are smooth function on the phase space $M = T^*Q$. M is a symplectic manifold. Following the example of the Hamiltonian function (which is a smooth

---

[6] This prescription is called the *inner product*: $i_X\alpha(Y_1, \dots, Y_k) \stackrel{\text{def}}{=} \alpha(X, Y_1, \dots, Y_k)$ is said to be the inner product of X with α. The indentity (5.94) then reads $L_X\alpha = i_X(d\alpha) + \mathrm{d}(i_X\alpha)$.

function on M, too), we can assign to f and g vector fields X_f and X_g, respectively, by means of the definition (5.91). As ω is nondegenerate, the vector fields are uniquely fixed by the equations

$$\omega(X_f, \bullet) = \mathrm{d}f \quad \text{and} \quad \omega(X_g, \bullet) = \mathrm{d}g\,. \tag{5.95}$$

The Poisson bracket of f and g is nothing but the expression

$$\{f, g\} \stackrel{\text{def}}{=} \omega(X_g, X_f)\,. \tag{5.96}$$

To see this, let us interpret (5.96) as a definition and let us verify that locally (i.e. in charts) it is the same as (2.122). From (5.95) we have the local representation of X_f,

$$X_f = \left(\frac{\partial f}{\partial \underset{\sim}{p}}, -\frac{\partial f}{\partial \underset{\sim}{q}}\right),$$

and an analogous one for X_g. Inserting these into ω, we find, according to (5.82), that

$$\omega(X_g, X_f) = \frac{\partial g}{\partial \underset{\sim}{p}}\left(-\frac{\partial f}{\partial \underset{\sim}{q}}\right) - \left(-\frac{\partial g}{\partial \underset{\sim}{q}}\right)\frac{\partial f}{\partial \underset{\sim}{p}} = \{f, g\}\,,$$

i.e. precisely the expression (2.122). While the latter form is formulated in charts, the definition (5.96) is free of coordinates on M.

The properties of Poisson brackets, well known to us from Chap. 2, can also be formulated and proved in a manner that is independent of coordinates. One has the following.

(i) The Poisson bracket can be expressed in terms of Lie derivatives, viz.

$$\{f, g\} = L_{X_f} g = \mathrm{d}g(X_f) = -L_{X_g} f = -\mathrm{d}f(X_g)\,. \tag{5.97}$$

(The reader should verify this in local form.)

Comparing this with the definition (5.93) of the Lie derivative yields assertions (ii) and (iii).

(ii) The quantity f is constant along the flow of X_g if and only if $\{f, g\} = 0$. The same statement holds with f and g interchanged. For example, let ψ_τ be the flow of X_g. Then, from (5.93)

$$\frac{\mathrm{d}}{\mathrm{d}\tau}(\Psi_\tau^* f) = \frac{\mathrm{d}}{\mathrm{d}\tau}(f \circ \Psi_\tau) = \Psi_\tau^* L_{X_g} f = -\Psi_\tau^*\{f, g\}\,.$$

This is zero if and only if the Poisson bracket vanishes.

(iii) Let Φ_t be the flow of the Hamiltonian vector field X_{H}, g being a dynamical quantity as above. In the same manner as in (ii) one shows that

$$\frac{\mathrm{d}}{\mathrm{d}t}(g \circ \Phi_t) = \{H, g \circ \Phi_t\}\,. \tag{5.98}$$

If g does not depend explicitly on time, this is identical with (2.128). As we know, the canonical equations themselves can be written in the form of (5.98), cf. (2.127). What we have gained compared to Chap. 2 is this: the definition (5.96), the expressions (5.97), and the equations of motion (5.98) are formulated in a way independent of coordinates (without charts). Furthermore, they are not restricted to finite dimensions.

There are many more properties of Poisson brackets that can be derived using the geometric formulation. As we studied them in some detail in Chap. 2, though using a local representation, we restrict the discussion to a few characteristic examples.

The smooth functions $\mathcal{F}(M)$ on the phase space (which form a real vector space), together with the Poisson brackets, generate a Lie algebra. In order to see this, we must verify that $\{f, g\}$ is bilinear, that $\{f, f\}$ vanishes, and that the Jacobi identity holds true, viz.

$$\{f, \{g, h\}\} + \{g, \{h, f\}\} + \{h, \{f, g\}\} = 0 . \tag{5.99}$$

In local form, this identity was obtained by direct calculation, cf. Sect. 2.32 (2.131). In a coordinate-free framework one proceeds as follows. Define a Poisson bracket for one-forms $\mathrm{d}f$, $\mathrm{d}g$ (instead of functions, as above), by

$$\{\mathrm{d}f, \mathrm{d}g\} \stackrel{\text{def}}{=} \omega([X_f, X_g], \bullet) . \tag{5.100}$$

This Poisson bracket is again a one-form and we have $\mathrm{d}\{f, g\} = \{\mathrm{d}f, \mathrm{d}g\}$. The last equation establishes the relation to the Poisson bracket of functions. (Abraham and Marsden (1981) provide a proof.) With this result, and on the basis of the definition (5.100) as well as (5.95), we conclude that the vector field $X_{\{f,g\}}$ defined by $\omega(X_{\{f,g\}}, \bullet) = \mathrm{d}\{f, g\}$ equals the commutator of X_f and X_g, $X_{\{f,g\}} = [X_f, X_g]$.

In a second step we write out the individual terms of (5.99), making use of (5.97):

$$\begin{aligned}
\{f, \{g, h\}\} &= L_{X_f}(L_{X_g} h) , \\
\{g, \{h, f\}\} &= -L_{X_g}(L_{X_f} h) , \\
\{h, \{f, g\}\} &= -L_{X_{\{f,g\}}} h = -[L_{X_f}, L_{X_g}] h .
\end{aligned}$$

In the last expression we made use of the property $L_{[v,w]} = [L_v, L_w]$ of the Lie derivative. Adding the three terms indeed yields the identity (5.99). We have only sketched this proof here, because we had something else in mind: for the definition (5.96) of the Poisson bracket, together with the definition (5.95) of the vector fields corresponding to the functions f and g, it was essential that the canonical two-form was *closed*. Finally, then, this is the reason the algebra of the smooth functions $\mathcal{F}(M)$, with the composition $\{,\}$, is a Lie algebra.

The following proposition is of interest in the light of the discussion in Sect. 2.32.

Proposition. Let (φ, U) be a chart taken from the atlas for the symplectic manifold (M, ω), chosen such that points $u \in U$ are represented by $q^1, \ldots, q^f, p_1, \ldots, p_f$.

This chart is symplectic (i.e. the canonical two-form becomes $\omega = \sum_{i=1}^{f} \mathrm{d}q^i \wedge \mathrm{d}p_i$) if and only if the following Poisson brackets are fulfilled:

$$\{q^i, q^j\} = 0 = \{p_i, p_j\}, \{p_j, q^i\} = \delta^i_j \ . \tag{5.101}$$

Proof. (a) If the chart is symplectic one verifies (5.101) by direct calculation. (b) We assume these equations to hold and determine the matrix representation $\Omega \equiv (\omega_{ik})$ of ω in the domain of this chart (φ, U). Ω is regular and hence has an inverse $(\sigma^{ik}) \equiv \Sigma$. From (5.97) and (5.96) we have

$$\{q^i, q^k\} = \mathrm{d}q^i(X_{q^k}) = (X_{q^k})^i = \sigma^{ik} \ , \quad i, k = 1, \ldots, f \ .$$

In a similar fashion one shows that $\{p_i, p_k\} = \sigma^{i+f,k+f}$ and $\{q^i, p_k\} = \sigma^{i,k+f} = -\sigma^{k+f,i}$. By assumption

$$\Sigma = \begin{pmatrix} 0 & -1\!\!1 \\ 1\!\!1 & 0 \end{pmatrix} = -\mathbf{J} = \mathbf{J}^{-1} \ ,$$

where $\mathbf{J}$ is defined as in (2.102). We conclude that $\Omega = \mathbf{J}$ and hence that the chart is symplectic. □

Finally, the invariance of Poisson brackets under canonical transformations (2.124) is rediscovered in the following form. Let F be a diffeomorphism connecting two symplectic manifolds, $F : (M, \omega) \to (N, \varrho)$. This mapping is symplectic precisely if it preserves the Poisson brackets of functions and/or one-forms, i.e.

$$\{F^* f, F^* g\} = F^*\{f, g\} \quad \text{for all} \quad f, g \in \mathcal{F}(N) \ .$$

In this case F^* preserves the Lie algebra structure on the vector space of the smooth function.

5.5.7 Time-Dependent Hamiltonian Systems

The preceding sections 5.5.1–6 gave an introduction to the mathematical foundations of the theory of Hamiltonian and Jacobi. They should be sufficient to study the theory of time-dependent systems as well, without any major difficulties. We restrict our discussion to a few remarks and refer to the more specialized, mathematical literature on mechanics for more details.

If the Hamiltonian function depends explicitly on time, $H : M \times \mathbb{R}_t \to \mathbb{R}$, then also the corresponding Hamiltonian vector field depends on time, i.e. assigns to each point (m, t) of the direct product of phase space and time axis, a tangent vector in $T_m M \times \mathbb{R}$. The manifold $M \times \mathbb{R}_t$ cannot be symplectic because its dimension is odd. However, the canonical two-form ω has maximal rank on $M \times \mathbb{R}_t$, namely $2f$, where $f = \dim Q$. In a local chart representation of $(m, t) \in U \times \mathbb{R}_t$, $U \subset M$, viz. $(q^1, \ldots q^f, p_1, \ldots, p_f, \tau)$, the canonical two-form reads

$$\omega|_U = \sum \mathrm{d}q^i \wedge \mathrm{d}p_i$$

according to Darboux's theorem, provided the chart is a symplectic one. As ω was given by $\omega = -\mathrm{d}\theta$, we have locally

$$\mathrm{d}(\theta - \sum p_i \mathrm{d}q^i) = 0 \ .$$

The form in parentheses is closed. Hence, locally, according to Poincaré's lemma, it can be written as the exterior derivative of a function, i.e.

$$\theta = \sum p_i \mathrm{d}q^i + \mathrm{d}\tau \ .$$

Note that the exterior product $\theta \wedge \mathrm{d}\theta \wedge \ldots \wedge \mathrm{d}\theta$, with f factors $\mathrm{d}\theta$, is a volume form on $M \times \mathbb{R}_t$.

It is not difficult to generalize the time-independent situation discussed in the previous sections to the case of time-dependent Hamiltonian vector fields. For every fixed $t \in \mathbb{R}$, such a vector field

$$X : M \times \mathbb{R} \to TM$$

is a vector field on M. One associates with it a vector field $\tilde{X}$ on $M \times \mathbb{R}$,

$$\tilde{X} : M \times \mathbb{R} \to T(M \times \mathbb{R}) \cong TM \times T\mathbb{R} \ ,$$

($\cong$ means isomorphic) through the assignment

$$(m, t) \to (X(m, t), (t, 1)) \ .$$

Regarding the integral curves of $\tilde{X}$, we can say the following. Let $\gamma : I \to M$ be an integral curve of X going through the point m. Then $\tilde{\gamma} : I \to M \times \mathbb{R}$ is the integral curve of $\tilde{X}$, passing through the point $(m, 0)$, precisely if $\tilde{\gamma}(t) = (\gamma(t), t)$. This is easily verified. Write

$$\tilde{\gamma}(t) = (\gamma(t), \tau(t)) \ .$$

This is an integral curve of $\tilde{X}$ provided

$$\tilde{\gamma}'(t) = (\gamma'(t), \tau'(t)) = \tilde{X}(\tilde{\gamma}(t)) \ ,$$

i.e. provided

$$\gamma'(t) = X(\gamma(t), t) \quad \text{and} \quad \tau'(t) = 1 \ .$$

However, as $\tau(0)$ should be equal to 0, we conclude that $\tau(t) = t$. The flux of $\tilde{X}$ is expressed in terms of the flux of X, viz.

$$\tilde{\Phi}_t(m, s) = ((t + s), \Phi_{t,s}(m)) \ .$$

Let M be the phase space and let H be a time-dependent Hamiltonian function on $M \times \mathbb{R}$. Then $H(m,t)$, for fixed t, is a function on M,

$$H_t(m) \stackrel{\text{def}}{=} H(m,t) : M \to \mathbb{R} \,,$$

whose vector field $X_{\rm H}$ is determined as before. Define the vector field

$$X_{\rm H} : M \times \mathbb{R} \to TM : (m,t) \mapsto X_{{\rm H}_t}(m)$$

as well as the corresponding vector field $\tilde{X}_{\rm H}$, to be constructed as above. The corresponding integral curves of $\tilde{X}_{\rm H}$ move across $M \times \mathbb{R}$, those of $X_{\rm H}$ move across M. The latter are identical with the phase portraits introduced in Chap. 1.

The canonical equations of motion hold in every symplectic chart. So $\gamma : I \to U$, with $I \subset \mathbb{R}$ and $U \subset M$, is an integral curve of $X_{\rm H}$ if and only if the equations

$$\left.\begin{aligned} \frac{\mathrm{d}}{\mathrm{d}t}[q^i(\gamma(t))] &= \partial H(\gamma(t),t)/\partial p_i \\ \frac{\mathrm{d}}{\mathrm{d}t}[p_i(\gamma(t))] &= -\partial H(\gamma(t),t)/\partial q_i \end{aligned}\right\} i = 1, \ldots f$$

are fulfilled.

5.6 Lagrangian Mechanics and Lagrange Equations

On the one hand, the Lagrangian function is defined as a smooth function on the tangent bundle TQ of the coordinate manifold Q, $L : TQ \to \mathbb{R}$. As we know from Chap. 2, on the other hand, it appears in the expressions for the Legendre transformation from Lagrangian mechanics, formulated on TQ, to Hamilton–Jacobi mechanics, which lives on T^*Q, and vice versa. The geometric approach shows very clearly that this is more than just a simple transformation of variables. The formulation of Hamilton and Jacobi is characteristic for the cotangent bundle T^*Q. The aim of this section is to show that Lagrangian mechanics is rather different from this, also as far as its geometric interpretation is concerned. The main difference is that on the tangent bundle one can define differential equations of second order (i.e. the Euler–Lagrange equations well known to us), in a natural way, while this is not possible on T^*Q.

5.6.1 The Relation Between the Two Formulations of Mechanics

When expressed in local coordinates, the first step of the Legendre transformation is the assignment

$$\Phi_{\rm L} : \{q^i, \dot{q}^j\} \to \left\{q^i, \frac{\partial L}{\partial \dot{q}^j} \stackrel{\text{def}}{=} p_j\right\} . \tag{5.102}$$

Going back from the charts to the original manifolds TQ and T^*Q, (5.102) says that we assign to an element of T_uQ, for fixed base point $u \in Q$, an element of T_u^*Q by means of derivatives of the Lagrangian function. In other words, the fibre T_uQ over $u \in Q$ of the tangent bundle TQ is mapped to the fibre T_u^*Q of the cotangent bundle over the same base point. This mapping is linear and makes use of the partial derivatives of the Lagrangian function within the fiber T_uQ (in charts: q is fixed, the derivatives are taken with respect to $\dot{q}$). Thus, let v_u be an element of T_uQ, the fibre of TQ in u. Denoting the restriction of the Lagrangian function to this fiber by L_u, the mapping Φ_L (5.102) corresponds to the assignment

$$\Phi_\mathrm{L} : T_uQ \to T_u^*Q : v_u \mapsto DL_u(v_u) , \tag{5.103}$$

where D denotes the derivatives of L. The precise definition of D on manifolds would lead us too far from our main subject. Therefore, the following, somewhat qualitative remarks that clarify matters in charts may be sufficient. Let (φ, U) be a chart taken from the atlas for Q and $(T\varphi, TU)$ the induced chart for TQ. $L^{(\varphi)}$ denotes the restriction of the Lagrangian function to the domains of these charts. Then $L^{(\varphi)} \circ T\varphi^{-1}$ is a function on $\mathbb{R}^f \times \mathbb{R}^f$, as shown schematically in Fig. 5.12.

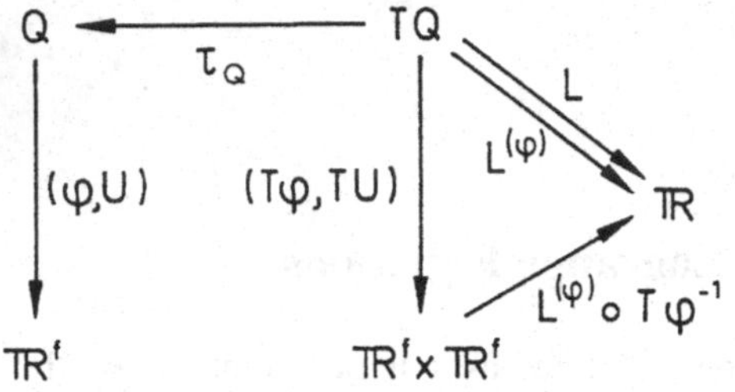

Fig. 5.12. The Lagrangian function is defined on the tangent bundle TQ (velocity space). Its representation in charts $L^{(\varphi)} \circ T\varphi^{-1}$ is the local form that one knows from Chap. 2

Denoting the derivatives with respect to the first and the second arguments by D_1 and D_2, respectively, we have

$$D_1 L^{(\varphi)} \circ T\varphi^{-1} = \left\{ \frac{\partial L}{\partial q^i} \right\} , \tag{5.104a}$$

$$D_2 L^{(\varphi)} \circ T\varphi^{-1} = \left\{ \frac{\partial L}{\partial \dot{q}^i} \right\} . \tag{5.104b}$$

The derivative DL_u of (5.103) leaves the base point u unchanged. Hence it is of the type (5.104b).

Φ_L being a mapping from TQ to T^*Q that is induced by the Lagrangian function, the canonical forms C1F (5.74) and C2F (5.79) can be pulled back from T^*Q to TQ. If Φ_L is a regular mapping[7], it is symplectic, so that canonical mechanics on T^*Q can be pulled back to TQ. If, furthermore, Φ_L is a diffeomorphism, then the two formulations of mechanics are completely equivalent. As we know from

[7] A mapping $\Phi : M \to N$ is said to be *regular* in the point $p \in M$ if the corresponding differential, or tangent, mapping from T_pM to $T\Phi_{(p)}N$ is surjective.

Chap. 2, this is true if and only if (in charts) the matrix of the second derivatives of L with respect to $\dot{q}$ is nowhere singular, i.e. if

$$\det\left(\frac{\partial^2 L}{\partial \dot{q}^k \partial \dot{q}^i}\right) \neq 0 \tag{5.105}$$

holds on the domain of definition of the problem. Strictly speaking, one should distinguish the cases where Φ_{L} is regular from those where, in addition, it is a diffeomorphism. In the first case, the condition (5.105) holds only locally while in the second it holds on the domain of *all* charts. For what follows we assume that L is chosen such that Φ_{L} is a diffeomorphism.

5.6.2 The Lagrangian Two-Form

The canonical two-form ω_0, defined by (5.79), can be pulled back to TQ by means of Φ_{L}. This yields what is called the Lagrangian two-form

$$\omega_{\mathrm{L}} \stackrel{\text{def}}{=} \Phi_{\mathrm{L}}^* \omega_0 \,. \tag{5.106}$$

The pull-back of ω_0, the canonical two-form on T^*Q, to ω_{L} on TQ is defined as described in Sect. 5.4.1 (5.41). Very much like ω_0 the form ω_{L} is closed,

$$\mathrm{d}\omega_{\mathrm{L}} = 0 \,.$$

This follows because the exterior derivative of the pull-back of a form $\mathrm{d}(F^*\omega)$ is equal to the pull-back $F^*(\mathrm{d}\omega)$ of the exterior derivative of the original form (see also Exercise 5.11).

Furthermore, the operation of pull-back commutes with the restriction to open neighborhoods on the manifold, on which a given form is defined. For $F: M \to N$ and $\overset{k}{\omega}$ an exterior k-form on N, one has

$$(F^* \overset{k}{\omega})|_{U\subset M} = F^*(\overset{k}{\omega}|_{F(U)\subset N}) \,.$$

Therefore, the expression of ω_{L} in charts can be computed from the local representation (5.80) of ω_0. Let U be the domain of a chart on Q and TU the corresponding domain on TQ. Then we have in the domain of the chart (φ, U)

$$\begin{aligned}
\omega_{\mathrm{L}}|_{TU} &= (\Phi_{\mathrm{L}}^*\omega_0)|_{TU} = \Phi_{\mathrm{L}}^*(\omega_0|_{T^*U}) \\
&= \Phi_{\mathrm{L}}^*\left(\sum \mathrm{d}q^i \wedge \mathrm{d}p_i\right) \\
&= \sum \mathrm{d}(\Phi_{\mathrm{L}}^* q^i) \wedge \mathrm{d}(\Phi_{\mathrm{L}}^* p_i) \,.
\end{aligned}$$

Here we have used the equality $F^*(\sigma \wedge \tau) = (F^*\sigma) \wedge (F^*\tau)$ for two exterior forms σ and τ, as well as the fact that the exterior derivative commutes with F^*. The last expression for ω_{L} contains the functions q^i and p_k, pulled back to TQ, for which we have

$$\Phi_L^* q^i = q^i \,, \quad \Phi_L^* p_k = \frac{\partial L}{\partial \dot{q}^k} \,.$$

Thus we find

$$\omega_L|_{TU} = \sum_i \mathrm{d}q^i \wedge \mathrm{d}\frac{\partial L}{\partial \dot{q}^i} \,.$$

The exterior derivative of the function $\partial L/\partial \dot{q}^k$ is easily calculated with the rules of Sect. 5.4.4. Thus, we obtain

$$\omega_L|_{TU} = \sum_{i,k} \left(\frac{\partial^2 L}{\partial q^k \partial \dot{q}^i} \mathrm{d}q^i \wedge \mathrm{d}q^k + \frac{\partial^2 L}{\partial \dot{q}^i \partial \dot{q}^k} \mathrm{d}q^i \wedge \mathrm{d}\dot{q}^k \right) . \tag{5.107}$$

The same result is obtained from the pull-back to TQ of thc canonical onc-form (5.74), $\theta_L \stackrel{\text{def}}{=} \Phi_L^* \theta_0$. In charts it reads

$$\theta_L|_{TU} = \sum \frac{\partial L}{\partial \dot{q}^i} \mathrm{d}q^i \,.$$

Taking the negative exterior derivative, $\omega_L = -\mathrm{d}\theta_L$, yields again the expression (5.107).

Thus, if the mapping Φ_L is regular, or even a diffeomorphism, then Φ_L is symplectic: it maps the symplectic manifold (T^*Q, ω_0) onto the symplectic manifold (TQ, ω_L).

5.6.3 Energy Function on *TQ* and Lagrangian Vector Field

In discussing the Legendre transformation in Chap. 2, we considered the function

$$E(\underset{\sim}{q}, \dot{\underset{\sim}{q}}, t) = \sum \dot{q}^i \frac{\partial L}{\partial \dot{q}^i} - L(\underset{\sim}{q}, \dot{\underset{\sim}{q}}, t) \,, \tag{5.108}$$

which led to the Hamiltonian function, after transformation to the variables $\underset{\sim}{q}$ and $\underset{\sim}{p}$ (taking account of the condition (5.105)). For autonomous systems this was the expression for the energy, the energy then being a constant of the motion. Given the Hamiltonian function and the canonical two-form ω_0, the Hamiltonian vector field was constructed following the definition HVF (5.91). A similar construction can be performed on TQ. For that purpose we first define the function E on the manifold TQ, its chart representation being given by (5.108) above. With $u \in Q$, $v_u \in TQ$, the first term on the right-hand side of (5.108) is given a coordinate-free meaning by the definition

$$W : TQ \to \mathbb{R} : v_u \mapsto \Phi_L(v_u) \cdot v_u \tag{5.109a}$$

According to (5.103), $\Phi_L(v_u)$ is a linear mapping from T_uQ to $\mathbb{R}$, i.e. it is an element of T_u^*Q, which acts on $v_u \in TQ$. One verifies easily that, in charts, W is indeed given by the first term on the right-hand side of (5.108). W is said to be the *action*.

The *energy function*, understood to be a smooth function on TQ, is then defined by

$$E \stackrel{\text{def}}{=} W - L \,. \tag{5.109b}$$

We follow the analogous construction on phase space, Sect. 5.5.5. We take the exterior derivative of E and define the Lagrangian vector field by means of the Lagrangian two-form ω_{L}, as follows.

LVF. Given the function $E = W - L$ on TQ, as well as the two-form $\omega_{\text{L}} = \Phi_{\text{L}}^{*}\omega_0$, with Φ_{L} being a regular mapping (or even a diffeomorphism), the *Lagrangian vector field* X_{E} is defined uniquely by

$$\omega_{\text{L}}(X_{\text{E}}, \bullet) = \text{d}E \,. \tag{5.110}$$

In local form E is given by (5.108) and therefore

$$\begin{aligned}\text{d}E|_{TU} &= \sum_{i,k}\left(\frac{\partial L}{\partial \dot{q}^i}\delta^{ik} + \dot{q}^i\frac{\partial^2 L}{\partial \dot{q}^k \partial \dot{q}^i}\right)\text{d}\dot{q}^k + \sum_{i,k}\dot{q}^i\frac{\partial^2 L}{\partial q^k \partial \dot{q}^i}\text{d}q^k \\ &\quad - \sum_k \frac{\partial L}{\partial q^k}\text{d}q^k - \sum_k \frac{\partial L}{\partial \dot{q}^k}\text{d}\dot{q}^k \\ &= \sum_{i,k}\dot{q}^i\frac{\partial^2 L}{\partial \dot{q}^k \partial \dot{q}^i}\text{d}\dot{q}^k + \sum_{i,k}\left(\dot{q}^i\frac{\partial^2 L}{\partial q^k \partial \dot{q}^i} - \frac{\partial L}{\partial q^i}\delta^{ik}\right)\text{d}q^k \,.\end{aligned}$$

It is instructive to write out explicitly the local form of (5.110) as well as the vector field X_{E}. For the sake of simplicity, we do this for the case of one degree of freedom, $f = 1$. The general case is no more difficult and will be dealt with in the next section. Let ∂ and $\bar{\partial}$ denote the base fields $\partial/\partial q^i$ and $\partial/\partial \dot{q}^i$, respectively. Then, in coordinates, the Lagrangian vector field is $X_{\text{E}} = v\partial + \bar{v}\bar{\partial}$, while another, arbitrary vector field reads $Y = w\partial + \bar{w}\bar{\partial}$. From (5.107) we have

$$\omega_{\text{L}}(X_{\text{E}}, Y) = \frac{\partial^2 L}{\partial \dot{q}^2}(v\bar{w} - \bar{v}w) \,,$$

while the action of $\text{d}E$ on Y gives the local result

$$\text{d}E(Y) = \left(\dot{q}\frac{\partial^2 L}{\partial q \partial \dot{q}} - \frac{\partial L}{\partial q}\right)w + \dot{q}\frac{\partial^2 L}{\partial \dot{q}^2}\bar{w} \,.$$

Inserting these expressions into the equation $\omega_{\text{L}}(X_{\text{E}}, Y) = \text{d}E(Y)$ and comparing the coefficients of w and $\bar{w}$, we obtain

$$v = \dot{q},\, \bar{v} = \left(\frac{\partial L}{\partial q} - \dot{q}\frac{\partial}{\partial q}\frac{\partial L}{\partial \dot{q}}\right) \Big/ \frac{\partial^2 L}{\partial \dot{q}^2} \,.$$

It is seen that the condition (5.105) is essential.

We now follow the pattern of (5.92) and try to determine the integral curve of the Lagrangian vector field X_{E},

$$\dot{c}(t) = (X_{\mathrm{E}})_{c(t)} \ .$$

Here $c : I \to \mathbb{R}$ is a curve on TQ. In charts $c(t)$ is $\binom{q(t)}{\dot{q}(t)}$ and obeys the differential equations

$$\dot{q}(t) = v = \dot{q} \ ,$$

$$\ddot{q}(t) = \bar{v} = \left(\frac{\partial L}{\partial q} - \dot{q} \frac{\partial}{\partial q} \frac{\partial L}{\partial \dot{q}} \right) \bigg/ \frac{\partial^2 L}{\partial \dot{q}^2} \ .$$

While the first of these just tells us that the time derivative of the first coordinate is equal to the second, the second equation has a somewhat surprising form. True, it is obtained from the Euler–Lagrange equation

$$\frac{\partial L}{\partial q} - \frac{\mathrm{d}}{\mathrm{d}t} \frac{\partial L}{\partial \dot{q}} = 0 \ ,$$

by taking the derivative with respect to t and by solving for $\ddot{q}$. However, it is a differential equation of *second* order and therefore, geometrically speaking, it is different from the canonical equations (5.92). Let us consider this new feature in more detail.

5.6.4 Vector Fields on Velocity Space TQ and Lagrange Equations

A smooth vector field X that is defined on the tangent bundle TQ of a manifold Q leads from TQ to $T(TQ)$, the tangent bundle of the tangent bundle,

$$X : TQ \to T(TQ) \ .$$

Let τ_Q denote the projection from TQ to Q and $T\tau_Q$ the corresponding tangent mapping. The composition $T\tau_Q \circ X$ maps TQ onto TQ, as shown in Fig. 5.13. If this composition produces just the identity on TQ, i.e. if $T\tau_Q \circ X = \mathrm{id}_{TQ}$, the

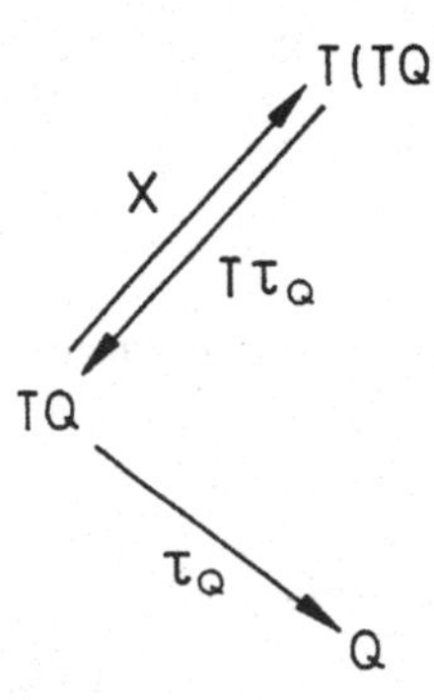

Fig. 5.13. A vector field on TQ generates an equation of second order if it fulfills the condition (5.111)

vector field X defines a differential equation of second order. This follows from the following proposition.

Proposition. The smooth vector field X has the property

$$T\tau_Q \circ X = \mathrm{id}_{TQ} \tag{5.111}$$

if and only if each integral curve $c : I \to TQ$ of X obeys the differential equation

$$(\tau_Q \circ c)^\bullet = c\,. \tag{5.112}$$

Proof. For each point $v_u \in TQ$ there is a curve going through that point such that $\dot{c}(\tau) = X(c(\tau))$, with $\tau \in I$. $T\tau_Q \circ X$ is the identity on TQ precisely if $T\tau_Q \circ \dot{c}(\tau) = c(\tau)$ holds true. Working out the left-hand side, we obtain

$$T\tau_Q \circ \dot{c}(\tau) = T\tau_Q \circ Tc(\tau, 1) = T(\tau_Q \circ c)(\tau, 1) = (\tau_Q \circ c)^\bullet(\tau)\,,$$

which proves (5.112). □

From a physicist's point of view the integral curves c of X are not exactly the solutions one is looking for. Rather, we are interested in the orbits γ on the base manifold Q itself. These are the physical orbits in the manifold of generalized coordinates (the ones one can "see"), i.e. the orbits that we denoted by $\Phi_{s,t}(q_0)$ in earlier sections. It is not difficult, however, to obtain these curves from c (integral curve of X on TQ) and from τ_Q (projection of TQ on Q). Indeed, $\gamma \stackrel{\text{def}}{=} \tau_Q \circ c : I \to Q$ is a curve on Q, since $c : I \to TQ$ and $\tau_Q : TQ \to Q$. A curve of this kind that is associated to the vector field X is said to be a *base integral curve*. The condition (5.112) can be written as $\dot{\gamma} = c$, which means the following: the vector field X defines a differential equation of second order if and only if each of its integral curves is equal to the derivative of its corresponding base integral curve $\gamma = \tau_Q \circ c$.

In charts the Lagrangian vector field X_{E} reads

$$X_{\mathrm{E}} = \sum v^i \partial_i + \sum \bar{v}^i \bar{\partial}_i$$

with $\partial_i \stackrel{\text{def}}{=} \partial/\partial q^i$, $\bar{\partial}_i \stackrel{\text{def}}{=} \partial/\partial \dot{q}^i$, cf. Sect. 5.6.3. Then $v^i = \dot{q}^i$, while the components $\bar{v}^i = \bar{v}^i(\underset{\sim}{q}, \underset{\sim}{\dot{q}})$ fulfill the differential equations

$$\frac{\mathrm{d}^2}{\mathrm{d}t^2} q^i(t) = \bar{v}^i(\underset{\sim}{q}(t)), \underset{\sim}{\dot{q}}(t))\,. \tag{5.113}$$

As before, $\begin{pmatrix} q^i \\ \dot{q}^i \end{pmatrix}$ is the local representation of the point $c(t)$ or $\dot{\gamma}(t)$. Of course, in charts one obtains the well-known Euler–Lagrange equations. In order to show this for the general case ($f > 1$), calculate $\omega_{\mathrm{L}}(X_{\mathrm{E}}, Y)$ as well as $\mathrm{d}E(Y)$, for an arbitrary vector field

$$Y = \sum w^i \partial_i + \sum \bar{w}^i \bar{\partial}_i \,.$$

One finds that

$$\omega_{\mathrm{L}}(X_{\mathrm{E}}, Y) = \sum_{i,k} \frac{\partial^2 L}{\partial q^k \partial \dot{q}^i}(v^i w^k - v^k w^i) + \sum_{i,k} \frac{\partial^2 L}{\partial \dot{q}^k \partial \dot{q}^i}(v^i \bar{w}^k - \bar{v}^k w^i) \tag{5.114}$$

and, similarly,

$$\mathrm{d}E(Y) = \sum_{i,k} \dot{q}^i \frac{\partial^2 L}{\partial \dot{q}^k \partial \dot{q}^i} \bar{w}^k + \sum_{i,k} \left(\dot{q}^i \frac{\partial^2 L}{\partial q^k \partial \dot{q}^i} - \frac{\partial L}{\partial q^i} \delta^{ik} \right) w^k \,. \tag{5.115}$$

We insert $v^i = \dot{q}^i$ in (5.114) and set it equal to (5.115). The terms in $\bar{w}^k$ cancel, while the comparison of the coefficients of w^k yields the equations

$$\frac{\partial L}{\partial q^k} - \sum_i \frac{\partial^2 L}{\partial q^i \partial \dot{q}^k} \dot{q}^i - \sum_i \frac{\partial^2 L}{\partial \dot{q}^i \partial \dot{q}^k} \bar{v}^i = 0 \,.$$

Finally, inserting the result (5.113), these equations become

$$\frac{\partial L}{\partial q^k} - \frac{\mathrm{d}}{\mathrm{d}t} \frac{\partial L}{\partial \dot{q}^k} = 0 \,,$$

i.e. the set of Euler–Lagrange equations, as expected.

5.6.5 The Legendre Transformation and the Correspondence of Lagrangian and Hamiltonian Functions

We had assumed the mapping Φ_{L} (5.103) from $T_u Q$ to $T_u^* Q$ to be a diffeomorphism. Locally this means that the condition (5.105) is satisfied everywhere. As we learnt in Chap. 2, Sect.2.15, we can then go over from Lagrangian mechanics to Hamilton–Jacobi mechanics and vice versa, as we wish. In this section we want to clarify this relationship using the geometric language.

With Φ_{L} a diffeomorphism, geometric objects can be transported between TQ and T^*Q at will. For example, if $X : TQ \to T(TQ)$ is a vector field on TQ, then

$$Y \stackrel{\mathrm{def}}{=} T\Phi_{\mathrm{L}} \circ X \circ \Phi_{\mathrm{L}}^{-1} : T^*Q \to T(T^*Q)$$

is a vector field on the manifold T^*Q. Here, $T\Phi_{\mathrm{L}}$ is the tangent mapping corresponding to Φ_{L}. It relates $T(TQ)$ with $T(T^*Q)$. As Φ_{L} is a diffeomorphism, $T\Phi_{\mathrm{L}}$ is an isomorphism. In this case one has the following results.

(i) Proposition. Let the Lagrangian function be such that Φ_L is a diffeomorphism. Let E be the function on TQ defined by (5.109b). Finally, define the function

$$H \stackrel{\text{def}}{=} E \circ \Phi_L^{-1} : T^*Q \to \mathbb{R}$$

on T^*Q. Then the Lagrangian vector field X_E and the vector field X_H, which corresponds to H, by the definition (5.91), are related by

$$T\Phi_L \circ X_E \circ \Phi_L^{-1} = X_H \,. \tag{5.116}$$

Φ_L maps the integral curves of X_E onto those of X_H. The two vector fields, X_E on TQ and X_H on T^*Q, have the same base integral curves (i.e. the same physical solutions on Q).

Proof. It is sufficient to establish the relation (5.116) because the remaining assertions all follow from it. Let $v \in TQ$, $w \in T_v(TQ)$, let v^* be the image of v by Φ_L, and w^* the image of w by the tangent mapping $T\Phi_L$, i.e. $w^* = T_v\Phi_L(w)$. At the point v we then have

$$\omega_0(T\Phi_L(X_E), w^*) = \omega_L(X_E, w) = \mathrm{d}E(w) = \mathrm{d}(H \circ \Phi_L)(w) \,.$$

On the other hand, at the point $v^* = \Phi_L(v)$, we know that

$$\omega_0(T\Phi_L(X_E), w^*) = \mathrm{d}H(w^*) = \omega_0(X_H, w^*) \,.$$

The assertion (5.116) now follows because w^* is arbitrary, $T\Phi_L$ is an isomorphism, and ω_0 is not degenerate. It is then also clear that the integral curves of X_E and X_H are related by Φ_L. Finally, denoting the projections from TQ and from T^*Q to Q by τ_Q and by τ_Q^*, respectively, we know that $\tau_Q = \tau_Q^* \circ \Phi_L$. Hence, the base integral curves are the same. □

(ii) The canonical one-form ω_0 (5.74) is closely related to the action W, (5.109a). With $H - E \circ \Phi_L^{-1}$, one has

$$\theta_0(X_H) = W \circ \Phi_L^{-1} \,. \tag{5.117a}$$

Conversely, if $\theta_L \stackrel{\text{def}}{=} \Phi_L^*\theta_0$ is the pull-back of the canonical one-form on TQ, then

$$\theta_0(X_E) = W \,. \tag{5.117b}$$

In charts this is easy to verify. For example, (5.117a) is equivalent to the statement that $\theta_0(X_H) \circ \Phi_L$ is equal to W. We have

$$\theta_0(X_H) = \sum_i p_i \frac{\partial H}{\partial p_i}$$

so that, indeed

$$\theta_0(X_\mathrm{H}) \circ \Phi_\mathrm{L} = \sum_i \frac{\partial L}{\partial \dot{q}^i} \dot{q}^i = W \, .$$

(iii) A transformation analogous to (5.103) can also be defined for the inverse direction, i.e. going from T^*Q to TQ. Let H be a smooth function on T^*Q. In analogy to the definition (5.103) let us define the transformation

$$\Phi_\mathrm{H} : T^*Q \to T^{**}Q \cong TQ \, . \tag{5.118}$$

If this mapping Φ_H is a diffeomorphism[8], one can define the quantities

$$E \stackrel{\mathrm{def}}{=} H \circ \Phi_\mathrm{H}^{-1} \, , \quad W \stackrel{\mathrm{def}}{=} \theta_0(X_\mathrm{H}) \circ \Phi_\mathrm{H}^{-1} \, , \quad L \stackrel{\mathrm{def}}{=} W - E \tag{5.119}$$

in analogy to (ii) above. This yields a Lagrangian system on TQ with L the Lagrangian function. For this L we again construct the mapping Φ_L (5.103). It then follows that $\Phi_\mathrm{L} = \Phi_\mathrm{H}^{-1}$, or $\Phi_\mathrm{L} \circ \Phi_\mathrm{H} = \mathrm{id}_{T^*Q}$ and $\Phi_\mathrm{H} \circ \Phi_\mathrm{L} = \mathrm{id}_{TQ}$. This leads to the following theorem.

Theorem. The Lagrangian functions on TQ for which the corresponding mappings Φ_L are diffeomorphisms, and the Hamiltonian functions for which the corresponding Φ_H are diffeomorphisms, correspond to each other in a bijective manner.

The proof, which is simple, makes use of the tools introduced above, see e.g. Abraham and Marsden (1981). Thus, under the assumptions stated above, there is a one-to-one correspondence between the two descriptions of mechanics. The relationship between them is illustrated once more in Fig. 5.14.

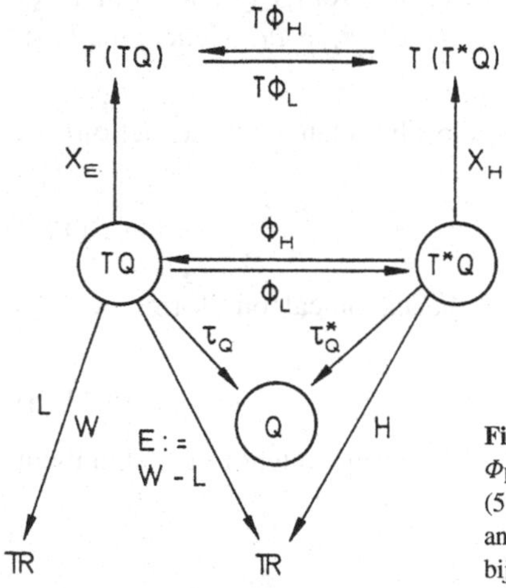

Fig. 5.14. If the condition (5.105) is fulfilled, the mapping Φ_L is a diffeomorphism. Its inverse Φ_H is defined by (5.118). A Lagrangian formulation of mechanics on TQ and a Hamiltonian formulation on T^*Q then correspond bijectively

---

8 As is easy to guess, this is true if $\det(\partial^2 H/\partial p_k \partial p_i)$ vanishes nowhere.

Note that most systems studied in nonrelativistic mechanics have this property. As a counterexample, however, we remind the reader of the relativistic description of a free particle, Sect. 4.11: in this case the Lagrangian function did not have the regularity required for Φ_{L} to be a diffeomorphism.

5.7 Riemannian Manifolds in Mechanics

A *Riemannian manifold* (M, g) is a differential manifold M equipped with a metric g. Differential or smooth manifolds are defined in Sect. 5.2.2; the metric is a smooth tensor field of type $\mathcal{T}_2^0(M)$, and its properties are summarized in definition (ME) in Sect. 5.5.1. As can be seen from (5.67b), or from (5.69), the metric defines a scalar product on T_pM, the tangent space attached to the point $p \in M$. This scalar product is often written in the "bra" and "ket" notation, i.e. by making use of the symbols $\langle \ldots |$ and $| \ldots \rangle$, such that

$$g_p(v, w) \equiv \langle v | w \rangle , \qquad v, w \in T_pM . \tag{5.120}$$

The phase space of a Hamiltonian system is a *symplectic* manifold, cf. the definition (S2) in Sect. 5.5.4. Symplectic manifolds are very different from Riemannian manifolds: While all symplectic manifolds look the same *locally*, this is not true for Riemannian manifolds. The first statement is the content of Darboux' theorem (Sect. 5.5.4) which may be expressed in more physical terms by the statement that locally and outside of equilibrium positions, any Hamiltonian vector field can be rectified (cf. Sect. 2.37.1)[9].

In this section we show that for certain systems of Lagrangian mechanics the coordinate manifold Q can be interpreted as a Riemannian manifold with the metric as defined by the kinetic energy; and that solutions of the Euler–Lagrange equations are nothing but *geodesics* of Q. In this way we discover another illustration and example of the geometrical nature of mechanics; at the same time we prepare the ground for general relativity, which is a geometrical theory, in an even deeper sense.

In what follows we first introduce the notions of parallel transport and affine connection that one needs in order to define parallel vector fields and to write down the geodesic equation. We then show that geodesics are solutions of Euler–Lagrange equations and conclude with a beautiful application of this somewhat formal chapter.

[9] The *global* properties of symplectic manifolds are the subject of an important research field of mathematics. The present state of the art is described in the book by Hofer and Zehnder (1994). This book should be readily accessible for the mathematically minded reader.

5.7.1 Affine Connection and Parallel Transport

To begin with, let M simply be a Euclidean space $\mathbb{R}^n$, equipped with the metric defined in (5.5) and (5.6). Let $W = \sum W^i \partial_i$ and $V = \sum V^i \partial_i$ be smooth vector fields on M, $V_p \in T_pM$ the local representative of V at the point p. We now ask the question of how W at p will change in the direction of V_p. The answer is simple in this case: At the point p we let V act on the functions (the components) $W^i(p)$ and use the result to construct the vector field $\sum V(W^i)\partial_i$. This is the local and natural expression for the *covariant derivative* of W with respect to V

$$D_V(W) = \sum_{i=1}^{n} V(W^i)\partial_i = \sum_{i,k=1}^{n} V^k \frac{\partial W^i}{\partial x^k} \partial_i \,. \tag{5.121}$$

Obviously this expression is linear in W. Regarding its dependence on V it is also possible to calculate the covariant derivative along the sum of two vector fields, viz $D_{V_1+V_2}W = D_{V_1}W + D_{V_2}W$, as well as along the vector field $f \cdot V$, where f is a smooth function on M, viz $D_{fV}W = f(D_V W)$. However, letting D_V act on the vector field $(f \cdot W)$ is a different matter; one finds

$$D_V(fW) = \sum_{i=1}^{n} V(fW^i)\partial_i = (Vf)\sum_{i=1}^{n} W^i\partial_i + f\sum_{i=1}^{n} V(W^i)\partial_i = (Vf)W + fD_V W \,.$$

This formula expresses a generalized product rule, or Leibniz rule.

In case of a smooth manifold M which is not $\mathbb{R}^n$ the formula (5.121) no longer holds, and there is no obvious and natural definition of a covariant derivative. Asking the question of how a vector field W changes along the direction of another vector field means that we have to compare elements of two distinct tangent spaces, say $W_p \in T_pM$ with $W_q \in T_qM$. In order to make such a comparison possible we first need to know how to transport W_p in a parallel fashion from T_pM to T_qM (by means of a vector space isomorphism). Only then can one compare the result of the parallel transport with W_q. As parallel transport, in general, is not given in a canonical way, an explicit rule is necessary. It needs to be constructed in a manner consistent with what we know from the flat space $\mathbb{R}^n$. Fixing the rule of parallel transport on a smooth manifold means choosing what is called a *connection*. The example studied above suggests the following defining properties of a connection D:

CONN. A connection D on a smooth manifold M is a mapping

$$D : \mathcal{V}(M) \times \mathcal{V}(M) \longrightarrow \mathcal{V}(M)\,, \tag{5.122}$$

which has the following properties:

(i) It is $\mathcal{F}(M)$-linear in the first argument, that is to say

$$D_{V_1+V_2}W = D_{V_1}W + D_{V_2}W\,, \tag{5.123a}$$

$$D_{fV}W = f(D_V W)\,; \tag{5.123b}$$

(ii) it is $\mathbb{R}$-linear in its second argument, that is

$$D_V\,(\lambda_1 W_1 + \lambda_2 W_2) = \lambda_1 D_V(W_1) + \lambda_2 D_V(W_2)\,,$$
$$\lambda_1, \lambda_2 \in \mathbb{R}\,; \tag{5.124}$$

(iii) it obeys the Leibniz rule

$$D_V(fW) = (Vf)W + f D_V W\,, \quad f \in \mathcal{F}(M)\,. \tag{5.125}$$

The vector field $D_V W$ is called the *covariant derivative* of W along V and with reference to the connection D.

Clearly, the parallel transport is fixed if its action on all base vectors is known. Therefore, if in a local chart we choose $V = \partial_i$ and $W = \partial_j$, the result is again a vector field which can be expanded along base fields,

$$D_{\partial_i}(\partial_j) = \sum_{k=1}^{n} \Gamma_{ij}^k \partial_k\,. \tag{5.126}$$

This equation defines the *Christoffel symbols* Γ_{ij}^k of the connection D. For example, if one computes the covariant derivative of a vector field W along the base field ∂_i, equations (5.125) and (5.126) yield the following local expression

$$D_{\partial_i}\left(\sum_k W^k \partial_k\right) = \sum_k \left\{\frac{\partial W^k}{\partial x^i} + \sum_j \Gamma_{ij}^k W^j\right\} \partial_k\,. \tag{5.127}$$

One of the central theorems of Riemannian geometry is the following: Among the set of connections on a Riemannian manifold M there is a special, uniquely determined connection which, in addition to (5.123–125) has the properties

$$[V, W] = D_V W - D_W V\,, \tag{5.128}$$

$$X\langle V|W\rangle = \langle D_X|W\rangle + \langle V|D_X W\rangle \text{ for all } X, V, W \in \mathcal{V}(M)\,. \tag{5.129}$$

This special connection is called the *Levi–Civita connection*.

The first of the additional properties (5.128) says that the commutator (5.32) of the vector fields V and W equals the difference of the covariant derivative of W along V and of V along W. By applying (5.128) to two base fields ∂_i and ∂_j we see that the Christoffel symbols are symmetric in their lower indices,

$$\Gamma^k_{ij} = \Gamma^k_{ji} \,. \tag{5.130}$$

Indeed the left-hand side vanishes because the base fields commute; the right-hand side, according to (5.126), gives (5.130)[10].

As it also possesses the property (5.129) the Levi–Civita connection is said to be *metric*. Indeed, the covariant derivative can also be applied to other smooth objects defined on M such as the metric g. One can show that (5.129) is equivalent to the condition $Dg = 0$, which says that the covariant derivative of the metric along any smooth vector field vanishes.

In local coordinates the Christoffel symbols can be expressed in terms of derivatives of the metric tensor g_{ik} as well as by its inverse g^{km}. We skip this calculation and simply quote the result

$$\Gamma^k_{ij} = \frac{1}{2} \sum_m g^{km} \left(\frac{\partial g_{jm}}{\partial x^i} + \frac{\partial g_{mi}}{\partial x^j} - \frac{\partial g_{ij}}{\partial x^m} \right) . \tag{5.131}$$

The symmetry (5.130) is obvious in this explicit formula.

5.7.2 Parallel Vector Fields and Geodesics

A smooth curve $\alpha : I \subset \mathbb{R}_\tau \to M$ on the manifold M is itself a smooth, one-dimensional manifold. Consider a smooth vector field $Z \in \mathcal{V}(\alpha)$ on this submanifold of M. Let τ be the parameter describing the curve, let the dot denote the derivative with respect to τ and let $\dot{\alpha}$ be its tangent vector field. The derivative of Z with respect to τ can then be computed as follows,

$$\dot{Z} = \sum_k \frac{\mathrm{d}Z^k}{\mathrm{d}\tau} \partial_k + \sum_k Z^k D_{\dot{\alpha}}(\partial_k) = \sum_k \left\{ \frac{\mathrm{d}Z^k}{\mathrm{d}\tau} + \sum_{lm} \Gamma^k_{lm} \frac{\mathrm{d}(x^l \circ \alpha)}{\mathrm{d}\tau} Z^m \right\} \partial_k \,.$$

This is the rate of change of Z as one moves along the curve. In particular, if $\dot{Z} = 0$ the vector field Z is said to be *parallel*. Given a tangent vector $z \in T_{\alpha(\tau_0)}M$ to the point $\alpha(\tau_0)$ on the curve we can now state precisely how to perform parallel transport of a given vector along the curve α. In particular, for every smooth curve $\alpha : I \to M$ there is a unique parallel vector field Z such that at $\tau = \tau_0$ it equals a given tangent vector, say $Z(\tau_0) = z$.

A case of special interest is when $Z = \dot{\alpha}$, i.e. where Z is the tangent vector field of a curve α. Obviously, $\dot{Z}$ is then none other than the acceleration $\ddot{\alpha}$. Geodesics, from the point of view of physics, describe motion of free fall on the manifold,

[10] The condition (5.128) expresses the fact that the Levi–Civita connection has vanishing torsion.

i.e. motion with vanishing acceleration. Geometrically speaking they are curves on the manifold which link arbitrary points p and q such that the length of the arc pq is extremal. An elementary example is provided by the unit sphere in three dimensions, $M = S^2$, where the geodesics are the great circles. The geodesic distance between any two points $A, B \in S^2$ is either a minimum (if the smaller segment of the great circle joining them is chosen), or a maximum (if the larger segment is chosen). If A and B are antipodes, the geodesic length corresponds to a saddle point (cf. Sect. 2.36).

These remarks illustrate the geometrical definition of geodesics.

Geodesics on a smooth Riemannian manifold are smooth curves $\gamma : I \to M$ whose tangent vector field $\dot{\gamma}$ is parallel.

This definition and our previous remarks allow us to write down a differential equation for geodesics in local coordinates. It reads

$$\frac{\mathrm{d}^2}{\mathrm{d}\tau^2}(x^l \circ \gamma) + \sum_{jk} \Gamma^l_{jk}(\gamma)\frac{\mathrm{d}}{\mathrm{d}\tau}(x^j \circ \gamma)\frac{\mathrm{d}}{\mathrm{d}\tau}(x^k \circ \gamma) = 0 . \tag{5.132a}$$

Here, the functions $(x^i \circ \gamma)$ are coordinate functions on the curve γ. As their meaning is obvious and as there is no real danger of confusion one simplifies the notation by writing just x^i for short. The geodesic equation then takes the simpler form

$$\ddot{x}^l + \sum_{jk} \Gamma^l_{jk}(\gamma)\dot{x}^j\dot{x}^k = 0 . \tag{5.132b}$$

5.7.3 Geodesics as Solutions of Euler–Lagrange Equations

As we have seen, geodesics describe force-free, unaccelerated motion on a given manifold. They are curves whose length is an extremum and, therefore, they are solutions of Euler–Lagrange equations. This is the content of the following theorem

Theorem on geodesics. Let (Q, g) be a Riemannian manifold and

$$L : TQ \longrightarrow \mathbb{R} , \quad L(v) = \frac{1}{2}\langle v|v\rangle$$

a Lagrangian function. A curve γ is a solution of the Euler–Lagrange equations if and only if it is geodesic on Q.

Proof: In local coordinates the Lagrangian function reads

$$L(v) = \frac{1}{2}\sum_{ij} g_{ij}(q)v^i v^j \equiv \frac{1}{2}\sum_{ij} g_{ij}(q)\dot{q}^i\dot{q}^j .$$

Lagrange's equations (2.18) yield

$$\frac{\mathrm{d}}{\mathrm{d}t}\left(\sum_j g_{ij}\dot{q}^j\right) - \frac{1}{2}\sum_{jk}\frac{\partial g_{jk}}{\partial q^i}\dot{q}^j\dot{q}^k = 0\,. \tag{5.133}$$

We calculate the time derivative in the first term

$$\frac{\mathrm{d}}{\mathrm{d}t}\left(\sum_j g_{ij}\dot{q}^j\right) = \sum_j g_{ij}\ddot{q}^j + \sum_{ij}\frac{\partial g_{ij}}{\partial q^k}\dot{q}^j\dot{q}^k\,,$$

multiply the entire equation from the left with the inverse g^{li} of the metric tensor, and sum over i to obtain the differential equation

$$\ddot{q}^l + \sum_{ijk} g^{li}\left(\frac{\partial g_{ij}}{\partial q^k} - \frac{1}{2}\frac{\partial g_{jk}}{\partial q^i}\right)\dot{q}^j\dot{q}^k =$$

$$\ddot{q}^l + \frac{1}{2}\sum_{ijk} g^{li}\left(\frac{\partial g_{ij}}{\partial q^k} + \frac{\partial g_{ik}}{\partial q^j} - \frac{\partial g_{jk}}{\partial q^i}\right)\dot{q}^j\dot{q}^k = 0\,.$$

In the second step we have written the first term of the expression within brackets twice by making use of its symmetry in j and k. In its second form, upon inserting the formula (5.131) for the Christoffel symbols, the differential equation becomes precisely the geodesic equation (5.132b). This proves the theorem.

Remark: With $L = T = g_{ik}\dot{q}^i\dot{q}^k/2$ and with T the kinetic energy, (5.133) shows that the geodesic equation has the form of (2.18). The integral

$$\lambda := \int_{\tau_1}^{\tau_2}\mathrm{d}\tau\,\sqrt{\sum_{ik} g_{ik}(q(\tau))\dot{q}^i\dot{q}^k} = \int_{\tau_1}^{\tau_2}\mathrm{d}\tau\,\sqrt{2T} \tag{5.134}$$

is the length of the curve with boundary values $\gamma(\tau_1) = a$ and $\gamma(\tau_2) = b$. As long as T does not vanish these geodesics are curves whose length λ is extremal because as $T \neq 0$

$$\frac{\mathrm{d}}{\mathrm{d}t}\frac{\partial\sqrt{T}}{\partial\dot{q}^i} - \frac{\partial\sqrt{T}}{\partial q^i} = 0 = \frac{1}{2\sqrt{T}}\left(\frac{\mathrm{d}}{\mathrm{d}t}\frac{\partial T}{\partial\dot{q}^i} - \frac{\partial T}{\partial q^i}\right)\,.$$

5.7.4 Example: Force-Free Asymmetric Top

We wish to conclude this chapter by illustrating these general results by means of a particularly beautiful example[11]: We show that Euler's equations (3.59) are geodesic equations on the Riemannian manifold $M = \mathrm{SO}(3)$, with the metric being determined by the inertia tensor $\mathbf{J}$.

We start by recalling that by using $\mathbf{S}(\varphi) = \sum_{i=1}^{3}\varphi_i\mathbf{J}_i$ the rotation matrix (3.45a) can be written as an exponential series in $\mathbf{S}$ and that the action of the

[11] V.I. Arnol'd: Ann. Inst. Fourier **16**, 319 (1966)

latter on any vector equals the cross product of φ with that vector (cf. Sect. 2.22). Thus, in symbols

$$\mathbf{R}(\varphi) = \exp\{-\mathbf{S}(\varphi)\} \quad \text{and} \quad \mathbf{S}(\varphi)\boldsymbol{x} = \boldsymbol{\varphi} \times \boldsymbol{x}\,.$$

The action of the matrix $\boldsymbol{\Omega}(\tau) = \dot{\mathbf{R}}^T(\tau)\mathbf{R}(\tau)$, eq. (3.53), on a vector in the laboratory system is that given in (3.56b). Clearly, these formulas can be rotated to the body-fixed system,

$$\overline{\boldsymbol{\Omega}}\,\overline{\boldsymbol{x}} = \overline{\boldsymbol{\omega}} \times \overline{\boldsymbol{x}}\,, \text{ where } \overline{\boldsymbol{\omega}} = \mathbf{R}\boldsymbol{\omega}\,,\ \overline{\boldsymbol{\Omega}} = \mathbf{R}\boldsymbol{\Omega}\mathbf{R}^T\,. \tag{5.135}$$

The Lagrangian function is equal to the kinetic energy expressed in the body-fixed system,

$$L = T = \frac{1}{2}\overline{\boldsymbol{\omega}} \cdot \mathbf{J} \cdot \overline{\boldsymbol{\omega}}\,. \tag{5.136}$$

Let $\mathbf{R}(\tau)$ be a smooth curve on the manifold $M = \mathrm{SO}(3)$ which assumes the boundary values $\mathbf{R}(\tau_1) = \mathbf{R}_1$ and $\mathbf{R}(\tau_2) = \mathbf{R}_2$ and which is such that the length (5.134) is an extremum. We show that any such geodesic obeys Euler's equations (3.58) for vanishing external torque.

Let $\mathbf{R}(\tau)$ be a geodesic, $\mathbf{R}_0 \in \mathrm{SO}(3)$ a *constant*, fixed rotation. We compute

$$\left[\frac{\mathrm{d}}{\mathrm{d}t}(\mathbf{R}_0\mathbf{R}(\tau))^{\mathrm{T}}\right](\mathbf{R}_0\mathbf{R}(\tau)) = \left[\frac{\mathrm{d}}{\mathrm{d}t}\mathbf{R}(\tau)\right]^{\mathrm{T}}\mathbf{R}_0^{\mathrm{T}}\mathbf{R}_0\mathbf{R}(\tau) = \dot{\mathbf{R}}^{\mathrm{T}}(\tau)\mathbf{R}(\tau)\,.$$

From this we conclude that $\boldsymbol{\Omega}$, and hence also $\boldsymbol{\omega}$ as well as $\overline{\boldsymbol{\omega}}$ remain unchanged. This means that if $\mathbf{R}(\tau)$ is a geodesic, so is $(\mathbf{R}_0\mathbf{R}(\tau))$. Therefore, it is sufficient to discuss the special geodesic which goes through $\mathbf{R}(\tau = 0) = \mathbb{1}$. In this case $\dot{\mathbf{R}}(0) = \boldsymbol{\Omega}(0)$. We compute $\overline{\boldsymbol{\Omega}}$ in the neighborhood of $\varphi = 0$ as follows

$$\begin{aligned}\overline{\boldsymbol{\Omega}} &= \mathbf{R}\boldsymbol{\Omega}\mathbf{R}^{-1} = \mathbf{R}(\tau)\dot{\mathbf{R}}^{\mathrm{T}}(\tau)\\ &= (\mathbb{1} - \mathbf{S} + \ldots)\left(\dot{\mathbf{S}} + \frac{1}{2}\dot{\mathbf{S}}\mathbf{S} + \frac{1}{2}\mathbf{S}\dot{\mathbf{S}} + \ldots\right)\\ &= \dot{\mathbf{S}} - \frac{1}{2}[\mathbf{S}, \dot{\mathbf{S}}] + \mathcal{O}(\varphi^2)\,,\end{aligned}$$

and use the identity (cf. Sect. 3.12)

$$[\mathbf{S}, \dot{\mathbf{S}}] \equiv [\mathbf{S}(\boldsymbol{\varphi}), \dot{\mathbf{S}}(\boldsymbol{\varphi})] = \mathbf{S}(\boldsymbol{\varphi} \times \dot{\boldsymbol{\varphi}})\,.$$

Note that $\boldsymbol{\varphi}$ and $\dot{\boldsymbol{\varphi}}$ here are independent variables and need not have the same direction. We conclude that

$$\overline{\boldsymbol{\Omega}} = \mathbf{S}(\dot{\boldsymbol{\varphi}}) - \frac{1}{2}\mathbf{S}(\boldsymbol{\varphi} \times \dot{\boldsymbol{\varphi}}) + \mathcal{O}(\varphi^2)\,.$$

On the other hand (5.135) implies that $\overline{\boldsymbol{\Omega}} = \mathbf{S}(\overline{\boldsymbol{\omega}})$ and we conclude that

$$\overline{\omega} = \dot{\varphi} - \frac{1}{2}\varphi \times \dot{\varphi} + \mathcal{O}(\varphi^2)\,. \tag{5.137}$$

Inserting (5.137) into (5.136) and keeping track of the symmetry of the inertia tensor we have

$$T = \frac{1}{2}\dot{\varphi} \cdot \mathbf{J} \cdot \dot{\varphi} - \frac{1}{2}\dot{\varphi} \cdot \mathbf{J} \cdot (\varphi \times \dot{\varphi}) + \mathcal{O}(\varphi^2)\,.$$

In calculating, in a next step, the derivatives with respect to $\boldsymbol{\varphi}$ and to $\dot{\boldsymbol{\varphi}}$ it is useful to rewrite the second term of T by using the identities $\boldsymbol{a}\cdot(\boldsymbol{b}\times\boldsymbol{c}) = \boldsymbol{b}\cdot(\boldsymbol{c}\times\boldsymbol{a}) = \boldsymbol{c}\cdot(\boldsymbol{a}\times\boldsymbol{b})$ with $\boldsymbol{a} = (\dot{\boldsymbol{\varphi}}\cdot\mathbf{J})^T = \mathbf{J}\cdot\dot{\boldsymbol{\varphi}}$, $\boldsymbol{b} = \boldsymbol{\varphi}$, and $\boldsymbol{c} = \dot{\boldsymbol{\varphi}}$. To first order we find

$$\frac{\partial T}{\partial \varphi} = -\frac{1}{2}\dot{\varphi} \times (\mathbf{J}\dot{\varphi}) + \mathcal{O}(\varphi) = -\frac{1}{2}\overline{\omega} \times (\mathbf{J}\,\overline{\omega}) + \mathcal{O}(\omega)\,.$$

In much the same way one finds

$$\frac{\mathrm{d}}{\mathrm{d}\tau}\frac{\partial T}{\partial \dot{\varphi}} = \mathbf{J}\ddot{\varphi} - \frac{1}{2}(\mathbf{J}\dot{\varphi}) \times \dot{\varphi} + \mathcal{O}(\varphi) = \mathbf{J}\,\dot{\overline{\omega}} + \frac{1}{2}\overline{\omega} \times (\mathbf{J}\,\overline{\omega}) + \mathcal{O}(\varphi)\,.$$

Thus, taking $\boldsymbol{\varphi} = 0$ one recovers the geodesic equation

$$\frac{\mathrm{d}}{\mathrm{d}\tau}\frac{\partial T}{\partial \dot{\varphi}} - \frac{\partial T}{\partial \varphi} = \mathbf{J}\,\dot{\overline{\omega}} + \overline{\omega} \times (\mathbf{J}\,\overline{\omega}) = 0\,. \tag{5.138}$$

This equation is identical to Euler's equations (3.58), with $\overline{D} = 0$. Thus, Euler's equations of motion have a simple geometrical interpretation which is helpful in visualizing their content:

The spinning top without external forces follows geodesics of the smooth manifold $SO(3)$.

6. Stability and Chaos

In this chapter we study a larger class of dynamical systems that include but go beyond Hamiltonian systems. We are interested, on the one hand, in *dissipative systems*, i.e. systems that lose energy through frictional forces or into which energy is fed from exterior sources, and, on the other hand, in discrete, or discretized, systems such as those generated by studying flows by means of the Poincaré mapping. The occurence of dissipation implies that the system is coupled to other, external systems, in a controllable manner. The strength of such couplings appears in the set of solutions, usually in the form of parameters. If these parameters are varied it may happen that the flow undergoes an essential and qualitative change, at certain critical values of the parameters. This leads rather naturally to the question of stability of the manifold of solutions against variations of the control parameters and of the nature of such a structural change. In studying these questions, one realizes that deterministic systems do not always have the well-ordered and simple behavior that we know from the integrable examples of Chap. 1, but that they may exhibit completely unordered, chaotic behavior as well. In fact, in contradiction with traditional views, and perhaps also with one's own intuition, chaotic behavior is not restricted to dissipative systems (turbulence of viscous fluids, dynamics of climates, etc.). Even relatively simple Hamiltonian systems with a small number of degrees of freedom exhibit domains where the solutions have strongly chaotic character. As we shall see, some of these are relevant for celestial mechanics.

6.1 Qualitative Dynamics

In the preceding chapters, we dealt primarily with fundamental properties of mechanical systems, with principles that allowed the construction of their equations of motion, and with general methods of solving these equations. The integrable cases, although a minority among the dynamical systems, were of special importance because they allowed us to follow specific solutions analytically, to appreciate the significance and the power of conservation laws, and to study the restrictions that the latter impose on the manifold of motions in phase space.

On the other hand, there are questions to which we have paid less attention so far; for example: What is the long-term behavior of a periodic motion that is subject to a small pertubation? What is the structure of the flow of a mechanical system (i.e. the set of *all* possible solutions) in the large? Are there structural, characteristic

F. Scheck, *Mechanics*, Graduate Texts in Physics, 5th ed.,
DOI 10.1007/978-3-642-05370-2_6,

properties of the flow that do not depend on the specific values of the constants appearing in the equations of motion? Can there be "ordered" and "unordered" types of motions, in a given system? If yes, can one define a quantitative measure for the lack of "order"? If a given system depends on external control parameters (strength of a perturbation, amplitude and frequency of a forced vibration, varying degree of friction, etc.), are there critical values of the parameters where the flow of the system changes its structure in the large?

These questions show that, here, we approach the analysis of mechanical systems in a somewhat different spirit. The equations of motion are assumed to be known (even though they may depend on control parameters that can be varied). We concentrate less on the individual solution but, instead, study the flow as a whole, its stability, its topological structure, and its behavior over long time periods. It is this kind of analysis we wish to call qualitative dynamics. Quite logically, it leads one to investigate the stability of equilibrium positions and of periodic orbits, to study attractors for dissipative systems (i.e. manifolds of lower dimension than the original phase space, to which the system tends, for large times, under the action of dissipation), to study bifurcations (i.e. structural changes of the flow at critical values of the control parameters), and to analyse the pattern of disordered motion if it occurs.

6.2 Vector Fields as Dynamical Systems

The dynamics of a very great variety of dynamical systems can be cast in the form of systems of first-order differential equations, viz.

$$\boxed{\frac{\mathrm{d}}{\mathrm{d}t}\underset{\sim}{x}(t) = \underset{\sim}{F}(\underset{\sim}{x}(t), t)}\ . \tag{6.1}$$

Here, t is the time variable, $\underset{\sim}{x}(t)$ is a point in the configuration space of the system, and $\underset{\sim}{F}$ is a vector field that is continuous and often also differentiable. The space of the variables $\underset{\sim}{x}$ may be the velocity space, described locally by generalized coordinates q^i and velocities $\dot{q}^i$, or the phase space that we describe locally by the q^i and the canonically conjugate momenta p_i. There are, of course, other cases where the $\underset{\sim}{x}$ live in some other manifold: an example is provided by the Eulerian angles that parametrize the rotational motion of rigid bodies.

As an example, let the equation of the motion be given in the form

$$\ddot{y} + f_1(y,t)\dot{y} + f_2(y,t) = 0\ .$$

It is easy to recast this in the form of (6.1), by taking

$$x_1(t) \stackrel{\text{def}}{=} y(t)\ , \quad x_2(t) \stackrel{\text{def}}{=} \dot{y}(t)\ ,$$

so that $\dot{x}_1 = x_2$, $\dot{x}_2 = -f_1 x_2 - f_2$. The pattern (6.1), of course, is not restricted to Lagrangian or Hamiltonian systems. It also describes systems with dissipation, that is systems where either mechanical energy is converted to other forms of energy or where energy from external sources is fed into the system. Thus, (6.1) describes a large class of dynamical systems, defined on the space of $\underset{\sim}{x}$ variables and the time axis $\mathbb{R}_t$. This equation is a local expression of the underlying physical laws. For instance, it relates the acceleration at every point of space and at each time t with the given field of forces. In this sense it determines the dynamics locally, "in the small". The temporal evolution of the system, starting from an arbitrary but fixed initial configuration, will be known only when we have the complete solution of the differential equation (6.1) that obeys this initial condition. As an example, consider the Kepler problem (Sect. 1.7.2) for a given initial position $\boldsymbol{r}_0$ and initial velocity $\dot{\boldsymbol{r}}_0$ of the relative coordinate. Take $T_0 = \mu\dot{\boldsymbol{r}}_0^2/2$ to be smaller than $|U(r_0)| \equiv A/r_0$ (where $A = Gm_1m_2$) and take $l \equiv |\boldsymbol{r}_0 \times \dot{\boldsymbol{r}}_0|\mu$ to be different from zero. The specific solution that assumes this initial configuration is the Keplerian ellipse with parameters $l^2/A\mu$ and

$$\varepsilon = \sqrt{1 + 2(T_0 + U(r_0))l^2/\mu A^2} \, .$$

This specific solution, though, gives little information on the general dynamics of mass points in the field of the gravitational force $\boldsymbol{F} = -\boldsymbol{\nabla} U$. Only when we know the solutions for *all* allowed initial configurations do we learn that, besides ellipses and circles, the Kepler problem also admits hyperbolas and parabolas as the typical scattering orbits. In other words, the diversity of the dynamics hidden in an equation such as (6.1) will come to light only if one knows and understands all solutions, i.e. the complete flow of the vector field $\underset{\sim}{F}$.

These remarks apply to a system whose law of motion is given once and for all. In the case of real physical systems, this assumption is true only in exceptional situations, for the following reasons.

(i) It may happen that the force law is not known exactly. Its explicit form may contain one or several parameters that one whishes to determine from the observed motions. Here is an example: if one doubts the long-range character of the Coulomb potential between two point charges e_1 and e_2, one might assume $U(\boldsymbol{r}) = e_1e_2/r^\alpha$ with $\alpha = 1 + \varepsilon$, with the idea of studying the dependence of the corresponding dynamics on the parameter ε (see also Practical Example 1.4).

(ii) The vector field $\underset{\sim}{F}$ on the right-hand side of (6.1) describes the influence of an external system that might be varied. An example is an oscillator that is coupled to an external oscillation of variable frequency and variable amplitude.

(iii) It may be that the differential equation (6.1) contains a predominant force field for which all physically allowed solutions are known. In addition, it contains further terms that describe the coupling of the system to other, external systems, the coupling being weak enough so that they may be taken to be small perturbations of the initial, soluble system. This is the situation that we studied in Sect. 2.38–2.40.

In all cases and examples quoted above, the vector field $\underset{\sim}{F}$ contains additional parameters that can be varied and that may have a decisive influence on the manifolds of solutions. For example, it may happen that the solutions of (6.1) change their structure completely once the parameters cross certain critical values. Stable solutions can turn into unstable ones, a periodic solution, by the bifurcation phenomenon, can double its frequency, etc.

From these remarks it is clear that the task of studying deterministic dynamic systems on the basis of their equation of motion (6.1) is a very ambitious one. When formulated this generally, this field of *differentiable dynamics*, by far, is not a closed subject. On the contrary, there are only relatively few rigorous results and a number of empirical results based on numerical studies. Therefore, studying this branch of mechanics leads one very quickly into the realm of modern research in this field.

6.2.1 Some Definitions of Vector Fields and Their Integral Curves

In this section we take up the tools introduced in Chap. 5 and discuss some concepts that are important for studying vector fields as dynamic systems. The local form of (6.1) is sufficient for an understanding of most of what follows in subsequent sections. Therefore, the reader who is not used to the geometrical language may skip this section. On the other hand, if one wishes to learn more about the subjects touched upon in this chapter, some knowledge of the content of Chap. 5 is mandatory, as the specialized literature and the research in this field make extensive use of the concepts and methods of topology and differential geometry.

In reality, (6.1) is a coordinate expression of the differential equation (5.42) for integral curves of a smooth vector field $\mathcal{F}$ on the manifold M. In physics, typically M is the phase space T^*Q or the velocity space TQ, i.e. the cotangent or tangent bundles of the coordinate manifold, respectively.

The curve $\Phi_m : I \to M$ is an integral curve of $\mathcal{F}$ if the tangent vector field $\dot{\Phi}_m$ coincides with $\mathcal{F}_{\Phi_m(t)}$, the restriction of $\mathcal{F}$ to points along the curve Φ_m,

$$\dot{\Phi}_m(t) = \mathcal{F}_{\Phi_m(t)} \,, \quad t \in I \subset \mathbb{R}_t \,, \quad \mathcal{F} \in \mathcal{X}(M) \,. \tag{6.2}$$

I is an open interval on the time axis $\mathbb{R}_t$ that contains the origin $t = 0$. The integral curve Φ_m is chosen such that it goes through m at time zero. (We adopt the notation of Sect. 1.19 because we shall use results from there.)

In the coordinates of the chart (φ, U) we obtain the differential equation (5.43), i.e.

$$\frac{\mathrm{d}}{\mathrm{d}t}(x^i \circ \Phi_m) = \mathcal{F}^i(x^k \circ \Phi_m, t) \,, \tag{6.3}$$

or, in a somewhat simplified notation, (6.1).

Somewhat more generally, we have the following. For each m_0 of M there is an open neigborhood V on M, an open interval I on the time axis containing the origin $t = 0$, and a smooth mapping

$$\Phi : V \times I \to M \,, \tag{6.4}$$

such that, for every fixed $m \in V$, the curve $\Phi(m, t)$ is the integral curve $\Phi_m(t) \equiv \Phi(m, t)$ of $\mathcal{F}$ that goes through m at time $t = 0$, $\Phi(m, t = 0) = m$. The theorem of Sect. 1.19 guarantees the existence and uniqueness of this integral curve. Φ is said to be the *local flow* of the vector field $\mathcal{F}$ and the integral curves $\Phi_m : I \to M$ are said to be the *flux* or *flow lines* of Φ. Keeping the time variable in $\Phi(m,t)$ fixed and letting m wander through V, we obtain the *flow fronts*

$$\Phi_t(m) \stackrel{\text{def}}{=} \Phi(m,t) \tag{6.5}$$

of the flow Φ. This local manifold of solutions may be visualized as shown in Fig. 6.1. Given a fixed time $t \in I$, each point of the domain V flows along a certain section of its integral curve Φ_m. The domain as a whole moves on to $\Phi_t(V)$.

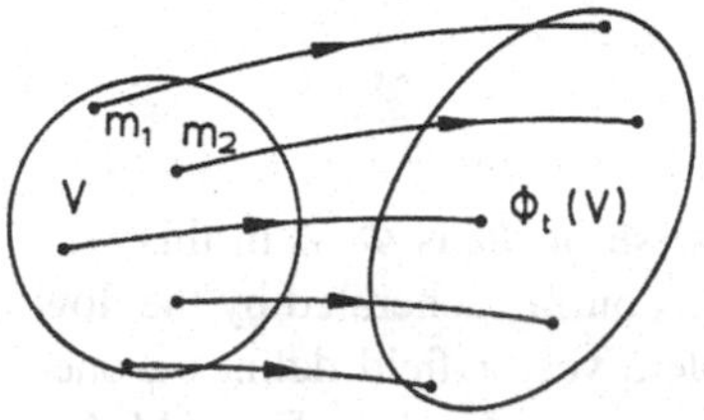

Fig. 6.1. During time t, the flow of a vector field transports a domain V to $V' = \Phi_t(V)$. The figure also shows the orbits along which the points of V move during this time

If $I_{(m)}$ is the maximal allowed interval on the time axis for which Φ_m exists, Φ_m is unique and is said to be the *maximal integral curve* through m. Applying this reasoning to every point of M yields a uniquely determined, open set $\Omega \subset M \times \mathbb{R}_t$, on which the *maximal flow* $\Phi : \Omega \to M$ of the vector field $\mathcal{F}$ is defined. This leads to the following.

Definition. A vector field $\mathcal{F}$ is said to be *complete* if $\Omega = M \times \mathbb{R}_t$, i.e. if its maximal flow is defined on the whole manifold and for all times.

The Hamiltonian vector field of the harmonic oscillator provides an example of a complete vector field,

$$(\mathcal{F}^i) \equiv (X^i_{\mathrm{H}}) = \left(\frac{\partial H}{\partial p}, -\frac{\partial H}{\partial q}\right) = (p, -q) \,. \tag{6.6}$$

Its maximal flow

$$\Phi(m \equiv (q,p), t) = \begin{pmatrix} \cos t & \sin t \\ -\sin t & \cos t \end{pmatrix} \begin{pmatrix} q \\ p \end{pmatrix}$$

is defined on the whole phase space (this is the example of Sect. 5.3.1 with $t_0 = 0$). In practice there are examples of vector fields that are not complete. For instance, in the Kepler problem the origin of the potential must be cut out of the orbit plane because of its singularity at this point. The corresponding Hamiltonian vector field then ceases to be complete on $\mathbb{R}^2$. Similarly, in relativistic mechanics and in general relativity there are vector fields (such as velocity fields of geodesics) that are not complete. For a complete vector field, Φ is a global flow,

$$\Phi : M \times \mathbb{R}_t \to M \,, \tag{6.7}$$

that may be interpreted in yet another way. As in (6.5) let us keep the time fixed. The flow (6.7) then generates a smooth mapping of M onto itself,

$$\Phi_t : M \to M : m \mapsto \Phi_t(m) \stackrel{\text{def}}{=} \Phi(m, t) \,, \tag{6.8}$$

which has the following properties. For $t = 0$ it is the identical mapping on M, $\Phi_0 = \mathrm{id}_M$. Taking the composition of (6.7) with itself, twice or several times, one obtains

$$\Phi_{t+s} = \Phi_t \circ \Phi_s \quad \text{for} \quad t, s \in \mathbb{R}_t \,.$$

For each t, Φ_t is a diffeomorphism of M. The inverse of Φ_t is Φ_{-t}. In this way we obtain a one-parameter group of diffeomorphisms on M, generated by the flow Φ and the assignment $t \mapsto \Phi_t$. Thus every complete vector field defines a one-parameter group of diffeomorphisms. Conversely, a group $\Phi : M \times \mathbb{R} \to M$ that depends on a real parameter defines a complete vector field.

6.2.2 Equilibrium Positions and Linearization of Vector Fields

Suppose the laws of motion of a physical system are described by the local equation (6.1) or, more generally, by an equation of the form of (6.2). A set of differential equations of this kind is called a *dynamical system* (although, strictly speaking, only the set of all solutions to these equations describes the dynamics). For what follows, we treat dynamical systems in the simplified form of (6.1), i.e. in the form of differential equations on $\mathbb{R}^n$. For manifolds that are not Euclidean spaces this means that we work in local charts. Exceptions regarding dynamical systems on more general, smooth manifolds will be mentioned explicitly.

A point $\underset{\sim}{x}_0$ is said to be an *equilibrium position* of the vector field $\underset{\sim}{F}$ if $\underset{\sim}{F}(\underset{\sim}{x}_0) = 0$. Equivalently, one also talks about a *singular* or *critical point* of the vector field. For an autonomous system, for example, (6.1) becomes

$$\dot{\underset{\sim}{x}}(t) = \underset{\sim}{F}(\underset{\sim}{x}(t)) \,. \tag{6.9}$$

At a critical point $\underset{\sim}{x}_0$ the velocity vector vanishes so that the system cannot move out of this point. However, as such, (6.9) says nothing about whether the configuration $\underset{\sim}{x}_0$ is stable or unstable against perturbations. One learns more about this if one linearizes (6.9) about the point $\underset{\sim}{x}_0$. For this purpose we introduce the following definitions.

(i) Linearization in the neighborhood of a critical point. In the terminology of Sect. 6.2.1 the linearization of a vector field at a critical point m_0 is defined to be the linear mapping

$$\mathcal{F}'(m_0) : T_{m_0}M \to T_{m_0}M \,,$$

which assigns to every tangent vector $v \in T_{m_0}M$ the derivative

$$\mathcal{F}'(m_0) \cdot v = \frac{\mathrm{d}}{\mathrm{d}t}(T\Phi_{m_0}(t) \cdot v)\big|_{t=0} \,.$$

Here Φ is the flow of $\mathcal{F}$ and $T\Phi$ is the corresponding tangent mapping.

In the simplified form of (6.9), valid on $\mathbb{R}^n$, linearization means simply that we expand about the point $\underset{\sim}{x} = \underset{\sim}{x}_0$. Thus, take $\underset{\sim}{y} = \underset{\sim}{x} - \underset{\sim}{x}_0$ and $\underset{\sim}{F}(\underset{\sim}{x}_0) = 0$. From (6.9) we obtain the differential equation

$$\dot{y}^i(t) = \sum_{k=1}^{n} \frac{\partial F^i}{\partial x^k}\bigg|_{\underset{\sim}{x}_0} y^k(t) \tag{6.10a}$$

or, in more compact notation,

$$\dot{\underset{\sim}{y}}(t) = \mathbf{D}\underset{\sim}{F}\big|_{\underset{\sim}{x}_0} \cdot \underset{\sim}{y}(t) \,. \tag{6.10b}$$

This is a differential equation of the type studied in Sect. 1.21. The symbol $\mathbf{D}F$ denotes the matrix of partial derivatives, very much as in Sect. 2.29.1. For an autonomous system (6.9) this matrix is independent of time. The linear system (6.10) obtained from it is homogeneous and autonomous.

The following case is more general (see Exercise 1.22 and the example of Sect. 1.26).

(ii) Linearization in the neighborhood of a given solution. Let $\underset{\sim}{\Phi}(t)$ be a solution of (6.1) and let $\underset{\sim}{y}(t) = \underset{\sim}{x}(t) - \underset{\sim}{\Phi}(t)$. Then, from (6.1),

$$\dot{\underset{\sim}{y}}(t) = \underset{\sim}{F}(\underset{\sim}{y}(t) + \underset{\sim}{\Phi}(t), t) - \dot{\underset{\sim}{\Phi}}(t) = \underset{\sim}{F}(\underset{\sim}{y}(t) + \underset{\sim}{\Phi}(t), t) - \underset{\sim}{F}(\underset{\sim}{\Phi}(t), t) \,.$$

Expanding the right-hand side in a Taylor series about the solution $\underset{\sim}{\Phi}(t)$ yields the linear and homogenous differential equation

$$\dot{y}^i(t) = \sum_{k} \frac{\partial F^i}{\partial x^k}\big(\underset{\sim}{x} = \underset{\sim}{\Phi}(t), t\big) y^k(t) \,, \tag{6.11}$$

where the partial derivatives of $\underset{\sim}{F}$ must be taken along the orbit $\underset{\sim}{\Phi}(t)$. Even if $\underset{\sim}{F}$ has no explicit time dependence, the linearized system (6.11) is not autonomous. It becomes autonomous only if the specific solution is chosen to be an equilibrium position, $\underset{\sim}{\Phi}(t) = \underset{\sim}{x}_0$, taking us back to the first case (6.10).

In the simpler case of linearizing an autonomous system about an equilibrium position we obtain the linear, homogeneous, and autonomous system (6.10), i.e.

$$\dot{\underset{\sim}{y}}(t) = \mathbf{A}\underset{\sim}{y}(t) \tag{6.12}$$

with the matrix $\mathbf{A}$ being given by

$$A_{ik} = \frac{\partial F^i}{\partial x^k}\bigg|_{\underset{\sim}{x}_0} \,.$$

The linear system (6.12) can be solved explicitly. The solution that fulfills the initial condition $\underset{\sim}{y}(s) = \underset{\sim}{y}_0$ is

$$\underset{\sim}{y}(t) \equiv \underset{\sim}{\Psi}_{t,s}(\underset{\sim}{y}_0) = \exp[(t-s)\mathbf{A}]\underset{\sim}{y}_0 \tag{6.13}$$

with $\underset{\sim}{\Psi}_{s,s}(\underset{\sim}{y}_0) = \underset{\sim}{y}_0$ and with

$$\exp[(t-s)\mathbf{A}] = \sum_{n=0}^{\infty} \frac{(t-s)^n}{n!}\mathbf{A}^n \,.$$

If $\mathbf{A}$ is given in diagonal form, this series becomes particularly simple. With α_i denoting the eigenvalues of $\mathbf{A}$ the exponential series also has diagonal form, its eigenvalues being $\exp(\lambda\alpha_i)$ with $\lambda = t - s$. For this reason, the eigenvalues of the matrix $\mathbf{A} = \mathbf{D}F$ are called *characteristic exponents* of the vector field $\underset{\sim}{F}$ at the point $\underset{\sim}{x}_0$.

For the sake of illustration we consider two examples. The first is the example of Sect. 1.21.1, which is understood to be the linearization of the plane pendulum at the point $\underset{\sim}{x} = 0$. From (1.46)

$$\mathbf{A} = \begin{pmatrix} 0 & 1/m \\ -m\omega^2 & 0 \end{pmatrix} .$$

The eigenvalues of $\mathbf{A}$ are easily found. From the characteristic equation $\det(\alpha\mathbb{1} - \mathbf{A}) = 0$ one finds $\alpha_1 = \mathrm{i}\omega$, $\alpha_2 = -\mathrm{i}\omega$, so that the diagonalized matrix is

$$\overset{0}{\mathbf{A}} = \begin{pmatrix} \mathrm{i}\omega & 0 \\ 0 & -\mathrm{i}\omega \end{pmatrix} . \tag{6.14}$$

In the second example we add a friction term to the plane pendulum, proportional to the velocity of the motion. Thus, in linearized form we obtain the differential equation

$$m\ddot{q} + 2\gamma m\dot{q} + m\omega^2 q = 0 \,, \tag{6.15}$$

where γ is a constant with the dimension of a frequency.

Using the notation of Sect. 1.18, $y^1 = q$, $y^2 = m\dot{q}$, (6.15) becomes

$$\begin{pmatrix} \dot{y}^1 \\ \dot{y}^2 \end{pmatrix} = \mathbf{A}\begin{pmatrix} y^1 \\ y^2 \end{pmatrix} \quad \text{with} \quad \mathbf{A} = \begin{pmatrix} 0 & 1/m \\ -m\omega^2 & -2\gamma \end{pmatrix} .$$

The eigenvalues of $\mathbf{A}$ are computed as in the previous example. For $\gamma^2 < \omega^2$ (this is the case of weak friction) one finds two, complex conjugate characteristic exponents

$$|\gamma| < \omega : \overset{0}{\mathbf{A}} = \begin{pmatrix} -\gamma + \mathrm{i}\sqrt{\omega^2 - \gamma^2} & 0 \\ 0 & -\gamma - \mathrm{i}\sqrt{\omega^2 - \gamma^2} \end{pmatrix} \tag{6.16a}$$

For $\gamma^2 > \omega^2$ (this is the aperiodic limit) one finds two real characteristic exponents both of which have the same sign as γ,

$$|\gamma| > \omega : \overset{0}{\mathbf{A}} = \begin{pmatrix} -\gamma + \sqrt{\gamma^2 - \omega^2} & 0 \\ 0 & -\gamma - \sqrt{\gamma^2 - \omega^2} \end{pmatrix} . \tag{6.16b}$$

In all cases, $\underset{\sim}{y}_0 = (y^1, y^2) = 0$ is an equilibrium position. In the case of damped motion, (6.16a) and (6.16b) show that all solutions (6.13) approach the origin exponentially as t goes to infinity. This point is certainly one of stable equilibrium. If, on the other hand, $\gamma < 0$, the oscillations are enhanced and every initial configuration except $\underset{\sim}{y}_0 = 0$ moves away from the origin, no matter how close to 0 it is chosen. In this situation the origin is certainly a point of unstable equilibrium.

In the case of purely harmonic oscillations (6.14) the origin is again stable but in a weaker sense than with positive damping. Indeed, if we perturb the oscillator a little from its position of rest, it becomes a stationary state of motion with small amplitude. It neither returns to zero nor moves away from it for large times. The origin is stable but, obviously, its stability is of a different character than for the damped oscillator. Let us study these different kinds of stability in more detail.

6.2.3 Stability of Equilibrium Positions

Let $\underset{\sim}{x}_0$ be a stable critical point of the vector field $\underset{\sim}{F}$, i.e. $\underset{\sim}{F}(\underset{\sim}{x}_0, t) = 0$ and $\underset{\sim}{x}_0$ is an equilibrium position of the dynamical system (6.1) or (6.9). The notion of stability of the critical point is qualified by the following definitions.

S1. The point $\underset{\sim}{x}_0$ is said to be *stable* (or *Liapunov stable*) if for every neighborhood U of $\underset{\sim}{x}_0$ there is a further neighborhood V of $\underset{\sim}{x}_0$ such that the integral curve, that, at time $t = 0$, goes through an arbitrary point $\underset{\sim}{x} \in V$, exists in the limit $t \to +\infty$ and never leaves the domain U. Thus, when expressed in symbols, we have for $\underset{\sim}{x} \in V$ and $\Phi_x(0) = \underset{\sim}{x}$, $\underset{\sim}{\Phi}_x(t) \in U$ for all $t > 0$.

S2. The point $\underset{\sim}{x}_0$ is said to be *asymptotically stable* if there is a neighborhood U of $\underset{\sim}{x}_0$ that is such that the integral curve $\underset{\sim}{\Phi}_x(t)$ through an arbitrary $\underset{\sim}{x} \in U$ is defined for $t \to +\infty$ and tends to $\underset{\sim}{x}_0$ as t goes to infinity. Thus, with $\underset{\sim}{\Phi}(\underset{\sim}{x}, t)$ denoting the flow,

$$\underset{\sim}{\Phi}(U, s) \subset \underset{\sim}{\Phi}(U, t) \subset U \quad \text{for} \quad s > t > 0 \quad \text{and}$$
$$\lim_{t \to +\infty} \Phi_x(t) = \underset{\sim}{x}_0 , \quad \text{for all} \quad x \in U .$$

In the first case orbits that belong to initial configurations close to $\underset{\sim}{x}_0$ stay in the neighborhood of that point, at all later times. In the second case they move toward the critical point as time increases. Clearly, S2 contains the situation defined in S1: a point that is asymptotically stable is also Liapunov stable.

The following proposition gives more precise information on how rapidly the points of the neighborhood U in S2 move towards $\underset{\sim}{x}_0$ as time increases.

Proposition I. Let $\underset{\sim}{x}_0$ be an equilibrium position of the dynamical system (6.1), which is approximated by the linearization (6.10b) in a neighborhood of $\underset{\sim}{x}_0$. Assume that for all eigenvalues α_i of $\mathbf{D}\underset{\sim}{F}|_{\underset{\sim}{x}_0}$ we have $\mathrm{Re}\{\alpha_i\} < -c < 0$. Then there is a neighborhood U of $\underset{\sim}{x}_0$ such that the flow of $\underset{\sim}{F}$ on U (i.e. which fulfills $\Phi(U, t = 0) = U$) is defined for all positive times, as well as a constant $\tilde{d}$ such that for all $\underset{\sim}{x} \in U$ and all $t > 0$ we have

$$\|\Phi_x(t) - \underset{\sim}{x}_0\| \le d\,\mathrm{e}^{-ct}\,\|\underset{\sim}{x} - \underset{\sim}{x}_0\| \,. \tag{6.17}$$

Here $\|\ldots\|$ denotes the distance function. The result (6.17) tells us that the orbit through $\underset{\sim}{x} \in U$ converges to $\underset{\sim}{x}_0$ uniformly and at an exponential rate.

A criterion for instability of an equilibrium position is provided by the following.

Proposition II. Let $\underset{\sim}{x}_0$ be an equilibrium position of the dynamical system (6.1). If $\underset{\sim}{x}_0$ is stable then none of the characteristic exponents (i.e. the eigenvalues of $\mathbf{D}\underset{\sim}{F}|_{\underset{\sim}{x}}$ of the linearization of (6.1)) has a positive real part.

We skip the proofs of these propositions and refer, for example, to Hirsch and Smale (1974). Instead, we wish to illustrate them by a few examples and to give the normal forms of the linearization (6.10) for the case of two dimensions.

One should note that definitions S1, S2 and propositions I and II apply to arbitrary smooth vector fields, and not only to linear systems. In general, the linearization (6.10) clarifies matters only in the immediate neighborhood of the critical point $\underset{\sim}{x}_0$. The question of the actual size of the domain around $\underset{\sim}{x}_0$ from which all integral curves converge to $\underset{\sim}{x}_0$, in the case of asymptotic stability, remains open, except for linear systems. We shall return to this below.

For a system with one degree of freedom, $f = 1$, the space on which the system (6.1) is defined has dimension 2. In the linearized form (6.10) it is

$$\begin{pmatrix} \dot{y}^1 \\ \dot{y}^2 \end{pmatrix} = \begin{pmatrix} a_{11} & a_{12} \\ a_{21} & a_{22} \end{pmatrix} \begin{pmatrix} y^1 \\ y^2 \end{pmatrix}$$

with $a_{ik} = (\partial F^i/\partial x^k)|_{\underset{\sim}{x}_0}$. The eigenvalues are obtained from the characteristic polynomial $\det(\alpha \mathbb{1} - \mathbf{A}) = 0$, i.e. from the equation

$$\alpha^2 - \alpha(a_{11} + a_{22}) + a_{11}a_{22} - a_{12}a_{21} = 0 \,,$$

which may be expressed by means of the trace $t = \mathrm{Tr}A$ and the determinant $d = \det A$ as follows:

$$\alpha^2 - t\alpha + d = 0 \,. \tag{6.18}$$

As is well known, the roots of this equation fulfill the relations

$$\alpha_1 + \alpha_2 = t \,, \quad \alpha_1\alpha_2 = d \,.$$

If the discriminant $D = t^2 - 4d$ is positive or zero, the solutions α_1 and α_2 are real. In this case we must distinguish the following possibilities.

(i) $\alpha_1 < \alpha_2 < 0$, i.e. $d > 0$ and $-2\sqrt{d} < t < 0$. For diagonal **A** the solutions are $y^1 = \exp(\alpha_1 t) y_0^1$, $y^2 = \exp(\alpha_2 t) y_0^2$ and we obtain the pattern shown in Fig. 6.2a. The origin is asymptotically stable: it is a node.

(ii) $\alpha_1 = \alpha_2 < 0$. This is a degenerate case contained in (i) and is shown in Fig. 6.2b.

(iii) $\alpha_2 < 0 < \alpha_1$, i.e. $d < 0$. Here the origin is unstable. The orbits show the typical pattern of a saddle point; see Fig. 6.2c: some orbits approach the origin, others leave it.

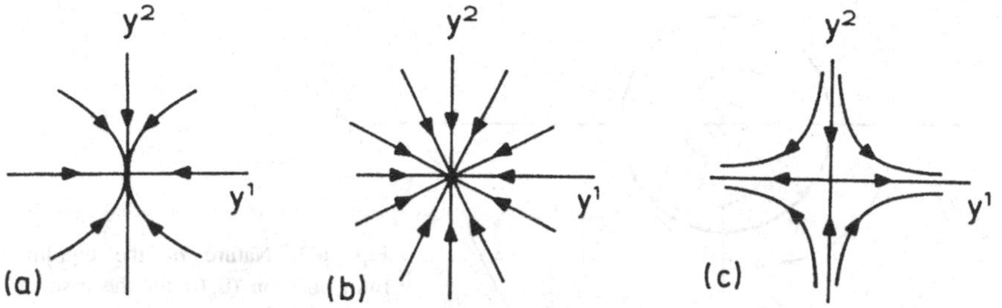

Fig. 6.2a–c. Typical behavior of a system with one degree of freedom in the neighborhood of an equilibrium position. In cases (**a**) and (**b**) the equilibrium is asymptotically stable. In case (**c**) it is unstable and has the structure of a saddle-point

If the discriminant D is negative, the characteristic exponents are complex conjugate numbers

$$\alpha_1 = \sigma + \mathrm{i}\varrho \,, \quad \alpha_2 = \sigma - \mathrm{i}\varrho \,,$$

with σ and ϱ real. Here $t = 2\sigma$, $d = \sigma^2 + \varrho^2$. The various cases that are possible here are illustrated in Fig. 6.3, which shows the examples of the damped, the excited, and the unperturbed oscillator (6.16a). The figure shows the solution with initial condition $y_0^1 = 1$, $y_0^2 = 0$ of (6.15), viz.

$$y^1(\tau) \equiv q(\tau) = \left[\cos\left(\tau\sqrt{1-g^2}\right) + \left(g/\sqrt{1-g^2}\right)\sin\left(\tau\sqrt{1-g^2}\right)\right]\mathrm{e}^{-g\tau} \,,$$
$$y^2(\tau) \equiv \dot{q}(\tau) = -\left(1/\sqrt{1-g^2}\right)\sin\left(\tau\sqrt{1-g^2}\right)\mathrm{e}^{-g\tau} \,. \qquad (6.19)$$

Here we have introduced $\tau \equiv \omega t$ and $g \equiv \gamma/\omega$. Curve A corresponds to $g = 0$, curve B to $g = 0.15$, and curve C to $g = -0.15$. In the framework of our analysis these examples tell us the following.

(iv) Curve A. Here $\sigma = 0$, $\varrho = \omega$, so that $t = 0$ and $d \geq 0$. The origin is stable but not asymptotically stable. It is said to be a *center*.

(v) Curve B. Here $\sigma = -\gamma < 0$, $\varrho = \sqrt{\omega^2 - \gamma^2}$, so that $t < 0$, $d \geq 0$. The origin is now asymptotically stable.

(vi) Curve C. Now $\sigma = -\gamma$ is positive while ϱ is as in (v); therfore $t > 0$, $d \geq 0$. The orbits move away from the origin like spirals. The origin is unstable for $t \to +\infty$.

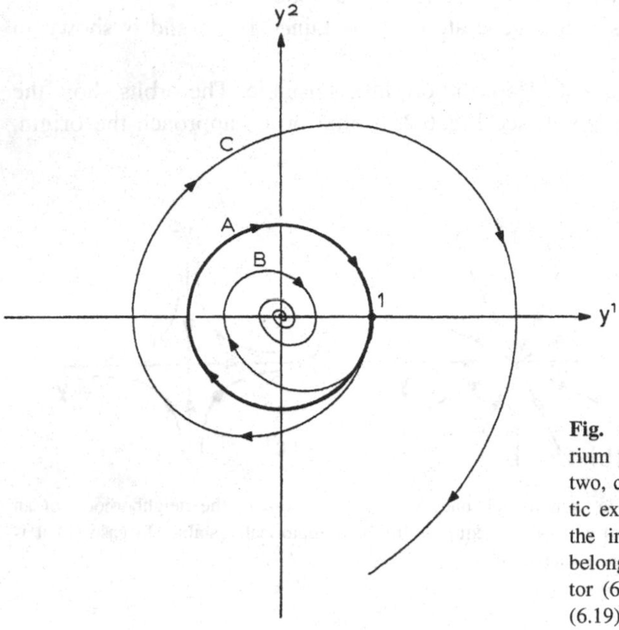

Fig. 6.3. Nature of the equilibrium position (0, 0) for the case of two, complex conjugate, characteristic exponents. All curves start from the initial configuration (1, 0) and belong to the example of the oscillator (6.15) with the explicit solution (6.19)

Our discussion shows the typical cases that occur. Figure 6.4 illustrates the various domains of stability in the plane of the parameters (t, d). The discussion is easily completed by making use of the real normal forms of the matrix $\mathbf{A}$ and by considering all possible cases, including the question of stability or instability as t tends to $-\infty$.

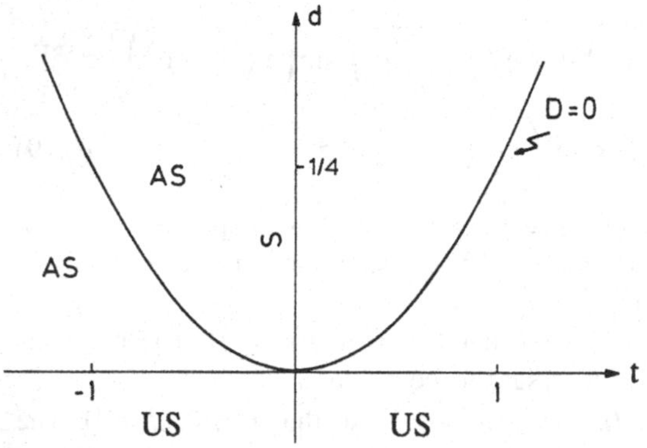

Fig. 6.4. For a system with $f = 1$ the various stability regions are determined by the trace t and the determinant d of the linearization A. AS means asymptotically stable (the equilibrium position is a node). S means stable (center) and US means unstable (saddle-point)

6.2.4 Critical Points of Hamiltonian Vector Fields

It is instructive to try the stability criteria developed above on canonical systems. These are governed by the canonical equations (2.99),

$$\dot{x} = \mathbf{J} H_{,x} \,, \tag{6.20}$$

where $\mathbf{J}$ is defined as in (2.102), its properties being

$$\det \mathbf{J} = 1 \,, \quad \mathbf{J}^T = \mathbf{J}^{-1} = -\mathbf{J}, \mathbf{J}^2 = -\mathbb{1} \,.$$

The system (6.20) having an equilibrium position at $\underset{\sim}{x}_0$, the Hamiltonian vector field X_{H} vanishes at that point. As $\mathbf{J}$ is regular, also the vector of partial derivatives $H_{,x}$ vanishes in $\underset{\sim}{x}_0$. Linearizing around $\underset{\sim}{x}_0$, i.e. setting $\underset{\sim}{y} = \underset{\sim}{x} - \underset{\sim}{x}_0$ and expanding the right-hand side of (6.20), we obtain the linear system

$$\dot{\underset{\sim}{y}} = \mathbf{A} \underset{\sim}{y}$$

with $\mathbf{A} = \mathbf{JB}$ and $\mathbf{B} = \{\partial^2 H / \partial x^k \partial x^i |_{x = x_0}\}$.

The matrix $\mathbf{B}$ is symmetric, $\mathbf{B} = \mathbf{B}^T$. Making use of the properties of $\mathbf{J}$ we have

$$\mathbf{A}^T \mathbf{J} + \mathbf{J}\mathbf{A} = 0 \,. \tag{6.21}$$

A matrix that obeys condition (6.21) is said to be *infinitesimally symplectic*. This name becomes clear if one considers a symplectic matrix $\mathbf{M}$ that differs only a little from $\mathbb{1}$,

$$\mathbf{M} = \mathbb{1} + \varepsilon \mathbf{A} + O(\varepsilon^2) \,.$$

The defining relation (2.113) then indeed yields (6.21), to first order in ε[1].

The following result applies to matrices which fulfill condition (6.21).

> **Proposition.** If α is an eigenvalue of the infinitesimally sympletic matrix $\mathbf{A}$, having multiplicity k, then also $-\alpha$ is an eigenvalue of $\mathbf{A}$ and has the same multiplicity. If $\alpha = 0$ is an eigenvalue then its multiplicity is even.

Proof. The proof makes use of the properties of $\mathbf{J}$, of the symmetry of $\mathbf{B}$ and of well-known properties of determinants. The eigenvalues are the zeros of the characteristic polynomial $P(\alpha) = \det(\alpha \mathbb{1} - \mathbf{A})$. Therefore, it is sufficient to show that $\det(\alpha \mathbb{1} - \mathbf{A}) = \det(\alpha \mathbb{1} + \mathbf{A})$. This is seen as follows

[1] In Sect. 5.5.4 symplectic transformations are defined without reference to coordinates, see definition SYT. If these are chosen to be infinitesimal, $F = id + \varepsilon \mathbf{A}$, then to first order in ε relation (6.21) is obtained in the coordinate-free form $\omega(Ae, e') + \omega(e, Ae') = 0$.

$$\begin{aligned} P(\alpha) &= \det(\alpha \mathbb{1} - \mathbf{A}) = \det(-\alpha \mathbf{J}^2 - \mathbf{J}\mathbf{B}) = \det \mathbf{J} \det(-\alpha \mathbf{J} - \mathbf{B}) \\ &= \det(-\alpha \mathbf{J} - \mathbf{B})^{\mathrm{T}} = \det(\alpha \mathbf{J} - \mathbf{B}) = \det(\alpha \mathbb{1} - \mathbf{J}^{-1}\mathbf{B}) \\ &= \det(\alpha \mathbb{1} + \mathbf{J}\mathbf{B}) = \det(\alpha \mathbb{1} + \mathbf{A}) . \end{aligned}$$

Thus, the proposition is proved. □

This result shows that the assumptions of proposition I (Sect. 6.2.3) can never be fulfilled for canonical systems. As a consequence, canonical systems cannot have asymptotically stable equilibria. Proposition II of Sect. 6.2.3, in turn, can be applied to canonical systems: it tells us that the equilibrium can only be stable if all characteristic exponents are purely imaginary. As an example consider the case of small oscillations about an absolute minimum q_0 of the potential energy described in Practical Example 2.1. Expanding the potential energy about q_0 up to second order in $(q - q_0)$, we obtain equations of motion that are linear. After we have transformed to normal coordinates the Hamiltonian function that follows from the Lagrangian function (A.8), Practical Example 2.1, reads

$$H = \frac{1}{2} \sum_{i=1}^{f} (P_i^2 + \Omega_i^2 Q_i^2) .$$

Setting $Q_i' = \sqrt{\Omega_i} Q_i$ and $P_i' = P_i / \sqrt{\Omega_i}$, H takes the form

$$H = \frac{1}{2} \sum_{i=1}^{f} \Omega_i (P_i'^2 + Q_i'^2) . \tag{6.22}$$

We calculate the matrix $\mathbf{A} = \mathbf{J}\mathbf{B}$ from this: $\mathbf{A}$ takes the standard form

$$\mathbf{A} = \left(\begin{array}{ccc|ccc} & & & \Omega_1 & & 0 \\ & 0 & & & \ddots & \\ & & & 0 & & \Omega_f \\ \hline -\Omega_1 & & 0 & & & \\ & \ddots & & & 0 & \\ 0 & & -\Omega_f & & & \end{array} \right) . \tag{6.23}$$

Generally, one can show that the linearization of a Hamiltonian system has precisely this standard form if the Hamiltonian function of its linearization is positive-definite. (Clearly, in the case of small oscillations about an absolute minimum of the potential energy, H does indeed have this property.) Diagonalizing the matrix (6.23) shows that the characteristic exponents take the purely imaginary values

$$\pm \mathrm{i}\Omega_1, \pm \mathrm{i}\Omega_2, \ldots, \pm \mathrm{i}\Omega_f .$$

Therefore, a system for which **A** has the standard form (6.23) does not contradict the criterion for stability of proposition II of Sect. 6.2.3. The point $\underset{\sim}{x}_0$ has a chance to be stable although this is not decided by the above propositions. In order to proceed one tries to find an auxiliary function $V(\underset{\sim}{x})$ that has the property that it vanishes in $\underset{\sim}{x}_0$ and is positive everywhere in a certain open neighborhood U of that point. In the example (6.22) this could be the energy function (with $x^i \equiv Q'_i$, $x^{i+f} = P'_i$, $i = 1, \ldots, f$)

$$V(\underset{\sim}{x}) \equiv E(\underset{\sim}{x}) = \frac{1}{2}\sum_{i=1}^{f} \Omega_i[(x^{i+f})^2 + (x^i)^2] .$$

We then take the time derivative of $V(\underset{\sim}{x})$ along orbits of the system. If this derivative is negative or zero everywhere in U, this means that no solution moves outward, away from $\underset{\sim}{x}_0$. Then the point $\underset{\sim}{x}_0$ is stable.

Remark. An auxiliary function that has these properties is called a *Liapunov function.*

The test for stability by means of a Liapunov function can also be applied to systems that are not canonical, and it may even be sharpened there. Indeed, if the derivative of $V(\underset{\sim}{x})$ along solutions is negative everywhere in U, then all orbits move inward, towards $\underset{\sim}{x}_0$. Therefore this point is asymptotically stable. Let us illustrate this by the example of the oscillator (6.15) with and without damping. The point $(q = 0, \dot{q} = 0)$ is an equilibrium position. A suitable Liapunov function is provided by the energy function,

$$V(q, \dot{q}) \stackrel{\text{def}}{=} \tfrac{1}{2}(\dot{q}^2 + \omega^2 q^2) \quad \text{with } V(0, 0) = 0 . \tag{6.24}$$

Calculate $\dot{V}$ along solution curves:

$$\dot{V} = \frac{\partial V}{\partial q}\dot{q} + \frac{\partial V}{\partial \dot{q}}\ddot{q} = \omega^2 q\dot{q} - 2\gamma\dot{q}^2 - \omega^2 q\dot{q} = -2\gamma\dot{q}^2 . \tag{6.25}$$

In the second step we have replaced $\ddot{q}$ by q and $\dot{q}$, using (6.15). For $\gamma = 0$, $\dot{V}$ vanishes identically. No solution moves outward or inward and therefore $(0, 0)$ is stable: it is a center. For positive γ, $\dot{V}$ is strictly negative along all orbits. The solutions move inward and therefore $(0, 0)$ is asymptotically stable.

6.2.5 Stability and Instability of the Free Top

A particularly beautiful example of a nonlinear system with stable and unstable equilibria is provided by the motion of a free, asymmetric rigid body. Following the convention of Sect. 3.13 (3.60), the principal axes are labeled such that $0 < I_1 < I_2 < I_3$. We set

$$x^i \stackrel{\text{def}}{=} \overline{\omega_i} \quad \text{and} \quad F^1 \stackrel{\text{def}}{=} \frac{I_2 - I_3}{I_1} x^2 x^3 \quad \text{(with cyclic permutations)} .$$

The Eulerian equations (3.59) take the form (6.1), viz.

$$\dot{x}^1 = -\frac{I_3 - I_2}{I_1} x^2 x^3 \,, \quad \dot{x}^2 = +\frac{I_3 - I_1}{I_2} x^3 x^1 \,, \quad \dot{x}^3 = -\frac{I_2 - I_1}{I_3} x^1 x^2 \,. \tag{6.26}$$

Here, we have written the right-hand sides such that all differences $I_i - I_k$ are positive. This dynamical system has three critical points (equilibria) whose stability we wish to investigate, viz.

$$\underset{\sim}{x}_0^{(1)} = (\omega, 0, 0) \,, \quad \underset{\sim}{x}_0^{(2)} = (0, \omega, 0) \,, \quad \underset{\sim}{x}_0^{(3)} = (0, 0, \omega) \,,$$

ω being an arbitrary positive constant. We set $\underset{\sim}{y} = \underset{\sim}{x} - \underset{\sim}{x}_0^{(i)}$ and linearize equations (6.26). For example, in the neighborhood of the point $\underset{\sim}{x}_0^{(1)}$ we obtain the linear system

$$\dot{\underset{\sim}{y}} \equiv \begin{pmatrix} \dot{y}^1 \\ \dot{y}^2 \\ \dot{y}^3 \end{pmatrix} = \begin{pmatrix} 0 & 0 & 0 \\ 0 & 0 & \omega \dfrac{I_3 - I_1}{I_2} \\ 0 & -\omega \dfrac{I_2 - I_1}{I_3} & 0 \end{pmatrix} \begin{pmatrix} y_1 \\ y_2 \\ y_3 \end{pmatrix} \equiv \mathbf{A} \underset{\sim}{y} \,.$$

The characteristic exponents follow from the equation $\det(\alpha \mathbb{1} - \mathbf{A}) = 0$ and are found to be

$$\alpha_1^{(1)} = 0 \,, \quad \alpha_2^{(1)} = -\alpha_3^{(1)} = \mathrm{i}\omega\sqrt{(I_2 - I_1)(I_3 - I_1)/I_2 I_3} \,. \tag{6.27a}$$

A similar analysis yields the following characteristic exponents at the points $\underset{\sim}{x}_0^{(2)}$ and $\underset{\sim}{x}_0^{(3)}$, respectively:

$$\alpha_1^{(2)} = 0 \,, \quad \alpha_2^{(2)} = -\alpha_3^{(2)} = \omega\sqrt{(I_3 - I_2)(I_2 - I_1)/I_1 I_3} \,, \tag{6.27b}$$

$$\alpha_1^{(3)} = 0 \,, \quad \alpha_2^{(3)} = -\alpha_3^{(3)} = \mathrm{i}\omega\sqrt{(I_3 - I_2)(I_3 - I_1)/I_1 I_2} \,. \tag{6.27c}$$

Note that in the case of (6.27b) one of the characteristic exponents has a positive real part. Proposition II of Sect. 6.2.3 tells us that $\underset{\sim}{x}_0^{(2)}$ cannot be a stable equilibrium. This confirms our conjecture of Sect. 3.14 (ii), which we obtained from Fig. 3.22: rotations about the axis corresponding to the intermediate moment of inertia cannot be stable.

Regarding the other two equilibria, the characteristic exponents (6.27a) and (6.27c) are either zero or purely imaginary. Therefore, $\underset{\sim}{x}_0^{(1)}$ and $\underset{\sim}{x}_0^{(3)}$ have a chance of being stable. This assertion is confirmed by means of the following Liapunov functions for the points $\underset{\sim}{x}_0^{(1)}$ and $\underset{\sim}{x}_0^{(3)}$, respectively:

$$V^{(1)}(\underset{\sim}{x}) \stackrel{\text{def}}{=} \tfrac{1}{2}[I_2(I_2 - I_1)(x^2)^2 + I_3(I_3 - I_1)(x^3)^2] \,,$$

$$V^{(3)}(\underset{\sim}{x}) \stackrel{\text{def}}{=} \tfrac{1}{2}[I_1(I_3 - I_1)(x^1)^2 + I_2(I_3 - I_2)(x^2)^2] \,.$$

$V^{(1)}$ vanishes at $\underset{\sim}{x} = \underset{\sim}{x}_0^{(1)}$; it is positive everywhere in the neighborhood of this point. Taking the time derivative of $V^{(1)}$ along solutions, we obtain, making use of the equations of motion (6.26),

$$\begin{aligned}\dot{V}^{(1)}(\underset{\sim}{x}) &= I_2(I_2 - I_1)x^2\dot{x}^2 + I_3(I_3 - I_1)x^3\dot{x}^3 \\ &= [(I_2 - I_1)(I_3 - I_1) - (I_3 - I_1)(I_2 - I_1)]x^1x^2x^3 = 0 \,.\end{aligned}$$

An analogous result is obtained for $V^{(3)}(\underset{\sim}{x})$. As a consequence the equilibria $\underset{\sim}{x}_0^{(1)}$ and $\underset{\sim}{x}_0^{(3)}$ are stable. However, they are not Liapunov stable, in the sense of the definition (St3) below.

6.3 Long-Term Behavior of Dynamical Flows and Dependence on External Parameters

In this section we investigate primarily dissipative systems. The example of the damped oscillator (6.15), illustrated by Fig. 6.3 may suggest that the dynamics of dissipative systems is simple and not very interesting. This impression is misleading. The behavior of dissipative systems can be more complex by far than the simple "decay" of the motion whereby all orbits approach exponentially an asymptotically stable point. This is the case, for instance, if the system also contains a mechanism that, on average, compensates for the energy loss and thus keeps the system in motion. Besides points of stability there can be other structures of higher dimension that certain subsets of orbits will cling to asymptotically. In approaching these *attractors* for $t \to +\infty$, the orbits will lose practically all memory of their initial condition, even though the dynamics is strictly deterministic. On the other hand, there are systems where orbits on an attractor with neighboring initial conditions, for increasing time, move apart exponentially. This happens in dynamical systems that possess what are called *strange attractors*. They exhibit the phenomenon of extreme sensitivity to initial conditions, which is one of the agents for deterministic chaos: two orbits whose distance, on average, increases exponentially, pertain to initial conditions that are indistinguishable from any practical point of view.

For this phenomenon to happen, there must be at least three dynamical variables. In point mechanics this means that the phase space must have dimensions 4, 6, or higher. Obviously, there is a problem in representing the flow of a dynamical system as a whole because of the large number of dimensions. On the other hand, if we deal with finite motions that stay in the neighborhood of a periodic orbit, it may be sufficient to study the intersection of the orbits with hypersurfaces of smaller dimension, perpendicular to the periodic orbit. This is the concept of Poincaré mapping. It leads to a discretization of the flow: e.g. one records the flow only at discrete times t_0, t_0+T, t_0+2T, etc., where T is the period of the reference orbit, or else, when it hits a given transversal hypersurface. The mapping of an m-dimensional flow on a hypersurface of dimension $(m-1)$, in general, may give a good impression of its topology. There may even be cases where it is sufficient to study a single variable at special, discrete points (e.g., maxima of a function). One then obtains a kind of return mapping in one dimension that one may think of as a stroboscopic observation of a one-dimensional system. If this mapping is suitably chosen it may give hints to the behavior of the flow as a whole.

In general, dynamical systems depend on one or several parameters that control the strength of external influences. An example is provided by forced oscillations, the frequency and amplitude of the the exciting oscillation being the control parameters. In varying these parameters one may hit critical values at which the structure of the flow changes qualitatively. Critical values of this kind are called *bifurcations*. Bifurcations, too, play an important role in the development of deterministic chaos.

This section is devoted to a more precise definition of the concepts sketched above. They are then discussed and illustrated by means of a number of examples.

6.3.1 Flows in Phase Space

Consider a connected domain U_0 of initial conditions in phase space that has the oriented volume V_0.

(i) For *Hamiltonian systems* Liouville's theorem tells us that the flow Φ carries this initial set across phase space as if it were a connected part of an incompressible fluid. Total volume and orientation are preseved; at any time t the image U_t of U_0 under the flow has the same volume $V_t = V_0$. Note, however, that this may be effected in rather different ways: for a system with $f = 2$ (i.e. four-dimensional phase space) let U_0 be a four-dimensional ball of initial configurations. The flow of the Hamiltonian vector field may be such that this ball remains unchanged or is deformed only slightly, as it wanders through phase space. At the other extreme, it may be such that the flow drives apart, at an exponential rate $\exp(\alpha t)$, points of one direction in U_0, while contracting points in a direction perpendicular to the first, at a rate $\exp(-\alpha t)$ so that the total volume is preserved.[2] Liouville's theorem is respected in either situation. In the former case orbits through U_0 possess a certain stability. In the latter case they are unstable in the sense that there are orbits with arbitrarily close initial conditions that nevertheless move apart at an exponential rate. Even though the system is deterministic, it is practically impossible to reconstruct the precise initial condition from an observation at a time $t > 0$.

(ii) For *dissipative systems* the volume V_0 of the initial set U_0 is not conserved. If the system loses energy, the volume will decrease monotonically. This may happen in such a way, that the initial domain shrinks more or less uniformly along all independent directions in U_0. There is also the possibility, however, that one direction spreads apart while others shrink at an increased rate such that the volume as a whole decreases.

A measure of constant increase or decrease of volume in phase space is provided by the Jacobian determinant of the matrix of partial derivatives $\mathbf{D}\Phi$ (2.119).

[2] In systems with $f = 1$, i.e. with a two-dimensional phase space, and keeping clear from saddle point equilibria, the deformation can be no more than linear in time, cf. Exercise 6.3.

If this determinant is 1, then Liouville's theorem applies. If it decreases as a function of time, the phase space volume shrinks. Whenever the Jacobian determinant is different from zero, the flow is invertible. If it has a zero at a point $\underset{\sim}{x}$ of phase space, the flow is irreversible at this point.

A simple phenomenological method of introducing dissipative terms into Hamiltonian systems consists in changing the differential equation for $\underset{\sim}{p}(t)$ in the following manner:

$$\dot{p}_j = -\frac{\partial H}{\partial q^j} - R_j(\underset{\sim}{q}, \underset{\sim}{p}) \, . \tag{6.28}$$

One calculates the time derivative of H along solutions of the equations of motion,

$$\frac{\mathrm{d}H}{\mathrm{d}t} = \sum \frac{\partial H}{\partial q^j}\dot{q}^i + \sum \frac{\partial H}{\partial p_j}\dot{p}_j = -\sum_{i=1}^{f} \dot{q}^i R_i(\underset{\sim}{q}, \underset{\sim}{p}) \, . \tag{6.29}$$

Depending on the nature of the dissipative terms R_i, the energy decreases either until the system has come to rest or the flow has reached a submanifold of lower dimension than $\dim \mathbb{P}$ on which the dissipative term $\sum \dot{q}^i R_i(\underset{\sim}{q}, \underset{\sim}{p})$ vanishes.

In the example (6.15) of the damped oscillator we have

$$H = (p^2/m + m\omega^2 q^2)/2 \quad \text{and} \quad R = 2\gamma m \dot{q} \, ,$$

so that

$$\frac{dH}{dt} = -2\gamma m \dot{q}^2 = -\frac{2\gamma}{m} p \, . \tag{6.30}$$

In this example the leakage of energy ceases only when the system has come to rest, i.e. when it has reached the asymptotically stable critical point $(0, 0)$.

6.3.2 More General Criteria for Stability

In the case of dynamical systems whose flow shows the behavior described above, the stability criteria of Sect. 6.2.3 must be generalized somewhat. Indeed, an orbit that tends to a periodic orbit, for $t \to +\infty$, can do so in different ways. Furthermore, as this concerns a local property of flows, one might ask whether there are subsets of phase space that are preserved by the flow, without "dissolving" for large times. The following definitions collect the concepts relevant to this discussion.

Let $\underset{\sim}{F}$ be a complete vector field on $\mathbb{R}^n$, or phase space $\mathbb{R}^{2f}$, or, more generally, on the manifold M, depending on the system one is considering. Let B be a subset of M whose points are possible initial conditions for the flow of the differential equation (6.1),

$$\Phi_{t=0}(B) = B \, .$$

For positive or negative times t this subset moves to $\Phi_t(B)$. The image $\Phi_t(B)$ can be contained in B, but it may also have drifted out of B, partially or completely. We sharpen the first possibility as follows.

(i) If the image of B under the flow is contained in B, for all $t \geq 0$,

$$\Phi_t(B) \subset B \, , \tag{6.31}$$

the set B is said to be *positively invariant.*

(ii) Similarly, if the condition (6.31) held in the past, i.e. for all $t \leq 0$, B is said to be *negatively invariant.*

(iii) Finally, B is said to be *invariant* if its image under the flow is contained in B for all t,

$$\Phi_t(t) \subset B \quad \text{for all } t \, . \tag{6.32}$$

(iv) If the flow has several, neighboring domains for which (6.32) holds, obviously, their union has the same property. For this reason one says that B is a *minimal set* if it is closed, nonempty, and invariant in the sense of (6.32), and if it cannot be decomposed into subsets that have the same properties.

A periodic orbit of a flow Φ_t has the property $\Phi_{t+T}(m) = \Phi_t(m)$, for all points m on the orbit, T being the period. Very much like equilibrium positions, closed, periodic orbits are generally exceptional in the diversity of integral curves of a given dynamical system. Furthermore, equilibrium positions may be understood as special, degenerate examples of periodic orbits. For this reason equilibria and periodic orbits are called *critical elements* of the vector field $\underset{\sim}{F}$ that defines the dynamical system (6.1). It is not difficult to verify that critical elements are minimal sets in the sense of definitions (iii) and (iv) above.

Orbits that move close to each other, for increasing time, or tend towards each other, can do so in different ways. This kind of "moving stability" leads us to the following definitions. We consider a reference orbit A, say, the orbit of a mass point m_A. This may, but need not, be a critical element. Let another mass point m_B move along a neighboring orbit B. At time $t = 0$ m_A starts from m_A^0 and m_B starts from m_B^0, their initial distance being smaller than a given $\delta > 0$,

$$\|m_B^0 - m_A^0\| < \delta \quad (t = 0) \, . \tag{6.33}$$

These orbits are assumed to be complete (or, at least, to be defined for $t \geq 0$), i.e. they should exist in the limit $t \to \pm\infty$ (or, at least, for $t \to +\infty$). Then orbit A is stable if

St1. for every test orbit B that fulfills (6.33) there is an $\varepsilon > 0$ such that for $t \geq 0$ orbit B, as a whole, never leaves the tube with radius ε around orbit A (*orbital stability*); or

St2. the distance of the actual position of $m_B(t)$ from orbit A tends to zero in the limit $t \to +\infty$ (*asymptotic stability*); or

St3. the distance of the actual positions of m_A and m_B at time t tends to zero as $t \to +\infty$ (*Liapunov stability*).

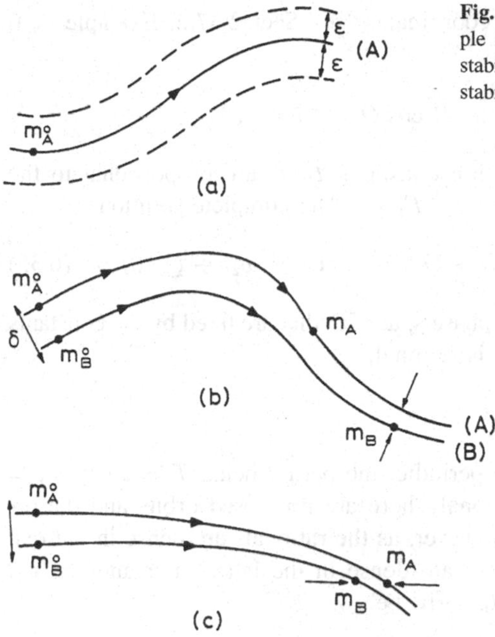

Fig. 6.5. Stability of an orbit (A) for the example of a system in two dimensions. (**a**) Orbital stability, (**b**) asymptotic stability, (**c**) Liapunov stability

In Fig. 6.5 we sketch the three types of stability for the example of a dynamical system in two dimensions. Clearly, analogous criteria can be applied to the past, i.e. the limit $t \to -\infty$. As a special case, m_A may be taken to be an equilibrium position in which case orbit A shrinks to a point. Orbital stability as defined by St1 is the weakest form and corresponds to case S1 of Sect. 6.2.3. The two remaining cases (St2 and St3) are now equivalent and correspond to S2 of Sect. 6.2.3.

Remarks: Matters become particularly simple for vector fields on two-dimensional manifolds. We quote the following propositions for this case.

Proposition I. Let $\underset{\sim}{F}$ be a vector field on the compact, connected manifold M (with $\dim M = 2$) and let B be a minimal set in the sense of definition (iv) above. Then B is either a critical point or a periodic orbit, or else $B = M$ and M has the structure of a two-dimensional torus T^2.

Proposition II. If, in addition, M is orientable and if the integral curve $\Phi_t(m)$ contains no critical points for $t \geq 0$, then either $\Phi_t(m)$ is dense in M (it covers all of M), which is T^2, or $\Phi_t(m)$ is a closed orbit.

For the proofs we refer to Abraham and Marsden (1981, Sect. 6.1).

As an example for motion on the torus T^2, let us consider two uncoupled oscillators

$$\dot{p}_1 + \omega_1^2 q_1 = 0 \,, \quad \dot{p}_2 + \omega_2^2 q_2 = 0 \,. \tag{6.34}$$

Transformation to action and angle coordinates (see Sect. 2.37.2, Example (vi)) gives

$$q_i = \sqrt{2P_i/\omega_i}\sin Q_i\,, \quad p_i = \sqrt{2\omega_i P_i}\cos Q_i\,, \quad i = 1, 2$$

with $P_i = I_i = \text{const}$. The integration constants I_1, I_2 are proportional to the energies of the individual oscillators, $I_i = E_i/\omega_i$. The complete solutions

$$P_1 = I_1\,, \quad P_2 = I_2\,, \quad Q_1 = \omega_1 t + Q_1(0)\,, \quad Q_2 = \omega_2 t + Q_2(0) \qquad (6.35)$$

lie on tori T^2 in the four-dimensional phase space $\mathbb{R}^4$ that are fixed by the constants I_1 and I_2. If the ratio of frequencies is rational,

$$\omega_2/\omega_1 = n_2/n_1\,, \quad n_i \in \mathbb{N}\,,$$

the combined motion on the torus is periodic, the period being $T = 2\pi n_1/\omega_1 = 2\pi n_2/\omega_2$. If the ratio ω_2/ω_1 is irrational, there are no closed orbits and the orbits cover the torus densely. (Note, however, as the rationals are dense in the real numbers, the orbits of the former case are dense in the latter.) For another and nonlinear example we refer to Sect. 6.3.3(ii) below.

6.3.3 Attractors

Let $\underset{\sim}{F}$ be a complete vector field on $M \equiv \mathbb{R}^n$ (or on another smooth manifold M, for that matter) that defines a dynamic system of the type of (6.1). A subset A of M is said to be an *attractor* of the dynamical system if it is closed and invariant (in the sense of definition 6.3.2(iii)) and if it obeys the following conditions.

(i) A is contained in an open domain U_0 of M that is positively invariant itself. Thus, according to definition 6.3.2(i), U_0 has the property

$$\Phi_t(U_0) \subset U_0 \quad \text{for} \quad t \geq 0\,.$$

(ii) For any neighborhood V of A contained entirely in U_0 (i.e. which is such that $A \subset V \subset U_0$), one can find a positive time $T > 0$ beyond which the image of U_0 by the flow ϕ_t of $\underset{\sim}{F}$ is contained in V,

$$\Phi_t(U_0) \subset V \quad \text{for all } t \geq T\,.$$

The first condition says that there should exist open domains of M that contain the attractor and that do not disperse under the action of the flow, for large positive times. The second condition says that, asymptotically, integral curves within such domains converge to the attractor. In the case of the damped oscillator, Fig. 6.3, the origin is a (pointlike) attractor. Here, U_0 can be taken to be the whole of $\mathbb{R}^2$ because any orbit is attracted to the point (0,0) like a spiral. It may happen that M contains several attractors (which need not be isolated points) and therefore that each individual attractor attracts the flow only in a finite subset of M. For this reason one defines the *basin of an attractor* to be the union of all neighborhoods of A that fulfill the two conditions (i) and (ii). Exercise 6.6 gives a simple example.

Regarding condition (ii) one may ask the question whether, for fixed U_0, one can choose the neighborhood V such that it does not drift out of U_0 under the action of the flow for positive times, i.e. whether $V_t \equiv \Phi_t(V) \subset U_0$ for all $t \geq 0$. If this latter condition is fulfilled, the attractor, A, is said to be *stable*. The following examples may help to illustrate the concept of attractor in more depth.

Example (i) Forced oscillations (Van der Pol's equation). The model (6.15) of a pendulum with damping or external excitation is physically meaningful only in a small domain, for several reasons. The equation of motion being a linear one, it tells us that if $q(t)$ is a solution, so is every $\bar{q}(t) = \lambda q(t)$, with λ an arbitrary real constant. Thus, by this simple rescaling, the amplitude and the velocity can be made arbitrarily large. The assumption that friction is proportional to $\dot{q}$ then cannot be a good approximation. On the other hand, if one chooses γ to be negative, then according to (6.30) the energy that is delivered to the system grows beyond all limits. It is clear that either extrapolation – rescaling or arbitrarily large energy supply – must be limited by nonlinear dynamical terms.

In an improved model one will choose the coefficient γ to depend on the amplitude in such a way that the oscillation is stabilized: if the amplitude stays below a certain critical value, we wish the oscillator to be excited; if it exceeds that value, we wish the oscillator to be damped. Thus, if $u(t)$ denotes the deviation from the state of rest, (6.15) shall be replaced by

$$m\ddot{u}(t) + 2m\gamma(u)\dot{u}(t) + m\omega^2 u(t) = 0\,, \tag{6.36a}$$

$$\text{where} \quad \gamma(u) \stackrel{\text{def}}{=} -\gamma_0(1 - u^2(t)/u_0^2) \tag{6.36b}$$

and where $\gamma_0 > 0$. u_0 is the critical amplitude beyond which the motion is damped. For small amplitudes $\gamma(u)$ is negative, i.e. the motion is enhanced.

We introduce the dimensionless variables

$$\tau \stackrel{\text{def}}{=} \omega t\,, \quad q(\tau) \stackrel{\text{def}}{=} (\sqrt{2\gamma_0}/u_0\sqrt{\omega})u(t)$$

and set $p = \dot{q}(\tau)$. The equation of motion can be written in the form of (6.28), where

$$H = \tfrac{1}{2}(p^2 + q^2) \quad \text{and} \quad R(q, p) = -(\varepsilon - q^2)p\,, \quad \varepsilon \stackrel{\text{def}}{=} 2\gamma_0/\omega\,,$$

and therefore

$$\begin{aligned} \dot{q} &= p\,, \\ \dot{p} &= -q + (\varepsilon - q^2)p\,. \end{aligned} \tag{6.36c}$$

Figure 6.6 shows three solutions of this model for the choice $\varepsilon = 0.4$ that are obtained by numerically integrating the system (6.36c) (the reader is invited to

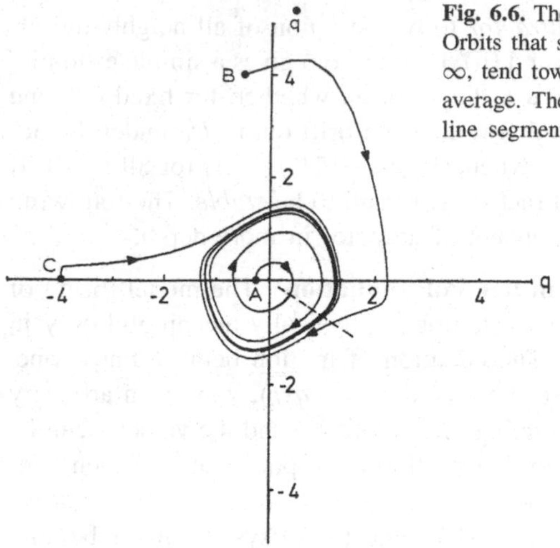

Fig. 6.6. The dynamical system (6.36a) has an attractor. Orbits that start inside or outside the attractor, for $t \to \infty$, tend towards the attractor at an exponential rate, on average. The control parameter is $\varepsilon = 0.4$. The dashed line segment is a transverse section

repeat this calculation). The figure shows clearly that the solutions tend rapidly to a *limit curve*, which is itself a solution of the system. (In Exercise 6.9 one is invited to find out empirically at what rate the solutions converge to the attractor.) Point A, which starts from the initial condition ($q_0 = -0.25$, $p_0 = 0$), initially moves outward and, as times goes on, clings to the attractor from the inside. Points B ($q_0 = -0.5$, $p_0 = 4$) and C ($q_0 = -4$, $p_0 = 0$) start outside and tend rapidly to the attractor from the outside. In this example the attractor seems to be a closed, and hence periodic, orbit. (We can read this from Fig. 6.6 but not what the dimension of the attractor is.) Figure 6.7 shows the coordinate $q(\tau)$ of the point A as a function of the time parameter τ. After a time interval of about twenty times the inverse frequency of the unperturbed oscillator it joins the periodic motion on the attractor. On the attractor the time average of the oscillator's energy $E = (p^2 + q^2)/2$ is conserved. This means that, on average, the driving term proportional to ε feeds in as much energy into the system as the latter loses through damping. From (6.29) we have

$$dE/d\tau = \varepsilon p^2 - q^2 p^2 .$$

Taking the time average, we have $\overline{dE/d\tau} = 0$, and hence

$$\overline{\varepsilon p^2} = \overline{q^2 p^2} , \tag{6.37}$$

the left-hand side being the average energy supply, the right-hand side the average loss through friction.

For $\varepsilon = 0.4$ the attractor resembles a circle and the oscillation shown in Fig. 6.7 is still approximately a harmonic one. If, instead, we choose ε appreciably larger, the limit curve gets strongly deformed and takes more the shape of a hysteresis curve. At the same time, $q(\tau)$ shows a behavior that deviates strongly from a sine

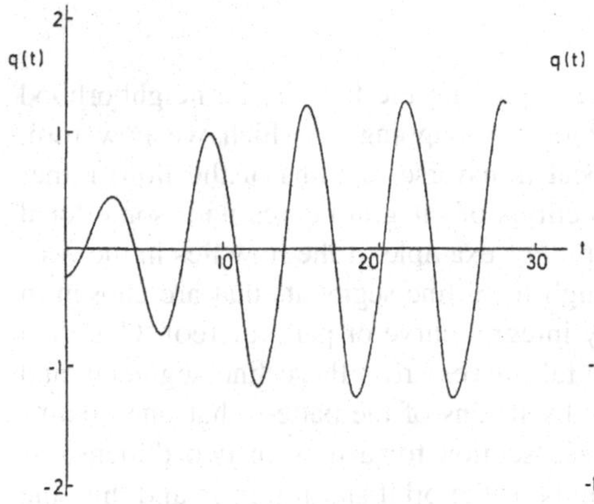

Fig. 6.7. Motion of the point A of Fig. 6.6, with initial condition ($q = -0.25$, $p = 0$) for $\varepsilon = 0.4$, as a function of time. It quickly joins the periodic orbit on the attractor

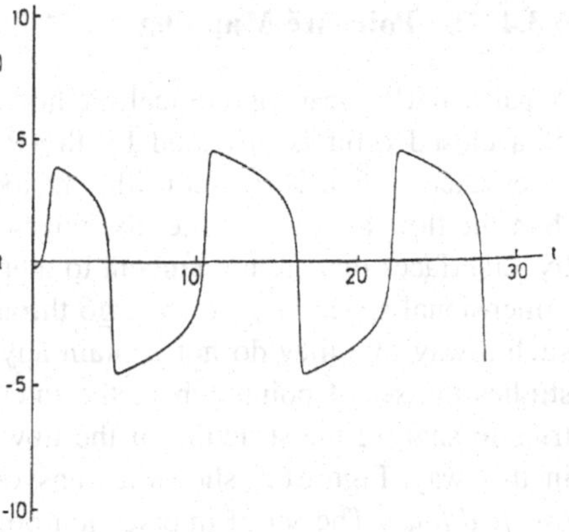

Fig. 6.8. Motion of the point A, with initial condition $(-0.25, 0)$ as in Fig. 6.7 but here with $\varepsilon = 5.0$

curve. Figure 6.8 shows the example $\varepsilon = 5.0$. The time variation of $q(\tau)$ shows cleary that it must contain at least two different scales.

Example (ii) Two coupled Van der Pol oscillators. The second example is directly related to, and makes use of the results of, the first. We consider two identical systems of the type (6.36c) but add to them a linear coupling interaction. In order to avoid resonances, we introduce an extra term into the equations that serves the purpose of taking the unperturbed frequencies out of tune. Thus, the equations of motion read

$$\begin{aligned} \dot{q}_i &= p_i \,, \quad i = 1, 2 \,, \\ \dot{p}_1 &= -q_1 + (\varepsilon - q_1^2)p_1 + \lambda(q_2 - q_1) \,, \\ \dot{p}_2 &= -q_2 - \varrho q_2 + (\varepsilon - q_2^2)p_2 + \lambda(q_1 - q_2) \,. \end{aligned} \tag{6.38}$$

Here ϱ is the detuning parameter while λ describes the coupling. Both are assumed to be small.

For $\lambda = \varrho = 0$ we obtain the picture of the first example, shown in Fig. 6.6, for each variable: two limit curves in two planes of $\mathbb{R}^4$ that are perpendicular to each other and whose form is equivalent to a circle. Their direct product defines a torus T^2, embedded in $\mathbb{R}^4$. This torus being the attractor, orbits in its neighborhood converge towards it, at an approximately exponential rate. For small perturbations, i.e. $\varrho, \lambda \ll \varepsilon$, one can show that the torus remains stable as an attractor for the coupled system (see Guckenheimer and Holmes 2001, Sect. 1.8). Note, however, the difference to the Hamiltonian system (6.35). There, for given energies E_1, E_2, the torus is the manifold of motions, i.e. all orbits start and stay on it, for all times. Here, the torus is the attractor to which the orbits tend in the limit $t \to +\infty$. The manifold of motions is four-dimensional but, as time increases, it "descends" to a submanifold of dimension two.

6.3.4 The Poincaré Mapping

A particularly clear topological method of studying the flow in the neighborhood of a closed orbit is provided by the Poincaré mapping to which we now turn. In essence, it consists in considering local transverse sections of the flow, rather than the flow as a whole, i.e. the intersections of integral curves with some local hypersurfaces that are not tangent to them. For example, if the flow lies in the two-dimensional space $\mathbb{R}^2$, we let it go through local line segments that are chosen in such a way that they do not contain any integral curve or parts thereof. One then studies the set of points where the integral curves cross these line segments and tries to analyze the structure of the flow by means of the pattern that one obtains in this way. Figure 6.6 shows a transverse section for a flow in two dimensions (*dashed line*). The set of intersection points of the orbit starting in A and this line section shows the average exponential approach to the attractor (see also Exercise 6.10).

A flow in three dimensions is cut locally by planes or other two-dimensional smooth surfaces that are chosen such that they do not contain any integral curves. An example is shown in Fig. 6.9: at every turn the periodic orbit Γ crosses the transverse section S at the same point, while a neighboring, nonperiodic orbit cuts the surface S at a sequence of distinct points. With these examples in mind the following general definition will be readily plausible.

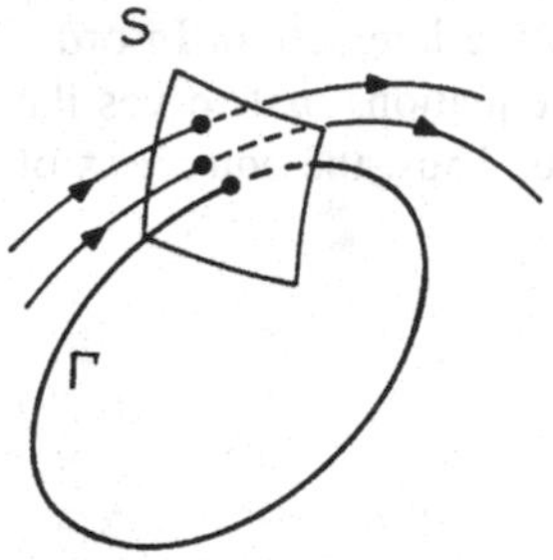

Fig. 6.9. Transverse section for a periodic orbit in $\mathbb{R}^3$

Definition. Let $\underset{\sim}{F}$ be a vector field on $M = \mathbb{R}^n$ (or on any other smooth manifold of dimension n). A local transverse section of $\underset{\sim}{F}$ at the point $\underset{\sim}{x} \in M$ is an open neighborhood on a hypersurface of dimension $\dim S = \dim M - 1 = n - 1$ (i.e. a submanifold of M) that contains $\underset{\sim}{x}$ and is chosen in such a way that, at none of the points $\underset{\sim}{s} \in S$ does the vector field $\underset{\sim}{F}(\underset{\sim}{s})$ lie in the tangent space $T_s S$.

The last condition makes sure that all flow lines going through points $\underset{\sim}{s}$ of S do indeed intersect with S and that none of them lies in S.

Consider a periodic orbit Γ with period T, and let S be a local transverse section at a point $\underset{\sim}{x}_0$ on Γ. Without restriction of generality we may take $\underset{\sim}{x}_0(t = 0) = 0$. Clearly, we have $\underset{\sim}{x}_0(nT) = 0$, for all integers n. As $\underset{\sim}{F}$ does not vanish in

$\underset{\sim}{x}_0$, there is always a transverse section S that fulfills the conditions of the definition above. Let S_0 be a neighborhood of $\underset{\sim}{x}_0$ that is contained in S. We ask the question at what time $\tau(\underset{\sim}{x})$ an arbitrary point $\underset{\sim}{x} \in S_0$ that follows the flow is taken back to the transverse section S for the first time. For $\underset{\sim}{x} = \underset{\sim}{x}_0$ the answer is simply $\tau(\underset{\sim}{x}_0) = T$ and $\Phi_T(\underset{\sim}{x}_0) = \Phi_0(\underset{\sim}{x}_0) = \underset{\sim}{x}_0$. However, points in the neighborhood of $\underset{\sim}{x}_0$ may return to S later or earlier than T or else may not return to the transverse section at all. The initial set of S_0, after one turn, is mapped onto a neighborhood S_1, i.e. into the set of points

$$S_1 = \{\Phi_{\tau(\underset{\sim}{x})}(\underset{\sim}{x}) | \underset{\sim}{x} \in S_0\} . \tag{6.39}$$

Note that different points of S_0 need different times for returning to S for the first time (if they escape, this time is infinite). Therefore, S_1 is not a front of the flow. The mapping generated in this way,

$$\pi : S_0 \to S_1 : \underset{\sim}{x} \mapsto \Phi_{\tau(\underset{\sim}{x})}(\underset{\sim}{x}) , \tag{6.40}$$

is said to be the *Poincaré mapping*. It describes the behavior of the flow, as a function of discretized time, on a submanifold S whose dimension is one less than the dimension of the manifold M on which the dynamical system is defined. Figure 6.10 shows a two-dimensional transverse section for a flow on $M = \mathbb{R}^3$.

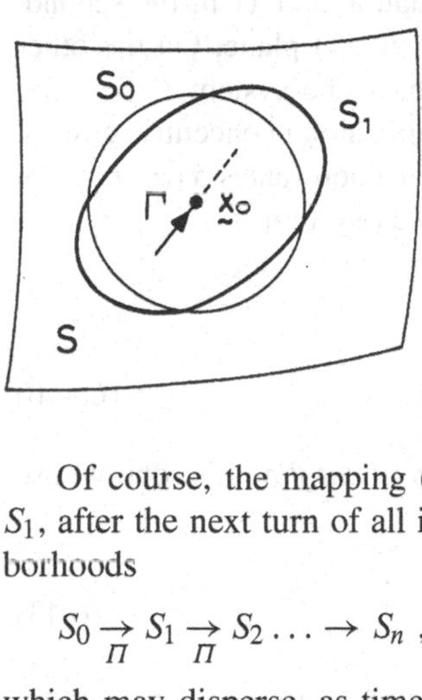

Fig. 6.10. Poincaré mapping of an initial domain S_0 in the neighborhood of a periodic orbit Γ. The point $\underset{\sim}{x}_0$, where Γ hits the transverse section, is a fixed point of the mapping

Of course, the mapping (6.40) can be iterated by asking for the image S_2 of S_1, after the next turn of all its points, etc. One obtains a sequence of open neighborhoods

$$S_0 \underset{\Pi}{\to} S_1 \underset{\Pi}{\to} S_2 \ldots \to S_n ,$$

which may disperse, as time goes to $+\infty$, or may stay more or less constant, or may shrink to the periodic orbit Γ. This provides us with a useful criterion for the investigation of the flow's long-term behavior in the neighborhood of a periodic orbit, or, more generally, in the neighborhood of an attractor. In particular, the Poincaré mapping allows for a test of stability of a periodic orbit or of an attractor.

In order to answer the question of stability in the neighborhood of the periodic orbit Γ, it suffices to linearize the Poincaré mapping at the point $\underset{\sim}{x}_0$. Thus, one considers the mapping

$$\mathbf{D}\Pi(0) = \{\partial \Pi^i / \partial x^k |_{\underset{\sim}{x}=0}\} . \tag{6.41}$$

(In the case of a general manifold M this is the tangent map $T\Pi$ at $\underset{\sim}{x}_0 \in M$.) The eigenvalues of the matrix (6.41) are called *characteristic multipliers* of the vector field $\underset{\sim}{F}$ at the periodic orbit Γ. They tell us whether there is stability or instability in a neighborhood of the closed orbit Γ. We have the following. Let Γ be a closed orbit of the dynamical system $\underset{\sim}{F}$ and let Π be a Poincaré mapping in $\underset{\sim}{x}_0 = 0$. If all characteristic multipliers lie strictly inside the unit circle, the flow will tend to Γ smoothly as $t \to +\infty$. This orbit is asymptotically stable. In turn, if the absolute value of one of the characteristic exponents exceeds one, the closed orbit Γ is unstable.

We study two examples. The first concerns flows in the plane for which transverse sections are one-dimensional. The second illustrates flows on the torus T^2 in $\mathbb{R}^4$, or in its neighborhood, in which case the transverse section may be taken to be a subset of planes that cut the torus.

Example (i) Consider the dynamical system

$$\begin{aligned} \dot{x}_1 &= \mu x_1 - x_2 - (x_1^2 + x_2^2)^n x_1 , \\ \dot{x}_2 &= \mu x_2 + x_1 - (x_1^2 + x_2^2)^n x_2 , \end{aligned} \tag{6.42a}$$

where the exponent n takes the values $n = 1, 2$, or 3 and where μ is a real parameter. Without the coupling terms $(-x_2)$ in the first equation and x_1 in the second, the system (6.42a) is invariant under rotations in the (x_1, x_2)-plane. On the other hand, without the nonlinearity and with $\mu = 0$, we have the system $\dot{x}_1 = -x_2$, $\dot{x}_2 = x_1$ whose solutions move uniformly about the origin, along concentric circles. One absorbs this uniform rotation by introducing polar coordinates $x_1 = r\cos\phi$, $x_2 = r\sin\phi$. The system (6.42a) becomes the decoupled system

$$\begin{aligned} \dot{r} &= \mu r - r^{2n+1} \equiv -\frac{\partial}{\partial r} U(r, \phi) , \\ \dot{\phi} &= 1 \equiv -\frac{\partial}{\partial \Phi} U(r, \phi) . \end{aligned} \tag{6.42b}$$

The right-hand side of the first equation (6.42b) is a *gradient flow* (i.e. one whose vector field is a gradient field, cf. Exercise 6.7), with

$$U(r, \phi) = -\frac{1}{2}\mu r^2 + \frac{1}{2n+2} r^{2n+2} - \phi . \tag{6.43}$$

The origin $r = 0$ is a critical point. Orbits in its neighborhood behave like spirals around (0,0) with radial dependence $r = \exp(\mu t)$. Thus, for $\mu < 0$, the point (0,0) is asymptotically stable. For $\mu > 0$ this point is unstable. At the same time, there appears a periodic solution

$$x_1 = R(\mu)\cos t , \quad x_2 = R(\mu)\sin t \quad \text{with} \quad R(\mu) = \sqrt[2n]{\mu} ,$$

which turns out to be an asymptotically stable attractor: solutions starting outside the circle with radius $R(\mu)$ move around it like spirals and tend exponentially towards the circle, for increasing time; likewise, solutions starting inside the circle move outward like spirals and tend to the circle from the inside. (The reader is invited to sketch the flow for $\mu > 0$.)

In this example it is not difficult to construct a Poincaré mapping explicitly. It is sufficient to cut the flow in the (x_1, x_2)-plane with the semi-axis $\phi = \phi_0 = \text{const.}$ Starting from (x_1^0, x_2^0) on this line, with $r_0 = \sqrt{(x_1^0)^2 + (x_2^0)^2}$, the flow hits the line again after the time $t = 2\pi$. The image of the starting point has the distance $r_1 = \Pi(r_0)$, where r_1 is obtained from the first equation (6.42b). Indeed, if Ψ_t denotes the flow of that equation, $r_1 = \Psi_{t=2\pi}(r_0)$.

Let us take the special case $n = 1$ and $\mu > 0$. Taking the time variable $\tau = \mu t$, the system (6.42b) becomes

$$\frac{\mathrm{d}r}{\mathrm{d}\tau} = r\left(1 - \frac{1}{\mu}r^2\right), \quad \frac{\mathrm{d}\phi}{\mathrm{d}\tau} = \frac{1}{\mu}. \tag{6.44}$$

With $r(\tau) = 1/\sqrt{\varrho(\tau)}$ we obtain the differential equation $\mathrm{d}\varrho/\mathrm{d}\tau = 2(1/\mu - \varrho)$, which can be integrated analytically. One finds $\varrho(c, \tau) = 1/\mu + c\exp(-2\tau)$, c being an integration constant determined from the initial condition $\varrho(\tau = 0) = \varrho_0 = 1/r_0^2$. Thus, the integral curve of (6.44) starting from (r_0, ϕ_0) reads

$$\Phi_\tau(r_0, \phi_0) = \left(1/\sqrt{\varrho(c, \tau)}, \phi_0 + \tau/\mu \bmod 2\pi\right) \tag{6.45}$$

with $c = 1/r_0^2 - 1/\mu$. Hence, the Poincaré mapping that takes (r_0, ϕ_0) to $(r_1, \phi_1 = \phi_0)$ is given by (6.45) with $\tau = 2\pi$, viz.

$$\Pi(r_0) = \left[1/\mu + \left(\frac{1}{r_0^2} - \frac{1}{\mu}\right)\mathrm{e}^{-4\pi}\right]^{-1/2}. \tag{6.46}$$

This has the fixed point $r_0 = \sqrt{\mu}$, which represents the periodic orbit. Linearizing in the neighborhood of this fixed point we find

$$\mathbf{D}\Pi(r_0 = \sqrt{\mu}) = \left.\frac{\mathrm{d}\Pi}{\mathrm{d}r_0}\right|_{r_0=\sqrt{\mu}} = \mathrm{e}^{-4\pi}.$$

The characteristic multiplier is $\lambda = \exp(-4\pi)$. Its absolute value is smaller than 1 and hence the periodic orbit is an asymptotically stable attractor.

Example (ii) Consider the flow of an autonomous Hamiltonian system with $f = 2$ for which there are two integrals of the motion. Suppose we have already found a canonical transformation to action and angle coordinates, i.e. one by which both coordinates are made cyclic, i.e.

$$\{q_1, q_2, p_1, p_2, H\} \to \{\theta_1, \theta_2, I_1, I_2, \tilde{H}\}, \tag{6.47}$$

and $\tilde{H} = \omega_1 I_1 + \omega_2 I_2$. An example is provided by the decoupled oscillators (6.34). As both θ_k are cyclic, we have

$$\dot{I}_i(q, p) = 0 \,, \quad \text{or} \quad I_i(q, p) = \text{const} = I_i(q_0, p_0)$$

along any orbit. Returning to the old coordinates for the moment, this means that the Poisson brackets

$$\{H, I_i\} \quad \text{and} \quad \{I_i, I_j\} \quad (i, j = 1, 2) \tag{6.48}$$

vanish.[3] In the new coordinates we have

$$\dot{\theta}_i = \partial\tilde{H}/\partial I_i = \omega_i \quad \text{or} \quad \theta_i(t) = \omega_i t + \theta_i^0 \,. \tag{6.49}$$

From (6.49) we see that the manifold of motions is the torus T^2, embedded in the four-dimensional phase space. For the transversal section of the Poincaré mapping it is natural to choose a part S of a plane that cuts the torus and is perpendicular to it. Let $\theta_1(t)$ be the angular variable running *along* the torus, and $\theta_2(t)$ the one running along a *cross section* of the torus. A point $s \in S_0 \subset S$ returns to S for the first time after $T = (2\pi/\omega_1)$. Without loss of generality we measure time in units of this period T, $\tau = t/T$, and take $\theta_1^0 = 0$. Then we have

$$\theta_1(\tau) = 2\pi\tau \,, \quad \theta_2(\tau) = 2\pi\tau\omega_2/\omega_1 + \theta_2^0 \,. \tag{6.49'}$$

Call C the curve of intersection of the torus and the transverse section of S. The Poincaré mapping maps points of C on the same curve. The points of intersection of the orbit (6.49′) with S appear, one after the other, at $\tau = 0, 1, 2, \ldots$ If the ratio of frequencies is rational , $\omega_2/\omega_1 = m/n$, the first $(n-1)$ images of the point $\theta_2 = \theta_2^0$ are distinct points on C, while the nth image coincides with the starting point. If, in turn, the ratio ω_2/ω_1 is irrational, a point s_0 on C is shifted, at each iteration of the Poincaré mapping, by the azimuth $2\pi\omega_2/\omega_1$. It never returns to its starting position. For large times the curve C is covered discontinuously but densely.

6.3.5 Bifurcations of Flows at Critical Points

In Example (i) of the previous section the flow is very different for positive and negative values of the control parameter. For $\mu < 0$ the origin is the only critical element. It turns out to be an asymptotically stable equilibrium. For $\mu > 0$ the flow has the critical elements $\{0, 0\}$ and $\{R(\mu)\cos t, R(\mu)\sin t\}$. The former is an unstable equilibrium position, the latter a periodic orbit that is an asymptotically stable attractor. If we let μ vary from negative to positive values, then, at $\mu = 0$, a stable, periodic orbit branches off from the previously stable equilibrium point $\{0, 0\}$. At the same time, the equilibrium position becomes unstable as shown in Fig. 6.11. Another way of expressing the same result is to say that the origin acts like a *sink* for the flow at $\mu < 0$. For $\mu > 0$ it acts like a *source* of the flow, while the periodic orbit with radius $R(\mu)$ is a sink. The structural change of the flow happens at the point $(\mu = 0, r = 0)$, in the case of this specific example. A point of this nature is said to be a bifurcation point.

---

[3] H, I_1, and I_2 are in involution, for definitions cf. Sect. 2.37.2

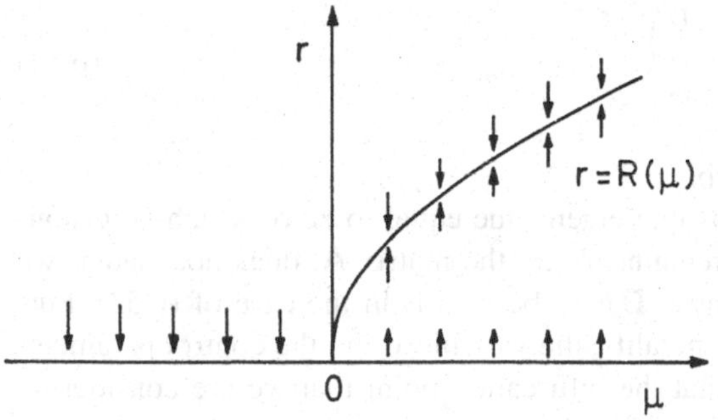

Fig. 6.11. For the system (6.42a) the point $r = 0$ is asymptotically stable for $\mu < 0$. At $\mu = 0$ a periodic solution (circle with radius $R(\mu)$) splits off and becomes an asymptotically stable attractor. At the same time the point $r = 0$ becomes unstable

The general case is that of the dynamical system

$$\dot{\underset{\sim}{x}} = \underset{\sim}{F}(\mu, \underset{\sim}{x}) \,. \tag{6.50}$$

whose vector field depends on a set $\underset{\sim}{\mu} = \{\mu_1, \mu_2, \ldots, \mu_k\}$ of k control parameters. The critical points $\underset{\sim}{x}_0(\mu)$ of the system (6.50) are obtained from the equation

$$\underset{\sim}{F}(\mu, \underset{\sim}{x}_0) = 0 \,. \tag{6.51}$$

The solutions of this implicit equation, in general, depend on the values of the parameters μ. They are smooth functions of μ if and only if the determinant of the matrix of partial derivatives $\mathbf{D}\underset{\sim}{F} = \{\partial F^i(\mu, \underset{\sim}{x})/\partial x^k\}$ does not vanish in $\underset{\sim}{x}_0$. This is a consequence of the theorem on implicit functions, which guarantees that (6.51) can be solved for $\underset{\sim}{x}_0$, provided that the condition is fulfilled. The points $(\underset{\sim}{\mu}, \underset{\sim}{x}_0)$ where this condition is *not* fulfilled, i.e. where $\mathbf{D}\underset{\sim}{F}$ has at least one vanishing eigenvalue, need special consideration. Here, several branches of differing stability may merge or split off from each other. By crossing this point, the flow changes its structure in a qualitative manner. Therefore, a point $(\underset{\sim}{\mu}, \underset{\sim}{x}_0)$ where the determinant of $\mathbf{D}\underset{\sim}{F}$ vanishes, or, equivalently, where at least one of its eigenvalues vanishes, is said to be a bifurcation point.

The general discussion of the solutions of (6.51) and the complete classification of bifurcations is beyond the scope of this book. A good account of what is known about this is given by Guckenheimer and Holmes (2001). We restrict our discussion to bifurcations of codimension 1.[4] Thus, the vector field depends on only one parameter μ, but is still a function of the n-dimensional variable $\underset{\sim}{x}$. If $(\mu_0, \underset{\sim}{x}_0)$ is a bifurcation point, the following two forms of the matrix of partial derivatives $\mathbf{D}\underset{\sim}{F}$ are typical (cf. Guckenheimer and Holmes 2001):

$$\mathbf{D}\underset{\sim}{F}(\mu, \underset{\sim}{x})|_{\mu_0, \underset{\sim}{x}_0} = \begin{pmatrix} 0 & 0 \\ 0 & \mathbf{A} \end{pmatrix} , \tag{6.52}$$

where $\mathbf{A}$ is a $(n-1) \times (n-1)$ matrix, as well as

[4] The codimension of a bifurcation is defined to be the smallest dimension of a parameter space $\{\mu_1, \ldots, \mu_k\}$ for which this bifurcation does occur.

$$\mathbf{D}\underset{\sim}{F}(\mu, \underset{\sim}{x})|_{\mu_0, \underset{\sim}{x}_0} = \begin{pmatrix} 0 & -\omega & 0 \\ \omega & 0 & 0 \\ 0 & 0 & \mathbf{B} \end{pmatrix} , \tag{6.53}$$

with $\mathbf{B}$ a $(n-2) \times (n-2)$ matrix.

In the first case (6.52) $\mathbf{D}\underset{\sim}{F}$ has one eigenvalue equal to zero, which is responsible for the bifurcation. As the remainder, i.e. the matrix $\mathbf{A}$, does not matter, we can take the dimension of the matrix $\mathbf{D}\underset{\sim}{F}$ to be $n = 1$, in the case of (6.52). Furthermore, without restriction of generality, the variable $\underset{\sim}{x}$ and the control parameter μ can be shifted in such a way that the bifurcation point that we are considering occurs at $(\mu = 0, \underset{\sim}{x}_0 = 0)$. Then the following types of bifurcations are contained in the general form (6.52).

(i) *The saddle-node bifurcation*:

$$\dot{x} = \mu - x^2 . \tag{6.54}$$

For $\mu > 0$ the branch $x_0 = \sqrt{\mu}$ is the set of stable equilibria and $x_0 = -\sqrt{\mu}$ the set of unstable equilibria, as shown in Fig. 6.12. These two branches merge at $\mu = 0$ and compensate each other because, for $\mu < 0$, there is no equilibrium position.

(ii) *The transcritical bifurcation*:

$$\dot{x} = \mu x - x^2 . \tag{6.55}$$

Here the straight lines $x_0 = 0$ and $x_0 = \mu$ are equilibrium positions. For $\mu < 0$ the former is asymptotically stable and the latter is unstable. For $\mu > 0$, on the other hand, the former is unstable and the latter is asymptotically stable, as shown in Fig. 6.13. The four branches coincide for $\mu = 0$, the semi-axes ($x_0 = 0$, or $x_0 = \mu, \mu < 0$) and ($x_0 = \mu$, or $x_0 = 0, \mu > 0$) exchange their character of stability; hence the name of the bifurcation.

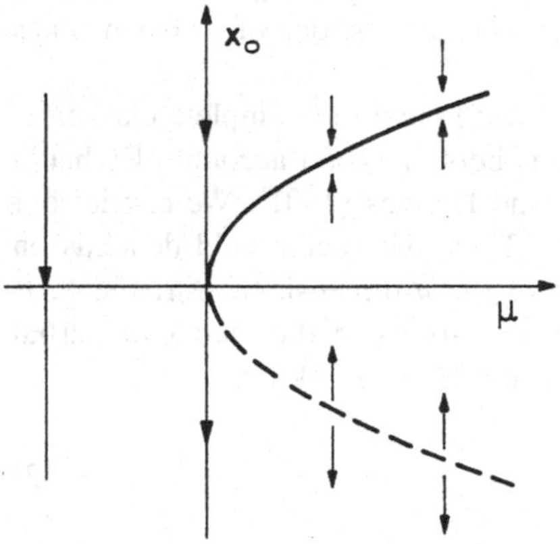

Fig. 6.12. Illustration of a saddle-node bifurcation at $x_0 = \mu = 0$. As in Fig. 6.11 the arrows indicate the direction of the flow in the neighborhood of the equilibria

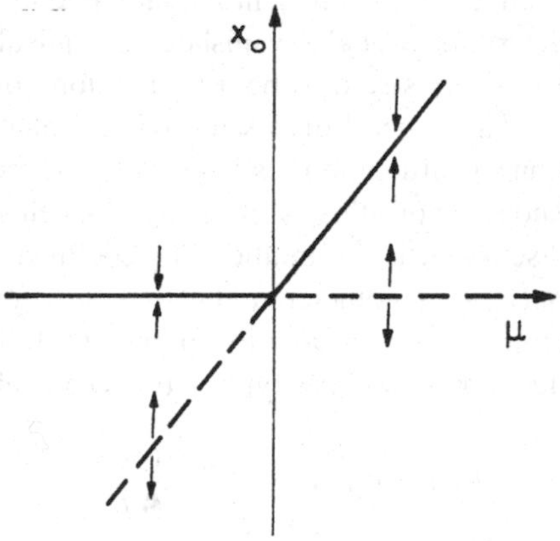

Fig. 6.13. The transcritical bifurcation. In crossing the point of bifurcation $\mu = 0$, the two semi-axes exchange their character of stability

(iii) *The pitchfork bifurcation*:

$$\dot{x} = \mu x - x^3 \,. \tag{6.56}$$

All the points of the straight line $x_0 = 0$ are critical points. These are asymptotically stable if μ is negative, but become unstable if μ is positive. In addition, for $\mu > 0$, the points on the parabola $x_0^2 = \mu$ are asymptotically stable equilibria, as shown in Fig. 6.14. At $\mu = 0$, the single line of stability on the left of the figure splits into the "pitchfork" of stability (the parabola) and the semi-axis of instability.

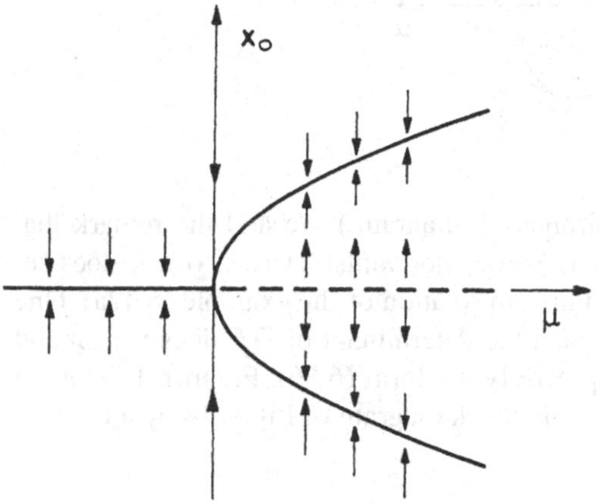

Fig. 6.14. The pitchfork bifurcation

In all examples and prototypes considered above the signs of the nonlinear terms are chosen such that they act against the constant or linear terms for $\mu > 0$, i.e. in such a way that they have a stabilizing effect as one moves from the line $x_0 = 0$ to positive x. The bifurcations obtained in this way are called *supercritical*. It is instructive to study the bifurcation pattern (6.54–6.56) for the case of the opposite sign of the nonlinear terms. The reader is invited to sketch the resulting bifurcation diagrams. The so-obtained bifurcations are called *subcritical*. In the case of the second normal form (6.53) the system must have at least two dimensions and $\mathbf{D}\underset{\sim}{F}$ must have (at least) two complex conjugate eigenvalues. The prototype for this case is the following.

(iv) *The Hopf bifurcation*:

$$\begin{aligned} \dot{x}_1 &= \mu x_1 - x_2 - (x_1^2 + x_2^2)x_1 \,, \\ \dot{x}_2 &= \mu x_2 + x_1 - (x_1^2 + x_2^2)x_2 \,. \end{aligned} \tag{6.57}$$

This is the same as the example (6.42a), with $n = 1$. We can take over the results from there and draw them directly in the bifurcation diagram $(\mu, \underset{\sim}{x}_0)$. This yields the picture shown in Fig. 6.15. (Here, again, it is instructive to change the sign of the nonlinear term in (6.57), turning the supercritical bifurcation into a subcritical

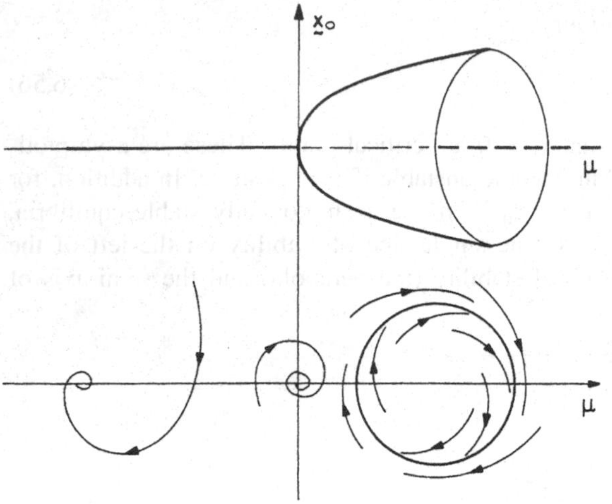

Fig. 6.15. The Hopf bifurcation in two dimensions. The lower part of the figure shows the behavior of the flow in the neighborhood of the asymptotically stable equilibrium and of the asymptotically stable periodic solution

one. The reader should sketch the bifurcation diagram.) We add the remark that here and in (6.53) the determinant of $\mathbf{D}\underset{\sim}{F}$ does not vanish at $(\mu_0, \underset{\sim}{x}_0)$. It does so, however, once we have taken out the uniform rotation of the example (6.42a). One then obtains the system (6.42b) for which the determinant of $\mathbf{D}\underset{\sim}{F}$ does vanish and whose first equation (for $n = 1$) has precisely the form (6.56). Figure 6.15 may be thought of as being generated from the pitchfork diagram of Fig. 6.14 by a rotation in the second x-dimension.

6.3.6 Bifurcations of Periodic Orbits

We conclude this section with a few remarks on the stability of closed orbits, as a function of control parameters. Section 6.3.5 was devoted exclusively to the bifurcation of *points* of equilibrium. Like the closed orbits, these points belong to the critical elements of the vector field. Some of the results obtained there can be translated directly to the behavior of periodic orbits at bifurcation points, by means of the Poincaré mapping (6.40) and its linearization (6.41).

A qualitatively new feature, which is important for what follows, is the bifurcation of a periodic orbit leading to period doubling. It may be described as follows. Stability or instability of flows in the neighborhood of closed orbits is controlled by the matrix (6.41), that is the linearization of the Poincaré mapping. The specific bifurcation in which we are interested here occurs whenever one of the characteristic multipliers (i.e. the eigenvalues of (6.41)) crosses the value -1, as a function of the control parameter μ. Let s_0 be the point of intersection of the periodic orbit Γ with a transverse section. Clearly, s_0 is a fixed point of the Poincaré mapping, $\Pi(s_0) = s_0$. As long as all eigenvalues of the matrix $\mathbf{D}\Pi(s_0)$ (6.41) are inside the unit circle (i.e. have absolute values smaller than 1), the distance from s_0 of another point s in the neighborhood of s_0 will decrease monotonically by successive iterations of the Poincaré mapping. Indeed, in linear approximation we have

$$\Pi^n(s) - s_0 = (\mathbf{D}\Pi(s_0))^n (s - s_0) \ . \tag{6.58}$$

Suppose the matrix $\mathbf{D}\Pi(s_0)$ to be diagonal. We assume the first eigenvalue to be the one that, as a function of the control parameter μ, moves outward from somewhere inside the unit circle, by crossing the value -1 at some value of μ. All other eigenvalues, for simplicity, are supposed to stay inside the unit circle. In this special situation it is sufficient to consider the Poincaré mapping only in the 1-direction on the transverse section, i.e. in the direction the eigenvalue λ_1 refers to. Call the coordinate in that direction u. If we suppose that $\lambda_1(\mu)$ is real and, initially, lies between 0 and -1, the orbit that hits the transverse section at the point s_1 of Fig. 6.16a appears in u_2 after one turn, in u_3 after two turns, etc. It approaches the point s_0 asymptotically and the periodic orbit through s_0 is seen to be stable. If, on the other hand, $\lambda_1(\mu) < -1$, the orbit through s_1 moves outward rapidly and the periodic orbit through s_0, obviously, is unstable.

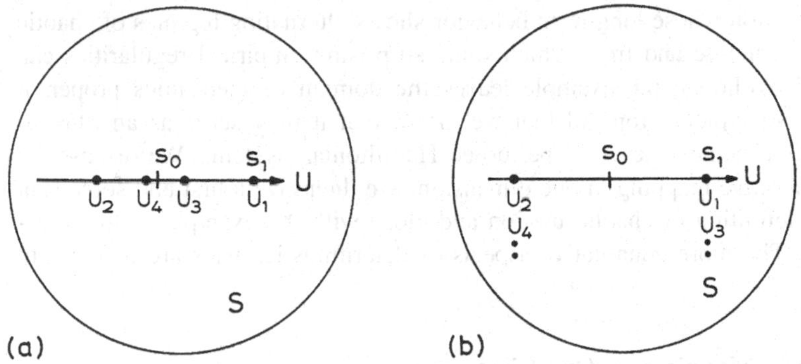

Fig. 6.16a,b. Poincaré mapping in the neighborhood of a periodic orbit, for the case where a characteristic multiplier approaches the value -1 from above (**a**), and for the case where it equals that value (**b**)

A limiting situation occurs if there is a value μ_0 of the control parameter for which $\lambda_1(\mu_0) = -1$. Here we obtain the pattern shown in Fig. 6.16b: after one turn the orbit through s_1 appears in $u_2 = -u_1$, after the second turn in $u_3 = +u_1$, then in $u_4 = -u_1$, then in $u_5 = +u_1$, etc. This applies to each s on the u axis, in a neighborhood of s_0. As a result, the periodic orbit Γ through s_0 has only a sort of saddle-point stability: orbits in directions other than the u-axis are attracted towards it, but orbits whose intersections with the transverse section lie on the u-axis will be caused to move away by even a small perturbation. Thus, superficially, the point (μ_0, s_0) seems to be a point of bifurcation having the character of the pitchfork of Fig. 6.14. A closer look shows, however, that there is really a new phenomenon. In a system as described by the bifurcation diagram of Fig. 6.14, the integral curve tends *either* to the point $x_0 = +\sqrt{\mu}$, *or* to the point $x_0 = -\sqrt{\mu}$, for positive μ. In the case shown in Fig. 6.16b, on the contrary, the orbit *alternates* between u_1 and $-u_1$. In other words, it is a periodic orbit Γ_2 with the period $T_2 = 2T$, T being the period of the original reference orbit.

If we take Γ_2 as the new reference orbit, the Poincaré mapping must be redefined in such a way that Γ_2 hits the transverse section for the first time after the time T_2. One can then study the stability of orbits in the neighborhood of Γ_2. By varying the control parameter further, it may happen that the phenomenon of period doubling described above happens once more at, say, $\mu = \mu_1$, and that we have to repeat the analysis given above. In fact, there can be a sequence of bifurcation points (μ_0, s_0), (μ_1, s_1), etc. at each of which the period is doubled. We return to this phenomenon in the next section.

6.4 Deterministic Chaos

This section deals with a particularly impressive and characteristic example for deterministic motion whose long-term behavior shows alternating regimes of chaotic and ordered structure and from which some surprising empirical regularities can be extracted. Although the example leaves the domain of mechanics proper, it seems to be so typical, from all that we know, that it may serve as an illustration for chaotic behavior even in perturbed Hamiltonian systems. We discuss the concept of iterative mapping in one dimension. We then give a first and somewhat provisional definition of chaotic motion and close with the example of the logistic equation. The more quantitative aspects of deterministic chaos are deferred to Sect. 6.5.

6.4.1 Iterative Mappings in One Dimension

In Sect. 6.3.6 we made use of the Poincaré mapping of a three-dimensional flow for investigating the stability of a closed orbit as a function of the control parameter μ. We found, in the simplest case, that the phenomenon of period doubling could be identified in the behavior of a single dimension of the flow, provided one concentrates on the direction for which the characteristic multiplier $\lambda(\mu)$ crosses the value -1 at some critical value of μ. The full dimension of the flow of $\underset{\sim}{F}(\mu, \underset{\sim}{x})$ did not matter. We can draw two lessons from this. Firstly, it may be sufficient to choose a single direction within the transverse section S (more generally, a one-dimensional submanifold of S) and to study the Poincaré mapping along this direction only. The picture that one obtains on this one-dimensional submanifold may already give a good impression of the flow's behavior in the large. Secondly, the restriction of the Poincaré mapping to one dimension reduces the analysis of the complete flow and of its full, higher-dimensional complexity, to the analysis of an iterative mapping in one dimension,

$$u_i \to u_{i+1} = f(u_i) . \tag{6.59}$$

In the example of Sect. 6.3.6, for instance, this iterative mapping is the sequence of positions of a point on the transverse section at times $0, T, 2T, 3T, \ldots$ Here the behavior of the full system (6.1) at a point of bifurcation is reduced to a difference equation of the type (6.59).

There is another reason one-dimensional systems of the form (6.59) are of interest. Strongly dissipative systems usually possess asymptotically stable equilibria and/or attractors. In this case a set of initial configurations filling a given volume of phase space will be strongly quenched, by the action of the flow and as time goes by, so that the Poincaré mapping quickly leads to structures that look like pieces of straight lines or arcs of curves. This observation may be illustrated by the example (6.38). Although the flow of this system is four dimensional, it converges to the torus T^2, the attractor, at an exponential rate. Therefore, considering the Poincaré mapping for large times, we see that the transverse section of the torus will show all points of intersection lying on a circle. This is also true if the torus is a strange attractor. In this case the Poincaré mapping shows a chaotic regime in a small strip in the neighborhood of the circle (see e.g. Bergé, Pomeau, Vidal 1987). Finally, iterative equations of the type (6.59) describe specific dynamical systems of their own that are formulated by means of difference equations (see e.g. Devaney 1989, Collet and Eckmann 1990). In Sect. 6.4.3 below we study a classic example of a discrete dynamical system (6.59). It belongs to the class of *iterative mappings on the unit interval*, which are defined as follows.

Let $f(\mu, x)$ be a function of the control parameter μ and of a real variable x in the interval $[0, 1]$. f is continuous, and in general also differentiable, and the range of μ is chosen such that the iterative mapping

$$x_{i+1} = f(\mu, x_i) , \quad x \in [0, 1] , \tag{6.60}$$

does not lead out of the interval $[0, 1]$. An equation of this type can be analyzed graphically, and particularly clearly, by comparing the graph of the function $y(x) = f(\mu, x)$ with the straight line $z(x) = x$. The starting point x_1 has the image $y(x_1)$, which is then translated to the straight line as shown in Fig. 6.17a. This yields the next value x_2, whose image $y(x_2)$ is again translated to the straight line, yielding the next iteration x_3, and so on. Depending on the shape of $f(\mu, x)$ and on the starting value, this iterative procedure may converge rapidly to the fixed point $\bar{x}$ shown in Fig. 6.17a. At this point the straight line and the graph of f intersect and we have

$$\bar{x} = f(\mu, \bar{x}) . \tag{6.61}$$

The iteration $x_1 \to x_2 \to \ldots \to \bar{x}$ converges if the absolute value of the derivative of the curve $y = f(\mu, x)$ in the point $y = x$ is smaller than 1. In this case $\bar{x}$ is an equilibrium position of the dynamical system (6.60), which is asymptotically stable. If the modulus of the derivative exceeds 1, on the other hand, the point $\bar{x}$

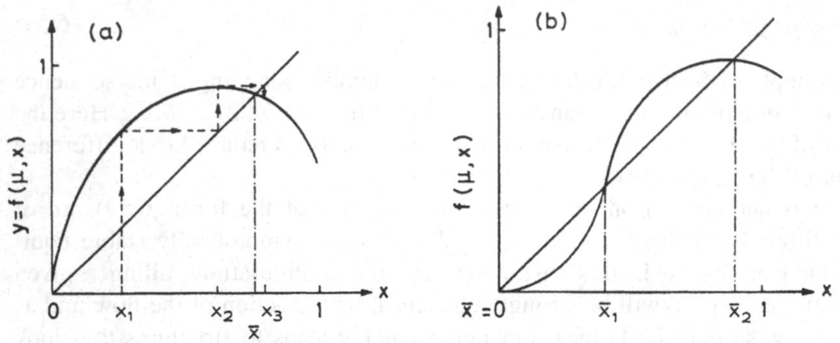

Fig. 6.17a,b. The iteration $x_{i+1} = f(\mu, x_i)$ converges to $\bar{x}$, provided $|\mathrm{d}f/\mathrm{d}x|_{\bar{x}} < 1$ **(a)**. In **(b)** both 0 and $\bar{x}_2$ are stable but $\bar{x}_1$ is unstable

is unstable. In the example shown in Fig. 6.17a $\bar{x} = 0$ is unstable. Figure 6.17b shows an example where $\bar{x}_0 = 0$ and $\bar{x}_2$ are stable, while $\bar{x}_1$ is unstable. By the iteration (6.60) initial values $x_1 < \bar{x}_1$ tend to $\bar{x}_0$, while those with $\bar{x}_1 < x_1 \leq 1$ tend to $\bar{x}_2$.

The nature and the position of the equilibria are determined by the control parameter μ. If we let μ vary within its allowed interval of variation, we may cross certain critical values at which the stability of points of convergence changes and, hence, where the structure of the dynamical system changes in an essential and qualitative way. In particular, there can be bifurcations of the type described in Sects. 6.3.5 and 6.3.6. We do not pursue the general discussion of iterated mappings (6.60) here and refer to the excellent monographs by Collet and Eckmann (1990) and Guckenheimer and Holmes (1990). An instructive example is given in Sect. 6.4.3 below. Also we strongly recommend working out the PC-assisted examples of Exercises 6.12–14, which provide good illustrations for iterative mappings and give an initial feeling for chaotic regimes.

6.4.2 Qualitative Definitions of Deterministic Chaos

Chaos and chaotic motion are intuitive concepts that are not easy to define in a quantitative and measurable manner. An example taken from daily life may illustrate the problem. Imagine a disk-shaped square in front of the main railway station of a large city, say somewhere in southern Europe, during rush hour. At the edges of the square busses are coming and going, dropping passengers and waiting for new passengers who commute with the many trains entering and leaving the station. Looking onto the square from the top, the motion of people in the crowd will seem to us nearly or completely chaotic. And yet we know that every single passenger follows a well-defined path: he gets off the train on platform 17 and makes his way through the crowd to a target well known to him, say bus no. 42, at the outer edge of the square.

Now image the same square on a holiday, on the day of a popular annual fair. People are coming from all sides, wandering between the stands, going here and there rather erratically and without any special purpose. Again, looking at the square from the top, the motion of people in the crowd will seem chaotic, at least to our intuitive conception. Clearly, in the second case, motion is more accidental and less ordered than in the first. There is more chaoticity in the second situation than in the first. The question arises whether this difference can be made quantitative. Can one indicate measurements that answer quantitatively whether a given type of motion is really unordered or whether it has an intrinsic pattern one did not recognize immediately?[5]

We give here two provisional definitions of chaos but return to a more quantitative one in Sect. 6.5 below. Both of them, in essence, define a motion to be chaotic whenever it cannot be predicted, in any practical sense, from earlier configurations of the same dynamical system. In other terms, even though the motion is strictly deterministic, predicting a state of motion from an initial configuration may require knowledge of the latter to a precision that is far beyond any practical possibility.

(i) The first definition makes use of Fourier analysis of a sequence of values $\{x_1, x_2, \dots, x_n\}$, which are taken on at the discrete times $t_\tau = \tau \cdot \Delta$, $\tau = 1, 2, \dots, n$. Fourier transformation assigns to this sequence another sequence of complex numbers $\{\tilde{x}_1, \tilde{x}_2, \dots, \tilde{x}_n\}$ by

$$\tilde{x}_\sigma \stackrel{\text{def}}{=} \frac{1}{\sqrt{n}} \sum_{\tau=1}^{n} x_\tau \mathrm{e}^{-\mathrm{i}2\pi\sigma\tau/n} , \quad \sigma = 1, 2, \dots, n . \tag{6.62}$$

While the former is defined over the time variable, the latter is defined over a frequency variable, as will be clear from the following. The sequence $\{x_i\}$ is recorded during the total time

$$T = t_n = n\Delta ,$$

or, if we measure time in units of the interval Δ, $T = n$. The sequence $\{x_\tau\}$ may be understood as a discretized function $x(t)$ such that $x_\tau = x(\tau)$ (with time in units of Δ). Then $F = 2\pi/n$ is the frequency corresponding to time T, and the sequence $\{\tilde{x}_\sigma\}$ is the discretization of a function $\tilde{x}$ of the frequency variable with $\tilde{x}_\sigma = \tilde{x}(\sigma \cdot F)$. Thus, time and frequency are conjugate variables.

Although the $\{x_\tau\}$ are real, the $\tilde{x}_\sigma$ of (6.62) are complex numbers. However, they fulfill the relations $\tilde{x}_{n-\sigma} = \tilde{x}_\sigma^*$ and thus do not contain additional degrees of freedom. One has the relation

$$\sum_{\tau=1}^{n} x_\tau^2 = \sum_{\sigma=1}^{n} |\tilde{x}_\sigma|^2$$

[5] In early Greek cosmology chaos meant "the primeval emptiness of the universe" or, alternatively, "the darkness of the underworld". The modern meaning is derived from Ovid, who defined chaos as "the original disordered and formless mass from which the maker of the Cosmos produced the ordered universe" (*The New Encyclopedia Britannica*). Note that the loan-word *gas* is derived from the word chaos. It was introduced by J.B. von Helmont, a 17th-century chemist in Brussels.

and the inverse transformation reads[6]

$$x_\tau = \frac{1}{\sqrt{n}} \sum_{\sigma=1}^{n} \tilde{x}_\sigma e^{i 2\pi\tau\sigma/n} \,. \tag{6.63}$$

The following correlation function is a good measure of the predictability of a signal at a later time, from its present value:

$$g_\lambda \stackrel{\text{def}}{=} \frac{1}{n} \sum_{\sigma=1}^{n} x_\sigma x_{\sigma+\lambda} \,. \tag{6.64}$$

g_λ is a function of time, $g_\lambda = g_\lambda(\lambda \cdot \Delta)$. If this function tends to zero, for increasing time, this means that any correlation to the system's past gets lost. The system ceases to be predictable and thus enters a regime of irregular motion.

One can prove the following properties of the correlation function g_λ. It has the same periodicity as x_τ, i.e. $g_{\lambda+n} = g_\lambda$. It is related to the real quantities $|\tilde{x}_\sigma|^2$ by the formula

$$g_\lambda = \frac{1}{n} \sum_{\sigma=1}^{n} |\tilde{x}_\sigma|^2 \cos(2\pi\sigma\lambda/n) \,, \quad \lambda = 1, 2, \ldots, n \,. \tag{6.65}$$

Hence, it is the Fourier transform of $|\tilde{x}_\sigma|^2$. Equation (6.65) can be inverted to give

$$\tilde{g}_\sigma \stackrel{\text{def}}{=} |\tilde{x}_\sigma|^2 = \sum_{\lambda=1}^{n} g_\lambda \cos(2\pi\sigma\lambda/n) \,. \tag{6.66}$$

The graph of $\tilde{g}_\sigma$ as a function of frequency gives direct information on the sequence $\{x_\tau\}$, i.e. on the signal $x(t)$. For instance, if $\{x_\tau\}$ was generated by a stroboscopic measurement of a singly periodic motion, then $\tilde{g}_\sigma$ shows a sharp peak at the corresponding frequency. Similarly, if the signal has a quasiperiodic structure, the graph of $\tilde{g}_\sigma$ contains a series of sharp frequencies, i.e. peaks of various strengths. Examples are given, for instance, by Bergé, Pomeau, and Vidal (1987). If, on the other hand, the signal is totally aperiodic, the graph of $\tilde{g}_\lambda$ will exhibit a practically continuous spectrum. When inserted in the correlation function (6.65) this means that g_λ will go to zero for large times. In this case the long-term behavior of the system becomes practically unpredictable. Therefore, the correlation function (6.65), or its Fourier transform (6.66), provides a criterion for the appearance of chaotic behavior: if g_λ tends to zero, after a finite time, or, equivalently, if $\tilde{g}_\lambda$ has a continuous domain, one should expect to find irregular, chaotic motion of the system.

---

[6] In proving this formula one makes use of the "orthogonality relation"

$$\frac{1}{n} \sum_{\sigma=1}^{n} e^{i 2\pi m\sigma/n} = \delta_{m0} \,,$$

$$m = 0, 1, \ldots, n-1 \,.$$

(see also Exercise 6.15)

(ii) The second definition, which is closer to the continuous systems (6.1), starts from the *strange* or *hyperbolic attractors*. A detailed description of this class of attractors is beyond the scope of this book (see, however, Devaney 1989, Bergé, Pomeau, Vidal 1987, and Exercise 6.14), and we must restrict our discussion to a few qualitative remarks. One of the striking properties of strange attractors is that they can sustain orbits that, on average, move apart exponentially (without escaping to infinity and, of course, without intersecting). In 1971 Newhouse, Takens, and Ruelle made the important discovery that flows in three dimensions can exhibit this kind of attractors[7]. Very qualitatively this may be grasped from Fig. 6.18, which shows a flow that strongly contracts in one direction but disperses strongly in the other direction. This flow has a kind of hyperbolic behavior. On the plane where the flow lines drive apart, orbits show extreme sensitivity to initial conditions. By folding this picture and closing it with itself one obtains a strange attractor on which orbits wind around each other (without intersecting) and move apart exponentially[8].

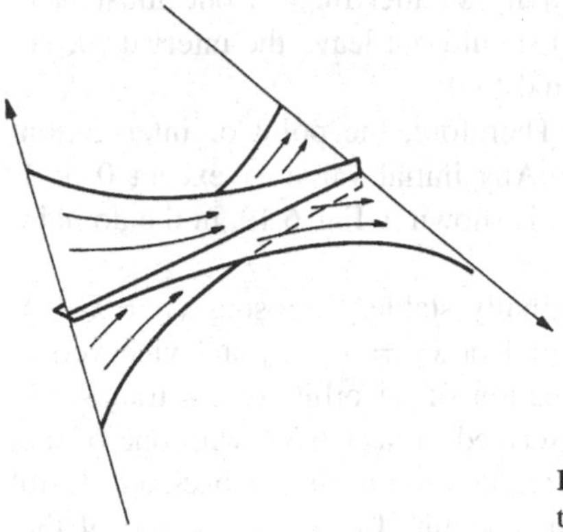

Fig. 6.18. Flow in $\mathbb{R}^3$ that can be bent and glued such that it generates a strange attractor

Whenever there is extreme sensitivity to initial conditions, the long-term behavior of dynamical systems becomes unpredictable, from a practical viewpoint, so that the motion appears to be irregular. Indeed, numerical studies show that there is deterministically chaotic behavior on strange attractors. This provides us with another plausible definition of chaos: flows of deterministic dynamical systems will exhibit chaotic regimes when orbits diverge strongly and, as a consequence, practically "forget" their initial configurations.

---

[7] Earlier it was held that chaotic motion would occur only in systems with very many degrees of freedom, such as gases in macroscopic vessels.

[8] See R.S. Shaw: "Strange attractors, chaotic behavior and information flow", Z. Naturforschung **A36**, (1981) 80.

6.4.3 An Example: The Logistic Equation[9]

An example of a dynamical system of the type (6.60) is provided by the logistic equation

$$x_{i+1} = \mu x_i(1 - x_i) \equiv f(\mu, x_i) \tag{6.67}$$

with $x \in [0, 1]$ and $1 < \mu \leq 4$. This seemingly simple system exhibits an extremely rich structure if it is studied as a function of the control parameter μ. Its structure is typical for systems of this kind and reveals several surprising and universal regularities. We illustrate this by means of numerical results for the iteration (6.67), as a function of the control parameter in the interval given above. It turns out that this model clearly exhibits all the phenomena described so far: bifurcations of equilibrium positions, period doubling, regimes of chaotic behavior, and attractors.

We analyze the model (6.67) as described in Sect. 6.4.1. The derivative of $f(\mu, x)$, taken at the intersection $\bar{x} = (\mu - 1)/\mu$ with the straight line $y = x$, is $f'(\mu, \bar{x}) = 2 - \mu$. In order to keep $|f'|$ initially smaller than 1, one must take $\mu > 1$. On the other hand, the iteration (6.67) should not leave the interval [0, 1]. Hence, μ must be chosen smaller than or equal to 4.

In the interval $1 < \mu < 3$, $|f'| < 1$. Therefore, the point of intersection $\bar{x} = (\mu - 1)/\mu$ is one of stable equilibrium. Any initial value x_1 except 0 or 1 converges to $\bar{x}$ by the iteration. The curve $\bar{x}(\mu)$ is shown in Fig. 6.19, in the domain $1 < \mu \leq 3$.

At $\mu = \mu_0 = 3$ this point becomes marginally stable. Choosing $x_1 = \bar{x} + \delta$ and linearizing (6.67), the image of x_1 is found in $x_2 = \bar{x} - \delta$, and vice versa. If we think of $x_1, x_2, \ldots$ as points of intersection of an orbit with a transverse section, then we have exactly the situation described in Sect. 6.3.6 with one of the characteristic multipliers crossing the value -1. The orbit oscillates back and forth between $x_1 = \bar{x} + \delta$ and $x_2 = \bar{x} - \delta$, i.e. it has acquired twice the period of the original orbit, which goes through $\bar{x}$. Clearly, this tells us that

$$(\mu_0 = 3\ ,\ \bar{x}_0 \equiv \bar{x}(\mu_0)) \tag{6.68}$$

is a bifurcation. In order to determine its nature, we investigate the behavior for $\mu > \mu_0$. As we just saw, the point $\bar{x} = (\mu - 1)/\mu$ becomes unstable and there is period doubling. This means that stable fixed points no longer fulfill the condition $\bar{x} = f(\mu, \bar{x})$ but instead return only after two steps of the iteration, i.e. $\bar{x} = f(\mu, f(\mu, \bar{x}))$. Thus, we must study the mapping $f \circ f$, that is the iteration

$$x_{i+1} = \mu^2 x_i(1 - x_i)[1 - \mu x_i(1 - x_i)]\ , \tag{6.69}$$

[9] This equation takes its name from its use in modeling the evolution of, e.g., animal population over time, as a function of fecundity and of the physical limitations of the surroundings. The former would lead to an exponential growth of the population, the latter limits the growth, the more strongly the bigger the population. If A_n is the population in the year n, the model calculates the population the following year by an equation of the form $A_{n+1} = rA_n(1 - A_n)$ where r is the growth rate, and $(1 - A_n)$ takes account of the limitations imposed by the environmental conditions. See e.g. `hypertextbook.com/chaos/42.shtml`.

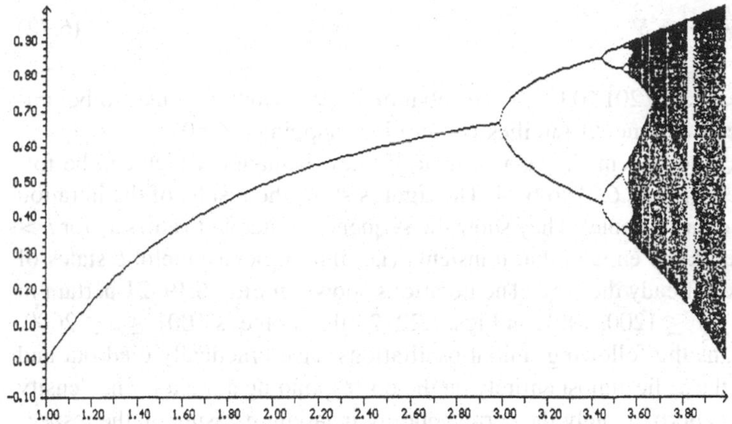

Fig. 6.19. Numerical results for large number of iterations of the logistic equation (6.67). The first bifurcation occurs at $(\mu_0 = 3, \bar{x}_0 = 2/3)$, the second at $\mu_1 = 1 + \sqrt{6}$, etc. The range of μ shown is $1 < \mu \leq 4$

and find its fixed points. Indeed, if one sketches the function $g = f \circ f$, one realizes immediately that it possesses two stable equilibria. This is seen also in Fig. 6.19, in the interval $3 \leq \mu < 1 + \sqrt{6} \simeq 3.449$. Returning to the function f, this tells us that the iteration (6.67) alternates between the two fixed points of $g = f \circ f$. If we interpret the observed pattern as described in Sect. 6.3.6 above, we realize that the bifurcation (6.68) is of the "pitchfork" type shown in Fig. 6.14.

The situation remains stable until we reach the value $\mu_1 = 1 + \sqrt{6}$ of the control parameter. At this value two new bifurcation points appear:

$$\left(\mu_1 = 1 + \sqrt{6}, \bar{x}_{1/2} = \tfrac{1}{10}(4 + \sqrt{6} \pm (2\sqrt{3} - \sqrt{2}))\right) . \tag{6.70}$$

At these points the fixed points of $g = f \circ f$ become marginally stable, while for $\mu > \mu_1$ they become unstable. Once more the period is doubled and one enters the domain where the function

$$h \stackrel{\text{def}}{=} g \circ g = f \circ f \circ f \circ f$$

possesses four stable fixed points. Returning to the original function f, this means that the iteration visits these four points alternately, in a well-defined sequence.

This process of period doublings $2T, 4T, 8T, \ldots$ and of pitchfork bifurcations continues like a cascade until μ reaches the limit point

$$\mu_\infty = 3.56994\ldots . \tag{6.71}$$

This limit point was discovered numerically (Feigenbaum 1979). The same is true for the pattern of successive bifurcation values of the control parameter, for which the following regularity was found empirically. The sequence

$$\lim_{i\to\infty} \frac{\mu_i - \mu_{i-1}}{\mu_{i+1} - \mu_i} = \delta \qquad (6.72)$$

has the limit $\delta = 4.669\,201\,609\ldots$ (Feigenbaum 1979), which is found to be universal for sufficiently smooth families of iterative mappings (6.60).

For $\mu > \mu_\infty$ the system shows a structurally new behavior, which can be followed rather well in Figs. 6.21 to 6.24. The figures show the results of the iteration (6.67) obtained on a computer. They show the sequence of iterated values x_i for $i > i_n$ with i_n chosen large enough that transients (i.e. initial, nonasymptotic states of oscillations) have already died out. The iterations shown in Figs. 6.19–21 pertain to the range $1001 \le i \le 1200$, while in Figs. 6.22–24 that range is $1001 \le i \le 2000$. This choice means the following: initial oscillations have practically died out and the sequence of the x_i lie almost entirely on the corresponding attractor. The density of points reflects approximately the corresponding invariant measure on the respective attractor. Figure 6.22 is a magnified section of Fig. 6.21 (the reader should mark in Fig. 6.21 the window shown in Fig. 6.22). Similarly, Fig. 6.23 is a magnified

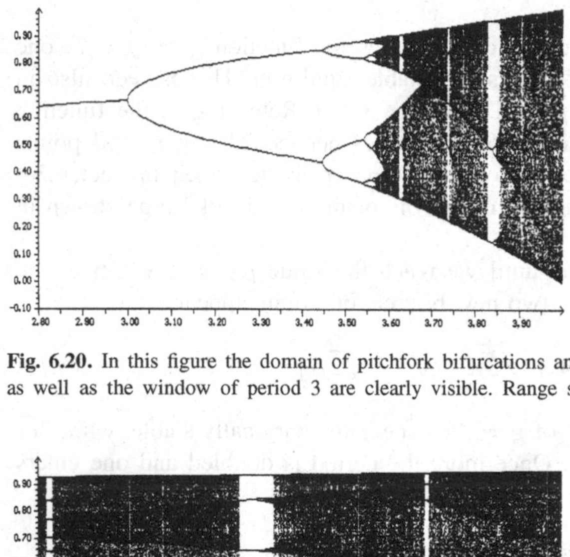

Fig. 6.20. In this figure the domain of pitchfork bifurcations and period doubling up to about $16\,T$ as well as the window of period 3 are clearly visible. Range shown: $2.8 \le \mu \le 4$

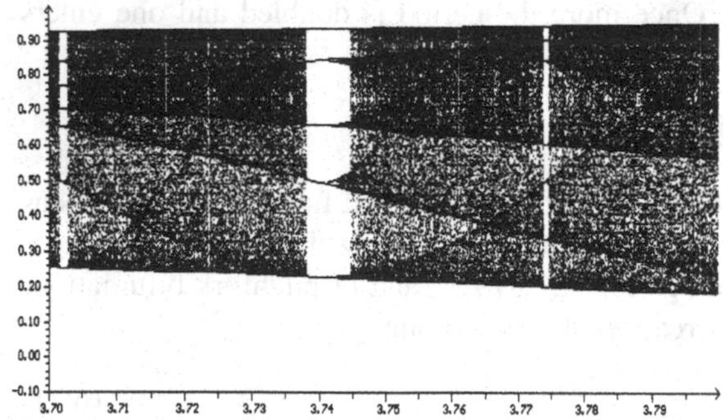

Fig. 6.21. Range shown is $3.7 \le \mu \le 3.8$ in a somewhat expanded representation. The window with period 5 is well visible

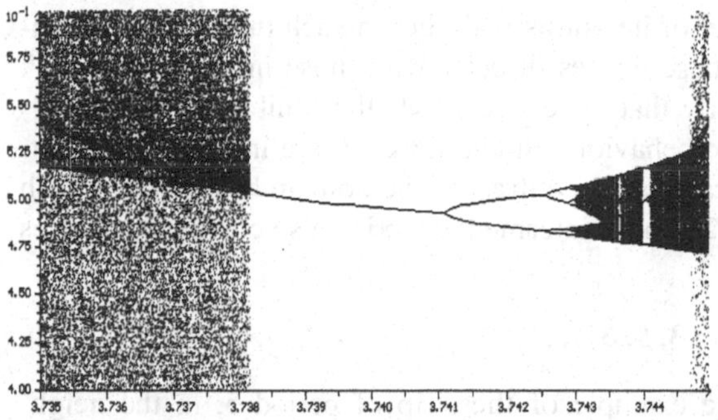

Fig. 6.22. A window of Fig. 6.21, with $0.4 \leq x \leq 0.6$ and $3.735 \leq \mu \leq 3.745$

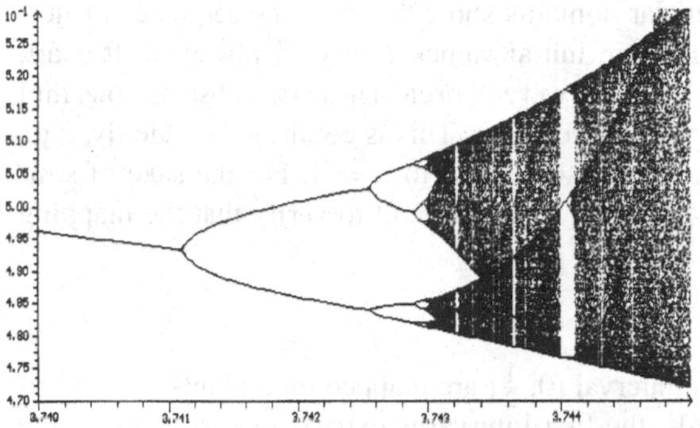

Fig. 6.23. The domain of bifurcations of Fig. 6.22 (itself cut out of Fig. 6.21) is shown in an enlarged representation with $0.47 \leq x \leq 0.53$ and $3.740 \leq \mu \leq 3.745$

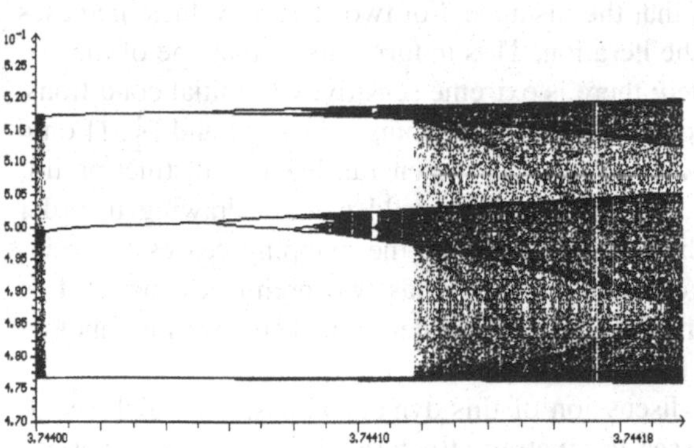

Fig. 6.24. Here one sees a magnification of the periodic window in the right-hand half of Fig. 6.23. The window shown corresponds to $0.47 \leq x \leq 0.53$ and $3.7440 \leq \mu \leq 3.7442$

section of Fig. 6.22. The number of iterations was chosen such that one may compare the average densities on these figures directly with those in Figs. 6.19–21[10].

The figures show very clearly that once μ exceeds the limit value μ_∞ (6.71) there appear domains of chaotic behavior, which, however, are interrupted repeatedly by strips with periodic attractors. In contrast to the domain below μ_∞, which shows only periods of the type 2^n, these intermediate strips also contain sequences of periods

$$p2^n, p3^n, p5^n \quad \text{with} \quad p = 3, 5, 6, \ldots .$$

Figures 6.22 and 6.23 show the example of the strip of period 5, in the neighborhood of $\mu = 3.74$. A comparison of Figs. 6.23 and 6.20 reveals a particularly startling phenomenon: the pattern of the picture in the large is repeated in a sectional window in the small.

A closer analysis of the irregular domains show that here the sequence of iterations $\{x_i\}$ never repeats. In particular, initial values x_1 and x_1' always drift apart, for large times, no matter how close they were chosen. These two observations hint clearly at the chaotic structure of these domains. This is confirmed explicitly, e.g., by the study of the iteration mapping (6.67) close to $\mu = 4$. For the sake of simplicity we only sketch the case $\mu = 4$. It is not difficult to verify that the mapping

$$f(\mu = 4, x) = 4x(1 - x)$$

has the following properties.

(i) The points $x_1 < x_2$ of the interval $[0, \frac{1}{2}]$ are mapped onto points $x_1' < x_2'$ of the interval $[0, 1]$. In other words, the first interval is expanded by a factor 2, the relative ordering of the preimages remains unchanged. Points $x_3 < x_4$ taken from $[\frac{1}{2}, 1]$ are mapped onto points x_3' and x_4' of the expanded interval $[0, 1]$. However, the ordering is reversed. Indeed, with $x_3 < x_4$ one finds $x_3' > x_4'$. The observed dilatation of the images tells us that the distance δ of two starting values increases exponentially, in the course of the iteration. This in turn tells us that one of the criteria for chaos to occur is fulfilled: there is extreme sensitivity to initial conditions.

(ii) The change of orientation between the mappings of $[0, \frac{1}{2}]$ and $[\frac{1}{2}, 1]$ onto the interval $[0, 1]$, tells us that an image x_{i+1}, in general, has two distinct preimages, $x_i \in [0, \frac{1}{2}]$ and $x_i' \in [\frac{1}{2}, 1]$. (The reader should make a drawing in order to convince him or herself.) Thus, if this happens, the mapping ceases to be invertible. x_{i+1} has two preimages, each of which has two preimages too, and so on. It is not possible to reconstruct the past of the iteration. Thus, we find another criterion for chaotic pattern to occur.

One can pursue further the discussion of this dynamical system, which seems so simple and yet which possesses fascinating structures. For instance, a classification of the periodic attractors is of interest that consists in studying the sequence in which the stable points are visited, in the course of the iteration. Fourier analysis and, specifically, the behavior of the correlation functions (6.65) and (6.66) in

[10] I thank Peter Beckmann for providing these impressive figures and for his advice regarding the presentation of this system.

the chaotic zones are particularly instructive. The (few) rigorous results as well as several conjectures for iterative mappings on the unit interval are found in the book by Collet and Eckmann (1990). For a qualitative and well-illustrated presentation consult Bergé, Pomeau, and Vidal (1987).

6.5 Quantitative Measures of Deterministic Chaos

6.5.1 Routes to Chaos

The transition from a regular pattern of the solution manifold of a dynamical system to regimes of chaotic motion, as a function of control parameters, can happen in various ways. One distinguishes the following routes to chaos.

(i) *Frequency doubling*. The phenomenon of frequency doubling is characteristic for the interval $1 < \mu \le \mu_\infty = 3.56994$ of the logistic equation (6.67). Above the limit value μ_∞ the iterations (6.67) change in a qualitative manner. A more detailed analysis shows that periodic attractors alternate with domains of genuine chaos, the chaotic regimes being characterized by the observation that the iteration $x_n \mapsto x_{n+1} = f(\mu, x_n)$ yields an infinite sequence of points that never repeats and that depends on the starting value x_1. This means, in particular, that sequences starting at neighboring points x_1 and x_1' eventually move away from each other. Our qualitative analysis of the logistic mapping with μ close to 4 in Sect. 6.4.3 (i) and (ii) showed how this happens. The iteration stretches the intervals $[0, \frac{1}{2}]$ and $[\frac{1}{2}, 1]$ to larger subintervals of $[0, 1]$ (for $\mu = 4$ this is the full interval). It also changes orientation by folding back the values that would otherwise fall outside the unit interval. As we saw earlier, this combination of *stretching* and *back-folding* has the consequence that the mapping becomes irreversible and that neighboring starting points, on average, move apart exponentially. Let x_1 and x_1' be two neighboring starting values for the mapping (6.67). If one follows their evolution on a calculator, one finds that after n iterations their distance is given approximately by

$$|x_n' - x_n| \simeq e^{\lambda n} |x_1' - x_1| . \tag{6.73}$$

The factor λ in the argument of the exponential is called the *Liapunov characteristic exponent*. Negative λ is characteristic for a domain with a periodic attractor: the points approach each other independently of their starting values. If λ is positive, on the other hand, neighboring points move apart exponentially. There is extreme sensitivity to initial conditions and one finds a chaotic pattern. Indeed, a numerical study of (6.67) gives the results (Bergé, Pomeau and Vidal 1987)

$$\begin{aligned} &\text{for } \mu = 2.8 \,, \quad \lambda \simeq -0.2 \,, \\ &\text{for } \mu = 3.8 \,, \quad \lambda \simeq +0.4 \,. \end{aligned} \tag{6.74}$$

(ii) *Intermittency*. In Sect. 6.3.6 we studied the Poincaré mapping at the transition from stability to instability for the case where one of the eigenvalues of

$D\Pi(s_0)$ crosses the unit circle at -1. There are other possibilities for the transition from stability to instability. (a) As a function of the control parameter, an eigenvalue can leave the interior of the unit circle at $+1$. (b) Two complex conjugate eigenvalues $c(\mu)e^{\pm i\phi(\mu)}$ leave the unit circle along the directions ϕ and $-\phi$. All three situations play their role in the transition to chaos. In case (a) one talks about *intermittency of type I*, in case (b) about *intermittency of type II*, while the first case above is also called *type-III intermittency*. We wish to discuss type I in a little more detail.

Figures 6.19 and 6.20 show clearly that at the value $\mu = \mu_c = 1+\sqrt{8} \simeq 3.83$ a new cycle with period 3 is born. Therefore, let us consider the triple iteration $h(\mu, x) = f \circ f \circ f$. Figure 6.25 shows that the graph of $h(\mu = \mu_c, x)$ is tangent to the straight line $y = x$ in three points $\bar{x}^{(1)}$, $\bar{x}^{(2)}$, $\bar{x}^{(3)}$. Thus, at these points, we have

$$h(\mu_c, \bar{x}^{(i)}) = \bar{x}^{(i)} , \qquad \frac{\mathrm{d}}{\mathrm{d}x} h(\mu_c, \bar{x}^{(i)}) = 1 .$$

In a small interval around μ_c and in a neighborhood of any one of the three fixed points, h must have the form

$$h(\mu, x) \simeq \bar{x}^{(i)} + (x - \bar{x}^{(i)}) + \alpha(x - \bar{x}^{(i)})^2 + \beta(\mu - \mu_c) .$$

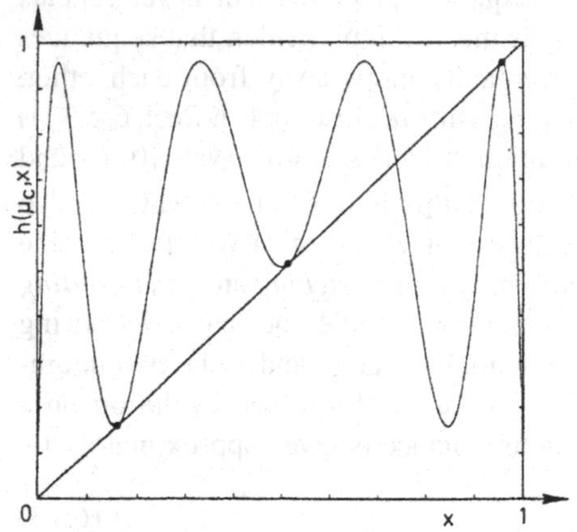

Fig. 6.25. Graph of the threefold iterated mapping (6.67) for $\mu = \mu_c = 1 + \sqrt{8}$. The function $h = f \circ f \circ f$ is tangent to the straight line in three points

We study the iterative mapping $x_{n+1} = h(\mu, x_n)$ in this approximate form, i.e. we register only every third iterate of the original mapping (6.67). We take $z = \alpha(x - \bar{x}^{(i)})$ and obtain

$$z_{n+1} = z_n + z_n^2 + \eta \qquad (6.75)$$

with $\eta = \alpha\beta(\mu - \mu_c)$. The expression (6.75) holds in the neighborhood of any of the three fixed points of $h(\mu, x)$. For negative η, (6.75) has two fixed points, at $z_- = -\sqrt{-\eta}$ and at $z_+ = \sqrt{-\eta}$, the first of which is stable, while the second is unstable. For $\eta = 0$ the two fixed points coincide and become marginally stable. For small, but positive η, a new phenomenon is observed as illustrated in Fig. 6.26. Iterations with a negative starting value of the variable z, move for a long time within

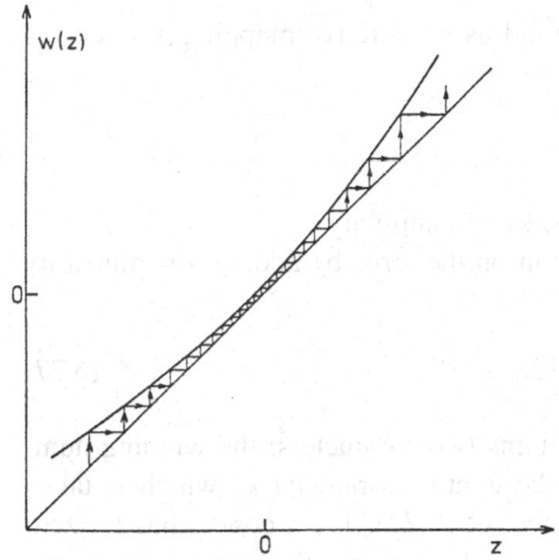

Fig. 6.26. The iterative mapping (6.75) with small positive η spends a long time in the narrow channel between the curve $w = z + z^2 + \eta$ and the straight line $w = z$

the narrow channel between the graph of the function $z + z^2 + \eta$ and the straight line $w = z$. As long as z is small the behavior is oscillatory and has nearly the same regularity as with negative values of η. This phase of the motion is said to be the *laminar phase*. When $|z|$ increases, the iteration quickly moves on to a chaotic or *turbulent phase*. However, the motion can always return to the first domain, i.e. to the narrow channel of almost regular behavior. Practical models such as the one by Lorenz (see e.g. Bergé, Pomeau, Vidal 1987) that contain this transition to chaos indeed show regular oscillatory behavior interrupted by bursts of irregular and chaotic behavior.

For small $|z|$ the iteration remains in the channel around $z = 0$ for some finite time. In this case successive iterates lie close to each other so that we can replace (6.75) by a differential equation. Replacing $z_{n+1} - z_n$ by dz/dn, we obtain

$$\frac{dz}{dn} = \eta + z^2 . \tag{6.76}$$

[Note that this is our equation (6.54) with a destabilizing nonlinearity, which describes then a subcritical saddle-node bifurcation.] Equation (6.76) is integrated at once,

$$z(n) = \sqrt{\eta}\,\mathrm{tg}\,(\sqrt{\eta}(n - n_0)) .$$

n_0 is the starting value of the iteration and may be taken to zero, without restriction. This explicit solution tells us that the number of iterations needed for leaving the channel is of the order of $n \sim \pi/2\sqrt{\eta}$. Hence, $1/\sqrt{\eta}$ is a measure of the time that the system spends in the laminar regime. Finally, one can show that the Liapunov exponent is approximately $\lambda \sim \sqrt{\eta}$, for small values of η.

(iii) *Quasiperiodic motion with nonlinear perturbation.* A third route to chaos may be illustrated by Example (ii) of Sect. 6.3.4. We consider quasiperiodic motion on the torus T^2, choosing the Poincaré section as described in (6.49′), i.e. we register the points of intersection of the orbit with the transverse section of the torus at $\theta_1 = 0 \,(\mathrm{mod}\, 2\pi)$. When understood as an iterative mapping, the second equation (6.49′) reads

$$\theta_{n+1} = \left(\theta_n + 2\pi \frac{\omega_2}{\omega_1}\right) \bmod 2\pi \;,$$

where we write θ instead of θ_2, for the sake of simplicity.

Let us perturb this quasiperiodic motion on the torus by adding a nonlinearity as follows:

$$\theta_{n+1} = \left(\theta_n + 2\pi \frac{\omega_2}{\omega_1} + \kappa \sin\theta_n\right) \bmod 2\pi \,. \tag{6.77}$$

This model, which is due to Arnol'd, contains two parameters: the winding number $\beta = \omega_2/\omega_1$, with $0 \le \beta < 1$, and the control parameter κ, which is taken to be positive. For $0 \le \kappa < 1$ the derivative of (6.77), $1 + \kappa \cos\theta_n$, has no zero and hence the mapping is invertible. For $\kappa > 1$, however, this is no longer true. Therefore, $\kappa = 1$ is a critical point where the behavior of the flow on T^2 changes in a qualitative manner. Indeed, one finds that the mapping (6.77) exhibits chaotic behavior for $\kappa > 1$. This means that in crossing the critical value $\kappa = 1$ from regular to irregular motion, the torus is destroyed. As a shorthand let us write (6.77) as follows: $\theta_{n+1} = f(\beta, \kappa, \theta_n)$. The *winding number* is defined by the limit

$$w(\beta, \kappa) \stackrel{\text{def}}{=} \lim_{n\to\infty} \frac{1}{2\pi n}\left[f^n(\beta, \kappa, \theta) - \theta\right] . \tag{6.78}$$

Obviously, for $\kappa = 0$ it is given by $w(\beta, 0) = \beta = \omega_2/\omega_1$. The chaotic regime above $\kappa = 1$ may be studied as follows. For a given value of κ we choose $\beta \equiv \beta_n(\kappa)$ such that the starting value $\theta_0 = 0$ is mapped to $2\pi p_n$, after q_n steps, q_n and p_n being integers,

$$f^{q_n}(\beta, \kappa, 0) = 2\pi p_n \;.$$

The winding number is then a rational number. $w(\beta, \kappa) = p_n/q_n \equiv r_n$, $r_n \in \mathbb{Q}$. This sequence of rationals may be chosen such that r_n tends to a given irrational number $\bar{r}$, in the limit $n \to \infty$. An example of a very irrational number is the *Golden Mean*. Let $r_n = F_n/F_{n+1}$, where the F_n are the *Fibonacci numbers*, defined by the recurrence relation $F_{n+1} = F_n + F_{n-1}$ and the initial values $F_0 = 0$, $F_1 = 1$. Consider

$$r_n = \frac{F_n}{F_{n+1}} = \frac{1}{1 + r_{n-1}}$$

in the limit $n \to \infty$. Hence $\bar{r} = 1/(1 + \bar{r})$. The positive solution of this equation is the Golden Mean $\bar{r} = (\sqrt{5} - 1)/2$[11].

[11] The Golden Mean is a well-known concept in the fine arts, in the theory of proportions. For example, a column of height H is divided into two segments of heights h_1 and h_2, with $H = h_1 + h_2$ such that the proportion of the shorter segment to the longer is the same as that of the longer to the column as a whole, i.e. $h_1/h_2 = h_2/H = h_2/(h_1 + h_2)$. The ratio $h_1/h_2 = \bar{r} = (\sqrt{5} - 1)/2$ is the Golden Mean. This very irrational number has a remarkable continued fraction representation: $r = 1/(1 + 1/(1 + \ldots$

With this choice, the winding numbers defined above, $w(\beta_n(\kappa), \kappa) = r_n$, will converge to $w = \bar{r}$. A numerical study of this system along the lines described above reveals remarkable regularities and scaling properties that are reminiscent of the logistic mapping (6.67) (see e.g. Guckenheimer and Holmes (1990), Sect. 6.8.3 and references quoted there).

6.5.2 Liapunov Characteristic Exponents

Chaotic behavior is observed whenever neighboring trajectories, on average, diverge exponentially on attractors. Clearly, one wishes to have a criterion at hand that allows one to measure the speed of this divergence. Thus, we consider a solution $\underset{\sim}{\Phi}(t, \underset{\sim}{y})$ of the equation of motion (6.1), change its initial condition by the amount $\delta \underset{\sim}{y}$, and test whether, and if yes at what rate, the solutions $\underset{\sim}{\Phi}(t, \underset{\sim}{y})$ and $\underset{\sim}{\Phi}(t, \underset{\sim}{y}+\delta \underset{\sim}{y})$ move apart. In linear approximation their difference obeys (6.11), i.e.

$$\begin{aligned} \delta \dot{\underset{\sim}{\Phi}} &\equiv \dot{\underset{\sim}{\Phi}}(t, \underset{\sim}{y}+\delta \underset{\sim}{y}) - \dot{\underset{\sim}{\Phi}}(t, \underset{\sim}{y}) = \Lambda(t)[\underset{\sim}{\Phi}(t, \underset{\sim}{y}+\delta \underset{\sim}{y}) - \underset{\sim}{\Phi}(t, \underset{\sim}{y})] \\ &\equiv \Lambda(t)\delta \underset{\sim}{\Phi} , \end{aligned} \tag{6.79}$$

the matrix $\Lambda(t)$ being given by

$$\Lambda(t) = \left(\frac{\partial F_i}{\partial \phi_k}\right)_{\underset{\sim}{\Phi}} .$$

Unfortunately, (6.79), in general, cannot be integrated analytically and one must resort to numerical algorithms, which allow the determination of the distance of neighboring trajectories as a function of time. Nevertheless, imagine we had solved (6.79). At $t = 0$ we have $\delta \underset{\sim}{\Phi} = \underset{\sim}{\Phi}(0, \underset{\sim}{y}+\delta \underset{\sim}{y}) - \underset{\sim}{\Phi}(0, \underset{\sim}{y}) = \delta \underset{\sim}{y}$. For $t > 0$ let

$$\delta \underset{\sim}{\Phi}(t) = \mathbf{U}(t) \cdot \delta \underset{\sim}{y} \tag{6.80}$$

be the solution of the differential equation (6.79). From (6.79) one sees that the matrix $\mathbf{U}(t)$ itself obeys the differential equation

$$\dot{\mathbf{U}}(t) = \Lambda(t)\mathbf{U}(t)$$

and therefore may be written formally as follows:

$$\mathbf{U}(t) = \exp\left\{\int_0^t dt' \Lambda(t')\right\}\mathbf{U}(0) , \quad \text{with} \quad \mathbf{U}(0) = 1\!\!1 . \tag{6.81}$$

Although this is generally not true, imagine the matrix to be independent of time. Let $\{\lambda_k\}$ denote its eigenvalues (which may be complex numbers) and use the basis system of the corresponding eigenvectors. Then $\mathbf{U}(t) = \{\exp(\lambda_k t)\}$ is also diagonal. Whether or not neighboring trajectories diverge exponentially depends on whether or not the real part of one of the eigenvalues $\mathrm{Re}\,\lambda_k = \frac{1}{2}(\lambda_k + \lambda_k^*)$ is positive. This can be tested by taking the logarithm of the trace of the product $\mathbf{U}^\dagger\mathbf{U}$,

$$\frac{1}{2t} \ln \mathrm{Tr}\,(\mathbf{U}^{\dagger}(t)\mathbf{U}(t)) = \frac{1}{2t} \ln \mathrm{Tr}\,(\exp\{(\lambda_k + \lambda_k^*)t\}) ,$$

and by letting t go to infinity. In this limit only the eigenvalue with the largest positive real part survives. With this argument in mind one defines

$$\mu_1 \stackrel{\text{def}}{=} \lim_{t\to\infty} \frac{1}{2t} \ln \mathrm{Tr}\,(\mathbf{U}^{\dagger}(t)\mathbf{U}(t)) . \tag{6.82}$$

The real number μ_1 is called the leading *Liapunov characteristic exponent*. It provides a quantitative criterion for the nature of the flow: whenever the leading Liapunov exponent is positive, there is (at least) one direction along which neighboring trajectories move apart, on average, at the rate $\exp(\mu_1 t)$. There is extreme sensitivity to initial conditions: the system exhibits chaotic behavior.

The definition (6.82) applies also to the general case where $\Lambda(t)$ depends on time. Although the eigenvalues and eigenvectors of $\Lambda(t)$ now depend on time, (6.82) has a well-defined meaning. Note, however, that the leading exponent depends on the reference solution $\Phi(t, \underset{\sim}{y})$.

The definition (6.82) yields only the leading Liapunov exponent. If one wishes to determine the next to leading exponent $\mu_2 \le \mu_1$, one must take out the direction pertaining to μ_1 and repeat the same analysis as above. Continuing this procedure yields all Liapunov characteristic exponents, ordered according to magnitude,

$$\mu_1 \ge \mu_2 \ge \ldots \ge \mu_f . \tag{6.83}$$

The dynamical system exhibits chaotic behavior if and only if the leading Liapunov exponent is positive.

For discrete systems in f dimensions, $\underset{\sim}{x}_{n+1} = \underset{\sim}{F}(\underset{\sim}{x}_n)$, $\underset{\sim}{x} \in \mathbb{R}^f$, the Liapunov exponents are obtained in an analogous fashion. Let $v^{(1)}$ be those vectors in the tangent space at an arbitrary point $\underset{\sim}{x}$ that grow at the fastest rate under the action of the linearization of the mapping $\underset{\sim}{F}$, i.e. those for which $|(\mathbf{D}\underset{\sim}{F}(\underset{\sim}{x}))^n v^{(1)}|$ is largest. Then

$$\mu_1 = \lim_{n\to\infty} \frac{1}{n} \ln |(\mathbf{D}\underset{\sim}{F}(\underset{\sim}{x}))^n v^{(1)}| . \tag{6.84}$$

In the next step, one determines the vectors $v^{(2)}$ that grow at the second fastest rate, leaving out the subspace of the vectors $v^{(1)}$. The same limit as in (6.84) yields the second exponent μ_2 etc. We consider two simple examples that illustrate this procedure.

(i) Take $\underset{\sim}{F}$ to be two-dimensional and let $\underset{\sim}{x}^0$ be a fixed point, $\underset{\sim}{x}^0 = \underset{\sim}{F}(\underset{\sim}{x}^0)$. $\mathbf{D}F(\underset{\sim}{x}^0)$ is diagonalizable,

$$\mathbf{D}F(\underset{\sim}{x}^0) = \begin{pmatrix} \lambda_1 & 0 \\ 0 & \lambda_2 \end{pmatrix} ,$$

with, say, $\lambda_1 > \lambda_2$. Choose $v^{(1)}$ from $\{\mathbb{R}^2 \backslash 2\text{-axis}\}$, i.e. in such a way that its 1-component does not vanish, and choose $v^{(2)}$ along the 2-axis,

$$v^{(1)} = \begin{pmatrix} a^{(1)} \\ b^{(1)} \end{pmatrix}, \quad a^{(1)} \neq 0 \; ; \quad v^{(2)} = \begin{pmatrix} a^{(2)} \\ b^{(2)} \end{pmatrix}, \quad a^{(2)} = 0 \, .$$

Then one obtains

$$\mu_i = \lim_{n\to\infty} \frac{1}{n} \ln \left| \begin{pmatrix} (\lambda_1)^n & 0 \\ 0 & (\lambda_2)^n \end{pmatrix} \begin{pmatrix} a^{(i)} \\ b^{(i)} \end{pmatrix} \right|$$

$$= \lim_{n\to\infty} \frac{1}{n} \ln |\lambda_i|^n = \ln |\lambda_i|$$

and $\mu_1 > \mu_2$.

(ii) Consider a mapping in two dimensions, $\underset{\sim}{x} = \binom{u}{v}$, that is defined on the unit square $0 \leq u \leq 1$, $0 \leq v \leq 1$, by the equations

$$u_{n+1} = 2u_n (\text{mod } 1) \, , \tag{6.85a}$$

$$v_{n+1} = \begin{cases} a v_n & \text{for } 0 \leq u_n < \frac{1}{2} \\ a v_n + \frac{1}{2} & \text{for } \frac{1}{2} \leq u_n \leq 1 \end{cases} \tag{6.85b}$$

with $a < 1$. Thus, in the direction of v this mapping is a contraction for $u < \frac{1}{2}$ and a contraction and a shift for $u \geq \frac{1}{2}$. In the direction of u its effect is stretching and back-bending whenever the unit interval is exceeded. (It is called the *baker's transformation* because of the obvious analogy to kneading, stretching, and back-folding of dough.) This dissipative system is strongly chaotic. This will become clear empirically if the reader works out the example $a = 0.4$ on a PC, by following the fate of the points on the circle with origin $(\frac{1}{2}, \frac{1}{2})$ and radius a, under the action of successive iterations. The original volume enclosed by the circle is contracted. At the same time horizontal distances (i.e. parallel to the 1-axis) are stretched exponentially because of (6.85a). The system possesses a strange attractor, which is stretched and folded back onto itself and which consists of an infinity of horizontal lines. Its basin of attraction is the whole unit square. Calculation of the Liapunov characteristic exponents by means of the formula (6.84) gives the result

$$\mu_1 = \ln 2 \, , \quad \mu_2 = \ln |a| \, ,$$

and thus $\mu_2 < 0 < \mu_1$.

6.5.3 Strange Attractors

Example (ii) of the preceding section shows that the system (6.85a) lands on a strangely diffuse object which is neither an arc of a curve nor a piece of a surface in the unit square, but somehow "something in between". This strange attractor does indeed have zero volume, but its geometric dimension is not an integer. Geometric structures of this kind are said to be *fractals*. Although a rigorous discussion of this concept and a detailed analysis of fractal-like strange attractors is beyond the scope of this book, we wish at least to give an idea of what such objects are.

Imagine a geometric object of dimension d embedded in a space $\mathbb{R}^n$, where d need not necessarily be an integer. Scaling all its *linear* dimensions by a factor λ, the object's volume will change by a factor $\kappa = \lambda^d$, i.e.

$$d = \frac{\ln \kappa}{\ln \lambda} .$$

Clearly, for points, arcs of curves, surfaces, and volumes in $\mathbb{R}^3$ one finds in this way the familiar dimensions $d = 0$, $d = 1$, $d = 2$ and $d = 3$, respectively. A somewhat more precise formulation is the following. A set of points in $\mathbb{R}^n$, which is assumed to lie in a finite volume, is covered by means of a set of elementary cells B whose diameter is ε. These cells may be taken to be little cubes of side length ε, or little balls of diameter ε, or the like. If $N(\varepsilon)$ is the minimal number of cells needed to cover the set of points completely, the so-called *Hausdorff dimension* of the set is defined to be

$$d_\mathrm{H} = \lim_{\varepsilon \to 0} \frac{\ln(N(\varepsilon)}{\ln(1/\varepsilon)} , \tag{6.86}$$

provided this limit exists. To cover a single point, one cell is enough, $N(\varepsilon) = 1$; to cover an arc of length L one needs at least $N(\varepsilon) = L/\varepsilon$ cells; more generally, to cover a p-dimensional smooth hypersurface F, $N(\varepsilon) = F/\varepsilon^p$ cells will be enough. In these cases, the definition (6.86) yields the familiar Euclidean dimensions $d_\mathrm{H} = 0$ for a point, $d_\mathrm{H} = 1$ for an arc, and $d_\mathrm{H} = p$ for the hypersurface F with $p \le n$.

For fractals, on the other hand, the Hausdorff dimension is found to be noninteger. A simple example is provided by the *Cantor set of the middle third*, which is defined as follows. From a line segment of length 1 one cuts out the middle third. From the remaining two segments $[0, \frac{1}{3}]$ and $[\frac{2}{3}, 1]$ one again cuts out the middle third, etc. By continuing this process an infinite number of times one obtains the middle third Cantor set. Taking $\varepsilon_0 = \frac{1}{3}$, the minimum number of intervals of length ε_0 needed to cover the set is $N(\varepsilon_0) = 2$. If we take $\varepsilon_1 = \frac{1}{9}$ instead, we need at least $N(\varepsilon_1) = 4$ intervals of length ε_1, etc. For $\varepsilon_n = 1/3^n$ the minimal number is $N(\varepsilon_n) = 2^n$. Therefore,

$$d_\mathrm{H} = \lim_{n \to \infty} \frac{\ln 2^n}{\ln 3^n} = \frac{\ln 2}{\ln 3} \simeq 0.631 .$$

Another simple and yet interesting example is provided by the so-called *snowflake set*, which is obtained by the following prescription. One starts from an equilateral triangle in the plane. To the middle third of each of its sides one adds another equilateral triangle, of one third the dimension of the original one, and keeps only the outer boundary. One repeats this procedure infinitely many times. The object generated in this way has infinite circumference. Indeed, take the side length of the initial triangle to be 1. At the nth step of the construction described above the side length of the last added triangles is $\varepsilon_n = 1/3^n$. Adding a triangle to the side of length ε_{n-1} breaks its up in four segments of length ε_n each. Therefore, the circumference is $C_n = 3 \times 4^n \times \varepsilon_n = 4^n/3^{n-1}$. Clearly, in the limit $n \to \infty$ this diverges, even though the whole object is contained in a finite portion of the plane. On the other hand, if one calculates the Hausdorff dimensions in the same way as for the middle third Cantor set one finds $d_{\mathrm{H}} = \ln 4/\ln 3 \simeq 1.262$.

There are further questions regarding chaotic regimes of dynamical systems, such as: If strange attractors have the structure of fractals, can one measure their generalized dimension? Is it possible to describe deterministically chaotic motion on the attractor quantitatively by means of a test quantity (some kind of entropy), which would tell us whether the chaos is rich or poor? These questions lead us beyond the range of the tools developed in this book. In fact, they are the subject of present-day research and no final answers have been given so far. We refer to the literature quoted in the Appendix for an account of the present state of knowledge.

6.6 Chaotic Motions in Celestial Mechanics

We conclude this chapter on deterministic chaos with a brief account of some fascinating results of recent research on celestial mechanics. These results illustrate in an impressive way the role of deterministically chaotic motion in our planetary system. According to the traditional view, the planets of the solar system move along their orbits with the regularity of a clockwork. To a very good approximation, the motion of the planets is strictly periodic, i.e. after one turn each planet returns to the same position, the planetary orbits are practically fixed in space relative to the fixed stars. From our terrestrial point of view no motion seems more stable, more uniform over very long time periods than the motion of the stars in the sky. It is precisely the regularity of planetary motion that, after a long historical development, led to the discovery of Kepler's laws and, eventually, to Newton's mechanics.

On the other hand, our solar system with its planets, their satellites, and the very many smaller objects orbiting around the sun is a highly complex dynamical system whose stability has not been established in a conclusive manner. Therefore, it is perhaps not surprising that there are domains of deterministically chaotic motion even in the solar system with observable consequences. It seems, for instance, that chaotic motion is the main reason for the formation of the *Kirkwood gaps* (these are gaps in the asteroid belts between Mars and Jupiter which appear at some rational ratios of the periods of revolution of the asteroid and Jupiter) and that chaotic motion also provides an important source for the transport of meteorites to the earth (Wisdom 1987).

In this section we describe an example of chaotic tumbling of planetary satellites which is simple enough that the reader may reproduce some of the figures on a PC. We then describe some recent results regarding the topics mentioned above.

6.6.1 Rotational Dynamics of Planetary Satellites

The moon shows us always the same face. This means that the period of its spin (its intrinsic angular momentum) is equal to the period of its orbital motion and that its axis of rotation is perpendicular to the plane of the orbit. In fact, this is its final stage, which was reached after a long-term evolution comprising two phases: a *dissipative phase*, or slowing-down phase, and the final, *Hamiltonian phase* that we observe today. Indeed, although we ignore the details of the moon's formation, it probably had a much faster initial rate of rotation and its axis of rotation was not perpendicular to the plane of the orbit. Through the action of friction by tidal forces, the rotation was slowed down, over a time period of the order of the age of the planetary system, until the period of rotation became equal to the orbital period. At the same time the axis of rotation turned upright such as to point along the normal to the orbit plane. These results can be understood on the basis of simple arguments regarding the action of tidal forces on a deformable body and simple energy considerations. In the synchronous phase of rotation (i.e. spin period equal to orbital period) the effect of tidal forces is minimal. Furthermore, for a given frequency of rotation, the energy is smallest if the rotation takes place about the principal axis with the largest moment of inertia. Once the satellite has reached this stage, the motion is Hamiltonian, to a very good approximation.

Thus, any satellite close enough to its mother planet that tidal forces can modify its motion in the way described above and within a time period comparable to the age of the solar system will enter this synchronous phase, which is stable in the case of our moon. This is not true, as we shall see below, if the satellite has a strongly asymmetric shape and if it moves on an ellipse of high eccentricity.

The *Voyager 1* and 2 space missions took pictures of Hyperion, one of the farthest satellites of Saturn, on passing close to Saturn in November 1980 and August 1981. Hyperion is an asymmetric top whose linear dimensions were determined to be

$$190\,\text{km} \times 145\,\text{km} \times 114\,\text{km}$$

with an uncertainty of about $\pm 15\,\text{km}$. The eccentricity of its elliptical orbit is $\varepsilon = 0.1$; its orbital period is 21 days. The surprising prediction is that Hyperion performs a chaotic tumbling motion in the sense that its angular velocity and the orientation of its axis of rotation are subject to strong and erratic changes within a few periods of revolution. This chaotic dance, which, at some stage, must have also occurred in the history of other satellites (such as Phobos and Deimos, the companions of Mars), is a consequence of the asymmetry of Hyperion and of the eccentricity of its orbit. This is what we wish to show within the framework of a simplified model.

The model is shown in Fig. 6.27. Hyperion H moves around Saturn S on an ellipse with semimajor axis a and eccentricity ε. We simulate its asymmetric shape by means of four mass points 1 to 4, that have the same mass m and are arranged in the orbital plane as shown in the figure. The line 2–1 (the distance between 2 and 1 is d) is taken to be the 1-axis, the line 4–3 (distance $e < d$) is taken to be the 2-axis. The moments of inertia are then given by

$$I_1 = \tfrac{1}{2}me^2 < I_2 = \tfrac{1}{2}md^2 < I_3 = \tfrac{1}{2}m(d^2 + e^2)\,. \tag{6.87}$$

As we said above, the satellite rotates about the 3-axis, i.e. the axis with the largest moment of inertia. This axis is perpendicular to the orbit plane (in Fig. 6.27 it points towards the reader). It is reasonable to assume that Hyperion's motion has no appreciable effect on Saturn, its mother planet, whose motion is very slow compared to that of Hyperion.

The gravitational field at the position of Hyperion is not homogeneous. As I_1 and I_2 are not equal, the satellite is subject to a net torque that depends on its position in the orbit. We calculate the torque for the pair (1,2). The result for the pair (3,4) will then follow immediately. We have

$$\boldsymbol{D}^{(1,2)} = \frac{\boldsymbol{d}}{2} \times (\boldsymbol{F}_1 - \boldsymbol{F}_2) \,,$$

where $\boldsymbol{F}_i = -GmM\boldsymbol{r}_i/r_i^3$ is the force acting on the mass point i, M being the mass of Saturn. The distance $d = |\boldsymbol{d}|$ being small compared to the radial distance r from Saturn we have, with the notations as in Fig. 6.27,

$$\frac{1}{r_i^3} = \frac{1}{r^3}\left(1 \pm \frac{d}{r}\cos\alpha + \frac{d^2}{4r^2}\right)^{-3/2} \simeq \frac{1}{r^3}\left(1 \mp \frac{3}{2}\frac{d}{r}\cos\alpha\right) .$$

(The upper sign holds for r_1, the lower sign for r_2.) Inserting this approximation as well as the cross product $\boldsymbol{r} \times \boldsymbol{d} = -rd\sin\alpha\,\hat{\boldsymbol{e}}_3$, one finds

$$\boldsymbol{D}^{(1,2)} \simeq (3d^2 mMG/4r^3)\sin 2\alpha\,\hat{\boldsymbol{e}}_3 = (3GMI_2/2r^3)\sin 2\alpha\,\hat{\boldsymbol{e}}_3 \,.$$

In the second step we inserted the expression (6.87) for I_2. The product GM can be expressed by the semimajor axis a and the orbital period T, using Kepler's third law (1.23). The mass of Hyperion (which in the model is $4m$) is small compared to M, and therefore it is practically equal to the reduced mass. So, from (1.23)

$$GM = (2\pi/T)^2 a^3 \,.$$

The calculation is the same for the pair (3,4). Hence, the total torque $\boldsymbol{D}^{(1,2)} + \boldsymbol{D}^{(3,4)}$ is found to be

$$\boldsymbol{D} \simeq \frac{3}{2}\left(\frac{2\pi}{T}\right)^2 (I_2 - I_1)\left(\frac{a}{r}\right)^3 \sin 2\alpha\,\hat{\boldsymbol{e}}_3 \,. \tag{6.88}$$

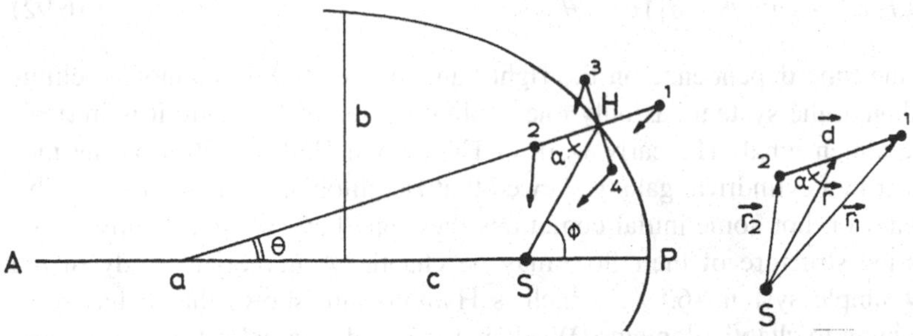

Fig. 6.27. A simple model for the asymmetric satellite Hyperion of the planet Saturn

This result remains valid if the satellite is described by a more realistic distribution of mass. It shows that the resulting torque vanishes if $I_1 = I_2$. With this result the equation of motion (3.52) for the rotational motion of the satellite reads

$$I_3\ddot{\theta} = \frac{3}{2}\left(\frac{2\pi}{T}\right)^2 (I_2 - I_1)\left(\frac{a}{r(t)}\right)^3 \sin 2\alpha \,. \tag{6.89}$$

Here, the angle θ describes the orientation of the satellite's 1-axis relative to the line SP (joining Saturn and Hyperion's perisaturnion) and Φ is the usual polar angle of Keplerian motion. As $\alpha = \Phi - \theta$, (6.89) reads

$$I_3\ddot{\theta} = -\frac{3}{2}\left(\frac{2\pi}{T}\right)^2 (I_2 - I_1)\left(\frac{a}{r(t)}\right)^3 \sin 2[\theta - \Phi(t)] \,. \tag{6.89$'$}$$

This equation contains only one explicit degree of freedom, θ, but its right-hand side depends on time because the orbital radius r and the polar angle Φ are functions of time. Therefore, in general, the system is not integrable. There is an exception, however. If the orbit is a circle, $\varepsilon = 0$ (cf. Sect. 1.7.2 (ii)) the average circular frequency

$$n \stackrel{\text{def}}{=} \frac{2\pi}{T} \tag{6.90}$$

is the true angular velocity, i.e. we have $\Phi = nt$ and with $\theta' = \theta - nt$ the equation of motion becomes

$$I_3\ddot{\theta}' = -\tfrac{3}{2}n^2(I_2 - I_1)\sin 2\theta' \,, \quad \varepsilon = 0 \,. \tag{6.91}$$

If we set

$$z_1 \stackrel{\text{def}}{=} 2\theta' \,, \quad \omega^2 \stackrel{\text{def}}{=} 3n^2\frac{I_2 - I_1}{I_3} \,, \quad \tau \stackrel{\text{def}}{=} \omega t \,,$$

(6.91) is recognized to be the equation of motion (1.40) of the plane pendulum, viz. $\mathrm{d}^2 z_1/\mathrm{d}\tau^2 = -\sin z_1$, which can be integrated analytically. The energy is an integral of the motion; it reads

$$E = \tfrac{1}{2}I_3\dot{\theta}'^2 - \tfrac{3}{4}n^2(I_2 - I_1)\cos 2\theta' \,. \tag{6.92}$$

If $\varepsilon \neq 0$, the time dependence on the right-hand side of (6.89′) cannot be eliminated. Although the system has only one explicit degree of freedom, it is intrinsically three-dimensional. The early work of Hénon and Heiles (1964) on the motion of a star in a cylindrical galaxy showed that Hamiltonian systems may exhibit chaotic behavior. For some initial conditions they may have regular solutions, but for others the structure of their flow may be chaotic. A numerical study of the seemingly simple system (6.89′), which is Hamiltonian, shows that it has solutions pertaining to chaotic domains (Wisdom 1987, and original references quoted there). One integrates the equation of motion (6.89′) numerically and studies the result on a transverse section (cf. the Poincaré mapping introduced in Sect. 6.3.4), which is chosen as follows. At every passage of the satellite at the point P of closest approach to the mother planet one records the momentary orientation of the satellite's 1-axis with respect to the line SP of Fig. 6.27. One then plots the relative change $(\mathrm{d}\theta/\mathrm{d}t)/n$ of the orientation at successive passages through P, for various initial conditions. One obtains figures of the kind shown in Figs. 6.28–30. We start by commenting on Fig. 6.28. One-dimensional manifolds, i.e. curves, correspond to quasiperiodic motion. If, on the other hand, the "measured" points fill a surface, this is a hint that there is chaotic motion. The scattered points in the middle part of the figure all pertain to the same, chaotic orbit. Also, the two orbits forming an "X" at about $(\frac{\pi}{2}, 2.3)$ are chaotic, while the islands in the chaotic zones correspond to states of motion where the ratio of the spin period and the period of the orbit are rational. For example, the island at (0, 0.5) is the remnant of the synchronous motion where Hyperion, on average, would always show the same face to the mother

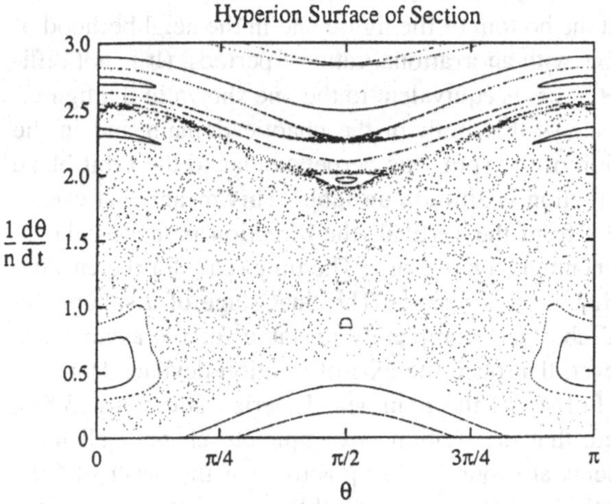

Fig. 6.28. Chaotic behavior of Hyperion, a satellite of Saturn. The picture shows the relative change of orientation of the satellite as a function of its orientation, at every passage in P, the point of closest approach to Saturn (from Wisdom 1987)

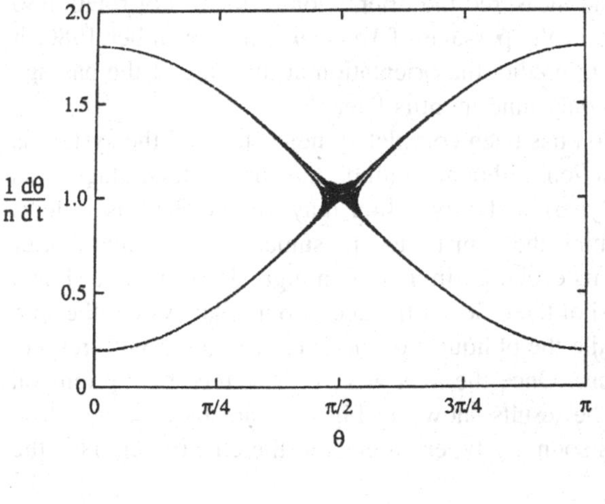

Fig. 6.29. Analogous result to the one shown in Fig. 6.28, for Deimos, a satellite of Mars whose asymmetry (6.93) is $\alpha = 0.81$ and whose orbital eccentricity is $\varepsilon = 0.0005$ (from Wisdom 1987)

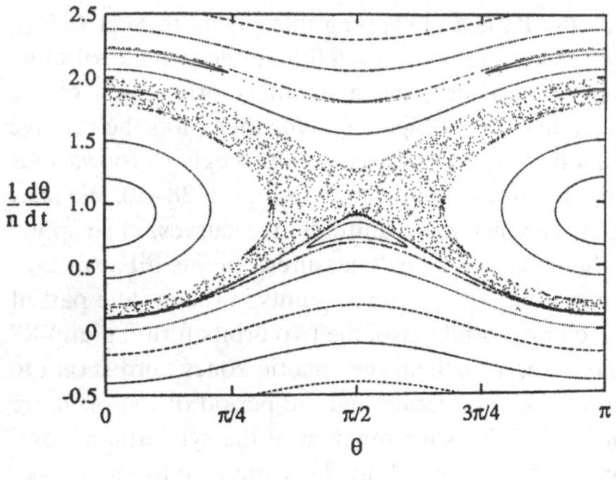

Fig. 6.30. Analogous result to those in Figs. 6.28 and 6.29 for Phobos, a satellite of Mars whose asymmetry (6.93) is $\alpha = 0.83$ and whose orbital eccentricity is $\varepsilon = 0.015$ (from Wisdom 1987)

planet. The synchronous orbit at $\theta = \pi$ would be the one where the satellite shows the opposite face. The curves at the bottom of the figure and in the neighborhood of $\theta = \frac{\pi}{2}$ are quasiperiodic motions with an irrational ratio of periods. (It is not difficult to see that the range $\pi \leq \theta \leq 2\pi$ is equivalent to the one shown in the figure.)

A more detailed analysis shows that both in the chaotic domain and in the synchronous state the orientation of the spin axis perpendicular to the orbit plane is unstable. One says that the motion is attitude unstable. This means that even a small deviation of the spin axis from the vertical (the direction perpendicular to the orbit plane) will grow exponentially, on average. The time scale for the ensuing tumbling is of the order of a few orbital periods. The final stage of a spherically symmetric moon, as described above, is completely unstable for the asymmetric satellite Hyperion. Note, however, that once the axis of rotation deviates from the vertical, one has to solve the full set of the nonlinear Eulerian equations (3.52). In doing this one finds, indeed, that the motion is completely chaotic: all three Liapunov characteristic exponents are found to be positive (of the order of 0.1). In order to appreciate the chaoticity of Hyperion's tumbling the following remark may be helpful. Even if one had measured the orientation of its axis of rotation to ten decimal places, at the time of the passage of *Voyager 1* in November 1980, it would not have been possible to predict the orientation at the time of the passage of *Voyager 2* in August 1981, only nine months later.

Up to this point tidal friction has been completely neglected and the system is exactly Hamiltonian. Tidal friction, although unimportant in the final stage, was important in the history of Hyperion. Its evolution may be sketched as follows (Wisdom 1987). In the beginning the spin period presumably was much shorter and Hyperion probably began its evolution in a domain high above the one shown in Fig. 6.28. Over a time period of the order of the age of our solar system the spin rotation was slowed down, while the obliquity of the axis of rotation with respect to the vertical decreased to zero. Once the axis was vertical, the assumptions on which the model (6.89′) and the results shown in Fig. 6.28 are based came close to being realized. However, as soon as Hyperion entered the chaotic regime, "the work of the tides over aeons was undone in a matter of days" (Wisdom 1987). It began to tumble erratically until this day[12].

In order to understand further the rather strange result illustrated by Fig. 6.28 for the case of Hyperion, we show the results of the same calculation for two satellites of Mars in Figs. 6.29 and 6.30: Deimos and Phobos. The asymmetry parameter

$$\alpha = \sqrt{\frac{3(I_2 - I_1)}{I_3}}, \tag{6.93}$$

which is the relevant quantity in the equation of motion (6.89′) and whose value is 0.89 in the case of Hyperion, is very similar for Deimos and Phobos: 0.81 and 0.83, respectively. However, the eccentricities of their orbits around Mars are much smaller than for Hyperion. They are 0.0005 for Deimos and 0.015 for Phobos. The synchronous phase at $(\theta = 0, (1/n)\mathrm{d}\theta/\mathrm{d}t = 1)$ that we know from our moon is still clearly visible in Figs. 6.29 and 6.30, while in the case of Hyperion it has drifted down in Fig. 6.28. Owing to the smallness of the eccentricities, the chaotic domains are correspondingly less developed. Even though today Deimos and Phobos no longer tumble, they must have gone through long periods of chaotic tumbling in the course of their history. One can estimate that Deimos' chaotic tumbling phase may have lasted about 100 million years, whereas Phobos' tumbling phase may have lasted about 10 million years.

6.6.2 Orbital Dynamics of Asteroids with Chaotic Behavior

As we learnt in Sect. 2.37 the manifold of motions of an integrable Hamiltonian system with f degrees of freedom is $\Delta^f \times T^f$, with $\Delta^f = \Delta_1 \times \Delta_2 \times \ldots \times \Delta_f$ being the range of the action variables $I_1, I_2, \ldots, I_f$ and T^f the f-dimensional torus spanned by the angle variables $\theta_1, \theta_2, \ldots, \theta_f$. Depending on whether or not the corresponding, fundamental frequencies are rationally dependent, one talks about resonant or nonresonant tori, respectively. These tori (the so-called KAM tori) and their stability with respect to small perturbations play an important role in perturbation theory of Hamiltonian systems, as explained in Sect. 2.39.

In the past it was held that the Kirkwood gaps referred to in the introduction were due to a breakdown of the KAM tori in the neighborhood of resonances. It seems that this rather qualitative explanation is not conclusive. Instead, recent investigations of the dynamics of asteroids, which are based on long-term calculations, seem to indicate that the Kirkwood gaps are due rather to chaotic behavior in a Hamiltonian system.

Here we wish to describe briefly one of the examples studied, namely the gap in the asteroid belt between Mars and Jupiter, which occurs at the ratio 3:1 of the periods of the asteroid and of Jupiter. Clearly, the integration of the equations of

12 The observations of *Voyager 2* are consistent with this prediction, since it found Hyperion in a position clearly out of the vertical. More recently, Hyperion's tumbling was positively observed from the earth (J. Klavetter et al., Science **246** (1989) 998, Astron. J. **98** (1989) 1855).

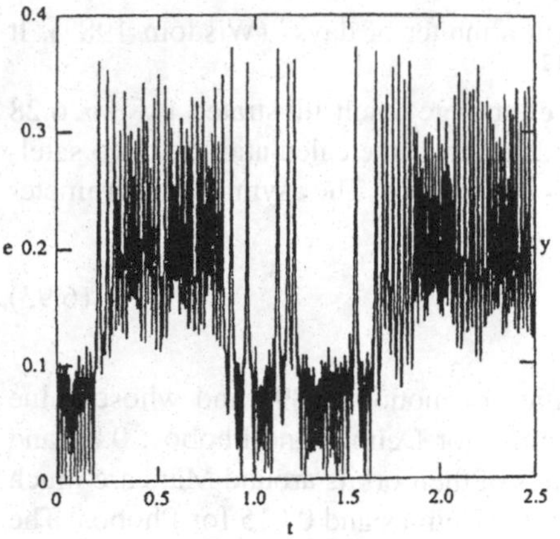

Fig. 6.31. Eccentricity of a typical orbit in the chaotic domain close to the 3:1 resonance, as a function of time measured in millions of years. From periods of small, though irregular, values of ε the orbit makes long-term excursions to large values of the eccentricity

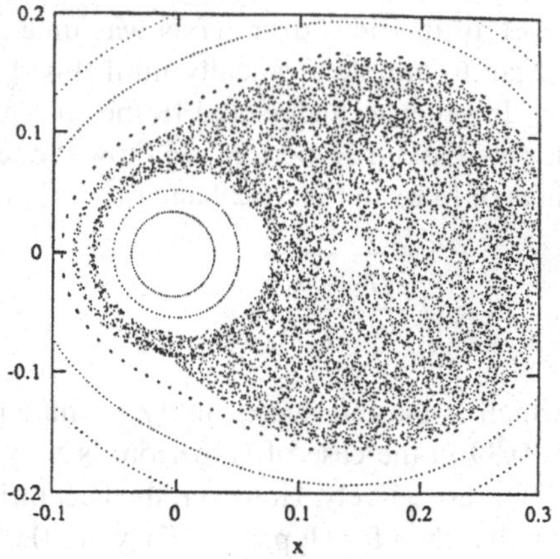

Fig. 6.32. Surface of section for the orbit shown in Fig. 6.31. The radial coordinate of the points shown is the eccentricity

motion over a time span of several millions of years is a difficult problem of applied mathematics for which dedicated methods had to be designed. We cannot got into these methods[13] and must restrict the discussion to a few characteristic results.

The main result of these calculations is that the orbits of asteroids in the neighborhood of the 3:1 resonance exhibit chaotic behavior in the following sense: the eccentricity of the asteroid's elliptic orbit varies in an irregular way, as a function of time, such that an asteroid with an initial eccentricity of, say, 0.1 makes long excursions to larger eccentricities. Figure 6.31 shows an example for a time interval of 2.5 million years, which was calculated for the planar system Sun–asteroid–Jupiter. The problem is formulated in terms of the coordinates (x, y) of the asteroid in the plane and in terms of the time dependence of the orbit parameters due to the motion of Jupiter along its orbit. Averaging over the orbital period yields an effective, two-dimensional system for which one defines effective coordinates

$$x = \varepsilon \cos(\bar{\omega} - \bar{\omega}_{\mathrm{J}}) , \quad y = \varepsilon \sin(\bar{\omega} - \bar{\omega}_{\mathrm{J}}) . \tag{6.94}$$

Here $\bar{\omega}$ and $\bar{\omega}_{\mathrm{J}}$ are the longitudes of the perihelia for the asteroid and for Jupiter, respectively. The quantities (6.94) yield a kind of Poincaré section if one records x and y each time a certain combination of the mean longitudes goes to zero. Figure 6.32 shows the section obtained in this way for the orbit shown in Fig. 6.31. These figures show clearly that orbits in the neighborhood of the 3:1 resonance are strongly chaotic. At the same time they provide a simple explanation for the observation that a strip of orbits in the neighborhood of this resonance is empty: all

[13] The reader will find hints to the original literature describing these methods in Wisdom's review (1987).

orbits with $\varepsilon > 0.4$ cross the orbit of Mars. As we know that orbits in the neighborhood of the resonance make long excursions to larger eccentricities, there is a finite probability for the asteroids to come close to Mars, or even to hit this planet, and thereby to be scattered out of their original orbit. Thus, deterministically chaotic motion played an important role in the formation of the 3:1 Kirkwood gap[14].

Another, very interesting observation, which follows from these investigations, is that irregular behavior near the 3:1 resonance may play an important role in the transport of meteorites from the asteroid belt to the earth. Indeed, the calculations show that asteroidal orbits starting at $\varepsilon = 0.15$ make long-term excursions to eccentricities $\varepsilon = 0.6$ and beyond. In this case they cross the orbit of the earth. Therefore, chaotic orbits in the neighborhood of the 3:1 gap can carry debris from collisions between asteroids directly to the surface of earth. In other words, deterministically chaotic motion may be responsible for an important transport mechanism of meteorites to earth, i.e. of objects that contain important information about the history of our solar system.

In this last section we returned to celestial mechanics, the point of departure of all of mechanics. Here, however, we discovered qualitatively new types of deterministic motion that are very different from the serene and smooth running of the planetary clockwork whose construction principles were investigated by Kepler. The solar system was always perceived as the prime example of a mechanical system evolving with great regularity and impressive predictability. We have now learnt that it contains chaotic behavior (tumbling of asymmetric satellites, chaotic variations of orbital eccentricities of asteroids near resonances, the chaotic motion of Pluto) very different from the harmony and regularity that, historically, one expected to find. At the same time, we have learnt that mechanics is not a closed subject that has disappeared in the dusty archives of physics. On the contrary, it is more than ever a lively and fascinating field of research, which deals with important and basic questions in many areas of dynamics.

---

[14] Analogous investigations of the 2:1 and 3:2 resonances indicate that there is chaotic behavior at the former while there is none at the latter. This is in agreement with the observation that there is a gap at 2:1 but not at 3:2.

7. Continuous Systems

A distinctive feature of the mechanical systems we have discussed so far is that their number of degrees of freedom is *finite* and hence countable. The mechanics of deformable macroscopic media goes beyond this framework. The reaction of a solid state to external forces, the flow behavior of a liquid in a force field, or the dynamics of a gas in a vessel cannot be described by means of finitely many coordinate variables. The coordinates and momenta of point mechanics are replaced by field quantities, i.e. functions or fields defined over space and time, which describe the dynamics of the system. The mechanics of continua is an important discipline of classical physics on its own and goes far beyond the scope of this book. In this epilog we introduce the important concept of dynamical field, generalize the principles of canonical mechanics to continuous systems, and illustrate them by means of some instructive examples. At the same time, this serves as a basis for electrodynamics, which is a typical and especially important field theory.

7.1 Discrete and Continuous Systems

Earlier we pointed out the asymmetry between the time variable on the one hand and the space variables on the other, which is characteristic for nonrelativistic physics, cf. Sects. 1.6 and 4.7. In a Galilei-invariant world, time has an absolute nature while space does not. In the mechanics of mass points and of rigid bodies there is still another asymmetry, which we also pointed out in Sect. 1.6 and which is this: time plays the role of a *parameter*, whereas the position $\boldsymbol{r}(t)$ of a particle, or, likewise, the coordinates $\{\boldsymbol{r}_s(t),\ \theta_k(t)\}$ of a rigid body, or, even more generally, the flow $\underset{\sim}{\boldsymbol{\Phi}}(t, t_0, \underset{\sim}{x}_0)$ in phase space are the genuine, *dynamical variables* that obey the mechanical equations of motion. Geometrically speaking, the latter are the "geometrical curves", while t is the orbit parameter (length of arc) that indicates in which way the system moves along its orbits.

This is different for the case of a continuous system, independent of whether it is to be described nonrelativistically or relativistically. Here, besides the time coordinate, also the space coordinates take over the role of parameters. Their previous role as dynamical variables is taken over by new objects, the *fields*. It is the fields that describe the state of motion of the system and obey a set of equations of motion. We develop this important new concept by means of a simple example.

F. Scheck, *Mechanics*, Graduate Texts in Physics, 5th ed.,
DOI 10.1007/978-3-642-05370-2_7,

Example. *Linear chain and vibrating string.* Let n mass points of mass m be joined by identical, elastic springs in such a way that their equilibrium positions are $x_1^0, x_2^0, \ldots, x_n^0$, cf. Fig. 7.1. As shown in part (a) of that figure we displace the mass points along the straight line joining them. The deviations from the equilibrium positions are denoted by

$$u_i(t) = x_i(t) - x_i^0 , \quad i = 1, 2, \ldots, n .$$

The kinetic energy is given by

$$T = \sum_{i=1}^{n} \frac{1}{2} m \dot{u}_i^2 . \tag{7.1}$$

The forces being harmonic the potential energy reads

$$U = \sum_{i=1}^{n-1} \frac{1}{2} k \left(u_{i+1} - u_i\right)^2 + \frac{1}{2} k \left(u_1^2 + u_n^2\right) . \tag{7.2}$$

The last two terms stem from the spring connecting particle 1 with the wall and from the one connecting particle n with the wall, at the other end of the chain. We ascribe the coordinate x_0^0 to the left suspension point and x_{n+1}^0 to the right suspension point of the chain and we require that their deviations and their velocities be zero at all times, i.e. $u_0(t) = u_{n+1}(t) = 0$. The potential energy can then be written as

$$U = \frac{1}{2} k \sum_{i=0}^{n} \left(u_{i+1} - u_i\right)^2 , \tag{7.2$'$}$$

and the natural form of the Lagrangian function reads

$$L = T - U , \tag{7.3}$$

with T as given by (7.1) and U by (7.2′). This Lagrangian function describes longitudinal motions of the mass points, i.e. motions along their line of connection. We obtain the same form of the Lagrangian function if we let the mass points move only transversely to that line, i.e. as shown in Fig. 7.1b. Let d be the distance between the equilibrium positions. The distance between neighboring mass points can be approximated as follows:

$$\sqrt{d^2 + \left(v_{i+1} - v_i\right)^2} \simeq d + \frac{1}{2} \frac{\left(v_{i+1} - v_i\right)^2}{d} ,$$

provided the differences of transverse amplitudes remain small compared to d. The force driving the mass points back is approximately transverse, its potential energy being given by

$$U = \frac{S}{2d} \sum_{i=0}^{n} \left(v_{i+1} - v_i\right)^2 , \tag{7.4}$$

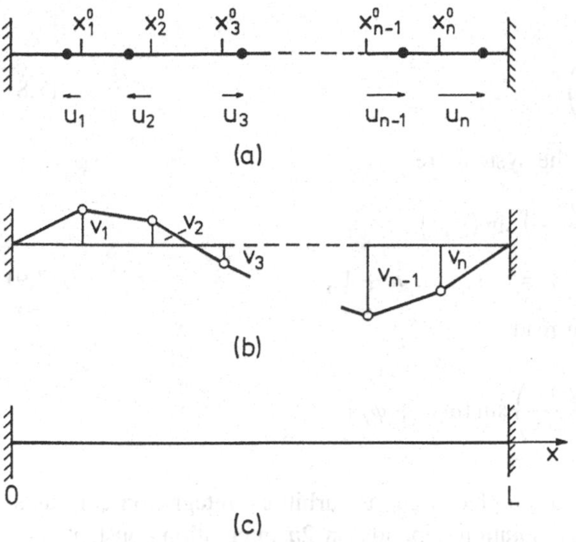

Fig. 7.1. A linear chain of finitely many mass points, which may oscillate (**a**) longitudinally or (**b**) transversely. (**c**) shows a vibrating string of the same length as the chain, for comparison

where S is the string constant. As before, we must take into account the condition $v_0(t) = v_{n+1}(t) = 0$ for the two points of suspension. In reality, the chain can perform longitudinal and transverse motions simultaneously and the two types of motion will be coupled. For the sake of simplicity, we restrict the discussion to purely transverse or purely longitudinal motions and do not consider mixed modes.

In the first case we set $\omega_0 = \sqrt{k/m}$ and $q_i(t) = u_i(t)$. In the second case we set $\omega_0 = \sqrt{S/md}$ and $q_i(t) = v_i(t)$. In either case the Lagrangian function reads

$$L = \frac{1}{2} m \sum_{j=0}^{n} \left\{ \dot{q}_j^2 - \omega_0^2 \left(q_{j+1} - q_j\right)^2 \right\} \tag{7.5}$$

with the conditions $q_0 = \dot{q}_0 = 0$, $q_{n+1} = \dot{q}_{n+1} = 0$. The equations of motion, which follow from (7.5), are

$$\ddot{q}_j = \omega_0^2 \left(q_{j+1} - q_j\right) - \omega_0^2 \left(q_j - q_{j-1}\right), \quad j = 1, \ldots, n. \tag{7.6}$$

We solve these equations by means of the following substitution:

$$q_j(t) = A \sin\left(j \frac{p\pi}{n+1}\right) e^{i\omega_p t}. \tag{7.7}$$

Obviously, we can let j run from 0 to $n+1$ because q_0 and q_{n+1} vanish for all times. In (7.7) p is a positive integer. The quantities ω_p are the eigenfrequencies of the coupled system (7.6) and could be determined by means of the general method developed in Practical Example 1 of Chap. 2. Here they may be obtained directly by inserting the substitution (7.7) into the equation of motion (7.6). One obtains

$$\omega_p^2 = 2\omega_0^2 \left(1 - \cos\left(\frac{p\pi}{n+1}\right)\right)$$

or

$$\omega_p = 2\omega_0 \sin\left(\frac{p\pi}{2(n+1)}\right) . \tag{7.8}$$

Hence, the normal modes of the system are

$$q_j^{(p)}(t) = A^{(p)} \sin\left(j\frac{p\pi}{n+1}\right) \sin\left(\omega_p t\right)$$
$$\text{with} \quad p = 1, \ldots, n\, ; \quad j = 0, 1, \ldots, n+1\, , \tag{7.9}$$

and the most general solution reads

$$q_j(t) = \sum_{p=1}^{n} A^{(p)} \sin\left(j\frac{p\pi}{n+1}\right) \sin\left(\omega_p t + \varphi_p\right) ,$$

where the amplitudes $A^{(p)}$ and the phases φ_p are arbitrary integration constants. (As expected, the most general solution depends on $2n$ integration constants.)

Let us compare these solutions, for the example of transverse oscillations, to the normal modes of a vibrating string spanned between the same end points as the chain (see Fig. 7.1c). The length of the string is $L = (n+1)d$. Its state of vibration for the pth harmonic is described by

$$\varphi(x,t) = A^{(p)} \sin\left(\frac{p\pi x}{L}\right) \sin\left(\omega_p t\right) \; ; \quad \omega_p = p\bar{\omega}_0 \, . \tag{7.10}$$

Here, ω_p is the p-fold of a basic frequency $\bar{\omega}_0$ that we may choose such that it coincide with the frequency (7.8) of the chain for $p = 1$, viz.

$$\bar{\omega}_0 = 2\omega_0 \sin\left(\frac{\pi}{2(n+1)}\right) . \tag{7.11}$$

The solution (7.10) is closely related to the solution (7.9); we shall work out the exact relationship in the next subsection. Here we wish to discuss a direct comparison of (7.9) and (7.10).,

At a fixed time t the amplitude of the normal mode (7.9), with a given p and with $1 \leqslant p \leqslant n$, has exactly the same shape as the amplitude of the vibration (7.10) at the points $x = jL/(n+1)$ on the string. Figure 7.2 shows the example $p = 2$ for $n = 7$ mass points. The full curve shows the first harmonic of the vibrating string; the points indicate the positions of the seven mass points according to the normal oscillation (7.9) with $p = 2$. (Note, however, that the frequencies ω_p and $p\bar{\omega}_0$ are not the same.)

The *discrete* system (7.9) has n degrees of freedom which, clearly, are countable. The dynamical variables are the coordinates $q_j^{(p)}(t)$ and the corresponding momenta $p_j^{(p)}(t) = m\dot{q}_j^{(p)}(t)$. Time pays the role of a parameter.

In the *continuous* system (7.10) we are interested in the local amplitude $\varphi(x,t)$, for fixed time and as a function of the continuous variable $x \in [0, L]$. Thus, the

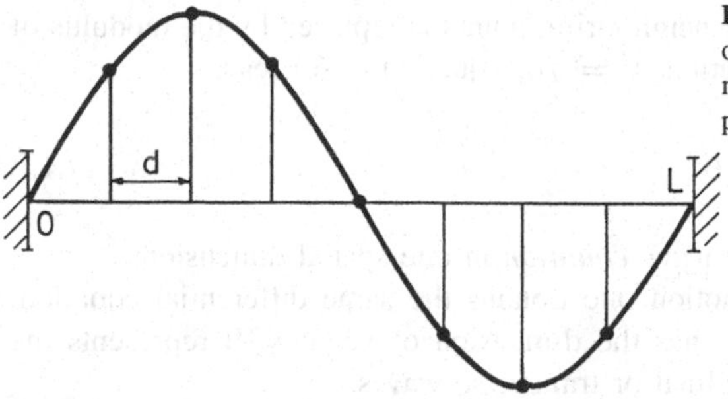

Fig. 7.2. Transversely oscillating chain of 7 mass points and the second harmonic of the vibrating string, for comparison

function φ over time t and position x on the string takes over the role of a dynamical variable.

If we suppose that the continuous system was obtained from the discrete system by letting the number of particles n become very large and their distance d correspondingly small, we realize that the variable x of the former takes over the role of the counting index j of the latter. This means, firstly, that the number of degrees of freedom has become infinite and that the degrees of freedom are not even countable. Secondly, the coordinate x, very much like the time t, has become a *parameter*. For given $t = t_0$ the function $\varphi(x, t_0)$ describes the shape of the vibration in the space $x \in [0, L]$; conversely, for fixed $x = x_0$, $\varphi(x_0, t)$ describes the motion of the string at that point, as a function of time.

7.2 Transition to the Continuous System

The transition from the discrete chain of mass points to the continuous string can be performed explicitly for the examples (7.2′) and (7.4) of longitudinal or transverse vibrations. Taking the number of mass points n to be very large and their distance d to be infinitesimally small (such that $(n+1)d = L$ stays finite), we have $q_j(t) \equiv \varphi(x = jL/(n+1), t)$ and

$$q_{j+1} - q_j \simeq \left.\frac{\partial \varphi}{\partial x}\right|_{x=jd+d/2} d \,, \quad q_j - q_{j-1} \simeq \left.\frac{\partial \varphi}{\partial x}\right|_{x=jd-d/2} d \,,$$

and therefore

$$\left(q_{j+1} - q_j\right) - \left(q_j - q_{j-1}\right) \simeq \left.\frac{\partial^2 \varphi}{\partial x^2}\right|_{x=jd} d^2 \,.$$

The equation of motion (7.6) becomes the differential equation

$$\frac{\partial^2 \varphi}{\partial t^2} \simeq \omega_0^2 d^2 \frac{\partial^2 \varphi}{\partial x^2} \,. \tag{7.6′}$$

In the case of longitudinal vibrations, $\omega_0^2 d^2 = kd^2/m$. In the limit $n \to \infty$ the ratio m/d becomes the mass density ϱ per unit length, while the product of the string constant k and the distance d of neighboring points is replaced by the modulus of elasticity $\eta = kd$. With the notation $v^2 = \eta/\varrho$ equation (7.6′) reads

$$\frac{\partial^2\varphi(x,t)}{\partial t^2} - v^2\frac{\partial^2\varphi(x,t)}{\partial x^2} = 0 \,. \tag{7.12}$$

This differential equation is the *wave equation* in one spatial dimension.

In the case of transverse motion one obtains the same differential equation, with $v^2 = S/\varrho$. The quantity v has the dimension of velocity. It represents the speed of propagation of longitudinal or transverse waves.

In a next step let us study the limit of the Lagrangian function obtained in performing the transition to the continuum. The sum over the mass points is to be replaced by the integral over x, the mass m by the product ϱd, and the quantity $m\omega_0^2(q_{j+1} - q_j)^2$ by

$$\varrho d\left(\frac{\partial\varphi}{\partial x}\right)^2 \omega_0^2 d^2 = \varrho d\left(\frac{\partial\varphi}{\partial x}\right)^2 v^2 \,.$$

The infinitesimal distance d is nothing but the differential dx. Thus, we obtain

$$L = \int_0^L \mathrm{d}x\,\mathcal{L} \,, \tag{7.13}$$

where

$$\mathcal{L} = \frac{1}{2}\varrho\left[\left(\frac{\partial\varphi}{\partial t}\right)^2 - v^2\left(\frac{\partial\varphi}{\partial x}\right)^2\right] \,. \tag{7.14}$$

the function $\mathcal{L}$ is called the *Lagrangian density.* In the general case, it depends on the *field* $\varphi(x,t)$, its derivatives with respect to space and time, and, possibly, also explicitly on t and x, i.e. it has the form

$$\mathcal{L} \equiv \mathcal{L}\left(\varphi, \frac{\partial\varphi}{\partial x}, \frac{\partial\varphi}{\partial t}, x, t\right) \,. \tag{7.15}$$

The analogy to the Lagrangian function of point mechanics is the following. The dynamical variable q is replaced by the field φ, $\dot{q}$ is replaced by the partial derivatives $\partial\varphi/\partial x$ and $\partial\varphi/\partial t$, and the time parameter is replaced by the space and time coordinates x and t. The spatial coordinate now plays the same role as the time coordinate, and therefore a certain symmetry between the two types of coordinates is restored.

We now turn to the question whether the equation of motion (7.12) can be obtained from the Lagrangian density (7.14) or, in the more general case, from (7.15).

7.3 Hamilton's Variational Principle for Continuous Systems

Let $\mathcal{L}(\varphi,\ \partial\varphi/\partial x,\ \partial\varphi/\partial t,\ x,\ t)$ be a Lagrangian density assumed to be at least C^1 in the field φ and in its derivatives. Let $L = \int \mathrm{d}x\, \mathcal{L}$ be the corresponding Lagrangian function. For the sake of simplitics we consider the example of one spatial dimension. The generalization to three dimensions is straightforward and can be guessed easily at the end.

As φ is the dynamical variable, Hamilton's variational principle now requires that the functional

$$I[\varphi] \stackrel{\text{def}}{=} \int_{t_1}^{t_2} \mathrm{d}t\, L = \int_{t_1}^{t_2} \mathrm{d}t \int \mathrm{d}x\, \mathcal{L} \tag{7.16}$$

assumes an extreme value if φ is a physically possible solution. Like in the mechanics of mass points one embeds the solution with given values $\varphi(x, t_1)$ and $\varphi(x, t_2)$ at the end points in a set of comparative fields. In other words, one varies the field φ such that its variation vanishes at the times t_1 and t_2, and requires $I[\varphi]$ to be an extremum. Let $\delta\varphi$ denote the variation of the field, $\dot{\varphi}$ the time derivative and φ' the space derivative of φ. Then

$$\delta I[\varphi] = \int_{t_1}^{t_2} \mathrm{d}t \int \mathrm{d}x \left\{ \frac{\partial \mathcal{L}}{\partial \varphi}\delta\varphi + \frac{\partial \mathcal{L}}{\partial \dot{\varphi}}\delta\dot{\varphi} + \frac{\partial \mathcal{L}}{\partial \varphi'}\delta\varphi' \right\} .$$

Clearly, the variation of a derivative is equal to the derivative of the variation,

$$\delta\dot{\varphi} = \frac{\partial}{\partial t}(\delta\varphi)\ , \quad \delta\varphi' = \frac{\partial}{\partial x}(\delta\varphi)\ .$$

Furthermore, the field φ shall be such that it vanishes at the boundaries of the integration over x. By partial integration of the second term with respect to t and of the third term with respect to x, and noting that $\delta\varphi$ vanishes at the boundaries of the integration, we obtain

$$\delta I[\varphi] = \int_{t_1}^{t_2} \mathrm{d}t \int \mathrm{d}x \left\{ \frac{\partial \mathcal{L}}{\partial \varphi} - \frac{\partial}{\partial t}\left(\frac{\partial \mathcal{L}}{\partial \dot{\varphi}}\right) - \frac{\partial}{\partial x}\left(\frac{\partial \mathcal{L}}{\partial \varphi'}\right) \right\} \delta\varphi\ .$$

The condition $\delta I[\varphi] = 0$ is to hold for all admissible variations $\delta\varphi$. Therefore, the expression in the curly brackets of the integrand must vanish. This yields the *Euler–Lagrange equation for continuous systems* (here in one space dimension),

$$\frac{\partial \mathcal{L}}{\partial \varphi} - \frac{\partial}{\partial t}\frac{\partial \mathcal{L}}{\partial(\partial\varphi/\partial t)} - \frac{\partial}{\partial x}\frac{\partial \mathcal{L}}{\partial(\partial\varphi/\partial x)} = 0\ . \tag{7.17}$$

We illustrate this equation by means of the example (7.14). In this example $\mathcal{L}$ does not depend on φ but only on $\dot{\varphi}$ and on φ'. $\mathcal{L}$ does not depend explicitly on x or t, either. The variable x is confined to the interval $[0, L]$. Both φ and $\delta\varphi$ vanish at the end points of this interval. We have

$$\frac{\partial \mathcal{L}}{\partial(\partial\varphi/\partial t)} = \varrho\frac{\partial\varphi}{\partial t}\,, \qquad \frac{\partial \mathcal{L}}{\partial(\partial\varphi/\partial x)} = -v^2\varrho\frac{\partial\varphi}{\partial x}\,,$$

and the equation of motion (7.17) yields the wave equation (7.12), as expected,

$$\frac{\partial^2\varphi}{\partial t^2} - v^2\frac{\partial^2\varphi}{\partial x^2} = 0\,.$$

Its general solutions are $\varphi_+(x,t) = f(x - vt)$, $\varphi_-(x,t) = f(x + vt)$, with $f(z)$ an arbitrary differentiable function of its argument $z = x \mp vt$. The first of these describes a wave propagating in the positive x direction, the second describes a wave propagating in the negative x direction. As the wave equation is linear in the field variable φ, any linear combination of two independent solutions is also a solution. As an example we consider two harmonic solutions (i.e. two pure sine waves) with wavelength λ and equal amplitude,

$$\varphi_+ = A\sin\left(\frac{2\pi}{\lambda}(x - vt)\right)\,, \qquad \varphi_- = A\sin\left(\frac{2\pi}{\lambda}(x + vt)\right)\,.$$

Their sum

$$\varphi = \varphi_+ + \varphi_- = 2A\sin\left(\frac{2\pi}{\lambda}x\right)\cos\left(\frac{2\pi}{\lambda}vt\right)$$

describes a *standing wave.* It has precisely the form of the solution (7.10) if

$$\frac{2\pi}{\lambda}x = \frac{p\pi x}{L} \quad \text{or} \quad \lambda = \frac{2L}{p}\,, \qquad p = 1, 2, \ldots \,.$$

The length L of the string must be an integer multiple of half the wavelength. The frequency of the vibration with wavelength λ is given by

$$\omega_p = \frac{2\pi v}{\lambda} = p\bar{\omega}_0\,, \quad \text{with} \quad \bar{\omega}_0 = \frac{\pi v}{L}\,. \tag{7.18}$$

Thus, the transverse oscillations of our original chain of mass points are standing waves. Note also that their frequency (7.11) takes on the correct continuum value (7.18). Indeed, when the number n of mass points is very large, the sine in (7.11) can be replaced by its argument,

$$\begin{aligned}\bar{\omega}_0 &= 2\omega_0\sin\left(\frac{\pi}{2(n+1)}\right) \\ &\simeq 2\omega_0\frac{\pi}{2(n+1)} = \frac{\pi}{L}\omega_0 d = \frac{\pi v}{L}\,,\end{aligned}$$

where we have set $L = (n+1)d$ and replaced $\omega_0 d$ by v.

We conclude this subsection with another example in one time and three space coordinates. Let $\varphi(\boldsymbol{x}, t)$ be a real field and let the Lagrangian density be given by

$$\mathcal{L} = \frac{1}{2}\left\{\frac{1}{v^2}\left(\frac{\partial\varphi}{\partial t}\right)^2 - \sum_{i=1}^{3}\left(\frac{\partial\varphi}{\partial x^i}\right)^2 - \mu^2\varphi^2\right\}, \tag{7.19}$$

where μ has the physical dimension of an inverse length. The generalization of (7.16) and (7.17) to three spatial dimensions is obvious. It yields the equation of motion

$$\frac{\partial\mathcal{L}}{\partial\varphi} - \frac{\partial}{\partial t}\frac{\partial\mathcal{L}}{\partial(\partial\varphi/\partial t)} - \sum_{i=1}^{3}\frac{\partial}{\partial x^i}\frac{\partial\mathcal{L}}{\partial(\partial\varphi/\partial x^i)} = 0\,. \tag{7.20}$$

In the example defined by (7.19) we obtain

$$\frac{1}{v^2}\frac{\partial^2\varphi}{\partial t^2} - \Delta\varphi + \mu^2\varphi = 0\,, \tag{7.21}$$

where $\Delta = \sum_{i=1}^{3}\partial^2/(\partial x^i)^2$ is the Laplacian operator.

For $\mu = 0$ (7.21) is the wave equation in three space dimensions. For the case $\mu \neq 0$ and the velocity v equal to the speed of light c, the differential equation (7.21) is called the *Klein–Gordon equation.*

7.4 Canonically Conjugate Momentum and Hamiltonian Density

The continuous field variable φ whose equation of motion is derived from the Lagrangian density $\mathcal{L}$ is the analog of the coordinate variables q_j of point mechanics. The canonically conjugate momenta are defined in (2.39) to be the partial derivatives of the Lagrangian function L with respect to $\dot{q}_j$. Following that analogy we define

$$\pi(x) \stackrel{\text{def}}{=} \frac{\partial\mathcal{L}}{\partial(\partial\varphi/\partial t)}\,. \tag{7.22}$$

For example, with $\mathcal{L}$ as given in (7.14) we find $\pi(x) = \varrho\dot{\varphi}(x)$. This is nothing but the local density of momentum for transverse vibrations of the string (or, likewise, longitudinal vibrations of a rubber band). Following the pattern of the definition (2.38) one constructs the function

$$\tilde{\mathcal{H}} = \dot{\varphi}\frac{\partial\mathcal{L}}{\partial(\partial\varphi/\partial t)} - \mathcal{L}$$

and, by means of Legendre transformation, the Hamiltonian density $\mathcal{H}$. In the example (7.14), for instance, one finds

$$\mathcal{H} = \frac{1}{2\varrho}\left[\pi^2(x) + \varrho^2 v^2\left(\frac{\partial\varphi}{\partial x}\right)^2\right]\,.$$

$\mathcal{H}$ describes the energy density of the vibrating system. Therefore, $H = \int_0^L \mathrm{d}x\, \mathcal{H}$ is the total energy of the system. For example, inserting the explicit solution (7.10) into the expression for $\mathcal{H}$, one easily finds that

$$H = \frac{1}{4} \varrho A^{(p)^2} \bar{\omega}_0^2 p^2 .$$

This is the total energy contained in the pth harmonic vibration.

7.5 Example: The Pendulum Chain

A generalization of the harmonic transverse oscillations of the n-point system of Sect. 7.1 is provided by the chain of pendulums shown in Fig. 7.3. It consists of n identical mathematical pendulums of length l and mass m which are suspended along a straight line and which swing in planes perpendicular to that line. They are coupled by means of harmonic forces in such a way that the torque acting between the ith and the $(i+1)$th pendulum is proportional to the difference of their deviations from the vertical, i.e. is given by $-k(\varphi_{i+1} - \varphi_i)$. The line of suspension may be thought of as being realized by a torsion bar. As the chain is fixed at its ends, we formally add two more, motionless pendulums at either end of the bar, to which we ascribe the numbers 0 and $(n+1)$. This means that the angles φ_0 and φ_{n+1} are taken to be zero at all times. The kinetic and potential energies of this system are given by (cf. Sect. 1.17.2)

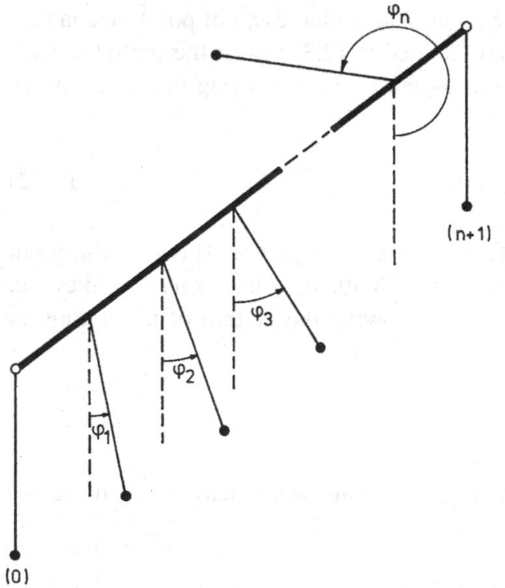

Fig. 7.3. A chain of pendulums, which are coupled by harmonic forces. While the first three show small deviations from the vertical, pendulum number n has made almost a complete turn

$$T = \frac{1}{2} m l^2 \sum_{j=0}^{n+1} \dot{\varphi}_j^2 ,$$

$$U = mgl \sum_{i=0}^{n+1} (1 - \cos\varphi_i) + \frac{1}{2} k \sum_{i=0}^{n} (\varphi_{i+1} - \varphi_i)^2 . \tag{7.23}$$

From the Lagrangian function in its natural form, $L = T - U$, and the Euler–Lagrange equations (2.28) we obtain the equations of motion

$$\ddot{\varphi}_i - \omega_0^2 \left[(\varphi_{i+1} - \varphi_i) - (\varphi_i - \varphi_{i-1}) \right] + \omega_1^2 \sin\varphi_i = 0 , \quad i = 1, \dots, n . \tag{7.24}$$

Here we introduced the following constants

$$\omega_0^2 = \frac{k}{ml^2} , \quad \omega_1^2 = \frac{g}{l} .$$

With $g = 0$ (7.24) is identical to (7.6). For $k = 0$ we recover the equation of motion of the plane pendulum that we studied in Sect. 1.17.2.

Let the horizontal distance of the pendulums be d so that the length of the chain is $L = (n+1)d$. We consider the transition to the corresponding continuous system by taking the limit $n \to \infty$, $d \to 0$. The countable variables $\varphi_1(t), \dots, \varphi_n(t)$ are replaced with the continuous variable $\varphi(x, t)$, x taking over the role of the counting index, which runs from 1 to n. While the discrete system had n degrees of freedom, the continuous system has an uncountably infinite number of degrees of freedom. Let $\varrho = m/d$ be the mass density. The constant in the harmonic force is set equal to $k = \eta/d$, η being proportional to the modulus of torsion of the bar. When d tends to zero, k must formally tend to infinity in such a way that the product $\eta = kd$ stays finite. At the same time the quantity

$$\omega_0^2 d^2 = \frac{d}{m} \frac{kd}{l^2} = \frac{\eta}{\varrho l^2} \equiv v^2$$

stays finite in that limit. In the same way as in Sect. 7.2, (7.24) becomes the equation of motion

$$\frac{\partial^2 \varphi(x,t)}{\partial t^2} - v^2 \frac{\partial^2 \varphi(x,t)}{\partial x^2} + \omega_1^2 \sin\varphi(x,t) = 0 . \tag{7.25}$$

This is the wave equation (7.12), supplemented by the nonlinear term $\omega_1^2 \sin\varphi$. Equation (7.25) is said to be the *Sine–Gordon equation.* In contrast to the wave equation (7.12) or to the Klein–Gordon equation (7.21) it is nonlinear in the field variable φ. It is the Euler–Lagrange equation (7.17) corresponding to the following Lagrangian density:

$$\mathcal{L} = \frac{1}{2} \varrho l^2 \left[\left(\frac{\partial\varphi}{\partial t} \right)^2 - v^2 \left(\frac{\partial\varphi}{\partial x} \right)^2 - 2\omega_1^2 (1 - \cos\varphi) \right] . \tag{7.26}$$

The latter may be obtained from (7.23) in the limit described above. Let us discuss solutions of (7.25) for two special cases.

(i) *The case of small deviations from the vertical.* In its discrete form (7.24) this coupled system of nonlinear equations of motions can only be solved in closed form for the case of small deviations from the vertical. Taking $\sin\varphi_1 \simeq \varphi_1$, we see that (7.24) becomes a linear system which may be solved along the lines of Practical Example 1 of Chap. 2, or following Sect. 7.1 above. In analogy to (7.9) we set

$$\varphi_j^{(p)}(t) = A^{(p)} \sin\left(\frac{jp\pi}{n+1}\right) \sin\left(\omega_p t\right) , \quad j = 0, 1, \ldots, n+1 \tag{7.27}$$

and obtain ω_p in terms of ω_1 and ω_0, as follows:

$$\begin{aligned} \omega_p^2 &= \omega_1^2 + 2\omega_0^2\left(1 - \cos\left(\frac{p\pi}{n+1}\right)\right) \\ &= \omega_1^2 + 4\omega_0^2 \sin^2\left(\frac{p\pi}{2(n+1)}\right) . \end{aligned} \tag{7.28}$$

The corresponding solution of the continuous system (7.25), assuming small deviations from the vertical, is obtained with $\sin\varphi(x,t) \simeq \varphi(x,t)$ and by making use of the results of Sect. 7.2. For large n (7.28) gives

$$\begin{aligned} \omega_p^2 &\simeq \omega_1^2 + 4\omega_0^2\left(\frac{p\pi}{2(n+1)}\right)^2 = \omega_1^2 + \omega_0^2 d^2\left(\frac{p\pi}{L}\right)^2 \\ &= \omega_1^2 + v^2\left(\frac{p\pi}{L}\right)^2 , \end{aligned}$$

so that (7.27) yields the pth harmonic oscillation

$$\varphi^{(p)}(x,t) = A^{(p)} \sin\left(\frac{p\pi x}{L}\right) \sin\left(\omega_p t\right)$$

with

$$\omega_p^2 = \omega_1^2 + \left(\frac{p\pi}{L}\right)^2 v^2 = \omega_1^2 + p^2\bar{\omega}_0^2$$

and with $\bar{\omega}_0$ as in (7.18).

(ii) *Soliton solutions.* For the continuous chain of infinite extension there are interesting and simple exact solutions of the equation of motion (7.25). Introducing the dimensionless variables

$$z \stackrel{\text{def}}{=} \frac{\omega_1}{v} x , \quad \tau \stackrel{\text{def}}{=} \omega_1 t ,$$

(7.25) takes the form

$$\frac{\partial^2\varphi(z,\tau)}{\partial\tau^2} - \frac{\partial^2\varphi(z,t)}{\partial z^2} + \sin\varphi(z,\tau) = 0 . \tag{7.25$'$}$$

Furthermore, we take $\varphi = 4\arctan f(z,\tau)$. With $f = \tan(\varphi/4)$ and using the well-known trigonometric formulae

$$\sin(2x) = \frac{2\tan x}{1+\tan^2 x}\,, \quad \tan(2x) = \frac{2\tan x}{1-\tan^2 x}\,,$$

we obtain

$$\sin\varphi = 4f\left(1-f^2\right)/\left(1+f^2\right)^2 .$$

From (7.25′) follows a differential equation for f

$$\left(1+f^2\right)\left(\frac{\partial^2 f}{\partial\tau^2} - \frac{\partial^2 f}{\partial z^2}\right) + f\left[1-f^2+2\left(\frac{\partial f}{\partial z}\right)^2 - 2\left(\frac{\partial f}{\partial\tau}\right)^2\right] = 0\,.$$

Finally, we set $y = (z+\alpha\tau)/\sqrt{1-\alpha^2}$, with α a real parameter in the interval $-1 < \alpha < 1$. If f is understood to be a function of y, the following differential equation is obtained:

$$\left(1+f^2\right)\frac{\mathrm{d}^2 f}{\mathrm{d}y^2} - f\left[1-f^2+2\left(\frac{\mathrm{d}f}{\mathrm{d}y}\right)^2\right] = 0\,.$$

It is not difficult to guess two simple solutions of the latter. They are $f_\pm = \mathrm{e}^{\pm y}$. Thus, the original differential equation (7.25′) has the special solutions

$$\varphi_\pm(z,\tau) = 4\arctan\left(\exp\left\{\pm\frac{z+\alpha\tau}{\sqrt{1-\alpha^2}}\right\}\right)\,. \tag{7.29}$$

As an example choose the positive sign, take $\alpha = -0.5$ and consider the time $\tau = 0$. For sufficiently large negative z the amplitude φ_+ is practically zero. For $z = 0$ it is $\varphi_+(0,0) = \pi$, while for sufficiently large positive z it is almost equal to 2π. In a diagram with z as the abscissa and $\varphi_+(z,\tau)$ the ordinate, this transition of the field from the value 0 to the value 2π propagates, with increasing time, in the positive z-direction and with the (dimensionless) velocity α. One may visualize the continuous pendulum chain as an infinitely long rubber belt whose width is l and which is suspended vertically. The process just described is then a flip-over of a vertical strip of the belt from $\varphi = 0$ to $\varphi = 2\pi$ which moves with constant velocity along the rubber belt. This strange and yet simple motion is characteristic of the nonlinear equation of motion (7.25). It is called a *soliton solution.* Expressing the results in terms of the original, dimensionful variables x and t, one sees that the soliton moves with velocity $v\alpha$ along the positive or the negative x-direction.

7.6 Comments and Outlook

So far we have studied continuous systems by means of examples taken from the mechanics of finitely many, say n, point particles, by letting n go to infinity. This limiting procedure is very useful for understanding the role of the fields $\varphi(\boldsymbol{x}, t)$ as the new dynamical variables which replace the coordinate functions $q(t)$ of the mechanics of point particles. This does not mean, however, that every continuous system can be obtained by or could be thought of as the limit $f \to \infty$ of a discrete system with f degrees of freedom. On the contrary, the set of classical, continuous systems is much richer than one might expect on the basis of the examples studied above. Continuous systems form the subject of *classical field theory*, an important branch of physics in its own right. Field theory, for which electrodynamics is a prominent example, goes beyond the scope of this book and we can do no more than add a few comments and an outlook here.

Let us suppose that the dynamics of N fields

$$\left\{\varphi^i(x) | i = 1, 2, \ldots, N\right\}$$

can be described by means of a Lagrange density $\mathcal{L}$ in such a way that the equations of motion that follow from it satisfy the postulate of special relativity (cf. Sect. 4.3), i.e., that they are form invariant with respect to Lorentz transformations. Assume, furthermore, that each of the fields $\varphi^i(x)$ is invariant under Lorentz transformations of space–time, viz.

$$\varphi'^i(x' = \Lambda x) = \varphi^i(x) \quad \text{with} \quad \Lambda \in L_+^\uparrow .$$

Fields possessing this simple transformation behavior are called *scalar fields*. The variational principle (7.16) is independent of the choice of coordinates $(\boldsymbol{x}, t)$. Indeed, the hypersurface $(\boldsymbol{x}, t_1 = \text{const.})$ and $(\boldsymbol{x}, t_2 = \text{const.})$ can be deformed into an arbitrary, smooth, three-dimensional, hypersurface Σ in space-time. In the action integral (7.16) one then integrates over the volume enclosed by Σ and chooses the variations $\delta\varphi^i$ of the fields such that they vanish on the hypersurface Σ. The form of the equations of motion (7.17) is always the same. This has an important consequence: Whenever the Lagrange density $\mathcal{L}$ is *invariant* under Lorentz transformations, the equations of motion (7.17) which follow from (7.16) are *form invariant*, i.e., they have the same form in every frame of reference.

The Lagrange density (7.14) may serve as an example for a Lorentz-invariant theory, provided we replace the parameter v by the velocity of light c. The equation of motion which follows from it

$$\frac{1}{c^2}\frac{\partial^2 \varphi}{\partial t^2} - \frac{\partial^2 \varphi}{\partial \boldsymbol{x}^2} = 0 \tag{7.30}$$

is form invariant. (This equation is the source-free wave equation.) With φ a scalar field it is even fully invariant itself. It is instructive to check this: With $x'^\mu = \Lambda^\mu{}_\nu x^\nu$ and $x_\mu = g_{\mu\nu}x^\nu$, and making use of the following simplified notation for the partial derivatives

$$\partial_\mu := \frac{\partial}{\partial x^\mu}\,, \qquad \partial^\mu := \frac{\partial}{\partial x_\mu} \tag{7.31}$$

one sees that (7.30) contains the expression $\partial_\mu \partial^\mu \varphi$. The transformation behavior of ∂_ν follows from the following calculation

$$\partial_\nu \equiv \frac{\partial}{\partial x^\nu} = \frac{\partial x'^\mu}{\partial x^\nu}\frac{\partial}{\partial x'^\mu} = \Lambda^\mu{}_\nu \frac{\partial}{\partial x'^\mu} \equiv \Lambda^\mu{}_\nu \partial'_\mu\,.$$

The transformation behavior of ∂^ν being the inverse of the above, the differential operator $\partial_\nu \partial^\nu$ is a Lorentz invariant operator. It is often called the Laplace operator in four dimensions and is denoted by the symbol $\Box$,

$$\Box := \partial_\nu \partial^\nu = \frac{1}{c^2}\frac{\partial^2}{\partial t^2} - \sum_{i=1}^{3} \frac{\partial^2}{(\partial x^i)^2} \equiv \frac{1}{c^2}\frac{\partial^2}{\partial t^2} - \Delta\,, \tag{7.32}$$

where Δ is the Laplace operator in three dimensions. Note that the derivative terms in (7.14) as well as in (7.19) (taking $v = c$ in either example) can be rewritten in the form of an invariant scalar product $(\partial_\mu \varphi)(\partial^\mu \varphi)$ of

$$\partial_\mu \varphi = \left(\frac{1}{c}\frac{\partial \varphi}{\partial t}, \nabla\varphi\right) \quad \text{and} \quad \partial^\mu \varphi = \left(\frac{1}{c}\frac{\partial \varphi}{\partial t}, -\nabla\varphi\right).$$

Thus, a Lorentz invariant theory of our fields φ^i could be designed by means of a Lagrange density of the form

$$\mathcal{L}\left(\varphi^i, \partial_\mu \varphi^i\right) = \frac{1}{2}\left\{\sum_{i=1}^{N} \left(\partial_\mu \varphi^i\right)\left(\partial^\mu \varphi^i\right) - \sum_{i=1}^{N} \lambda_i \left[\varphi^i(x)\right]^2 - U\left(\varphi^i(x)\right)\right\}, \tag{7.33}$$

with $U(\varphi^i)$ a Lorentz scalar function of the fields. The first term on the right-hand side of (7.33) is the analog of the kinetic energy in the mechanics of point particles, the last term is the analog of the potential. The second term, which is new, is called *mass term* because in the quantized version of the theory it does indeed contain the rest masses of the particles which are described by the fields. Of course, it could equally well be considered as part of the potential U.

In this discussion one recognizes, though in a sketchy manner only, an important building principle for classical field theories: Very much like in mechanics of point particles, symmetries and invariances can be read off, or can be built into, the Lagrange density $\mathcal{L}$. Above we considered the example of form invariance with respect to Lorentz transformations. As the next step in a deeper and more detailed analysis one would derive the theorem of Emmy Noether, in its form adapted to field theory, which states that the energy, the momentum, or the angular momentum are conserved quantities whenever $\mathcal{L}$ is invariant under translations in time, translations in space, or under rotations, respectively. A new feature is the appearence of a local energy *density* (cf. the example studied in Sect. 7.4), and, analogously, momentum and angular momentum *densities*. Noether's theorem concerns *local* quantities. If the energy density changes locally, i.e., if it changes in a finite domain of space and time, then there must be a continuity equation which guarantees that the total energy (i.e., the integral of the energy density over space) remains unchanged. Analogous statements apply to momentum and angular momentum densities.

Finally, $\mathcal{L}$ may possess further, *inner*, symmetries which have to do with transformations on the fields. In this case there are additional conservation laws, or continuity equations, as shown by the following simple example.

Given two real scalar fields and a Lagrange density of the form (7.33) which is such that $\lambda_1 = \lambda_2 \equiv \lambda$ and where U depends on the sum of the squares of the fields only,

$$\mathcal{L}\left(\varphi^i, \partial_\mu \varphi^i\right) = \frac{1}{2}\left\{\sum_{i=1}^{2}\left(\partial_\mu \varphi^i\right)\left(\partial^\mu \varphi^i\right) - \lambda \sum_{i=1}^{2}\left[\varphi^i(x)\right]^2 \right.$$
$$\left. -U\left(\sum_{i=1}^{2}\left(\varphi^i(x)\right)^2\right)\right\} . \tag{7.34}$$

In addition to being invariant under Lorentz transformations in space and time $\mathcal{L}$ is obviously invariant under orthogonal transformations of the fields as a whole, of the kind

$$\begin{aligned} \varphi'^1(x) &= \varphi^1(x)\cos\alpha - \varphi^2(x)\sin\alpha ,\\ \varphi'^2(x) &= \varphi^1(x)\sin\alpha + \varphi^2(x)\cos\alpha \end{aligned} \tag{7.35}$$

with $\alpha \in [0, 2\pi]$. Equations (7.35) describe a formal rotation in the two-dimensional, inner, space which is spanned by the independent fields φ^1 and φ^2. In particular, if we choose the angle α to be infinitesimal, $\alpha = \varepsilon$, then (7.35) becomes

$$\delta\varphi^1 := \varphi'^1 - \varphi^1 = -\varepsilon\varphi^2 , \quad \delta\varphi^2 := \varphi'^2 - \varphi^2 = \varepsilon\varphi^1 . \tag{7.36}$$

As these changes in the fields are special cases of variations, one can calculate the corresponding change of $\mathcal{L}$. Let us write (7.36) as $\delta\varphi^i = \sum_{k=1}^{2} \varepsilon_{ik}\varphi^k$ with $\varepsilon_{11} = \varepsilon_{22} = 0$ and $-\varepsilon_{12} = \varepsilon_{21} = \varepsilon$. Then we find

$$\begin{aligned} \delta\mathcal{L} &= \sum_{i=1}^{2}\left(\frac{\partial\mathcal{L}}{\partial\varphi^i}\delta\varphi^i + \frac{\partial\mathcal{L}}{\partial(\partial_\mu\varphi^i)}\delta\partial_\mu\varphi^i\right)\\ &= \partial_\mu\left[\sum_{i,k=1}^{2}\frac{\partial\mathcal{L}}{\partial(\partial_\mu\varphi^i)}\varepsilon_{ik}\varphi^k\right] \equiv \partial_\mu j^\mu(x) , \end{aligned}$$

where we have replaced $\partial\mathcal{L}/\partial\varphi^i$ in the first term by

$$\frac{\partial\mathcal{L}}{\partial\varphi^i} = \partial_\mu \frac{\partial\mathcal{L}}{\partial(\partial_\mu\varphi^i)}$$

making use of the equations of motion. The right-hand side of the above equation is nothing but a divergence in four dimensions, the quantity εj^μ being defined by the expression in square brackets. The left-hand side vanishes because the change of the Lagrange density is zero, $\delta\mathcal{L} = 0$. It is not difficult to show that j^μ is a four-vector with respect to Lorentz transformations. In the concrete example considered here one calculates the explicit form of this vector from the Lagrange density (7.34) and making use of the formulae (7.36). The result is

$$j^\mu(x) = \left(\partial^\mu \varphi^2(x)\right)\varphi^1(x) - \left(\partial^\mu \varphi^1(x)\right)\varphi^2(x)\,. \tag{7.37}$$

The statement that the four-divergence of the quantity j^μ vanishes, in fact, is a continuity equation. The time component j^0 and the space components $\boldsymbol{j}$ have the same physical dimension. Therefore, if $\boldsymbol{j}$ is a current density, that is, if it has dimension, e.g., charge × velocity per unit of volume, then j^0 is not yet a density, which, in our example, should have dimension charge per unit volume. However, $\varrho(x, t) = j^0/c$ is a density with the correct physical dimension. Therefore, in a given frame of reference, we set $j^\mu = (c\varrho, \boldsymbol{j})$ so that the continuity equation becomes

$$\partial_\mu j^\mu = \frac{\partial \varrho(\boldsymbol{x}, t)}{\partial t} + \nabla \cdot \boldsymbol{j}(\boldsymbol{x}, t) = 0\,. \tag{7.38}$$

When the density ϱ in a given, finite, space volume increases or decreases, this change is compensated by a flow of charge into this volume, or out of this volume. The total charge contained in the fields is given by the integral of the density ϱ over the entire space. Provided the fields and, therefore, also the current density $\boldsymbol{j}$ vanish sufficiently fast at infinity, equation (7.38) implies that the total charge $Q := \int \mathrm{d}^3 x \varrho(\boldsymbol{x}, t)$ is a constant of the motion,

$$\frac{\mathrm{d}}{\mathrm{d}t} Q = \frac{\mathrm{d}}{\mathrm{d}t} \int \mathrm{d}^3 x \varrho(\boldsymbol{x}, t) = - \int \mathrm{d}^3 x \nabla \cdot \boldsymbol{j}(\boldsymbol{x}, t) = 0\,. \tag{7.39}$$

Indeed, the right-hand side of this equation vanishes because the volume integral of the divergence over space equals the surface integral of the radial component of $\boldsymbol{j}$ over the surface at infinity[1].

In the example (7.34) invariance with respect to the transformations (7.35) leads to the conservation law (7.38), or (7.39), with $j^\mu(x)$ as given by the expression (7.37). It is useful to replace the real fields φ^1 and φ^2 by a complex field and its complex conjugate, through the definitions

$$\phi(x) = \frac{1}{\sqrt{2}}\left(\varphi^1(x) + i\varphi^2(x)\right)\,, \quad \phi^*(x) = \frac{1}{\sqrt{2}}\left(\varphi^1(x) - i\varphi^2(x)\right)\,.$$

[1] One shows, furthermore, that the charge Q is a Lorentz invariant quantity, i.e., that its value does not depend on the frame of reference in which it is calculated. This holds if and only if $\partial_\mu j^\mu(x) = 0$.

The Lagrange density (7.34) then takes a simpler form, namely

$$\mathcal{L} = \left(\partial_\mu \phi^*\right)\left(\partial^\mu \phi\right) - \lambda \phi^* \phi - U\left(\phi^* \phi\right) . \tag{7.40}$$

Similarly, the transformation (7.35) simplifies to

$$\phi'(x) = \phi(x)\,\mathrm{e}^{\mathrm{i}\alpha} , \quad \phi'^*(x) = \phi^*(x)\,\mathrm{e}^{-\mathrm{i}\alpha} , \tag{7.41}$$

and the quantity (7.37) becomes

$$j^\mu(x) = -\mathrm{i}\left[\phi^*(x)\partial^\mu\phi(x) - \left(\partial^\mu\phi^*(x)\right)\phi(x)\right] . \tag{7.42}$$

In quantum physics one learns that, indeed, this expression is a suitable candidate for the description of the electric charge and current densities of a scalar particle.

Exercises

Chapter 1: Elementary Newtonian Mechanics

1.1 Under the assumption that the orbital angular momentum $\boldsymbol{l} = \boldsymbol{r} \times \boldsymbol{p}$ of a particle is conserved show that its motion takes place in a plane spanned by $\boldsymbol{r}_0$, the initial position, and $\boldsymbol{p}_0$, the initial momentum. Which of the orbits of Fig. 1 are possible in this case? (O denotes the origin of the coordinate system.)

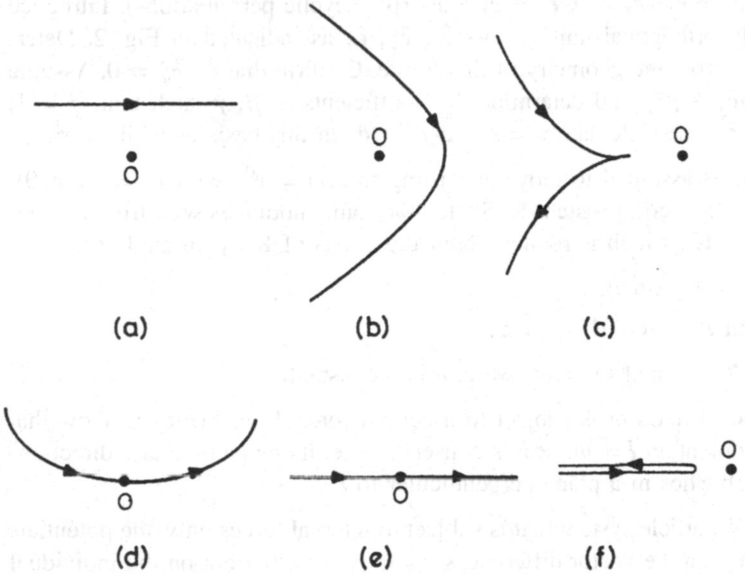

Fig. 1.

1.2 In the plane of motion of Exercise 1.1 introduce polar coordinates $\{r(t), \varphi(t)\}$. Calculate the line element $(\mathrm{d}s)^2 = (\mathrm{d}x)^2 + (\mathrm{d}y)^2$, as well as $\boldsymbol{v}^2 = \dot{x}^2 + \dot{y}^2$ and $\boldsymbol{l}^2$, in the polar coordinates. Express the kinetic energy in terms of $\dot{r}$ and $\boldsymbol{l}^2$.

1.3 For the description of motions in $\mathbb{R}^3$ one may use Cartesian coordinates $\boldsymbol{r}(t) = \{x(t), y(t), z(t)\}$, or spherical coordinates $\{r(t), \theta(t), \varphi(t)\}$. Calculate the infinitesimal line element $(\mathrm{d}s)^2 = (\mathrm{d}x)^2 + (\mathrm{d}y)^2 + (\mathrm{d}z)^2$ in spherical coordinates. Use this result to derive the square of the velocity $\boldsymbol{v}^2 = \dot{x}^2 + \dot{y}^2 + \dot{z}^2$ in these coordinates.

F. Scheck, *Mechanics*, Graduate Texts in Physics, 5th ed.,
DOI 10.1007/978-3-642-05370-2, © Springer-Verlag Berlin Heidelberg 2010

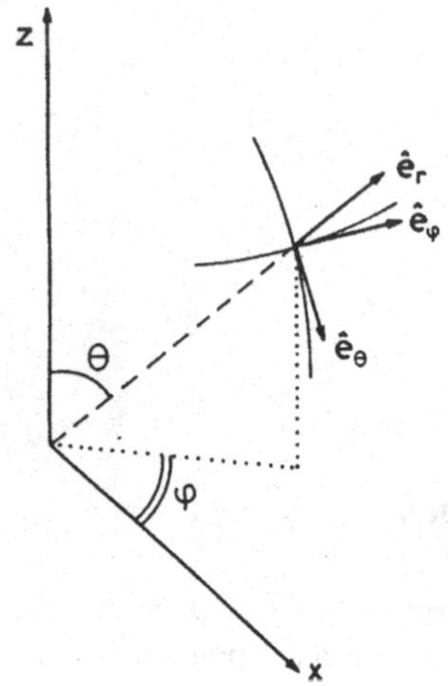

Fig. 2.

1.4 Let $\hat{e}_x$, $\hat{e}_y$, $\hat{e}_z$ be Cartesian unit vectors. They then fulfill $\hat{e}_x^2 = \hat{e}_y^2 = \hat{e}_z^2 = 1$, $\hat{e}_x \cdot \hat{e}_y = \hat{e}_x \cdot \hat{e}_z = \hat{e}_y \cdot \hat{e}_z = 0$, $\hat{e}_z = \hat{e}_x \times \hat{e}_y$ (plus cyclic permutations). Introduce three, mutually orthogonal unit vectors $\hat{e}_r$, $\hat{e}_\varphi$, $\hat{e}_\theta$ as indicated in Fig. 2, Determine $\hat{e}_r$ and $\hat{e}_\varphi$ from the geometry of this figure. Confirm that $\hat{e}_r \cdot \hat{e}_\varphi = 0$. Assume $\hat{e}_\theta = \alpha\hat{e}_x + \beta\hat{e}_y + \gamma\hat{e}_z$ and determine the coefficients α, β, γ such that $\hat{e}_\theta^2 = 1$, $\hat{e}_\theta \cdot \hat{e}_\varphi = 0 = \hat{e}_\theta \cdot \hat{e}_r$. Calculate $\boldsymbol{v} = \dot{\boldsymbol{r}} = \mathrm{d}(r\hat{e}_r)/\mathrm{d}t$ in this basis as well as $\boldsymbol{v}^2$.

1.5 A particle is assumed to move according to $\boldsymbol{r}(t) = \boldsymbol{v}^0 t$ with $\boldsymbol{v}^0 = \{0, v, 0\}$, with respect to the inertial system **K**. Sketch the same motion as seen from another reference frame **K**′, which is rotated about the z-axis of **K** by an angle Φ,

$$x' = x\cos\Phi + y\sin\Phi,$$
$$y' = -x\sin\Phi + y\cos\Phi,\quad z' = z,$$

for the cases $\Phi = \omega$ and $\Phi = \omega t$, were ω is a constant.

1.6 A particle of mass m is subject to a central force $\boldsymbol{F} = F(r)\boldsymbol{r}/r$. Show that the angular momentum $\boldsymbol{l} = m\boldsymbol{r} \times \dot{\boldsymbol{r}}$ is conserved (i.e. its magnitude and direction) and that the orbit lies in a plane perpendicular to $\boldsymbol{l}$.

1.7 (i) In an N-particle system that is subject to internal forces only, the potentials V_{ik} depend only on the vector differences $\boldsymbol{r}_{ik} = \boldsymbol{r}_i - \boldsymbol{r}_k$, but not on the individual vectors $\boldsymbol{r}_i$. Which quantities are conserved in this system?
(ii) If V_{ik} depends only on the modulus $|\boldsymbol{r}_{ik}|$ the force acts along the straight line joining i to k. There is one more integral of the motion.

1.8 Sketch the one-dimensional potential

$$U(q) = -5q\mathrm{e}^{-q} + q^{-4} + 2/q \quad \text{for} \quad q \geq 0$$

and the corresponding phase portraits for a particle of mass $m = 1$ as a function of energy and initial position q_0. In particular, find and discuss the two points of equilibrium. Why are the phase portraits symmetric with respect to the abscissa?

1.9 Study two identical pendula of length l and mass m, coupled by a harmonic spring, the spring being inactive when both pendulums are at rest. For small deviations from the vertical the energy reads

$$E = \frac{1}{2m}(x_2^2 + x_4^2) + \frac{1}{2}m\omega_0^2(x_1^2 + x_3^2) + \frac{1}{2}m\omega_1^2(x_1 - x_3)^2$$

with $x_2 = m\dot{x}_1$, $x_4 = m\dot{x}_3$. Identify the individual terms of this equation. Derive from it the equations of motion in phase space,

$$\frac{\mathrm{d}\boldsymbol{x}}{\mathrm{d}t} = \mathbf{M}\boldsymbol{x}\,.$$

The transformation

$$\boldsymbol{x} \to \boldsymbol{u} = \mathbf{A}\boldsymbol{x} \quad \text{with} \quad \mathbf{A} = \frac{1}{\sqrt{2}}\begin{pmatrix} \mathbb{1} & \mathbb{1} \\ \mathbb{1} & -\mathbb{1} \end{pmatrix} \quad \text{and}$$

$$\mathbb{1} \equiv \begin{pmatrix} 1 & 0 \\ 0 & 1 \end{pmatrix}$$

decouples these equations. Write the equations obtained in this way in dimensionless form and solve them.

1.10 The one-dimensional harmonic oscillator satisfies the differential equation

$$m\ddot{x}(t) = -\lambda x(t)\,, \tag{1.1}$$

with m the inertial mass, λ a positive constant, and $x(t)$ the deviation from equilibrium. Equivalently, (1.1) can be written as

$$\ddot{x} + \omega^2 x = 0, \quad \omega^2 \stackrel{\text{def}}{=} \lambda/m\,. \tag{1.2}$$

Solve the differential equation (1.2) by means of $x(t) = a\cos(\mu t) + b\sin(\mu t)$ for the initial condition

$$x(0) = x_0 \quad \text{and} \quad p(0) = m\dot{x}(0) = p_0\,. \tag{1.3}$$

Let $x(t)$ be the abscissa and $p(t)$ the ordinate of a Cartesian coordinate system. Draw the graph of the solution with $\omega = 0.8$ that goes through the point $(x_0 = 1, p_0 = 0)$.

1.11 Adding a weak friction force to the system of Exercise 1.10 yields the equation of motion

$$\ddot{x} + \kappa\dot{x} + \omega^2 x = 0\,.$$

"Weak" means that $\kappa < 2\omega$. Solve the differential equation by means of

$$x(t) = \mathrm{e}^{at}[x_0\cos\tilde{\omega}t + (p_0/m\tilde{\omega})\sin\tilde{\omega}t]\,.$$

Draw the graph $(x(t), p(t))$ of the solution with $\omega = 0.8$ which goes through $(x_0 = 1, p_0 = 0)$.

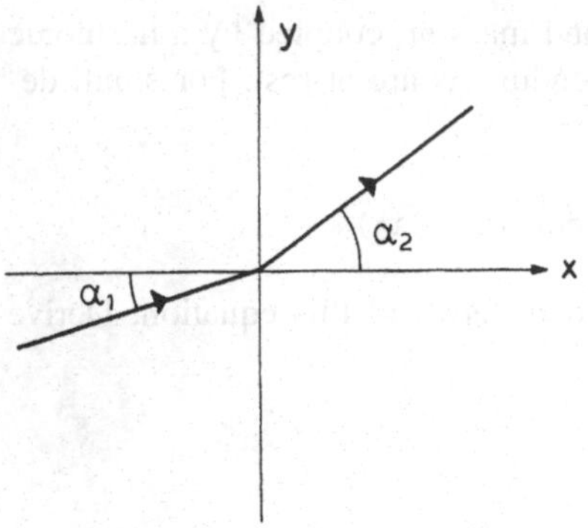

Fig. 3.

1.12 A mass point of mass m moves in the piecewise constant potential (see Fig. 3)

$$U = \begin{cases} U_1 & \textit{for } x < 0 \\ U_2 & \textit{for} x > 0. \end{cases}$$

In crossing from the domain $x < 0$, where its velocity was v_1, to the domain $x > 0$, it changes its velocity (modulus and direction). Express U_2 in terms of the quantities U_1, v_1, α_1, and α_2. What is the relation of α_1 to α_2 when (i) $U_1 < U_2$ and (ii) $U_1 > U_2$? Work out the relationship to the law of refraction of geometrical optics.

Hint: Make use of the principle of energy conservation and show that one component of the momentum remains unchanged in crossing from $x < 0$ to $x > 0$.

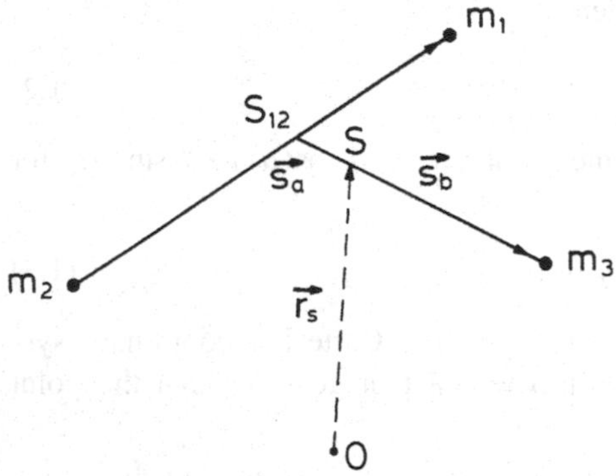

Fig. 4.

1.13 In a system of three mass points m_1, m_2, m_3 let S_{12} be the center-of-mass of 1 and 2 and S the center-of-mass of the whole system. Express the coordinates $\boldsymbol{r}_1$, $\boldsymbol{r}_2$, $\boldsymbol{r}_3$ in terms of $\boldsymbol{r}_s$, $\boldsymbol{s}_a$, and $\boldsymbol{s}_b$, as defined in Fig. 4. Calculate the total kinetic energy in terms of the new coordinates and interpret the result. Write the total angular momentum in terms of the new coordinates and show that $\sum_i \boldsymbol{l}_i = \boldsymbol{l}_s + \boldsymbol{l}_a + \boldsymbol{l}_b$, where $\boldsymbol{l}_s$ is the angular momentum of the center-of-mass and $\boldsymbol{l}_a$ and $\boldsymbol{l}_b$ are relative angular momenta. By considering a Galilei transformation $\boldsymbol{r}' = \boldsymbol{r} + \boldsymbol{\omega} t + \boldsymbol{a}$, $t' = t + s$ show that $\boldsymbol{l}_s$ depends on the choice of the origin, while $\boldsymbol{l}_a$ and $\boldsymbol{l}_b$ do not.

1.14 *Geometric similarity*. Let the potential $U(\boldsymbol{r})$ be a homogeneous function of degree α in the coordinates (x, y, z), i.e. $U(\lambda\boldsymbol{r}) = \lambda^\alpha U(\boldsymbol{r})$.

(i) Show by making the replacements $\boldsymbol{r} \to \lambda\boldsymbol{r}$ and $t \to \mu t$, and choosing $\mu = \lambda^{1-\alpha/2}$, that the energy is modified by a factor λ^α and that the equation of motion remains unchanged.

The consequence is that the equation of motion admits solutions that are geometrically similar, i.e. the time differences $(\Delta t)_a$ and $(\Delta t)_b$ of points that correspond to each other on geometrically similar orbits (a) and (b) and the corresponding linear dimensions L_a and L_b are related by

$$\frac{(\Delta t)_b}{(\Delta t)_a} = \left(\frac{L_b}{L_a}\right)^{1-\alpha/2} .$$

(ii) What are the consequences of this relationship for

- the period of harmonic oscillation?
- the relation between time and height of free fall in the neighborhood of the earth's surface?
- the relation between the periods and the semimajor axes of planetary ellipses?

(iii) What is the relation of the energies of two geometrically similar orbits for

- the harmonic oscillation?
- the Kepler problem?

1.15 *The Kepler problem.* (i) Show that the differential equation for $\Phi(r)$, in the case of finite orbits, has the following form:

$$\frac{\mathrm{d}\Phi}{\mathrm{d}r} = \frac{1}{r}\sqrt{\frac{r_\mathrm{P} r_\mathrm{A}}{(r - r_\mathrm{P})(r_\mathrm{A} - r)}}, \qquad (1.4)$$

where r_P and r_A denote the perihelion and the aphelion, respectively. Calculate r_P and r_A and integrate (1.4) with the boundary condition $\Phi(r_\mathrm{P}) = 0$.

(ii) Change the potential to $U(r) = (-A/r) + (B/r^2)$ with $|B| \ll l^2/2\mu$. Determine the new perihelion r'_P and the new aphelion r'_A and write the differential equation for $\Phi(r)$ in a form analogous to (1.4). Integrate this equation as in (i) and determine two successive perihelion positions for $B > 0$ and for $B < 0$.

Hint:

$$\frac{\mathrm{d}}{\mathrm{d}x} \arccos\left(\frac{\alpha}{x} + \beta\right) = \frac{\alpha}{x}\frac{1}{\sqrt{x^2(1-\beta^2) - 2\alpha\beta x - \alpha^2}} .$$

1.16 The most general solution of the Kepler problem reads, in terms of polar coordinates r and Φ,

$$r(\Phi) = \frac{p}{1 + \varepsilon \cos(\phi - \phi_0)}\,.$$

The parameters are given by

$$p = \frac{l^2}{A\mu}\,, \quad (A = Gm_1m_2)\,,$$

$$\varepsilon = \sqrt{1 + \frac{2El^2}{\mu A^2}}\,, \quad \left(\mu = \frac{m_1 m_2}{m_1 + m_2}\right).$$

What values of the energy are possible if the angular momentum is given? Calculate the semimajor axis of the earth's orbit under the assumption $m_{\text{Sun}} \gg m_{\text{Earth}}$;

$$G = 6.672 \times 10^{-11}\ \text{m}^3\,\text{kg}^{-1}\,\text{s}^{-2}\,,$$
$$m_{\text{S}} = 1.989 \times 10^{30}\ \text{kg}\,,$$
$$m_{\text{E}} = 5.97 \times 10^{24}\ \text{kg}\,.$$

Calculate the semimajor axis of the ellipse along which the sun moves about the center-of-mass of the sun and the earth and compare the result to the solar radius ($6.96 \times 10^8\ m$).

1.17 Determine the interaction of two electric dipoles $\boldsymbol{p}_1$ and $\boldsymbol{p}_2$ (example for noncentral potential force).

Hints: Calculate the potential of a single dipole $\boldsymbol{p}_1$, making use of the following approximation. The dipole consists of two charges $\pm e_1$ at a distance $\boldsymbol{d}_1$. Let e_1 tend to infinity and $|\boldsymbol{d}_1|$ to zero, in such a way that their product $\boldsymbol{p}_1 = \boldsymbol{d}_1 e_1$ stays constant. Then calculate the potential energy of a finite dipole $\boldsymbol{p}_2$ in the field of the first and perform the same limit $e_2 \to \infty$, $|\boldsymbol{d}_2| \to 0$, with $\boldsymbol{p}_2 = \boldsymbol{d}_2 e_2$ constant, as above. Calculate the forces that act on the two dipoles.

Answer:

$$W(1,2) = (\boldsymbol{p}_1 \cdot \boldsymbol{p}_2)/r^3 - 3(\boldsymbol{p}_1 \cdot \boldsymbol{r})(\boldsymbol{p}_2 \cdot \boldsymbol{r})/r^5\,,$$

$$\begin{aligned}\boldsymbol{F} &= -\nabla_1 W = \big[3(\boldsymbol{p}_1 \cdot \boldsymbol{p}_2)/r^5 \\ &\quad -15(\boldsymbol{p}_1 \cdot \boldsymbol{r})(\boldsymbol{p}_2 \cdot r)/r^7\big]\boldsymbol{r} \\ &\quad +3\big[\boldsymbol{p}_1(\boldsymbol{p}_2 \cdot \boldsymbol{r}) + \boldsymbol{p}_2(\boldsymbol{p}_1 \cdot \boldsymbol{r})\big]/r^5 = -\boldsymbol{F}_{12}\,.\end{aligned}$$

1.18 Let the motion of a point mass be governed by the law

$$\dot{\boldsymbol{v}} = \boldsymbol{v} \times \boldsymbol{a}\,, \quad \boldsymbol{a} = \text{const}\,. \tag{1.5}$$

Show that $\dot{\boldsymbol{r}} \cdot \boldsymbol{a} = \boldsymbol{v}(0) \cdot \boldsymbol{a}$ holds for all t and reduce (1.5) to an inhomogeneous differential equation of the form $\ddot{\boldsymbol{r}} + \omega^2 \boldsymbol{r} = f(t)$. Solve this equation by means of the substitution $\boldsymbol{r}_{\text{inhom}}(t) = \boldsymbol{c}t + \boldsymbol{d}$. Express the integration constants in terms of the initial values $\boldsymbol{r}(0)$ and $\boldsymbol{v}(0)$. Describe the curve $\boldsymbol{r}(t) = \boldsymbol{r}_{\text{hom}}(t) + \boldsymbol{r}_{\text{inhom}}(t)$.

Hint:

$$\boldsymbol{a}_1 \times (\boldsymbol{a}_2 \times \boldsymbol{a}_3) = \boldsymbol{a}_2(\boldsymbol{a}_1 \cdot \boldsymbol{a}_3) - \boldsymbol{a}_3(\boldsymbol{a}_1 \cdot \boldsymbol{a}_2) \,.$$

1.19 An iron ball falls vertically onto a horizontal plane from which it is reflected. At every bounce it loses the nth fraction of its kinetic energy. Discuss the orbit $x = x(t)$ of the bouncing ball and derive the relation between x_{max} and t_{max}.

Hint: Study the orbit between two successive bounces and sum over previous times.

1.20 Consider the following transformations of the coordinate system:

$$\{t, \boldsymbol{r}\} \underset{\text{E}}{\rightarrow} \{t, \boldsymbol{r}\}, \quad \{t, \boldsymbol{r}\} \underset{\text{P}}{\rightarrow} \{t, -\boldsymbol{r}\}, \quad \{t, \boldsymbol{r}\} \underset{\text{T}}{\rightarrow} \{-t, \boldsymbol{r}\},$$

as well as the transformation P·T that is generated by performing first T and then P. Write these transformations in the form of matrices that act on the four-component vector $\binom{t}{\boldsymbol{r}}$. Show that {E, P, T, PT} form a group.

1.21 Let the potential $U(r)$ of a two-body system be C^2 (twice continuously differentiable). For fixed relative angular momentum, under which additional condition on $U(r)$ are there circular orbits? Let E_0 be the energy of such an orbit. Discuss the motion for $E = E_0 + \varepsilon$ for small positive ε. Study the special cases

$$U(r) = r^n \quad \text{and} \quad U(r) = \lambda / r \,.$$

1.22 Following the methods explained in Sect. 1.26 show the following.

(i) In the northern hemisphere a falling object experiences a *southward* deviation of second order (in addition to the first-order eastward deviation).

(ii) A stone thrown vertically upward falls down *west* of its point of departure, the deviation being four times the eastward deviation of the falling stone.

1.23 Let a two-body system be subject to the potential

$$U(r) = -\frac{\alpha}{r^2}$$

in the relative coordinate r, with positive α. Calculate the scattering orbits $r(\Phi)$. For fixed angular momentum what are the values of α for which the particle makes one (two) revolutions about the center of force? Follow and discuss an orbit that collapses to $r = 0$.

1.24 A pointlike comet of mass m moves in the gravitational field of a sun with mass M and radius R. What is the total cross section for the comet to crash on the sun?

1.25 Solve the equations of motion for the example of Sect. 1.21.2 (Lorentz force with constant fields) for the case

$$\boldsymbol{B} = B\hat{\boldsymbol{e}}_z\,, \quad \boldsymbol{E} = E\hat{\boldsymbol{e}}_z\,.$$

1.26 Kepler problem and Hodograph: Let p_x and p_y be the components of the momentum in the plane of motion of the Kepler problem. Show: In momentum space, spanned by (p_x, p_y), all bound orbits are circles. Give the position and the radius of these circles. The curve described by the tip of the velocity, or momentum, vector is called *hodograph.*

Chapter 2: The Principles of Canonical Mechanics

2.1 The energy $E(q, p)$ is an integral of a finite, one-dimensional, periodic motion. Why is the portrait symmetric with respect to the q-axis? The surface enclosed by the periodic orbit is

$$F(E) = \oint p\,\mathrm{d}q = 2\int_{q_{\min}}^{q_{\max}} p\,\mathrm{d}q\,.$$

Show that the change of $F(E)$ with E equals the period T of the orbit, $T = \mathrm{d}F(E)/\mathrm{d}E$, Calculate F and T for the example

$$E(q, p) = p^2/2m + m\omega^2 q^2/2\,.$$

2.2 A weight glides without friction along a plane inclined by the angle α with respect to the horizontal. Study this system by means of d'Alembert's principle.

2.3 A ball rolls without friction on the inside of a circular annulus. The annulus is put upright in the earth's gravitational field. Use d'Alembert's principle to derive the equation of motion and discuss its solutions.

2.4 A mass point m that can only move along a straight line is tied to the point A by means of a spring. The distance of A to the straight line is l (cf. Fig. 5). Calculate (approximately) the frequency of oscillation of the mass point.

2.5 Two equal masses m are connected by means of a (massless) spring with spring constant x. They move without friction along a rail, their distance being l when the spring is inactive. Calculate the deviations $x_1(t)$ and $x_2(t)$ from the equilibrium positions, for the following initial conditions:

$$x_1(0) = 0\,, \quad \dot{x}_1(0) = v_0,$$
$$x_2(0) = l\,, \quad \dot{x}_2(0) = 0.$$

2.6 Given a function $F(x_1, \ldots, x_f)$ that is homogeneous and of degree N in its f variables, show that

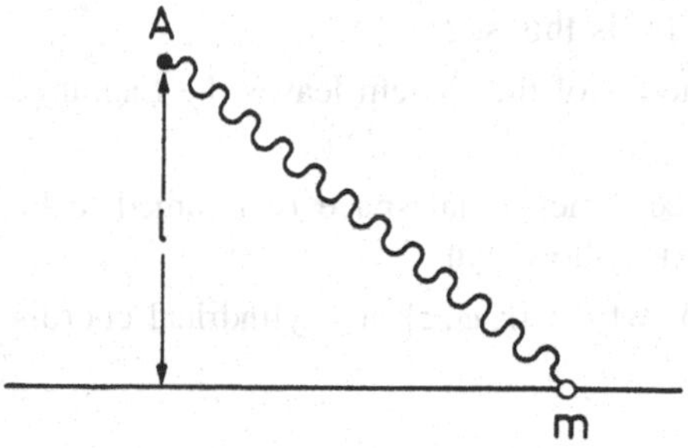

Fig. 5.

$$\sum_{i=1}^{f} \frac{\partial F}{\partial x_i} x_i = NF \,.$$

2.7 If in the integral

$$I[y] = \int_{x_1}^{x_2} \mathrm{d}x \; f(y, y')$$

f does not depend explicitly on x, show that

$$y' \frac{\partial f}{\partial y'} - f(y, y') = \text{const} \,.$$

Apply this result to $L(q, q) = T - U$ and identify the constant. T is assumed to be a homogeneous quadratic form in $\dot{q}$.

2.8 Solve the following two problems (whose solutions are well known) by means of variational calculus:

(i) the shortest connection between two points (x_1, y_1) and (x_2, y_2) in the Euclidean plane;

(ii) the shape of a homogeneous, fine-grained chain suspended at its end points (x_1, y_1) and (x_2, y_2) in the gravitational field.

Hints: Make use of the result of Exercise 2.7. The equilibrium shape of the chain is determined by the lowest position of its center of mass. The line element is given by

$$\mathrm{d}s = \sqrt{(\mathrm{d}x)^2 + (\mathrm{d}y)^2} = \sqrt{1 + y'^2}\, \mathrm{d}x \,.$$

2.9 Two coupled pendula can be described by means of the Lagrangian function

$$L = \tfrac{1}{2} m \left(\dot{x}_1^2 + \dot{x}_2^2\right) - \tfrac{1}{2} m \omega_0^2 \left(x_1^2 + x_2^2\right) - \tfrac{1}{4} m \left(\omega_1^2 - \omega_0^2\right) (x_1 - x_2)^2 .$$

(i) Show that the Lagrangian function

$$\begin{aligned} L' &= \tfrac{1}{2} m (\dot{x}_1 - \mathrm{i}\omega_0 x_1)^2 + \tfrac{1}{2} m (\dot{x}_2 - \mathrm{i}\omega_0 x_2)^2 \\ &\quad - \tfrac{1}{4} m \left(\omega_1^2 - \omega_0^2\right) (x_1 - x_2)^2 \end{aligned}$$

leads to the same equations of motion. Why is this so?

(ii) Show that transforming to the eigenmodes of the system leaves the Lagrange equations form invariant.

2.10 The force acting on a body in three-dimensional space is assumed to be axially symmetric with respect to the z-axis. Show that

(i) its potential has the form $U = U(r, z)$, where $\{r, \varphi, z\}$ are cylindrical coordinates,

$$x = r\cos\varphi\,, \quad y = r\sin\varphi\,, \quad z = z\,;$$

(ii) the force always lies in a plane containing the z-axis.

2.11 With respect to an inertial system $\mathbf{K}_0$ the Lagrangian function of a particle is

$$L_0 = \tfrac{1}{2}m\dot{\boldsymbol{x}}_0^2 - U(\boldsymbol{x}_0)\,.$$

The frame of reference $\mathbf{K}$ has the same origin as $\mathbf{K}_0$ but rotates about the latter with constant angular velocity $\boldsymbol{\omega}$. Show that the Lagrangian with respect to $\mathbf{K}$ reads

$$L = \frac{m\dot{\boldsymbol{x}}^2}{2} + m\dot{\boldsymbol{x}}\cdot(\omega \times \boldsymbol{x}) + \frac{m}{2}(\boldsymbol{\omega}\times\boldsymbol{x})^2 - U(\boldsymbol{x})\,.$$

Derive the equations of motion of Sect. 1.25 from this.

2.12 A planar pendulum is suspended such that its point of suspension glides without friction along a horizontal axis. Construct the kinetic and potential energies and a Lagrangian function for this problem.

2.13 A pearl of mass m glides (without friction) along a planar curve $s = s(\Phi)$ put up vertically. s is the length of arc and Φ the angle between the tangent to the curve and the horizontal line (see Fig. 6).

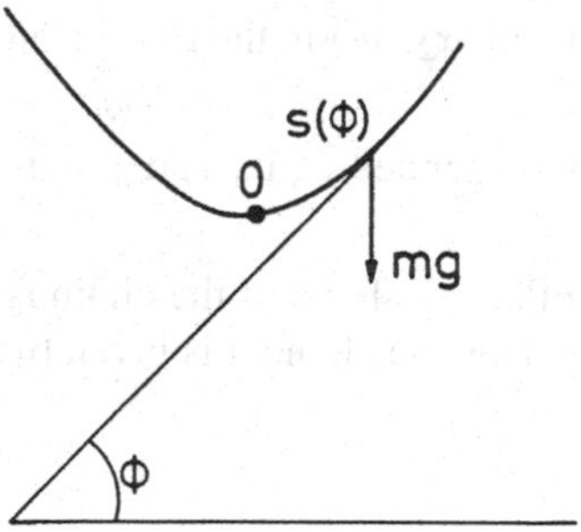

Fig. 6.

(i) Derive the equation for $s(t)$ for harmonic oscillations.

(ii) What is the relation between $s(t)$ and $\Phi(t)$? Discuss this relation and the motion that follows from it. What happens in the limit where s can reach its maximal amplitude?

(iii) From the explicit solution calculate the force of constraint and the effective force that acts on the pearl.

2.14 *Geometrical interpretation of the Legendre transformation.* Given $f(x)$ with $f''(x) > 0$. Construct $(\mathcal{L}f)(x) = xf'(x) - f(x) = xz - f(x) \equiv F(x, z)$, where $z = f'(x)$. The inverse $x = x(z)$ of the latter exists and so does the Legendre transform of $f(x)$, which is $zx(z) - f(x(z)) = \mathcal{L}f(z) = \Phi(z)$.

(i) Comparing the graphs of the functions $y = f(x)$ and $y = zx$ (for fixed z) one sees with

$$\frac{\partial F(x, z)}{\partial x} = 0$$

that $x = x(z)$ is the point where the vertical distance between the two graphs is maximal (see Fig. 7).

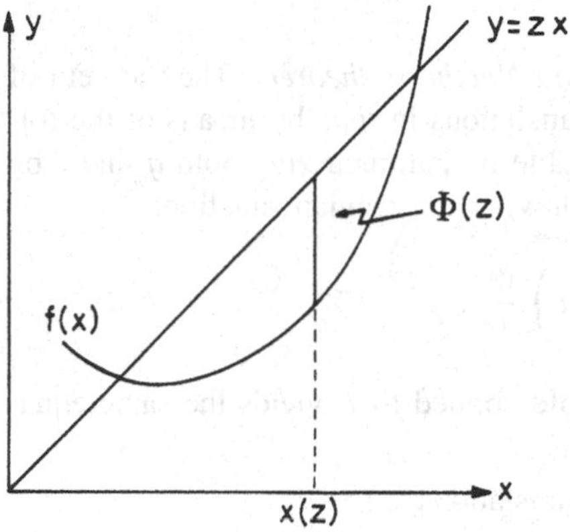

Fig. 7.

(ii) Take the Legendre transform of $\Phi(z)$, i.e. $(\mathcal{L}\Phi)(z) = z\Phi'(z) - \Phi(z) = zx - \Phi(z) \equiv G(z, x)$ with $\Phi'(z) = x$. Identify the straight line $y = G(z, x)$ for fixed z and with $x = x(z)$ and show that one has $G(z, x) = f(x)$. Sketch the picture that one obtains if one keeps $x = x_0$ fixed and varies z.

2.15 (i) Let

$$L(q_1, q_2, \dot{q}_1, \dot{q}_2, t) = T - U \quad \text{with}$$

$$T = \sum_{i,k=1}^{2} c_{ik}\dot{q}_i\dot{q}_k + \sum_{k=1}^{2} b_k\dot{q}_k + a\,.$$

Under what condition can one construct $H(\underset{\sim}{q}, \underset{\sim}{p}, t)$ and what are p_1, p_2, and H? Confirm that the Legendre transform of H is again L and that

$$\det\left(\frac{\partial^2 L}{\partial\dot{q}_k\partial\dot{q}_i}\right)\det\left(\frac{\partial^2 H}{\partial p_n\partial p_m}\right) = 1\,.$$

Hint: Take $d_{11} = 2c_{11}$, $d_{12} = d_{21} = c_{12} + c_{21}$, $d_{22} = 2c_{22}$, $\pi_i = p_i - b_i$.

(ii) Assume now that $L = L(x_1 \equiv \dot{q}_1,\ x_2 \equiv \dot{q}_2,\ q_1,\ q_2,\ t) \equiv L(x_1, x_2, \boldsymbol{u})$ with $\boldsymbol{u} \stackrel{\text{def}}{=} (q_1, q_2, t)$ to be an arbitrary Lagrangian function. We expect the momenta $p_i = p_i(x_1, x_2, \boldsymbol{u})$ derived from L to be independent functions of x_1 and x_2, i.e. that there is no function $F(p_1(x_1, x_2, \boldsymbol{u}),\ p_2(x_1, x_2, \boldsymbol{u}))$ that vanishes identically.

Show that, if p_1 and p_2 were dependent, the determinant of the second derivatives of L with respect to the x_i would vanish.

Hint: Consider $\mathrm{d}F/\mathrm{d}x_1$ and $\mathrm{d}F/\mathrm{d}x_2$.

2.16 A particle of mass m is described by the Lagrangian function

$$L = \frac{1}{2}m(\dot{x}^2 + \dot{y}^2 + \dot{z}^2) + \frac{\omega}{2}l_3 ,$$

where l_3 is the z-component of angular momentum and ω is a constant frequency. Find the equations of motion, write them in terms of the complex variable $x + \mathrm{i}y$ and of z, and solve them. Construct the Hamiltonian function and find the *kinematical* and *canonical* momenta. Show that the particle has only kinetic energy and that the latter is conserved.

2.17 *Invariance under time translations and Noether's theorem.* The theorem of E. Noether can be applied to the case of translations in *time* by means of the following trick. Make t a coordinate-like variable by parametrizing both q and t by $q = q(\tau)$, $t = t(\tau)$ and by defining the following Lagrangian function:

$$\bar{L}\left(\boldsymbol{q}, t, \frac{\mathrm{d}\boldsymbol{q}}{\mathrm{d}\tau}, \frac{\mathrm{d}t}{\mathrm{d}\tau}\right) \stackrel{\text{def}}{=} L\left(\boldsymbol{q}, \frac{1}{\mathrm{d}t/\mathrm{d}\tau}\frac{\mathrm{d}\boldsymbol{q}}{\mathrm{d}\tau}, t\right)\frac{\mathrm{d}t}{\mathrm{d}\tau} .$$

(i) Show that Hamilton's variational principle applied to $\bar{L}$ yields the same equations of motion as for L.

(ii) Assume L to be invariant under time translations

$$h^s(\boldsymbol{q}, t) = (\boldsymbol{q}, t + s) . \qquad (2.1)$$

Apply Noether's theorem to $\bar{L}$ and find the constant of the motion corresponding to the invariance (2.1).

2.18 A mass point is scattered elastically on a sphere with center P and radius R (see Fig. 8). Show that the physically possible orbit has *maximal* length.

Hints: Show first that the angles α and β must be equal and construct the action integral. Show that any other path $AB'\Omega$ would be shorter than for those points

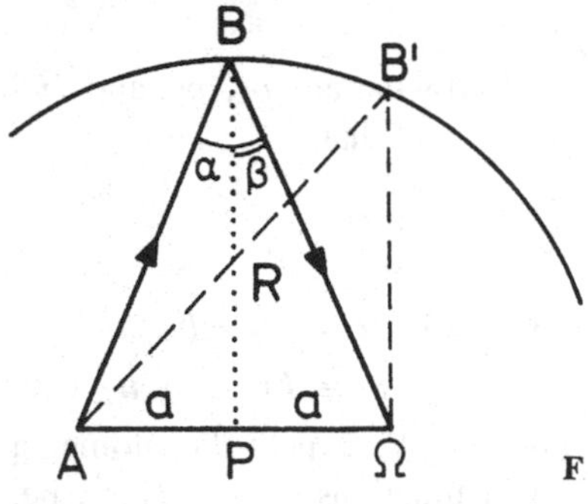

F **Fig. 8.**

where the sum of the distances to A and Ω is constant and equal to the length of the physical orbit.

2.19 (i) Show that canonical transformations leave the physical dimension of the *product* $p_i q_i$ unchanged, i.e. $[P_i Q_i] = [p_i q_i]$. Let Φ be the generating function for a canonical transformation. Show that

$$[p_i q_i] = [P_k Q_k] = [\Phi] = [H \cdot t]\,,$$

where H is the Hamiltonian function and t the time.

(ii) In the Hamiltonian function $H = p^2/2m + m\omega^2 q^2/2$ of the harmonic oscillator introduce the variables

$$x_1 \stackrel{\text{def}}{=} \omega\sqrt{m}q\,, \quad x_2 \stackrel{\text{def}}{=} p/\sqrt{m}\,, \quad \tau \stackrel{\text{def}}{=} \omega t\,,$$

thus obtaining $H = (x_1^2 + x_2^2)/2$. What is the generating function $\hat{\Phi}(x_1, y_1)$ for the canonical transformation $x \underset{\hat{\Phi}}{\rightarrow} y$ that corresponds to the function $\Phi(q, Q) = (m\omega q^2/2)\cot Q$? Calculate the matrix $M_{ik} = \partial x_i / \partial y_k$ and confirm $\det \mathbf{M} = 1$ and $\mathbf{M}^T \mathbf{J} \mathbf{M} = \mathbf{J}$.

2.20 The group Sp_{2f} is particularly simple for $f = 1$, i.e. in two dimensions.
(i) Show that every matrix

$$\mathbf{M} = \begin{pmatrix} a_{11} & a_{12} \\ a_{21} & a_{22} \end{pmatrix}$$

is symplectic if and only if $a_{11}a_{22} - a_{12}a_{21} = 1$.
(ii) Therefore, the orthogonal matrices

$$\mathbf{O} = \begin{pmatrix} \cos\alpha & \sin\alpha \\ -\sin\alpha & \cos\alpha \end{pmatrix}$$

and the symmetric matrices

$$\mathbf{S} = \begin{pmatrix} x & y \\ y & z \end{pmatrix} \quad \text{with} \quad xz - y^2 = 1$$

belong to Sp_{2f}. Show that every $M \in Sp_{2f}$ can be written as a product

$$\mathbf{M} = \mathbf{S} \cdot \mathbf{O}$$

of a symmetric matrix $\mathbf{S}$ with determinant 1 and an orthogonal matrix $\mathbf{O}$.

2.21 (i) Evaluate the following Poisson brackets for a single particle:

$$\{l_i, r_k\}\,, \quad \{l_i, p_k\}\,, \quad \{l_i, \boldsymbol{r}\}\,, \quad \{l_i, \boldsymbol{p}^2\}\,.$$

(ii) If the Hamiltonian function in its natural form $H = T + U$ is invariant under rotations, what quantities can U depend on?

2.22 Making use of the Poisson brackets show that for the system $H = T + U(r)$ with $U(r) = \gamma/r$ and γ a constant, the vector

$$\boldsymbol{A} = \boldsymbol{p} \times \boldsymbol{l} + \boldsymbol{x} m\gamma/r$$

is an integral of the motion (*Lenz' vector* or *Hermann–Bernoulli–Laplace vector*).

2.23 The motion of a particle of mass m is described by

$$H = \frac{1}{2m}\left(p_1^2 + p_2^2\right) + m\alpha q_1 , \quad \alpha = \text{const} .$$

Construct the solution of the equations of motion for the initial conditions

$$q_1(0) = x_0 , \quad q_2(0) = y_0 , \quad p_1(0) = p_x , \quad p_2(0) = p_y ,$$

making use of Poisson brackets.

2.24 For a three-body system with masses m_i, coordinates $\boldsymbol{r}_i$, and momenta $\boldsymbol{p}_i$ introduce the following coordinates (*Jacobian coordinates*[2]):

$$\varphi_1 \stackrel{\text{def}}{=} \boldsymbol{r}_2 - \boldsymbol{r}_1 \quad \text{(relative coordinate of particles 1 and 2)},$$

$$\varphi_2 \stackrel{\text{def}}{=} \boldsymbol{r}_3 - \frac{m_1\boldsymbol{r}_1 + m_2\boldsymbol{r}_2}{m_1 + m_2} \quad \text{(relative coordinate of particle 3 and the center of mass of the first two)},$$

$$\varphi_3 \stackrel{\text{def}}{=} \frac{m_1\boldsymbol{r}_1 + m_2\boldsymbol{r}_2 + m_3\boldsymbol{r}_3}{m_1 + m_2 + m_3} \quad \text{(center of mass of the three particles)},$$

$$\pi_1 \stackrel{\text{def}}{=} \frac{m_1\boldsymbol{p}_2 - m_2\boldsymbol{p}_1}{m_1 + m_2} ,$$

$$p_2 \stackrel{\text{def}}{=} \frac{(m_1 + m_2)\boldsymbol{p}_3 - m_3(\boldsymbol{p}_1 + \boldsymbol{p}_2)}{m_1 + m_2 + m_3} ,$$

$$\pi_3 \stackrel{\text{def}}{=} \boldsymbol{p}_1 + \boldsymbol{p}_2 + \boldsymbol{p}_3 .$$

(i) What is the physical interpretation of the momenta $\boldsymbol{\pi}_1$, $\boldsymbol{\pi}_2$, $\boldsymbol{\pi}_3$?

(ii) How would you define such coordinates for four or more particles?

(iii) Show in at least two (equivalent) ways that the transformation

$$\{\boldsymbol{r}_1, \boldsymbol{r}_2, \boldsymbol{r}_3, \boldsymbol{p}_1, \boldsymbol{p}_2, \boldsymbol{p}_3\} \to \{\varphi_1, \varphi_2, \varphi_3, \pi_1, \pi_2, \pi_3\}$$

is canonical.

2.25 Given a Lagrangian function L for which $\partial L/\partial t = 0$, study only those variations of the orbits $q_k(t, \alpha)$ which belong to a fixed energy $E = \sum_k \dot{q}_k(\partial L/\partial \dot{q}_k) - L$

---

[2] Jacobi, C.G.J., *Sur l'élimination des noeuds dans le problème des trois corps*, Crelles Journal für reine und angewandte Mathematik, XXVI (1843) 115

and whose end points are kept fixed irrespective of the time $(t_2 - t_1)$ that the system needs to move from the initial to the end point, i.e.

$$q_k(t,\alpha) \quad \text{with} \quad \begin{cases} q_k(t_1(\alpha),\alpha) = q_k^{(1)} \\ q_k(t_2(\alpha),\alpha) = q_k^{(2)} \end{cases} \quad \text{for all } \alpha\,. \tag{2.1}$$

Thus, initial and final times are also varied, $t_i = t_i(\alpha)$.

(i) Calculate the variation of $I(\alpha)$,

$$\delta I = \left.\frac{\mathrm{d}I(\alpha)}{\mathrm{d}\alpha}\right|_{\alpha=0} \mathrm{d}\alpha = \int_{t_1(\alpha)}^{t_2(\alpha)} L(q_k(t,\alpha),\ \dot{q}_k(t,\alpha))\,\mathrm{d}t\,. \tag{2.2}$$

(ii) Show that the variational principle

$$\delta K = 0 \quad \text{with} \quad K \stackrel{\text{def}}{=} \int_{t_1}^{t_2} (L + E)\,\mathrm{d}t$$

together with the prescriptions (2.1) is equivalent to the Lagrange equations (*the Principle of Euler and Maupertuis*).

2.26 The kinetic energy

$$T = \sum_{i,k=1}^{f} q_{ik}\dot{q}_i\dot{q}_k = \frac{1}{2}(L + E)$$

is assumed to be a positive symmetric quadratic form in the $\dot{q}_i$. The orbit in the space spanned by the q_k is described by the length of arc s such that $T = (\mathrm{d}s/\mathrm{d}t)^2$. With $E = T + U$ the integral K of Exercise 2.25 can be replaced with an integral over s. Show that the integral principle obtained in this way is equivalent to Fermat's principle of geometric optics,

$$\delta \int_{x_1}^{x_2} n(\boldsymbol{x},\nu)\,\mathrm{d}s = 0$$

(n: index of refraction, ν: frequency).

2.27 Let $H = p^2/2 + U(q)$, where the potential is such that it has a local minimum at q_0. Thus, in an interval $q_1 < q_0 < q_2$ the potential forms a potential well. Sketch a potential with this property and show that there is an interval $U(q_0) < E \leq E_{\max}$ where there are periodic orbits. Consider the characteristic equation of Hamilton and Jacobi (2.154). If $S(q, E)$ is a complete integral then $t - t_0 = \partial S/\partial E$. Take the integral

$$I(E) \stackrel{\text{def}}{=} \frac{1}{2\pi} \oint_{\Gamma_E} p\,\mathrm{d}q$$

over the periodic orbit Γ_E with energy E (this is the surface enclosed by Γ_E). Write $I(E)$ as an integral over time and show that

$$\frac{\mathrm{d}I}{\mathrm{d}E} = \frac{T(E)}{2\pi}\,.$$

2.28 In Exercise 2.27 replace $S(q, E)$ by $\bar{S}(q, I)$ with $I = I(E)$ as defined there. $\bar{S}$ generates the canonical transformation $(q, p, H) \to (\theta, I, \bar{H} = E(\Omega))$. What are the canonical equations in the new variables? Can they be integrated?

2.29 Let $H^0 = p^2/2+q^2/2$. Calculate the integral $I(E)$ defined in Exercise 2.27. Solve the characteristic equation of Hamilton and Jacobi (2.154) and write the solution as $\bar{S}(q, I)$. Then $\theta = \partial\bar{S}/\partial I$. Show that (q, p) and (θ, I) are related by the canonical transformation (2.95) of Sect. 2.24 (ii).

2.30 We assume that the Lagrangian of a mechanical system with one degree of freedom does not depend explicitly on time. In Hamilton's variational principle we make a *smooth* change of the end points q^a and q^b, as well as of the running time $t = t_2 - t_1$, in the sense that the solution $\varphi(t)$ for the values (q^a, q^b, t) and the solution $\phi(s, t)$ which belongs to the values (q'^a, q'^b, t') are related in a smooth manner: $\varphi(t) \mapsto \phi(s, t)$ such that $\phi(s, t)$ is differentiable in s and $\phi(s = 0, t) = \varphi(t)$.

Show that the corresponding change of the action integral I_0 into which the physical solution is inserted (this function is called *Hamilton's principal function*), is given by the following expression

$$\delta I_0 = -E\,\delta t + p^b\,\delta q^b - p^a\,\delta q^a\ .$$

2.31 The vector $\boldsymbol{A}$ that is introduced in Exercise 2.22 lies in the plane perpendicular to $\boldsymbol{\ell}$. Calculate $|\boldsymbol{A}|$ as a function of the energy. When does this vanish? Let ϕ denote the angle between $\boldsymbol{x}$ (orbit vector) and $\boldsymbol{A}$. Calculate $\boldsymbol{x} \cdot \boldsymbol{A}$ and show that this yields the orbit's equation in the form $r = r(\phi)$. Determine the modulus and the direction of $\boldsymbol{A}$, calculate the cross product $\boldsymbol{\ell} \times \boldsymbol{A}$ and from there the quantity

$$\left(\boldsymbol{p} - \frac{1}{\ell^2}\boldsymbol{\ell} \times \boldsymbol{A}\right)^2 .$$

This calculation yields an alternative solution of Exercise 1.26.

Chapter 3: The Mechanics of Rigid Bodies

3.1 Let two systems of reference **K** and $\bar{\mathbf{K}}$ be fixed in the center of mass of a rigid body, the axes of the former being fixed in *space*, those of the latter fixed in the body. If **J** is the inertia tensor with respect to **K** and $\bar{\mathbf{J}}$ the one as calculated in $\bar{\mathbf{K}}$, show that (i) **J** and $\bar{\mathbf{J}}$ have the same eigenvalues. (Use the characteristic polynomial.)

(ii) $\bar{\mathbf{K}}$ is now assumed to be a system of principal axes of inertia. What is the form of $\bar{\mathbf{J}}$? Calculate **J** for the case of rotation of the body about the 3-axis.

3.2 Two particles with masses m_1 and m_2 are held by a rigid but massless straight connection with length l. What are the principal axes and what are the moments of inertia?

3.3 The inertia tensor of a rigid body is found to have the form

$$I_{ik} = \begin{pmatrix} I_{11} & I_{12} & 0 \\ I_{21} & I_{22} & 0 \\ 0 & 0 & I_{33} \end{pmatrix}, \quad I_{21} = I_{12}.$$

Determine the three moments of inertia and consider the following special cases.
(i) $I_{11} = I_{22} = A$, $I_{12} = B$. Can I_{33} be arbitrary?

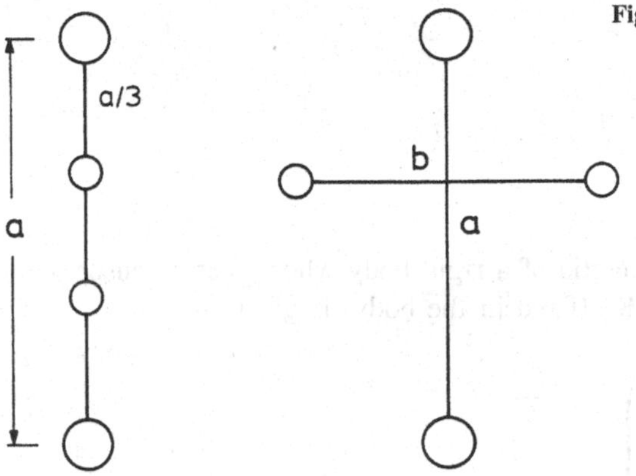

Fig. 9.

(ii) $I_{11} = A$, $I_{22} = 4A$, $I_{12} = 2A$. What can you say about I_{33}? What is the shape of the body in this example?

3.4 Construct the Lagrangian function for general, force-free motion of a conical top (height h, mass M, radius of base circle R). What are the equations of motion? Are there integrals of the motion and what is their physical interpretation?

3.5 Calculate the moments of inertia of a torus filled homogeneously with mass. Its main radius is R; the radius of its section is r.

3.6 Calculate the moment of inertia I_3 for two arrangements of four balls, two heavy (radius R, mass M) and two light (radius r, mass m) with homogeneous mass density, as shown in Fig. 9. As a model of a dancer's pirouette compare the angular velocity for the two arrangements, with L_3 fixed and equal in the two cases.

3.7 (i) Let the boundary of a homogeneous body be defined by the formula (in spherical coordinates)

$$R(\theta) = R_0(1 + \alpha \cos\theta),$$

i.e. $\varrho(r, \theta, \Phi) = \varrho_0 = \text{const}$ for $r \leq R(\theta)$ and all θ and Φ, and $\varrho(r, \theta, \Phi) = 0$ for $r > R(\theta)$. If M is the total mass, calculate ϱ_0 and the moments of inertia.

(ii) Perform the same calculation for a homogeneous body whose shape is given by

$$R(\theta) = R_0(1 + \beta Y_{20}(\theta))$$

with $Y_{20}(\theta) = \sqrt{5/16\pi}(3\cos^2\theta - 1)$ being the spherical harmonic with $l = 2$, $m = 0$. In both examples sketch the body.

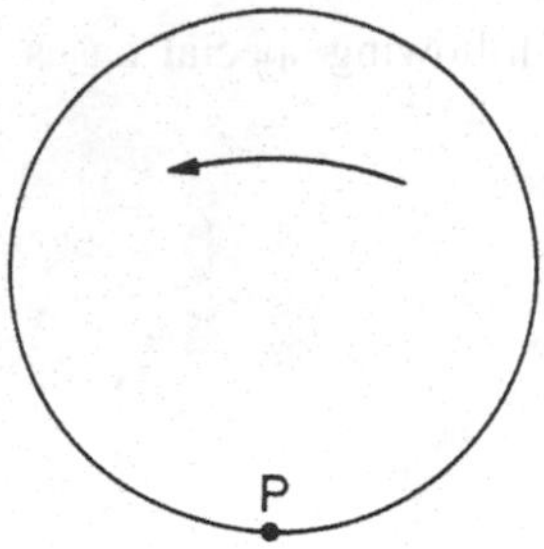

Fig. 10.

3.8 Determine the moments of inertia of a rigid body whose inertia tensor with respect to a system of reference $\mathbf{K}_1$ (fixed in the body) is given by

$$\mathbf{J} = \begin{pmatrix} \frac{9}{8} & \frac{1}{4} & \frac{-\sqrt{3}}{8} \\ \frac{1}{4} & \frac{3}{2} & \frac{-\sqrt{3}}{4} \\ \frac{-\sqrt{3}}{8} & \frac{-\sqrt{3}}{4} & \frac{11}{8} \end{pmatrix} .$$

Can one indicate the relative position of the principal inertia system $\mathbf{K}_0$ relative to $\mathbf{K}_1$?

3.9 A ball with radius a is filled homogeneously with mass such that the density is ϱ_0. The total mass is M.

(i) Write the mass density ϱ with respect to a body-fixed system centered in the center of mass and express ϱ_0 in terms of M. Let the ball rotate about a point P on its surface (see Fig. 10).

(ii) What is the same density function $\varrho(\boldsymbol{r}, t)$ as seen from a space-fixed system centered on P ?

(iii) Give the inertia tensor in the body-fixed system of (i). What is the moment of inertia for rotation about a tangent to the ball in P ?

Hint: Use the step function $\Theta(x) = 1$ for $x \geq 0$, $\Theta(x) = 0$ for $x < 0$.

3.10 A homogeneous circular cylinder with length h, radius r, and mass m rolls along an inclined plane in the earth's gravitational field.

(i) Construct the full kinetic energy of the cylinder and find the moment of inertia relevant to the described motion.

(ii) Construct the Lagrangian function and solve the equation of motion.

3.11 *Manifold of motions of the rigid body.* A rotation $R \in SO(3)$ can be determined by a unit vector $\hat{\varphi}$ (the direction about which the rotation takes place) and an angle φ.

(i) Why is the interval $0 \le \varphi \le \pi$ sufficient for describing every rotation?

(ii) Show that the parameter space $(\hat{\boldsymbol{\varphi}}, \varphi)$ fills the interior of a sphere with radius π in $\mathbb{R}^3$. This ball is denoted by D^3. Confirm that antipodal points on the ball's surface represent the same rotation.

(iii) There are two types of closed orbit in D^3, namely those which can be contracted to a point and those which connect two antipodal points. Show by means of a sketch that every closed curve can be reduced by continuous deformation to either the former or the latter type.

3.12 Calculate the Poisson brackets (3.92–95).

Chapter 4: Relativistic Mechanics

4.1 (i) A neutral π meson (π^0) has constant velocity v_0 along the x^3-direction. Write its energy-momentum vector. Construct the special Lorentz transformation that leads to the particle's rest system.

(ii) The particle decays *isotropically* into two photons, i.e. with respect to its rest system the two photons are emitted in all directions with equal probability. Study their decay distribution in the laboratory system.

4.2 The decay $\pi \to \mu + \nu$ (cf. Example (i) of Sect. 4.9.2) is isotropic in the pion's rest system. Show that above a certain fixed energy of the pion in the laboratory system there is a maximal angle beyond which no muons are emitted. Calculate that energy and the maximal emission angle as a function of m_π and m_μ (see Fig. 11). Where do muons go in the laboratory system that in the pion's rest system were emitted forward, backward, or transversely with respect to the pion's velocity in the laboratory?

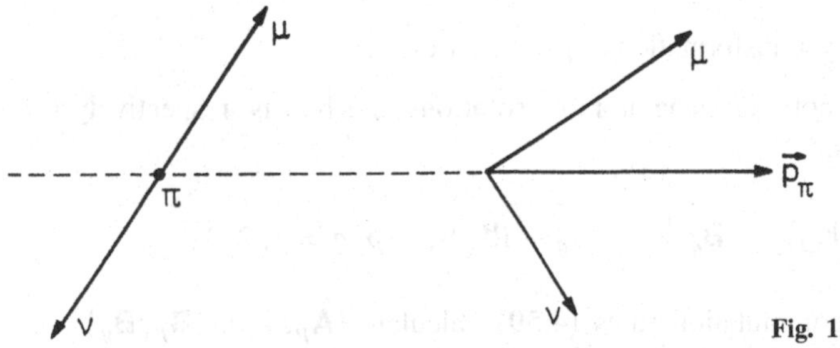

Fig. 11.

4.3 Consider a two-body reaction $A + B \to A + B$ for which the relative velocity of A (the *projectile*) and B (the *target*) is not small compared to the speed of light.

Examples are

$$e^- + e^+ \to e^- + e^- , \quad \nu + e \to e + \nu , \quad p + p \to p + p .$$

Denoting the four-momenta before and after the scattering by q_A, q_B and q'_A, q'_B the following quantities are Lorentz scalars, i.e. they have the same values in every system of reference,

$$s \stackrel{\text{def}}{=} c^2(q_A + q_B)^2 , \quad t \stackrel{\text{def}}{=} c^2(q_A - q'_A)^2 .$$

Conservation of energy and momentum requires $q'_A + q'_B = q_A + q_B$. Furthermore, we have $q_A^2 = q_A'^2 = (m_A c^2)^2$, $q_B^2 = q_B'^2 = (m_B c^2)^2$.

(i) Express s and t in terms of the energies and momenta of the particles in the center-of-mass frame. Denoting the modulus of the 3-momentum by q^* and the scattering angle by θ^*, write s and t in terms of these variables.

(ii) Define $u = c^2(q_A - q'_B)^2$ and show that

$$s + t + u = 2\left(m_A^2 + m_B^2\right) c^4 .$$

4.4 Calculate the variables s and t (as defined in Exercise 4.3) in the laboratory system, i.e. in that system where B is at rest before the scattering. What is the relation between the scattering angle θ in the laboratory system and θ^* in the center-of-mass frame? Compare to the nonrelativistic expression (1.80).

4.5 In its rest system the electron's spin is described by the 4-vector $s^\alpha = (0, \boldsymbol{s})$. What is the form of this vector in a frame where the electron has the momentum p? Calculate the scalar product $(s \cdot p) = s^\alpha p_\alpha$.

4.6 Show that

(i) every lightlike vector z ($z^2 = 0$) can be brought to the form (1,1,0,0) by means of Lorentz transformations;

(ii) every spacelike vector can be transformed to the form $(0, z^1, 0, 0)$, where $z^1 = \sqrt{-z^2}$.

Indicate the necessary transformations in both cases.

4.7 If $\mathbf{J}_i$ and $\mathbf{K}_i$ denote the generators of rotations and boosts, respectively (cf. Sect. 4.5.2 (iii)) define

$$\mathbf{A}_p \stackrel{\text{def}}{=} \frac{1}{2}\left(\mathbf{J}_p + \mathrm{i}\mathbf{K}_p\right) , \quad \mathbf{B}_q \stackrel{\text{def}}{=} \frac{1}{2}\left(\mathbf{J}_q - \mathrm{i}\mathbf{K}_q\right) , \quad p, q = 1, 2, 3 .$$

Making use of the commutation rules (4.59) calculate $[\mathbf{A}_p, \mathbf{A}_q]$, $[\mathbf{B}_p, \mathbf{B}_q]$, and $[\mathbf{A}_p, \mathbf{B}_q]$ and compare to (4.59).

4.8 Study the behavior of $\mathbf{J}_i$ and $\mathbf{K}_j$ with respect to space inversion, i.e. determine $\mathbf{P}\mathbf{J}_i\mathbf{P}^{-1}$, $\mathbf{P}\mathbf{K}_j\mathbf{P}^{-1}$.

4.9 In quantum theory one prefers to use the quantities

$$\hat{\mathbf{J}}_i \stackrel{\text{def}}{=} \mathrm{i}\mathbf{J}_i\,,\quad \hat{\mathbf{K}}_j \stackrel{\text{def}}{=} -\mathrm{i}\mathbf{K}_j\,.$$

What are the commutators (4.59) for these matrices? Show that the matrices $\hat{\mathbf{J}}_i$ are Hermitian, i.e. that $(\hat{\mathbf{J}}_i^{\mathrm{T}})^* = \hat{\mathbf{J}}_i$.

4.10 A muon decays predominantly into an electron and two nearly massless neutrinos, $\mu^- \to \mathrm{e}^- + \nu_1 + \nu_2$. If the muon is at rest, show that the electron assumes its maximal momentum whenever the neutrinos are emitted parallel to each other. Calculate the maximal and minimal energies of the electron as functions of m_μ and m_e.

Answer:

$$E_{\max} = \frac{m_\mu^2 + m_\mathrm{e}^2}{2m_\mu}c^2\,,\qquad E_{\min} = m_\mathrm{e}c^2\,.$$

Draw the corresponding momenta in the two cases.

4.11 A particle of mass M is assumed to decay into three particles (1,2,3) with masses m_1, m_2, m_3. Determine the maximal energy of particle 1 in the rest system of the decaying particle as follows. Set

$$\boldsymbol{p}_1 = -f(x)\hat{\boldsymbol{n}}\,,\quad \boldsymbol{p}_2 = xf(x)\hat{\boldsymbol{n}}\,,\quad \boldsymbol{p}_3 = (1-x)f(x)\hat{\boldsymbol{n}}\,,$$

where $\hat{\boldsymbol{n}}$ is a unit vector and x is a number between 0 and 1. Find the maximum of $f(x)$ from the principle of energy conservation.

Examples:

(i) $\mu^- \to \mathrm{e}^- + \nu_1 + \nu_2$ (cf. Exercise 4.10),

(ii) Neutron decay: $\mathrm{n} \to \mathrm{p} + \mathrm{e} + \nu$.

What is the maximal energy of the electron? What is the value of $\beta = |\boldsymbol{v}|/c$ for the electron? $m_\mathrm{n} - m_\mathrm{p} = 2.53 m_\mathrm{e}$, $m_\mathrm{p} = 1836 m_\mathrm{e}$.

4.12 Pions π^+, π^- have the mean lifetime $\tau \simeq 2.6 \times 10^{-8}$ s and decay predominantly into a muon and a neutrino. Over what distance can they fly, on average, before decaying if their momentum is $p_\pi = x \cdot m_\pi c$ with $x = 1$, 10, or 1000? ($m_\pi \simeq 140\,\mathrm{MeV/c^2} = 2.50 \times 10^{-28}\,\mathrm{kg}$).

4.13 The free neutron is unstable. Its mean lifetime is $\tau \simeq 900$ s. How far can a neutron fly on average if its energy is $E = 10^{-2}\, m_\mathrm{n}c^2$ or $E = 10^{14}\, m_\mathrm{n}c^2$?

4.14 Show that a free electron cannot radiate a single photon, i.e. the process

$$\mathrm{e} \to \mathrm{e} + \gamma$$

cannot take place because of energy and momentum conservation.

4.15 The following transformation

$$\mathcal{I} : x^\mu \mapsto \bar{x}^\mu = \frac{R^2}{x^2} x^\mu$$

implies the relation $\sqrt{x^2}\sqrt{\bar{x}^2} = R^2$. This is an obvious generalization of the well-known inversion at the circle of radius R, $r \cdot \bar{r} = R^2$. Show that the sequence of transformations: inversion $\mathcal{I}$ of x^μ, translation $\mathcal{T}$ of the image by the vector $R^2 c^\mu$, and another inversion of the result, i.e.,

$$x' = (\mathcal{I} \circ \mathcal{T} \circ \mathcal{I})x$$

is precisely the special conformal transformation (4.102).

4.16 Consider the following Lagrangian

$$L = \frac{1}{2} m \left(\psi \dot{\boldsymbol{q}}^2 - c_0^2 \frac{(\psi - 1)^2}{\psi} \right) \equiv L(\dot{\boldsymbol{q}}, \psi)$$

which contains the additional, dimensionless, degree of freedom ψ. The parameter c_0 has the physical dimension of a velocity. Show: The extremum of the action integral yields a theory obeying special relativity for which c_0 is the maximal velocity, in other words, one obtains the Lagrangian (4.97) with the velocity of light c replaced by c_0. Consider the limit $c_0 \to \infty$.

Chapter 5: Geometric Aspects of Mechanics

5.1 Let $\overset{k}{\omega}$ be an exterior k-form, $\overset{l}{\omega}$ an exterior l-form. Show that their exterior product is symmetric if k and/or l are even and antisymmetric if both are odd, i.e.

$$\overset{k}{\omega} \wedge \overset{l}{\omega} = (-1)^{k \cdot l} \, \overset{l}{\omega} \wedge \overset{k}{\omega} \, .$$

5.2 Let x_1, x_2, x_3 be local coordinates in the Euclidean space $\mathbb{R}^3$, $\mathrm{d}s^2 = E_1 \, \mathrm{d}x_1^2 + E_2 \, \mathrm{d}x_2^2 + E_3 \, \mathrm{d}x_3^2$ the square of the line element, and $\hat{\boldsymbol{e}}_1$, $\hat{\boldsymbol{e}}_2$, $\hat{\boldsymbol{e}}_3$ unit vectors along the coordinate directions. What is the value of $\mathrm{d}x_i(\hat{\boldsymbol{e}}_j)$, i.e. of the action of the one-form $\mathrm{d}x_i$ on the unit vector $\hat{\boldsymbol{e}}_j$?

5.3 Let $\boldsymbol{a} = \sum_{i=1}^3 a_i(x) \hat{\boldsymbol{e}}_i$ be a vector field with $a_i(x)$ smooth functions on M. To every such vector field we associate a one-form $\overset{1}{\omega}_{\boldsymbol{a}}$ and a two-form $\overset{2}{\omega}_{\boldsymbol{a}}$ such that

$$\overset{1}{\omega}_{\boldsymbol{a}}(\boldsymbol{\xi}) = (\boldsymbol{a} \cdot \boldsymbol{\xi}) \, , \quad \overset{2}{\omega}_{\boldsymbol{a}}(\boldsymbol{\xi}, \boldsymbol{\eta}) = (\boldsymbol{a} \cdot (\boldsymbol{\xi} \times \boldsymbol{\eta})) \, .$$

Show that

$$\overset{1}{\omega}_a = \sum_{i=1}^{3} a_i(x)\sqrt{E_i}\,\mathrm{d}x_i\,,$$

$$\overset{2}{\omega}_a = a_i(x)\sqrt{E_2 E_3}\,\mathrm{d}x_2 \wedge \mathrm{d}x_3 + \text{cyclic permutations}\,,$$

5.4 Making use of the results of Exercise 5.3 determine the components of ∇f in the basis $\{\hat{\boldsymbol{e}}_1, \hat{\boldsymbol{e}}_2, \hat{\boldsymbol{e}}_3\}$

Answer:

$$\nabla f = \sum_{i=1}^{3} \frac{1}{\sqrt{E_i}} \frac{\partial f}{\partial x^i} \hat{\boldsymbol{e}}_i\,.$$

5.5 Determine the functions E_i for the case of Cartesian, cylindrical, and spherical coordinates. In each case give the components of ∇f.

5.6 To the force $\boldsymbol{F} = (F_1, F_2)$ in the plane we associate the one-form $\omega = F_1\,\mathrm{d}x^1 + F_2\,\mathrm{d}x^2$. When we apply ω onto a displacement vector, $\omega(\boldsymbol{\xi})$ is the work done by the force. What is the dual $*\omega$ of the form ω? What is its interpretation?

5.7 The Hodge star operator assigns to every k-form ω the $(n-k)$-form $*\omega$. Show that

$$*(*\omega) = (-1)^{k\cdot(n-k)}\omega\,.$$

5.8 Let $\boldsymbol{E} = (E_1, E_2, E_3)$ and $\boldsymbol{B} = (B_1, B_2, B_3)$ be electric and magnetic fields that in general depend on x and t. We assign the following exterior forms to them:

$$\varphi \stackrel{\text{def}}{=} \sum_{i=1}^{3} E_i \mathrm{d}x^i\,,$$

$$\omega \stackrel{\text{def}}{=} B_1\,\mathrm{d}x^2 \wedge \mathrm{d}x^3 + B_2\,\mathrm{d}x^3 \wedge \mathrm{d}x^1 + B_3\,\mathrm{d}x^1 \wedge \mathrm{d}x^2\,.$$

Write the homogeneous Maxwell equation $\operatorname{curl}\boldsymbol{E} + \dot{\boldsymbol{B}}/c = 0$ as an equation between the forms φ and ω.

5.9 If d denotes the exterior derivatives and $*$ the Hodge star operator, the codifferential δ is defined by

$$\delta \stackrel{\text{def}}{=} *\,\mathrm{d}\,*\,.$$

Show that $\Delta \stackrel{\text{def}}{=} \mathrm{d}\circ\delta + \delta\circ\mathrm{d}$, when applied to functions, is the Laplacian operator

$$\Delta = \sum_i \frac{\partial^2}{\partial x^{i2}}\,.$$

5.10 Let

$$\overset{k}{\omega} = \sum_{i_1<\dots<i_k} \omega_{i_1\dots i_k}(\underset{\sim}{x})\, \mathrm{d}x^{i_1} \wedge \dots \wedge \mathrm{d}x^{i_k}$$

be an exterior k-form over a vector space W. Let $F : V \to W$ be a smooth mapping of the vector space V onto W. Show that the pull-back $F^*(\overset{k}{\omega} \wedge \overset{l}{\omega})$ of the exterior product of two such forms is equal to the exterior product of the pull-back of the individual forms $(F^* \overset{k}{\omega}) \wedge (F^* \overset{l}{\omega})$.

5.11 With the same assumptions as in Exercise 5.10 show that the exterior derivative and the pull-back commute,

$$\mathrm{d}(F^*\omega) = F^*(\mathrm{d}\omega)\,.$$

5.12 Let x and y be Cartesian coordinates in $\mathbb{R}^2$, $V = y\partial_x$ and $W = x\partial_y$ two vector fields on $\mathbb{R}^2$. Calculate the Lie bracket $[V, W]$. Sketch the vector fields V, W, and $[V, W]$ along circles about the origin.

5.13 Prove the follow assertions.

(i) The set of all tangent vectors to the smooth manifold M at the point $p \in M$ form a real vector space, denoted by T_pM, whose dimension is $n = \dim M$.

(ii) If M is $\mathbb{R}^n$, T_pM is isomorphic to that space.

5.14 The canonical two-form for a system with two degrees of freedom reads $\omega = \sum_{i=1}^{2} \mathrm{d}q^i \wedge \mathrm{d}p_i$. Calculate $\omega \wedge \omega$ and confirm that this product is proportional to the oriented volume element in phase space.

5.15 Let $H^{(1)} = p^2/2 + (1 - \cos q)$ and $H^{(2)} = p^2/2 + q(q^2 - 3)/6$ be the Hamiltonian functions for two systems with one degree of freedom. Construct the corresponding Hamiltonian vector fields and sketch them along some of the solution curves.

5.16 Let $H = H^0 + H'$ with $H^0 = (p^2 + q^2)/2$ and $H' = \varepsilon q^3/3$. Construct the Hamiltonian vector fields X_{H^0} and X_H and calculate $\omega(X_H, X_{H^0})$.

5.17 Let L and L' be two Lagrangian functions on TQ for which Φ_L and $\Phi_{L'}$ are regular. The corresponding vector fields and canonical two-forms are X_E, $X_{E'}$, ω_L, and $\omega_{L'}$. Show that each of the following assertions implies the other:

(i) $L' = L + \alpha$, where α: $TQ \to \mathbb{R}$ is a closed one-form, i.e. $\mathrm{d}\alpha = 0$;

(ii) $X_E = X_{E'}$ and $\omega_L = \omega_{L'}$.

Show that in local coordinates this is the result obtained in Sect. 2.10.

Chapter 6: Stability and Chaos

6.1 Study the two-dimensional linear system $\dot{\underset{\sim}{y}} = \mathbf{A}\underset{\sim}{y}$, where $\mathbf{A}$ has one of the Jordan normal forms

$$\text{(i)} \quad \mathbf{A} = \begin{pmatrix} \lambda_1 & 0 \\ 0 & \lambda_2 \end{pmatrix}, \quad \text{(ii)} \quad \mathbf{A} = \begin{pmatrix} a & b \\ -b & a \end{pmatrix}, \quad \text{(iii)} \quad \mathbf{A} = \begin{pmatrix} \lambda & 0 \\ 1 & \lambda \end{pmatrix}.$$

In all three cases determine the characteristic exponents and the flow (6.13) with $s = 0$. Suppose the system is obtained by linearizing a dynamical system in the neighborhood of an equilibrium position. (i) corresponds to the situations shown in Figs. 6.2a–c. Draw the analogous pictures for (ii) for $(a = 0,\ b > 0)$ and $(a < 0,\ b > 0)$, and for (iii) with $\lambda < 0$.

6.2 The variables α and β on the torus $T^2 = S^1 \times S^1$ define the dynamical system

$$\dot{\alpha} = a/2\pi\,, \quad \dot{\beta} = b/2\pi\,, \quad 0 \le \alpha,\ \beta \le 1\,,$$

where a and b are real constants. Cutting the torus at $(\alpha = 1,\ \beta)$ and at $(\alpha,\ \beta = 1)$ yields a square of length 1. Draw the solutions with initial condition $(\alpha_0,\ \beta_0)$ in this square for b/a rational and irrational.

6.3 Show that in an autonomous Hamiltonian system with one degree of freedom (and hence two-dimensional phase space) neighboring trajectories can diverge at most linearly with increasing time as long as one keeps clear from saddle points.

Hint: Make use of the characteristics equation (2.154) of Hamilton and Jacobi.

6.4 Study the system

$$\begin{aligned} \dot{q}_1 &= -\mu q_1 - \lambda q_2 + q_1 q_2 \\ \dot{q}_2 &= \lambda q_1 - \mu q_2 + \left(q_1^2 - q_2^2\right)/2\,, \end{aligned}$$

where $0 \le \mu \ll 1$ is a damping term and λ with $|\lambda| \ll 1$ is a detuning parameter. Show that if $\mu = 0$ the system is Hamiltonian. Find a Hamiltonian function for this case. Draw the projection of its phase portraits for $\lambda > 0$ onto the (q_1, q_2)-plane and determine the position and the nature of the critical points.

Show that the picture obtained above is structurally unstable when μ is chosen to be different from zero and positive, by studying the change of the critical points for $\mu \ne 0$.

6.5 Given the Hamiltonian function on $\mathbb{R}^4$

$$H(q_1, q_2, p_1, p_2) = \frac{1}{2}\left(p_1^2 + p_2^2\right) + \frac{1}{2}\left(q_1^2 + q_2^2\right) + \frac{1}{3}\left(q_1^3 + q_2^3\right)$$

show that this system possesses two independent integrals of the motion and sketch the structure of its flow.

6.6 Study the flow of the equations of motion $p = \dot{q}$, $\dot{p} = q - q^3 - p$ and determine the position and the nature of its critical points. Two of these are attractors. Determine their basin of attraction by means of the Liapunov function $V = p^2/2 - q^2/2 + q^4/4$.

6.7 Dynamical systems of the type

$$\dot{\underset{\sim}{x}} = -\partial U/\partial \underset{\sim}{x} \equiv -U_{,x}$$

are called *gradient flows*. They are quite different from the flows of Hamiltonian systems. Making use of a Liapunov function show that if U has an isolated minimum at $\underset{\sim}{x}_0$, then $\underset{\sim}{x}_0$ is an asymptotically stable equilibrium position. Study the example

$$\dot{x}_1 = -2x_1(x_1 - 1)(2x_1 - 1)\,, \quad \dot{x}_2 = -2x_2\,.$$

6.8 Consider the equations of motion

$$\dot{q} = p\,, \quad \dot{p} = \frac{1}{2}(1 - q^2)$$

of a system with $f = 1$. Sketch the phase portrait of typical solutions with given energy. Study its critical points.

6.9 By numerical integration find the solutions of the Van der Pool equation (6.36) for initial conditions close to (0,0) and for various values of ε in the interval $0 < \varepsilon \leq 0.4$. Draw $q(t)$ as a function of time, as in Fig. 6.7. Use the result to find out empirically at what rate the orbit approaches the attractor.

6.10 Choose the straight line $p = q$ as the transverse section for the system (6.36), Fig. 6.6. Determine numerically the points of intersection of the orbit with initial condition (0.01,0) with that line and plot the result as a function of time.

6.11 The system in $\mathbb{R}^2$

$$\dot{x}_1 = x_1\,, \quad \dot{x}_2 = -x_2 + x_1^2$$

has a critical point in $x_1 = 0 = x_2$. Show that for the linearized system the line $x_1 = 0$ is a stable submanifold and the line $x_2 = 0$ an unstable one. Find the corresponding manifolds for the exact system by integrating the latter.

6.12 Study the mapping $x_{i+1} = f(x_i)$ with $f(x) = 1 - 2x^2$. Substitute $u = (4\pi)\arcsin\sqrt{(x+1)/2}$ and show that there are no stable fixed points. Calculate numerically 50 000 iterations of this mapping for various initial values $x_1 \neq 0$ and plot the histogram of the points that land in one of the intervals $[n/100, (n+1)/100]$ with $n = -100, -99, \ldots, +99$. Follow the development of two close initial values x_1, x_1', and verify that they diverge in the course of the iteration. (For a discussion see Collet, Eckmann 1990.)

6.13 Study the flow of Roessler's model

$$\dot{x} = -y - z\,, \quad \dot{y} = x + ay\,, \quad \dot{z} = b + xz - cz$$

for $a = b = 0.2$, $c = 5.7$ by numerical integration. The graphs of x, y, z as functions of time and their projections onto the (x, y)-plane and the $(x, \dot{x})$-plane are particularly interesting. Consider the Poincaré mapping for the transverse section $y + z = 0$. As $\dot{x} = 0$, x has an extremum on the section. Plot the value of the extremum x_{i+1} as a function of the previous extremum x_i (see also Bergé, Pomeau, Vidal 1984 and references therein).

6.14 Although this is more than an exercise, the reader is strongly encouraged to study the system known as Hénon's attractor. It provides a good illustration of chaotic behavior and extreme sensitivity to initial conditions (see also, Bergé, Pomeau, Vidal 1984, Sect. 3.2 and Devaney 1989, Sect. 2.6, Exercise 10).

6.15 Show that

$$\sum_{\sigma=1}^{n} \exp\left[\mathrm{i}\frac{2\pi}{n}\sigma m\right] = n\delta_{m0}\,, \quad (m = 0, \ldots, n-1)\,.$$

Use this result to prove (6.63), (6.65), and (6.66).

6.16 Show that by a linear substitution $y = \alpha x + \beta$ the system (6.67) can be transformed to $y_{i+1} = 1 - \gamma y_i^2$. Determine γ in terms of μ and show that y lies in the interval $(-1, 1]$ and γ in $(0, 2]$ (cf. also Exercise 6.12 above). Making use of this transformed equation derive the values of the first bifurcation points (6.68) and (6.70).

Solution of Exercises

Cross-references to a specific section or equation in the main text of the book are marked with a capital M preceding the number of that section or equation. For instance, Sect. M3.7 refers to Chap. 3, Sect. 7, of the main text, while (M4.100) refers to eq. (4.100) in Chap. 4. Cross references within this set of solutions should be fairly obvious.

Chapter 1: Elementary Newtonian Mechanics

1.1 The time derivative of angular momentum is $\dot{\boldsymbol{l}} = \dot{\boldsymbol{r}} \times \boldsymbol{p} + \boldsymbol{r} \times \dot{\boldsymbol{p}} = m\dot{\boldsymbol{r}} \times \dot{\boldsymbol{r}} \times \boldsymbol{r} + \boldsymbol{r} \times \boldsymbol{F}$. By assumption this is zero which implies that the force $\boldsymbol{F}$ must be proportional to $\boldsymbol{r}$, $\boldsymbol{F} = \alpha \boldsymbol{r}$, $\alpha \in \mathbb{R}$. If we decompose the velocity into a component along $\boldsymbol{r}$ and a component perpendicular to it, then $\boldsymbol{F}$ will change only the former. Therefore, the motion takes place in a spatially fixed *plane* perpendicular to the angular momentum $\boldsymbol{l} = m\boldsymbol{r}(t) \times \dot{\boldsymbol{r}}(t) = m\boldsymbol{r}_0 \times \boldsymbol{v}_0$, itself a constant. Motion along (a), (b), (e), and (f) is possible. Motion along (c) is not possible because $\boldsymbol{l}$ would vanish at the turning point but would be different from zero beforeand after passing through that point.

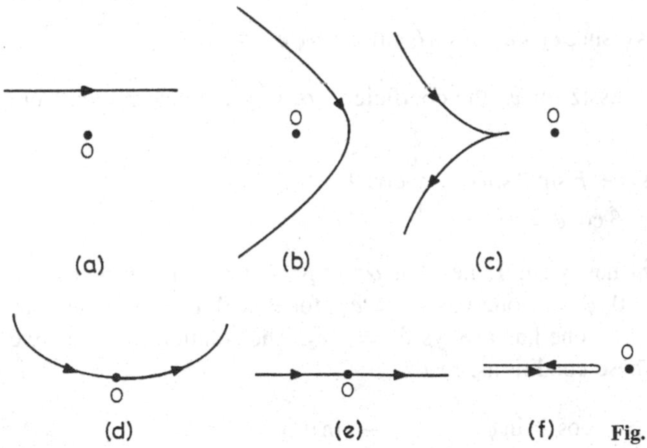

Fig. 1.

Similarly (d) is not possible because $\boldsymbol{l}$ would vanish in O but not before and after.

1.2 We note that $x(t) = r(t)\cos\varphi(t)$, $y(t) = r(t)\sin\varphi(t)$ and, hence $dx = dr\cos\varphi - rd\varphi\sin\varphi$, $dy = dr\sin\varphi + rd\varphi\cos\varphi$. In taking $(ds)^2 = (dx)^2 + (dy)^2$ the mixed terms cancel so that $(ds)^2 = (dr)^2 + r^2(d\varphi)^2$. Thus, the velocity is $\boldsymbol{v}^2 = \dot{r}^2 + r^2\dot{\varphi}^2$. As neither $\boldsymbol{r}$ nor $\boldsymbol{v}$ have a z-component, the x- and y-components of $\boldsymbol{l} = m\boldsymbol{r} \times \boldsymbol{v}$ vanish. The z-component is

$$
\begin{aligned}
l_z &= m(xv_z - yv_x) \\
&= mr(\dot{r}\sin\varphi\cos\varphi + r\dot{\varphi}\cos^2\varphi - \dot{r}\cos\varphi\sin\varphi + r\dot{\varphi}\sin^2\varphi) \\
&= mr^2\dot{\varphi}\,.
\end{aligned}
$$

Thus one finds

$$
\boldsymbol{v} = \dot{r}^2 + \frac{\boldsymbol{l}^2}{m^2r^2} \quad \text{and} \quad T = \frac{1}{2}m\dot{r}^2 + \frac{\boldsymbol{l}^2}{2mr^2}\,.
$$

If $\boldsymbol{l}$ is constant this means that the product $r^2\dot{\varphi} = \text{const.}$, thus correlating the angular velocity $\dot{\varphi}$ with the radial distance, cf. the examples (a), (b), (e), and (f), of Exercise 1.1. A motion of type (d) could only be possible if, on approaching O, $\dot{\varphi}$ were to go to infinity in such a way that the product $r^2\dot{\varphi}$ stays finite. But then the shape of the orbit would be different, see Exercise 1.23.

1.3 In analogy to the solution of the previous exercise one finds $(ds)^2 = (dr)^2 + r^2(d\theta)^2 + r^2\sin^2\theta(d\varphi)^2$. Thus, $\boldsymbol{v}^2 = \dot{r}^2 + r^2\dot{\theta}^2 + r^2\sin^2\theta\dot{\varphi}^2$.

1.4 Having solved Exercise 1.3 one first reads off $\hat{\boldsymbol{e}}_r$ from Fig. 2: $\hat{\boldsymbol{e}}_r = \hat{\boldsymbol{e}}_x\sin\theta\cos\varphi + \hat{\boldsymbol{e}}_y\sin\theta\sin\varphi + \hat{\boldsymbol{e}}_z\cos\theta$. At the point with azimuth φ, $\hat{\boldsymbol{e}}_\varphi$ is tangent to a great cirlce, see Fig. 3. Hence, $\hat{\boldsymbol{e}}_\varphi = -\hat{\boldsymbol{e}}_x\sin\varphi + \hat{\boldsymbol{e}}_y\cos\varphi$ (check the special cases $\varphi = 0$ and $\pi/2$!). One verifies that

$$
\hat{\boldsymbol{e}}_r\cdot\hat{\boldsymbol{e}}_\varphi = -\sin\theta\cos\varphi\sin\varphi\hat{\boldsymbol{e}}_x\cdot\hat{\boldsymbol{e}}_x + \sin\theta\sin\varphi\cos\varphi\hat{\boldsymbol{e}}_y\cdot\hat{\boldsymbol{e}}_y = 0\,.
$$

Starting from the given ansatz for $\hat{\boldsymbol{e}}_\theta$ the coefficients α, β, γ are determined from the equations

$$
\begin{aligned}
\hat{\boldsymbol{e}}_\theta\cdot\hat{\boldsymbol{e}}_r &= \alpha\sin\theta\cos\varphi + \beta\sin\theta\sin\varphi + \gamma\cos\theta = 0\,, \\
\hat{\boldsymbol{e}}_0\cdot\hat{\boldsymbol{e}}_\varphi &= -\alpha\sin\varphi + \beta\cos\varphi = 0\,,
\end{aligned}
$$

keeping in mind that $\hat{\boldsymbol{e}}_\theta$ has norm 1, i.e. that $\alpha^2 + \beta^2 + \gamma^2 = 1$. Furthermore, from Fig. 2 and for $\theta = 0$, $\varphi = 0$ one has $\hat{\boldsymbol{e}}_\theta = \hat{\boldsymbol{e}}_x$, for $\theta = 0$, $\varphi = \pi/2$ one has $\hat{\boldsymbol{e}}_\theta = \hat{\boldsymbol{e}}_y$, while for $\theta = \pi/2$ one has always $\hat{\boldsymbol{e}}_\theta = -\hat{\boldsymbol{e}}_z$. The solution of the above equation which meets these conditions, reads

$$
\alpha = \cos\theta\cos\varphi\,,\quad \beta = \cos\theta\sin\varphi\,,\quad \gamma = -\sin\theta\,.
$$

In this basis we find

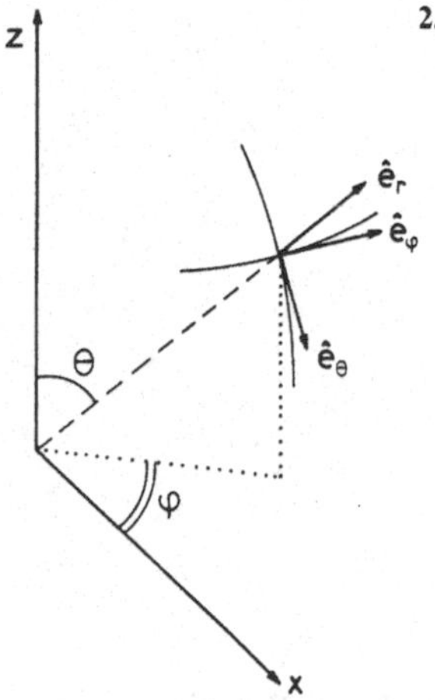

2.

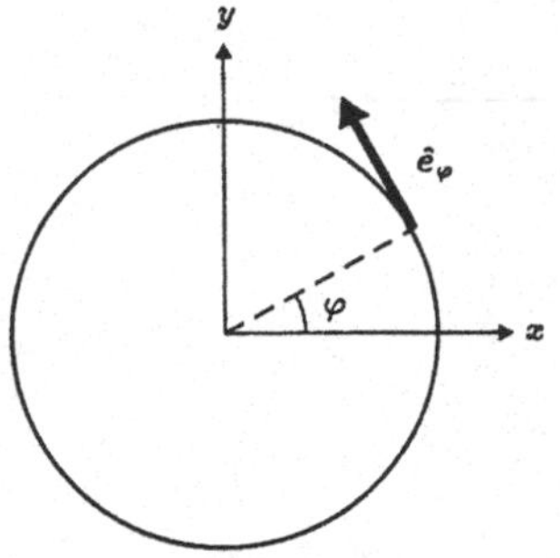

Fig. 3.

$$
\begin{aligned}
\boldsymbol{v} = \dot{\boldsymbol{r}} &= \dot{r}\hat{\boldsymbol{e}}_r + r\dot{\hat{\boldsymbol{e}}}_r \\
&= \dot{r}\hat{\boldsymbol{e}}_r + r((\dot{\theta}\cos\theta\cos\varphi - \dot{\varphi}\sin\theta\sin\varphi)\hat{\boldsymbol{e}}_x \\
&\quad +(\dot{\theta}\cos\theta\sin\varphi + \dot{\varphi}\sin\theta\cos\varphi)\hat{\boldsymbol{e}}_y - \dot{\theta}\sin\theta\hat{\boldsymbol{e}}_z) \\
&= \dot{r}\hat{\boldsymbol{e}}_r + r(\dot{\theta}\hat{\boldsymbol{e}}_\theta + \dot{\varphi}\sin\varphi\hat{\boldsymbol{e}}_\varphi)\,,
\end{aligned}
$$

from which follows the result $\boldsymbol{v}^2 = \dot{r}^2 + r^2(\dot{\theta}^2 + \dot{\varphi}^2\sin^2\theta)$ that we found in the previous exercise.

1.5 With respect to the frame **K**, $\boldsymbol{r}(t) = vt\hat{\boldsymbol{e}}_y$, i.e., $x(t) = 0 = z(t)$ and $y(t) = vt$. In the rotating frame

$$
\begin{aligned}
\dot{x}' &= \dot{x}\cos\phi + \dot{y}\sin\phi + \dot{\phi}(-x\sin\phi + y\cos\phi) \\
\dot{y}' &= -\dot{x}\sin\phi + \dot{y}\cos\phi - \dot{\phi}(x\cos\phi + y\sin\phi) \\
\dot{z}' &= \dot{z} = 0\,.
\end{aligned}
$$

In the first case, $\phi = \omega = \text{const.}$, the particle moves uniformly along a straight line with velocity $\boldsymbol{v}' = (v\sin\omega, v\cos\omega, 0)$. In the second case, $\phi = \omega t$, $\dot{x}' = v\sin\omega t + \omega vt\cos\omega t$, $\dot{y}' = v\cos\omega t - \omega vt\sin\omega t$. Integrating over time, $x'(t) = vt\sin\omega t$, $y'(t) = vt\cos\omega t$, and $z'(t) = 0$. The apparent motion as seen by an observer in the accelerated frame **K**$'$, is sketched in Fig. 4.

1.6 The equation of motion of the particle reads

$$
m\ddot{\boldsymbol{r}} = \boldsymbol{F} = f(r)\frac{\boldsymbol{r}}{r}\,.
$$

Take the time derivative of the angular momentum, $\dot{\boldsymbol{l}} = m\dot{\boldsymbol{r}} \times \dot{\boldsymbol{r}} + m\boldsymbol{r} \times \ddot{\boldsymbol{r}}$. The first term is always zero. The second term vanishes because, by the equation of motion, the acceleration is proportional to $\boldsymbol{r}$. Hence, $\dot{\boldsymbol{l}} = 0$, which means that the magnitude and the direction of the angular momentum are conserved. As $\boldsymbol{l}$ is perpendicular to $\boldsymbol{r}$ and the velocity $\dot{\boldsymbol{r}}$ this proves the assertion.

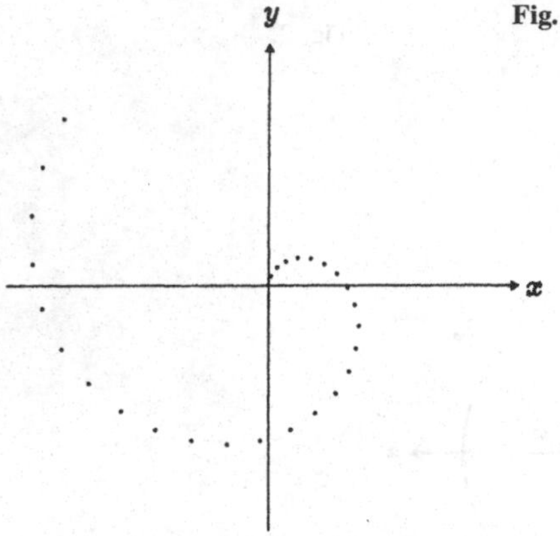

Fig. 4.

1.7 (i) By Newton's third law the forces between two bodies fulfill $\boldsymbol{F}_{ik} = -\boldsymbol{F}_{ki}$ or $-\nabla_i V_{ik}(\boldsymbol{r}_i, \boldsymbol{r}_k) = \nabla_k V_{ik}(\boldsymbol{r}_i, \boldsymbol{r}_k)$. Hence, V can only depend on $(\boldsymbol{r}_i - \boldsymbol{r}_k)$. Constants of the motion are: total momentum $\boldsymbol{P}$, energy E; furthermore, we have for the center-of-mass motion

$$\boldsymbol{r}_S(t) - \boldsymbol{P}/Mt = \boldsymbol{r}_S(0) = \text{const.}\,.$$

(ii) When V_{ij} depends only on the modulus $|\boldsymbol{r}_i - \boldsymbol{r}_k|$, we have

$$\begin{aligned} \boldsymbol{F}_{ji} &= -\nabla_i V_{ij}(|\boldsymbol{r}_i - \boldsymbol{r}_k|) = -V'_{ij}(|\boldsymbol{r}_i - \boldsymbol{r}_k|)\nabla_i |\boldsymbol{r}_i - \boldsymbol{r}_k| \\ &= -V'_{ij}(|\boldsymbol{r}_i - \boldsymbol{r}_k|)\frac{\boldsymbol{r}_i - \boldsymbol{r}_k}{|\boldsymbol{r}_i - \boldsymbol{r}_k|}\,. \end{aligned}$$

In this case the total angular momentum is another constant of the motion.

1.8 For $q \to 0$ the potential goes to infinity like $1/q^4$, while for $q \to \infty$ it tends to zero. Between these points it has two extrema as sketched in Fig. 5. As the energy $E = p^2/2 + U(q)$ is conserved, the phase portraits are given by $p = [2(E - U(q))]^{1/2}$. The figure shows a few examples. The minimum at $q = 2$ is a stable equilibrium point, the maximum just beyond $q = 6$ is unstable. The orbits with $E \approx 0.2603$ are separatrices. The phase portraits are symmetric with respect to reflection in the q-axis because $(q, p = +\sqrt{\ldots})$ and $(q, p = -\sqrt{\ldots})$ belong to the same portrait.

1.9 The term $(x_2^2 + x_4^2)/(2m)$ is the total kinetic energy while $U(x_1, x_3) = m(\omega_0^2(x_1^2 + x_3^2) + \omega_1^2(x_1 - x_3)^2)/2$ is the potential energy. The forces acting on pendula 1 and 2 are, respectively, $-\partial U/\partial x_1$, and $-\partial U/\partial x_3$. Thus, the equations of motion are

$$\begin{pmatrix} \dot{x}_1 \\ \dot{x}_2 \\ \dot{x}_3 \\ \dot{x}_4 \end{pmatrix} = \begin{pmatrix} 0 & 1/m & 0 & 0 \\ -m(\omega_0^2 + \omega_1^2) & 0 & m\omega_1^2 & 0 \\ 0 & 0 & 0 & 1/m \\ m\omega_1^2 & 0 & -m(\omega_0^2 + \omega_1^2) & 0 \end{pmatrix} \begin{pmatrix} x_1 \\ x_2 \\ x_3 \\ x_4 \end{pmatrix},$$

Fig. 5

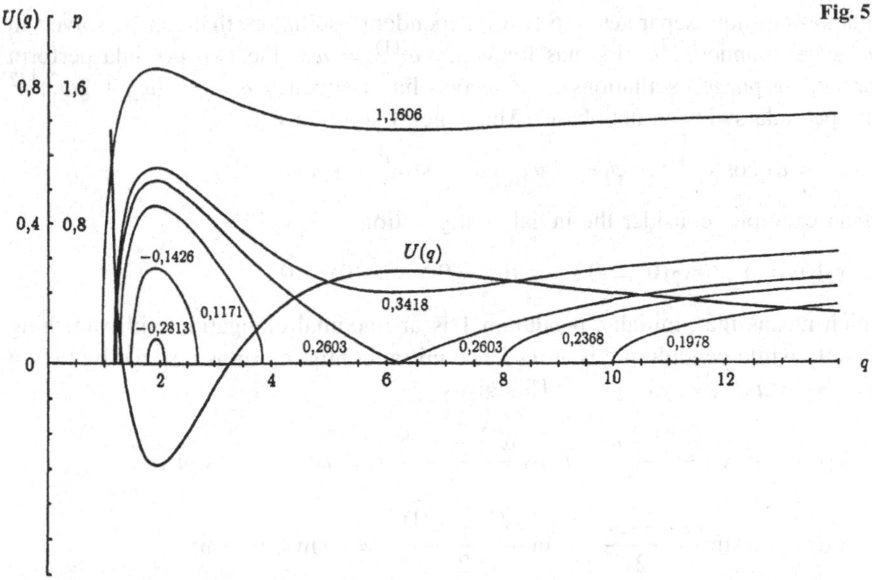

or, for short, $\dot{x} = Mx$. The transformation as given above

$$u_1 = \frac{1}{\sqrt{2}}(x_1 + x_3)\ , \quad u_2 = \frac{1}{\sqrt{2}}(x_2 + x_4)\ ,$$
$$u_3 = \frac{1}{\sqrt{2}}(x_1 - x_3)\ , \quad u_4 = \frac{1}{\sqrt{2}}(x_2 - x_4)$$

leads to sums and differences of the original coordinates and momenta. We note that the matrix M has the structure

$$M = \left(\begin{array}{c|c} B & C \\ \hline C & B \end{array}\right)$$

where B and C are 2×2 matrices. Furthermore the transformation A is invertible and, in fact, the inverse equals A. Thus

$$\frac{du}{dt} = AMA^{-1}u \quad \text{with} \quad A^{-1} = A\ .$$

It is useful to note that one can do the calculations in terms of the 2×2 submatrices as if these were (possibly noncommuting) numbers. For example,

$$AMA^{-1} = AMA = \left(\begin{array}{c|c} B + C & 0 \\ \hline 0 & B - C \end{array}\right) \quad \text{with}$$
$$B + C = \begin{pmatrix} 0 & 1/m \\ -m\omega_0^2 & 0 \end{pmatrix} \quad \text{and}$$
$$B - C = \begin{pmatrix} 0 & 1/m \\ -m(\omega_0^2 + 2\omega_1^2) & 0 \end{pmatrix}\ .$$

This system now separates into two independent oscillators that can be solved in the usual manner. The first has frequency $\omega^{(1)} = \omega_0$ (the two pendula perform parallel, in-phase oscillations); the second has frequency $\omega^{(2)} = (\omega_0^2 + 2\omega_1^2)^{1/2}$ (the pendula swing in antiphase). The general solution is

$$u_1 = a_1 \cos(\omega^{(1)} t + \varphi_1)\ , \quad u_3 = a_2 \cos(\omega^{(2)} t + \varphi_2)\ .$$

As an example, consider the initial configuration

$$x_1(0) = a\ , \quad x_2(0) = 0\ , \quad x_3(0) = 0\ , \quad x_4(0) = 0\ ,$$

which means that, initially, pendulum 1 is at maximal elongation with vanishing velocity while pendulum 2 is at rest. The initial configuration is realized by taking $a_2 = a_1 = a\sqrt{2}$, $\varphi_1 = \varphi_2 = 0$. This gives

$$x_1(t) = a \cos \frac{\omega^{(1)} + \omega^{(2)}}{2} t \cos \frac{\omega^{(2)} - \omega^{(1)}}{2} t = a \cos \Omega t \cos \omega t\ ,$$

$$x_3(t) = a \sin \frac{\omega^{(1)} + \omega^{(2)}}{2} t \sin \frac{\omega^{(2)} - \omega^{(1)}}{2} t = a \sin \Omega t \sin \omega t\ ,$$

Where $\Omega := (\omega^{(1)} + \omega^{(2)})/2$, $\omega := (\omega^{(2)} - \omega^{(1)})/2$. If $\Omega/\omega = p/q$ with $p, q \in \mathbb{Z}$ and $p > q$, hence rational, the system returns to its initial configuration after time $t = 2\pi p/\Omega = 2\pi q/\omega$. For earlier times one has $t = \pi p/(2\Omega)$: $x_1 = 0$, $x_3 = a$ (pendulum 1 at rest, pendulum 2 has maximal elongation); $t = \pi p/\Omega$: $x_1 = -a$, $x_3 = 0$; $t = 3\pi p/(2\Omega)$: $x_1 = 0$, $x_3 = -a$. The oscillation moves back and forth between pendulum 1 and pendulum 2. If Ω/ω is not rational, the system will come close, at a later time, to the initial configuration but will never assume it exactly (cf. Exercise 6.2). In the example considered here, this will happen if $\Omega t \approx 2\pi n$ and $\omega t \approx 2\pi m$ (with $m, n \in \mathbb{Z}$), i.e. if Ω/ω can be approximated by the ratio of two integers. It may happen that these integers are large so that the "return time" becomes very large.

1.10 As the differential equation is linear, the two terms are solutions precisely when $\mu = \omega$; a and b are integration constants which are fixed by the initial condition as follows

$$\begin{aligned} x(t) &= a \cos \omega t + b \sin \omega t\ , \\ p(t) &= -am\omega \sin \omega t + mb\omega \cos \omega t\ . \end{aligned}$$

$x(0) = x_0$ gives $a = x_0$, $p(0) = p_0$ gives $b = p_0/(m\omega)$. The solution with $\omega = 0.8$, $x_0 = 1$, $p_0 = 0$ reads $x(t) = \cos(0.8t)$.

1.11 From the ansatz one has

$$\begin{aligned} \dot{x}(t) &= \alpha x(t) + e^{\alpha t}(-\tilde{\omega} x_0 \sin \tilde{\omega} t + p_0/m \cos \tilde{\omega} t) \\ \ddot{x}(t) &= \alpha^2 x(t) + 2\alpha e^{\alpha t}(-\tilde{\omega} x_0 \sin \tilde{\omega} t + p_0/m \cos \tilde{\omega} t) \\ &\quad - e^{\alpha t}\tilde{\omega}^2 (x_0 \cos \tilde{\omega} t + p_0/m\tilde{\omega} \sin \tilde{\omega} t) \\ &= -\alpha^2 x + 2\alpha \dot{x} - \tilde{\omega}^2 x\ . \end{aligned}$$

Inserting and comparing coefficients one finds

$$\alpha = -\frac{\kappa}{2}\,, \quad \tilde{\omega} = \sqrt{\omega^2 - \alpha^2} = \sqrt{\omega^2 - \kappa^2/4}\,.$$

The special solution $x(t) = \mathrm{e}^{-\kappa t/2}\cos(\sqrt{0.64 - \kappa^2/4}\,t)$, approaches the origin in a spiraling motion as $t \to \infty$.

1.12 Energy conservation formulated for the two domains yields

$$\frac{m}{2}\boldsymbol{v}_1^2 + U_1 = E = \frac{m}{2}\boldsymbol{v}_2^2 + U_2\,.$$

As the potential energy U depends on x only there can be no force perpendicular to the x-axis. Therefore, the component of the momentum along the direction perpendicular to that axis cannot change in going from $x < 0$ to $x > 0$: $v_{1\perp} = v_{2\perp}$. The law of conservation of energy hence reads

$$\frac{m}{2}\boldsymbol{v}_{1\perp}^2 + \frac{m}{2}\boldsymbol{v}_{1\|}^2 + U_1 = \frac{m}{2}\boldsymbol{v}_{2\perp}^2 + \frac{m}{2}\boldsymbol{v}_{2\|}^2 + U_2\,, \quad \text{or}$$
$$\frac{m}{2}\boldsymbol{v}_{1\|}^2 + U_1 = \frac{m}{2}\boldsymbol{v}_{2\|}^2 + U_2\,.$$

from which follows

$$\sin^2\alpha_1 = \frac{\boldsymbol{v}_{1\perp}^2}{\boldsymbol{v}_1^2}\,, \quad \sin^2\alpha_2 = \frac{\boldsymbol{v}_{2\perp}^2}{\boldsymbol{v}_2^2}\,, \quad \text{directly yielding}$$
$$\frac{\sin\alpha_1}{\sin\alpha_2} = \frac{|\boldsymbol{v}_2|}{|\boldsymbol{v}_1|}\,.$$

For $U_1 < U_2$ we find $|\boldsymbol{v}_1| > |\boldsymbol{v}_2|$, hence $\alpha_1 < \alpha_2$. For $U_1 < U_2$ all inequalities are reversed.

1.13 Let $M = m_1 + m_2 + m_3$ be the total mass and $m_{12} = m_1 + m_2$. From the figure one sees that $\boldsymbol{r}_2 + \boldsymbol{s}_a = \boldsymbol{r}_1$, $\boldsymbol{s}_{12} + \boldsymbol{s}_b = \boldsymbol{r}_3$, where $\boldsymbol{s}_{12}$ is the center-of-mass coordinate of particles 1 and 2. Solving for $\boldsymbol{r}_1, \boldsymbol{r}_2, \boldsymbol{r}_3$ we find

$$\boldsymbol{r}_1 = \boldsymbol{r}_S - \frac{m_3}{M}\boldsymbol{s}_b + \frac{m_2}{m_{12}}\boldsymbol{s}_a\,,$$
$$\boldsymbol{r}_2 = \boldsymbol{r}_S - \frac{m_3}{M}\boldsymbol{s}_b - \frac{m_1}{m_{12}}\boldsymbol{s}_a\,,$$
$$\boldsymbol{r}_3 = \boldsymbol{r}_S + \frac{m_{12}}{M}\boldsymbol{s}_b\,.$$

Inserting these into the kinetic energy all mixed terms cancel. The result contains only terms quadratic in $\dot{\boldsymbol{r}}_S, \dot{\boldsymbol{s}}_a, \dot{\boldsymbol{s}}_b$

$$T = \underbrace{\frac{1}{2}M\dot{\boldsymbol{r}}_S^2}_{T_S} + \underbrace{\frac{1}{2}\mu_a\dot{\boldsymbol{s}}_a^2}_{T_a} + \underbrace{\frac{1}{2}\mu_b\dot{\boldsymbol{s}}_b^2}_{T_b} \quad \text{with} \quad \mu_a = \frac{m_1 m_2}{m_{12}}\,, \quad \mu_b = \frac{m_{12}m_3}{M}\,.$$

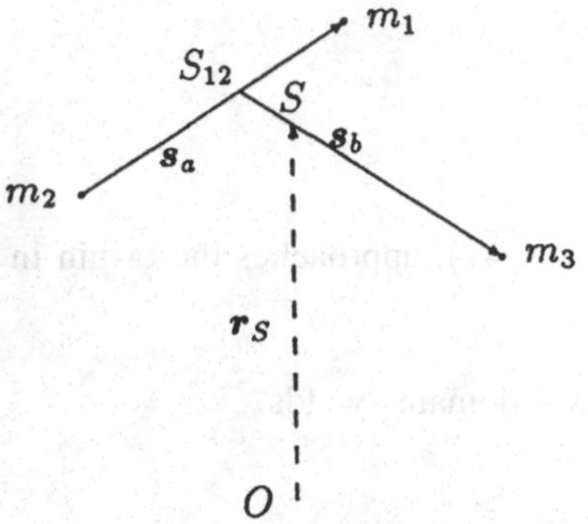

Fig. 6.

T_S is the kinetic energy of the center-of-mass motion, μ_a is the reduced mass of the subsystem consisting of particles 1 and 2. μ_b is the reduced mass of the subsystem consisting of particle 3 and the center-of-mass S_{12} of particles 1 and 2, T_b is the kinetic energy of the relative motion of particle 3 and S_{12}.

In an analogous way, the angular momentum is found to be

$$\boldsymbol{L} = \sum_i \boldsymbol{l}_i = \underbrace{M\boldsymbol{r}_S \times \dot{\boldsymbol{r}}_S}_{\boldsymbol{l}_S} + \underbrace{\mu_a \boldsymbol{s}_a \times \dot{\boldsymbol{s}}_a}_{\boldsymbol{l}_a} + \underbrace{\mu_b \boldsymbol{s}_b \times \dot{\boldsymbol{s}}_b}_{\boldsymbol{l}_b} ,$$

all mixed terms having cancelled.

By a special (and proper) Galilei transformation, $\boldsymbol{r}_S \to \boldsymbol{r}'_S = \boldsymbol{r}_S + \boldsymbol{w}t + \boldsymbol{a}$, $\dot{\boldsymbol{r}}_S \to \dot{\boldsymbol{r}}'_S = \dot{\boldsymbol{r}}_S + \boldsymbol{w}$, $\boldsymbol{s}_a \to \boldsymbol{s}_a$, $\boldsymbol{s}_b \to \boldsymbol{s}_b$ and, hence,

$$\boldsymbol{l}'_S = \boldsymbol{l}_S + M(\boldsymbol{a} \times (\dot{\boldsymbol{r}}_S + \boldsymbol{w}) + (\boldsymbol{r}_S - t\dot{\boldsymbol{r}}_S) \times \boldsymbol{w}) ,$$

while $\boldsymbol{l}'_a = \boldsymbol{l}_a$, $\boldsymbol{l}'_b = \boldsymbol{l}_b$ remain unchanged.

1.14 (i) With $U(\lambda r) = \lambda^\alpha U(r)$ and $r' = \lambda r$ the forces from $\tilde{U}(r') := U(\lambda r)$ and from $U(r)$, respectively, differ by the factor $\lambda^{\alpha-1}$. Indeed

$$\boldsymbol{F}' = -\nabla_{r'}\tilde{U} = -\frac{1}{\lambda}\nabla_r \tilde{U} = -\lambda^{\alpha-1}\nabla_r U = \lambda^{\alpha-1}\boldsymbol{F} .$$

Integrating $\boldsymbol{F}' \cdot d\boldsymbol{r}'$ over a path in $\boldsymbol{r}'$ space and comparing with the corresponding integral over $\boldsymbol{F} \cdot d\boldsymbol{r}$, the work done in the two cases differs by the factor λ^α. Changing t to $t' = \lambda^{1-\alpha/2} t$,

$$\left(\frac{d\boldsymbol{r}'}{dt'}\right)^2 = \lambda^2 \lambda^{\alpha-2} \left(\frac{d\boldsymbol{r}}{dt}\right)^2 ,$$

which means that the kinetic energy

$$T = \frac{1}{2} m \left(\frac{d\boldsymbol{r}'}{dt'}\right)^2$$

differs from the original one by the same factor λ^α. Thus, this holds for the total energy, too, $E' = \lambda^\alpha E$. The indicated relation between time differences and linear dimensions of geometrically similar orbits follows.

(ii) For harmonic oscillation the assumption holds with $\alpha = 2$. The ratio of the periods of two geometrically similar orbits is $T_a/T_b = 1$, independently of the linear dimensions.

In the homogeneous gravitational field $U(z) = mgz$ and, hence, $\alpha = 1$. Times of free fall and initial height H are related by $T \propto H^{1/2}$.

In the case of the Kepler problem $U = -A/r$ and, hence, $\alpha = -1$. Two geometrically similar ellipses with semimajor axes a_a and b_b have circumference U_a and U_b, respectivley, such that $U_a/U_b = a_a/a_b$. Therefore the ratio of the periods T_a and T_b is $T_a/T_b = (U_a/U_b)^{3/2}$ from which follows $(T_a/T_b)^2 = (a_a/a_b)^3$, Kepler's third law.

(iii) The general relation is $E_a/E_b = (L_a/L_b)^\alpha$. If A_i denotes the amplitude of harmonic oscillation, $E_a/E_b = A_a^2/A_b^2$. In the case of Kepler motion $E_a/E_b = a_b/a_a$: the energy is inversely proportional to the semimajor axis.

1.15 (i) From the equations of Sect. M1.24

$$r_P = \frac{p}{1+\varepsilon} = -\frac{A}{2E}\frac{1-\varepsilon^2}{1+\varepsilon} = -\frac{A}{2E}(1-\varepsilon)\;;$$

$$r_A = -\frac{A}{2E}(1+\varepsilon)\;.$$

From these we calculate

$$r_P + r_A = -\frac{A}{E}\,, \quad r_P \cdot r_A = \frac{A^2}{4E^2}(1-\varepsilon^2) = \frac{l^2}{-2\mu E}\;.$$

Inserting this into the differential equation we obtain

$$\frac{d\phi}{dr} = \frac{l}{r^2\sqrt{2\mu\left(E + \frac{A}{r} - \frac{l^2}{2\mu r^2}\right)}}\;.$$

This is precisely eq. (M1.67) with $U_{\text{eff}} = -A/r + l^2/2\mu r^2$. Integration of eq. (1.4) with the boundary condition as indicated implies

$$\phi(r) - \phi(r_P) = \int_{r_P}^{r} \frac{1}{r}\left(\frac{r_P r_A}{(r-r_P)(r_A-r)}\right)^{1/2} dr\;.$$

We make use of the indicated formula with

$$\alpha = 2\frac{r_A r_P}{r_A - r_P}\,, \quad \beta = -\frac{r_A + r_P}{r_A - r_P}\,,$$

and obtain

$$\phi(r) = \arccos \frac{2r_A r_P - (r_A + r_P)r}{(r_A - r_P)r} \,.$$

(ii) There are two possibilities for solving this equation: (a) the new equations are obtained by replacing l^2 with $\bar{l}^2 = l^2 + 2\mu B$. For the remainder, the solution is exactly the same as for the Kepler problem. If $B > 0 (B < 0)$, then $\bar{l} > l (\bar{l} < l)$, i.e., in the case of repulsion (attraction) the orbit becomes larger (smaller). (b) With $U(r) = U_0(r) + B/r^2$, $U_0(r) = -A/r$, the differential equation for $\phi(r)$ is written in the same form as above

$$\frac{d\phi}{dr} = \frac{\sqrt{r_A r_P}}{r\sqrt{(r - r'_P)(r'_A - r)}} \,,$$

where r'_P, r'_A denote perihelion and aphelion, respectively, for the perturbed potential. They are obtained from the formula $(r - r_P)(r_A - r) + B/E = (r - r'_P)(r'_A - r)$. Multiplying the differential equation by $((r'_P r'_A)/(r_P r_A))^{1/2}$ and integrating as before

$$\phi(r) = \sqrt{\frac{r_P r_A}{r'_P r'_A}} \arccos \frac{2r'_A r'_P - r(r'_A + r'_P)}{r(r'_A - r'_P)} \,.$$

From this solution follows $r(\phi) = 2r'_P r'_A / [r'_P + r'_A + (r'_A - r'_P) \cos \sqrt{r'_P r'_A / r_P r_A}\, \phi]$. The first passage through perihelion is set to $\phi_{P1} = 0$. The second is $\phi_{P2} = 2\pi((r_P r_A)/(r'_P r'_A))^{1/2} = 2\pi l / \sqrt{l^2 + 2\mu B} \approx 2\pi(1 - \mu B / l^2)$. The perihelion precession is $(\phi_{P2} - 2\pi)$. It is independent of the energy E. For $B > 0$ (additional repulsion) the motion lags behind, and for $B < 0$ (additional attraction) the motion advances as compared to the Kepler case.

1.16 For fixed l, the energy must fulfill $E \geq -\mu A^2/(2l^2)$. The lower limit is assumed for circular orbits with radius $r_0 = l^2/\mu A$. The semimajor axis (in relative motion) follows from Kepler's third law $a^3 = G_N(m_E + m_S)T^2/(4\pi^2)$. This gives $a = 1.495 \times 10^{11}$ m ($T = 1$ y $= 3.1536 \times 10^7$ s). This is approximately equal to a_E, the semimajor axis of the earth in the center-of-mass system. The sun moves on an ellipse with semimajor axis

$$a_S = \frac{m_E}{m_E + m_S} a \approx 449 \,\text{km} \,.$$

This is far within the sun's radius $R_S \approx 7 \times 10^5$ km.

1.17 We arrange the two dipoles as sketched in Fig. 7. The potential created by the first dipole at a point situated at r is

$$\Phi_1 = e_1 \left(\frac{1}{|\boldsymbol{r} - \boldsymbol{d}_1|} - \frac{1}{|\boldsymbol{r}|} \right) \approx e_1 \left(\frac{1}{r} + \frac{\boldsymbol{r} \cdot \boldsymbol{d}_1}{r^3} - \frac{1}{r} \right) = \frac{\boldsymbol{r} \cdot (e_1 \boldsymbol{d}_1)}{r^3} \,.$$

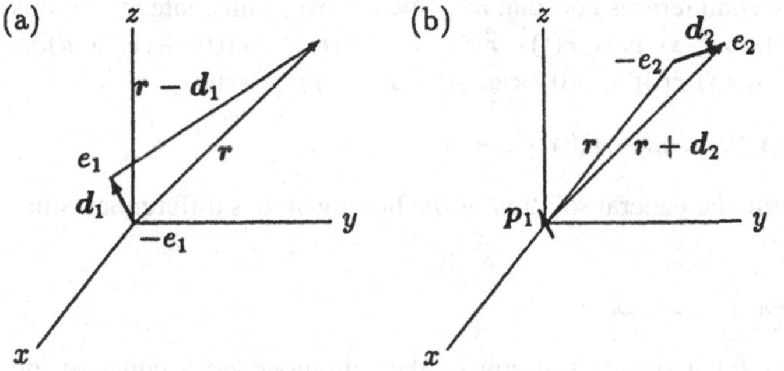

Fig. 7.

Here, we have expanded

$$\frac{1}{|\boldsymbol{r}-\boldsymbol{d}_1|}=\frac{1}{\sqrt{r^2+d_1^2-2\boldsymbol{r}\cdot\boldsymbol{d}_1}}$$

up to the term linear in $\boldsymbol{d}_1$. In the limit we obtain $\Phi_1=\boldsymbol{r}\cdot\boldsymbol{p}_1/r^3$. The potential energy of the second dipole in the field of the first reads

$$W=e_1(\Phi_1(\boldsymbol{r}+\boldsymbol{d}_2)-\Phi_1(\boldsymbol{r}))=e_2\left(\frac{\boldsymbol{p}_1\cdot(\boldsymbol{r}+\boldsymbol{d}_2)}{|\boldsymbol{r}+\boldsymbol{d}_2|^3}-\frac{\boldsymbol{p}_1\cdot\boldsymbol{r}}{r^3}\right).$$

Expanding again up to terms linear in $\boldsymbol{d}_2$

$$W\approx e_2\left(\frac{\boldsymbol{p}_1\cdot\boldsymbol{r}}{r^3}\left(1-3\frac{\boldsymbol{r}\cdot\boldsymbol{d}_2}{r^2}\right)+\frac{\boldsymbol{p}_1\cdot\boldsymbol{d}_2}{r^3}-\frac{\boldsymbol{p}_1\cdot\boldsymbol{r}}{r^3}\right).$$

Finally, taking the limit $e_2\to\infty$, $\boldsymbol{d}_2\to 0$, with $e_2\boldsymbol{d}_2=\boldsymbol{p}_2$ finite, this yields

$$W(1,2)=\frac{\boldsymbol{p}_1\cdot\boldsymbol{p}_2}{r^3}-3\frac{(\boldsymbol{p}_1\cdot\boldsymbol{r})(\boldsymbol{p}_2\boldsymbol{r})}{r^5}.$$

From this expression one calculates the components of $\boldsymbol{F}_{21}=-\nabla_1 W=-\boldsymbol{F}_{12}$, making use of relations such as

$$\frac{\partial}{\partial x_1}=\frac{\partial r}{\partial x_1}\frac{\partial}{\partial r}=\frac{x_1-x_2}{r}\frac{\partial}{\partial r}, \quad \text{etc.}\,.$$

So, for example

$$\frac{\partial W(1,2)}{\partial x_1}=-(\boldsymbol{p}_1\cdot\boldsymbol{p}_2)\frac{3}{r^4}\frac{x_1-x_2}{r}-\frac{3}{r^5}\left(p_1^x(\boldsymbol{p}_2\cdot\boldsymbol{r})\right.$$
$$\left.+(\boldsymbol{p}_1\cdot\boldsymbol{r})p_2^x\right)+(\boldsymbol{p}_1\cdot\boldsymbol{r})(\boldsymbol{p}_2\cdot\boldsymbol{r})\frac{15}{r^6}\frac{x_1-x_2}{r}.$$

1.18 Take the time derivative of $\dot{\boldsymbol{r}}\cdot\boldsymbol{a}$,

$$\frac{d}{dt}\dot{\boldsymbol{r}}\cdot\boldsymbol{a}=\ddot{\boldsymbol{r}}\cdot\boldsymbol{a}=\dot{\boldsymbol{v}}\cdot\boldsymbol{a}=(\boldsymbol{v}\times\boldsymbol{a})\cdot\boldsymbol{a}=0\,.$$

Thus, $\dot{\boldsymbol{r}}\cdot\boldsymbol{a}$ is constant in time and the indicated relation holds for all times. Taking the time derivative of (5) and inserting $\dot{\boldsymbol{v}}$, we find $\ddot{\boldsymbol{v}}=\dot{\boldsymbol{v}}\times\boldsymbol{a}=(\boldsymbol{v}\times\boldsymbol{a})\times\boldsymbol{a}=$

$-\boldsymbol{a}^2\boldsymbol{v}+(\boldsymbol{v}\cdot\boldsymbol{a})\boldsymbol{a}$. The second term is constant as shown above. Thus, integrating this equation over t from 0 to t, we have $\ddot{\boldsymbol{r}}(t)-\ddot{\boldsymbol{r}}(0)=-\omega^2(\boldsymbol{r}(t)-\boldsymbol{r}(0))+(\boldsymbol{v}(0)\cdot\boldsymbol{a})\boldsymbol{a}t$, where $\omega^2:=\boldsymbol{a}^2$. By eq. (5) $\ddot{\boldsymbol{r}}(0)=\boldsymbol{v}(0)\times\boldsymbol{a}$, so that we may write

$$\ddot{\boldsymbol{r}}(t)+\omega^2\boldsymbol{r}(t)=(\boldsymbol{v}(0)\cdot\boldsymbol{a})\boldsymbol{a}t+\boldsymbol{v}(0)\times\boldsymbol{a}+\omega^2\boldsymbol{r}(0)\ .$$

This is the desired form, the general solution of the homogeneous differential equations is

$$\boldsymbol{r}_{\text{hom}}(t)=\boldsymbol{c}_1\sin\omega t+\boldsymbol{c}_2\cos\omega t\ .$$

With the given ansatz for a special solution of the inhomogeneous equation the constants are found to be

$$\begin{aligned}
\boldsymbol{c}_1&=\frac{1}{\omega^3}\left(\boldsymbol{a}^2\boldsymbol{v}(0)-(\boldsymbol{v}(0)\cdot\boldsymbol{a})\boldsymbol{a}\right)=\frac{1}{\omega^3}(\boldsymbol{a}\times(\boldsymbol{v}(0)\times\boldsymbol{a}))\\
\boldsymbol{c}_2&=-\frac{1}{\omega^2}\boldsymbol{v}(0)\times\boldsymbol{a}\\
\boldsymbol{c}&=\frac{1}{\omega^2}(\boldsymbol{v}(0)\cdot\boldsymbol{a})\boldsymbol{a}\\
\boldsymbol{d}&=\frac{1}{\omega^2}\boldsymbol{v}(0)\times\boldsymbol{a}+\boldsymbol{r}(0)\ .
\end{aligned}$$

The solution therefore reads

$$\begin{aligned}
\boldsymbol{r}(t)=&\frac{1}{\omega^3}\boldsymbol{a}\times(\boldsymbol{v}(0)\times\boldsymbol{a})\sin\omega t+\frac{1}{\omega^2}(\boldsymbol{v}(0)\cdot\boldsymbol{a})\boldsymbol{a}t\\
&+\frac{1}{\omega^2}\boldsymbol{v}(0)\times\boldsymbol{a}(1-\cos\omega t)+\boldsymbol{r}(0)\ .
\end{aligned}$$

It represents a helix winding around the vector $\boldsymbol{a}$.

1.19 The ball falls from initial height h_0. It hits the plane for the first time at $t_1=\sqrt{2h_0/g}$, the velocity then being $u_1=-\sqrt{2h_0g}=-gt_1$. Furthermore, with $\alpha:=\sqrt{(n-1)/n}$

$$v_i=-\alpha u_i\ ,\quad u_{i+1}=-v_i\ ,\quad t_{i+1}-t_i=\frac{2v_i}{g}\ .$$

The first two equations give $v_1=\alpha g t_1$ and $v_i=\alpha^i g t_1$. The third equation yields

$$t_i^0-t_i=\frac{v_i}{g}=t_{i+1}-t_i^0\quad\text{and}\quad t_{i+1}^0-t_{i+1}=\frac{v_{i+1}}{g}\ ,$$

and, from there, $t_{i+1}^0-t_i^0=(v_{i+1}+v_i)/g=t_1(\alpha+1)\alpha^i$. With $t_0^0=0$ we have at once

$$t_i^0=t_1(1+\alpha)\sum_{\nu=0}^{i-1}\alpha^\nu\ .$$

From $h_i=v_i^2/(2g)$, finally, $h_i=\alpha^{2i}h_0$.

1.20 The answer is contained in the following table giving the products of the elements

	E	P	T	$P\cdot T$
E	E	P	T	$P\cdot T$
P	P	E	$P\cdot T$	T
T	T	$P\cdot T$	E	P
$P\cdot T$	$P\cdot T$	T	P	E

1.21 Let R and E_0 denote the radius and the energy of a circular orbit, respectively. The differential equation for the radial motion reads

$$\frac{dr}{dt}=\sqrt{\frac{2}{\mu}}\sqrt{E_0-U_{\text{eff}}(r)}\,,\quad U_{\text{eff}}(r)=U(r)+\frac{l^2}{2\mu r^2}\,.$$

From this follows $E_0=U_{\text{eff}}(R)$, $U'_{\text{eff}}\big|_{r=R}=0$, $U''_{\text{eff}}\big|_{r=R}>0$ or, for $U(r)$,

$$U'(R)=\frac{l^2}{\mu}\frac{1}{R^3}\quad\text{and}\quad U''(R)>-\frac{3l^2}{\mu}\frac{1}{R^4}\,.$$

If $E=E_0+\varepsilon$,

$$\frac{dr}{dt}=\sqrt{\frac{2}{\mu}}\sqrt{\varepsilon-\frac{1}{2}(r-R)^2U''_{\text{eff}}(R)}\,.$$

Setting $\kappa:=U''_{\text{eff}}(R)$ we obtain, choosing $\varsigma=r'-R$,

$$t-t_0=\sqrt{\frac{\mu}{\kappa}}\int_{r_0-R}^{r-R}\frac{d\varsigma}{\sqrt{2\varepsilon/\kappa-\varsigma^2}}=\sqrt{\frac{\mu}{\kappa}}\arcsin\left((r-R)\sqrt{\frac{\kappa}{2\varepsilon}}\right)\,.$$

Solving for $r-R$ yields

$$r-R=\sqrt{\frac{2\varepsilon}{\kappa}}\sin\sqrt{\frac{\kappa}{\mu}}(t-t_0)\,.$$

Thus, the radial distance oscillates around the value R. More specifically, one finds (i) $U(r)=r^n$, $U'(r)=nr^{n-1}$, $U''(r)=n(n-1)r^{n-2}$. This yields the equation

$$nR^{n-1}=\frac{l^2}{\mu R^3}\Rightarrow R=\sqrt[n+2]{\frac{l^2}{\mu n}}\,,$$

$$\kappa=n(n-1)R^{n-2}+\frac{3l^2}{\mu R^4}>0\Leftrightarrow \underbrace{n(n-1)R^{n+2}}_{l^2/(\mu n)}+\frac{3l^2}{\mu}=\frac{(n+2)l^2}{\mu}>0\,.$$

(ii) $U(r)=\lambda/r$, $U'(r)=-\lambda/r^2$, $U''(r)=2\lambda/r^3$. From this $R=-l^2/(\mu\lambda)$, $\kappa=-\lambda/R^3$. This is greater than zero of λ is negative.

1.22 (i) The eastward deviation follows from the formula given in Sect. M1.26, $\Delta \approx (2\sqrt{2}/3)g^{-1/2}H^{3/2}\omega\cos\varphi$. With $\omega = 2\pi/(1\,\text{day}) = 7.27 \times 10^{-5}\,\text{s}^{-1}$ and $g = 9.81\,\text{ms}^{-2}$ one finds $\Delta \approx 2.2\,\text{cm}$.

(ii) We proceed as in Sect. M1.26 (b) and determine the eastward deviation u from the linearized ansatz $\boldsymbol{r}(t) = \boldsymbol{r}^{(0)}(t) + \omega\boldsymbol{u}(t)$, inserting here the unperturbed solution, $\boldsymbol{r}^0(t) = gt(T - \frac{1}{2}t)\hat{\boldsymbol{e}}_v$. This gives $(d'^2/dt^2)\boldsymbol{u}(t) \approx 2g\cos\varphi(t - T)\hat{\boldsymbol{e}}_v$. Integrating twice,

$$\boldsymbol{u}(t) = \frac{1}{3}g\cos\varphi(t^3 - 3Tt^2)\hat{\boldsymbol{e}}_0 .$$

The stone returns to the surface of the earth after time $t = 2T$. The eastward deviation is found to be *negative*, $\Delta \approx -\frac{4}{3}g\omega\cos\varphi T^3$, which means, in reality, that it is a westward deviation. Its magnitude is four times larger than in case (i).

(iii) Denote the eastward deviation by $\boldsymbol{u}$ as before (directed from west to east), the southward deviation by $\boldsymbol{s}$ (directed from north to south). A local, earth-bound, coordinate system is given by $(\hat{\boldsymbol{e}}_1, \hat{\boldsymbol{e}}_0, \hat{\boldsymbol{e}}_v)$, $\hat{\boldsymbol{e}}_1$ defining the direction N–S, $\hat{\boldsymbol{e}}_0$ and $\hat{\boldsymbol{e}}_v$ being defined as in Sect. M1.26 (b). Thus, $\boldsymbol{u} = u\hat{\boldsymbol{e}}_0$, $\boldsymbol{s} = s\hat{\boldsymbol{e}}_1$. The equation of motion (M1.74′), together with $\boldsymbol{\omega} = \omega(-\cos\varphi, 0, \sin\varphi)$, implies

$$\ddot{s} = 2\omega^2 \sin\varphi \dot{u} .$$

Inserting the approximate solution $u \approx \frac{1}{3}gt^3\cos\varphi$ and integrating over time twice, one obtains

$$s(t) = \frac{1}{6}\omega^2 g\sin\varphi\cos\varphi t^4 .$$

1.23 For $E > 0$ all orbits are scattering orbits. If $l^2 > 2\mu\alpha$,

$$\begin{aligned}\phi - \phi_0 &= \frac{l}{\sqrt{2\mu E}}\int_{r_0}^{r}\frac{dr'}{\sqrt{r'^2 - (l^2 - 2\mu\alpha)/(2\mu E)}}\\ &= r_p^{(0)}\int_{r_0}^{r}\frac{dr'}{r'\sqrt{r'^2 - r_\text{P}^2}} ,\end{aligned} \tag{1}$$

where μ is the reduced mass, $r_\text{P} = \sqrt{(l^2 - 2\mu\alpha)/(2\mu E)}$ the perihelion and $r_\text{P}^{(0)} = l/\sqrt{2\mu E}$. The particle is assumed to come from infinity, traveling parallel to the x-axis. Then the solution is $\phi(r) = l/\sqrt{l^2 - 2\mu\alpha}\,\arcsin(r_\text{P}/r)$. If $\alpha = 0$, the corresponding solution is $\phi^{(0)}(r) = \arcsin(r_\text{P}^{(0)}/r)$; the particle moves along a straight line parallel to the x-axis, at the distance $r_\text{P}^{(0)}$. For $\alpha \neq 0$

$$\phi(r = r_\text{P}) = \frac{l}{\sqrt{l^2 - 2\mu\alpha}}\frac{\pi}{2} ,$$

that is, after the scattering and asymptotically, the particle moves in the direction $l/\sqrt{l^2 - 2\mu\alpha}\,\pi$. Before that it travels around the center of force n times if the condition

$$\frac{l}{\sqrt{l^2 - 2\mu\alpha}} \left(\arcsin \frac{r_P}{\infty} - \arcsin \frac{r_P}{r_P} \right) = \frac{r_P^{(0)}}{r_P} \left(\pi - \frac{\pi}{2} \right) > n\pi$$

is fulfilled. The number

$$n = \left[\frac{r_P^{(0)}}{2r_P} \right]$$

is independent of energy E.

In the case $l^2 < 2\mu\alpha$ eq. (1) can also be integrated. With the same initial condition one obtains

$$\phi(r) = \frac{r_P^{(0)}}{b} \ln \frac{b + \sqrt{b^2 + r^2}}{r} ,$$

where $b = ((2\mu\alpha - l^2)/(2\mu E))^{1/2}$. The particle travels around the force center on a spiral-like orbit, towards the center. As the radial distance tends to zero, the angular velocity increases in such a way as to respect Kepler's second law (M1.22).

1.24 Let the comet and the sun approach each other with energy E. Long before the collision the relative momentum has the magnitude $q = \sqrt{2\mu E}$, with μ the reduced mass, the angular momentum has the magnitude $l = qb$. The comet crashes when the perihelion r_P of its hyperbola is smaller or equal R, i.e., when $b \le b_{\max}$ with $b_{\max}$ following from the condition $r_P = R$, viz.

$$\frac{p}{1+\varepsilon} = R \quad \text{with} \quad p\frac{l^2}{A\mu} = \frac{q^2b^2}{A\mu} , \quad \varepsilon - \sqrt{1 + \frac{2Eq^2b^2}{\mu A^2}}$$

and $A = GmM$. One finds $b_{\max} = \sqrt{1 + A/(ER)}$ and, hence,

$$\sigma = \int_0^{b_{\max}} 2\pi b\, db = \pi R^2 \left(1 + \frac{A}{ER} \right) .$$

For $A = 0$ this is the area of the sun seen by the comet. With increasing gravitational attraction ($A > 0$) this surface increases by the ratio (potential energy at the sun's edge)/(energy of relative motion).

1.25 As explained in Sect. M1.21.2 the equation of motion reads

$$\dot{\boldsymbol{x}} = A\boldsymbol{x} + \boldsymbol{b} ,$$

with A as given in eq. (M1.50), and

$$A = \begin{bmatrix} 0 & 0 & 0 & 1/m & 0 & 0 \\ 0 & 0 & 0 & 0 & 1/m & 0 \\ 0 & 0 & 0 & 0 & 0 & 1/m \\ 0 & 0 & 0 & 0 & K & 0 \\ 0 & 0 & 0 & -K & 0 & 0 \\ 0 & 0 & 0 & 0 & 0 & 0 \end{bmatrix} , \quad \boldsymbol{b} = e \begin{bmatrix} 0 \\ 0 \\ 0 \\ E_x \\ E_y \\ E_z \end{bmatrix} .$$

The last of the six equations is integrated immediately, giving $x_6 = eEt + C_1$. Inserting this into the third and integrating yields

$$x_3 = z = \frac{eE_z}{2m} t^2 + C_1 t + C_2 \, .$$

The initial conditions $z(0) = z^{(0)}$, $\dot{z}(0) = v_z^{(0)}$ give $C_2 = z^{(0)}$, $C_1 = v_z^{(0)}$. The remaining equations are coupled equations. Taking the time derivative of the fourth and replacing $\dot{x}_5$ by the right-hand side of the fifth gives $\ddot{x}_4 = -K^2 x_4 + eKE_y$ which is integrated to $x_4 = C_3 \sin Kt + C_4 \cos Kt + eE_y/K$. Making use of the fifth equation once more yields $x_3 = C_3 \cos Kt - C_4 \sin Kt + C_5$. Also the fourth equation yields the condition $C_5 = -eE_x/K$. These two expressions are inserted into the first and second equations so that these can be integrated yielding

$$x_1 = -\frac{C_3}{Km} \cos Kt + \frac{C_4}{Km} \sin Kt + \frac{e}{Km} E_y t + C_6$$

$$x_2 = +\frac{C_3}{Km} \sin Kt + \frac{C_4}{Km} \cos Kt - \frac{e}{Km} E_x t + C_7 \, .$$

Upon insertion of the initial conditions $x(0) = x^{(0)}$, $y(0) = y^{(0)}$, $\dot{x}(0) = v_x^{(0)}$, $\dot{y}(0) = v_y^{(0)}$ we finally obtain

$$C_3 = mv_y^{(0)} + \frac{e}{K} E_x \, , \quad C_4 = mv_x^{(0)} - \frac{e}{K} E_y \, ,$$

$$C_6 = x^{(0)} + \frac{v_y^{(0)}}{K} + \frac{e}{mK^2} E_x \, , \quad C_7 = y^{(0)} - \frac{v_x^{(0)}}{K} + \frac{e}{mK^2} E_y \, .$$

If the electric field points along the z-direction, $\boldsymbol{E} = E\hat{\boldsymbol{e}}_z$, then the motion is the superposition of a uniformly accelerated motion along the z-direction and a circular motion in the (x, y)-plane. That is to say the particle runs along a spiral.

1.26 Using Cartesian coordinates in the plane of the motionand allowing for an arbitrary initial position of the perihelion, the solution eq. (M1.21) reads

$$x(t) = \frac{p}{1+\varepsilon\cos(\phi-\phi_0)} \cos(\phi - \phi_0) \, ,$$

$$y(t) = \frac{p}{1+\varepsilon\cos(\phi-\phi_0)} \sin(\phi - \phi_0) \, .$$

Differentiate these formulae with respect to the time variable and replace the derivative $\dot{\phi}$ by $\ell/(\mu r^2)$, by means of eq. (M1.19a). Inserting $p = \ell^2/(A\mu)$ one obtains

$$p_x(t) = \mu\dot{x} = -\frac{A\mu}{\ell} \sin(\phi - \phi_0) \, ,$$

$$p_y(t) = \mu\dot{y} = \frac{A\mu}{\ell} \{\cos(\phi - \phi_0) + \varepsilon\} \, .$$

This equation describes a circle with radius $A\mu/\ell$ whose center has the coordinates

$$\left(0, \varepsilon\frac{A\mu}{\ell}\right) = \left(0, \sqrt{(A\mu/\ell)^2 + 2\mu E}\right) .$$

See also Exercise 2.31 below.

Chapter 2: The Principles of Canonical Mechanics

2.1 We take the derivative of $F(E)$ with respect to E

$$\frac{dF}{dE} = 2\frac{d}{dE}\int_{q_{\min}(E)}^{q_{\max}(E)} \sqrt{2m(E-U(q))}\,dq = 2\int_{q_{\min}(E)}^{q_{\max}(E)} \frac{m}{\sqrt{2m(E-U(q))}}\,dq$$
$$+2\sqrt{2m\underbrace{(E-U(q_{\max}))}_{=0}}\,\frac{dq_{\max}}{dE} - 2\sqrt{2m\underbrace{(E-U(q_{\min}))}_{=0}}\,\frac{dq_{\min}}{dE}\,.$$

To find T we must calculate the time integral over one period. In doing so we note that

$$m\frac{dq}{dt} = p = \sqrt{2m(E-U(q))}\,, \quad \text{and hence}\,,$$
$$dt = \frac{m\,dq}{\sqrt{2m(E-U(q))}}\,.$$

Therefore,

$$T = 2\int_{q_{\min}(E)}^{q_{\max}(E)} \frac{m}{\sqrt{2m(E-U(q))}}\,dq\,.$$

This, however, is precisely the expression calculated above. For the example of the oscillator with $q = q_0 \sin\omega t$, $p = m\omega q_0 \cos\omega t$, one finds $F = m\omega\pi q_0^2 = (2\pi/\omega)E$ and $T = 2\pi/\omega$.

2.2 Choose the plane as sketched in Fig. 8. D'Alembert's principle $(\boldsymbol{F}-\dot{\boldsymbol{p}})\cdot\delta\boldsymbol{r} = 0$, with $\boldsymbol{F} = -mg\hat{\boldsymbol{e}}_3$, admits virtual displacements along the line of intersection of the inclined plane and the (1,3)-plane as well as long the 2-axis. Denoting the two independent variables by q_1, q_2, this means that $\delta\boldsymbol{r} = \delta q_1\hat{\boldsymbol{e}}_\alpha + \delta q_2\hat{\boldsymbol{e}}_2$ with

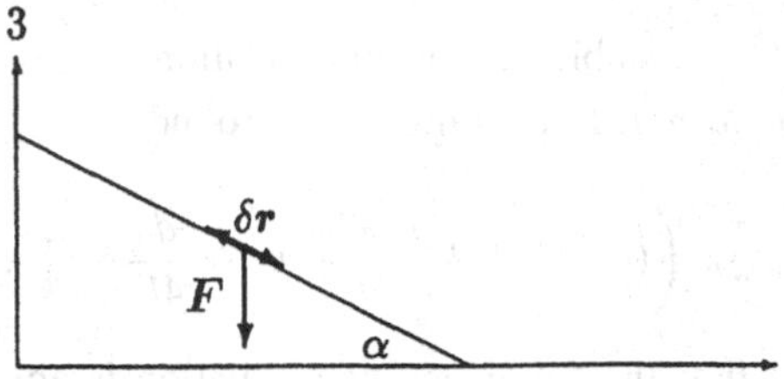

Fig. 8.

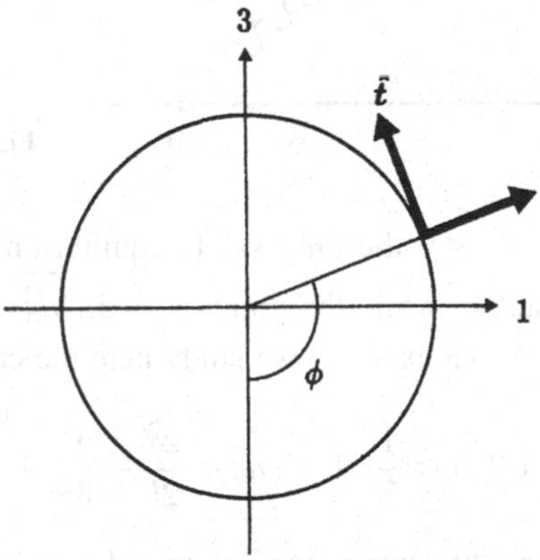

Fig. 9.

$\hat{\boldsymbol{e}}_\alpha = \hat{\boldsymbol{e}}_1 \cos\alpha - \hat{\boldsymbol{e}}_3 \sin\alpha$. Inserting this yields the equations of motion $\ddot{q}_1 = g\sin\alpha$, $\ddot{q}_2 = 0$ whose solutions read

$$q_1(t) = (g\sin\alpha)\frac{t^2}{2} + v_1 t + a_1 , \quad q_2(t) = v_2 t + a_2 .$$

2.3 Choose the (1,3)-plane to coincide with the plane of the annulus and take its center to be the origin. Choosing the unit vectors $\hat{\boldsymbol{t}}$ and $\hat{\boldsymbol{n}}$ as shown in Fig. 9, viz.

$$\hat{\boldsymbol{t}} = \hat{\boldsymbol{e}}_1 \cos\phi + \hat{\boldsymbol{e}}_3 \sin\phi , \quad \hat{\boldsymbol{n}} = \hat{\boldsymbol{e}}_1 \sin\phi - \hat{\boldsymbol{e}}_3 \cos\phi$$

we find

$$\delta\boldsymbol{r} = \hat{\boldsymbol{t}} R\delta\phi , \quad \dot{\boldsymbol{r}} = R\dot{\phi}\hat{\boldsymbol{t}} , \quad \ddot{\boldsymbol{r}} = R\ddot{\phi}\hat{\boldsymbol{t}} - R\dot{\phi}^2\hat{\boldsymbol{n}} ,$$

the force acting on the system being $\boldsymbol{F} = -mg\hat{\boldsymbol{e}}_3$.

D'Alembert's principle $(\boldsymbol{F} - \dot{\boldsymbol{p}}) \cdot \delta\boldsymbol{r} = 0$ yields the equation of motion $\ddot{\phi} + g\sin\phi/R = 0$. This is the equation of motion of the planar pendulum that was studied in Sect. M1.17.

2.4 Let d_0 be the length of the spring in its rest state and let κ be the string constant. When the mass point is at the position x the length of the string is $d = \sqrt{x^2 + l^2}$. The corresponding potential energy is

$$U(x) = \frac{1}{2}\kappa(d - d_0)^2 .$$

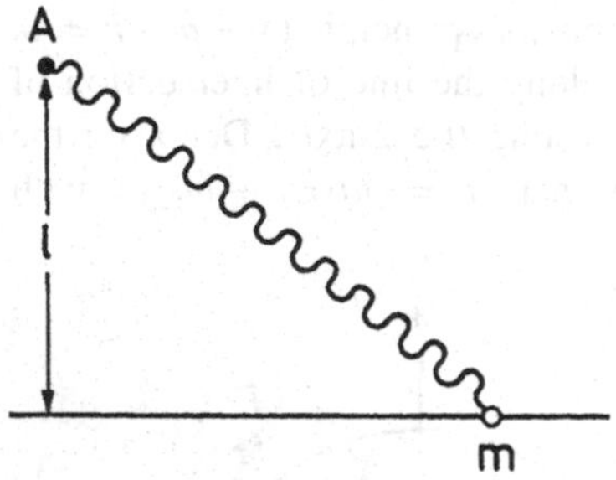

Fig. 10.

For $d_0 \leq l$ the only stable equilibrium position is $x = 0$. For $d_0 > l$, $x = 0$ is unstable, while the points $x = \pm\sqrt{d_0^2 - l^2}$ are stable equilibrium positions.

As an example we study here the case $d_0 \leq l$. Expanding $U(x)$ around $x = 0$,

$$U(x) \approx \frac{1}{2}\kappa\left(l - d_0 + \frac{x^2}{2l} - \frac{x^4}{8l^3}\right)^2 \approx \frac{1}{2}\kappa\left((l - d_0)^2 + \frac{l - d_0}{l}x^2 + \frac{d_0}{4l^3}x^4\right) .$$

From this expression we would conclude that the frequency of oscillation is approximately

$$\omega = \sqrt{\frac{\kappa}{m}\frac{l-d_0}{l}} \; .$$

However, this does not hold for all values of d_0. For $d_0 = l$ the quadratic term vanishes, and x^4 is the leading order. In the other extreme, $d_0 = 0$, we have $U(x) = \kappa(x^2 + l^2)/2$, i.e. a purely harmonic potential (the constant terms in the potential are irrelevant). Thus, the approximation is acceptable only when d_0 is small compared to l.

2.5 A suitable Lagrangian function for this system reads

$$L = \underbrace{\frac{1}{2}m(\dot{x}_1^2 + \dot{x}_2^2)}_{T} - \underbrace{\frac{1}{2}\kappa(x_1 - x_2)^2}_{U} \; .$$

Introduce the following coordinates: $u_1 := x_1 + x_2$, $u_2 := x_1 - x_2$. Except for a factor 1/2 these are the center-of-mass and relative coordinates, respectively. The Lagrangian becomes $L = m(\dot{u}_1^2 + \dot{u}_2^2)/4 - \kappa u_2^2/2$. The equations of motion that follow from it are

$$\ddot{u}_1 = 0 \,, \quad m\ddot{u}_2 + 2\kappa u_2 = 0 \; .$$

The solutions are $u_1 = C_1 t + C_2$, $u_2 = C_3 \sin\omega t + C_4 \cos\omega t$, with $\omega = \sqrt{2\kappa/m}$. It is not difficult to rewrite the initial conditions in the new coordinates, viz.

$$\begin{aligned} u_1(0) &= +l \qquad & \dot{u}_1(0) &= v_0 \\ u_2(0) &= -l \qquad & \dot{u}_2(0) &= v_0 \; . \end{aligned}$$

The constants are determined from these so that the final solution is

$$\begin{aligned} x_1(t) &= \frac{v_0}{2}\left(t + \frac{1}{\omega}\sin\omega t\right) - \frac{l}{2}(1 - \cos\omega t) \\ x_2(t) &= \frac{v_0}{2}\left(t - \frac{1}{\omega}\sin\omega t\right) + \frac{l}{2}(1 + \cos\omega t) \; . \end{aligned}$$

2.6 By hypothesis $F(\lambda x_1, \ldots, \lambda x_n) = \lambda^N F(x_1, \ldots, x_N)$. We take the first derivative of this equation with respect to λ and set $\lambda = 1$. The left-hand side is

$$\frac{d}{d\lambda} F(\lambda x_1, \ldots, \lambda x_n)\bigg|_{\lambda=1} = \sum_{i=1}^{n} \frac{\partial F}{\partial x_i}\frac{d(\lambda x_i)}{d\lambda}\bigg|_{\lambda=1} = \sum_{i=1}^{n} \frac{\partial F}{\partial x_i} x_i \; .$$

The same operation on the right-hand side gives NF.

2.7 In the general case the Euler-Lagrange equation reads

$$\frac{\partial f}{\partial y} = \frac{d}{dx}\frac{\partial f}{\partial y'} \; .$$

Multiply this equation by y' and add the term $y''\partial f/\partial y'$ on both sides. The right-hand side is combined to

$$y'\frac{\partial f}{\partial y} + y''\frac{\partial f}{\partial y'} = \frac{d}{dx}\left(y'\frac{\partial f}{\partial y'}\right) .$$

If f does not depend explicitly on x then the left-hand side is $df(y, y')/dx$. The whole equation can be integrated directly and yields the desired relation. Applying this result to $L(\dot{q}, q) = T(\dot{q}) - U(q)$ gives

$$\sum_i \dot{q}_i \frac{\partial T(\dot{q})}{\partial \dot{q}_i} - T + U = \text{const.} .$$

If T is a homogeneous, quadratic form in $\dot{q}$ the solution to Exercise 2.6 tells us that the first term equals $2T$. Therefore, the constant is the energy $E = T + U$.

2.8 (i) We must minimize the arc length

$$L = \int ds = \int_{x_1}^{x_2} \sqrt{1 + y'^2}\, dx$$

i.e. we must choose $f(y, y') = \sqrt{1 + y'^2}$. Applying the result of the preceding exercise we obtain

$$y' \frac{\sqrt{y'}}{\sqrt{1 + y'^2}} - \sqrt{1 + y'^2} = \text{const.} ,$$

or $y' = \text{const.}$ Thus, $y = ax + b$. Inserting the boundary conditions $y(x_1) = y_1$, $y(x_2) = y_2$ gives

$$y(x) = \frac{y_2 - y_1}{x_2 - x_1}(x - x_1) + y_1 .$$

(ii) The position of the center of mass is determined by the equation

$$M r_S = \int r\, dm ,$$

M denoting the mass of the chain, dm the mass element. If λ is the mass per unit length, $dm = \lambda\, ds$. As the x-coordinate of the center of mass is irrelevant, the problem is to find the shape for which its y-coordinate is lowest. Thus we have to minimize the functional

$$\int y\, ds = \int_{x_1}^{x_2} y\sqrt{1 + y'^2}\, dx .$$

The result of the preceding exercise leads to

$$\frac{yy'^2}{\sqrt{1+y'^2}} - y\sqrt{1+y'^2} = -\frac{\sqrt{y}}{\sqrt{1+y'^2}} = C \, .$$

This equation can be solved for y',

$$y' = \sqrt{Cy^2 - 1} \, .$$

This is a separable differential equation whose general solution is

$$y(x) = \frac{1}{\sqrt{C}} \cosh\left(\sqrt{C}\, x + C'\right) .$$

The constants C and C', finally, must be chosen such that the boundary conditions $y(x_1) = y_1$, $y(x_2) = x_2$ are fulfilled.

2.9 (i) In either case the equations of motion read

$$\ddot{x}_1 = -m\omega_0^2 x_1 - \frac{1}{2} m(\omega_1^2 - \omega_0^2)(x_1 - x_2)$$
$$\ddot{x}_2 = -m\omega_0^2 x_2 + \frac{1}{2} m(\omega_1^2 - \omega_0^2)(x_1 - x_2) \, .$$

The reason for this result becomes clear when we calculate the difference $L' - L$:

$$L' - L = -\mathrm{i}\omega_0 m(x_1\dot{x}_1 + x_2\dot{x}_2) = -\frac{i}{2}\omega_0 m \frac{d}{dt}(x_1^2 + x_2^2) \, .$$

The two Lagrangian functions differ by the total time derivative of a function which depends on the coordinates only. By the general considerations of Sects. M2.9 and M2.10 such an addition does not alter the equations of motion.

(ii) The transformation to eigenmodes reads

$$z_1 = \frac{1}{\sqrt{2}}(x_1 + x_2) \, , \quad z_2 = \frac{1}{\sqrt{2}}(x_1 - x_2) \, .$$

This transformation

$$\begin{pmatrix} x_1 \\ x_2 \end{pmatrix} \underset{F}{\longrightarrow} \begin{pmatrix} z_1 \\ z_2 \end{pmatrix}$$

is one-to-one. Both F and F^{-1} are differentiable. Thus, F, being a diffeomorphism, leaves the Lagrange equations invariant.

2.10 The axial symmetry of the force suggests the use of cylindrical coordinates. In these coordinates the force must not have a component along the unit vector $\hat{e}_\phi$. Furthermore, since

$$\nabla U(r, \varphi, z) = \frac{\partial U}{\partial r}\hat{\boldsymbol{e}}_r + \frac{1}{r}\frac{\partial U}{\partial \varphi}\hat{\boldsymbol{e}}_\varphi + \frac{\partial U}{\partial z}\hat{\boldsymbol{e}}_z ,$$

U must not depend on φ. The unit vectors $\hat{\boldsymbol{e}}_r$ and $\hat{\boldsymbol{e}}_z$ span a plane that contains the z-axis.

2.11 By a (passive) infinitesimal rotation we have

$$\boldsymbol{x} \approx \boldsymbol{x}_0 - (\hat{\boldsymbol{\varphi}} \times \boldsymbol{x}_0)\,\varepsilon \quad \text{or} \quad \boldsymbol{x}_0 \approx \boldsymbol{x} + (\hat{\boldsymbol{\varphi}} \times \boldsymbol{x})\,\varepsilon .$$

Here $\hat{\boldsymbol{\varphi}}$ is the direction about which the rotation takes place, ε is the angle of rotation, so that $\hat{\boldsymbol{\varphi}}\varepsilon = \boldsymbol{\omega}\,dt$. Thus $\dot{\boldsymbol{x}}_0 = \dot{\boldsymbol{x}} + (\boldsymbol{\omega} \times \boldsymbol{x})$, the dot denoting the time derivative in the system of reference that one considers. Inserting this into the kinetic energy one finds

$$T = m\left(\dot{\boldsymbol{x}}^2 + 2\dot{\boldsymbol{x}} \cdot (\boldsymbol{\omega} \times \boldsymbol{x}) + (\boldsymbol{\omega} \times \boldsymbol{x})^2\right)/2 .$$

Meanwhile $U(\boldsymbol{x}_0)$ becomes $U(\boldsymbol{x}) = \bar{U}(R^{-1}(t)\boldsymbol{x})$. We calculate

$$\frac{\partial L}{\partial \dot{x}_i} = m\dot{x}_i + m(\boldsymbol{\omega} \times \boldsymbol{x})_i$$

$$\frac{\partial L}{\partial x_i} = -\frac{\partial \bar{U}}{\partial x_i} + m(\dot{\boldsymbol{x}} \times \boldsymbol{\omega})_i + m((\boldsymbol{\omega} \times \boldsymbol{x}) \times \boldsymbol{\omega})_i .$$

This leads to the equation of motion

$$m\ddot{\boldsymbol{x}} = -\nabla U - 2m(\boldsymbol{\omega} \times \dot{\boldsymbol{x}}) - m\boldsymbol{\omega} \times (\boldsymbol{\omega} \times \boldsymbol{x}) - m(\dot{\boldsymbol{\omega}} \times \boldsymbol{x}) .$$

2.12 Let the coordinates of the point of suspension be $(x_A, 0)$, φ the angle between the pendulum and the vertical, with $-\pi \le \varphi \le \pi$. The coordinates of the mass point are

$$x = x_A + l\sin\varphi , \quad y = -l\cos\varphi ,$$

m denoting the mass, l the length of the pendulum. Inserting these into

$$L = \frac{m}{2}(\dot{x}^2 + \dot{y}^2) - mg(y + l)$$

gives the answer

$$L = \frac{m}{2}(\dot{x}_A^2 + l^2\dot{\varphi}^2 + 2l\cos\varphi\,\dot{x}_A\dot{\varphi}) + mgl(\cos\varphi - 1) .$$

2.13 (i) If the oscillation is to be harmonic $s(t)$ must obey the following equation

$$\ddot{s} + \kappa^2 s = 0 \Rightarrow s(t) = s_0 \sin\kappa t .$$

(ii) The Lagrangian function reads

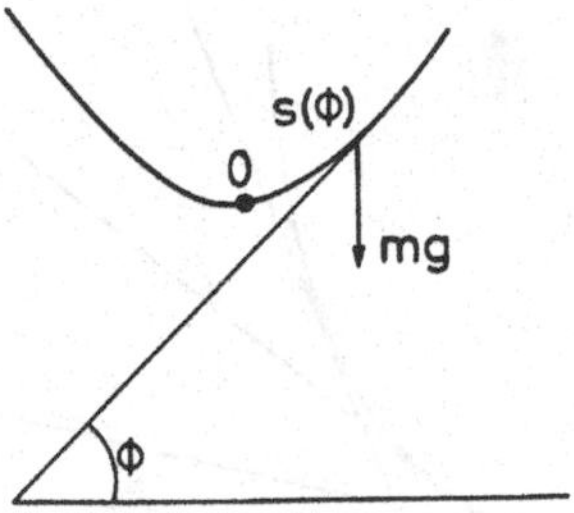

Fig. 11.

$$L = \frac{m}{2}\dot{s}^2 - U$$

where the potential energy is given by (see Fig. 11)

$$U = mgy = mg\int_0^s \sin\phi\, ds\,.$$

The Euler-Lagrange equation reads $m\ddot{s}+mg\sin\phi = 0$. Inserting the above relation for $s(t)$ we obtain the equation $s_0\kappa^2 \sin\kappa t = g\sin\phi$.

Since the absolute value of the sine function is always smaller than, or equal to, 1 one has

$$\lambda := s_0\frac{\kappa^2}{g} \le 1\,.$$

Thus we obtain the equation $\phi(t) = \arcsin(\lambda\sin\kappa t)$ whose derivatives are

$$\dot{\phi} = \frac{\lambda\kappa\cos\kappa t}{\sqrt{1-\lambda^2\sin^2\kappa t}}\,, \qquad \ddot{\phi} = \frac{-\lambda\kappa^2(1-\lambda^2)\sin\kappa t}{(1-\lambda^2\sin^2\kappa t)^{3/2}}\,.$$

In the limit $\lambda \to 1$, ϕ goes to zero and $\dot{\phi}$ goes to κ, except if $\kappa t = (2n+1)/2\pi$ where they are singular.

(iii) The force of constraint is the one perpendicular to the orbit. It is

$$\boldsymbol{Z}(\phi) = mg\cos\phi\begin{pmatrix}-\sin\phi\\ \cos\phi\end{pmatrix}.$$

The effective force is then

$$\boldsymbol{E} = -mg\begin{pmatrix}0\\1\end{pmatrix} + \boldsymbol{Z}(\phi) = -mg\sin\phi\begin{pmatrix}\cos\phi\\ \sin\phi\end{pmatrix}.$$

2.14 (i) The condition $\partial F(x,z)/\partial x = 0$ implies $z - \partial f/\partial z = 0$, i.e., $z = f'(x)$. Therefore, $x = x(z)$ is the point where the *vertical* distance between $y = zx$ (with z fixed) and $y = f(x)$ is largest (see Fig. 12).

(ii) The figure shows that $(\mathcal{L}\Phi)(z) = zx - \Phi(z) \equiv G(x,z)$, z fixed, is tangent to $f(x)$ at the point $x = x(z)$ (the derivative being z).

Keeping $x = x_0$ fixed and varying z yields the picture shown in Fig. 13. For fixed z $y = G(x,z)$ is the tangent to $f(x)$ in $x(z)$. $G(x_0,z)$ is the ordinate of the

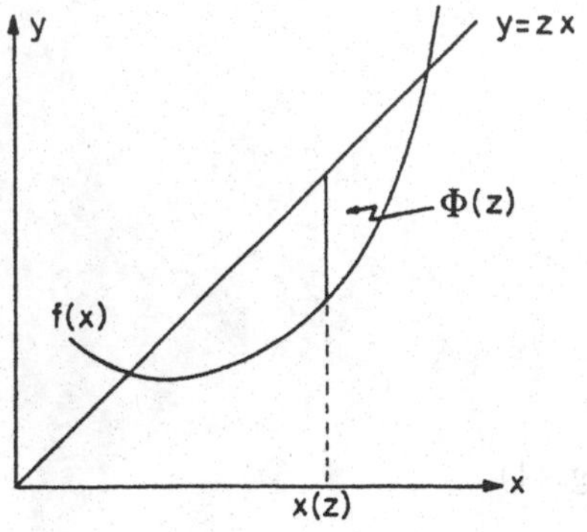

Fig. 12.

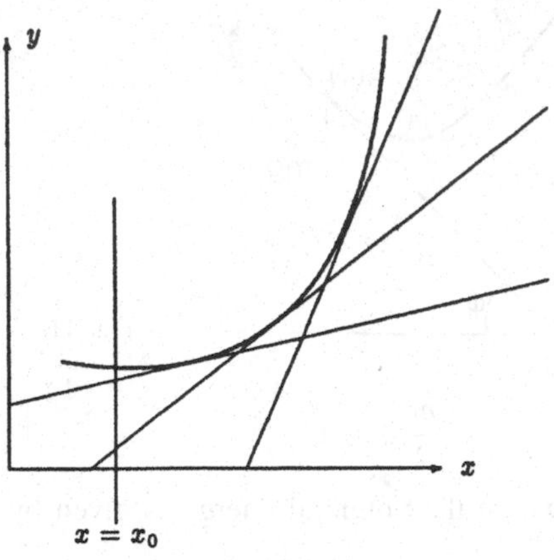

Fig. 13.

intersection point of that tangent with the straight line $x = x_0$. The maximum is at $x_0 = x(z)$, i.e. $z(x_0) = f'(x)|_{x=x_0}$. As $f'' > 0$ all tangents are below the curve. The envelope of this set of straight lines is the curve $y = f(x)$.

2.15 (i) In a first step we determine the canonically conjugate momenta

$$p_1 = \frac{\partial L}{\partial \dot{q}_1} = 2c_{11}\dot{q}_1 + (c_{12} + c_{21})\dot{q}_2 + b_1$$

$$p_2 = \frac{\partial L}{\partial \dot{q}_2} = (c_{12} + c_{21})\dot{q}_1 + 2c_{22}\dot{q}_2 + b_2 .$$

Using the given abbreviations this can be written in the form

$$\pi_1 = d_{11}\dot{q}_1 + d_{12}\dot{q}_2 , \quad \pi_2 = d_{21}\dot{q}_1 + d_{22}\dot{q}_2 .$$

For this to be solvable in terms of $\dot{q}_i$ the determinant

$$D := d_{11}d_{22} - d_{12}d_{21} = \det\left(\frac{\partial L}{\partial \dot{q}_i\, \partial \dot{q}_k}\right) \neq 0$$

must be different from zero. The $\dot{q}_i$ can then be expressed in terms of the π_i:

$$\dot{q}_1 = \frac{1}{D}(d_{22}\pi_1 - d_{12}\pi_2) , \quad \dot{q}_2 = \frac{1}{D}(-d_{21}\pi_1 - d_{11}\pi_2) .$$

We construct the Hamiltonian function and obtain

$$H = p_1\dot{q}_1 + p_2\dot{q}_2 - L = \frac{1}{D}\left(c_{22}\pi_1^2 - (c_{12} + c_{21})\pi_1\pi_2 + c_{11}\pi_2^2\right) - a + U .$$

The above determinant is found to be

$$\det\left(\frac{\partial H}{\partial p_i\,\partial p_k}\right) = \det\left(\frac{\partial^2 H}{\partial \pi_i\,\partial \pi_k}\right)$$
$$= \frac{1}{D^2}\begin{vmatrix} d_{22} & (d_{12}+d_{21})/2 \\ (d_{12}+d_{21})/2 & d_{11} \end{vmatrix} = \frac{1}{D}\,.$$

The inverse transformation, the construction of L from H, proceeds along the same lines.

(ii) Assume that there exists a function

$$F(p_1(x_1, x_2, \boldsymbol{u}), p_2(x_1, x_2, \boldsymbol{u}))$$

which vanishes identically in the domain of definition of the x_i, $\boldsymbol{u}$ fixed. Take the derivatives

$$0 = \frac{dF}{dx_1} = \frac{\partial F}{\partial p_1}\frac{\partial p_1}{\partial x_1} + \frac{\partial F}{\partial p_2}\frac{\partial p_2}{\partial x_1}$$
$$0 = \frac{dF}{dx_2} = \frac{\partial F}{\partial p_1}\frac{\partial p_1}{\partial x_2} + \frac{\partial F}{\partial p_2}\frac{\partial p_2}{\partial x_2}$$

By assumption the partial derivatives of F with respect to p_i do not vanish (otherwise the system of equations would be trivial). Therefore, the determinant

$$D = \det\begin{pmatrix} \dfrac{\partial p_1}{\partial x_1} & \dfrac{\partial p_2}{\partial x_1} \\ \dfrac{\partial p_1}{\partial x_2} & \dfrac{\partial p_2}{\partial x_2} \end{pmatrix} = \det\left(\frac{\partial^2 L}{\partial \dot{x}_i\,\partial \dot{x}_k}\right)$$

must be different from zero. This proves the assertion.

2.16 We introduce the complex variable $w := x + \mathrm{i}y$. Then

$$x = (w + w^*)/2\,, \quad y = -\mathrm{i}(w - w^*)/2\,, \quad \dot{x}^2 + \dot{y}^2 = \dot{w}\dot{w}^*\,.$$

l_3 is calculated to be

$$l_3 = m(x\dot{y} - y\dot{x}) = \frac{m}{2\mathrm{i}}(\dot{w}w^* - w\dot{w}^*)\,.$$

Expressed in the new coordinates the Lagrangian function reads

$$L = \frac{m}{2}(\dot{w}\dot{w}^* + \dot{z}^2) - \frac{\mathrm{i}m\omega}{4}(\dot{w}w^* - w\dot{w}^*)\,.$$

The equations of motion are

$$\frac{m}{2}\ddot{w}^* - \frac{\mathrm{i}m\omega}{4}\dot{w}^* = \frac{\mathrm{i}m\omega}{4}\dot{w}^*\,, \quad m\ddot{z} = 0\,.$$

The first of these is written in terms of the variable $u := \dot{w}^*$. It becomes $\dot{u} = \mathrm{i}\omega u$, its solution being $u = \mathrm{e}^{\mathrm{i}\omega t}$. w^* is the time integral of this function, viz.

$$w^* = -\frac{\mathrm{i}}{\omega}\,\mathrm{e}^{\mathrm{i}\omega t} + C\,,$$

where C is a complex constant. Take the complex conjugate

$$w = \frac{\mathrm{i}}{\omega}\,\mathrm{e}^{-\mathrm{i}\omega t} + C^*\,,$$

from which follow the solutions for x and y

$$x = \frac{1}{\omega}\sin\omega t + C_1\,, \quad y = \frac{1}{\omega}\cos\omega t + C_2\,,$$

where $C_1 = \Re\{C\}$, $C_2 = -\Im\{C\}$.

The solution for the z coordinate is simple: $z = C_3 t + C_4$, i.e., uniform motion along a straight line. The canonically conjugate momenta are

$$p_x = m\dot{x} - \frac{m}{2}\,\omega y\,, \quad p_y = m\dot{y} + \frac{m}{2}\,\omega x\,, \quad p_z = m\dot{z}\,,$$

while the kinetic momenta are given by $p_{\text{kin}} = m\dot{x}$. In order to construct the Hamiltonian function the velocities are expressed in terms of the canonical momenta

$$\dot{x} = \frac{1}{m}\,p_x + \frac{\omega}{2}\,y\,, \quad \dot{y} = \frac{1}{m}\,p_y - \frac{\omega}{2}\,x\,, \quad \dot{z} = \frac{1}{m}\,p_z\,.$$

Then H is found to be

$$H = \boldsymbol{p}\cdot\dot{\boldsymbol{x}} - L = \frac{1}{2m}\,p_{\text{kin}}^2\,.$$

2.17 (i) Hamilton's variational principle, when applied to $\bar{L}$, requires

$$\bar{I} := \int_{\tau_1}^{\tau_2} \bar{L}\,d\tau$$

to be an extremum. Now, since

$$\int_{\tau_1}^{\tau_2} \bar{L}\,d\tau = \int_{t_1}^{t_2} L\,dt \quad \text{with} \quad t_i = t(\tau_i)\,,\ i = 1, 2\,,$$

the action integral $\bar{I}$ is extremal if and only if the Lagrange equations that follow from L are fulfilled.

(ii) We define $q = (q_1, \ldots, q_f)$, $t = q_{f+1}$. From Noether's theorem the quantity

$$I = \sum_{i=1}^{f+1} \frac{\partial\bar{L}}{\partial\dot{q}_i}\frac{d}{ds}\,h^s(q_1, \ldots, q_{f+1})|_{s=0}$$

is an integral of the motion provided $\bar{L}$ is invariant under $(q_1, \ldots, q_{f+1}) \to h^s(q_1, \ldots, q_{f+1})$, i.e., in the case considered here, under $(q_1, \ldots, q_{f+1}) \to (q_1, \ldots, q_{f+1} + s)$. Here

$$\left.\frac{dh^s}{ds}\right|_{s=0} = (0, \ldots, 0, 1)$$

and

$$\frac{\partial \bar{L}}{\partial \dot{q}_{f+1}} = \frac{\partial \bar{L}}{\partial (dt/d\tau)} = L + \sum_{i=1}^{f} \frac{\partial L}{\partial \dot{q}_i}\left(-\frac{1}{(dt/d\tau)^2}\right)\frac{dq_i}{d\tau}\frac{dt}{d\tau} = L - \sum_{i=1}^{f} \frac{\partial L}{\partial \dot{q}_i}\frac{dq_i}{dt} .$$

The integral of the motion is

$$I = L - \sum_{i=1}^{f} \frac{\partial L}{\partial \dot{q}_i}\frac{dq_i}{dt} .$$

Except for a sign this is the expression for the energy.

2.18 The points for which the sum of their distances to A and to Ω is constant lie on the ellipsoid with foci A, Ω, semi-major axis $\sqrt{R^2 + a^2}$, and semi-minor axis R. The reflecting sphere lies *inside* that ellipsoid and is tangent to it in B. Thus, any other path than the one shown in Fig. 14 would be shorter than the one through B for which $\alpha = \beta$.

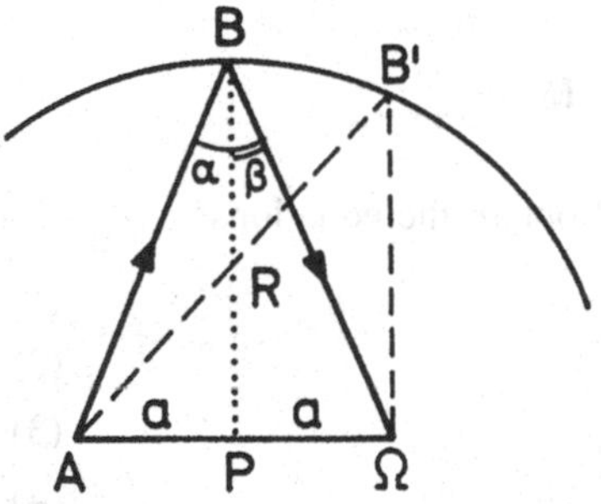

Fig. 14.

2.19 (i) As usual we set $x_\alpha = (q_1, \ldots, q_f; p_1, \ldots, p_f)$, and $y_\beta = (Q_1, \ldots, Q_f; P_1, \ldots, P_f)$, as well as $M_{\alpha\beta} = \partial x_\alpha / \partial y_\beta$. We have

$$\mathbf{M}^T \mathbf{J} \mathbf{M} = \mathbf{J} \quad \text{and} \quad \mathbf{J} = \begin{pmatrix} 0_{f\times f} & 1\!\!1_{f\times f} \\ -1\!\!1_{f\times f} & 0_{f\times f} \end{pmatrix} . \tag{1}$$

The equation always relates $\partial P_k/\partial p_i$ to $\partial q_i/\partial Q_k$, $\partial Q_j/\partial p_l$ to $\partial q_l/\partial P_j$ etc. From this follows that $[P_k \cdot Q_k] = [p_j \cdot q_j]$. Let $\Phi(x, y)$ be generating function of the canonical transformation. As $\tilde{H} = H + \partial\Phi/\partial t$, the function Φ has the dimension of the product $H \cdot t$. The assertion then follows from the canonical equations.

(ii) With the canonical transformation Φ and using $\tau := \omega t$, H goes over into $\tilde{H} = H + \partial\Phi/\partial\tau$. Hence $[\Phi] = [H] = [x_1 x_2] = [\omega][pq]$. The new generalized coordinate $y_1 = Q$ has no dimension. As $y_1 y_2$ has the same dimension as $x_1 x_2$, y_2 must have the dimension of H, or $\tilde{H}$, that is, y_2 must equal ωP. Therefore,

$$\hat{\Phi}(x_1, y_1) = \frac{1}{2} x_1^2 \cot y_1 \, .$$

From this one calculates

$$x_2 = \frac{\partial\hat{\Phi}}{\partial x_1} = x_1 \cot y_1 \, , \quad y_2 = -\frac{\partial\hat{\Phi}}{\partial y_1} = \frac{x_1^2}{2\sin^2 y_1}$$

or,

$$x_1 = \sqrt{2y_2}\sin y_1 \, , \quad x_2 = \sqrt{2y_2}\cos y_1 \, .$$

Using these formulas one finds

$$\mathbf{M}_{\alpha\beta} = \frac{\partial x_\alpha}{\partial y_\beta} = \begin{pmatrix} (2y_2)^{1/2}\cos y_1 & (2y_2)^{-1/2}\sin y_1 \\ -(2y_2)^{1/2}\sin y_1 & (2y_2)^{-1/2}\cos y_1 \end{pmatrix} .$$

One easily verifies the conditions $\det\mathbf{M} = 1$ and $\mathbf{M}^T\mathbf{J}\mathbf{M} = \mathbf{J}$.

2.20 (i) For $f = 1$ the condition $\det\mathbf{M} = 1$ is necessary and sufficient because, quite generally,

$$\mathbf{M}^T\mathbf{J}\mathbf{M} = \begin{pmatrix} 0 & 1 \\ -1 & 0 \end{pmatrix}(a_{11}a_{22} - a_{12}a_{21}) = \mathbf{J}\det\mathbf{M} \, .$$

(ii) We calculate $\mathbf{S}\cdot\mathbf{O}$, set it equal to $\mathbf{M}$, and obtain the equations

$$x\cos\alpha - y\sin\alpha = a_{11} \qquad (1)$$

$$x\sin\alpha + y\cos\alpha = a_{12} \qquad (2)$$

$$y\cos\alpha - z\sin\alpha = a_{21} \qquad (3)$$

$$y\sin\alpha + z\cos\alpha = a_{22} \qquad (4)$$

From the combination $((2) - (3))/((1) + (4))$ of the equations

$$\tan\alpha = \frac{a_{12} - a_{21}}{a_{11} + a_{22}} \, .$$

This allows us to calculate $\sin\alpha$ and $\cos\alpha$, so that the subsystems $((1), (2))$ and $((3), (4))$ can be solved for x, y, and z. One finds $x = a_{11}\cos\alpha + a_{12}\sin\alpha$, $z = a_{22}\cos\alpha - a_{21}\sin\alpha$, $y^2 = xz - 1$.

There is a special case, however, that must be studied separately: This is when $a_{11} + a_{22} = 0$. If $a_{12} \neq a_{21}$ we take the reciprocal of the above relation

$$\cot\alpha = \frac{a_{11}+a_{22}}{a_{12}-a_{21}}\ .$$

If, however, $a_{12} = a_{21}$ the matrix **M** is symmetric and **O** can be taken to be the unit matrix, i.e., $\alpha = 0$.

2.21 (i) Using the product rule one has $\{fg, h\} = f\{g, h\} + g\{f, h\}$. Hence

$$\{l_i, r_k\} = \{\varepsilon_{imn} r_m p_n, r_k\} = \varepsilon_{imn} r_m \{p_n, r_k\} + \varepsilon_{imn} p_n$$
$$\{r_m, r_k\} = \varepsilon_{imn} r_m \delta_{nk} = \varepsilon_{imk} r_m$$

and, in a similar fashion, $\{l_i, p_k\} = \varepsilon_{ikm} p_m$. In calculating the third Poisson bracket we note that

$$\begin{aligned}\{l_i, r\} &= \{\varepsilon_{imn} r_m p_n, r\} = \varepsilon_{imn} r_m \{p_n, r\} + \varepsilon_{imn} p_n \{r_m, r\}\\ &= \varepsilon_{imn} r_m \frac{\partial r}{\partial r_n} = \varepsilon_{imn} r_m r_n \frac{1}{r} = 0\ .\end{aligned}$$

Finally, we have

$$\begin{aligned}\left\{l_i, \boldsymbol{p}^2\right\} &= \{\varepsilon_{imn} r_m p_n, p_k p_k\} = \varepsilon_{imn} r_m \{p_n, p_k p_k\} + \varepsilon_{imn} p_n \{r_m, p_k p_k\}\\ &= -2\varepsilon_{imn} p_n p_k \delta_{mk} = -\varepsilon_{imn} p_n p_m = 0\ .\end{aligned}$$

(ii) U can only depend on r.

2.22 The vector A is a constant of the motion precisely when the Poisson bracket of each of its components with the Hamiltonian function vanishes. Therefore, we calculate

$$\begin{aligned}\{H, A_k\} &= \left\{\frac{1}{2m}\boldsymbol{p}^2 + \frac{\gamma}{r}, \varepsilon_{klm} p_l l_m + \frac{m\gamma}{r} r_k\right\}\\ &= \frac{1}{2m} e_{klm}\left\{\boldsymbol{p}^2, p_l l_m\right\} + \gamma\varepsilon_{klm}\{1/r, p_l l_m\}\\ &\quad + \frac{\gamma}{2}\left\{\boldsymbol{p}^2, r_k/r\right\} + m\gamma^2\{1/r, r_k/r\}\ .\end{aligned}$$

The fourth bracket vanishes. The first three are calculated as follows

$$\left\{\boldsymbol{p}^2, p_l l_m\right\} = \left\{\boldsymbol{p}^2, p_l\right\} l_m + \left\{\boldsymbol{p}^2, l_m\right\} p_l = 0$$
$$\{1/r, p_l l_m\} = \{1/r, p_l\} l_m + \{1/r, l_m\} p_l = r_l/r^3 l_m$$
$$\left\{\boldsymbol{p}^2, r_k/r\right\} = 1/r\left\{\boldsymbol{p}^2, r_k\right\} + r_k\left\{\boldsymbol{p}^2, 1/r\right\} = 2p_k/r - 2r_k\boldsymbol{p}\cdot\boldsymbol{x}/r^3\ .$$

Inserting these results we obtain

$$\{H, A_k\} = \gamma\varepsilon_{klm} r_l/r^3 l_m + \gamma\left(p_k/r - r_k\boldsymbol{p}\cdot\boldsymbol{x}/r^3\right) = 0\ .$$

This vector is often called Lenz vector or, in the German literature, Lenz-Runge vector, although apparently neither H.F.E. Lenz nor C. Runge claimed priority for it. Its discovery is due to Jakob Hermann (published in Giornale dei Letterati d'Italia, vol. 2 (1710) p. 447). The conservation of this vector was also known to Joh. I Bernoulli and to P.-S. de Laplace, see H. Goldstein, Am. J. Phys. **44** (1976) No. 11. very much in the spirit of linear algebra.

2.23 Calculation of the Poisson brackets yields differential equations which are solved taking proper account of the initial conditions as follows:

$$\begin{aligned}
\dot{p}_1 &= \{H, p_1\} = -m\alpha \Rightarrow p_1 = -m\alpha t + p_x \,, \\
\dot{p}_2 &= \{H, p_2\} = 0 \Rightarrow p_2 = p_y \,, \\
\dot{q}_1 &= \{H, q_1\} = \frac{1}{m}\, p_1 \Rightarrow q_1 = -\frac{1}{2}\alpha t^2 + \frac{p_x}{m}\, t + x_0 \,, \\
\dot{q}_2 &= \{H, q_2\} = \frac{1}{m}\, p_2 \Rightarrow q_2 = \frac{p_y}{m}\, t + y_0 \,.
\end{aligned}$$

2.24 (i) Let $\mu_1 = m_1 m_2/(m_1+m_2)$ and $\mu_2 = (m_1+m_2)m_3/(m_1+m_2+m_3)$ be the reduced masses of the two two-body systems (1, 2) and ((center-of-mass of 1 and 2), 3), respectively. Then $\pi_1 = \mu_1 \dot{\varrho}_1$ and $\pi_2 = \mu_2 \dot{\varrho}_2$. This explains the meaning of these two momenta. The momentum π_3 is the center-of-mass momentum.

(ii) We define

$$M_j := \sum_{i=1}^{j} m_i \,,$$

i.e., M_j is the total mass of particles $1, 2, \ldots, j$. We can then write

$$\begin{aligned}
\boldsymbol{\varrho}_j &= \boldsymbol{r}_{j+1} - \frac{1}{M_j} \sum_{i=1}^{j} m_i \boldsymbol{r}_i \,, \quad j = 1, \ldots, N-1 \\
\boldsymbol{\varrho}_N &= \frac{1}{M_N} \sum_{i=1}^{N} m_i \boldsymbol{r}_i \,, \\
\boldsymbol{\pi}_j &= \frac{1}{M_{j+1}} \left(M_j \boldsymbol{p}_{j+1} - m_{j+1} \sum_{i=1}^{j} \boldsymbol{p}_i \right) \,, \quad j = 1, \ldots, N-1 \\
\boldsymbol{\pi}_N &= \sum_{i=1}^{N} \boldsymbol{p}_i \,.
\end{aligned}$$

(iii) We choose the following possibilities:

a) As the Poisson bracket of $\boldsymbol{r}_i$ and $\boldsymbol{p}_k$ ist $\{\boldsymbol{p}_k, \boldsymbol{r}_i\} = 1\!\!1_{3\times 3}\delta_{ik}$ we must also have $\{\boldsymbol{\pi}_k, \boldsymbol{\varrho}_i\} = 1\!\!1_{3\times 3}\delta_{ik}$. We use this suggestive short hand notation for $\{(\boldsymbol{p}_k)_m, (\boldsymbol{q}_i)_n\} = \delta_{ik}\delta_{mn}$ with $(\cdot)_m$ denoting the mth Cartesian coordinate. Using the former brackets one calculates the latter brackets from the defining formulas. For instance, with $m_{12} := m_1 + m_2$, $M := m_1 + m_2 + m_3$,

$$\{\pi_1, \varrho_1\} = \left(\frac{m_1}{m_{12}} + \frac{m_2}{m_{12}}\right)\mathbb{1} = \mathbb{1} \,,$$
$$\{\pi_2, \varrho_1\} = \left(\frac{m_3}{M} - \frac{m_3}{M}\right)\mathbb{1} = 0 \,, \quad \text{etc.}$$

b) In the 18-dimensional phase space introduce the variables $x = (\boldsymbol{r}_1, \boldsymbol{r}_2, \boldsymbol{r}_3, \boldsymbol{p}_1, \boldsymbol{p}_2, \boldsymbol{p}_3)$ and $y = (\boldsymbol{\varrho}_1, \boldsymbol{\varrho}_2, \boldsymbol{\varrho}_3, \boldsymbol{\pi}_1, \boldsymbol{\pi}_2, \boldsymbol{\pi}_3)$, calculate the matrix $M_{\alpha\beta} := \partial y_\alpha / \partial x_\beta$ and verify that this matrix is symplectic, i.e. that it satisfies eq. (M2.113). This calculation can be simplified by noting that **M** has the form

$$\begin{pmatrix} A & 0 \\ 0 & B \end{pmatrix}$$

such that

$$\mathbf{M}^T\mathbf{J}\mathbf{M} = \begin{pmatrix} 0 & \mathbf{A}^T\mathbf{B} \\ -\mathbf{B}^T\mathbf{A} & 0 \end{pmatrix}$$

Thus, it suffices to verify that $\mathbf{A}^T\mathbf{B} = \mathbb{1}_{9\times 9}$. One finds

$$\mathbf{A} = \begin{bmatrix} -\mathbb{1} & \mathbb{1} & 0 \\ -m_1/m_{12}\mathbb{1} & -m_2/m_{12}\mathbb{1} & \mathbb{1} \\ m_1/M\mathbb{1} & m_2/M\mathbb{1} & m_3/M\mathbb{1} \end{bmatrix} ,$$
$$\mathbf{B} = \begin{bmatrix} -m_2/m_{12}\mathbb{1} & m_1/m_{12}\mathbb{1} & 0 \\ -m_3/M\mathbb{1} & -m_3/M\mathbb{1} & m_{12}/M\mathbb{1} \\ \mathbb{1} & \mathbb{1} & \mathbb{1} \end{bmatrix} ,$$

the entries being themselves 3×3 matrices. In a next step one calculates $(\mathbf{A}^T\mathbf{B})_{ik} = \sum_l A_{li}B_{lk}$. For instance, one finds

$$(\mathbf{A}^T\mathbf{B})_{11} = \frac{m_2}{M_{12}} + \frac{m_1 m_3}{m_{12}M} + \frac{m_1}{M} = 1 \,, \quad \text{etc.}$$

and verifies, eventually, that $\mathbf{A}^T\mathbf{B} = \mathbb{1}_{9\times 9}$.

2.25 (i) In the situation described in the exercise the variation of $I(\alpha)$ is

$$\begin{aligned} \delta I &= \left.\frac{dI(\alpha)}{d\alpha}\right|_{\alpha=0} d\alpha \\ &= L\left(q_k(t_2(0),0), \dot{q}_k(t_2(0),0)\right) \left.\frac{dt_2(\alpha)}{d\alpha}\right|_{\alpha=0} d\alpha \\ &\quad - L\left(q_k(t_1(0),0), \dot{q}_k(t_1(0),0)\right) \left.\frac{dt_1(\alpha)}{d\alpha}\right|_{\alpha=0} d\alpha \\ &\quad + \int_{t_1(0)}^{t_2(0)} \left(\sum_k \frac{\partial L}{\partial q_k} \left.\frac{\partial q_k(t,\alpha)}{\partial \alpha}\right|_{\alpha=0} d\alpha + \sum_k \frac{\partial L}{\partial \dot{q}_k} \left.\frac{\partial \dot{q}_k(t,\alpha)}{\partial \alpha}\right|_{\alpha=0} d\alpha \right) dt \,. \end{aligned}$$

We define, as usual

$$\left.\frac{\partial q_k}{\partial \alpha}\right|_0 d\alpha = \delta q_k \quad \text{and} \quad \left.\frac{\partial \dot{q}_k}{\partial \alpha}\right|_0 d\alpha = \delta \dot{q}_k = \frac{d}{dt}\delta q_k \,,$$

and, in addition, $dt_i(\alpha)/d\alpha|_0 d\alpha = \delta t_i$, $i = 1, 2$. The time derivative $d\delta\, q_k/dt$, by partial integration, is shifted onto $\partial L/\partial \dot{q}_k$. Here, however, the terms at the boundaries do not vanish because the variations δt_i do not vanish. One has

$$\int_{t_1(0)}^{t_2(0)} \frac{\partial L}{\partial \dot{q}_k}\frac{d}{dt}\delta q_k\, dt = \left[\frac{\partial L}{\partial \dot{q}_k}\delta q_k\right]_{t_1(0)}^{t_2(0)} - \int_{t_1(0)}^{t_2(0)} \left(\frac{d}{dt}\frac{\partial L}{\partial \dot{q}_k}\right)\delta q_k\, dt \,.$$

The end points are kept fixed which means that

$$\left.\frac{dq_k(t_i(\alpha), \alpha)}{d\alpha}\right|_{\alpha=0} = 0 \,, \quad i = 1, 2 \,.$$

Taking the derivative with respect to α this implies

$$\begin{aligned} \left.\frac{dq_k(t_i(\alpha), \alpha)}{d\alpha}\right|_0 &= \left.\frac{\partial q_k}{\partial t}\right|_{t=t_i} \left.\frac{dt_i(\alpha)}{d\alpha}\right|_{\alpha=0} d\alpha + \left.\frac{\partial q_k}{\partial \alpha}\right|_{t=t_i,\alpha=0} d\alpha \\ &\equiv \dot{q}_k(t_i)\delta t_i + \delta q_k|_{t=t_i} = 0 \end{aligned}$$

Inserting this into δI one obtains the result

$$\delta I = \left[\left(L - \sum_k \frac{\partial L}{\partial \dot{q}_k}\dot{q}_k\right)\delta t_i\right]_{t_1(0)}^{t_2(0)} + \int_{t_1(0)}^{t_2(0)} dt \sum_k \left(\frac{\partial L}{\partial q_k} - \frac{d}{dt}\frac{\partial L}{\partial \dot{q}_k}\right)\delta q_k \,.$$

(ii) One calculates δK in exactly the way, viz.

$$\begin{aligned} \delta K &= K \int_{t_1}^{t_2} (L + E)\, dt \\ &= \left[\left(L - \sum_k \frac{\partial L}{\partial \dot{q}_k}\dot{q}_k\right)\delta t_i\right]_{t_1}^{t_2} \\ &\quad + \int_{t_1}^{t_2} dt \sum_k \left(\frac{\partial L}{\partial q_k} - \frac{d}{dt}\frac{\partial L}{\partial \dot{q}_k}\right)\delta q_k + [E\delta t_i]_{t_1}^{t_2} = 0 \,. \end{aligned}$$

Now, by assumption $E = \sum_k \dot{q}_k(\partial L/\partial \dot{q}_k) - L$ is constant. As a consequence the first and third terms of the equation cancel. As the variations δq_k are independent one finds indeed the implication

$$\delta K \stackrel{!}{=} 0 \Leftrightarrow \frac{\partial L}{\partial q_k} - \frac{d}{dt}\frac{\partial L}{\partial \dot{q}_k} = 0 \,, \quad k = 1, \ldots, f \,.$$

2.26 We write

$$T = \sum g_{ik}\dot{q}_i\dot{q}_k = \left(\frac{ds}{dt}\right)^2 = \frac{1}{2}(L + E) = E - U$$

and obtain $T\,dt = (ds/dt)ds = \sqrt{E-U}\,ds$. The principle of Euler and Maupertuis, $\delta K = 0$, requires

$$\delta \int_{q^1}^{q^2} \sqrt{E-U}\,ds = 0\,.$$

On the other hand, Fermat's principle states the following: A light pulse traverses the path ds in the time $dt = (n(x,\nu)/c)ds$. The path it chooses is such that the integral $\int dt$ is an extremum, i.e. that $\delta \int n(x,\nu)ds = 0$. The analogy is established if we associate with the particle an "index of refraction" which is given by the dimensionless quantity $((E-U)/mc^2)^{1/2}$; (see also Exercise 1.12).

2.27 For $U(q_0) < E \le E_{\max}$ the points of intersection q_1 and q_2 of the curves $y = U(q)$ and $y = E$ are turning points, $q(t)$ oscillates periodically between q_1 and q_2, cf. Fig. 15. Write the characteristic equation of Hamilton and Jacobi (M2.154) as

$$H\left(q, \frac{\partial S(q,P)}{\partial q}\right) = E\,. \tag{1}$$

We know that the transformed momentum obeys the differential equation $\dot{P} = 0$, i.e., that $P = \alpha = $ const. We are free to choose this constant to be the energy, $P = E$. Taking the derivative of eq. (1) with respect to $P = E$,

$$\frac{\partial H}{\partial p}\frac{\partial^2 S}{\partial q\,\partial P} = 1\,.$$

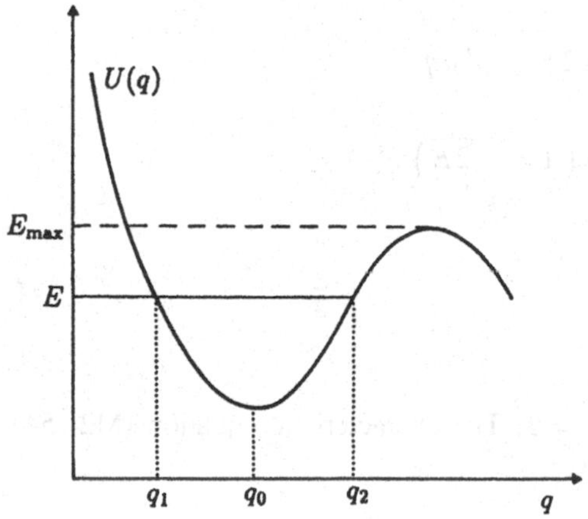

Fig. 15.

If $\partial H/\partial p \neq 0$ (this holds locally if E is larger than $U(q_0)$), then $(\partial^2 S)/(\partial q\,\partial P) \neq 0$. Thus, the equation $Q = \partial S(q,P)/\partial P$ can be solved locally for $q = q(Q,P)$. This yields

$$H\left(q(Q,P), \frac{\partial S}{\partial q}(q(Q,P),P)\right) \equiv \tilde{H}(Q,P) = E \equiv P \,.$$

From this we conclude

$$\dot{Q} = \frac{\partial \tilde{H}}{\partial P} = 1 \,, \quad \dot{P} = -\frac{\partial \tilde{H}}{\partial Q} = 0 \Rightarrow Q = t - t_0 = \frac{\partial S}{\partial E} \,.$$

The integral $I(E)$ becomes

$$I(E) = \frac{1}{2\pi} \oint_{\Gamma_E} p\, dq = \frac{1}{2\pi} \int_{t_0}^{t_0+T(E)} p \cdot \dot{q}\, dt$$

so that $dI(E)/dE = T(E)/(2\pi) \equiv \omega(E)$, in agreement with Exercise 2.1.

2.28 The function $\bar{S}(q, I)$, with I as in Exercise 2.27, generates the transformation from (q, p) to the action and angle variables (θ, I),

$$p = \frac{\partial \bar{S}(q,I)}{\partial q} \,, \quad \theta = \frac{\partial \bar{S}(q,I)}{\partial I} \,, \quad \text{with} \quad \tilde{H} = E(I) \,.$$

We then have $\dot{\theta} = \partial E/\partial I = \text{const.}$, $\dot{I} = 0$, which are integrated to $\theta(t) = (\partial E/\partial I)t + \theta_0$, $I = \text{const.}$ Call the circular frequency $\omega(E) := \partial E/\partial I$ so that $\theta(t) = \omega t + \theta_0$, $I = \text{const.}$

2.29 We calculate the integral $I(E)$ of Exercise 2.27 for the case $H = p^2/2 + q^2/2$: With $p = \sqrt{2E - q^2}$

$$\begin{aligned} I(E) &= \frac{1}{2\pi} \oint_{\Gamma_E} p\, dq = \frac{1}{2\pi} \oint_{\Gamma_E} \sqrt{2E - q^2}\, dq \\ &= \frac{1}{\pi} \int_{-A}^{+A} \sqrt{A^2 - q^2}\, dq \,, \quad (A = \sqrt{2E}) \,. \end{aligned}$$

Using

$$\int_{-A}^{A} \sqrt{A^2 - x^2}\, dx = \frac{\pi A^2}{2}$$

one finds $I(E) = A^2/2 = E$, i.e., $H = I$. The characteristic equation (M2.154) reads in the present example

$$\frac{1}{2}\left(\frac{\partial S}{\partial q}\right)^2 + \frac{1}{2} q^2 = E \,.$$

Its solution can be written as an indefinite integral $S = \int \sqrt{2E - q'^2}\, dq'$, or $\bar{S}(q, I) = \int \sqrt{2I - q'^2}\, dq'$. The angle variable follows from this

$$\theta = \frac{\partial \bar{S}}{\partial I} = \int \frac{1}{\sqrt{2I - q'^2}} dq' = \arcsin \frac{q}{\sqrt{2I}} ,$$

giving $q = \sqrt{2I} \sin \theta$. In a similar fashion one calculates

$$p = \frac{\partial \bar{S}}{\partial q} = \sqrt{2I - q^2} = \sqrt{2I} \cos \theta .$$

These are identical with the formulas that follow from the canonical transformation $\Phi(q, Q) = (q^2/2) \cot Q$, cf. eq. (M2.95).

2.30 Following Exercise 2.17 we take the time variable t to be a generalized coordinate $t = q_{f+1}$ and introduce in its place a new variable τ such that

$$\bar{L}\left(q, t, \frac{dq}{d\tau}, \frac{dt}{d\tau}\right) := L\left(q, \frac{1}{(dt/d\tau)} \frac{dq}{d\tau}\right) \frac{dt}{d\tau} .$$

By assumption $f = 1$, i.e., $q_1 = q$ and $q_2 \equiv q_{f+1} = t$. The action integral (the principal function) with the boundary conditions modified accordingly, reads

$$I_0^s = \int_{\tau_1'}^{\tau_2'} d\tau \, \bar{L}(\phi_1(s, \tau), \phi_{f+1}(s, \tau), \phi_1'(s, \tau), \phi'_{f+1}(s, \tau)) ,$$

the prime denoting the derivative with respect to τ. We now take the derivative of I_0^s with respect to s, at $s = 0$,

$$\frac{d}{ds} I_0^s \bigg|_{s=0} = \int_{\tau_1}^{\tau_2} d\tau \times \left\{ \frac{\partial \bar{L}}{\partial \phi_1} \frac{d\phi_1}{ds} + \frac{\partial \bar{L}}{\partial \phi_1'} \frac{d\phi_1'}{ds} + \frac{\partial \bar{L}}{\partial \phi_{f+1}} \frac{d\phi_{f+1}}{ds} + \frac{\partial \bar{L}}{\partial \phi'_{f+1}} \frac{d\phi'_{f+1}}{ds} \right\} . \quad (1)$$

Replacing $\partial \bar{L}/\partial \phi_1$ in the first term by $(d/d\tau)(\partial \bar{L}/\partial \phi_1')$, through the equations of motion, the first two terms of the curly brackets in eq. (1) can be combined to a total derivative with respect to τ. The integral over τ can be rewritten as an integral over t, so that the first two terms give the contribution

$$\int_{t_1}^{t_2} dt \frac{d}{dt} \left(\frac{\partial L}{\partial \dot{\phi}_1} \frac{d\phi_1}{ds} \right) = p^b \frac{dq^b}{ds} - p^a \frac{dq^a}{ds} .$$

(Note that here the dot means the derivative with respect to t, as before.) In the third term we replace $\partial \bar{L}/\partial \phi_{f+1}$ by $(d/d\tau)(\partial \bar{L}/\partial \phi'_{f+1})$, (again by virtue of the equation of motion). In this term and in the last term of the curly brackets in eq. (1) we make use of the solution to Exercise 2.17, viz.

$$\frac{\partial \bar{L}}{\partial(\partial \phi_{f+1}/\partial \tau)} = \frac{\partial \bar{L}}{\partial(\partial t/\partial \tau)} = L - \frac{\partial L}{\partial \dot{\phi}_1} \frac{d\phi_1}{dt}$$

so that these terms can also be combined to a total derivative

$$\frac{d}{d\tau}\left(\left(L-\frac{\partial L}{\partial\dot{\phi}_1}\frac{d\phi_1}{dt}\right)\frac{d\phi_{f+1}}{ds}\right) . \tag{2}$$

As above, in doing the integral one replaces τ by the variable t. The inner bracket in eq. (2) is the energy (to within the sign) so that we obtain the difference of the term $-E(d\phi_{f+1}/ds)$ at the two boundary points, that is, $(-E)$ times the derivative of the time $t = t_2 - t_1$ by s. Summing up one indeed obtains the result of the assertion. Finally, the generalization to $f > 1$ is obvious.

2.31 Start from the Hamiltonian function $H = \boldsymbol{p}^2/(2m) + U(r)$ with $U(r) = \gamma/r$, and from $\boldsymbol{A} = \boldsymbol{p} \times \boldsymbol{\ell} + mU(r)\boldsymbol{x}$. Clearly, $\boldsymbol{A} \cdot \boldsymbol{\ell} = 0$, the vector $\boldsymbol{A}$ is perpendicular to $\boldsymbol{\ell}$ and, hence, lies in the plane of the orbit. Making use of the formula $\boldsymbol{x} \cdot (\boldsymbol{p} \times \boldsymbol{\ell}) = \boldsymbol{\ell} \cdot (\boldsymbol{x} \times \boldsymbol{p}) = \ell^2$ one calculates (with $\ell = |\boldsymbol{\ell}|$)

$$\begin{aligned}\boldsymbol{A}^2 &= (\boldsymbol{p} \times \boldsymbol{\ell})^2 + 2mU(r)\,\boldsymbol{x} \cdot (\boldsymbol{p} \times \boldsymbol{\ell}) + m^2\gamma^2 \\ &= \ell^2\big(\boldsymbol{p}^2 + 2mU(r)\big) + m^2\gamma^2 \\ &= m^2\gamma^2 + 2m\ell^2 H = m^2\gamma^2 + 2m\ell^2 E .\end{aligned}$$

This vanishes only if the energy E and hence also γ are negative. In the case of the Kepler problem $\gamma = -Gm_1m_2 \equiv -A$ (notation as in Sect. 1.7.2), $m \equiv \mu$ is the reduced mass. The vector $\boldsymbol{A}$ vanishes for $E = -\mu A^2/(2\ell^2)$ which is the case of the *circular* orbit.

Calculate the scalar product $\boldsymbol{x} \cdot \boldsymbol{A} = \boldsymbol{x} \cdot (\boldsymbol{p} \times \boldsymbol{\ell}) - \mu A r = \ell^2 - \mu A r$, set this equal to $r|\boldsymbol{A}|\cos\phi$ to obtain

$$\begin{aligned}r(\phi) &= \ell^2/|\boldsymbol{A}|\cos\phi + \mu A \\ &= \ell^2/(\mu A)/1 + \sqrt{1 + 2E\ell^2/(\mu A^2)}\cos\phi \equiv p/1 + \varepsilon\cos\phi .\end{aligned}$$

This is the solution given in Sect. 1.21, with $\phi_0 = 0$. One concludes that $\boldsymbol{A}$ points along the 1-axis in the orbital plane, from the center of force to the perihelion, its modulus being $|\boldsymbol{A}| = \varepsilon\mu A$.

The cross product is $\boldsymbol{\ell} \times \boldsymbol{A} = \ell^2\boldsymbol{p} - (\mu A/r)\,\boldsymbol{x} \times \boldsymbol{\ell}$ from which one finds

$$\left(\boldsymbol{p} - 1/\ell^2\,\boldsymbol{\ell} \times \boldsymbol{A}\right)^2 = \mu^2 A^2 1/\ell^2 .$$

Noting that $\boldsymbol{A} = \varepsilon\mu A\hat{\boldsymbol{e}}_1$ and $\boldsymbol{\ell} \times \boldsymbol{A} = \ell\varepsilon\mu A\hat{\boldsymbol{e}}_3 \times \hat{\boldsymbol{e}}_1 = \ell\varepsilon\mu A\hat{\boldsymbol{e}}_2$, and decomposing $\boldsymbol{p} = p_1\hat{\boldsymbol{e}}_1 + p_2\hat{\boldsymbol{e}}_2$ one obtains

$$p_1^2 + (p_2 - \varepsilon\mu A/\ell)^2 = (\mu A)^2/\ell^2 .$$

This is the equation of the hodograph of Exercise 1.26.

Chapter 3: The Mechanics of Rigid Bodies

3.1 (i) As $\mathbf{K}$ and $\bar{\mathbf{K}}$ differ by a time-dependent rotation, $\mathbf{J}$ is related to $\bar{\mathbf{J}}$ by $\mathbf{J} = \mathbf{R}(t)\mathbf{J}\mathbf{R}^{-1}(t)$, with $\mathbf{R}(t)$ the rotation matrix that describes the relative rotation of the two coordinate systems. The characteristic polynomial of $\mathbf{J}$ is invariant under similarity transformations. Indeed, by the multiplication law for determinants,

$$\begin{aligned}\det|\mathbf{J} - \lambda\mathbb{1}| &= \det\left|\mathbf{R}(t)\bar{\mathbf{J}}\mathbf{R}^{-1}(t) - \lambda\mathbb{1}\right| \\ &= \det\left|\mathbf{R}(t)(\bar{\mathbf{J}} - \lambda\mathbb{1})\mathbf{R}^{-1}(t)\right| \\ &= \det\left|\bar{\mathbf{J}} - \lambda\mathbb{1}\right| .\end{aligned}$$

The characteristic polynomials of $\mathbf{J}$ and $\bar{\mathbf{J}}$ are the same. Hence, their eigenvalues are pairwise equal.

(ii) If $\bar{K}$ is a principal-axes system, $\bar{\mathbf{J}}$ has the form

$$\bar{\mathbf{J}} = \begin{pmatrix} I_1 & 0 & 0 \\ 0 & I_2 & 0 \\ 0 & 0 & I_3 \end{pmatrix} .$$

A rotation about the 3-axis reads

$$\mathbf{R}(t) = \begin{pmatrix} \cos\phi(t) & \sin\phi(t) & 0 \\ -\sin\phi(t) & \cos\phi(t) & 0 \\ 0 & 0 & 1 \end{pmatrix} .$$

This allows us to compute $\mathbf{J}$ with the result

$$\mathbf{J} = \begin{pmatrix} I_1\cos^2\phi + I_2\sin^2\phi & (I_2 - I_1)\sin\phi\cos\phi & 0 \\ (I_2 - I_1)\sin\phi\cos\phi & I_1\sin^2\phi + I_2\cos^2\phi & 0 \\ 0 & 0 & I_3 \end{pmatrix} .$$

3.2 The straight line connecting the two atoms is a principal axis. The remaining axes are chosen perpendicular to the first and perpendicular to each other, as sketched in Fig. 16. Using the notation of that figure we have $m_1 a_1 = m_2 a_2$, $a_1 + a_2 = l$, and therefore

$$a_1 = \frac{m_2}{m_1 + m_2} l , \quad a_2 = \frac{m_1}{m_1 + m_2} l ,$$

with μ denoting the reduced mass.

$$I_1 = I_2 = m_1 a_1^2 + m_2 a_2^2 = \frac{m_1 m_2}{m_1 + m_2} l^2 = \mu l^2 ,$$

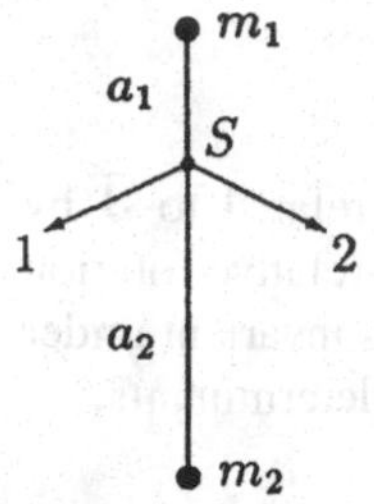

Fig. 16.

3.3 The moments of inertia are determined from the equation

$$\det(\mathbf{J} - \lambda \mathbb{1}) = \begin{vmatrix} I_{11} - \lambda & I_{12} & 0 \\ I_{21} & I_{22} - \lambda & 0 \\ 0 & 0 & I_{33} - \lambda \end{vmatrix} = 0 \,,$$

whose solutions are

$$I_{1,2} = \frac{I_{11} + I_{22}}{2} \pm \sqrt{\frac{(I_{11} - I_{22})^2}{4} + I_{12} I_{21}} \,, \quad I_3 = I_{33} \,.$$

(i) $I_{1,2} = A \pm B$. Thus it follows that $B \leq A$ and $A + B \geq 0$. Since $I_1 + I_2 \geq I_3$ we also have $I_3 \leq 2A$, i.e., $A \geq 0$.

(ii) $I_1 = 5A$, $I_2 = 0$. From $I_1 + I_2 \geq I_3$ and $I_2 + I_3 \geq I_1$ follows $I_3 = 5A$. The body is axially symmetric with respect to the 2-axis.

3.4 The motion being free we choose a principal-axes system attached to the center-of-mass, letting the 3-axis coincide with the symmetry axis. The moments of inertia are easily calculated, the result being

$$I_1 = I_2 = \frac{3}{20} M \left(R^2 + \frac{1}{4} h^2 \right) \,, \quad I_3 = \frac{3}{10} M R^2 \,.$$

A Lagrangian function is

$$L = T_{\text{rot}} = \frac{1}{2} \sum_{i=1}^{3} I_i \bar{\omega}_i^2 \,,$$

where $\bar{\omega}_i$ are the components of the angular velocity in the body fixed system. They are related to the Eulerian angles and their time derivatives by the formulae (M3.82),

$$\begin{aligned} \bar{\omega}_1 &= \dot{\theta} \cos \psi + \dot{\phi} \sin \theta \sin \psi \\ \bar{\omega}_2 &= -\dot{\theta} \sin \psi + \dot{\phi} \sin \theta \cos \psi \\ \bar{\omega}_3 &= \dot{\phi} \cos \theta + \dot{\psi} \,. \end{aligned}$$

Inserting these into L and noting that $I_1 = I_2$, we have

$$L(\phi,\theta,\psi,\dot{\phi},\dot{\theta},\dot{\psi}) = \frac{1}{2} I_1 \left(\dot{\theta}^2 + \dot{\phi}^2 \sin^2\theta\right) + \frac{1}{2} I_3 \left(\dot{\psi} + \dot{\phi}\cos\theta\right)^2 .$$

The variables ϕ and ψ are cyclic, hence

$$p_\phi = \frac{\partial L}{\partial \dot{\phi}} = I_1 \dot{\phi} \sin^2\theta + I_3 \left(\dot{\psi} + \dot{\phi}\cos\theta\right)\cos\theta ,$$
$$p_\psi = \frac{\partial L}{\partial \dot{\psi}} = I_3 \left(\dot{\psi} + \dot{\phi}\cos\theta\right) ,$$

are conserved. Furthermore, the energy $E = T_{\mathrm{rot}} = L$ is conserved. Note that $p_\phi = I_1(\bar{\omega}_1 \sin\psi + \bar{\omega}_2 \cos\psi)\sin\theta + I_3\bar{\omega}_3 \cos\theta$. This is the scalar product $\boldsymbol{L}\cdot\hat{\boldsymbol{e}}_{3_0}$ of the angular momentum and the unit vector in the 3-direction of the laboratory system, i.e. $p_\phi = L_3$. Regarding p_ψ we have $p_\psi = I_3\bar{\omega}_3 = \bar{L}_3$. The equations of motion read

$$\frac{d}{dt} p_\phi = \frac{d}{dt}\left(I_1(\bar{\omega}_1 \sin\psi + \bar{\omega}_2 \cos\psi)\sin\theta + I_3\bar{\omega}_3\cos\theta\right) = 0 \qquad (1)$$
$$\frac{d}{dt} p_\psi = I_3 \frac{d}{dt}\left(\dot{\psi} + \dot{\phi}\cos\theta\right) = 0 \qquad (2)$$
$$\frac{d}{dt}\frac{\partial L}{\partial \dot{\theta}} - \frac{\partial L}{\partial \theta} = I_1\ddot{\theta} - I_1\dot{\phi}^2 \sin\theta\cos\theta + I_3\left(\dot{\psi} + \dot{\phi}\cos\theta\right)\dot{\phi}\sin\theta = 0 . \qquad (3)$$

From the first of these $\dot{\phi} = (L_3 - \bar{L}_3\cos\theta)/(I_1 \sin^2\theta)$. Inserting this into the Lagrangian function gives

$$L = \frac{1}{2} I_1 \dot{\theta}^2 + \frac{1}{2 I_1 \sin^2\theta}\left(L_3 - \bar{L}_3\cos\theta\right)^2 + \frac{1}{2 I_3}\bar{L}_3^2 = E = \text{const.} \qquad (4)$$

If, on the other hand, $\dot{\phi}$ is inserted into the third equation of motion (3), one obtains

$$I_1\ddot{\theta} - \frac{\cos\theta}{I_1 \sin^3\theta}\left(L_3 - \bar{L}_3\cos\theta\right)^2 + \frac{1}{I_1\sin\theta}\bar{L}_3\left(L_3 - \bar{L}_3\cos\theta\right) = 0 ,$$

which is nothing but the time derivative of eq. (4).

3.5 We choose the 3-axis to be the symmetry axis of the torus. Let (r', ϕ) be polar coordinates in a section of the torus and ψ be the azimuth in the plane of the torus as sketched in Fig. 17. The coordinates (r', ψ, ϕ) are related to the cartesian coordinates by

$$x_1 = (R + r'\cos\phi)\cos\psi , \quad x_2 = (R + r'\cos\phi)\sin\psi , \quad x_3 = r'\sin\phi .$$

The Jacobian is

$$\frac{\partial(x_1, x_2, x_3)}{\partial(r', \psi, \phi)} = r'(R + r'\cos\phi) .$$

The volume of the torus is calculated to be

$$V = \int_0^r dr' \int_0^{2\pi} d\psi \int_0^{2\pi} d\phi\, r'(R + r'\cos\phi) = 2\pi^2 r^2 R ,$$

so that the mass density is $\varrho_0 = M/(2\pi^2 r^2 R)$. Thus

$$I_3 = \int d^3x \varrho_0 \left(x_1^2 + x_2^2\right) = \varrho_0 \int_0^{2\pi} d\psi \int_0^{2\pi} d\phi \int_0^r dr' \, r'(R + r' \cos\phi)^3$$
$$= M\left(R^2 + \frac{3}{4}r^2\right) ,$$

$$I_1 = \int d^3x \varrho_0 \left(x_2^2 + x_3^2\right) = \varrho_0 \int_0^{2\pi} d\psi \int_0^{2\pi} d\phi \int_0^r dr' \, r'(R + r' \cos\phi)$$
$$\cdot \left((R + r' \cos\phi)^2 \sin^2\psi + r'^2 \sin^2\phi\right) = \frac{1}{2}M\left(R^2 + \frac{5}{4}r^2\right) .$$

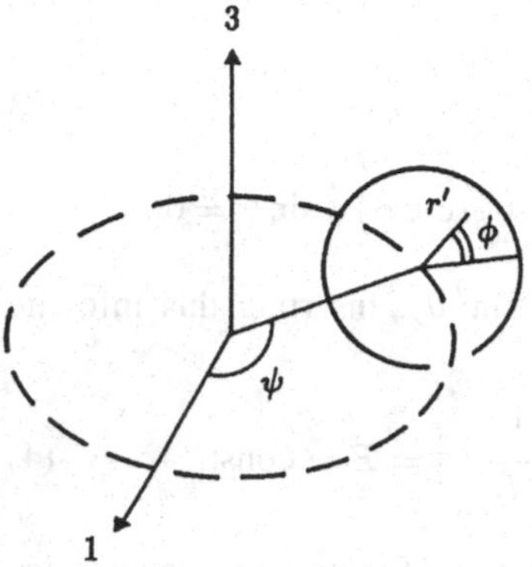

Fig. 17.

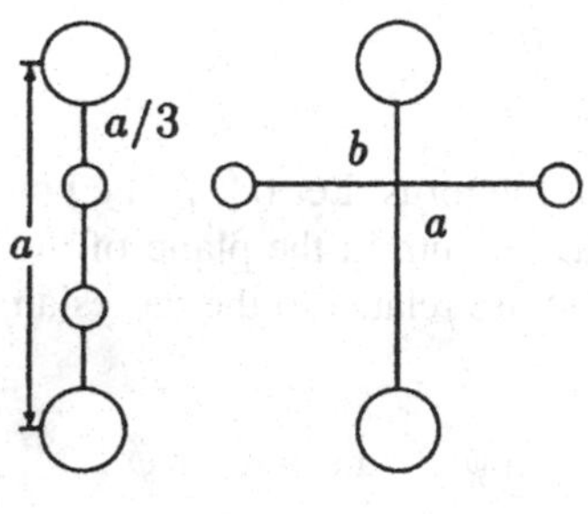

Fig. 18.

3.6 In the first position we have $I_3^{(a)} = 2(2/5)(MR^2 + mr^2)$ and $\omega_3^{(a)} = L_3/I_3^{(a)}$. In the second position the contribution of the two smaller balls is calculated by means of Steiner's theorem, $I_i' = I_i + m(\boldsymbol{a}^2 - a_i^2)$, now with $I_3 = 2mr^2/5$, $\boldsymbol{a} = \pm(b/2)\hat{\boldsymbol{e}}_1$:

$$I_3^{(b)} = 2\left((2/5)(MR^2 + mr^2) + mb^2/4\right) .$$

It follows that $\omega_3^{(b)} = L_3/I_3^{(b)}$, and $\omega_3^{(a)}/\omega_3^{(b)} = 1 + mb^2/(2I_3^{(a)})$. One rotates faster if the arms are close to one's body than if they are stretched out. Making use of Steiner's theorem once more we finally calculate $I_1 = I_2$. For the two arrangements one obtains the result

$$I_1^{(a)} = I_3^{(a)} + \frac{1}{2}a^2\left(M + \frac{1}{9}m\right), \quad I_1^{(b)} = \frac{4}{5}\left(MR^2 + mr^2\right) + \frac{5}{9}Ma^2 .$$

3.7 The relation between density and mass reads

$$M = \int \varrho(r)\, d^3r = \int_0^{2\pi} d\phi \int_0^{\pi} \sin\theta\, d\theta \int_0^{\infty} r^2\, dr\, \varrho(r,\theta,\phi) .$$

In our case, where ϱ depends on θ only, this means

$$M = \int_0^{2\pi} d\phi \int_0^{\pi} \sin\theta\, d\theta \int_0^{R(\theta)} r^2\, dr\, \varrho_0 = \frac{2\pi}{3}\varrho_0 \int_0^{\pi} \sin\theta\, d\theta\, R(\theta)^3 .$$

The moments of inertia I_3, $I_1 = I_2$ are calculated from the formulas

$$I_3 = \int \mathrm{d}^3x\, \varrho r^2(1-\cos^2\theta) , \quad I_1 + I_2 + I_3 = 2\int \mathrm{d}^3x\, \varrho r^2 ,$$

(i) Integration gives the results

$$M = \frac{4\pi}{3}\varrho_0 R_0^3(1+\alpha)^2 , \quad \text{i.e.,} \quad \varrho_0 = \frac{3}{4\pi}\frac{M}{R_0^3(1+\alpha^2)} .$$

$$I_1 = I_2 = \frac{2MR_0^2}{5(1+\alpha^2)}\left\{1 + 4\alpha^2 + \frac{9}{7}\alpha^4\right\} , \quad I_3 = \frac{2MR_0^2}{5(1+\alpha^2)}\left\{1 + 2\alpha^2 + \frac{3}{7}\alpha^4\right\} .$$

(ii) Substituting $z = \cos\theta$ the integrals are easily evaluated. With the abbreviation $\gamma := \sqrt{5/16\pi}\,\beta$ we obtain

$$M = \frac{4\pi}{3}\varrho_0 R_0^3\left(\frac{16}{35}\gamma^3 + \frac{12}{5}\gamma^2 + 1\right),$$

that is

$$\varrho_0 = \frac{3}{4\pi}\frac{M}{R_0^3}\left(\frac{16}{35}\gamma^3 + \frac{12}{5}\gamma^2 + 1\right)^{-1} .$$

$$I_1 = I_2 = \frac{2MR_0^2}{5}\cdot\frac{1 + \gamma + 64\gamma^2/7 + 8\gamma^3 + 688\gamma^4/77 + 2512\gamma^5/1001}{1 + 12\gamma^2/5 + 16\gamma^3/35} ,$$

$$I_3 = \frac{2MR_0^2}{5}\cdot\frac{1 - 2\gamma + 40\gamma^2/7 - 16\gamma^3/7 + 208\gamma^4/77 - 32\gamma^5/1001}{1 + 12\gamma^2/5 + 16\gamma^3/35} ,$$

3.8 The (principal) moments of inertia are the eigenvalues of the given tensor and, hence, are the roots of the characteristic polynomial $\det(\lambda \mathbb{1} - \mathbf{J})$. Calculating this determinant we are led to the cubic equation $\lambda^3 - 4\lambda^2 + 5\lambda - 2 = 0$. Its solutions are $\lambda_1 = \lambda_2 = 1$, $\lambda_3 = 2$. The inertia tensor in diagonal form reads

$$\mathring{J} = \begin{pmatrix} 1 & 0 & 0 \\ 0 & 1 & 0 \\ 0 & 0 & 2 \end{pmatrix} .$$

We write $\mathbf{J} = \mathbf{R}\mathring{J}\mathbf{R}^T$ and decompose the rotation matrix according to $\mathbf{R}(\psi, \theta, \phi) = \mathbf{R}_3(\psi)\mathbf{R}_2(\theta)\mathbf{R}_3(\phi)$. As the factor $\mathbf{R}_3(\phi)$ leaves $\mathring{J}$ invariant we can choose $\phi = 0$. With

$$\mathbf{R}_2(\theta) = \begin{pmatrix} \cos\theta & 0 & -\sin\theta \\ 0 & 1 & 0 \\ \sin\theta & 0 & \cos\theta \end{pmatrix} \quad \text{and} \quad \mathbf{R}_3(\psi) = \begin{pmatrix} \cos\psi & \sin\psi & 0 \\ -\sin\psi & \cos\psi & 0 \\ 0 & 0 & 1 \end{pmatrix}$$

we calculate

$$\mathbf{R}\mathring{J}\mathbf{R}^T = \mathbb{1} + \begin{pmatrix} \cos^2\psi \sin^2\theta & -\sin\psi\cos\psi\sin^2\theta & -\cos\psi\cos\theta\sin\theta \\ -\sin\psi\cos\psi\sin^2\theta & \sin^2\psi\sin^2\theta & \sin\psi\cos\theta\sin\theta \\ -\cos\psi\cos\theta\sin\theta & \sin\psi\cos\theta\sin\theta & \cos^2\theta \end{pmatrix} .$$

If this is set equal to $\mathbf{J}$ as given, we find $\cos^2\theta = 3/8$, and, with the following choice of signs for θ : $\cos\theta = \sqrt{3}/(2\sqrt{2})$ and $\sin\theta = \sqrt{5}/(2\sqrt{2})$, the result $\cos\psi = 1/\sqrt{5}$, $\sin\psi = -\sqrt{4/5}$.

3.9 (i) $\varrho(\boldsymbol{r}) = \varrho_0\theta(a - |\boldsymbol{r}|)$. The total mass is equal to the volume integral of $\varrho(\boldsymbol{r})$, viz.

$$M = \frac{4\pi}{3} a^3 \varrho_0 \Rightarrow \varrho_0 = \frac{3M}{4\pi a^3} .$$

(ii) We choose the 3-axis to be the axis of rotation. Let the coordinate in the body fixed system be (x, y, z), the coordinates in the space fixed system are

$$x' = x\cos\omega t - (y + a)\sin\omega t , \quad y' = x\sin\omega t + (y + a)\cos\omega t , \quad z' = z .$$

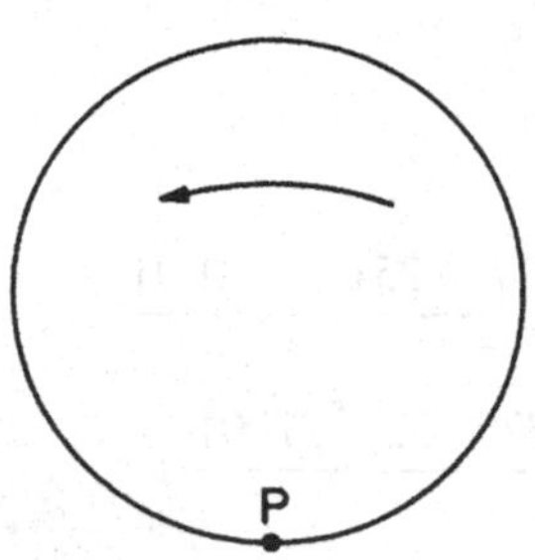

Fig. 19.

Inverting these equations we have

$$x = +x' \cos \omega t + y' \sin \omega t \ , \quad y = -x' \sin \omega t + y' \cos \omega t - a \ ,$$

whence

$$x^2 + y^2 = x'^2 + y'^2 + a^2 + 2a(x' \sin \omega t - y' \cos \omega t) \ .$$

From this we get

$$\varrho(\boldsymbol{r}', t) = \varrho_0 \theta \left(a - \sqrt{r'^2 + a^2 + 2a(x' \sin \omega t - y' \cos \omega t)}\right) .$$

(iii) In the case of a homogeneous sphere the inertia tensor is diagonal, all three moments of inerta are equal, $I_1 = I_2 = I_3 \equiv I$. Hence,

$$3I = I_1 + I_2 + I_3 = 2 \int d^3 r \varrho(r) r^2 = \frac{6Ma^2}{5} \ .$$

Making use of Steiner's theorem (M3.23) we find

$$I'_3 = I_3 + M\left(a^2 \delta_{33} - a_3^2\right) = \frac{7Ma^2}{5} \ .$$

3.10 (i) The volume of the cylinder is $V = \pi r^2 h$, hence the mass density is $\varrho_0 = m/(\pi r^2 h)$. The moment of inertia relevant for rotations about the symmetry axis is best calculated using cylindrical coordinates,

$$I_3 = \varrho_0 \int_0^{2\pi} d\phi \int_0^h dz \int_0^r \varrho^3 \, d\varrho = \frac{1}{2} m r^2 \ .$$

Call $q(t)$ the projection of the center-of-mass' orbit onto the inclined plane. When the center-of-mass moves by an amount dq, the cylinder rotates by an angle $d\phi = dq/r$. Therefore, the total kinetic energy is

$$T = \frac{1}{2} m \dot{q}^2 + \frac{1}{2} I_3 \frac{\dot{q}^2}{r^2} = \frac{3}{4} m \dot{q}^2 \ .$$

(ii) A Lagrangian function is

$$L = T - U = 3m\dot{q}^2/4 - mg(q_0 - q) \sin \alpha \ ,$$

where q_0 is the length of the inclined plane and α its angle of inclination. The equation of motion reads $3m\ddot{q}/2 = mg \sin \alpha$, the general solution being $q(t) = q(0) + v(0)t + (g \sin \alpha) t^2/3$.

3.11 (i) The rotation $R(\phi \cdot \hat{\boldsymbol{\phi}})$ is a right-handed rotation by an angle ϕ about the direction $\hat{\boldsymbol{\phi}}$, with $0 \le \phi \le \pi$. Any desired position is reached by means of rotations about $\hat{\boldsymbol{\phi}}$ and $-\hat{\bar{\boldsymbol{\phi}}}$.

(ii) With $\hat{\boldsymbol{\phi}}$ an arbitrary direction in $\mathbb{R}$, and with ϕ between 0 and π, the parameter space $(\phi, \hat{\boldsymbol{\phi}})$ is the ball D^3 (surface and interior of the unit sphere in $\mathbb{R}^3$). Every point $p \in D^3$ represents a rotation, the direction $\hat{\boldsymbol{\phi}}$ being given by the polar coordinates of p, and the angle ϕ being given by its distance from center. Note, however, that $A : (\hat{\boldsymbol{\phi}}, \phi = \pi)$ and $B : (-\hat{\boldsymbol{\phi}}, \phi = \pi)$ represent the same rotation.

(iii) There are two types of closed curves in D^3: curves of the type of C_1 as shown in Fig. 20, which can be contracted, by a continuous deformation, to a point, and curves such as C_2, which do not have this property. C_2 connects the antipodes A and B. As these points represent the same rotation, C_2 is a closed curve. Any continuous deformation of C_2 which shifts A to A' also shifts B to B', the antipodal point of A'.

While C_1 contains no jumps between antipodes, C_2 contains one such jump. One easily convinces oneself, by means of a drawing, that any closed curve with an *even* number of antipodal jumps can be deformed continuously into C_1 or, equivalently, into a point.

Take the example of a closed curve with two such jumps as shown in Fig. 21. One can let A_1 move to B_2 in such a way that the arc $B_1 A_2$ goes to zero, the sections $A_1 B_1$ and $A_2 B_2$ become equal and opposite so that the curve $A_1 B_2$ becomes like C_1 in Fig. 20. In a similar fashion one shows that all closed curves with an *odd* number of jumps can be continuously deformed into C_2. (One says that the two types of curves form homotopy classes.)

Fig. 20.

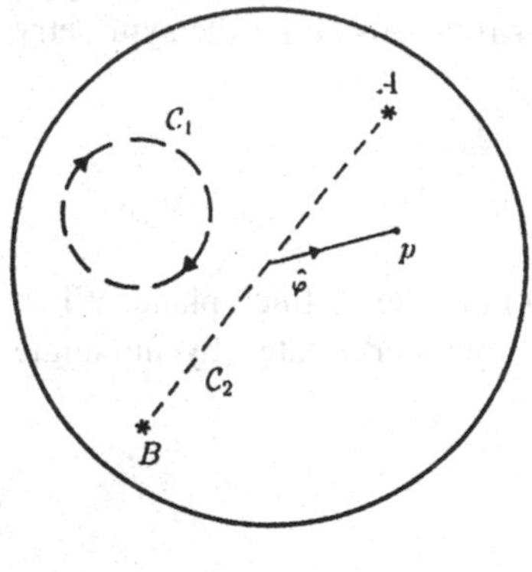

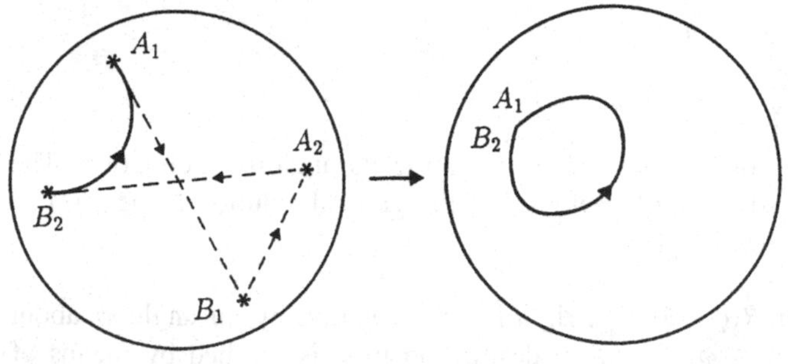

Fig. 21.

3.12 We do the calculation for the example of (M3.93). Equations (M3.89) yield expressions for the components of angular momentum in the body fixed system.

Making use of the relation $\{p_i, f(q_j)\} = \delta_{ij} f'(q_j)$ which follows from the definition of the Poisson brackets, we calculate readily

$$
\begin{aligned}
\{\bar{L}_1, \bar{L}_2\} &= \left\{ p_\phi \frac{\sin\psi}{\sin\theta} - p_\psi \sin\psi \cot\theta + p_\theta \cos\psi, \right. \\
&\qquad \left. p_\phi \frac{\cos\psi}{\sin\theta} - p_\psi \cos\psi \cot\theta - p_\theta \sin\psi \right\} \\
&= p_\theta \frac{\cos\theta}{\sin^2\theta} \left(- \{\sin\psi, p_\psi \cos\psi\} - \{p_\psi \sin\psi, \cos\psi\} \right) \\
&\quad + p_\phi \left(-\sin^2\psi \left\{ \frac{1}{\sin\theta}, p_\theta \right\} + \cos^2\psi \left\{ p_\theta, \frac{1}{\sin\theta} \right\} \right) \\
&\quad + \cot^2 \left(p_\psi \sin\psi, p_\psi \cos\psi \right\} + \left(\sin\psi \left\{ p_\psi \cot\theta, p_\theta \sin\psi \right\} \right. \\
&\quad \left. - \cos\psi \left\{ p_\theta \cos\psi, p_\psi \cot\theta \right\} \right) \\
&= + p_\phi \frac{\cos\theta}{\sin^2\theta} - p_\phi \frac{\cos\theta}{\sin^2\theta} - p_\psi \cot^2\theta + p_\psi \frac{1}{\sin^2\theta} \\
&= p_\psi = \bar{L}_3 \,.
\end{aligned}
$$

Chapter 4: Relativistic Mechanics

4.1 (i) Let the neutral pion fly in the 3-direction with velocity $\boldsymbol{v} = v_0 \hat{\boldsymbol{e}}_3$. The full energy–momentum vector of the pion is

$$
q = \left(\frac{1}{c} E_q, \boldsymbol{q} \right) = (\gamma_0 m_\pi c, \gamma_0 m_\pi \boldsymbol{v}) = \gamma_0 m_\pi c \left(1, \beta_0 \hat{\boldsymbol{e}}_3 \right) ,
$$

where $\beta_0 = v_0/c$, $\gamma_0 = (1 - \beta_0^2)^{-1/2}$. The special Lorentz transformation which takes us to the rest system of the pion, is

$$
\mathbf{L}_{-v} = \begin{pmatrix} \gamma_0 & 0 & 0 & -\gamma_0\beta_0 \\ 0 & 1 & 0 & 0 \\ 0 & 0 & 1 & 0 \\ -\gamma_0\beta_0 & 0 & 0 & \gamma_0 \end{pmatrix} .
$$

Indeed, $\mathbf{L}_{-v} q = q^* = (m_\pi c, 0)$.

(ii) In the rest system (Fig. 22, left-hand side) the two photons have the four-momenta $k_i^* = (E_i^*/c, \boldsymbol{k}_i^*)$, $i = 1, 2$. Conservation of energy and momentum requires $q^* = k_1^* + k_2^*$, i.e. $E_1^* + E_2^* = m_\pi c^2$ and $\boldsymbol{k}_1^* + \boldsymbol{k}_2^* = 0$. As photons are massless, $E_i^* = |\boldsymbol{k}_i^*| c$, and, as $\boldsymbol{k}_1^* = -\boldsymbol{k}_2^*$, one has $E_1^* = E_2^*$. Denote the absolute value of the spatial momenta by κ^*. Then $|\boldsymbol{k}_1^*| = |\boldsymbol{k}_2^*| = \kappa^* = m_\pi c/2$.

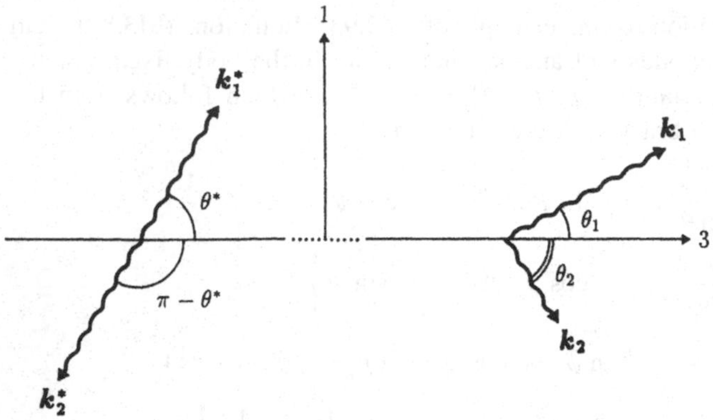

Fig. 22.

In the rest system the decay is isotropic. In the laboratory system only the direction $\hat{e}_3$ of the pion's momentum is singled out. Therefore, in this system the decay distribution is symmetric with respect to the 3-axis. We first study the situation in the $(1, 3)$-plane and then obtain the complete answer by rotation about the 3-axis. We have $(\boldsymbol{k}_1^*)_3 = \kappa^* \cos\theta^* = -(\boldsymbol{k}_2^*)_3$, $(\boldsymbol{k}_1^*)_1 = \kappa^* \sin\theta^* = -(\boldsymbol{k}_2^*)_1$, while the 2-components vanish. In the laboratory system we have $k_i = \mathbf{L}_{v_0} k_i^*$, viz.

$$
\begin{aligned}
&\frac{1}{2} E_1 = \gamma_0 \kappa^* \left(1 + \beta_0 \cos\theta^*\right) , \quad \frac{1}{c} E_2 = \gamma_0 \kappa^* \left(1 - \beta_0 \cos\theta^*\right) , \\
&(\boldsymbol{k}_1)_1 = \left(\boldsymbol{k}_1^*\right)_1 = \kappa^* \sin\theta^* , \quad (\boldsymbol{k}_2)_1 = \left(\boldsymbol{k}_2^*\right)_1 = -\kappa^* \sin\theta^* , \\
&(\boldsymbol{k}_1)_3 = \gamma_0 \kappa^* \left(\beta_0 + \cos\theta^*\right) , \quad (\boldsymbol{k}_2)_3 = \gamma_0 \kappa^* \left(\beta_0 - \cos\theta^*\right) , \\
&(\boldsymbol{k}_1)_2 = 0 = (\boldsymbol{k}_2)_2 .
\end{aligned}
$$

In the laboratory system we then find

$$
\tan\theta_1 = \frac{(\boldsymbol{k}_1)_1}{(\boldsymbol{k}_1)_3} = \frac{\sin\theta^*}{\gamma_0(\beta_0 + \cos\theta^*)} ; \quad \tan\theta_2 = \frac{-\sin\theta^*}{\gamma_0(\beta_0 - \cos\theta^*)} .
$$

Examples:

a) $\theta^* = 0$ (forward emission of one photon, backward emission of the other): From the formulas above one finds $E_1 = m_\pi c^2 \gamma_0 (1 + \beta_0)/2$, $E_2 = m_\pi c^2 \gamma_0 (1 - \beta_0)/2$, $\boldsymbol{k}_1 = m_\pi c \gamma_0 (\beta_0 + 1) \hat{e}_3 / 2$, $\boldsymbol{k}_2 = m_\pi c \gamma_0 (\beta_0 - 1) \hat{e}_3 / 2$, and, as $\beta_0 \leq 1$, $\theta_1 = 0$, $\theta_2 = \pi$.

b) $\theta^* = \pi/2$ (transverse emission): In this case $E_1 = E_2 = m_\pi c^2 \gamma_0 / 2$, $\boldsymbol{k}_1 = m_\pi c (\hat{e}_1 + \gamma_0 \beta_0 \hat{e}_3)/2$, $\boldsymbol{k}_2 = m_\pi c (-\hat{e}_1 + \gamma_0 \beta_0 \hat{e}_e)/2$, $\tan\theta_1 = \tan\theta_2 = 1/(\gamma_0 \beta_0)$.

c) $\theta^* = \pi/4$ and $\beta_0 = 1/\sqrt{2}$, i.e, $\gamma_0 = \sqrt{2}$: In this case one finds $E_1 = 3 m_\pi c^2 \gamma_0 / 4$, $E_2 = m_\pi c^2 \gamma_0 / 4$, $\boldsymbol{k}_1 = m_\pi c (\hat{e}_1 - 2\sqrt{2} \hat{e}_3)/(2\sqrt{2})$, $\boldsymbol{k}_2 = -m_\pi c \hat{e}_1 / (2\sqrt{2})$, $\theta_1 = \arctan(1/(2\sqrt{2})) \approx 0.108\pi$, $\theta_2 = \pi/2$.

In the rest system the decay distribution is isotropic, which means that the differential probability $d\Gamma$ for $\boldsymbol{k}_1^*$ to lie in the interval $d\Omega^* = \sin\theta^* \, d\theta^* \, d\phi^*$ is independent of θ^* and of ϕ^*. (To see this enclose the decaying pion by a unit sphere. If one considers a large number of decays then photon 1 will hit every surface element $d\Omega^*$ on that sphere with equal probability.) Thus,

$$d\Gamma = \Gamma_0 \, d\Omega^* \quad \text{with} \quad \Gamma_0 = \text{const.}$$

In the *laboratory system* the analogous distribution is no longer isotropic. It is distorted in the direction of flight but is still axially symmetric about that axis. We have

$$\frac{1}{\Gamma_0} d\Gamma = \left| \frac{d\Omega^*}{d\Omega} \right| d\Omega \quad \text{where} \quad \frac{d\Omega^*}{d\Omega} = \frac{\sin\theta^*}{\sin\theta} \frac{d\theta^*}{d\theta} .$$

The factor $\sin\theta^*/\sin\theta$ is calculated from the above formula for $\tan\theta_1$, viz.

$$\sin\theta = \frac{\tan\theta}{\sqrt{1+\tan^2\theta}} = \frac{\sin\theta^*}{\gamma_0(\beta_0 + \cos\theta^*)} ,$$

and the derivative $d\theta/d\theta^*$ is obtained from $\theta = \arctan(\sin\theta^*/\gamma_0(\beta_0 + \cos\theta^*))$ by making use of the relation $\gamma_0^2\beta_0^2 = \gamma_0^2 - 1$. One finds

$$\frac{d\Omega^*}{d\Omega} = \gamma_0^2 \left(1 + \beta_0 \cos\theta^*\right)^2 .$$

The cosine of θ^* is expressed in terms of the corresponding laboratory angle by the formula

$$\cos\theta^* = \frac{\cos\theta - \beta_0}{1 - \beta_0 \cos\theta} .$$

The shift of the angular distribution is well illustrated by the graph of the function

$$F(\theta) := \frac{d\Omega^*}{d\Omega} = \gamma_0^2 \left(1 + \beta_0 \frac{\cos\theta - \beta_0}{1 - \beta_0\cos\theta}\right)^2$$

for different values of β_0. Quite generally we have $dF/d\theta|_{\theta=0} = 0$. For $\beta_0 \to 1$ the value $F(0) = (1+\beta_0)/(1-\beta_0)$ tends to infinity. For small argument $\theta = \varepsilon \ll 1$, on the other hand,

$$F(\varepsilon) \approx \frac{1+\beta_0}{1-\beta_0} \left(1 - \frac{\varepsilon^2}{1-\beta_0}\right) ,$$

which means that for $\varepsilon^2 \approx (1 - \beta_0)$ F becomes very small. Therefore, when $\beta_0 \to 1$ the function $F(\theta)$ falls off very quickly with increasing θ. Figure 23 shows the examples $\beta_0 = 0$, $\beta_0 = 1/\sqrt{2}$, and $\beta_0 = 11/13$.

4.2 Let the energy-momentum four-vectors of π, μ, and ν be q, p, and k, respectively. We have always $q = p + k$. In the pion's rest system

$$q = (m_\pi c, \boldsymbol{q} = 0) , \quad p = \left(\frac{1}{c} E_p^*, \boldsymbol{p}^*\right) , \quad k = \left(\frac{1}{c} E_k^*, -\boldsymbol{p}^*\right) .$$

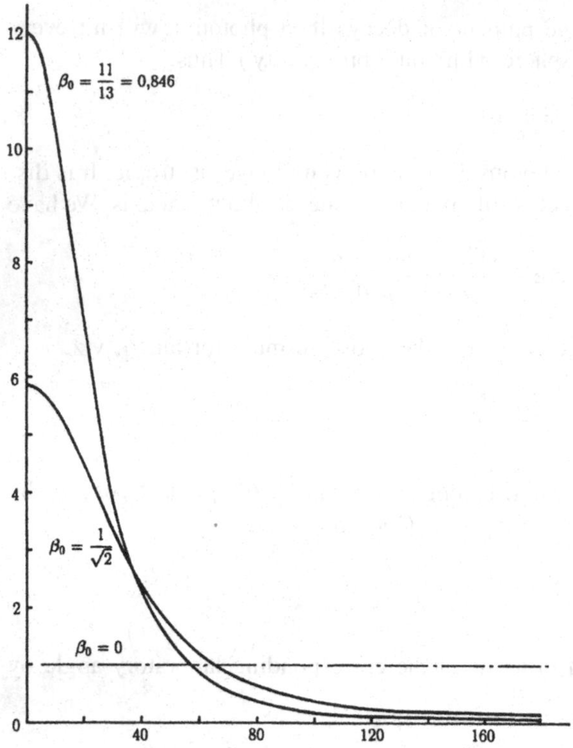

Fig. 23.

If $\kappa^* := |\boldsymbol{p}^*|$ denotes the magnitude of the momentum of the muon and of the neutrino, then

$$E_k^* = \kappa^* c = \frac{m_\pi^2 - m_\mu^2}{2m_\pi} c^2 \,, \quad E_p^* = \sqrt{(\kappa^* c)^2 + (m_\mu c)^2} = \frac{m_\pi^2 + m_\mu^2}{2m_\pi} c^2 \,.$$

In the laboratory system the situation is as follows: The pion has velocity $\boldsymbol{v}_0 = v_0 \hat{\boldsymbol{e}}_3$ and, therefore,

$$q = (E_q/c, \boldsymbol{q}) = (\gamma_0 m_\pi c, \gamma_0 m_\pi \boldsymbol{v}_0) = \gamma_0 m_\pi c(1, \beta_0 \hat{\boldsymbol{e}}_3)$$

with $\beta_0 = v_0/c$, $\gamma_0 = 1/\sqrt{1-\beta_0^2}$. It is sufficient to study the kinematics in the $(1,3)$-plane. The transformation from the pion's rest system to the laboratory system yields

$$\frac{1}{c} E_p = \gamma_0 \left(\frac{1}{c} E_p^* + \beta_0 p^{*3} \right) = \gamma_0 \left(\frac{1}{c} E_p^* + \beta_0 \kappa^* \cos\theta^* \right) \,,$$

$$p^1 = p^{*1} \,,$$

$$p^2 = p^{*2} = 0 \,,$$

$$p^3 = \gamma_0 \left(\frac{1}{c} \beta_0 E_0^* + p^{*3} \right) = \gamma_0 \left(\frac{1}{c} \beta_0 E_p^* + \kappa^* \cos\theta^* \right)$$

and, therefore, the relation between the angles of emission θ^* and θ is (cf. Fig. 24)

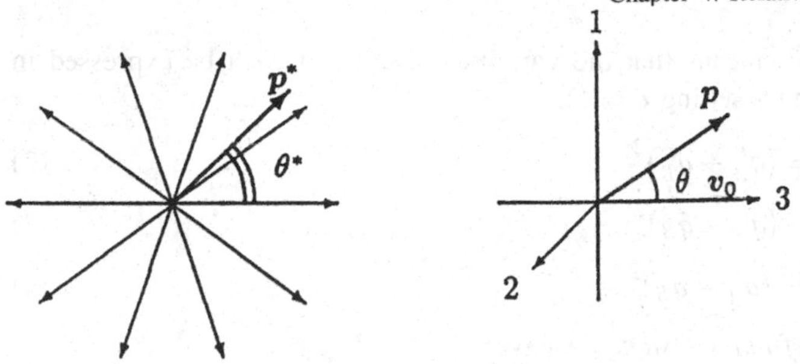

Fig. 24.

$$\tan\theta = \frac{p^1}{p^3} = \frac{\kappa^* \sin\theta^*}{\gamma_0(\beta_0 E_p^*/c + \kappa^* \cos\theta^*)}\,, \tag{1}$$

or

$$\tan\theta = \frac{\left(m_\pi^2 - m_\mu^2\right)\sin\theta^*}{\gamma_0\left(\beta_0\left(m_\pi^2 + m_\mu^2\right) + \left(m_\pi^2 - m_\mu^2\right)\cos\theta^*\right)}\,. \tag{2}$$

Making use of $\beta^* := c\kappa^*/E_p^* = (m_\pi^2 - m_\mu^2)/(m_\pi^2 + m_\mu^2)$ (this is the beta factor of the muon in the rest system of the pion), eq. (1) is rewritten

$$\tan\theta = \frac{\beta^* \sin\theta^*}{\gamma_0(\beta_0 + \beta^* \cos\theta^*)}\,. \tag{3}$$

There exists a maximal angle θ if the muons which are emitted backwards in the pion's rest system ($\theta^* = \pi$), have momenta

$$p^3 = \gamma_0 E_p^*/c\left(\beta_0 + \beta^* \cos\theta^*\right) = \gamma_0 E_p^*/c\left(\beta_0 - \beta^*\right) > 0\,,$$

i.e., if $\beta_0 > \beta^*$. The magnitude of the maximal angle is obtained from the condition $d\tan\theta/d\theta^* = 0$ which gives $\cos\beta^* = -\beta^*/\beta_0$ and, finally,

$$\tan\theta_{\max} = \frac{\beta^*\sqrt{\beta_0^2 - \beta^{*2}}}{\gamma_0(\beta_0^2 - \beta^{*2})} = \frac{\beta^*}{\gamma_0\sqrt{\beta_0^2 - \beta^{*2}}} = \frac{\beta^*\sqrt{1-\beta_0^2}}{\sqrt{\beta_0^2 - \beta^{*2}}}\,. \tag{4}$$

4.3 The variables s and t are the squared norms of four-vectors and are thus invariant under Lorentz transformations. The same is true for $u = c^2(q_A - q_B')^2$. For our calculations it is convenient to choose units such that $c = 1$. It is not difficult to re-insert the constant c in the final results. (This is important if we wish to expand in terms of v/c.) To reconstruct those factors one must keep in mind that terms like (mass times c^2) and (momentum times c) have the physical dimension of energy.

Conservation of energy and momentum means that the four equations

$$q_A + q_B = q_A' + q_B' \tag{1}$$

must be satisfied. This means that the variables s, t, u can each be expressed in two different ways (now setting $c = 1$):

$$s = (q_A + q_B)^2 = \left(q'_A + q'_B\right)^2 \tag{2}$$

$$t = \left(q_A - q'_A\right)^2 = \left(q'_B - q_B\right)^2 \tag{3}$$

$$u = \left(q_A - q'_B\right)^2 = \left(q'_A - q_B\right)^2 . \tag{4}$$

(i) In the *center-of-mass system* we have

$$q_A = \left(E^*_A, \boldsymbol{q}^*\right) , \quad q_B = \left(E^*_B, -\boldsymbol{q}^*\right) ,$$
$$q'_A = \left(E'^*_A, \boldsymbol{q}'^*\right) , \quad q'_B = \left(E'^*_B, -\boldsymbol{q}'^*\right) , \tag{5}$$

where $E^*_A = \sqrt{m_A^2 + (\boldsymbol{q}^*)^2}$ etc., with $q^* = |\boldsymbol{q}^*|$. Like in the nonrelativistic case energy conservation requires the magnitudes of the three-momenta in the center-of-mass system to be equal. However, the simple nonrelativistic formula

$$(q^*)_{\text{n.r.}} = \frac{m_B}{m_A + m_B} \left|\boldsymbol{q}_A^{\text{lab}}\right|$$

that follows from eq. (M1.79a) no longer holds. This is so because neither the nonrelativistic energy

$$T_r = \frac{m_A + m_B}{2m_A m_B} (q^*)^2_{\text{n.r.}}$$

nor the quantity

$$\frac{\left(\boldsymbol{q}_A + \boldsymbol{q}_B\right)^2}{2(m_A + m_B)}$$

are conserved. We have

$$s = \left(E^*_A + E^*_B\right)^2 = m_A^2 + m_B^2 + 2(q^*)^2$$
$$+2\sqrt{\left((q^*)^2 + m_A^2\right)\left((q^*)^2 + m_B^2\right)} . \tag{6}$$

Thus, s is the square of the total energy in the center-of-mass system. Reintroducing the velocity of light,

$$s = m_A^2 c^4 + m_B^2 c^4 + 2(q^*)^2 c^2 + 2\sqrt{\left((q^*)^2 c^2 + m_A^2 c^4\right)\left((q^*)^2 c^2 + m_B^2 c^4\right)} .$$

In a first step we check that s, when expanded in terms of $1/c$, gives the correct nonrelativistic kinetic energy T_{r} of relative motion (except for the rest masses, of course)

$$s \approx \left(m_A c^2 + m_B c^2\right)^2 \left(1 + \frac{1}{m_A m_B} (q^*)^2/c^2 + O\left(\frac{(q^*)^4}{m^4 c^4}\right)\right) ,$$

and, thus,

$$\sqrt{s} \approx m_A c^2 + m_B c^2 + \frac{m_A + m_B}{2m_A m_B}(q^*)^2 + O\left(\frac{(q^* c)^4}{(mc^2)^4}\right) .$$

The magnitude of the center-of-mass momentum is obtained from (6)

$$q^*(s) = \frac{1}{2\sqrt{s}}\sqrt{\left(s-(m_A+m_B)^2\right)\left(s-(m_A-m_B)^2\right)} . \qquad (7)$$

Clearly, the reaction can take place only if s is at least equal to the square of the sum of the rest energies,

$$s \geq s_0 := (m_A + m_B)^2 \mathrel{\widehat{=}} \left(m_A c^2 + m_B c^2\right)^2 .$$

s_0 is called the *threshold* of the reaction. For $s = s_0$ the momentum vanishes, which means that at threshold the kinetic energy of relative motion vanishes.

The variable t is expressed in terms of q^* and the scattering angle θ^* as follows:

$$t = \left(q_A - q'_A\right)^2 = q_A^2 + q_A'^2 - 2q_A \cdot q'_A = 2m_A^2 - 2E_A^* E_A'^* + 2\boldsymbol{q}^* \cdot \boldsymbol{q}'^* .$$

As the magnitudes of $\boldsymbol{q}^*$ and of $\boldsymbol{q}'^*$ are equal, $E_A^* = E_B^*$. Therefore,

$$t = -2(q^*)^2(1 - \cos\theta^*) . \qquad (8)$$

Except for the sign, t is the square of the momentum transfer $(\boldsymbol{q}^* - \boldsymbol{q}'^*)$ in the center-of-mass system. For fixed $s \geq s_0$, t varies as follows

$$-4(q^*)^2 \leq t \leq 0 .$$

Examples:

a) $e^- + e^- \to e^- + e^-$

$$s \geq s_0 = 4\left(m_e c^2\right)^2 , \qquad -(s - s_0) \leq t \leq 0 .$$

b) $\nu + e^- \to e^- + \nu$

$$s \geq s_0 = \left(m_e c^2\right)^2 , \qquad -\frac{1}{s}(s - s_0)^2 \leq t \leq 0 .$$

(ii) Calculating $s+t+u$ from the formulas (2)–(4) and making use of (1), we find $s+t+u = 2(m_A^2 + m_B^2)c^4$. More generally, for the reaction $A + B \to C + D$, one finds

$$s + t + u = \left(m_A^2 + m_B^2 + m_C^2 + m_D^2\right) c^4 .$$

4.4 In the *laboratory system*

$$q_A = (E_A, \boldsymbol{q}_A) \;, \quad q_B = \left(m_B c^2, 0\right) \;,$$
$$q'_A = (E'_A, \boldsymbol{q}'_A) \;, \quad q'_B = (E'_B, \boldsymbol{q}'_B) \;, \tag{1}$$

The scattering angle θ is the angle between the three-vectors $\boldsymbol{q}_A$ and $\boldsymbol{q}'_A$. From eq. (3) of the solution to Exercise 4.3 above (using $c = 1$),

$$t = q_A^2 + q_A'^2 - 2q_A q'_A = 2m_A^2 - 2E_A E'_A + 2\left|\boldsymbol{q}_A\right| \left|\boldsymbol{q}'_A\right| \cos\theta \;. \tag{2}$$

Equation (8) of Exercise 4.3 above gives an alternative expression for t. The aim is now to express the laboratory quantities E_A, E'_A, $\left|\boldsymbol{q}_A\right|$, $\left|\boldsymbol{q}'_A\right|$ in terms of the invariants s and t. Using (1) s is found to be, in the laboratory system, $s = m_A^2 + m_B^2 + 2E_A m_B$, that is,

$$E_A = \frac{1}{2m_B}\left(s - m_A^2 - m_B^2\right) \;. \tag{3}$$

From this, using $\boldsymbol{q}_A^2 = E_A^2 - m_A^2$,

$$\left|\boldsymbol{q}_A\right| = \frac{1}{2m_B}\sqrt{\left(s - (m_A + m_B)^2\right)\left(s - (m_A - m_B)^2\right)} = \frac{1}{m_B} q^* \sqrt{s} \tag{4}$$

with q^* as given by eq. (7) of Exercise 4.3 above. We now calculate $t = (q_B - q'_B)^2$ in the laboratory system and find $E'_B = (2m_B^2 - t)/(2m_B)$ and from this $E'_A = E_A + m_B - E'_B$ to be equal to

$$E'_A = \frac{1}{2m_B}\left(s + t - m_A^2 - m_B^2\right) = E_A + \frac{t}{2m_B} \;, \tag{5}$$

and, eventually, from $\boldsymbol{q}_A'^2 = E_A'^2 - m_A^2$

$$\left|\boldsymbol{q}'_A\right| = \frac{1}{2m_B}\sqrt{\left(s + t - (m_A + m_B)^2\right)\left(s + t - (m_A - m_B)^2\right)} \;. \tag{6}$$

From (2)

$$\cos\theta = \left(E_A E'_A - m_A^2 + \frac{t}{2}\right)\frac{1}{\left|\boldsymbol{q}_A\right| \left|\boldsymbol{q}'_A\right|} \;.$$

This is used to calculate $\sin\theta$ and $\tan\theta$, replacing all quantities which are not invariants by the expressions (3)–(6). With the abbreviations $\Sigma := (m_A + m_B)^2$ and $\Delta := (m_A - m_B)^2$ we find

$$\tan\theta = \frac{2m_B\sqrt{-t\,(st + (s - \Sigma)(s - \Delta))}}{(s - \Sigma)(s - \Delta) + t\left(s - m_A^2 + m_B^2\right)} \;.$$

Finally, $\cos\theta^*$ and $\sin\theta^*$ can also be expressed in terms of s and t, starting from eqs. (8) and (7) of Exercise 4.3 above,

$$\cos\theta^* = \frac{2st + (s-\Sigma)(s-\Delta)}{(s-\Sigma)(s-\Delta)} \,,$$

$$\sin\theta^* = 2\sqrt{s}\,\frac{\sqrt{-t\,(st + (s-\Sigma)(s-\Delta))}}{(s-\Sigma)(s-\Delta)} \,.$$

Replace the square root in the numerator of $\tan\theta$ by $\sin\theta^*$ and insert t in the denominator, as a function of $\cos\theta^*$, to obtain the final result

$$\boxed{\tan\theta = \frac{2m_B\sqrt{s}}{s - m_A^2 + m_B^2}\,\frac{\sin\theta^*}{\cos\theta^* + \dfrac{s + m_A^2 - m_B^2}{s - m_A^2 + m_B^2}}} \tag{7}$$

For $s \approx (m_A + m_B)^2$ one recovers the nonrelativistic result (M1.80). The case of two equal masses is particularly interesting. With $m_A = m_B \equiv m$

$$\tan\theta = \frac{2m}{\sqrt{s}}\tan\frac{\theta^*}{2} \,.$$

As $\sqrt{s} \geq 2m$ the scattering angle θ is always smaller than in the nonrelativistic situation.

4.5 If we wish to go from the rest system of a particle to another system where its four-momentum is $p = (E/c, \boldsymbol{p})$, we have to apply a special Lorentz transformation $\mathbf{L}(\boldsymbol{v})$ with $\boldsymbol{v}$ related to $\boldsymbol{p}$ by $\boldsymbol{p} = m\gamma\boldsymbol{v}$. Solving for $\boldsymbol{v}$,

$$\boldsymbol{v} = \frac{\boldsymbol{p}c}{\sqrt{\boldsymbol{p}^2 + m^2c^2}} = \frac{\boldsymbol{p}c^2}{E} \,.$$

Insertion into eq. (M4.41) and application to the vector $(0, \boldsymbol{s})$ gives

$$s = \mathbf{L}(\boldsymbol{v})(0, \boldsymbol{s}) = \left(\frac{\gamma}{c}\boldsymbol{s}\cdot\boldsymbol{v},\, \boldsymbol{s} + \frac{\gamma^2}{c^2(1+\gamma)}\,\boldsymbol{v}\cdot\boldsymbol{s}\boldsymbol{v}\right) .$$

As $s_\alpha p^\alpha$ is a Lorentz scalar, and hence is independent of the frame of reference that one uses, this quantity may be evaluated most simply in the rest system. It is found to vanish there and, hence, in any frame of reference.

4.6 In either case the coordinate system can be chosen such that the y- and z-components of the four-vector vanish and the x-component is positive, i.e., $z = (z^0, z^1, 0, 0)$ with $z^1 > 0$, If z^0 is smaller than zero we apply the time reversal operation (M4.30) so that, from here on, we assume $z^0 > 0$, without loss of generality.

(i) A light-like vector has $z^2 = 0$ and, hence, $z^0 = z^1$. We apply a boost with parameter λ along the x-direction, cf. (M4.39). In order to obtain the desired form of the four-vector we must have

$$z^0\cosh\lambda - z^0\sinh\lambda = 1 \quad \text{or} \quad z^0 e^{-\lambda} = 1 \,,$$

from which follows $\lambda = \ln z^0$.

(ii) For a space-like vector $z^2 = (z^0)^2 - (z^1)^2 < 0$, i.e., $0 < z^0 < z^1$. Applying a boost with parameter λ it is transformed to

$$\left(z^0 \cosh\lambda - z^1 \sinh\lambda, z' \cosh\lambda - z^0 \sinh\lambda, 0, 0\right) .$$

For the time component to vanish, one must have $\tanh\lambda = z^0/z^1$. Calculating $\sinh\lambda$ and $\cosh\lambda$ from this yields the assertion $z^1 = \sqrt{-z^2}$.

4.7 The commutation relations (M4.59) can be summarized as follows, making use of the Levi-Civita symbol:

$$\begin{aligned} \left[\mathbf{J}_p, \mathbf{J}_q\right] &= \varepsilon_{pqr}\mathbf{J}_r , \\ \left[\mathbf{K}_p, \mathbf{K}_q\right] &= -\varepsilon_{pqr}\mathbf{J}_r , \\ \left[\mathbf{J}_p, \mathbf{K}_q\right] &= \varepsilon_{pqr}\mathbf{K}_r . \end{aligned}$$

From this one obtains

$$\begin{aligned} \left[\mathbf{A}_p, \mathbf{A}_q\right] &= \varepsilon_{pqr}\mathbf{A}_r , \\ \left[\mathbf{B}_p, \mathbf{B}_q\right] &= \varepsilon_{pqr}\mathbf{B}_r , \\ \left[\mathbf{A}_p, \mathbf{B}_q\right] &= 0 . \end{aligned}$$

4.8 The explicit calculation gives

$$\mathbf{P}\mathbf{J}_i\mathbf{P}^{-1} = \mathbf{J}_i , \quad \mathbf{P}\mathbf{K}_j\mathbf{P}^{-1} = -\mathbf{K}_j .$$

This corresponds to the fact that space inversion does not alter the sense of rotation but reverses the direction of motion.

4.9 The commutators (M4.59) read in this basis

$$\begin{aligned} \left[\hat{J}_i, \hat{J}_j\right] &= \mathrm{i}\varepsilon_{ijk}\hat{J}_k , \\ \left[\hat{J}_i, \hat{K}_j\right] &= \mathrm{i}\varepsilon_{ijk}\hat{K}_k , \\ \left[\hat{K}_i, \hat{K}_j\right] &= -\mathrm{i}\varepsilon_{ijk}\hat{J}_k . \end{aligned}$$

The matrix $\mathbf{J}_i$ and $\mathbf{K}_j$ being real and skew-symmetric, we have for instance

$$\left(\hat{J}_i^T\right)^* = -(\mathrm{i}\mathbf{J}_i)^* = \mathrm{i}\mathbf{J}_i = \hat{J}_i .$$

4.10 This exercise is a special case of Exercise 4.11. The result is obtained from there by taking $m_2 = 0 = m_3$.

4.11 Energy conservation implies (again taking $c = 1$)

$$M = E_1 + E_2 + E_3 = \sqrt{m_1^2 + f^2} + \sqrt{m_2^2 + x^2 f^2} + \sqrt{m_3^2 + (1-x)^2 f^2}$$
$$= M(xf(x)) \,.$$

The maximum of $f(x)$ is found from the equation

$$0 \stackrel{!}{=} \frac{df}{dx} = -\frac{\partial M/\partial x}{\partial M/\partial f} = -fE_1 \frac{xE_3 - (1-x)E_2}{E_2E_3 + x^2E_1E_3 + (1-x)^2E_1E_2} \,,$$

or $xE_3 = (1-x)E_2$. Squaring this equation gives $x^2(m_3^2 + (1-x)^2 f^2) = (1-x)^2(m_2^2 + x^2 f^2)$, and from this the condition

$$x \stackrel{!}{=} \frac{m_2}{m_2 + m_3} \,.$$

Taking into account the condition

$$E_3 = \frac{1-x}{x} E_2 = \frac{m_3}{m_2} E_2 \,,$$

one obtains

$$M - E_1 = \frac{m_2 + m_3}{m_2} E_2 \,.$$

The square of this yields

$$M^2 - 2ME_1 + m_1^2 + f^2 = \frac{(m_2 + m_3)^2}{m_2^2} \left(m_2^2 + \frac{m_2^2}{(m_2 + m_3)^2} f^2 \right)$$

and from this

$$(E_1)_{\max} = \frac{1}{2M} \left(M^2 + m_1^2 - (m_2 + m_3)^2 \right) .$$

Examples:

(i) $\mu \to \mathrm{e} + \nu_1 + \nu_2 : m_2 = m_3 = 0,\ M = m_\mu,\ m_1 = m_\mathrm{e}$. Thus

$$(E_\mathrm{e})_{\max} = \frac{1}{2m_\mu} \left(m_\mu^2 + m_\mathrm{e}^2 \right) c^2 \,.$$

With $m_\mu / m_\mathrm{e} \approx 206.8$ one finds $(E_\mathrm{e})_{\max} \approx 104.4 m_\mathrm{e} c^2$.

(ii) $\mathrm{n} \to \mathrm{p} + \mathrm{e} + \nu$; $M = m_\mathrm{n}$, $m_1 = m_\mathrm{e}$, $m_2 = m_\mathrm{p}$, $m_3 = 0$. Therefore one obtains

$$(E_\mathrm{e})_{\max} = \frac{1}{2m_\mathrm{n}} \left(m_\mathrm{n}^2 + m_\mathrm{e}^2 - m_\mathrm{p}^2 \right) c^2 = \frac{1}{2m_\mathrm{n}} \left((2m_\mathrm{n} - \Delta)\Delta + m_\mathrm{e}^2 \right) c^2 \,,$$

where $\Delta := m_\mathrm{n} - m_\mathrm{p}$. Inserting numerical values yields $(E_\mathrm{e})_{\max} \approx 2.528 m_e c^2$. Thus $\gamma_{\max} = 2.528$ and $\beta_{\max} = \sqrt{\gamma_{\max}^2 - 1}/\gamma_{\max} = 0.918$. The electron is highly relativistic at the maximal energy.

4.12 The apparent lifetime $\tau^{(v)}$ in the laboratory system is related to the real lifetime $\tau^{(0)}$ by $\tau^{(v)} = \gamma\tau^{(0)}$. During this time the particle, on average, travels a distance

$$L = v\tau^{(v)} = \beta\gamma\tau^{(0)}c\,.$$

Now, the product $\beta\gamma$ equals $|\boldsymbol{p}|c/(mc^2)$, cf. eq. (M4.83), so that for $|\boldsymbol{p}| = xmc$ there follows the relation

$$L = x\tau^{(0)}c\,.$$

For pions, for instance, one has $\tau_\pi^{(0)}c \approx 780$ cm.

4.13 From the results of the preceding exercise we obtain $\tau_{\text{n}}^{(0)}c \approx 2.7\times 10^{13}$ cm. For $E = 10^{-2}m_{\text{n}}c^2$ one has $x = \sqrt{\gamma^2 - 1} = 0.142$, while for $E = 10^{14}m_{\text{n}}c^2$ one has $x \approx 10^{14}$.

4.14 Let p_1, p_2 be the energy–momentum four-vectors of the incoming and outgoing electron, respectively, and k that of the photon. Energy and momentum conservation in the reaction $\text{e} \to \text{e} + \gamma$ requires $p_1 = p_2 + k$. Squaring this relation and making use of

$$p_1^2 = m_{\text{e}}c^2 = p_2^2\,, \quad k^2 = 0\,,$$

one deduces $p_2 \cdot k = 0$. As k is a light-like four-vector this relation can only hold if p_2 is light-like, too, i.e., if $p_2^2 = 0$. This is in contradiction with the outgoing electron being on its mass shell, $p_2^2 = m_{\text{e}}^2$. Hence, the reaction cannot take place.

4.15 The first inversion leads from x^μ to $(R^2/x^2)x^\mu$, the translation that follows leads to $R^2(x^\mu/x^2 + c^\mu)$, and the second inversion, finally, to

$$x'^\mu = \frac{R^4\left(x^\mu + x^2c^\mu\right)x^4}{x^2R^4\left(x + x^2c\right)^2} = \frac{x^\mu + x^2c^\mu}{1 + 2(c\cdot x) + c^2x^2}\,.$$

The inversion $\mathcal{J}$ leaves invariant the two halves of the time-like hyperboloid $x^2 = R^2$, but interchanges those of the space-like hyperboloid $x^2 = -R^2$. The image of the light-cone by the inversion is at infinity. The light-cone as a whole stays invariant under the combined transformation $\mathcal{J}\circ\mathcal{T}\circ\mathcal{J}$.

4.16 As L does not depend on q, the equation of motion for this variable reads

$$\frac{d}{dt}\frac{\partial L}{\partial\dot{\boldsymbol{q}}} = m\frac{d}{dt}(\psi\dot{\boldsymbol{q}}) = 0\,. \tag{1}$$

In turn, L is independent of $\dot{\psi}$. The condition for the action integral to be extremal leads to the following equation for ψ,

$$\frac{\partial L}{\partial \psi} = \frac{1}{2} m \left(\dot{q}^2 - c_0^2 \frac{\psi^2 - 1}{\psi^2} \right) = 0 \, .$$

The solutions of this equations are

$$\psi_1 = \frac{c_0}{\sqrt{c_0^2 - \dot{q}^2}} \, , \quad \psi_2 = -\frac{c_0}{\sqrt{c_0^2 - \dot{q}^2}} \, .$$

Insertion of ψ_1 into the Lagrangian function yields

$$L \left(\dot{q}, \psi = \psi_1 \right) = \frac{1}{2} m \Big(-2c_0 \sqrt{c_0^2 - \dot{q}^2} + 2c_0^2 \Big) = -m c_0^2 \sqrt{1 - \dot{q}^2/c_0^2} + m c_0^2 \, .$$

This is nothing but eq. (M4.97), with c replaced by c_0, to which the constant energy $m c_0^2$ is added.

If we let c_0 go to infinity, ψ_1 trends to 1 and the Lagrangian function becomes $L_{\text{nr}} = m \dot{q}^2/2$, well-known from nonrelativistic motion. One verifies easily that (1) is the correct equation of motion in either case.

The second solution ψ_2 must be excluded. Obviously, the additional term

$$\frac{1}{2} m (\psi - 1) \left(\dot{q}^2 - c_0^2 \frac{\psi - 1}{\psi} \right) \, .$$

which is added to the Lagrangian function L_{nr} takes care of the requirement that the velocity $\dot{q}$ should not exceed the value c_0.

Chapter 5: Geometric Aspects of Mechanics

5.1 We make use of the decomposition (M5.52) for $\overset{k}{\omega}$ and $\overset{l}{\omega}$

$$\overset{k}{\omega} \wedge \overset{l}{\omega} \sum_{i_1 < \dots < i_k} \omega_{i_1 \dots i_k} \sum_{j_1 < \dots < j_l} \omega_{j_1 \dots j_l} dx^{i_1} \wedge \ldots \wedge dx^{i_k} \wedge dx^{j_1} \wedge \ldots \wedge dx^{j_l} \, .$$

The analogous decomposition of $\overset{l}{\omega} \wedge \overset{k}{\omega}$ (k and l interchanged) is obtained from this by shifting first dx^{j_1}, then dx^{j_2}, and so forth up to dx^{j_l}, across the product $dx^{i_1} \wedge dx^{i_2} \wedge \ldots \wedge dx^{i_k}$ from the right to the left. Each one of these operations gives rise to a factor $(-)^k$, so that one obtains the total factor $(-)^{kl}$.

5.2 We calculate

$$ds^2 \left(\hat{\boldsymbol{e}}_i, \hat{\boldsymbol{e}}_j \right) = \sum_{k=1}^{3} E_k \, dx^k(\hat{\boldsymbol{e}}_i) \, dx^k(\hat{\boldsymbol{e}}_j) = \sum_{k=1}^{3} E_k a_i^k a_j^k \, ,$$

where we have set $a_i^k := dx^k(\hat{\boldsymbol{e}}_i)$. As $ds^2(\hat{\boldsymbol{e}}_i, \hat{\boldsymbol{e}}_k) = \delta_{ik}$, we must have $a_i^k = b_i^k/\sqrt{E_k}$, where $\{b_i^k\}$ is an orthogonal matrix. This matrix must be orthogonal because the coordinate axes were chosen orthogonal. Therefore $dx^k(\hat{\boldsymbol{e}}_i) = \delta_i^k/\sqrt{E_k}$.

5.3 Consider

$$\overset{1}{\omega}_a = \sum_{i=1}^{3} \omega_i(x)\, dx^i \ ,$$

$$\overset{2}{\omega}_a = b_1(x)\, dx^2 \wedge dx^3 + \text{cyclic permutations}$$

whose coefficients, $\omega_i(x)$ and $b_i(x)$, are to be determined.

a) We calculate

$$\overset{1}{\omega}_a(\xi) = \sum_i \omega_i(x)\, dx^i(\xi) = \sum_i \omega_i(x)\, dx^i\left(\sum_k \xi^k \hat{e}_k\right) = \sum_i \omega_i(x) \xi^i \frac{1}{\sqrt{E_i}} \ .$$

Since, on the other hand,

$$\overset{1}{\omega}_a(\xi) = \boldsymbol{a} \cdot \boldsymbol{\xi} = \sum_i a_i(x) \xi^i$$

we deduce

$$\omega_i(x) = a_i(x) \sqrt{E_i} \ .$$

b) We calculate

$$\begin{aligned}\overset{2}{\omega}_a(\boldsymbol{\xi}, \boldsymbol{\eta}) &= b_1(x)(dx^2(\boldsymbol{\xi})\, dx^3(\boldsymbol{\eta}) - dx^2(\boldsymbol{\eta})\, dx^3(\boldsymbol{\xi})) + \text{cycl. perms.} \\ &= b_1(x)(\xi^2 \eta^3 - \eta^2 \xi^3)/\sqrt{E_2 E_3} + \text{cycl. perms.} \ .\end{aligned}$$

Comparing this with the scalar product of $\boldsymbol{a}$ and $\boldsymbol{\xi} \times \boldsymbol{\eta}$ yields

$$b_1(x) = \sqrt{E_2 E_3}\, a_1(x) \quad \text{(cyclic permutations)} \ .$$

5.4 Denote by $(\nabla f)_i$ the components of ∇f with respect to the orthogonal basis that we consider. We then have, according to the solution of Exercise 5.3,

$$\overset{1}{\omega}_{\nabla f} = \sum_i (\nabla f)_i \sqrt{E_i}\, dx^i \ .$$

With $\hat{\boldsymbol{\xi}} = \sum_i \xi^i \hat{\boldsymbol{e}}_i$ a unit vector, the function $\overset{1}{\omega}_{\nabla f}(\hat{\boldsymbol{\xi}}) = \sum_i (\nabla f)_i \xi^i$ is the directional derivative of f along the direction $\hat{\boldsymbol{\xi}}$. This quantity can be calculated alternatively from the total differential

$$df = \sum_i \frac{\partial f}{\partial x^i}\, dx^i$$

to be

$$df(\hat{\boldsymbol{\xi}}) = \sum_{i,k} \frac{\partial f}{\partial x^i} \xi^k \, dx^i \, (\hat{\boldsymbol{e}}_k) = \sum_i \frac{1}{\sqrt{E_i}} \frac{\partial f}{\partial x^i} \xi^i \, .$$

Comparing the two expressions yields the result

$$(\boldsymbol{\nabla} f)_i = \frac{1}{\sqrt{E_i}} \frac{\partial f}{\partial x^i} \, .$$

5.5 For *cartesian coordinates* we have $E_1 = E_2 = E_3 = 1$.

For *cylindrical coordinates* $(\hat{\boldsymbol{e}}_\varrho, \hat{\boldsymbol{e}}_\phi, \hat{\boldsymbol{e}}_z)$ we have $ds^2 = d\varrho^2 + \varrho^2 d\phi^2 + dz^2$, i.e., $E_1 = E_3 = 1$, $E_2 = \varrho^2$ and, therefore,

$$\boldsymbol{\nabla} f = \left(\frac{\partial f}{\partial \varrho}, \frac{1}{\varrho} \frac{\partial f}{\partial \phi}, \frac{\partial f}{\partial z} \right) \, .$$

For *spherical coordinates* $(\hat{\boldsymbol{e}}_r, \hat{\boldsymbol{e}}_\theta, \hat{\boldsymbol{e}}_\phi)$ we have $ds^2 = dr^2 + r^2 d\theta^2 + r \sin^2\theta \, d\phi^2$, which means that $E_1 = 1$, $E_2 = r^2$, $E_3 = r^2 \sin^2\theta$ and, hence,

$$\boldsymbol{\nabla} f = \left(\frac{\partial f}{\partial r}, \frac{1}{r} \frac{\partial f}{\partial \theta}, \frac{1}{r \sin\theta} \frac{\partial f}{\partial \phi} \right) \, .$$

5.6 The defining equation (M5.58) can be written alternatively as follows

$$(*\omega)(\hat{\boldsymbol{e}}_{i_{k+1}}, \dots, \hat{\boldsymbol{e}}_{i_n}) = \varepsilon_{i_1 \dots i_k i_{k+1} \dots i_n} \omega(\hat{\boldsymbol{e}}_{i_1}, \dots, \hat{\boldsymbol{e}}_{i_k}) \, .$$

Here $\varepsilon_{i_1 \dots i_n}$ is the totally antisymmetric Levi-Civita symbol. It equals $+1(-1)$ if $(i_1 \dots i_n)$ is an even (odd) permutation of $(1, \dots, n)$, and vanishes whenever two of its indices are equal. Thus, for $n = 2$, $*dx^1 = dx^2$, $*dx^2 = -dx^1$, and $*\omega = F_1 dx^2 - F_2 dx^1$. Therefore, $\omega(\boldsymbol{\xi}) = \boldsymbol{F} \cdot \boldsymbol{\xi}$, while $*\omega(\boldsymbol{\xi}) = \boldsymbol{F} \times \boldsymbol{\xi}$. If $\boldsymbol{\xi}$ is a displacement vector $\boldsymbol{\xi} = \boldsymbol{r}_A - \boldsymbol{r}_B$, $\boldsymbol{F}$ a constant force, $\omega(\boldsymbol{\xi})$ is the work of the force along that displacement. In turn, $*\omega(\boldsymbol{\xi})$ describes the change of the external torque.

5.7 For any base k-form $dx^{i_1} \wedge \dots \wedge dx^{i_k}$ with $i_1 < \dots < i_k$

$$*(dx^{i_1} \wedge \dots \wedge dx^{i_k}) = \varepsilon_{i_1 \dots i_k i_{k+1} \dots i_n} dx^{i_{k+1}} \wedge \dots \wedge dx^{i_n} \, .$$

Here we have assumed the indices on the right-hand side to be ordered, too, viz. $i_{k+1} < \cdot < i_n$. The dual of this form is again a k-form and is given by

$$** \, (dx^{i_1} \wedge \dots \wedge dx^{i_k}) = \varepsilon_{i_1 \dots i_k i_{k+1} \dots i_n} \varepsilon_{i_{k+1} \dots i_n j_1 \dots j_k} dx^{j_1} \wedge \dots \wedge dx^{j_k} \, .$$

All indices $i_1 \dots i_n$ must be different. Therefore, the set $(j_1 \dots j_k)$ must be a permutation of $(i_1 \dots i_k)$. If we choose the ordering $j_1 < \dots < j_k$, then $j_1 = i_1, \dots, j_k = i_k$. In the second ε-symbol interchange the group of indices $(i_1, \dots, i_k)$ with the group $(i_{k+1}, \dots, i_n)$. For i_1 this requires exactly $(n - k)$ exchanges of neighbors. The same holds true for i_2 up to i_k. This gives k times a sign factor $(-)^{n-k}$. As $(\varepsilon_{i_1 \dots i_n})^2 = 1$ for all indices different, we conclude

$$**\,(dx^{i_1}\wedge\ldots\wedge dx^{i_k}) = (-)^{k(n-k)}dx^{i_1}\wedge\ldots\wedge dx^{i_k}\,.$$

5.8 The exterior derivative of ϕ is calculated following the rule (CD3) of Sect. M5.4.4, viz.

$$\begin{aligned} d\phi &= \left(-\frac{\partial E_1}{\partial x^2}+\frac{\partial E_2}{\partial x^1}\right)dx^1\wedge dx^2 + \text{cyclic permutations} \\ &= (\text{curl}\,E)_3\,dx^1\wedge dx^2+\ldots\,. \end{aligned}$$

This yields the result $d\phi+\dot\omega/c=0$.

5.9 With f a smooth function $df=\sum(\partial f/\partial x^i)\,dx^i$, thus

$$\begin{aligned} *df &= \left(\partial f/\partial x^1\right)\,dx^2\wedge dx^3 + \text{cyclic permutations}\,, \\ d(*df) &= \left(\partial^2 f/(\partial x^i)^2\right)\,dx^1\wedge dx^2\wedge dx^3 + \text{cyclic permutations}\,, \end{aligned}$$

and

$$*d(*df) = \sum_i^3 \partial^2 f/(\partial x^i)^2\,.$$

Furthermore, $*f = f\,dx^1\wedge dx^2\wedge dx^3$ and $d(*f)=0$.

5.10 If $\overset{k}{\omega}$ is a k-form which is applied to k vectors $(\hat{\boldsymbol e}_1,\ldots\hat{\boldsymbol e}_k)$, then, by the definition of the pull-back (special case, for vector spaces, of (M5.41), Sect. 5.4.1) $F^*\overset{k}{\omega}(\hat{\boldsymbol e}_1,\ldots\hat{\boldsymbol e}_k)=\overset{k}{\omega}(F(\hat{\boldsymbol e}_1),\ldots,F(\hat{\boldsymbol e}_k))$. Then

$$F^*(\overset{k}{\omega}\wedge\overset{l}{\omega})(\hat{\boldsymbol e}_1,\ldots,\hat{\boldsymbol e}_{k+l}) = (\overset{k}{\omega}\wedge\overset{l}{\omega})(F(\hat{\boldsymbol e}_1),\ldots,F(\hat{\boldsymbol e}_{k+l}))\,,$$

which, in turn, equals $(F^*\overset{k}{\omega})\wedge(F^*\overset{l}{\omega})$.

5.11 This exercise is solved in close analogy to the solution of Exercise 5.10 above.

5.12 With $V:=y\partial_x$ and $W:=x\partial_y$ we find readily $Z:=[V,W]=(y\partial_x)(x\partial_y)-(x\partial_y)(y\partial_x)=y\partial_x-x\partial_x$.

5.13 Let v_1 and v_2 be elements of T_pM. Addition of vectors and multiplication by real numbers being defined as in (M5.20), it is clear that both v_1+v_2 and av_i with $a\in\boldsymbol R$ belong to T_pM, too. The dimension of T_pM is $n=\dim M$. T_pM is a vector space. In the case $M=\mathbb{R}^n$, T_pM isomorphic to M.

5.14 We have

$$\omega \wedge \omega = \sum_{i=1}^{2} \sum_{j=1}^{2} dq^i \wedge dp_i \wedge dq^j \wedge dp_j = -2dq^1 \wedge dq^2 \wedge dp_1 \wedge dp_2 \,.$$

This is so because we interchanged dp_i and dq^j, and because the terms ($i = 1, j = 2$) and ($i = 2, j = 1$) are equal.

5.15 $H^{(1)} = p^2/2 + 1 - \cos q$ is the Hamiltonian function that describes the planar mathematical pendulum. The corresponding Hamiltonian vector field reads

$$X_H^{(1)} = \frac{\partial H}{\partial p} \partial_q - \frac{\partial H}{\partial q} \partial_p = p\partial_q - \sin q \partial_p \,.$$

A sketch of this vector field will yield the vectors tangent to the curves of Fig. M1.10. The neighborhood of the point ($p = 0, q = \pi$) is particularly interesting as this represents an unstable equilibrium. For

$$H^{(2)} = \frac{1}{2} p^2 + \frac{1}{6} q(q^2 - 3)$$

the Hamiltonian vector field is

$$X_H^{(2)} = p\partial_q - \frac{1}{2} (q^2 - 1)\partial_p \,.$$

This vector field has two equilibrium points, ($p = 0, q = +1$) and ($p = 0, q = -1$). A sketch of $X_H^{(2)}$ will show that the former is a stable equilibrium (center), while the latter is unstable (saddle point). Linearization in the neighborhood of $q = +1$ means that we set $u := q - 1$ and keep up to linear terms in u only. Then $X_H^{(2)} \approx p\partial_u - u\partial_p$. This is the vector field of the harmonic oscillator or, equivalently, the vector field $X_H^{(1)}$ above, for small values of q.

Linearization of the system in the neighborhood of ($p = 0, q = -1$), in turn, means setting $u := q + 1$ so that $X_H^{(2)} \approx p\partial_u + u\partial_p$. Here the system behaves like the mathematical pendulum (described by $X_H^{(1)}$ above) in the neighborhood of its unstable equilibrium ($p = 0, q = \pi$) where $\sin q = -\sin(q - \pi) \approx -(q - \pi)$. (See also Exercise 6.8.)

5.16 One finds $X_{H^0} = p\partial_q - q\partial_p$, $X_H = p\partial_q - (q + \varepsilon q^2)\partial_p$, and, finally

$$\omega\left(X_H, X_{H^0}\right) = dH\left(X_{H^0}\right) = \varepsilon p q^2 = \left\{H^0, H\right\} \,.$$

5.17 For the proof consult for example Sect. 3.5.18 of Abraham and Marsden (1981).

Chapter 6: Stability and Chaos

6.1 (i) **A** is already diagonal. The flux is

$$\exp(t\mathbf{A}) = \begin{pmatrix} e^{t\lambda_1} & 0 \\ 0 & e^{t\lambda_2} \end{pmatrix} .$$

(ii) The characteristic exponents (i.e., the eigenvalues of **A**) are $\lambda_1 = a + ib$, $\lambda_2 = a - ib$ so that in the diagonalized form the flux reads as follows: With

$$\underset{\sim}{y} \to \underset{\sim}{u} = \mathbf{U}\underset{\sim}{y}\,, \quad \overset{\circ}{A} = \mathbf{UAU}^{-1} = \begin{pmatrix} a+ib & 0 \\ 0 & a-ib \end{pmatrix} ,$$

we have

$$\underset{\sim}{u}(t) = \exp(r\overset{\circ}{A})\underset{\sim}{u}(0) = \begin{pmatrix} e^{t(a+ib)} & 0 \\ 0 & e^{t(a-ib)} \end{pmatrix} \underset{\sim}{u}(0) .$$

For $a = 0$, $b > 0$ we find a (stable) center. For $a < 0$, $b > 0$ we find an (asymptotically stable) node.

(iii) The characteristic exponents are equal, $\lambda_1 = \lambda_2 \equiv \lambda$. For $\lambda < 0$ we again find a node.

6.2 The flux of this system is

$$\left(\alpha(\tau) = \frac{a}{2\pi}\tau + \alpha_0(\mod 1), \beta(\tau) = \frac{b}{2\pi}\tau + \beta_0(\mod 1) \right) .$$

If the ratio b/a is *rational*, i.e., $b/a = m/n$ with $m, n \in \mathbb{Z}$, the system returns to its initial position after the time $\tau = T$ where T follows from $\alpha_0 + aT/(2\pi) = \alpha_0$ (mod 1) and $\beta_0 + bT/(2\pi) = \beta_0$ (mod 1), i.e., $T = 2\pi n/a = 2\pi m/b$. We study the example ($a = 2/3, b = 1$) with initial condition ($\alpha_0 = 1/2, \beta_0 = 0$). This yields the results shown in Table 1 and in Fig. 25. In the figure the sections of the orbit are numbered in the order in which they appear.

Table 1

$\varrho = \frac{\tau}{2\pi}$	0	$\frac{3}{4}$	1	2	$\frac{9}{4}$	3
α	$\frac{1}{2}$	1	$\frac{1}{6}$	$\frac{5}{6}$	1	$\frac{1}{2}$
β	0	$\frac{3}{4}$	1	1	$\frac{1}{4}$	1

If the ratio b/a is *irrational* then the flux will cover the torus, or the square of Fig. 25, densely. As an example one may choose a "out of tune" at the value $a = 1/\sqrt{2} \approx 0.7071$, keeping $b = 1$ fixed, and plot the flux in the square. A

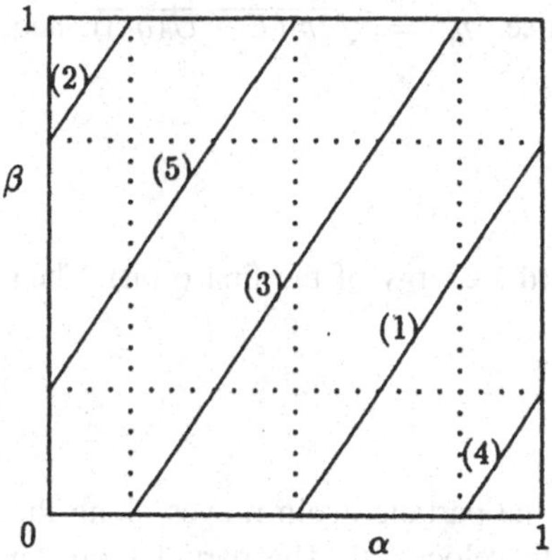

Fig. 25.

specific example is provided by two coupled oscillators, cf. Exercises 1.9 and 2.9. The modes of the system obey the differential equation $\ddot{u}_i + \omega^{(i)2} u_i = 0$, $i = 1, 2$. These are rewritten in terms of action and angle variables I_i and Θ_i, respectively, by means of the canonical transformation (M2.95). They become $\dot{I}_i = 0$, $\dot{\Theta}_i = \omega^{(i)}$, $i = 1, 2$. Embedded in the phase space which is now four-dimensional, we find two two-dimensional tori which are determined by the given values of $I_i = I_i^0 = \text{const.}$ Each of these tori carries the flux described above.

6.3 The Hamiltonian function has the form $H = p^2/(2m) + U(q)$. The characteristic equation

$$\frac{1}{2m}\left(\frac{\partial S_0(q,\alpha)}{\partial q}\right)^2 + U(q) = E_0$$

is integrated by quadrature:

$$S_0(q,\alpha) = \int_{q_0}^{q} \sqrt{2m\,(E_0 - U(q'))}\, dq' \,.$$

We have $p = m\dot{q} = \partial S_0/\partial q = \sqrt{2m(E_0 - U(q))}$ and hence

$$t(q) - t(q_0) = \int_{q_0}^{q} \frac{m}{\sqrt{2m\,(E_0 - U(q'))}}\, dq' = \frac{\partial S_0}{\partial E_0} \,.$$

(This holds away from equilibrium positions, in case the system possesses any equilibria.) Choose $P \equiv \alpha = E_0$. Then $Q = \partial S_0/\partial E_0 = t - t_0$, or, alternatively, $(\dot{P} = 0,\ \dot{Q} = 1)$. Thus, we have achieved rectification of the Hamiltonian vector field: In the coordinates (P, Q) the particle moves on the straight line $P = E_0$ with velocity $\dot{Q} = 1$.

Consider now an energy in the vicinity of E_0, $E = \beta E_0$, with β not far from 1. Let the particle travel from q_0 to a point q' in such a way that the time

$t(q') - t(q_0)$ is the same as for E_0. Of course, $p'_0 = \sqrt{2m(E - U(q_0))}$, and $p' = \sqrt{2m(E - U(q'))}$, and

$$t - t_0 = \int_{q_0}^{q} \frac{m\,dx}{\sqrt{2m(E - U(x))}} .$$

The new momentum is chosen to be $P = E_0$ (the energy of the first orbit). Then one has

$$Q = \frac{\partial S(q, E)}{\partial E_0} = \beta \frac{\partial S(q, E)}{\partial E} = \beta(t - t_0) ,$$

i.e., ($\dot{P} = 0$, $\dot{Q} = \beta$). In the new coordinates the particle again moves along the straight line $P = E_0$, this time, however, with velocity β. The particles on the orbits to be compared move apart linearly in time. If $U(q)$ is such that in some region of phase space all orbits are periodic, one should transform to action and angle variables, $I(E) = \text{const.}$, $\Theta = \omega(E)t + \Theta_0$. Also in this situation one sees that the orbits separate at most linearly in time. In the case of the oscillator, where ω is independent of E or I, their distance remains constant.

The integration described above is possible only if E is *larger* than the maximum of $U(q)$. For $E = U_{\max}(q)$ the running time goes to infinity logarithmically (cf. Sect. M1.23). The statement of this exercise does not apply when one of the trajectories is a separatrix.

6.4 For $\mu = 0$ the system becomes $\dot{q}_1 = \partial H/\partial q_2$ and $\dot{q}_2 = -\partial H/\partial q_1$, with $H = -\lambda(q_1^2 + q_2^2)/2 + (q_1 q_2^2 - q_1^3/3)/2$. The critical points (where the Hamiltonian vector field vanishes) are obtained from the system of equations $-\lambda q_2 + q_1 q_2 = 0$, $\lambda q_1 + (q_1^2 - q_2^2)/2 = 0$. One finds the following solutions: $P_0 = (q_1 = 0, q_2 = 0)$, $P_{1/2} = (q_1 = \lambda, q_2 = \pm\sqrt{3}\,\lambda)$, $P_3 = (q_1 = -2\lambda, q_2 = 0)$. Linearization in the neighborhood of P_0 leads to $\dot{q}_1 \approx -\lambda q_2$, $\dot{q}_2 \approx \lambda q_1$. Thus P_0 is a center. Linearization in the neighborhood of P_1 is achieved by the transformation $u_1 := q_1 - \lambda$, $u_2 := q_2 - \sqrt{3}\,\lambda$, whereby the differential equations become.

$$\dot{u}_1 = \sqrt{3}\,\lambda u_1 + u_1 u_2 \approx \sqrt{3}\,\lambda u_1 ;$$
$$\dot{u}_2 = 2\lambda u_1 - \sqrt{3}\,\lambda u_2 + \left(u_1^2 - u_2^2\right)/2 \approx 2\lambda u_1 - \sqrt{3}\,\lambda u_2 .$$

The flux tends to P_1 along $u_1 = 0$ but tends away from it along $u_2 = 0$. Thus P_1 is a saddle point. The same is true for P_2 and P_3. One easily verifies that these three points belong to the same energy $E = H(P_i) = -2\lambda^3/3$ and that they are pairwise connected by separatrices. Indeed, the straight lines $q_2 = \pm(q_1 + 2\lambda)/\sqrt{3}$ and $q_1 = \lambda$ are curves with constant energy $E = -2\lambda^3/3$ and build up the triangle (P_1, P_2, P_3).

If one switches on the damping terms by means of $1 \gg \mu > 0$, P_0 is still an equilibrium point because in the neighborhood of $(q_1 = 0, q_2 = 0)$ we have

$$\begin{pmatrix} \dot{q}_1 \\ \dot{q}_2 \end{pmatrix} \approx \begin{pmatrix} -\mu & -\lambda \\ \lambda & -\mu \end{pmatrix} \begin{pmatrix} q_1 \\ q_2 \end{pmatrix} \equiv \mathbf{A} \begin{pmatrix} q_1 \\ q_2 \end{pmatrix} .$$

From the equation $\det(x\mathbb{1} - \mathbf{A}) = 0$ one finds the characteristic exponents to be $x_{1/2} = -\mu \pm \mathrm{i}\lambda$. Thus P_0 becomes a node (a sink). The points P_1, P_2, and P_3, however, are no longer equilibrium positions. The lines which connect them are broken up.

6.5 We write H in two equivalent forms

$$\text{(i)}\ \ H = I_1 + I_2 \quad \text{with} \quad I_1 = \left(p_1^2 + q_1^2\right)/2 + q_1^3/3\,,$$

$$I_2 = \left(p_2^2 + q_2^2\right)/2 - q_2^3/3\,,$$

$$\text{(ii)}\ \ H = \left(p_1^2 + p_2^2\right)/2 + U(q_1, q_2) \quad \text{with} \quad U = \left(q_1^2 + q_2^2\right)/2 + \left(q_1^3 - q_2^3\right)/3$$

$$= \left(\Sigma^2 + \Delta^2\right)/4 + \Sigma^2\Delta/4 + \Delta^3/12\,,$$

$$\text{where} \quad \Sigma := q_1 + q_2\,, \quad \Delta := q_1 - q_2\,.$$

Then the equations of motion are

$$\dot{q}_1 = p_1\,, \quad \dot{q}_2 = p_2\,,$$

$$\dot{p}_1 = -q_1 - q_1^2\,, \quad \dot{p}_2 = -q_2 + q_2^2\,.$$

The critical points of this system are $P_0 : (q_1 = 0, q_2 = 0, p_1 = 0, p_2 = 0)$, $P_1 : (0, 1, 0, 0)$, $P_2 : (-1, 0, 0, 0)$, $P_3 : (-1, 1, 0, 0)$. One easily verifies that $dI_i/dt = 0$, $i = 1, 2$, i.e., I_1 and I_2 are independent integrals of the motion. The points P_1 and P_2 lie on two equipotential lines, viz. the straight line $q_1 - q_2 = -1$, and the ellipse $3(q_1+q_2)^2 + (q_1-q_2)^2 + 2(q_1-q_2) - 2 = 0$. In either case $U = 1/6$. As an example and using these results, one may sketch the projection of the flux onto the plane (q_1, q_2).

6.6 The critical points of the system $\dot{q} = p$, $\dot{p} = q - q^3 - p$ are $P_0 : (q = 0, p = 0)$, $P_1 : (1, 0)$, and $P_2 = (-1, 0)$. Linearization around P_0 gives

$$\begin{pmatrix} \dot{q} \\ \dot{p} \end{pmatrix} \approx \begin{pmatrix} 1 & 1 \\ 0 & -1 \end{pmatrix} \begin{pmatrix} q \\ p \end{pmatrix} = \mathbf{A} \begin{pmatrix} q \\ p \end{pmatrix}.$$

The eigenvalue of A are $\lambda_{1/2} = (-1 \pm \sqrt{5})/2$, hence $\lambda_1 > 0$ and $\lambda_2 < 0$ which means that P_0 is a saddle point. Linearizing in the neighborhood of P_1 and introducing the variables $u := q - 1$, $v := p$, the system becomes

$$\begin{pmatrix} \dot{u} \\ \dot{v} \end{pmatrix} \approx \begin{pmatrix} 0 & 1 \\ -2 & -1 \end{pmatrix} \begin{pmatrix} u \\ v \end{pmatrix}.$$

The characteristic exponents are $\mu_{1/2} = (-1 \pm \mathrm{i}\sqrt{7})/2$. The same values are found for the system linearized in the neighborhood of P_2. This means that both P_1 and P_2 are sinks.

The Liapunov function $V(q, p)$ has the value 0 in P_0, and the value $-1/4$ in P_1 and P_2. One easily verifies that P_1 and P_2 are minima and that V increases monotonically in a neighborhood of these points. For instance, close to P_1 take $u :=$

$q-1$, $v := p$. Then $\Phi_1(u,v) := V(q=u+1, p=v)+1/4 = v^2/2+u^2+u^3+u^4/4$. Indeed, at the point P_1 we find $\Phi_1(0,0)=0$ while Φ_1 is positive in a neighborhood of P_1.

Along solutions the function $V(q,p)$, or, equivalently, $\Phi_1(u,v)$, decreases monotonically. Let us check this for V:

$$\frac{dV}{dt} = \frac{dV}{\partial p}\dot{p} + \frac{\partial V}{\partial q}\dot{q} = \frac{\partial V}{\partial p}(q-q^3-p) + \frac{\partial V}{\partial q}p = -p^2 \, .$$

In order to find out towards which of the two sinks a given initial configuration will tend, one has to calculate the two separatrices that end in P_0. They form the boundaries of the basins of P_1 and P_2 as indicated in Fig. 26 by the blank and dotted areas, respectively.

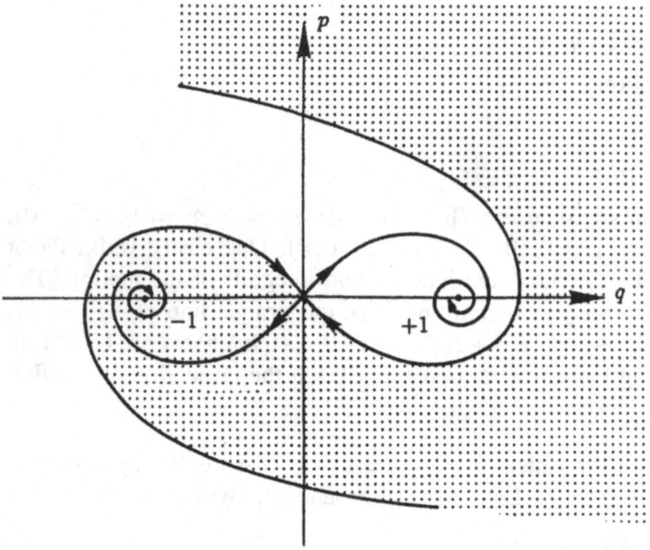

Fig. 26.

6.7 As x_0, by assumption, is an isolated minimum, a Liapunov function is chosen as follows: $V(x) := U(x) - U(x_0)$. In a certain neighborhood M of x_0, $V(x)$ is positive semi-definite and we have

$$\frac{d}{dt}V(\underset{\sim}{x}) = \sum_{i=1}^{n} \frac{\partial V}{\partial x_i}\dot{x}_i = -\sum_{i=1}^{n}\left(\frac{\partial U}{\partial x_i}\right)^2 \, .$$

If we follow a solution in the domain $M\backslash\{x_0\}$, $V(x)$ decreases, i.e., all solution curves tend "inwards", towards x_0. Thus, this point is asymptotically stable.

In the example $U(x_1,x_2) = x_1^2(x_1-1)^2 + x_2^2$. The points $x_0 = (0,0)$ and $x_0' = (1,0)$ are isolated minima and, hence, are asymptotically stable equilibria.

6.8 This system is Hamiltonian. A Hamiltonian function is $H = p^2/2 + q(q^2 - 3)/6$. The phase portraits are obtained by drawing the curves $H(q,p) = E =$

const. The Hamiltonian vector field $v_H = (p, (1 - q^2)/2)$ has two critical points whose nature is easily identified by linearizing in their neighborhoods. One finds

P_1 : $(q = -1, p = 0)$ and, with $u := q + 1$, $v := p$: $(\dot{u} \approx v, \dot{v} \approx u)$. Thus, P_1 is a saddle point.

P_2 : $(q = 1, p = 0)$ and, with $\dot{u} := q - 1$, $v := p$: $(\dot{u} \approx v, \dot{v} \approx -u)$. Thus, P_2 is a center. In the neighborhood of P_2 there will be harmonic oscillations with period 2π (see also Exercise 5.15).

6.9 The differential equation $\ddot{q} = f(q, \dot{q})$, with $f(q, \dot{q}) = -q + (\varepsilon - q^2)\dot{q}$, is solved numerically by means of a Runge–Kutta procedure as follows

$$q_{n+1} = q_n + h\left(\dot{q}_n + \frac{1}{6}(k_1 + k_2 + k_3)\right) + O(h^5)$$

$$\dot{q}_{n+1} = \dot{q}_n + \frac{1}{6}(k_1 + 2k_2 + 2k_3 + k_4) \ ,$$

h being the integration step in the time variable, and the auxiliary quantities k_i being defined by

$$k_1 = hf(q_n, \dot{q}_n) \ ,$$

$$k_2 = hf\left(q_n + \frac{h}{2}\dot{q}_n + \frac{h}{8}k_1, \dot{q}_n + \frac{1}{2}k_1\right) \ ,$$

$$k_3 = hf\left(q_n + \frac{h}{2}\dot{q}_n + \frac{h}{8}k_1, \dot{q}_n + \frac{1}{2}k_2\right) \ ,$$

$$k_4 = hf\left(q_n + h\dot{q}_n + \frac{h}{2}k_3, \dot{q}_n + k_3\right) \ .$$

One lets the dimensionless time variable $\tau = \omega t$ run from 0 to, say, 6π, in steps of 0.1, or 0.05, or 0.01. This will produce pictures of the type shown in Figs. (M6.6)–(M6.8). Alternatively, one may follow the generation of these figures on the screen of a PC. One will notice that all of them tend quickly to the attractor.

6.10 The program developed in Exercise 6.9 may be used to print out, for a given initial configuration, the time τ and the distance from the origin d, each time the orbit crosses the line $p = q$. One finds the following result:

$p = q > 0$:	τ	5.46	11.87	18.26	24.54	30.80	37.13	43.47
	d	0.034	0.121	0.414	1.018	1.334	1.375	1.378
$p = q < 0$:	τ	2.25	8.86	15.07	21.42	27.66	33.96	40.30
	d	0.018	0.064	0.227	0.701	1.238	1.366	1.378

Plotting $\ln d$ versus τ shows that this function increases approximately linearly (with slope ≈ 0.1) until it has reached the attractor. Thus, the point of intersection of the orbit and the straight line $p = q$ wanders towards the attractor at an approximately exponential rate. One finds a similar result for orbits which approach the attractor from the outside.

6.11 The motion manifold of this system is $\mathbb{R}^2$. For the linearized system ($\dot{x}_1 = x_1, \dot{x}_2 = -x_2$) the straight line $U_{\text{stab}} = (x_1 = 0, x_2)$ is a stable submanifold. Indeed, the velocity field points towards the equilibrium point $(0, 0)$ and the characteristic exponent is -1. The straight line $U_{\text{unst}} = (x_1, x_2 = 0)$, in turn, is an unstable submanifold: The velocity field points away from $(0, 0)$ and the characteristic exponent is $+1$. The full system can be transformed to

$$\ddot{x}_2 - \dot{x}_2 - 2x_2 = 0 \,, \quad x_1^2 = x_2 + \dot{x}_2 \,,$$

whose general solution is

$$x_2(t) = a \exp(2t) + b \exp(-t) \,, \quad x_1(t) = \sqrt{3a} \exp t \,,$$

or, equivalently,

$$x_2 = \frac{1}{3} x_1^2 + b\sqrt{3a} \frac{1}{x_1} \equiv \frac{1}{3} x_1^2 + \frac{c}{x_1} \,.$$

Among this set of solutions the orbit with $c = 0$ goes through the point (0,0) and is tangent to U_{unst} in that point. On the submanifold $V_{\text{unst}} = (x_1, x_2 = x_1^2/3)$ the velocity field moves away from $(0, 0)$.

The corresponding stable submanifold of the full system coincides with U_{stab} because, with $a = 0$, $x_1(t) = 0$, $x_2(t) = b \exp(-t)$ which means that $V_{\text{stab}} = (x_1 = 0, x_2)$.

6.12 We have $x_{n+1} = 1 - 2x_n^2$ and $y_i = 4/\pi \arcsin \sqrt{(x_i + 1)/2} - 1$. With $-1 \le x_i \le 0$ also $-1 \le y_i \le 0$, and with $0 \le x_i \le 1$ also $0 \le y_i \le 1$. We wish to know the relation between y_{n+1} and y_n. First, the relation $x_n \to y_{n+1}$ is $y_{n+1} = 4/\pi \arcsin(1 - x_n^2)^{1/2} - 1$. Using the addition theorem $\arcsin u + \arcsin v = \arcsin(u\sqrt{1 - v^2} + v\sqrt{1 - u^2})$ and setting $u = v = (1 + x)/2$ one shows

$$\arcsin \sqrt{1 - x^2} = 2 \arcsin \sqrt{\frac{x + 1}{2}} \quad \text{for} \quad -1 \le x \le 0 \,,$$

$$\arcsin \sqrt{1 - x^2} = \pi - 2 \arcsin \sqrt{\frac{x + 1}{2}} \quad \text{for} \quad 0 \le x \le 1 \,.$$

In the first case $y_n \le 0$ and $y_{n+1} = 1 + 2y_n$, in the second case $y_n \ge 0$ and $y_{n+1} = 1 - 2y_n$. These can be combined to $y_{n+1} = 1 - 2|y_n|$. The derivative of this iterative mapping is ± 2; its magnitude is larger than 1. There are no stable fixed points.

6.15 If, for instance, $m = 1$, then $z_\sigma := \exp(\mathrm{i}2\pi\sigma/n)$ are the roots of the equation $z^n - 1 = (z - z_1) \dots (z - z_n) = 0$. They lie on the unit circle in the complex plane and neighboring roots are separated by the angle $2\pi/n$. Expanding the product $(z - z_1) \dots (z - z_n) = z^n - z \sum_{\sigma=1}^{n} z_\sigma + \dots$, one sees that, indeed, $\sum_{\sigma=1}^{n} z_\sigma = 0$.

For the values $m = 2, \ldots, n-1$ we renumber the roots and obtain the same result. For $m = 0$ and for $m = n$, however, the sum is equal to 0. Multiplying $\tilde{x}_\sigma = \sum_{\tau=1}^{n} x_\tau \exp(-2\mathrm{i}\pi\sigma\tau/n)/\sqrt{n}$ by $1/\sqrt{n}\exp(2\mathrm{i}\pi\sigma\lambda/n)$ and summing over σ one obtains

$$\frac{1}{\sqrt{n}}\sum_{\sigma=1}^{n}\tilde{x}_\sigma \mathrm{e}^{2\mathrm{i}\pi\sigma\lambda/n} = \frac{1}{n}\sum_{\tau=1}^{n} x_\tau \sum_{\sigma=1}^{n}\mathrm{e}^{2\mathrm{i}\pi\sigma(\tau-\lambda)/n} = \sum_{\tau=1}^{n} x_\tau\delta_{\tau\lambda} = x_\lambda\ .$$

This is used to calculate

$$g_\lambda = \frac{1}{n}\sum_{\sigma=1}^{n} x_\sigma x_{\sigma+\lambda} = \frac{1}{n^2}\sum_{\mu,\nu}\tilde{x}_\mu\tilde{x}^*_{n-\nu}\sum_{\sigma}\mathrm{e}^{2\mathrm{i}\pi/n(\sigma(\mu+\nu)+\lambda\nu)}\ .$$

The orthogonality relation implies $\mu+\nu = 0\ (\mathrm{mod}\ n)$, and we have used $\tilde{x}^*_{n-\nu} = \tilde{x}_\nu$. Furthermore, $\tilde{x}_{\mu \bmod n} = \tilde{x}_\mu$ and, finally, $\tilde{x}_\mu$ and $\tilde{x}_{n-\mu}$ have the same modulus. It follows that $g_\lambda = 1/n\sum_{\mu=1}^{n}|\tilde{x}_\mu|^2\cos(2\pi\lambda\mu/n)$. The inverse of this formula $|\tilde{x}_\sigma|^2 = \sum_{\lambda=1}^{n} g_\lambda\cos(2\pi\sigma\lambda/n)$ is obtained in the same way.

6.16 If we set $y = \alpha x + \beta$, i.e., $x = y/\alpha - \beta/\alpha$, the relation

$$x_{i+1} = \mu x_i(1-x_i) = \mu\left(\frac{1}{\alpha}y_i - \frac{\beta}{\alpha}\right)\left(1+\frac{\beta}{\alpha}-\frac{1}{\alpha}y_i\right)$$

will take the desired form provided α and β are chosen such that they fulfill the equations $\alpha + 2\beta = 0$, $\beta(1-\mu(\alpha+\beta)/\alpha) = 1$. These give $\alpha = 4/(\mu-2)$, $\beta = -\alpha/2$, and, therefore, $\gamma = \mu(\mu-2)/4$. From $0 \le \mu < 4$ follows $0 \le \gamma < 2$. One then sees easily that $y_i \in [-1,+1]$ is mapped onto y_{i+1} in the same interval. Let $h(y,\gamma) := 1-\gamma y^2$. The first bifurcation occurs when $h(y,\gamma) = y$ and $\partial h(y,\gamma)/\partial y = 1$, i.e., when $\gamma_0 = 3/4$, $y_0 = 2/3$, or, correspondingly, $\mu_0 = 3/4$, $x_0 = 2/3$. Take then $k := h\circ h$, i.e., $k(y,\gamma) = 1-\gamma(1-\gamma y^2)^2$. The second bifurcation occurs at $\gamma_1 = 5/4$. The corresponding value y_1 of y is calculated from the system

$$k\left(y,\frac{5}{4}\right) = -\frac{1}{4}+\frac{25}{8}y^2-\frac{125}{64}y^4 = y\ , \tag{1}$$

$$\frac{\partial k}{\partial y}\left(y,\frac{5}{4}\right) = \frac{25}{4}y\left(1-\frac{5}{4}y^2\right) = -1\ . \tag{2}$$

Combining these equations according to $y\cdot(2)$–$4\cdot(1)$, one finds the quadratic equation

$$y^2 - \frac{4}{5}y - \frac{4}{25} = 0\ ,$$

whose solutions are $y_{1/2} = 2(1\pm\sqrt{2})/5$. From $\gamma_1 = 5/4$ we have $\mu_1 = 1+\sqrt{6}$, from this and from $y_{1/2}$, finally

$$x_{1/2} = \frac{1}{10}(4+\sqrt{6}\pm(2\sqrt{3}-\sqrt{2})) = 0.8499 \quad \text{and} \quad 0.4400\ .$$

Appendix

A. Some Mathematical Notions

"Order" and "Modulo". The symbol $\mathrm{O}(\varepsilon^n)$ stands for terms of the order ε^n and higher powers thereof that are being neglected.

Example. If in a Taylor series one wishes to take acount only of the first three terms, one writes

$$f(x) = f(0) + f'(0)x + \frac{1}{2!}f''(0)x^2 + \mathrm{O}(x^3). \tag{A.1}$$

This means that the right-hand side is valid up to terms of order x^3 and higher.

The notation $y = x \pmod a$ means that x and $x + na$ should be identified, n being any integer. Equivalently, this means that one should add to x or subtract from it the number a as many times as are necessary to have y fall in a given interval.

Example. Suppose two angles α and β are defined in the interval $[0, 2\pi]$. The equation $\alpha = f(\beta) \pmod{2\pi}$ means that one must add to the value of the function $f(\beta)$, or subtract from it, an integer number of terms 2π such that α does not fall outside its interval of definition.

Mappings. A mapping f that maps a set A onto a set B is denoted as follows:

$$f\colon A \to B\colon a \mapsto b. \tag{A.2}$$

It assigns to a given element $a \in A$ the element $b \in B$. The element b is said to be the *image* and a its *preimage*.

Examples. (i) The real function $\sin x$ maps the real x-axis onto the interval $[-1, +1]$,

$$\sin : \mathbb{R} \to [-1, +1]\colon x \mapsto y = \sin x .$$

(ii) The curve $\gamma\colon x = \cos\omega t,\ y = \sin\omega t$ in $\mathbb{R}^2$ is a mapping from the real t-axis onto the unit circle S^1 in $\mathbb{R}^2$,

$$\gamma : \mathbf{R}_t \to S^1 : t \mapsto (x = \cos\omega t,\ y = \sin\omega t) .$$

A mapping is called *surjective* if $f(A) = B$, i.e. if B is covered completely. The mapping is called *injective* if two distinct elements in A also have distinct images in B, in other words, if every $b \in B$ has at most one original $a \in A$. If it has both properties it is said to be *bijective*. In this case every element of A has exactly one image in B, and for every element of B there is exactly one preimage in A. In other words the mapping is then one-to-one.

Examples: (i) The mapping

$$f : \mathbb{R} \to \mathbb{R} : a \mapsto b = f(a) = a^3$$

is injective. Indeed, $b_1 = f(a_1) = f(a_2) = b_2$ implies the equality $a_1 = a_2$. The mapping is also surjective: For any $b \in \mathbb{R}$ the preimage is $a = b^{1/3}$ if b is positive, and $a = -b^{1/3}$ if b is negative.

(ii) The mapping

$$f : \mathbb{R} \to \mathbb{R} : a \mapsto b = f(a) = a^2$$

is not injective because $a_1 = 1$ and $a_2 = -1$ have the same image.

The *composition* $f \circ g$ of two mappings f and g means that g should be applied first and then f should be applied to the result of the first mapping, viz.

$$\text{If } g\colon A \to B \quad \text{and} \quad f\colon B \to C\,, \quad \text{then} \quad f \circ g\colon A \to C\,. \tag{A.3}$$

Example: Suppose f and g are functions over the reals. Then, with $y = g(x)$ and $z = f(y)$ we have $z = (f \circ g)(x) = f(g(x))$.

The *identical mapping* is often denoted by "id", i.e.

$$\text{id}\colon A \to A\colon a \mapsto a\,.$$

Special Properties of Mappings. A mapping $f\colon A \to B$ is said to be *continuous at the point* $u \in A$ if for every neighborhood V of its image $v = f(u) \in B$ there exists a neighborhood U of u such that $f(U) \subset V$. The mapping is *continuous* if this property holds at *every* point of A.

Homeomorphisms are bijective mappings $f\colon A \to B$ that are such that both f and its inverse f^{-1} are continuous.

Diffeomorphisms are differentiable, bijective mappings f that are such that both f and f^{-1} are smooth (i.e. differentiable, C^∞).

Derivatives. Let $f(x^1, x^2, \ldots, x^n)$ be a function over the space $\mathbb{R}^n$, $\{\hat{\boldsymbol{e}}_1, \ldots, \hat{\boldsymbol{e}}_n\}$, a set of orthogonal unit vectors. The *partial derivative* with respect to the variable x^i is defined as follows:

$$\frac{\partial f}{\partial x^i} = \lim_{h \to 0} \frac{f(x + h\hat{\boldsymbol{e}}_i) - f(x)}{h}\,. \tag{A.4}$$

Thus one takes the derivative with respect to x^i while keeping all other arguments $x^1, \ldots, x^{i-1}, x^{i+1}, \ldots, x^n$ fixed.

Collecting all partial derivatives yields a vector field called the *gradient*,

$$\nabla f = \left(\frac{\partial f}{\partial x^1}, \ldots, \frac{\partial f}{\partial x^n} \right) . \tag{A.5}$$

Since any direction $\hat{\boldsymbol{n}}$ in $\mathbb{R}^n$ can be decomposed in terms of the basis vectors $\hat{\boldsymbol{e}}_1, \ldots, \hat{\boldsymbol{e}}_n$, one can take the *directional derivative* of f along that direction, viz.

$$\frac{\partial f}{\partial \hat{\boldsymbol{n}}} = \sum_{i=1}^{n} \hat{\boldsymbol{n}}^i \frac{\partial f}{\partial x^i} \equiv \hat{\boldsymbol{n}} \cdot \nabla f . \tag{A.6}$$

The *total differential* of the function $f(x^1, \ldots, x^n)$ is defined as follows:

$$\mathrm{d}f = \frac{\partial f}{\partial x^1}\mathrm{d}x^1 + \frac{\partial f}{\partial x^2}\mathrm{d}x^2 + \cdots + \frac{\partial f}{\partial x^n}\mathrm{d}x^n . \tag{A.7}$$

Examples. (i) Let $f(x, y) = \frac{1}{2}(x^2 + y^2)$ and let $(x = r\cos\phi,\ y = r\sin\phi)$ with fixed r and $0 \le \phi \le 2\pi$ be a circle in $\mathbb{R}^2$. The normalized vector tangent to the circle at the point (x, y) is given by $v_t = (-\sin\phi,\ \cos\phi)$. Similarly, the normalized normal vector at the same point is given by $v_n = (\cos\phi,\ \sin\phi)$. The total differential of f is $\mathrm{d}f = x\mathrm{d}x + y\mathrm{d}y$ and its directional derivative along $\boldsymbol{v}_t$ is $\boldsymbol{v}_t \cdot \nabla f = -x\sin\phi + y\cos\phi = 0$; the directional derivative along $\boldsymbol{v}_n$ is given by $\boldsymbol{v}_n \cdot \nabla f = x\cos\phi + y\sin\phi = r$. For an arbitrary unit vector $\boldsymbol{v} = (\cos\alpha, \sin\alpha)$ we find $\boldsymbol{v} \cdot \nabla f = r(\cos\phi\,\cos\alpha + \sin\phi\,\sin\alpha)$. For fixed ϕ the absolute value of this real number becomes maximal if $\alpha = \phi\ (\mathrm{mod}\,\pi)$. Thus, the gradient defines the direction along which the function f grows or falls fastest.

(ii) Let $U(x, y) = xy$ be a potential in the plane $\mathbb{R}^2$. The curves along which the potential U is constant (they are called *equipotential lines*) are obtained by taking $U(x, y) = c$, with c a constant real number. They are given by $y = c/x$, i.e. by hyperbolas whose center of symmetry is the origin. Along these curves $\mathrm{d}U(x, y) = y\mathrm{d}x + x\mathrm{d}y = 0$ because $\mathrm{d}y = -(c/x^2)\mathrm{d}x = -(y/x)\mathrm{d}x$. The gradient is given by $\boldsymbol{\nabla} U = (\partial U/\partial x,\ \partial U/\partial y) = (y, x)$ and is perpendicular to the curves $U(x, y) = c$ at any point (x, y). It is the tangent vector field of another set of curves that obey the differential equation

$$\frac{\mathrm{d}y}{\mathrm{d}x} = \frac{x}{y} .$$

These latter curves are given by $y^2 - x^2 = a$.

Differentiability of Functions. The function $f(x^1, \ldots, x^i, \ldots, x^n)$ is said to be C^r with respect to x^i if it is r-times continuously differentiable in the argument x^i. A function is said to be C^∞ in some or all of its arguments if it is differentiable an infinite number of times. It is then also said to be *smooth*.

Variables and Parameters. Physical quantities often depend on two kinds of arguments, the *variables* and the *parameters*. This distinction is usually made on the basis of a physical picture and, therefore, is not canonical. Generally, variables are dynamical quantities whose time evolution one wishes to study. Parameters, on the other hand, are given numbers that define the system under consideration. In the example of a forced and damped oscillator, the deviation $x(t)$ from the equilibrium position is taken to be the dynamical variable while the spring constant, the damping factor, and the frequency of the driving source are parameters.

Lie Groups. The definition assumes that the reader is familiar with the notion of differentiable manifold, cf. Sect. 5.2. A Lie group is a finite dimensional, smooth, manifold G which in addition is a group and for which the product "·",

$$G \times G \to G : (g, g') \mapsto g \cdot g' ,$$

as well as the transition to the inverse,

$$G \to G : g \mapsto g^{-1}$$

are smooth operations.

G being a group means that it fulfills the group axioms, cf. e.g., Sect. 1.13: There exists an associative product; G contains the unit element e; for every $g \in G$ there exists an inverse $g^{-1} \in G$. Loosely speaking, smoothness means that the group elements depend differentiably on parameters, which may be thought of as angles for instance, and that group elements can be deformed in a continuous and differentiable manner.

A simple example is the unitary group

$$U(1) = \left\{ e^{i\alpha} \,|\, \alpha \in [0, 2\pi] \right\} .$$

This is an Abelian group (i.e., a commutative group). Further examples are provided by the rotation group SO(3) in three real dimensions, and the unitary group SU(2) which are dealt with in Sects. 2.21 and 5.2.3 (iv). The Galilei group is defined in Sect. 1.13, the Lorentz group is discussed in Sects. 4.4 and 4.5.

B. Historical Notes

There follow some biographical notes on scientists who made important contributions to mechanics. Some of these are marked by an asterisk and are treated in somewhat more detail, though without striving for completeness, because their impact on the understanding and the development of mechanics was particularly important.

***d'Alembert, Jean-Baptiste:** born 17 November 1717 in Paris, died 29 October 1783 in Paris. Writer, philosopher, and mathematician. Co-founder of the *Encyclopédie*. Important contributions to mathematics, mathematical physics and astronomy. His principal work "Traité de dynamique" contains the principle which bears his name.

Arnol'd, Vladimir Igorevich: 1937– , Russian mathematician.

Cartan, Elie: 1869–1951, French mathematician.

Coriolis, Gustave-Gaspard: 1792–1843, French mathematician.

Coulomb, Charles Augustin: 1736–1806, French physicist.

***Descartes, René (Cartesius):** born 31 March 1596 in La Haye (Touraine), died 11 February 1650 in Stockholm. French philosopher, mathematician and natural philosopher. In spite of the fact that Descartes' contributions to mechanics were not too successful (for instance, he proposed incorrect laws of collision), he contributed decisively to the development of analytic thinking without which modern natural science would not be possible. In this regard his book *Discours de la Méthode pour bien conduire sa Raison*, published in 1637, was particularly important. Also his imaginative conceptions – ether whirls carrying the planets around the sun; God having given eternal motion to the atoms relative to the atoms of the ether which span our space; the state of motion of atoms being able to change only by collisions – inspired the amateur researchers of the 17th century considerably. It was in this community where the real scholars of science found the resonance and support which they did not obtain from the scholastic and rigid attitude of the universities of their time.

***Einstein, Albert:** born 14 March 1879 in Ulm (Germany), died 18 April 1955 in Princeton, N.J. (U.S.A.). German-Swiss physicist, 1940 naturalized in the U.S.A. His most important contribution to mechanics is Special Relativity which he published between 1905 and 1907. In his General Relativity, published between 1914 and 1916, he succeeded in establishing a (classical) description of gravitation as one of the fundamental interactions. While Special Relativity is based on the assumption that space-time is the flat space $\mathbb{R}^4$, General Relativity is a dynamical and geometric field theory which allows one to determine the metric on space-time from the sources, i.e., from the given distribution of masses in the universe.

***Euler, Leonhard:** born 15 April 1707 in Basel (Switzerland), died 18 September 1783 in St. Petersburg (Russia). Swiss mathematician. Professor initially of physics, then of mathematics at the Academy of Sciences in St. Petersburg (from 1730 until 1741, and again from 1766 onwards), and, at the invitation of Frederick the Great, member of the Berlin Academy (1741–1766). Among his gigantic scientific work particularly relevant for mechanics: development of variational calculus; law of conservation of angular momentum as an independent principle; equations of motion for the top. He also made numerous contributions to continuum mechanics.

Fibonacci, Leonardi: ∼1175–∼1240. Italian mathematician who introduced the Indian-Arabic system of numbers. See also *Fibonacci numbers* in Sect. 6.5.

***Galilei, Galileo:** born 15 February 1564 in Pisa (Italy), died 8 January 1642 in Arcetri near Florence (Italy). Italian mathematician, natural philosopher, and philosopher, who belongs to the founding-fathers of natural sciences in the modern sense. Professor in Pisa (1589–1592), in Padua (1592–1610), mathematician and physicist at the court of the duke of Florence (1610–1633). From 1633 on under confinement to his house in Arcetri, as a consequence of the conflict with Pope Urban VIII and the Inquisition, which was caused by his defence of the Copernican, heliocentric, planetary system. Galilei made important contributions to the mechanics of simple machines and to observational astronomy. He developed the laws of free fall.

***Hamilton, William Rowan:** born 4 August 1805 in Dublin (Ireland), died 2 September 1865 in Dunsink near Dublin. Irish mathematician, physicist, and astronomer. At barely 22 years of age he became professor at the university of Dublin. Important contributions to optics and to dynamics. Developed the variational principle which was cast in its later and more elegant form by C.G.J. Jacobi.

***Huygens, Christiaan:** born 14 April 1629 in The Hague (Netherlands), died 8 July 1695 in The Hague. Dutch mathematician, physicist and astronomer. From 1666 until 1681 member of the Academy of Sciences in Paris. Although Huygens is not mentioned explicitly in this book he made essential contributions to mechanics: among others the correct laws for elastic, central collisions and, building on Galilei's discoveries, the classical principle of relativity.

Jacobi, Carl Gustav Jakob: 1804–1851, German mathematician.

***Kepler, Johannes:** born 27 December 1571 in Weil der Stadt (Germany), died 15 November 1630 in Regensburg (Germany). German astronomer and mathematician. Led a restless live, in part due to numerous misfortunes during the turbulent times before and during the Thirty Years War, but also for reasons to be found in his character. Of greatest importance for him was his acquaintance with the Danish astronomer Tyge (Tycho) Brahe, in the year 1600 in Prague, whose astronomical data were the basis for Kepler's most important works. Succeeding T. Brahe, Kepler became mathematician and astrologer at the imperial court, first under emperor Rudolf II, later under Mathias. Finally, from 1628 until his death in 1630, he was astrologer of the duke of Friedland and Sagan, A. von Wallenstein. Kepler's first two laws are contained in his *Astronomia nova* (1609), the third is contained in his main work *Harmonices Mundi* (1619). They were not generally accepted, however, until Newton's work who gave them a new, purely mechanical foundation. Kepler's essential achievement was to overcome the ancient opposition between celestial mechanics (where the circle was believed to be the natural inertial type of movement) on one hand and terrestial mechanics on the other (where the straight line is the inertial motion).

Kolmogorov, Andrei Nikolaevic: 1903–1987, Russian mathematician.

***Lagrange, Joseph Louis:** born 25 January 1736 in Torino, died 10 April 1813 in Paris. Italian-French mathematician. At the early age of 19 professor for mathematics at the Royal School of Artillery in Torino, from 1766 member of the Berlin

Academy, succeeding d'Alembert, then from 1786 member of the French Academy of Sciences, professor at Ecole Normale, Paris, in 1795 and at Ecole Polytechnique in 1797. His major work *Mécanique Analytique* which appeared in 1788, after Newton's *Principia* (1688) and Euler's *Mechanica* (1736), is the third of the historically important treatises of mechanics. Of special relevance for mechanics were his completion of variational calculus which he used to derive the Euler-Lagrange equations of motion, and his contributions to celestial perturbation theory.

*__Laplace, Pierre Simon de:__ born 28 March 1749 in Beaumont-en-Auge (France), died 5 March 1827 in Paris. French mathematician and physicist. From 1785 on member of the Academy, he became professor for mathematics at Ecole Normale (Paris) in 1794. He must have been rather flexible because he survived four political systems without harm. Under Napoleon he was for a short time Secretary of the Interior. Besides important contributions to celestial mechanics on the basis of which the stability of the planetary system was rendered plausible, he developed potential theory, along with Gauß and Poisson. Other important publications of his deal with the physics of vibrations and with thermodynamics.

Legendre, Adrien Marie: 1752–1833, French mathematician.

Leibniz, Gottfried Wilhelm: 1646–1716, German natural philosopher and philosopher.

Lie, Marius Sophus: 1842–1899, Norwegian mathematician.

Liapunov, Aleksandr Mikhailovich: 1857–1918, Russian mathematician.

Liouville, Joseph: 1809–1882, French mathematician.

Lorentz, Hendrik Antoon: 1853–1928, Dutch physicist.

*__Maupertuis, Pierre Louis Moreau de:__ born 28 September 1698 in St. Malo (Britanny, France), died 27 July 1759 in Basel (Switzerland). French mathematician and natural philosopher. From 1731 paid member of the Academy of Sciences of France, in 1746 he became the first president of the Prussian *Académie Royale des Sciences et Belles Lettres*, newly founded by Frederick the Great in Berlin who called many important scientists to the academy, notably L. Euler. In 1756, seriously ill, Maupertuis returned first to France but then joined his friend Joh. II Bernoulli in Basel where he died in 1759. Along with Voltaire, Maupertuis was a supporter of Newton's theory of gravitation which he had come to know while visiting London in 1728, and fought against Descartes' ether whirls. Of decisive importance for the development of mechanics was his principle of least action, formulated in 1747, although his own formulation was still somewhat vague (the principle was formulated in precise form by Euler and Lagrange). A widely noticed dispute of priority started by the Swiss mathematician Samuel König who attributed the principle to Leibniz, was eventually decided in favor of Maupertuis. This dispute alienated Maupertuis from the Prussian Academy and contributed much to his bad state of health.

Minkowski, Hermann: 1864–1909, German mathematician.

Moser, Jürgen: 1928–1999, German mathematician.

***Newton, Isaac:** born 24 December 1642 in Whoolsthorpe (Lincolnshire, England), died 20 March 1726 in Kensington (London), (both dates according to the Julian calender which was used in England until 1752). Newton, who had studied theology at Trinity College Cambridge, learnt mathematics and natural sciences essentially by himself. His first great discoveries, differential calculus, dispersion of light, and the law of gravity which he communicated to a small circle of experts in 1669, so impressed Isaac Barrow, then holding the "Lucasian Chair" of mathematics at Trinity College, that he renounced his chair in favor of Newton. In 1696 Newton was called at the Royal Mint and became its director in 1699. The Royal Society of London elected him president in 1703. Venerated and admired as the greatest English natural philosopher, Newton was buried in Westminster Abbey.

His principal work, with regard to mechanics, is the three-volume *Philosophiae Naturalis Principia Mathematica* (1687) which he wrote at the instigation of his pupil Halley and which was edited by Halley. Until this day the Principia have been studied and completely understood only by very few people. The reason for this is that Newton's presentation uses a highly geometrical language, divides matters into "definitions" and "axioms" which mutually complete and explain one another, following examples from antiquity, in a manner difficult to understand for us, and because he presupposes notions of scholastic and Cartesian philosophy we are normally not familiar with. Even during Newton's lifetime it took a long time before his contemporaries learned to appreciate this difficult and comprehensive work which, in addition to Newton's laws, deals with a wealth of problems in mechanics and celestial mechanics, and which contains Newton's original ideas about space and time that have stimulated our thinking ever since. Newton completed a long development that began during antiquity and was initiated by astronomy, by showing that celestial mechanics is determined only by the principle of inertia and the gravitational force and, hence, that it follows the same laws as the mechanics of our everyday world. At the same time he laid the foundation for a development which to this day is not concluded.

Noether, Emmy Amalie: 1882–1935, German mathematician. Belongs to the great scientific personalities in the mathematics of the 20th century. Her seminal work *Invariante Variationsprobleme*, published in 1918, contains two theorems which ever since we refer to as the "Noether theorems" and which provide important keys to various parts of theoretical physics, notably mechanics and classical field theory. Barely any other mathematical publication has had such a profound and lasting impact on theoretical physics in the 20th century.

***Poincaré, Jules Henri:** born 29 April 1854 in Nantes (France), died 17 July 1912 in Paris. French mathematician, professor at Sorbonne university in Paris.

Poincaré, very broad and extraordinarily productive, made important contributions to the many-body problem in celestial mechanics for which he received a prize donated by King Oscar II of Sweden. (Originally the prize was announced for the problem of convergence of the celestial perturbation series, the famous problem of the "small denominators" that was solved much later by Kolmogorov, Arnol'd, and Moser.) Poincaré may be considered the founder of qualitative dynamics and of the modern theory of dynamical systems.

Poisson, Siméon Denis: 1781–1840, French mathematician.

Bibliography

Mechanics Textbooks and Exercise Books

Fetter, A.L., Walecka, J.D.: *Theoretical Mechanics of Particles and Continua* (McGraw Hill, New York 1980)
Goldstein, H.: *Classical Mechanics* (Addison-Wesley, Reading 1980)
Honerkamp, J., Römer, H.: *Theoretical Physics* (Springer, Berlin, Heidelberg 1993)
Landau, L.D., Lifshitz, E.M.: *Mechanics* (Pergamon, Oxford and Addison-Wesley, Reading 1971)
Spiegel, M.R.: *General Mechanics, Theory and Applications* (McGraw Hill, New York 1976)

Some General Mathematical Texts

Arnol'd, V.I.: *Ordinary Differential Equations* (Springer, New York, Berlin, Heidelberg 1992)
Arnol'd, V.I.: *Geometrical Methods in the Theory of Ordinary Differential Equations* (Springer, New York, Berlin, Heidelberg 2nd ed. 1988)
Flanders, H.: *Differential Forms* (Academic, New York 1963)
Hamermesh, M.: *Group Theory and Its Applications to Physical Problems* (Addison-Wesley, Reading 1962)

Special Functions and Mathematical Formulae

Abramowitz, M., Stegun, I.A.: *Handbook of Mathematical Functions* (Dover, New York 1965)
Gradshteyn, I.S., Ryzhik, I.M.: *Tables of Integrals, Series and Products* (Academic, New York 1965)

Chapter 2

Boccaletti, D., Pucacco, G.: Theory of Orbits 1+2 (Springer, Berlin, Heidelberg 1996, 1999)
Rüssmann, H.: *Konvergente Reihenentwicklungen in der Störungstheorie der Himmelsmechanik*, Selecta Mathematica V (Springer, Berlin, Heidelberg 1979)
Rüssmann, H.: *Non-degeneracy in the Perturbation Theory of Integrable Dynamical Systems*, in Dodson, M.M., Vickers, J.A.G. (eds.) *Number Theory and Dynamical Systems*, London Mathematical Society Lecture Note Series 134 (Cambridge University Press 1989)

Chapter 4

Ellis, G.F.R., Williams, R.M.: *Flat and Curved Space-Times* (Clarendon Press, Oxford, 1994)
Hagedorn, R.: *Relativistic Kinematics* (Benjamin, New York 1963)
Jackson, J.D.: *Classical Electrodynamics* (Wiley, New York 1975)
Sexl, R.U., Urbantke, H.K.: *Relativität, Gruppen, Teilchen* (Springer, Berlin, Heidelberg 1976)
Weinberg, S.: *Gravitation and Cosmology, Principles and Applications of the General Theory of Relativity* (Wiley, New York 1972)

Chapter 5

Abraham, R., Marsden, J.E.: *Foundations of Mechanics* (Benjamin Cummings, Reading 1981)
Arnol'd, V.I.: *Mathematical Methods of Classical Mechanics* (Springer, New York, Berlin, Heidelberg 2nd ed. 1989)
Guillemin, V., Sternberg, S.: *Symplectic Techniques in Physics* (Cambridge University Press, New York 1990)
Hofer, H., Zehnder, E.: *Symplectic Invariants and Hamiltonian Dynamics* (Birkhäuser, Basel, 1994)
Marsden, J.E., Ratiu, T.S.: *Introduction to Mechanics and Symmetry* (Springer, New York, 1994)
O'Neill, B.: *Semi-Riemannian Geometry, With Applications to Relativity* (Academic, New York 1983)
Thirring, W.: *A Course in Mathematical Physics*, Vol. 1: *Classical Dynamical Systems* (Springer, Berlin, Heidelberg 2nd ed. 1992)

Chapter 6

Arnol'd, V.I.: *Catastrophe Theory* (Springer, Berlin, Heidelberg 1986)
Bergé, P., Pomeau, Y., Vidal, C.: *Order within Chaos; Towards a Deterministic Approach to Turbulence* (Wiley, New York 1986) French original (Hermann, Paris 1984)
Chirikov, B.V.: *A Universal Instability of Many-Dimensional Oscillator Systems*, Physics Reports **52**, 263 (1979)
Collet, P., Eckmann, J.P.: *Iterated Maps on the Interval as Dynamical Systems* (Birkhäuser, Boston 1990)
Devaney, R.L.: *An Introduction to Chaotic Dynamical Systems* (Benjamin Cummings, Reading 1989)
Feigenbaum, M.: J. Stat. Phys. **19**, 25 (1978) and **21**, 669 (1979)
Guckenheimer, J., Holmes, P.: *Nonlinear Oscillations, Dynamical Systems, and Bifurcations of Vector Fields* (Springer, Berlin, Heidelberg 1990)
Hénon, M., Heiles, C.: Astron. Journ. **69**, 73 (1964)
Hirsch, M.W., Smale, S.: *Differential Equations, Dynamical Systems and Linear Algebra* (Academic, New York 1974)
Palis, J., de Melo, W.: *Geometric Theory of Dynamical Systems* (Springer, Berlin, Heidelberg 1982)
Peitgen, H.O., Richter, P.H.: *The Beauty of Fractals, Images of Complex Dynamical Systems* (Springer, Berlin, Heidelberg 1986)
Ruelle, D.: *Chance and Chaos* (Princeton University Press, New Jersey, 1991)
Ruelle, D.: D.: *Elements of Differential Dynamics and Bifurcation Theory* (Academic, New York 1989)
Schuster, H.G.: *Deterministic Chaos, An Introduction* (Physik-Verlag, Weinheim 1987)
Wisdom, J.: *Chaotic Behaviour in the Solar System*, Nucl. Phys. B (Proc. Suppl.) **2**, 391 (1987)

Index

About the Author

Florian Scheck

Born 1936 in Berlin, son of Gustav Scheck, flutist. Studied physics at University of Freiburg im Breisgau, Germany. Physics diploma 1962, PhD in theoretical physics 1964, both at Freiburg University.

Guest scientist at the Weizmann Institute of Science, Rehovoth, Israel, 1964–1966; research assistant at University of Heidelberg, Germany, 1966–1968; research fellow at CERN, Geneva, 1968–1970.

Habilitation at the University of Heidelberg 1968. From 1970 until 1976 head of theory group at the Swiss Institute for Nuclear Research (PSI), Villigen, as well as lecturer and titular professor at ETH, Zurich.

Since 1976 professor of theoretical physics at Johannes Gutenberg University in Mainz, Germany. Numerous visits worldwide as guest scientist or guest professor.

Principal field of activity: theoretical elementary particle physics.